Handbook of Green Building Design, and Construction

LEED®, BREEAM®, and Green Globes®

Sam Kubba, Ph.D., LEED AP

AMSTERDAM • BOSTON • HEIDELBERG • LONDON
NEW YORK • OXFORD • PARIS • SAN DIEGO
SAN FRANCISCO • SINGAPORE • SYDNEY • TOKYO

Butterworth-Heinemann is an imprint of Elsevier

Butterworth-Heinemann is an imprint of Elsevier
225 Wyman Street, Waltham, MA 02451, USA
The Boulevard, Langford Lane, Kidlington, Oxford, OX5 1GB, UK

Notices
Knowledge and best practice in this field are constantly changing. As new research and experience
broaden our understanding, changes in research methods, professional practices, or medical
treatment may become necessary.

Practitioners and researchers must always rely on their own experience and knowledge in
evaluating and using any information, methods, compounds, or experiments described herein. In
using such information or methods they should be mindful of their own safety and the safety of
others, including parties for whom they have a professional responsibility.

To the fullest extent of the law, neither the Publisher nor the authors, contributors, or editors,
assume any liability for any injury and/or damage to persons or property as a matter of products
liability, negligence or otherwise, or from any use or operation of any methods, products,
instructions, or ideas contained in the material herein.

Library of Congress Cataloging-in-Publication Data
Kubba, Sam.
    Handbook of green building design and construction : LEED®, BREEAM®,
and Green Globes® / Sam Kubba, PhD, LEED AP.
    pages   cm
    ISBN 978-0-12-385128-4 (hardback)
    1. Sustainable buildings—Design and construction—Handbooks,
manuals, etc. 2. Buildings—Specifications—Handbooks, manuals, etc.
3. Sustainable construction—Handbooks, manuals, etc. 4. Sustainable
construction—Standards. I. Title.
  TH880.K8395 2012
  720'.47–dc23

British Library Cataloguing-in-Publication Data
A catalogue record for this book is available from the British Library.

For information on all Butterworth–Heinemann publications
visit our website at *http://store.elsevier.com*

Working together to grow
libraries in developing countries

www.elsevier.com | www.bookaid.org | www.sabre.org

ELSEVIER   BOOK AID International   Sabre Foundation

Transferred to Digital Printing in 2012

To my friends and colleagues everywhere,
without whom life would be meaningless and hollow and
To my wife and four children whose love and affection
continues to inspire me through the years

Contents

4. Green Project Cost Monitoring and Closeout

5. Building Information Modeling

6. Green Building Materials and Products

7. Indoor Environmental Quality

8. Water Efficiency and Sanitary Waste

9. Impact of Energy and Atmosphere

10. Green Design and Building Economics

11. Green Project Commissioning

12. Project Cost Analysis

13. Green Specifications and Documentation

14. Types of Building Contract Agreements

15. Green Business Development

16. Building Green Litigation and Liability Issues

Appendices

Foreword

When I was an Executive Editor at GreenBiz.com, I spent a lot of time trying to reduce the number of words used to make a point. I dream of the day when editors can edit out the word "green" from the title of Dr. Kubba's thorough Handbook. At that point, there will be no such thing as green buildings, only buildings that operate comfortably and economically in harmony with the natural flows of the Earth—buildings that not only nourish and restore the occupants but also the planet itself.

Confronted by overwhelming scientific evidence that our current linear make-use-toss-repeat way of life is fast rendering the planet unfit for human habitation, we urgently need to take measures to come back into balance with the beautiful and finely tuned web that supports all species'—including our—life. Green buildings are beginning to take us in this direction, and it is vital that the growing amount of knowledge to make buildings better be disseminated widely.

Recognizing that a chain can break at its weakest link, Dr. Kubba's *Handbook of Green Building Design, and Construction* covers all of the key elements of successfully executing a sustainable building from concept to operation, from permitting to (avoiding) litigation and liability. Drawing extensively from research conducted for his other two books on LEED® certification and managing green projects this book is chock-full of practical project advice, as well as numerous proof sources of both the benefits of green buildings and resources for debunking the most common myths that stand in the way of wider dissemination of building green.

When I was leading the development of LEED and was showing the draft system to design professionals, a few of them said, "I don't get it, this is just good design." Actually, they got it exactly: a green building is a good building. If a building isn't green, it's not a good building. Period. End of story. Dr. Kubba's book and his work to promote good building around the world make an important contribution to the betterment of humankind.

Robert (Rob) Watson, LEED AP (BD&C)
CEO, EcoTech International

Acknowledgments

It would not have been possible for me to produce a book of this size and scope without the active and passive support of many friends and colleagues who have contributed greatly to my thinking and insights during the writing of it and who were in many ways instrumental in the crystallization and formulation of my thoughts on the subjects and issues discussed within. To them I am heavily indebted, as I am to the innumerable people and organizations that have contributed ideas, comments, and illustrations, that have helped make this book a reality.

I must also unequivocally mention that without the unfailing fervor, encouragement, and wisdom of Mr. Kenneth McCombs, Senior Acquisitions Editor, Elsevier Science and Technology, this book would still be on the drawing board. It is always a great pleasure working with the company. I would like to thank Mr. Michael Joyce, Editorial Project Manager, who helped smooth many of the "bumps" along the way. Likewise, I must acknowledge the wonderful work of Ms. Marilyn E. Rash, Project Manager, who coordinated the day-to-day details of the *Handbook* and who saw it through production. I also wish to thank her for her unwavering commitment and support. I also wish to thank Ms. Samantha Graham, a highly valued and dedicated member of the Elsevier's support team, for proofreading the first pages. I would especially like to salute and express my deepest appreciation to all of the Elsevier team for seeing the book through production, and to SPi in India for formatting the pages, and Eric DeCicco for the excellent cover design.

I am particularly indebted to the U.S. Green Building Council (USGBC) and its staff for their assistance, continuous updates, and support on the new LEED® 2009 Version 3 Rating System, and to Ms. Anica Landreneau of HOK for reviewing Chapter 5. I also wish to express my appreciation to Mr. Rob Watson, CEO of EcoTech International for reviewing the Introduction and Chapters 1 and 2, in addition to his informative comments. Last but not least, I wish to record my gratitude to all those who came to my rescue during the final stretch of this work—the many nameless colleagues, architects, engineers, and contractors who kept me motivated with their ardent enthusiasm, support, and technical expertise. To these wonderful professionals, I can only say, "Thank you." I relied on them in so many ways, and while no words can reflect the depth of my gratitude to all of them for their assistance and advice, in the final analysis, I alone must bear responsibility for any mistakes, omissions, or errors that may have found their way into this book.

The Green Movement—Myths, History, and Overview

One of the hottest topics over the last decade in the field of property development is the concept of sustainable development and green building. Yet, it is not easy to give a precise definition of what makes a building green. One definition offered by the Office of the Federal Environmental Executive (OFEE) for green building is:

[T]he practice of (1) increasing the efficiency with which buildings and their sites use energy, water, and materials, and (2) reducing building impacts on human health and the environment, through better siting, design, construction, operation, maintenance, and removal—the complete building life cycle.

The EPA defines green building as, "the practice of creating structures and using processes that are environmentally responsible and resource-efficient throughout a building's life cycle from siting to design, construction, operation, maintenance, renovation, and deconstruction."

So essentially when correctly applied, green building is meant to improve design and construction practices so that the buildings we build last longer, cost less to operate, and facilitate increased productivity and better working environments for workers or residents. But even more than that, it is also about protecting our natural resources and improving the built environment so that the planet's ecosystems, people, enterprises, and communities can live a healthier and more prosperous life.

The general perception of the green movement has been considerably transformed since its early formative days and is today sweeping across the United States and much of the world. Furthermore, sustainable development principles are taking on an increasingly important role in real estate applications, particularly by forward-looking developers. In fact, many contractors are now seeking green certification and, with this in mind, the Associated Builders and Contractors, Inc. (ABC) has recently initiated a program that would certify "Green Contractors." Nevertheless, some developers refuse to jump on the environment-friendly, or "green" building, bandwagon mainly due to the misplaced notion that green buildings cost more or that they are impractical to construct.

GREEN BUILDING: MYTHS AND REALITIES

There are many myths about sustainability floating in the ether. One example is the myth that sustainability costs more, which ignores recent research as well as the reality that for any society to thrive and prosper, it must seek to create a healthy balance between its environmental, social, and economic dimensions. Sustainability is not just about building green but about building a healthy community and sustaining a quality way of life. As reminded often by President Obama and his cabinet, as a community we cannot afford to continue delaying the pursuit of new sources of energy such as wind, solar, and geothermal. With the state of the economy being what it is, these efforts would help create new jobs, attract new businesses, reduce our energy costs, and create a healthy environment. Although green building has made tremendous strides in the past few years, there remain many who are still unconvinced of its benefits due to the numerous myths and misconceptions floating around the mainstream construction and real estate industries, as described next.

Myth 1: Green/sustainable buildings cost much more than conventional buildings

Reality check: This is a very common misconception that continues to linger on even though it has been debunked many times over. Although on a price per square foot basis, building green may incur marginally greater upfront costs, in the long run a green home is more affordable and cost effective because the operational costs are lower when compared with conventional buildings. It is surprising therefore that some developers still believe that building with green materials or renovating to green specifications is cost-prohibitive. In addition to this, there are various strategies and approaches that can be employed to achieve inexpensive green building. These include reducing waste, optimal value engineering, right-sizing the structure to using solar panels, low-e windows, and energy-saving appliances, and more—all of which can help qualify the project for federal tax credits. Moreover, when green thinking becomes an integral part of the initial building plans, it is easier to design and incorporate green elements into the project.

Myth 2: It's just another fad and therefore not particularly important

Reality check: Over the last decade, we have witnessed an increasing interest in sustainability and a continuous growth in green building and green building certification—so much so that it has now become more than an integral part of the mainstream in the construction industry, and it is becoming the preferred building method. Furthermore, creating a healthy environment where green building does not exist cannot be considered a fad.

Myth 3: Green buildings are often "unattractive" or "ugly" and lack the aesthetic quality of conventional buildings

Reality check: A green/sustainable building doesn't have to look any different from a conventional building. In fact many of today's green buildings are virtually indistinguishable from traditional buildings. Moreover, green renovations of existing buildings should respect its character and if well designed, most likely won't be noticeable from either the interior or exterior. Thus, wood certified by the Forest Stewardship Council (FSC) looks essentially the same as other types of wood, and when using a vegetated roof, for example, it would not typically be visible from ground level. Moreover, one does not have to mount continuous rows of unattractive solar panels to be green or be obligated to go with solar power, although there are numerous ways to creatively integrate photovoltaic (PV) panels into a project that are both attractive and effective. Likewise, eco-friendly shingles are actually more attractive than the common asphalt versions and some renovations are actually invisible (e.g., extra insulation or a new energy-efficient HVAC system).

Myth 4: Green building is essentially about eco-friendly material selection

Reality check: Not at all. Green building is mainly concerned with how you design and orient your building, site selection, water conservation, energy performance, window location, and so on. However, making smart decisions regarding eco-friendly building materials (e.g., those possessing a high recycled content, low embodied energy, minimal VOCs) is an important aspect of green building, but they are only a small part of the overall equation. Alex Wilson, president of BuildingGreen Inc. and executive editor of *Environmental Building News*, says: "People are beginning to gain a greater understanding that green building is a systems approach to the entire construction process."

Myth 5: Green buildings do not fetch higher rental rates or capital compared with traditional buildings

Reality check: Recent surveys consistently show that there is a growing market demand for green buildings because they achieve much higher rentals, thus capital, as a result of reduced operation costs and higher productivity of employees. For example, a recent Building Owners and Managers Association (BOMA) survey in Seattle concluded that 61% of real estate leaders opine that green buildings enhance their corporate image and more than two-thirds of those surveyed believe that over the next five years tenants will make the "greenness" of property a significant factor in choosing space. Tenants and developers therefore do care about green and healthier environments and are willing to pay for it.

This trend is already particularly evident in high-end residential projects and flagship corporate office projects and is very likely to become widespread.

Myth 6: Green buildings do not provide the comfort levels that many of today's tenants demand

Reality check: On the contrary, green buildings are typically more comfortable and healthier than conventional buildings. In fact, one of the chief characteristics of sustainable design is to support the well-being of building occupants by reducing indoor air pollution from exposure to contaminants (e.g., asbestos, radon, and lead), therefore avoiding complaints such as sick building syndrome (SBS) and building-related illness (BRI). This can normally be achieved by selecting materials with low off-gassing potential; proper ventilation strategies; adequate access to daylight and views; and optimum comfort through control of lighting, humidity and temperature levels. This is not the case with traditional building environs.

Myth 7: Green building products are often difficult to find

Reality check: This may have been true a decade or so ago when it may sometimes have been difficult to find eco-friendly or energy-saving materials at a reasonable price; today, green building materials are more popular than ever and have become much more accessible. Where green building products are not readily accessible, it may be because they are not manufactured nationwide or they may be difficult to find in certain parts of the country; in such cases, it is usually possible to find satisfactory alternatives. Indeed, the number of green products and systems that are now readily available on the market has dramatically increased during recent years and is growing continually. So much so that green building products are now in the thousands and have become part of the mainstream. Much information—including performance data and contact details—can also be obtained from the various green product directories on the market such as the two comprehensive directories published by BuildingGreen Inc. (*GreenSpec® Directory* and *Green Building Products*).

Myth 8: Green building uses traditional tools and techniques and not cutting-edge technology

Reality check: The most successful green building design projects generally use a multidisciplinary and integrated design approach, where a number of consultants and the owner's representative participate as a team and the architect typically takes on the role of team leader rather than sole decision maker. In most cases, locally available materials and techniques are used in addition to the latest technology. This is reinforced by the U.S. Environmental Protection Agency's website, which clearly states that "green building research is being done by national laboratories, private companies, universities, and industry." According to a recent U.S. Green Building Council (USGBC®) report, in excess of 70% of the green building research is focused on energy and atmosphere research.

Myth 9: Green building products don't work as well as traditional ones

Reality check: Examples of typical products that frequently get a bad rap include double-flush and low-flow toilets. It may be true that when first introduced, low-flow toilets did not function that well, and some people are still of the opinion that 1.6 gallon-per-flush toilets don't work as well as traditional toilets, even though these fixtures have been mandated for all new construction for more than a decade. Moreover, recent surveys show that when customers are asked to comment on their satisfaction with their new 1.6-gal., high-efficiency toilet fixtures, the majority say they double-flush the same number of times or fewer with their new efficient fixture than with their old water waster. The "don't work as well" myth was reinforced with the introduction of compact fluorescent light bulbs (CFLs), which gave off harsh color, didn't last as long as claimed, and took too long to light up. Another green building product myth often cited relates to fiberglass insulation in that inhaling fiberglass fibers can lead to cancer, which is obviously false.

It is therefore important to research unfamiliar products and seek accurate information to back up any efficiency claims prior to formulating a final opinion regarding its suitability or lack thereof. However, generally speaking, most modern new green products work as well, if not better than traditional ones, and green products have been vastly improved in recent years. It should be noted that green materials like traditional building materials also have to meet strict quality-control standards, and as the green market grows, new upgrades will undoubtedly take place to improve quality and reliability.

Myth 10: Building green is too difficult and complicated

Reality check: Nothing is further from the truth; in fact many builders consider green building to be very easy and compares favorably with conventional building. Building green is a business that can be simple, uses common sense, and does not require a rocket sciencist to implement. Basically, build it smaller, use quality materials chosen for sustainability and efficiency, not for the fad of the month.

Myth 11: It is not possible to build a high-rise green building

Reality check: Green concepts do not generally inhibit or restrict building design or space usability. Furthermore, all modern techniques that apply to conventional building can be employed when building *green*. A good example of this is the Condé Nast Building (officially 4 Times Square) located in Midtown

Manhattan. The building boasts 48 stories and rises to 809 feet (247 m). It is environmentally friendly with gas-fired *absorption chillers*, and a high-performing insulating and shading curtain wall, which keeps the building's energy costs down by not requiring heating or cooling for most of the year. In addition, the building uses solar and fuel-cell technology, making it the first project of its size to incorporate these features in construction.

Myth 12: It is difficult or not possible to convert existing conventional buildings into energy efficient buildings

Reality check: It is not really difficult to convert existing buildings into green/sustainable buildings. Actually, there are numerous scientific ratings and checklists that builders can use to redesign and realign traditional buildings to meet modern green standards. Likewise, many rating systems, such as Leadership in Energy and Environmental Design (LEED®) for existing buildings, Canada's Go Green Plus, and the Japanese CASBEE certification system, all encourage such conversions. To this end, President Obama after becoming president committed his administration to retrofitting 75% of all existing federal buildings. It is important therefore to increase public awareness of how baseless these myths are and to do all that is possible to eliminate them.

Myth 13: Building green requires signing up for a green program or third-party certification

Reality check: This is definitely not a normal requirement for building green, although certification programs, such as Green Globes® and the U.S. Green Building Council's LEED, are excellent vehicles for increasing exposure and furthering the green movement and, it must be said, that without third-party certification, much of the value of "green" is lost. In addition, keep in mind that the LEED Rating System is, in most cases, a totally voluntary program: You pay your fees, follow the LEED guidelines, and ultimately receive a plaque or certificate stating your building has achieved a Silver, Gold, or whichever status. More important, however, remember that there are many financial and other government incentives to attain certification. Moreover, building owners and developers can reap the financial benefits of the "greenness" of their building projects by taking advantage of the various tax credits and private and public non-tax financial incentives available, as well as tenant monetization of reduced operations and maintenance costs and carbon and renewable energy tradable credits.

Myth 14: Going green is an all-or-nothing proposition

Reality check: Many developers and construction professionals have the misconception that going green with existing buildings involves large-scale remodeling. In fact, the degree and scale of incorporating green into a building is wholly up to the owner, depending on the individual lifestyle and budget. Many builders and designers often use green concepts and green products intuitively without being fully aware of them. This is rapidly changing with increased awareness and demand for green products, and many manufacturers and the construction industry find themselves moving in this direction.

GREEN BUILDING AND THE GREEN MOVEMENT: ITS HISTORY

For an in-depth and more comprehensive understanding of the modern green movement, it helps to try and trace its origins back to the beginning. However, it is almost impossible to determine precisely when a movement may have started. Long before the arrival of the industrial revolution and electrically powered heating and cooling systems, ancient and primitive populations were compelled to improvise using basic tools and natural materials to construct buildings that protected them from the harsh elements and extremes in temperature. Particularly, as the ancients had few other options at their disposal, these builders incorporated passive design that took advantage of the resources provided by nature, namely the sun and climate to heat, cool, and light their buildings. The Babylonians and Egyptians, for example, used adobe as their prime building material and built *badgeer* (wind shafts) into their palaces and houses. They took advantage of courtyards and narrow alleyways for shade. These are simple examples of how the ancients overcame the many challenges of climate that faced them.

More recently however, we find scholars like Mark Wilson who believe that the concept of green building first appeared in America more than a century ago. According to Wilson:

The revolutionary design philosophy known as First Bay Tradition had its roots in the San Francisco Bay Area in the 1890s. Indeed, the leading practitioners of this environmentally sensitive organic movement, Bernard Maybeck and Julia Morgan, developed a design philosophy that incorporated most of the concepts that are embraced by today's green movement in architecture.

Some historians associate its beginning with Rachel Carson's (1907−1964) book, *Silent Spring,* and the legislative fervor of the 1970s, or with Henry David Thoreau who in his book, *Marine Woods,* advocates for the respecting of nature and also for an awakening to the need for conservation and federal preservation of virgin forests. Many believe that the green movement had its roots in the energy crises of the 1970s and the creative approaches to saving energy that emanated from it, such as smaller building envelopes and the use of active and passive solar design.

When the 1973 OPEC oil crisis erupted, it brought the cost of energy into sharp focus and reminded us that our future prosperity and security was in the hands of a very small number of petroleum-producing countries. This catalyzing event effectively highlighted the need for diversified sources of energy and encouraged corporate and government investment in solar, wind, water, and geothermal sources of power. The energy crises artificially created by the imposition of an oil embargo by OPEC in 1973 caused an upward spike in gasoline prices and, for the first time, long lines of vehicles at gas stations around the country. This had a dramatic effect on a small group of enlightened and forward-thinking architects, environmentalists, and ecologists, who began questioning the wisdom of conventional building techniques and inspired them to seek new solutions to the problem of sustainability.

This nascent "environmental movement," which was partly inspired by Victor Olgyay's *Design with Climate*, Ralph Knowles's *Form and Stability*, and Rachel Carson's *Silent Spring*, served notice of the emergence of a new era in environmental design. It also captured the attention and imagination of the general public and caused many to clamor for a broader reexamination of the wisdom of our reliance on fossil fuels for transportation and buildings. Indeed, a number of legislative steps were initiated to clean up the environment, including the Clean Air Act, the National Environmental Policy Act, the Water Pollution Control Act, the banning of DDT, the Endangered Species Act, and the institution of Earth Day.

The response of the American Institute of Architects (AIA) to the energy crisis of 1973 was to form an energy task force to study energy-efficient design strategies, and in 1977 President Carter's administration founded what became the U.S. Department of Energy; one of its principal tasks was to focus on energy usage and conservation. The energy task force was later to become the AIA Committee on Energy. The energy committee prepared several papers, including "A Nation of Energy Efficient Buildings," which became effective AIA tools for lobbying Capitol Hill. Among the more active committee members in the late 1970s were Donald Watson, FAIA, and Greg Franta, FAIA, when the AIA was also advocating building energy research. The committee also collaborated with government and other organizations for more than a decade.

According to committee member Dan Williams, the Committee on Energy formed two main groups: the first researched mainly passive systems (e.g., reflective roofing materials and environmentally beneficial siting of buildings) to achieve its goal of energy savings. The second group primarily concentrated on solutions employing new technologies such as the use of triple-glazed windows. This was transformed into a more broadly scaled AIA Committee on the Environment (COTE) in 1989, and the following year the AIA (through COTE), and the AIA Scientific Advisory Committee on the Environment, managed to obtain funding from the then recently created U.S. Environmental Protection Agency (EPA) to embark on the development of a building products guide, which was published in 1992, based on life cycle analysis.

As the energy concerns began to subside in the years that followed, partially due to lower energy prices, the momentum for green building and energy-related issues, in general, also gradually weakened but was not stamped out due to the dedication of a core group of pioneering architects who continued to advance their green building energy conservation concept. Several notable buildings were constructed during the 1970s that utilized green design concepts: the Willis Faber and Dumas Headquarters in England, with a grass roof, daylighted atrium, and mirrored windows, and the Gregory Bateson Building in California, with energy-sensitive photovoltaic (solar cells), under-floor rock-store cooling systems, and area climate control devices).

During the 1980s, we witnessed numerous oil spills (e.g., the *Exxon Valdez* in 1989, among others), and while the industry presented significant opposition against environmental strictures, the various energy-related Acts remained in force. We also witnessed during the 1980s and early 1990s global conservation efforts by sustainability proponents such as Robert Berkebile who was a driving force in starting the AIA Committee on the Environment (COTE); William McDonough (Ford Motor Company's River Rouge Plant in Michigan), Sim Van der Ryn (Gregory Bateson Building in Sacramento, CA), and Sandra Mendler (World Resources Institute Headquarters Office, Washington, DC) in the United States. Other countries' proponents include Thomas Herzog of Germany (Design Center in Linz, Austria), British architects Norman Foster (Commerzbank Headquarters in Frankfurt, Germany) and Richard Rogers (The Pompidou Centre in Paris, France), and Malaysian architect Kenneth Yeang (Menara Mesiniaga in Kuala Lumpur, Malaysia). In 1987, the United Nations World Commission on Environment and Development, under Norwegian Prime Minister Gro Harlem Bruntland, suggested a definition for the term "sustainable development," as that which "meets the needs of the present without compromising the ability of future generations to meet their own needs."

In 1991, President George H.W. Bush issued a National Energy Policy, and AIA president James Lawler convened an advisory group to issue a response and resolution. The resolution, which the board passed a month later, called on all AIA members to undertake environmental reforms within their practices, including the immediate cessation of ozone-depleting refrigerants.

A United Nations Conference on Environment and Development (also known as the "Earth Summit"), hosted by Brazil in Rio de Janeiro in 1992, proved to be a spectacular success, drawing 17,000 attendees and boasting delegations from 172 governments and 2400 representatives of nongovernmental organizations. The conference witnessed the passage of Agenda 21 that provided a blueprint for achieving global sustainability. This resulted in the Rio Declaration on Environment and Development, the Statement of Forest Principles, the United Nations Framework Convention on Climate Change, and the United Nations Convention on Biological Diversity. Following on the heels of the Rio de Janeiro summit, the American Institute of Architects chose sustainability as its theme for the June 1993 UIA/AIA World Congress of Architects held in Chicago; an estimated 10,000 architects and design professionals from around the world attended the event. Today, this convention is recognized as a milestone in the history of the green building movement.

Encouraged by Bill Clinton's election to the presidency in November 1992, a number of proponents of sustainability began to circulate the grandiose idea of "greening" the White House itself. On Earth Day, April 21, 1993, President Bill Clinton announced his ambitious plans for "greening the White House" and to make the presidential mansion "a model for efficiency and waste reduction." To put this plan into effect, the President's Council on Environmental Quality assembled a team of experts that included members of the AIA, the U.S. Department of Energy's Federal Energy Management Program (FEMP), the EPA, the General Services Administration, the National Parks Service, the White House Office of Administration, and the Potomac Electric Power Company.

The "Greening the White House" initiative created a saving of more than $1.4 million in its first six years, primarily from improvements made to the lighting, heating, air conditioning, water sprinklers, insulation, and energy and water consumption reduction. Among other things, the initiative included a 600,000 sq. ft. Old Executive Office Building that was located across from the White House. Likewise, there was an energy audit by the Department of Energy (DOE), an environmental audit led by the EPA, and a series of well-attended design charettes consisting of design professionals, engineers, government officials, and environmentalists; the aim was to formulate sustainable energy-conservation strategies using available technologies. Within three years, the strategies resulted in significant improvements to the nearly 200-year-old mansion, including reducing its annual atmospheric emissions by an estimated 845 metric tons of carbon and an estimated $300,000 in annual energy and water savings.

Bill Browning, Hon. AIA says that "the process pioneered by the Greening of the White House charette has become an integral part of the green building movement." However, the deluge of federal greening projects was among several forces that drove the sustainability movement in the 1990s. To accelerate this process, former President Clinton issued a number of executive orders, the first being in September 1998, that directed the federal government to improve its use of recycled and eco-friendly products, including building products. A second executive order was issued in June 1999 to encourage government agencies to improve energy management and reduce emissions in federal buildings through the application of better design, construction, and operation techniques. Clinton issued a third executive order in April 2000 requiring federal agencies to integrate environmental accountability into their daily decision making as well as into their long-term planning. The team assembled by the President's Council on Environmental Quality produced important recommendations to preserve the historical presence of the structure as well as to maintain and improve comfort and productivity.

George W. Bush followed in his father's footsteps and during the eight years of his presidency, greening the White House was taken a little further with the installation of three solar systems, including a thermal setup on the pool cabana to heat water for the pool and showers, and photovoltaic panels to supplement the mansion's electrical supply. The White House greening approaches fit under several main headings:

- *Building Envelope*: Realizing that a significant amount of energy is lost through building elements, such as the roof and windows, an effort was made to analyze these and find solutions to increase their efficiency.
- *Lighting*: Energy-saving light bulbs were used wherever possible and the use of natural light was maximized. Steps were also taken to ensure lights were turned off in empty rooms.
- *Heating, Ventilation, and Air-Conditioning (HVAC)*: HVAC measures were used to reduce the amount of energy needed to heat and cool the buildings while simultaneously increasing occupant comfort. Correct ventilation is necessary to help achieve this.
- *Plug Loads*: The installation of energy-saving office equipment and replacement of refrigerators and coolers with more energy-efficient models.
- *Waste*: Initiation of a comprehensive recycling program for aluminum, glass, paper, newsprint, furniture, fluorescent lamps, paint solvents, batteries, laser printer cartridges, and organic yard waste.
- *Vehicles*: A program was initiated to lease vehicles that accept cleaner-burning alternative fuels, and the White House participates in a pilot program to test electric vehicles. Many employees are encouraged to use public transportation to decrease the use of automobiles.
- *Landscaping*: White House upgrades include methods to reduce unnecessary water and pesticide use, and the increased use of organic fertilizers on the grounds of the complex were studied.

The greening of the White House proved to be such a success that it created an underlying demand to green other properties in the extensive federal portfolio—for example, the Pentagon, the Presidio, and the DOE headquarters, as well as three national parks: Grand Canyon, Yellowstone, and Alaska's Denali. In 1996, the AIA/COTE and the DOE signed a Memorandum of Understanding to cooperate on research and development, the objective being to formulate a program consisting of a series of road maps for the construction and development of sustainable buildings during the twenty-first century.

The onslaught of green activity facilitated individual federal departments to also make significant headway. Thus, the Navy became emboldened and undertook eight pilot projects, including the Naval Facilities Engineering Command (NAVFAC) headquarters at the Washington Navy Yard. In 1997, the Navy also initiated development of an online resource, the *Whole Building Design Guide (WBDG)*, the main mission of which is to incorporate sustainability requirements into mainstream specifications and guidelines. A number of other federal agencies have now joined this project, which is now managed by the National Institute of Building Sciences (NIBS).

The green movement's emergence as a significant force was mainly a consequence of many forward-looking individuals and groups from all walks of life. As mentioned earlier, visionaries and innovative thinkers (e.g., Robert K. Watson—the father of LEED) have for decades recognized the challenges and need for serious changes in how we react to and treat our environment. The championing of green issues by forward-thinking politicians and celebrities played a pivotal role in addressing some of the environmental concerns that captivated the public's imagination during the early years of this century. Hollywood celebrities like Robert Redford were among the earlier true believers; he has been promoting solar energy since the 1970s. Redford has spent some 30 years on the board of the Natural Resources Defense Council (NRDC), which is described by *The New York Times* as "one of the nation's most powerful environmental groups." Redford also avidly lobbied Congress in support of environmental legislation and has energetically campaigned on behalf of local initiatives to address climate change and wilderness preservation. Other Hollywood environment-friendly celebrities who embraced green building and environmental causes include Brad Pitt, Daryl Hannah, Ed Begley Jr., Ed Norton, Cameron Diaz, and Leonardo DiCaprio. This has led to a wide array of stars following suit and who make it an avocation to champion their favorite environmental and green causes.

The green movement was further helped by "green politicians" from mayors to governors to heads of state in the United States and worldwide. An excellent example of this is former Vice President Al Gore whose release in May 2006 of his academy award-winning documentary film, *An Inconvenient Truth,* is credited with projecting global warming and climate change into the popular consciousness; it raised public awareness of many issues including that our quality of life is endangered, that our water is contaminated with toxic chemicals, and our natural resources are running out. Another eco-friendly politician is former California governor Arnold Schwarzenegger who made the state a global leader on climate change when he signed into law the historical milestone Global Warming Solutions Act of 2006; it commited the state to reduce its greenhouse-gas emissions by 80% below the 1990 levels by 2050. Other eco-friendly politicians include Ralph Nadar, former presidential candidate and a key leader of the U.S. Green Party; the left-wing mayor of London, Ken Livingstone; Angela Merkel, German chancellor and current leader of the G8, former environment minister, and an outspoken advocate for action against climate change; New Zealand's Prime Minister, Helen Clark; former European Union environment minister Margot Wallström (1999–2004); and Xie Zhenhua, China's vice minister of state development and reform and former environment minister.

When President-elect, Barack Obama was always an outspoken, vocal advocate for sustainability with regard to both the environment and the economic stimulus. He also frequently stresses the need to build a green economy to maintain America's competitive edge in the global labor market, while reducing our impact on the environment. For example, investments in a smart electric grid and energy-efficient homes, offices, and appliances will go a long way to reducing the overall energy consumption as a nation. This partially explains why, after taking office, President Obama put green building at the forefront of his sustainability agenda and proposed expanding federal grants that assist states and municipalities in building LEED-certified public buildings. Jerry Yudelson, a well-known green activist, believes that, "the impact of the Obama administration on green building is going to be to make it a permanent part of the economic, cultural, and financial landscape."

The President is making great strides toward changing our energy future. One of his first acts was endorsing the American Recovery and Reinvestment Act of 2009. This was to pump more than $825 billion into the U.S. economy via tax cuts, publicly funded investments in infrastructure, and work force development. This ambitious Recovery Act, which Obama constituted, is an unprecedented and historic investment in the clean energy economy, and is primarily designed to bolster clean technologies. President Obama believes that investments in clean energy today will lead to the industries of the future and will help put America back in the lead of the global clean energy economy in addition to creating millions of new green jobs.

A new report by from climate change consultants ICF International (ICFI) commissioned by Greenpeace, reports that the proposed "Green New Deal" environmental measures included in President Obama's $800 billion economic stimulus package is calculated to deliver minimum greenhouse-gas emissions savings of 61 million tonnes a year, which if correct is very significant as it is the equivalent of taking approximately 13 million cars off the road—and possibly more. Well-known green movement activist Lindsay McDuff says:

When politicians create or formulate policies, the business industries are consequently affected. With the rise in green policy, business executives from every arena are jumping on the green movement bandwagon, basically out of the growing market demand. Being green has become a selling advantage in the business world, and eager companies are starting to jump at the chance to get ahead.

According to an NPR report, the construction industry was expected to spend an estimated $10 billion dollars in 2006 on office buildings, apartment buildings, and smaller homes that are certified to be environmentally sound. The green movement today has become global and consists of miscellaneous individuals, activist groups, and diverse organizations seeking eco-friendly solutions to environmental concerns plaguing the planet.

GREEN BUILDING: AN OVERVIEW

Some scholars consider the *green building* movement to be primarily a reaction to the energy crises that came into being as a result of this, nurtured by efforts to make buildings more efficient and revamp the way energy, water, and materials are used. It should be noted that "green building" and "sustainable architecture" are relatively new terms in our vocabulary; they essentially represent a whole-systems approach incorporating a building's siting, design, construction, and operation in a manner that enhances the well-being of a building's occupants, while preserving the environment for future generations by conserving natural resources and safeguarding air and water quality. The core message, therefore, is essentially to improve conventional design and construction practices and standards so that the buildings we build today will last longer, be more efficient, cost less to operate, increase productivity, and contribute to healthier living and working environments for their occupants.

The arrival of green building concepts denotes a fundamental change in our approach to how we design and construct buildings. It is clear that the green building phenomenon has, during the last two decades, significantly impacted both the U.S. and global construction markets. Various environmental studies have consistently shown that buildings in the United States consume roughly one-third of all primary energy produced and nearly two-thirds of electricity produced. Research also shows that roughly 30% of all new and renovated U.S. buildings contain inferior indoor environmental quality as a result of an unacceptable level of noxious emissions, pathogens, and emittance of harmful substances found to exist in building materials. Continuing efforts are in place to address these environmental impacts, including the implementation of sustainability practices in construction project objectives.

The incorporation of sustainable practices into traditional design and construction procedures, however, is an approach that would require redefining and reassessing the current roles played by project participants in the design and construction process to help guarantee effective contribution to a sustainable project's objectives. One of the primary characteristic of a successful sustainable design is to apply a multidisciplinary and integrated "total" team approach that incorporates the various project members and stakeholders into the decision-making process, particularly during the early design phases. This holistic team approach helps ensure that the project will be a more productive, energy-efficient, and healthier building for both occupants and owner and will have less of a negative impact on the environment.

The 1990s saw the introduction of important new environmental rating systems for buildings. As international awareness of green issues increased, various international conferences were taking place such as the Green Building Challenge (GBC) held in Vancouver in October 1993; it was led by CANMET Energy Technology Centre of Natural Resources Canada. This event was a well-attended affair with representatives from 14 nations. The goal of such conferences is to create an international environmental rating system for buildings that takes into account regional and national environmental, economic, and social equity conditions.

The green building movement encouraged other parallel efforts to take shape. For example, in 1990 in the United Kingdom the Building Research Establishment introduced its own environmental building rating system, known as BREEAM®. In the United States in 1998, we witnessed the founding of the

U.S. Green Building Council, which developed the Leadership in Energy and Environmental Design Green Building Rating System. LEED has become the leading and most widely accepted green building rating system in the United States as evidenced by its dramatic growth during recent years. In 2004, the Green Building Initiative (GBI) introduced the Green Globes rating system into the United States (from Canada).

Over the past few years, we have seen a dramatic increase in the number of projects seeking LEED certification from the USGBC, which tends to confirm the significant inroads green building is making into the mainstream design and construction industry. While many builders were reluctant to participate in or encourage the green movement during its formative stages, this reluctance has rapidly diminished recently as more and more developers and clients jump on the green building bandwagon so that the construction industry too is now making serious efforts to embrace the initiative.

Since its inception, the LEED Rating System has grown to become an international forum and now encompasses more than 14,000 projects in the United States and some 40 countries, including Canada, India, China, the United Arab Emirates, Italy, and Israel. One of the Indian Green Building Council's declared objectives, for example, is to achieve 1 sq. ft. of green building for every Indian by 2012. The council's chairman, Prem C. Jain, recently stated that India already has an estimated 240 million sq. ft. of green buildings in place.

Although there has been enormous interest regarding green building issues, the amount of money allocated to research has been minimal at best, and presently constitutes a mere 0.2% of all federally funded research; this comes to roughly $193 million annually. This amount compares to a bare 0.02% of the estimated $1 trillion value of U.S. buildings constructed annually, while the building construction industry represents more than 10% of the U.S. GDP. The amount of funding allocated for green building is relatively minimal compared to funding for other research topics and this needs to be corrected. The federal government is one of several relevant funding sources that should be encouraged to provide appropriate financial support to research programs that have readily attainable strategies. We cannot progress toward achieving sustainability unless we can significantly improve green building practices. Failing to do so will have tragic consequences and generate an unduly negative impact on our ecosystem for years to come.

The enormous challenges we currently face about critical issues, such as global warming, water shortages, indoor environmental quality issues, and destruction of our ecosystem, are sobering. It has been clearly documented that conventionally constructed buildings contribute substantially to the environmental problems that are emerging in industrialized countries like the United States and China. For example, it has been estimated that current U.S. building operations account for about 38% of its carbon dioxide emissions and 71% of its electricity use. Likewise, the Environmental Information Administration (EIA) in 2008 estimated that building operations accounted for almost 40% of total energy use; the latter number increases to an estimated 48% if the energy required making building materials and constructing buildings are included.

It is further estimated that buildings annually consume about 13.6% of the country's potable water, and according to EPA estimates, waste from demolition, construction, and remodeling amounts to 136 million tons of landfill additions annually. In addition, construction and remodeling of buildings accounts for 3 billion tons or roughly 40% of raw material used globally each year. As the population in the United States continues to grow at its current pace from 306 million in 2009 to an estimated 370 million by 2030, the pressure and negative impact on the environment will also continue to increase unless we take urgent measures to appropriately adjust consumption patterns to take into account the limited natural resource available.

A typical example of the significant impact that green building research has had and is having can be seen by the impact of carbon emissions on global warming, which continues to receive national attention. This has resulted in several organizations (e.g., the AIA, ASHRAE, USGBC, and the Construction Specifications Institute) collectively adopting what has become known as the "2030 Challenge." This essentially consists of a series of goals and benchmarks for the architectural and engineering community to compare each building's design against the carbon footprint of similar buildings. The main goal is that all new construction will have net-zero carbon emissions by the year 2030, and that an equivalent amount of existing square footage will be renovated to use half of its previous energy use.

The 2030 Challenge applies the Commercial Buildings Energy Consumption Survey (CBECS) to benchmark energy use in kBtu per square foot; this allows a generalized correlation to the reduction of each building's carbon footprint. This goal will add a new dimension to the use of energy analysis as a tool to predict a building's carbon footprint, and this carbon footprint analysis will likely encourage increased measurement and verification to determine the status of each building upon completion. Should a building not perform according to design expectations, energy modeling and commissioning groups can diagnose prevailing operation issues in order to correct them.

The United Nations came out with a report in March 2007, "Buildings and Climate Change: Status, Challenges and Opportunities," that clearly reaffirms buildings' role in global warming. Achim Steiner, UN Under-Secretary General and UNEP Executive Director says:

Energy efficiency, along with cleaner and renewable forms of energy generation, is one of the pillars upon which a de-carbonized world will stand or fall. The savings that can be made right now are potentially huge and the costs to implement them relatively low if sufficient numbers of governments, industries, businesses, and consumers act.

He goes on to say:

This report focuses on the building sector. By some conservative estimates, the building sector worldwide could deliver emission reductions of 1.8 billion tonnes (1 tonne = 1000 *kilograms = 2025 pounds) of CO_2. A more aggressive energy-efficiency policy might deliver over 2 billion tonnes or close to three times the amount scheduled to be reduced under the Kyoto Protocol.*

It is no secret that to meet the 2030 Challenge, a dramatic change in our current approach and knowledge of building energy issues will be necessary.

The construction industry today is facing unprecedented and growing pressures originating from a global economic crisis, rising material costs, an increase in natural disasters, and the dramatic impact of the green consumer among other things. Together these trends have motivated the industry to increasingly reevaluate and revise its position by adopting sustainable design and construction methods in a serious effort to construct more efficient buildings designed to conserve energy and water, improve building operations, enhance the health and well-being of the general population, and minimize negative impacts on the environment.

The market share of green building will continue to develop and increase, partly due to growing public awareness, in addition to the unprecedented level of state and local government interest and initiatives such as the application of various incentive-based techniques to encourage green building practices. Regrettably, these efforts have not been totally successful because of certain obstacles and challenges encountered along the way, particularly the high cost of the new incentive programs and issues and stumbling blocks related to implementation and the lack of adequate resources. In an effort to assist communities in overcoming obstacles, the AIA commissioned a report, "Local Leaders in Sustainability—Green Incentives," which defines and explains these various programs, scrutinizes the main challenges that must be overcome to succeed, and highlights examples of best practices.

THE U.S. BUILT ENVIRONMENT

In 2008, Department of Commerce statistics estimated that the construction market accounted for 13.4% of the $13.2 trillion U.S. GDP, of which it projected that the value of green building construction would increase to $60 billion by 2010 ("McGraw-Hill Construction, 2008—Key Trends in the European and U.S. Construction Marketplace: SmartMarket Report"). It further forecasts that 82% of corporate America will be greening at least 16% of their real estate portfolios in 2009, and 18% of these corporations were expected to be greening more than 60% of their real estate portfolios. However, the volatile economic conditions of the current recession and the bad state of the U.S. economy have negatively impacted these forecasts. When the economy recovers, the green building industry is expected to witness significant growth in many sectors but particularly office, education, healthcare, government, hospitality, and industrial facilities. Of these, the three largest segments for nonresidential green building construction—office, education, and health care—could account for more than 80% of total green construction.

Green projects generally represent a diverse cross-section of the construction industry. For example, roofing companies are increasingly making the determination to focus on green technologies that allow their customers to harness the energy rooftop solutions can provide. Most of these companies are aware, however, that if they are to be successful in meeting the challenges that green technology presents, it will require reexamining how their company is to operate. This is in addition to making a serious commitment financially, in terms of manpower, green technologies, equipment, training, and education in green fundamentals.

The positive economic impact of new green technologies is also clearly evident in the plumbing industry; it is spurring economic growth for contractors around the country. Plumbing contractors have started to take an active role and create actions to take advantage of sustainable opportunities such as pushing for the installation of water and energy-efficient systems, and through the installation and use of green technologies, to promote energy efficiency and water conservation.

From the very beginning, it was the project designers and property owners/developers as stakeholders who have played a pivotal role in pursuing sustainable design and green construction practices, and it was them who became the driving force of the built environment concept. With both the source (designer/consultant) and the end user (owner) readily adopting sustainable design practices, and with the belief that green building will grow, it became obvious that the contractor/builder had to take on a modified role if green building projects were to be successfully executed. The end user had to become an active member of the project team along with the architect, mechanical/electrical/civil/structural engineer, landscape designer, and so on. Experienced builders have much to offer in terms of input on various aspects such as specifications, system performance, material selection, and minimizing construction waste. The contractor can also assist in the achieving of a green project's overall objectives by streamlining construction and applying value engineering methods, among other things.

As to be expected, much of the latest research conducted on the costs and benefits of green buildings come to the conclusion that energy and water savings on their own outweigh the initial cost premium in most green buildings, and the median increase that green buildings may incur (if at all) is less than 2% when compared with constructing conventional non-green buildings. This should dispel the myth and public perception that green buildings are much more expensive than conventional buildings. An international study that was published in 2008, "Greening Buildings and Communities: Costs and Benefits," concluded:

Most green buildings cost 0% to 4% more than conventional buildings, with the largest concentration of reported "green premiums" between 0% to 1%. Green premiums increase with the level of greenness but most LEED buildings, up through Gold level, can be built for the same cost as conventional buildings.

This report according to Henry Kelly, president of the Federation of American Scientists, "provides the first large-scale data resource on the cost and benefits of green buildings and sustainable community designs." Finally, Greg Kats, the preceding study's lead author and a managing director of Good Energies—one of the study's main supporters—says, "the deep downturn in real estate has not reduced the rapid growth in demand for and construction of green buildings," which "suggests a flight to quality as buyers express a market preference for buildings that are more energy efficient, more comfortable, and healthier." This is reaffirmed by the study, which determined that productivity and health benefits are a major motivating factor for building green.

Finally, a new national green building code has been approved. This new International Green Construction Code (IgCC), available from the International Code Council®, applies to all new and renovated commercial and residential buildings more than three stories high. It is a historic code that sets mandatory baseline standards for all aspects of building design and construction, including energy and water efficiency, site impacts, building waste, and materials. The new code differs from LEED in several ways. For example, the new code creates a mandatory "floor" that stipulates enforceable minimum standards that now must be reached on all aspects of building design and construction. LEED certification, on the other hand, is voluntary, and many building owners do not aspire to achieve it. Therefore, unlike LEED and Green Globes certifications, the new U.S. green codes will raise the standards for all buildings. It should be noted that California's building commission also adopted the first statewide green building code in January 2010. Although most environmental groups welcome the new standards, which mandate water use reductions and waste recycling in new buildings, there has been some criticism of its rating system.

Chapter 1

Green Concepts and Vocabulary

1.1 THE GREEN BUILDING MOVEMENT TODAY

The construction industry and the architectural/engineering professions have witnessed fundamental changes over recent years in the promotion of environmentally responsible buildings. Since the 1973 oil crisis, the green building movement has continued to gain momentum across all sectors of industry and "green" construction has become the norm on many new construction projects. Architects, designers, builders, and building owners are increasingly jumping on the green building bandwagon. National and local programs advancing green building principles are flourishing throughout the nation as well as globally.

Indeed, the green movement has penetrated most areas of our society, including the construction and home-building industries. Still, according to Achim Steiner, executive director of the United Nations Environment Programme (UNEP), "If targets for greenhouse gas (GHG) emissions reduction are to be met, decision-makers must unlock the potential of the building sector with much greater seriousness and vigor than they have to date and make mitigation of building-related emissions a cornerstone of every national climate change strategy." Steiner goes on to say:

Public policy is vital in triggering investment in energy efficient building stock, achieving energy and cost savings, reducing emissions, and creating millions of quality jobs. In developing countries where more than 50 percent of households (up to 80 percent in rural Africa) have no access to electricity, affordable, energy efficient, low-carbon housing helps address energy poverty.

Green construction remains in its relative infancy and is continuously developing. Moreover, although the practices and technologies used in green building construction continue to evolve and develop, and vary from region to region and from one country to the next, there remain certain fundamental principles that apply to all green projects: siting, structure design efficiency, energy efficiency, water efficiency, materials selection, indoor environmental quality (IEQ) operations and maintenance, and waste and toxics reduction. In today's world, national and global economic conditions, political pressure, and good environmental stewardship dictate that our built environment be sustainable.

At local and state levels, government is increasingly mandating that projects be built to green standards of construction, and this is driving our industry toward making sustainable projects for our clients and communities a priority.

With respect to building green and sustainability, architects and project teams should concentrate on designing and erecting buildings that are energy efficient, that use natural or reclaimed materials in their construction, and that are in tune with the environments in which they exist. Building green means being more efficient in the use of valuable resources such as energy, water, materials, and land than conventional building that simply adheres to code, which is why green buildings are more sympathetic to the environment and provide indoor spaces that occupants typically find to be healthier, more comfortable, and more productive. This is supported by a recent CoStar Group study finding that sustainable "green" buildings outperform their peer non-green assets in the key areas of occupancy, sale price, and rental rates, sometimes by wide margins.

Studies clearly show that buildings are primary contributors to environmental impacts—both during the Construction Phase and through their operation—which is why they have become a focus of green investment dollars. Studies also show that buildings are the world's prime consumers of natural resources, which is why today we see a flurry of architects, engineers, contractors, and builders reevaluating how residential and commercial buildings are being built. Additionally, we now see various incentive programs around the country and internationally to encourage and sometimes stipulate that developers and federal agencies go green. It should be noted, however, that while sustainable or green building is basically a strategy for creating healthier and more energy-efficient buildings—that is, environmentally optimal buildings—it has been found that buildings designed and operated with their life-cycle impacts taken into consideration provide significantly greater environmental, economic, and social benefits.

Moreover, the incorporation of green strategies and materials during the early Design Phase is the ideal approach to increase a project's potential market value. Sustainable buildings amass a vast array of practices and techniques to reduce and ultimately eliminate their negative impacts on the environment and on human health. For example, the EPA states that as many as 500 out of the 4100 or so commercial buildings that have earned the federal government's ENERGY STAR® rating use a full 50% less energy than average buildings. And many of those efficiency practices, such as upgrading light bulbs or office equipment, pay for themselves in energy cost savings.

Most green building programs typically focus on a number of environmentally related categories that emphasize taking advantage of renewable resources, such as natural daylight and sunlight, through active and passive solar as well as photovoltaic techniques and the innovative use of plants to produce green roofs and reduce rainwater runoff. But, as mentioned earlier, sustainability is best achieved when an integrated team approach is used in the building design and construction process. In fact, in today's high-tech world an integrated team approach to green building has become pivotal to a project's success; this means that all aspects of a project, from site selection to the structure, to interior finishes, are carefully considered from the outset.

Architects and property developers have come to realize that focusing on only one aspect of a building can have a severe negative impact on the project as a whole. For example, the design and construction of an inefficient building envelope can adversely affect indoor environmental quality in addition to increasing energy costs, whereas a proper sustainable envelope can help lower operating costs over the life of a building by increasing productivity and utilizing less energy and water. As mentioned earlier, sustainable developments can also provide tenants and occupants with a healthier and more productive working environment as a result of improved indoor air quality. This means that exposure to materials such as asbestos, lead, and formaldehydes, which may contain high volatile organic compound (VOC) emissions, are less likely in a green building and so potential health problems such as "sick building syndrome" (SBS) are avoided.

The main objective of most designers who engage in green building is to achieve both ecological and aesthetic harmony between a structure and its surrounding environment. Helen Brown, former board director of the U.S. Green Building Council (USGBC®) and a fellow of the Post Carbon Institute, echoes the sentiment of many green proponents:

Viewed through a green building lens, conventionally built buildings are rather poor performers. They generate enormous material and water waste as well as indoor and outdoor air pollution. As large containers and collection points of human activity, buildings are especially prodigious consumers of energy. They depend on both electricity and on-site fossil fuel use to support myriad transactions: transporting and exchanging water, air, heat, material, people, and information.

Brown also believes that the green building movement, which is now in its second decade, reduces (and eventually eliminates) the negative impacts buildings have on local and global ecosystems.

According to Rob Watson, author of the "Green Building Impact Report" issued in November 2008:

The construction and operation of buildings require more energy than any other human activity. The International Energy Agency (IEA) estimated in 2006 that buildings used 40 percent of primary energy consumed globally, accounting for roughly a quarter of the world's greenhouse gas emissions [Figure 1.1]. Commercial buildings comprise one-third of this total. Urbanization trends in developing countries are accelerating the growth of the commercial building sector relative to residential buildings, according to the World Business Council on Sustainable Development (WBCSD).

Additionally, it is estimated that buildings account for about 71% of all electricity consumed in America and 40% of global carbon dioxide emissions.

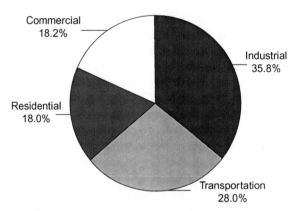

FIGURE 1.1   U.S. total greenhouse gas emissions in 2005. The Energy Information Administration (EIA) typically breaks down U.S. energy consumption into four end-use categories: industry, transportation, residential, and commercial. Almost all residential greenhouse emissions are CO_2, and are strongly related to energy consumption. *(Source: Adapted from Paul Emrath and Helen Fei Liu, the National Association of Home–Builders, Special Studies, Residential Greenhouse Gas Emissions, April 30, 2007.)*

The impact of building on the U.S. economy is clearly evident from the use of construction materials. For example, it is estimated that infrastructure supplies, building construction, and road building make up about 60% of the total flow of materials (excluding fuel) through the U.S. economy. Likewise, studies show that building construction and demolition waste accounts for roughly 60% of all nonindustrial waste. As for water usage, it is estimated that building occupants consume about 50 billion gallons per day (i.e., more than 12% of U.S. potable water consumption). This amount of water consumption is mainly to support municipal, agricultural, and industrial activities, which have more than tripled since 1950. Construction also impacts the indoor levels of air pollutants and VOCs in buildings, which can be two to five times higher than outdoor levels.

For all of these reasons, sustainable/green building strategies and best practices present a unique opportunity to create environmentally sound and resource-efficient buildings. By applying an integrated design approach from the beginning, this can be achieved especially by having the stakeholders—architects, engineers, land planners, and building owners and operators, as well as members of the construction industry—work together as a team to design a project. Indeed, architects and urban engineers around the world are building cities designed to cope with a future of growing populations, increasingly scarce resources, and the need to reduce carbon emissions. We see examples of future cities debuting in Great Britain, China, and the United Arab Emirates.

At the forefront of the U.S. green building offensive is the federal government, the nation's largest landlord. The General Services Administration

(GSA) was one of the first adopters of LEED for New Construction (LEED-NC)—to be discussed momentarily—and is committed to incorporating principles of sustainable design and energy efficiency into all of its building projects. It is the GSA's intent to integrate sustainable design as effortlessly as possible into existing design and construction processes. In this regard it recently announced that it will apply more stringent green building standards to its $12 billion construction portfolio, which presently includes more than 361 million square feet of space in 9600 federally owned and leased facilities occupied by more than 1.2 million federal employees. This portfolio consists of post offices, courthouses, border stations, and other buildings.

The GSA decided to use the USGBC's Leadership in Energy and Environmental Design (LEED) Green Building Rating System as its tool for evaluating and measuring achievements in its sustainable design programs. In keeping with the spirit of sustainability, the GSA recently increased its minimum standard requirement for new construction and substantial renovation of federally owned facilities by adopting the LEED Gold standard, which is the second highest level of certification (just below Platinum). Until recently, the GSA had only required a LEED Silver certification. In justifying this move, Robert Peck, former GSA commissioner of public buildings, stated, "Sustainable, better-performing federal buildings can significantly contribute to reducing the government's environmental footprint" and "this new requirement is just one of the many ways we're greening the federal real estate inventory to help deliver on President Obama's commitment to increase sustainability and energy efficiency across government."

CEO and founding chairman of the USGBC, Richard Fedrizzi, echoed the federal government's lead in adopting green building practices when he said: "The Federal government has been at the forefront of the sustainable building movement since its inception, providing resources, pioneering best practices and engaging multiple Federal agencies in the mission of transforming the built environment." A first ever White House Summit on Federal Sustainable Buildings held in January 2006 attracted over 150 federal facility managers and decision makers in addition to 21 government agencies, coming together to formulate and witness the signing of the "Federal Leadership in High Performance and Sustainable Buildings Memorandum of Understanding (MOU)." Signatories to this MOU committed to federal leadership in the design, construction, and operation of high-performance and sustainable buildings. The MOU highlights the sense of urgency felt by green building proponents and represents a significant accomplishment by the federal government through its collective effort to define common strategies and guiding principles. The signatory agencies now need to work with others in the private and public sectors to consolidate these goals.

The initiatives just described are clear indications that the gap between green and conventional construction is narrowing, and they signal that green construction has come of age, especially when we learn that there are more than 80 green building programs operating in the United States alone, and even more in numerous other countries such as Canada, Japan, China, India, Australia, and the United Kingdom. A measure of the growth of green building programs and their success is reflected by the number of cities that have established or adopted them. For example, the American Institute of Architects reports that by 2008 92 cities with populations greater than 50,000 had established green building programs—up from 22 in 2004—which represents a 318% increase.

Many programs in the United States are city, county, or state operated; there are also three that are national in scope: USGBC's LEED program, the Green Globes program of the Green Building Institute (GBI)—designed by the U.K. Building Research Establishment—and the National Green Building Standard of the National Association of Home Builders (NAHB). In the United States, LEED is the most widely recognized, but all of these programs were developed and operate outside of government.

The USGBC recently announced that, as of November 2010, the footprint of LEED-certified commercial space in the United States had surpassed 1 billion square feet. This is in addition to 6 billion square feet of projects around the world that are registered and working toward certification. It should be noted that, while not all projects that register with LEED achieve certification, this milestone remains significant. "The impact of these one billion square feet resonates around the world," according to Peter Templeton, president of the Green Building Certification Institute, which certifies LEED projects.

Chicago's recent hosting, for the second time, of the USGBC annual Greenbuild International Conference and Expo proved to be a great success in uniting many people from different countries, different backgrounds, and different professions around a single common cause: building a better, healthier, more sustainable world. Following the Chicago event, Rick Fedrizzi said,

For years, we've asked ourselves: Can we build it taller? Can we build it faster? Can we build it cheaper? At the USGBC, we ask a different question: Can we build it better? Can we build in ways that are more sustainable, more energy efficient and that provide clean air and good lighting? In ways that can create jobs, restore our economy, and build healthier, more livable communities?

1.2 GREEN BASICS—WHAT MAKES A BUILDING GREEN?

The term "green building," or "sustainable building," is relatively new to our language, and a precise definition is not easy. Green/sustainable building is also known as "high-performance building." The California Department of Resources Recycling and Recovery (CalRecycle), for example, defines it this way:

A structure that is designed, built, renovated, operated, or reused in an ecological and resource-efficient manner. Green buildings are designed to meet certain objectives such as protecting occupant health; improving employee productivity; using energy, water, and other resources more efficiently; and reducing the overall impact to the environment.

The EPA defines it as, "The practice of creating structures and using processes that are environmentally responsible and resource-efficient throughout a building's life cycle from siting to design, construction, operation, maintenance, renovation and deconstruction." This practice expands and complements the classical building design concerns of economy, utility, durability, and comfort. Still another definition of sustainable development was offered at the Gothenburg European Council meeting of June 2001: "a means of meeting the needs of the present generation without compromising those of the future."

However one wishes to define the term, green building or sustainable development has had a profound impact on the U.S. and global construction market over the last two decades, although it may be some years before we can ascertain its full impact on the building construction industry and its suppliers. Since there is no uniform definition of green, it is essential that every "green" term be specifically defined and that agreed-to objective standards of performance be established in a contract. For example, explicit energy efficiency requirements must be set forth in a carefully drafted and technically correct and verifiable standard.

While the United States remains the undisputed global leader in the construction of green buildings, we are witnessing a sharp increase around the world in investment in sustainability and green building practices. In this respect, the European Union (EU) agreed on a new sustainable development strategy in June 2006 that has the potential to determine how the EU economy evolves in the coming decades. In addition to the USGBC's LEED Rating System, there are many other green building assessment systems currently in use in many countries. Examples are the U.K. Building Research Establishment's Environmental Assessment Method (BREEAM), the Comprehensive Assessment System for Building Environmental Efficiency (CASBEE), the U.S. Green Globes®, the Qatar Sustainability Assessment System (QSAS), and the Green Building Council of Australia (GBCA) Green Star Rating Tool (Mago and Syal 2007).

As previously outlined, green building strategies relate primarily to land use, building design, construction, and operation, which together help minimize or mitigate a building's overall impact on the environment. The chief objectives of green buildings are therefore to increase the efficiency with which buildings utilize available natural resources such as energy, water, and materials, and simultaneously to minimize a building's adverse impact on human health and the environment. There are numerous strategies and approaches that can be used in green construction of a new building designed for long-term operation and maintenance savings. Moreover, the United States has a vast reservoir of existing buildings that can be made greener and more efficient; studies indicate that many property owners have shown considerable interest in exploring that possibility.

First of all, however, it is important to dispel the myths and misinformation that surround sustainability and green design. Only then will a number of pertinent strategies become apparent that will help achieve the goals and desires of building green. As discussed in the Introduction and echoed by Leah B. Garris, senior associate editor at *Buildings* magazine: "Myth and misinformation surround the topic of sustainability, clouding its definition and purpose, and blurring the lines between green fact and fiction." Remarking on aspects of aesthetics, the well-known green building proponent Alan Scott, principal of Green Building Services in Portland, Oregon, says, "You can have a green building that doesn't really 'look' any different than any other building."

Ralph DiNola, also a principal with Green Building Services, affirms this statement, believing that a level of sustainability can easily be achieved by designing a green building that looks "normal": "People don't really talk about the value of aesthetics in terms of the longevity of a building. A beautiful building will be preserved by a culture for a greater length of time than an ugly building." Thus a building's potential longevity is one of sustainability's principle characteristics, and aesthetics is a pivotal factor in achieving it.

Sustainability is really about understanding nature and working in harmony with it, not against it. It is not about building structures that purport to be environmentally responsible but that in reality sacrifice tenant/occupant comfort. This is not to suggest that purchasing green products or recycling assets at the end of their useful lives is not sustainable—it is. It is also appropriate for the environment and for the health of a building's occupants. However, before making a final determination, a developer or building owner should first take the time to research the various options that are most appropriate for the project and that offer the best possible return on investment. Failing to take the time to research the various sustainability options may lead to making incorrect decisions.

Many green professionals believe that sustainability starts with a thorough understanding of climate and that the primary reason green strategies are considered green is that they work in harmony with, not against, surrounding climatic and geographic conditions. This necessitates a true understanding of the environment in which a project is being designed in order to fully apply these conditions to a project's advantage. Most architects and designers who specialize in green building are fully aware of the need for familiarity with year-round weather conditions, such as temperature, rainfall, humidity, site topography, prevailing winds, indigenous plants, and so forth, to succeed in sustainable design. Although climate impacts sustainability in various ways, partly depending on a project's location, a measure of success in achieving sustainability can be made by comparing a project's performance to a baseline condition that relates to the microclimate and environmental conditions of where it is located.

To successfully achieve sustainability, it is also necessary to identify and minimize a building's need for resources that are in short supply or locally unavailable and to encourage the use of readily available resources such as sun, rain, and wind. A thorough understanding of the microclimate where the project is located is imperative because it reflects an understanding of what is and is not readily available, such as sun for heating and lighting, wind for ventilation, and rainwater for irrigation and other water requirements. CalRecycle, for example, cites the main elements of green buildings and sustainability as the following.

Siting. This includes selecting a suitable site that takes advantage of mass transit availability, and protecting and retaining existing landscape and natural features. Plants should be selected that have low water and pesticide needs and that generate minimum plant trimmings.

Water efficiency. This can be achieved by applying certain water efficiency strategies that, according to CalRecycle, include designing "for dual plumbing to use recycled water for toilet flushing or a gray water system that recovers rainwater or other nonpotable water for site irrigation" and "Minimize wastewater by using ultra low-flush toilets, low-flow showerheads, and other water conserving fixtures." In addition, CalRecycle suggests recirculating systems for centralized hot water distribution and the installation of point-of-use water-heating systems for more distant locations. The landscape should be metered separately from the buildings, and micro-irrigation should be used to supply water in nonturf areas. Whenever possible, state-of-the-art irrigation controllers and self-closing hose nozzles should be used.

Energy efficiency. To achieve optimum energy performance and energy efficiency, a number of passive strategies should be employed such as utilizing a building's size, shape, and orientation; passive solar design; and natural lighting. Alternative sources of energy should be considered such as photovoltaics and fuel cells, which are now widely used and readily available. Renewable energy sources are a sign of emerging technologies. Computer modeling has also become part of the mainstream and is a helpful tool in energy calculations, and optimizing the design of electrical and mechanical systems and the building envelope. (These are discussed in greater detail in later chapters.)

Materials efficiency and resource conservation. Selection of construction materials and products should be based on key characteristics such as reused and recycled content, zero or low off-gassing of harmful air emissions, zero or low toxicity, sustainably harvested materials, high recyclability, durability, longevity, and local production. Likewise, the incorporation of dimensional planning and other material efficiency strategies increases sustainability as well as the reuse of recycled construction and demolition materials.

Environmental air quality. Studies show that buildings with good overall indoor air quality can reduce the rate of respiratory disease, allergy, asthma, and sick building symptoms, and increase worker productivity. In addition to adequate ventilation, construction materials and interior finish products should be chosen with zero or low emissions to improve indoor air quality. Many building materials and cleaning/maintenance products emit toxic gases, such as VOCs and formaldehyde. These gases can have a harmful impact on occupants' health and productivity.

Building operation and maintenance. Commissioning of green buildings on completion ensures that they perform according to the design goals that were intended. Commissioning includes testing and adjusting the mechanical, electrical, and plumbing systems to certify that all equipment meets design criteria. It also requires staff instruction on the operation and maintenance of equipment. Proper maintenance allows a building to continue to perform at optimum levels, as designed and commissioned.

Both water conservation and energy efficiency rely heavily on climate, whereas indoor environment quality and material and resource conservation are largely independent of it. And although site sustainability depends on climate to some degree, and more specifically on the specifications and micro-elements that are particular to a site, it is important to note that different regions or locations may encounter different climates—hot, arid, humid, freezing, and windy. Therefore, understanding a region's climate and readily available resources can help avoid the use of inappropriate techniques on a project that may have an adverse impact and invariably increase the project's costs and therefore its viability.

1.3   GOING GREEN: INCENTIVES, BARRIERS, AND BENEFITS

Since the oil crises of the 1970s, but particularly over the last two decades, architects, designers, builders, and building owners have increasingly taken an interest in green building. The green building movement is flourishing throughout the nation as well as globally mainly because of increasing demand (as a result of public awareness of the benefits of green building) and because of the many national and local programs offering various incentives. Thousands of projects have been constructed over recent years that provide tangible evidence of what green building can accomplish in terms of resource efficiency, improved comfort levels, aesthetics, and energy efficiency.

Some of the primary benefits of building green, which are not always easily quantifiable and therefore not typically adequately considered in cost analysis, include

- Reduced energy consumption
- Reduced pollution
- Protection of ecosystems
- Improved occupant health and comfort
- Increased productivity
- Reduced landfill waste

The Dutch economist Nils Kok has published what is reportedly the most comprehensive statistical analysis to date on the relative value of green and conventional buildings. His September 2010 study concludes that U.S. buildings certified under LEED or the ENERGY STAR program charge 3% higher rent, have greater occupancy rates, and sell for 13% more than comparable properties. According to Kok, "Labeled buildings have effective rents (rent multiplied by occupancy rate) that are almost 8% higher than those of otherwise identical nearby non-rated buildings."

As for residential buildings, McGraw-Hill Construction's 2008 *SmartMarket Report* issue, "The Green Home Consumer," says that 70% of homebuyers are more or much more inclined to buy a green home over a conventional home in a depressed housing market. That number is 78% for those earning less than $50,000 a year; moreover, the report shows that 56% of respondents who bought green homes in 2008 earned less than $75,000 per year and 29% earned less than $50,000.

It is also interesting to note that studies have also shown that buildings' operating costs often represent only 10% or less of an organization's cost structure whereas personnel usually constitute the remaining 90%. This lends strong credence to the view that even minor improvements in worker comfort can result in substantial dividends in performance and productivity. Likewise, there is substantial evidence linking high-performance buildings with improved working conditions, which in turn typically lead to reduced employee turnover and absenteeism, increased productivity, improved health, and other benefits. This in turn has become a major contributing factor to the growth of building efficiency, particularly with respect to a building's occupants and tenants.

Even as the global economic recession continues to dominate year-end headlines, we see a cascade of newly released studies and reports that point to green building as one of the growing bright spots for the U.S. economy. With regard to existing buildings, more than 80% of commercial building owners have allocated funds to green initiatives according to "2008 Green Survey: Existing Buildings," a survey jointly funded by Incisive Media's Real Estate Forum and GlobeSt.com, the Building Owners and Managers Association (BOMA) International, and the USGBC, released at Greenbuild 2008. The survey concluded that nearly 70% of commercial building owners have already implemented some form of energy-monitoring system, and it confirmed that energy conservation is the most widely employed green program in commercial buildings, followed by recycling and water conservation. In addition, according to the survey, 45% of respondents planned to increase sustainability investments in 2009 and 60% of commercial building owners offered education programs to assist tenants in implementing green programs in their space, a number up 49.4% from the previous year. Pike Research has predicted that comprehensive efficiency retrofits will likely more than triple in annual revenue to $6.6 billion by the year 2013.

A study recently conducted by Henley University of Reading in the United Kingdom concluded that commercial building owners can reap higher rental premiums, of about 6%, for green buildings if the buildings enjoy LEED or ENERGY STAR certification. It also concluded that the more highly rated a building is, the higher the rental premium, and suggested a sales price premium of about 35% based on 127 price observations of LEED-rated buildings and a

31% premium based on 662 price observations of ENERGY STAR-rated buildings. Andrew Florance, president and CEO of CoStar, echoes these findings: "Green buildings are clearly achieving higher rents and higher occupancy, they have lower operating costs, and they're achieving higher sale prices."

In Turner Construction's "Green Building Barometer" survey, 84% of respondents report that their green buildings have resulted in lower energy costs and 68% report lower overall operating costs. These figures are perhaps not as high as one might expect from truly sustainable buildings. However, nearly 65% of those who have built green buildings claim that their investments have already produced a positive return on investment.

The constricted supply of green buildings, which still accounts for only a small percentage of the total U.S. building stock, appears to be one of the factors contributing to "green" premiums, particularly since the number of green-certified buildings continues to grow even as the supply fails to keep pace with demand. Most developers and property owners generally agree that among the more tangible benefits of attaining a green certification for a building (e.g., LEED, Green Globes, ENERGY STAR) is the use of this accomplishment as a marketing tool, and designers and contractors who have certified buildings in their portfolios typically find that they have a greater competitive marketing edge. Tenants and employees continue to show a clear preference for living and working in certifiably green buildings, resulting in a greater demand and in a greater capability to attract quality tenants and thus higher rents.

1.3.1   Tax Deductions and Incentives

In addition to the health and environmental benefits of living and working in a green building, there are various tax incentives for homeowners and businesses that purchase and install energy-efficient equipment or make energy efficiency improvements to existing structures. The federal government directly participates in cost-shared research by offering tax incentives to encourage consumers and businesses to develop and adopt energy-efficient technologies and products. Today many local and state governments, utility companies, and other entities nationwide are offering rebates, tax breaks, and other incentives to encourage the incorporation of eco-friendly elements in proposed building projects. In fact, the majority of large cities in the United States now do so. Recent estimates show that more than 65 local governments have already made a commitment to LEED standards in building construction, with some reducing the entitlement process by up to a year in addition to offering tax credits. Energy costs have become a major office building expense, although this can be reduced by as much as 30% and even more with the development of new technologies.

The American Recovery and Reinvestment Act (ARRA) of 2009 extended many energy efficiency and renewable energy tax incentives originally introduced in the Energy Policy Act (EPAct) of 2005, and ARRA extended similar provisions in the Emergency Economic Stabilization Act of 2008 (P.L. 110-343). It should be noted, however, that a tax credit is generally more useful than an equivalent tax deduction because the former reduces tax dollar for dollar whereas the latter only removes a certain percentage of the tax that is owed. Consumers should itemize all purchases on their federal income tax form in order to lower the total amount of tax they owe.

Tax Incentives Available for Commercial Buildings

Some of the tax incentives available under ARRA include (from the DOE website) those described in the following subsections.

Deduction of the Cost of Energy-Efficient Property Installed in Commercial Buildings

A tax deduction of up to $1.80 per square foot is available for buildings that save at least 50% of the heating and cooling energy of a building that meets American Society of Heating, Refrigeration and Air-Conditioning Engineers (ASHRAE) Standard 90.1-2001. Partial deductions of up to $0.60 per square foot can be taken for measures affecting the building envelope, lighting, or heating and cooling systems. This deduction has been extended through December 31, 2013.

Buildings must be within the scope of the ASHRAE standard, including addenda 90.1a-2003, 90.1b-2002, 90.1c-2002, 90.1d-2002, and 90.1 k-2002 (in effect as of April 2, 2003), and within the control of the building designer. Retrofit of existing buildings is also eligible for the tax deduction.

Extension of the Energy Investment Tax Credits

The 30% investment tax credits (ITC) for solar energy and qualified fuel cell properties are extended to January 1, 2017, and now also apply to qualified small wind energy properties. The cap for qualified fuel cells increases to $1500 per half kilowatt of capacity, and a new 10% ITC is available for combined heat and power systems and geothermal heat pumps.

Accelerated Depreciation for Smart Meters and Smart Grid Systems

Currently taxpayers generally recover the cost of smart electric meters and smart electric grid equipment over a 20-year period. The ARRA legislation allows them to recover the cost of this property over a 10-year period, unless it already qualifies for a shorter recovery schedule. The DOE website offers information about tax deductions available for relating the purchase and installation of energy-efficient products and the construction of new energy-efficient homes. As mentioned, the American Recovery and Reinvestment Act of 2009 offers tax credits for residential energy efficiency measures and renewable energy systems. Many of these credits were introduced in EPAct 2005 and amended in the Emergency Economic Stabilization Act of 2008.

Tax Incentives Available for Residential Buildings

Some incentives available under the ARRA legislation (from the DOE website) are described in the following subsections.

Energy Efficiency Tax Credits for Existing Homes

Homeowners are eligible for a tax credit of 30% of the cost of improvements to windows, roofing, insulation, and heating and cooling equipment. These improvements must be in service from January 1, 2009 through December 31, 2010 (i.e., now expired), and there is a limit of $1500 for all products. Improvements made in 2008 are not eligible for a tax credit. See the ENERGY STAR website (www.energystar.gov/index.cfm?c=tax_credits.tx_index) for latest updates of eligible improvements.

Renewable Energy Tax Credits for Existing or New Homes

Homeowners can receive a tax credit of 30% of the cost of the following renewable energy technologies with no upper limit: geothermal heat pumps, photovoltaic systems, solar water heaters, and small wind energy systems. Fuel cells are also eligible for a tax credit with a cap. These technologies must be placed in service by December 31, 2016. Again, see the ENERGY STAR website for detailed information.

The Internal Revenue Service offers information on tax incentives. One example is the Tax Incentives Assistance Project (TIAP), which offers a flyer with more information about these tax credits. TIAP is sponsored by a coalition of public interest nonprofit groups, government agencies, and other organizations in the energy efficiency field. The TIAP program is essentially designed to give consumers and businesses the information they need to make use of the federal income tax incentives for energy-efficient products and technologies. It was passed by Congress as part of the Energy Policy Act of 2005 and subsequently amended on a number of occasions. It is important for readers to visit the relevant websites for the latest updates because incentive programs change from time to time. Moreover, it is always wise to consult a tax professional on questions for specific situations. For example, credits received by builders for energy-efficient homes that are substantially completed after August 8, 2005 and purchased for use as a residence from January 1, 2006 through December 31, 2009 have now expired.

In California there is increasing evidence that the state's homebuyer tax credit, enacted at the beginning of 2009, is helping to generate new-home sales and, in turn, job-creating home construction. Various links to funding sources for green building that are available to homeowners, industry, government organizations, and nonprofits in the form of grants, tax credits, loans, and other sources can readily be found on the EPA's website (www.epa.gov).

Also, the Database of State Incentives for Renewables & Efficiency (DSIRE), a nonprofit project funded by the DOE through the North Carolina

Solar Center and the Interstate Renewable Energy Council, contains much information on its website (www.dsireusa.org/) regarding local, state, federal, and utility incentives for switching to renewable or efficient energy use. Other ways to obtain federal tax credits include the use of energy-efficient products such as those proposed by the ENERGY STAR program. The DOE likewise provides a list of qualified software programs that commercial building owners can use to calculate energy and power cost savings meeting federal tax incentive requirements.

1.3.2 Green Building Programs

Numerous cities throughout the United States are now promoting the use of various external green building programs. One excellent example of this is Seattle, which is one of the top cities in the nation for LEED facilities and is the largest single owner of LEED facilities in the world. This achievement was spurred by the city's adoption, in 2000, of its Sustainable Building Policy, which requires all new city-funded projects and renovations that contain in excess of 5000 square feet of occupied space to achieve a LEED Silver rating. The policy affects all city departments involved with construction, including the Department of Planning and Development (DPD), which monitors the policy's implementation.

According to the DPD website: "The City of Seattle currently has 32 projects—either completed, under construction, or planned—that are targeted for LEED certification. These represent capital improvement projects within six departments. So far the City has completed 20 projects; of these, 17 are certified, with many pending certification." Seattle currently promotes a number of green building programs:

- *Built Green*™*:* An environmentally friendly, nonprofit, residential building program of the Master Builders Association of King and Snohomish Counties, developed in partnership with King County, Snohomish County, and local environmental groups in Washington State. The Built Green programs are New Home Building, Remodeling, Multifamily Development, and Communities.
- *ENERGY STAR Homes:* A program for new homes that was created by the EPA and the Department of Energy. In Seattle 5995 ENERGY STAR-qualified homes have been built to date; in Washington State, that number is 14,673.
- *LEED for Homes:* A residential rating system recently created by the USGBC, which describes it as "a consensus-developed, third-party–verified, voluntary rating system which promotes the design and construction of high-performance green homes." It should be noted that the city of Cincinnati currently offers a sizable tax incentive for new and renovated homes that are certified under the LEED green building rating system. The local Cincinnati chapter of the National Association of Home Builders is currently

seeking adoption of its residential green building rating system, known as the National Green Building Standard™ (NGBS), for the same tax incentive.

- *Multifamily:* A green program for apartments, townhomes, and condominiums offering incentives for building efficiency and renewable energy (see www.seattle.gov/dpd/greenbuilding for more details).

In addition to Seattle, there are a large number of U.S. cities that promote green building programs. These include, among others, Phoenix, AZ; San José, CA; Minneapolis, MN; Portland, OR; and Pittsburgh, PA.

1.3.3 Defining Sustainable Communities

Interest in sustainability and sustainable communities arose out of a desire to increase quality of life and provide opportunities that economic development can bring, but in a manner that preserves the environment for present and future generations. However, the concept of sustainable communities remains somewhat elusive, perhaps even complicated, and a precise definition may vary from source to source. Community planners around the country have started to formulate a perception, or vision, of how such a community would grow to embrace the sustainability of its citizens' core values, which include community, social equity, economic prosperity, environmental stewardship, security, and opportunity.

Numerous cities, including Seattle, have already started to adopt comprehensive plans that include goals and policies designed to help guide development toward a more sustainable and environmentally friendly future. This new forward-looking "green urbanism" seeks to apply leading-edge tools, models, strategies, and technologies to help cities achieve eco-friendly sustainability goals and policies. The application of an integrated, whole-systems design approach to the planning of communities or neighborhoods puts a city in a stronger position to achieve increased environmental protection. Among the compelling inducements for building owners and property developers to invest in green buildings are the LEED certification program, which offers the financial benefits of operating a more efficient and less expensive facility. Adherence to LEED guidelines goes a long way toward ensuring that facilities are designed, constructed, and operated more effectively, mainly because LEED encourages project teams to concentrate on operating life-cycle costs rather than on initial construction costs.

As previously mentioned, many states are now offering various incentives in the form of tax benefits for green building and LEED compliance. An excellent example of this is New York, where Governor George Pataki, in May 2001, signed into law the nation's first Green Building Tax Credit (GBTC) program. This is a $25 million income tax credit created to promote the funding of concepts and ideas that encourage green building practices, particularly for owners and tenants of buildings that meet specific criteria for energy, indoor air quality, water conservation, materials, commissioning, appliances, and size as set out in the regulations. The program is adminstered by the state's Department of Environmental Conservation.

GBTC and other programs led to the building of the first high-rise green office building as well as the first high-rise green residential building in the United States. Governor Pataki also established New York's leading brownfield program encouraging increased development in cities across the state by creating a $200 million fund to support the redevelopment of contaminated sites and instituting a $135 million tax credit program to stimulate public/private brownfields investment.

It is surprising that until recently no single organization had the vision and foresight to move toward bringing green construction to the American home market. Any residential green programs that did previously exist were often sponsored by local homebuilder associations (HBAs), nonprofit organizations, or municipalities. Because this situation was unacceptable and unlikely to last, the National Association of Homebuilders and its research center (NAHB RC) took preemptive action and produced its Model Green Home Building Guidelines and various other utility programs. However, while these programs may have provided many of the answers to the nation's residential building market, with respect to commercial construction LEED was until recently the only viable certification program available.

This has changed, and many states, such as Oregon, now stipulate that "the building must meet an established standard set by the U.S. Green Building Council's LEED *or be rated by a comparable program approved by the Oregon Department of Energy.*" Likewise, in early 2007 New York City adopted broad sustainable rules for school construction. The School Construction Authority (SCA) created a new Green Schools Rating System, giving it a robust LEED equivalent standard and new green guidelines that represent a sweeping redefinition of the rules and immediately raised the bar for future construction projects even higher than required by the LEED law.

But LEED's competition on a national scale in the United States has come from Canada in the form of Green Globes. This system provides a green management tool that includes an assessment protocol, a rating system, and a guide for integrating environmentally friendly design in commercial buildings. Green Globes, which hopes to offer the U.S. commercial construction industry a simpler, less expensive method for assessing and rating a building's environmental performance, is a web-based auditing tool developed by a Toronto environmental consulting firm, Energy and Environment Canada. The greatest strength of it being web-based is purported to be its rapid and economical method for assessing and rating the environmental performance of new and existing buildings. Rights to market the program in the United States were purchased by the GBI, which has budgeted more than $800,000 as a first step in promoting national awareness of Green Globes as a viable alternative to LEED throughout the construction and development community, and in capturing a significant percentage of LEED's market share. Green Globes is discussed in greater detail in Chapter 2.

The building codes of California and New York are among the nation's leaders in sustainable development. Upon taking office in 2003, former California governor Arnold Schwarzenegger made it a priority to develop a self-sustaining solar industry for his state. The solar initiatives he introduced, including the Million Solar Roofs Initiative and $2.9 billion in incentives to homeowners and building owners who install solar electric systems, were pivotal in motivating and creating a solar industry in California, which has become the nation's largest solar market. On December 14, 2004, Schwarzenegger signed Executive Order S-20-04, which requires the design, construction, and operation of all new and renovated state-owned facilities to be LEED Silver-certified.

New York City's Local Law 86 (also known as "The LEED Law") took effect in January 2007. It basically requires that many of the city's new municipal buildings, as well as additions and renovations to its existing buildings, achieve standards of sustainability that meet various LEED criteria. Mayor Michael Bloomberg also announced his Greener, Greater Buildings Plan in 2009, in which he set a target of 30% reduction of greenhouse gas emissions by 2017. In 2007 New Mexico passed its own major green building tax credit, and Oregon followed with a 35% tax credit for the employment of solar energy systems. With these tax incentives, green building tenant attraction and retention continue to grow and become stronger, thereby making green building a sound investment.

1.3.4   Potential Risks of Building Green

On "navigating the potentially litigious waters of a new and expanding industry," Judah Lifschitz, co-president of the legal firm Shapiro, Lifschitz & Schram, says:

The potential liability associated with taking on a green project without proper preparation is huge. Potential causes of legal action include claims for misrepresentation, fraud, negligence, negligence per se and breach of contract. Some factors that will likely contribute to an increase in green-building litigation include:

1. *The volume of inexperienced parties attempting to build green*
2. *A lack of understanding and defining the term "green"*
3. *A lack of understanding of the Leadership in Energy and Environmental Design (LEED®) certification requirements*
4. *Unintentionally guaranteeing an outcome that does not occur*
5. *Failing to draft green building contracts to appropriately account for the unique risks inherent in green building projects.*

Thus, it is essential that before undertaking to work on a green project, you think through all aspects of the project and fully understand what a green building project constitutes and requires.

Lifschitz advises against promising more than can be delivered. He also warns that if stakeholders are "to avoid the prospect of costly claims and litigation, green-project participants must be proactive at the outset of a project and pay careful attention to potential pitfalls when drafting and negotiating contract documents."

1.4 ESTABLISHING MEASURABLE GREEN CRITERIA

With the green movement continuing to increase in strength and vitality, an urgent need has arisen to establish measurable green criteria. Rating systems in the United States and globally, such as LEED and Green Globes, are making a serious attempt to define the qualitative and/or quantitative measures of sustainability and the data needed to implement and assess them. These efforts are of paramount importance because they help in determining whether a building is having the impact on human health and the environment its designers contemplated, and in calculating the estimated cost of or saving from this achievement.

Early recognition of the urgent need to address sustainability problems, and the necessity for establishing measurable green criteria to facilitate this, is evidenced by the creation of the World Commission on Environment and Development (WCED) in December 1983 by the United Nations, with the main intent of addressing growing concerns "about the accelerating deterioration of the human environment and natural resources and the consequences of that deterioration for economic and social development." The establishment of WCED was a clear recognition early on by the General Assembly that the environmental problems we faced are global. The UN determined that it was in the best interest of all nations to establish common policies for sustainable development (see *Report of the World Commission on Environment and Development: Our Common Future*; www.un-documents.net/wced-ocf.htm).

After the formation of WCED came the Brundtland Commission in 1987, which produced the "Brundtland Report" (also known as "Our Common Future") that same year. The significance of this report is that it alerted the world to the urgency of making progress toward economic development that could be sustained without depleting natural resources or harming the environment. The report highlighted three essential components to achieve sustainable development: environmental protection, economic growth, and social equity. Some have found rather troubling the findings of this report, which include the following:

The "greenhouse effect," one such threat to life support systems, springs directly from increased resource use. [See Figure 1.2.] The burning of fossil fuels and the cutting and burning of forests release carbon dioxide (CO_2). The accumulation in the atmosphere of CO_2 and certain other gases traps solar radiation near the earth's surface, causing global warming. This could cause sea level rises over the next 45 years large enough to inundate many low lying coastal cities and river deltas. It could also drastically upset national and international agricultural production and trade systems.

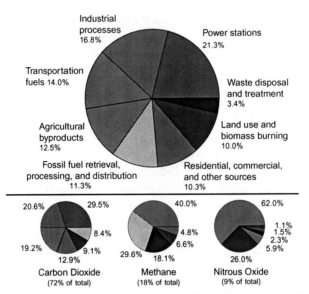

FIGURE 1.2 Global anthropogenic greenhouse gas emissions broken down into eight sectors for the year 2000. Concentrations of several greenhouse gases have increased over time, and human activity may be enhancing the greenhouse effect through release of carbon dioxide. *(Source: Robert A. Rohde, Wikipedia: Greenhouse gas—http://en.wikipedia.org/wiki/Greenhouse_gas.)*

Another threat arises from the depletion of the atmospheric ozone layer by gases released during the production of foam and the use of refrigerants and aerosols. A substantial loss of such ozone could have catastrophic effects on human and livestock health and on some life forms at the base of the marine food chain. The 1986 discovery of a hole in the ozone layer above the Antarctic suggests the possibility of a more rapid depletion than previously suspected.

The report goes on to say:

A variety of air pollutants are killing trees and lakes and damaging buildings and cultural treasures, close to and sometimes thousands of miles from points of emission. The acidification of the environment threatens large areas of Europe and North America. Central Europe is currently receiving more than one gram of sulfur on every square meter of ground each year. The loss of forests could bring in its wake disastrous erosion, siltation, floods, and local climatic change. Air pollution damage is also becoming evident in some newly industrialized countries.

Figure 1.3 is a graphic illustration of the greenhouse effect.

In trying to establish green measuring and performance criteria, we are immediately faced with several significant challenges, both conceptual and practical. On the conceptual side, we are challenged with the need to determine precisely what "performance" is. For example, it can be understood to mean that a building, as built, exhibits or embraces characteristics that are green or sustainable, and that building upgrades, renovations, and reconfigurations are sustainable. In some cases, green criteria measure the environmental results

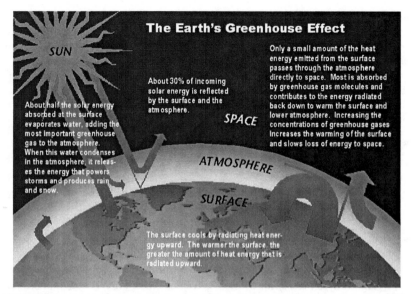

The Earth's Greenhouse Effect

SUN

About 30% of incoming
solar energy is reflected
by the surface and the
atmosphere.

About half the solar energy
absorbed at the surface
evaporates water, adding the
most important greenhouse
gas to the atmosphere.
When this water condenses
in the atmosphere, it releases the energy that powers
storms and produces rain
and snow.

SPACE

Only a small amount of the heat
energy emitted from the surface
passes through the atmosphere
directly to space. Most is absorbed
by greenhouse gas molecules and
contributes to the energy radiated
back down to warm the surface and
lower atmosphere. Increasing the
concentrations of greenhouse gases
increases the warming of the surface
and slows loss of energy to space.

ATMOSPHERE

SURFACE

The surface cools by radiating heat energy upward. The warmer the surface, the
greater the amount of heat energy that is
radiated upward.

FIGURE 1.3   Greenhouse gases, in the order of relative abundance, are water vapor, carbon dioxide, methane, nitrous oxide, and ozone. They come from natural sources and human activity. *(Source: Darren Samuelsohn,* Earth News, *2007; www.earthportal.org.)*

and consequences of sustainable strategies in terms of resource consumption and environmental impacts; in others, they measure the ultimate savings and costs realized from a sustainable building.

As for practical challenges, these mainly revolve around actual versus modeled performance. Actual performance data are generally preferable, but are not always easy to obtain, in which case models or estimations must be used. Where models are necessary, it is preferable when possible to use any relevant existing data as they better reflect operating as opposed to design performance. For example, it is difficult to obtain relevant data to support performance on campuses where buildings are not separately metered for energy or water use, which means that extra individual effort may be required to gather the needed data. But even when measured data are available, there still remains the question of quality and relia- bility. For the data to be really useful, we need to apply benchmarks for compar- ison to determine the level of performance and compare them to a typical well-documented building in a similar climate, with the same occupancies. Bench- marks can be a building's performance measured over time or based on external yardsticks such as LEED, Green Globes, or other acceptable organizations.

In a *Whole Building Design Guide* (WBDG) article entitled "Measuring Performance of Sustainable Buildings," Joel Todd, environmental consultant, and Kim Fowler, senior research engineer at Pacific Northwest National Laboratory, have this to say:

In recent years, building owners and designers, researchers, and others have begun performing studies related to the costs and benefits of sustainable design. Some of these studies attempt to address the full impact of sustainable design, while others emphasize the economic aspects, the environmental impacts, and the social aspects separately. Other differences in the studies include whether or not the data is measured, modeled, or some combination of both, whether the information is based on a single building or multiple buildings and the differences in how the baseline or benchmark is being used.

Some of these studies include:

- "The Costs and Financial Benefits of Green Buildings: A Report to California's Sustainable Building Task Force," G. Kats, L. Alevantis, A. Berman, E. Mills, J. Perlman, 2003
- *The Business Case for Sustainable Design in Federal Facilities*, the Department of Energy, Federal Energy Management Program (FEMP), August 2003
- LEED® Cost Study and LEED® Applications Guide, Steven Winter Associates for the General Services Adminstration, 2004
- *The Human Factors of Sustainable Building Design: Post-Occupancy Evaluation of the Philip Merrill Environmental Center, Annapolis, MD*, Judith Heerwagen and Leah Zagreus, prepared for the Department of Energy, 2005
- "Life-Cycle Cost Analysis (LCCA)", Sieglinde Fuller, National Institute of Standards and Technology, in WBDG, updated 2008
- "The Cost of Green Revisited," Lisa Fay Matthiessen and Peter Morris, 2007

It is problematic that we experienced a substantial building boom in recent years that was often underpinned by inferior design and construction strategies, among them highly inefficient HVAC systems that put buildings at the top of the list of contributors to global warming. Federal and private organizations are making serious attempts to address these problems with some success; due partly to these efforts, we are now witnessing a surge of interest in green concepts and sustainability. Many developers and project owners have become aware of the numerous benefits of green strategies and are increasingly seeking green—particularly LEED—certification for their buildings. The principal objectives of the green rating systems are to create incentives for producing high-performance buildings and to increase demand for sustainable construction. Green buildings have been shown to be economically viable and ecologically benign; their operation and maintenance have proven viable and sustainable over the long term.

Viable and sustainable green building has encouraged the collaboration between the Partnership for Achieving Construction Excellence and the Pentagon Renovation and Construction Program Office on the recently issued *Field Guide for Sustainable Construction*. This guide consists of ten chapters:

Chapter 1: Procurement. Specific procurement strategies to ensure sustainable construction requirements.

Chapter 2: Site/Environment. Methods to reduce the environmental impact of construction on the project site and surrounding environment.

Chapter 3: Material Selection. Environmentally friendly building materials as well as harmful and toxic materials that should be avoided.

Chapter 4: Waste Prevention. Methods to reduce and eliminate waste on construction projects.

Chapter 5: Recycling. Materials to recycle at each phase of construction and methods to support onsite recycling efforts.

Chapter 6: Energy. Methods to ensure and improve a building's energy performance, to reduce energy consumed during construction, and to identify opportunities for renewable energy source use.

Chapter 7: Building and Material Reuse. Reusable materials and methods to facilitate future reuse of facilities, systems, equipment, products, and materials.

Chapter 8: Construction Technologies. Technologies that can be used during construction to improve efficiency and reduce waste (especially paper).

Chapter 9: Health and Safety. Methods to improve quality of life for construction workers are identified.

Chapter 10: Indoor Environmental Quality. Methods to ensure that indoor environmental quality measures during construction are properly managed and executed.

The object of the *Field Guide* is to educate construction field workers, as well as supervisors and managers and other stakeholders, in making appropriate decisions to meet sustainable project goals. Most of the topics discussed are incorporated in LEED and other rating systems.

There are a number of other sustainability programs that outline important green criteria. DOE, for example, has an Environmental Protection Program, the goals and objectives of which are these (see www.hss.doe.gov/nuclearsafety/nfsp/fire/guidance/o4501admc1.pdf):

To implement sound stewardship practices that are protective of the air, water, land, and other natural and cultural resources impacted by DOE operations and by which DOE cost effectively meets or exceeds compliance with applicable environmental; public health; and resource protection laws, regulations, and DOE requirements. This objective must be accomplished by implementing Environmental Management Systems (EMSs) at DOE sites. An EMS is a continuing cycle of planning, implementing, evaluating, and improving processes and actions undertaken to achieve environmental goals.

Architects, designers, property developers, contractors, and other stakeholders must have a clear understanding of the certification programs currently available, and the need for certification if they are to remain competitive in an increasingly green market. Certification implies independent verification that a building has met accepted guidelines in these areas, as outlined for example, in LEED's Green Building Rating System. LEED certification of a project has become a recognized testimonial to its quality and environmental stewardship, especially since the rating system is now widely accepted by public and private owners—not only in the United States but in many countries around the world.

Because of the major impact on mainstream design and construction of the LEED Rating System, contractors and property developers are realizing that it is in their best interest to contribute to a project's achievement of green objectives. Success is not difficult to achieve once the LEED process is understood and its specific role in achieving LEED credits. This must be followed by involvement (preferably beginning in the planning phase) in all project phases through a team approach in an integrated design process. Measurable benchmarks are necessary to enable verification and to confirm a building's satisfactory performance. It must be pointed out here that ASHRAE places the responsibility of defining design intent requirements squarely on the shoulders of the project owner.

Normally, in the practice of architecture and engineering for building design, the owner hires an architect and/or engineer to develop the design intent in the project's Design Phase. To do this, it is imperative that he or she have in place all necessary information. Otherwise, it is difficult if not impossible to correctly measure a building or project's performance against the criteria on which the project's design and execution were based. For this reason, a project's plans, specifications, and so forth, are prepared in a manner that can provide measurable results to determine whether it has met the specified objectives and original design intent of the owner.

Likewise, before measurable sustainability criteria can be established, it is necessary to first agree on what is understood precisely by "green construction" and to clearly articulate what the finished product is to consist of. If not, in many building construction projects points of dispute will arise over how a building, product, or system is evaluated prior to, during, and after construction. Where disputes cannot be resolved through standard meeting minutes or interpretation (i.e., requests for information, RFIs), the parties may end up in mediation, arbitration, or trial.

USGBC currently promotes its LEED Rating System by emphasizing its simplicity in addition to its other benefits. However, the uniqueness of the LEED certification system is that it typically mandates performance over process. Moreover, USGBC, through its widely circulated and recently updated LEED V3 scoring system and other efforts, has compelled many contractors and their subcontractors to change the way they operate. LEED V3 has also improved its rating system by taking into account the impact of microclimate and by incorporating Regional Priority as a rating category. This is discussed in greater detail in Chapter 2.

Organizations such as NAHB have also put forward a set of green home-building guidelines. NAHB states that its guidelines "should be viewed as a dynamic document that will change and evolve as new information becomes available, improvements are made to existing techniques and technologies, and new research tools are developed." The NAHB Model Green Home Building

Guidelines were written to help move environmentally friendly home-building concepts further into the mainstream marketplace; they represent one of two rating systems that make up NAHBGreen, the National Green Building Program.

Unlike LEED's rating system, which has four levels (Certified, Silver, Gold, and Platinum) available to builders using the guidelines to rate their projects, NAHB's point program contains only three levels (Bronze, Silver, and Gold). It is available to builders wishing to implement NAHB's guidelines to rate their projects. "At all levels, there are a minimum number of points required for each of the seven guiding principles to assure that all aspects of green building are addressed and that there is a balanced, whole-systems approach. After reaching the thresholds, an additional 100 points must be achieved by implementing any of the remaining line items." It should be mentioned that Green Globes, which is very popular in Canada, uses the NAHB standards for evaluating residential buildings. Table 1.1 outlines the necessary points needed for the three NAHB threshold levels.

While the general appearance of a green building may be similar to that of conventional building forms, the conceptual design approach is fundamentally different because it revolves around a concern for the building's potential impact on the environment. It also endeavors to extend the life span of natural resources and to improve human comfort and well-being, as well as security, productivity, and energy efficiency. That sustainably designed buildings will result in reduced operating costs, including those of energy and water, as well as other intangible benefits, is now globally recognized. In this regard, the Indian Green Building Council (IGBC), which administers LEED's India rating system, highlights a number of salient green building attributes:

- Minimal disturbance to landscapes and site condition
- Use of recycled and environmentally friendly building materials

TABLE 1.1 NAHB Three-Tier Point System

Category	Bronze	Silver	Gold
Lot design, preparation, and development	8	10	12
Resource efficiency	44	60	77
Energy efficiency	37	62	100
Water efficiency	6	13	19
Indoor environmental quality	32	54	72
Operation, maintenance, and homeowner education	7	7	9
Global impact	3	5	6
Additional points from sections of applicant's choice	100	100	100

- Use of nontoxic and recycled/recyclable materials
- Efficient use of water and water recycling
- Use of energy-efficient and eco-friendly equipment
- Use of renewable energy
- Indoor air quality for human safety and comfort
- Effective controls and building management systems

Other programs, such as the *Whole Building Design Guide* in the United States, set out certain rules and principles relating to sustainable design. *WBDG*'s are outlined here as follows.

Objectives
- Avoid resource depletion of energy, water, and raw material
- Prevent environmental degradation caused by facilities and infrastructure throughout their life cycles
- Create livable, comfortable, safe, and productive built environments

Principles
- Optimize site potential
- Optimize energy use
- Protect and conserve water
- Use environmentally preferred products
- Enhance indoor environmental quality
- Optimize operations and maintenance procedures

According to James Woods (2008), executive director of the Building Diagnostics Research Institute:

Building performance is a set of facts and not just promises. If the promises are achieved and verified through measurement, beneficial consequences will result and risks will be managed. However, if the promises are not achieved, adverse consequences are likely to lead to increased risks to the occupants and tenants, building owners, designers and contractors; and to the larger interests of national security and climate change.

As Alan Bilka, a sustainable design expert with ICC Technical Services, correctly points out:

Over time, more and more "green" materials and methods will appear in the codes and/ or have an effect on current code text. But the implications of green and sustainable building are so wide and far-reaching that their effects will most certainly not be limited to one single code or standard. On the contrary, they will affect virtually all codes and will spill beyond the codes. Some green building concepts may become hotly contested political issues in the future, possibly requiring the creation of new legislation and/or entirely new government agencies.

The U.S. DOE's National Renewable Energy Laboratory has created a High-Performance Buildings (HPB) database to help improve building performance by showcasing examples of green buildings and providing a standardized format for displaying performance results. DOE is also working on standardizing methods for reporting building performance by collecting relevant data on sustainable topics such as land use, energy, materials, water conservation, and indoor environmental quality. The HPB database presents information at various levels of detail. An "overview" level, for example, describes key information, including a project's function and its most pertinent green features. More detailed information on the project is divided into a series of modules on process, performance, and results.

1.5   EMERGING DIRECTIONS

We live today in both challenging and exciting times in a volatile design and construction industry. Sustainable design has provided us with a way to efficiently use our resources and make a minimal impact on our environment, preserving it for ourselves and for future generations. However, as Keith Fox, president of McGraw-Hill Construction, states: "The one constant in a volatile industry has been our ability to help construction professionals make the right decisions to drive their businesses," and "during these tough economic times, gaining insight and intelligence about where our industry is headed and what role we will each need to play to be successful is extremely important."

Most design professionals as well as developers, contractors, manufacturers, and federal, state, and local governments are enthusiastically embracing this emerging green phenomenon. Moreover, the world of building design and construction has over the last couple of decades become increasingly green and today is an integral part of our global culture. The enthusiastic embrace of green and increased public awareness have brought growing pressure to bear on the construction industry to fundamentally change how it does business and executes projects. This is a significant development.

In addition, according to a 2008 Green Building Market Barometer online survey of commercial real estate executives conducted by New York City-based Turner Construction, even the 2008 credit collapse failed to adversely affect the desire of property developers to go green. This debunks the myth that green building is a fad and endorses the reality that it is global and here to stay. In fact, McGraw-Hill Construction, in *Construction Outlook*, predicted an increase in overall U.S. construction starts for 2011 of 8%, to $445.5 billion, following the 2% decline predicted for 2010. The report further predicted, based on research and analysis of macro trends, significant advances for each construction sector as follows:

- Single-family housing in 2011 will climb 27% in dollars, corresponding to a 25% increase in the number of units to 565,000 (McGraw-Hill Construction basis).
- Multifamily housing will rise 24% in dollars and 23% in units, continuing to move gradually upward.

- Commercial buildings will increase 16%, following a three-year decline, which lowered contracting by 62% in dollar terms. The level of activity expected for stores, warehouses, offices, and hotels in 2011 will still be quite weak by historical standards.
- The institutional building market will slip an additional 1% in 2011, retreating for the third straight year. The difficult fiscal climate for states and localities will continue to dampen school construction, although healthcare facilities should see moderate growth.
- Manufacturing buildings will increase 9% in dollars and 11% in square feet.
- Public works construction will drop 1%, given the fading benefits of the federal stimulus act for highway and bridge construction.
- Electric utilities will slide 10%, falling for the third year in a row.

Increased public environmental alertness has become an integral component of the corporate mainstream and of the general global awareness of the impact of humans on the environment; it has also caused an increase in consumer demand for sustainable products and services that is creating new challenges and opportunities for businesses in all areas of the construction industry. Enlightened corporations have responded to these challenges by becoming more mindful of environmental impacts. "Green" organizations, such as USGBC and Green Globes, play a pivotal role in raising corporate awareness and in encouraging increased participation in the green movement.

Moreover, LEED and other rating systems can be found the world over, in countries as diverse as Britain, Mexico, Australia, Spain, Canada, India, China, the United Arab Emirates, Israel, and Japan, to name but a few, where the green movement is well under way. What is even more profound is that green buildings are, by codification, becoming the law of the land. For some organizations, this will just mean business as usual, but for others it will be cataclysmic. The new codes will invariably mean increased expectations from designers and contractors and possibly increased litigation arising from greater standards of care.

The concept and practice of sustainability continue to have a profound impact on building construction and design and are helping to fundamentally transform the building market and change our perception of how we design, inhabit, and operate our buildings. In fact, among the primary factors that are accelerating the push toward building green are increased demand for green construction, particularly in the residential sector, increasing levels of government initiatives, and improvements in the quality and availability of environmentally friendly building materials. The growing demand for sustainability has forced many businesses to seek new ways to become more sustainable, mainly by focusing their efforts on improving their buildings' energy efficiency and interior environmental air quality.

Spurred by this growing demand for building projects that employ environmentally friendly and energy-efficient materials, a strong green movement in the construction industry is emerging. With this in mind, we witness a number of forward-looking companies, such as DPR Construction, jumping on the green bandwagon and being well prepared and well placed to deliver successful green projects. According to DPR, there are LEED-trained and -accredited professionals in every one of its offices across the country. It boasts a portfolio of more than $1 billon in green building projects, and claims to have trained more than 500 professionals in overall sustainability and green building above and beyond the available LEED programs. Furthermore, 27% of DPR's professionals have acquired LEED accreditation; this is reportedly the highest percentage of LEED-accredited professionals in the nation among general contractors.

President Obama inherited a depressed economy, yet even with the downturn in the economy and the construction industry, the number of "green buildings" being built in the United States is estimated to be valued in excess of $10 billion. According to the Department of Commerce (2008), the construction market constitutes about 13.4% of the $13.2 trillion U.S. GDP; this includes all commercial, residential, industrial, and infrastructure construction. Commercial and residential building construction on its own accounts for about 6.1% of GDP (figures from the Department of Commerce, 2008). Furthermore, as of 2006 the USGBC's LEED system had certified 775 million square feet of commercial office space as green. This represents a mere 2% of total U.S. commercial office space, but is expected to increase exponentially, with green buildings potentially accounting for 5 to 10% of the U.S. commercial construction market by 2012. Still, according to Howard Birnberg, executive director of the Association for Project Managers:

Whatever the condition of the economy, technology continues to advance. While it remains to be seen if Building Information Modeling (BIM) will be a game changer for the industry, the ability to integrate new technology is an expensive and endless challenge. Training of design and construction staff in new technology and important subjects such as project management has been widely neglected during the downturn. When workloads improve, many organizations will need to play catch-up on their staff training.

At the annual Greenbuild Conference held in Chicago in 2010, the USGBC announced that it had achieved a major milestone in the certification of more than 1 billion square feet of commercial real estate through its LEED Rating System. It further reported that another 6 billion square feet of projects around the world are registered and seeking to achieve LEED certification.

There is growing evidence that the shift toward green construction is truly a global trend, with more countries putting more resources into improving efficiency and sustainability. In 2008, for example, McGraw-Hill's "Global Green Building Trends" stated that 67% of global construction firms reported at least 16% of their projects as green. The study also projected that by 2013 the percentage of firms going green would be 94%. If this forecast is correct, it is a clear signal that green construction not only has become part of the mainstream but can expect a significant share of the construction industry's $4.7 trillion global market.

Europe has reportedly achieved the highest level of sustainable building activity in the international arena, with an unprecedented 44% of European construction firms building green on at least 60% of their projects. Next come North America and Australia, but the gaps are gradually closing. However, Asia has the greatest potential for increasing its market share, where the number of firms dedicated to green construction is projected to increase threefold in the coming years.

Although the American Institute of Architects (AIA) has concluded that buildings are the leading source of greenhouse gas emissions in the United States, a recent online survey conducted by Harris Interactive showed that only 4% of U.S. adults were aware of this fact. However, the quantity and quality of much of the data that generally relates to business and the environment remains inadequate, to say the least. It is unfortunate that many government agencies, corporations, nonprofit groups, and academic institutions have often maintained a lethargic approach and have produced relatively little to quantify or assess simple measures of business environmental impact.

Frank Hackett, an energy conservation sales consultant for Mayer Electric Supply, says that one of the simplest things a business can do to improve its efficiency and reduce costs is to update or retrofit its lighting system. One example he gives is modifying and updating existing lighting fixtures to use the more energy-efficient T-5 or T-8 fluorescent lamps in place of the T-12 models that are widely used at present. Replacement of the magnetic ballasts in lighting can also help increase the system's energy efficiency. Recent DOE estimates show that a significant percentage of a business's normal energy bill consists of lighting costs; reducing these costs can have a favorable economic impact. Employing automatic control systems that take advantage of natural light and automatically switch off when no one is present should also be considered as an energy-saving option.

According to a recent ruling issued by DOE, states must now certify that their building codes meet the requirements in the 2004 American Society of Heating, Refrigerating and Air-Conditioning Engineers/Illuminating Engineering Society of North America (ASHRAE/IESNA) energy efficiency standard. ASHRAE is moving forward on development of the nation's first standard for high-performance, green commercial buildings (Standard 189.1P). This standard will require buildings to be significantly more efficient than required by the current Standard 90.1-2007 (Energy Standard for Buildings except Low-Rise Residential Buildings).

The USGBC has reaffirmed its commitment to the development of Standard 189.1P, which when completed will be America's first national standard for use as a green building code. Standard 189.1P is being developed as an American National Standards Institute (ANSI) standard, created specifically for adoption by states, localities, and other building code jurisdictions that are ready to require a minimum level of green performance for all commercial buildings. According to ICC chief executive officer Richard P. Weiland:

The emergence of green building codes and standards is an important next step for the green building movement, establishing a much-needed set of baseline regulations for green buildings that is adoptable, usable and enforceable by jurisdictions. . . . The IgCC provides a vehicle for jurisdictions to regulate green for the design and performance of new and renovated buildings in a manner that is integrated with existing codes as an overlay, allowing all new buildings to reap the rewards of improved design and construction practices.

On October 3, 2008, former President Bush signed into law H.R. 1424 and extended the Energy Efficient Commercial Building Tax Deduction as part of the Emergency Economic Stabilization Act of 2008. This is not a tax credit; rather, an amount can be subtracted from gross taxable income and cannot be directly subtracted from tax owed. Still, it can offer benefits to the taxpayer and be used as an incentive to choosing energy-efficient building systems. Green building initiatives have been popping up on all sides of the equation.

California passed its Revised Title 24 Code in October 2005 in response to a legislative mandate to reduce California's energy consumption. The state's Energy Commission later adopted the 2008 standards on April 23, 2008, and its Building Standards Commission approved them for publication on September 11, 2008. Additionally, a new law that took effect on January 1, 2009, mandates that owners of all nonresidential properties in California make available to tenants, lenders, and potential buyers the energy consumption of their buildings as part of the state's participation in the ENERGY STAR program. These data are transmitted to the EPA's ENERGY STAR portfolio manager, who benchmarks the information under ENERGY STAR standards. On assembling these data, beginning in 2010, building owners are required to disclose them and the building's ratings. The compelling reasons that California's Energy Commission adopted changes to the Building Energy Efficiency Standards in 2008 include the following:

- To provide California with an adequate, reasonably priced, and environmentally sound supply of energy.
- To respond to Assembly Bill 32, the Global Warming Solutions Act of 2006, which mandates that California reduce its greenhouse gas emissions to 1990 levels by 2020.
- To pursue California energy policy that energy efficiency be the resource of first choice for meeting California's energy needs.
- To act on the findings of California's Integrated Energy Policy Report (IEPR) that standards are the most cost-effective means to achieve energy efficiency. The Building Energy Efficiency Standards will continue to be upgraded over time to reduce electricity and peak demand, and the role of the standards in reducing energy related to meeting California's water needs and in reducing greenhouse gas emissions will be recognized.

- To meet the West Coast Governors' Global Warming Initiative commitment to include aggressive energy efficiency measures in updates of state building codes.
- To meet the Executive Order in the Green Building Initiative to improve nonresidential buildings' energy efficiency through aggressive standards.

The majority of major U.S. cities have initiated some form of energy efficiency standards for new construction and existing buildings. For example, in April 2005 Washington State began requiring that all state-funded construction projects having more than 5000 square feet be built green. In May 2006, Seattle moved forward and approved a plan offering incentives to encourage site-appropriate packages of greening possibilities that include green roofs, exterior vertical landscaping, interior green walls, air filtration, and stormwater runoff management. Seattle can also boast of becoming the first municipality in the United States to adopt LEED Silver certification for its own major construction ventures.

The state that boasts the second-highest number of LEED-certified buildings in the nation is Pennsylvania, which currently has 83 and is just behind California. Pennsylvania has put into place four state funds for green projects, including a $20 million Sustainable Energy Fund that offers grants and loans for energy efficiency and renewable energy projects. The city of Philadelphia also recently enacted a "Green Roofs Tax Credit" to encourage the installation of roofs that support living vegetation, and has also proposed a "Sustainable Zoning" ordinance that mandates buildings that occupy a minimum of 90,000 square feet or more to incorporate green roofs in their design.

In April 2007, the Baltimore City Planning Commission voted to require developers to incorporate green building standards into their projects by 2010. Boston also amended its zoning code to require all public and private development projects in excess of 50,000 square feet to be constructed to green building standards. When Washington, D.C.'s Green Building Act of 2006 went into effect in March 2007, the district became the first major U.S. city to require LEED compliance for private projects. These new green building standards became mandatory in 2009 for privately owned, nonresidential construction projects with 50,000 square feet or more; compliance by public projects is now required as well.

The USGBC says that, as of September 2010:

Various LEED initiatives including legislation, executive orders, resolutions, ordinances, policies, and incentives are found in 45 states, including 442 localities (384 cities/towns and 58 counties), 35 state governments (including the Commonwealth of Puerto Rico), 14 federal agencies or departments, and numerous public school jurisdictions and institutions of higher education across the United States.

Furthermore, with the increasing move of building green into the mainstream, it seems that soon green or sustainable building will cease to be an option and become a requirement. In April 2008, Stacey Richardson, a product specialist with the Tremco Roofing & Building Maintenance Division, says in "The Green Movement Sweeps Eastward":

It is the way of the future, and industry developments in new green technology will provide building owners increasing access to energy-saving, environmentally-friendly systems and materials. Everything from bio-based adhesives and sealants, low-VOC or recycled-content building products, to the far-reaching capabilities of nanotechnology—the movement of building "renewable" and "energy-efficient" will only continue to strengthen.

Even colleges and universities (e.g., Harvard, Pennsylvania State, the University of Florida, the University of South Carolina, the University of California-Merced, and others) have jumped on the green bandwagon.

A November 2008 study, "Greening Buildings and Communities: Costs and Benefits" by Landmark International, purported to be the largest international study of its kind, is based on extensive financial and technical analysis of 150 green buildings built between 1998 and 2008 in 33 U.S. states and 10 countries worldwide. It provides the most detailed and reliable findings on the costs and financial benefits of building green. Some of its key findings are as follows:

- Most green buildings cost 0 to 4% more than conventional buildings, with the largest concentration of reported "green premiums" between 0 to 1%. Green premiums increase with the level of greenness, but most LEED buildings, up through the Gold level, can be built for the same cost as for conventional buildings. This stands in contrast to a common misperception that green buildings are much more expensive than conventional buildings.
- Energy savings alone make green building cost-effective, outweighing any initial cost premium in most cases. The present value of 20 years of energy savings in a typical green office ranges from $7 per square foot (Certified) to $14 per square foot (Platinum), which is more than the average additional cost of $3 to $8 per square foot for building green.
- Green building design goals are associated with improved health and with enhanced student and worker performance. Health and productivity benefits remain a major motivating factor for green building owners, but are difficult to quantify. Occupant surveys generally demonstrate greater comfort and productivity in green buildings.
- Green buildings create jobs by shifting spending from fossil fuel-based energy to domestic energy efficiency, construction, renewable energy, and other green jobs. A typical green office creates roughly one-third of a permanent job per year, equal to $1 per square foot of value in increased employment, compared to a similar nongreen building.
- Green buildings are seeing increased market value (higher sales/rental rates, increased occupancy, and lower turnover) compared to comparable conventional buildings. In a March 2008 study, CoStar, for example, reports an

average increased sales price from building green of more than $20 per square foot, providing a strong incentive to build green even for speculative builders.

- Roughly 50% of green buildings in the study's data set see the initial "green premium" paid back by energy and water savings in five or fewer years. Significant health and productivity benefits mean that over 90% of green buildings pay back an initial investment in five or fewer years.
- Green community design (e.g., LEED-ND) provides a distinct set of benefits to owners, residents, and municipalities, including reduced infrastructure costs, transportation and health savings, and increased property values. Green communities and neighborhoods have a greater diversity of uses, housing types, job types, and transportation options, and appear to better retain value in a market downturn than conventional sprawl.
- Annual gas savings in walkable communities can be as much as $1,000 per household. Annual health savings (from increased physical activity) can be more than $200 per household. CO_2 emissions can be reduced by 10 to 25%.
- Upfront infrastructure development costs in conservation developments can be reduced by 25%, approximately $10,000 per home.
- Religious and faith groups build green for ethical and moral reasons. Financial benefits are not the main motivating factor for many places of worship, religious educational institutions, and faith-based nonprofits. A survey of faith groups building green found that the financial cost-effectiveness of green building makes it a practical way to enact the ethical/moral imperative to care for the Earth and communities. Building green has also been found to energize and galvanize faith communities.

Green building to this day remains a relatively small, although burgeoning, market despite its impressive growth and the tremendous boom in green construction. But although sustainability on a large scale and attaining a corresponding market share remain elusive, with LEED-registered projects today representing just over 5% of the total square footage in U.S. new construction, it is estimated that sustainable construction projects will contribute $554 billion to the U.S. GDP by 2013.

Despite the fact that the nonprofit USGBC was founded in 1993, only in the last few years has it become a significant driving force in the green building construction movement. By the end of 2010, its membership consisted of roughly 18,500 companies and organizations. USGBC's important leadership role was achieved mainly through the early development of its LEED commercial building rating system. The process for earning LEED certification typically starts in the early planning stage, when interested stakeholders make a determination to pursue certification. This is followed by registering the project and paying the required fee. Once the project is completed and commissioned and all required numbers are handed in with supporting documentation, the project is submitted for evaluation and certification. This is discussed in greater detail in Chapter 2.

It is evident that USGBC has had a very significant impact on green building and has emerged as a clear leader in fostering and furthering green building efforts throughout the world. In the United States, the LEED Rating System is increasingly becoming the national standard for green building; it is also internationally recognized as a major tool for the design and construction of high-performance buildings and sustainable projects. With its eye clearly focused on the future, USGBC recently issued its updated LEED V3. It has also put in place a strategic plan for the period 2009–2013 in which it outlines the key strategic issues that face the green building community:

- *Shift in emphasis from individual buildings toward the built environment and broader aspects of sustainability, including a more focused approach to social equity;*
- *Need for strategies to reduce contribution of the built environment to climate change;*
- *Rapidly increasing activity of government in [the] green building arena;*
- *Lack of capacity in the building trades to meet the demand for green building;*
- *Lack of data on green building performance;*
- *Lack of education about how to manage, operate, and inhabit green buildings; and*
- *Increasing interest in and need for green building expertise internationally.*

One of the primary indicators reflecting international interest in the USGBC and LEED is the large annual attendance and the increasing number of countries represented at Greenbuild, the USGBC's International Conference and Expo. In 2008, nearly 30,000 people from 85 countries attended the Greenbuild conference in Boston (USGBC, "Green Building Facts") as compared to only 4200 who attended the 2002 event in Austin. In 2010 the conference was held in Chicago and was attended by over 28,000 visitors, with representatives from all 50 states, 114 countries, and 6 continents. The slightly lower than expected attendance in 2010 was mainly due to the downturn in the global economy. The decline in construction activity during the first two years of the Obama administration was broader, steeper, and faster than many economists anticipated as private nonresidential building markets succumbed to the credit crunch and many public markets waited for stimulus funding to be delivered. Nevertheless, the substantial attendance reflects the international importance of the annual Greenbuild event and once again shows that Greenbuild has become an important forum for international leaders in green building where ideas and information can be exchanged.

Attending the 2008 Greenbuild conference were many international groups, including a high-level delegation from China headed by the vice minister of construction Qiu Baoxing. This is significant because over the past decade China's economy has been expanding at a phenomenal rate and some forecasts predict it will become the largest in the world by 2020. In the wake of such growth, however, has come a series of potentially severe environmental challenges. China has been able to make substantial inroads in overcoming these challenges and reversing many of these environmental trends. To further this goal in March 2010, it announced the initiation of a new energy efficiency strategy, of which green building is a primary component. This was followed by the signing of a Memorandum of Understanding (MOU) between the Chinese Ministry of Construction and the USGBC in which points of mutual interest were identified for collaboration in the advancement of environmentally responsible construction in the United States and China.

Project teams around the world are today applying the LEED Rating System as developed in the United States. The USGBC has been quick to recognize, however, that certain criteria, processes, or technologies may not always be appropriate for all countries, and that successful strategies for encouraging and practicing green building vary from one country to another, depending on local conditions, traditions, and practices. This reality was addressed by the USGBC in its sanctioning of other countries to license LEED and allowing them to adapt the rating system to their specific needs—on the understanding that LEED's high standards would not be compromised. Various countries worldwide have expressed an interest in LEED licensing, and several, such as Canada and India, now have their own LEED licensing programs.

On the international stage, USGBC works through the World Green Building Council (WorldGBC), which was formed in 1999 by David Gottfried. WorldGBC defines itself as a union of national green building councils from around the world, making it the largest international organization influencing the green building marketplace. It is currently operating in nearly 70 countries. One of WorldGBC's primary goals is to help countries establish their own councils and find a way to work effectively with policy makers and local industry. The WorldGBC is devoted to transforming the global property industry to sustainability, as it states on its website: "WorldGBC draws on the support of its partnerships to support the work of Green Building Councils around the world and to further drive the transition towards market transformation of the global property industry. Key partnerships have been made with private sector companies, governmental and non-governmental organizations, and academic institutions."

The main mission of WorldGBC is to serve as a forum for knowledge transfer between green building councils and to support and promote individual council members. It has the additional mission of recognizing global green building leadership and encouraging the adoption and development of market-based environmental rating systems that meet local needs for each country, but the WorldGBC does not promote any particular system or methodology as a global standard.

Responding to terrorist attacks such as those on the World Trade Center in New York and in Mumbai, many architects and building owners are now demanding that their facilities be designed for greater blast resistance and to better withstand the effects of violent tornadoes and hurricanes—for example, by the use of blast-resistant windows with protective glazing. This is of particular importance with high-rise buildings. Federal buildings are now required to incorporate windows that provide protection against such potential threats. Likewise, there are increasing demands from governments and the public for structures to be sustainable and to meet general environmental requirements.

USGBC has emerged as a driving force in an industry that is projected to contribute $554 billion to the U.S. GDP through 2013. It leads a diverse following of professional designers and engineers, builders and environmentalists, federal agencies, corporations, and nonprofit organizations, as well as teachers and students. Moreover, USGBC now comprises some 80 local affiliates, 18,000 member companies and organizations, and more than 155,000 LEED Professional Credential holders. This unprecedented growth is further evidenced by the dramatic increase in the number of certified and registered projects since LEED was first launched in 2000. By the end of 2007, the square footage of U.S. office and commercial space registered or certified under LEED had reached 2.3 billion, an increase of more than 500% from two years earlier.

Certified projects are those that have been completed and verified through USGBC's process, while registered projects are those that are still in design or construction. "Green Building by the Numbers," a report published in April 2009 by USGBC, lists 2476 certified projects and 19,524 registered projects in more than 90 countries. Altogether, commercial building space with LEED certification amounts to more than 5 billion square feet. This astonishing increase in LEED project registration for new construction is significant as a clear indicator of future prospects for the green industry.

Bob Schroeder, industry director (Americas) for Dow Corning's construction division, has this to say: "Today, sustainable design has been recognized by the industry and the public as critical factors in achieving high quality architecture and benefiting the building owners—the companies that occupy these structures and the wider community." It has been shown that many design, construction, and consulting firms have laid off a significant percentage of their staff during the past three years, basically retaining top-performing staff who are able to quickly and effectively respond to client needs. Likewise, a substantial number of mergers and acquisitions have taken place, allowing firms to reposition themselves in emerging markets and gain expertise while developing new relationships. Even in a depressed economy, green design has become very important and is here to stay. It no longer suffices for owners, institutions, and agencies to just talk about building green. There is a strong incentive to implement elements of green design into new and existing construction projects.

While most design professionals consider great architecture to be a delicate balance of form and function, we find high-rise buildings being constructed on a global scale, with increasing ferocity, and with little concern for due diligence, the environment, or aesthetics. However, what is transpiring from the green building upheaval is the emergence of several interesting trends such as the building of spectacular landmarks, as exemplified by the Sydney Opera House (Figure 1.4) and Burj Khalifa in Dubai (Figure 1.5), which is the highest building in the world and was recognized by the Council on Tall Buildings and Urban Habitat (CTBUH) with its 2010 highest award,

FIGURE 1.4   Australia's Sydney Opera House, which overlooks the harbor, is considered to be one of the most recognizable buildings in the world and has become the city's landmark. It consists of 14 freestanding sculptures of spherical roofs and sail-like shells sheathed in white ceramic tiles. The original commission to build the opera house was won in competition by the late Danish architect John Utzon in 1957, but because his vision and design were too advanced for the architectural and engineering capabilities at the time, it wasn't until 1973 that the building was finally opened.

FIGURE 1.5   The Dubai Burj Khalifa, which was inaugurated on January 4, 2010, is currently the tallest skyscraper in the world in all three categories recognized by the CTBUH. *(Source: Skidmore, Owings & Merrill.)*

"Global Icon and Best Tall Building." Bill Baker, chief structural engineer on the project, says:

Burj Khalifa is a game changer. This incredible team of architects, engineers, consultants, and contractors has been able to create something that goes far beyond what has been done before. We are extremely grateful to the CTBUH for creating this prize for the project and recognizing the Burj Khalifa's impact on the art of tall buildings.

CTBUH ranks buildings on the basis of height to architectural top, highest occupied floor, and height to tip. Burj Khalifa was designed by the Chicago office of Skidmore, Owings & Merrill and constructed by Samsung/BESIX/Arabtec Corporation for Emaar Properties, UAE. Turner Construction International was the project and construction manager. The completed tower is 2717 feet (828 meters) high and was built at a cost of about $1.5 billion. It reportedly contains 160 habitable floors, 57 elevators, apartments, shops, swimming pools, spas, corporate suites, Italian fashion designer Giorgio Armani's 160-room hotel, and an observation platform on the 124th floor.

The towering Burj Khalifa skyscraper is the centerpiece of a large-scale, mixed-use development comprising residential, commercial, hotel, entertainment, shopping, and leisure outlets with open green spaces, water features, pedestrian boulevards, a shopping mall, and a tourist-oriented old town. The design of the tower combines historical and cultural influences with cutting-edge technology to achieve high performance. Its massing is manipulated in the vertical dimension to induce maximum vortex shedding and minimize the impact of wind on its movement.

We often find that the main driving force behind the creation of national landmarks is twofold: The primary desire is for important recognizable symbols to foster local and national pride; the secondary desire is essentially economic—for example, to increase tourism.

Chapter 2

Components of Sustainable Design and Construction

2.1 INTRODUCTION

Green building and contracting has become a hot selling point with home and business customers and can add tangible value to your business for years to come. An increasing number of designers, builders, and building owners are getting involved in green building practices. Specializing in green building and sustainability basically means incorporating environmentally friendly techniques and sustainable practices into a business' operations. National and local programs instigating and promoting green building are growing and reporting increasing success, while hundreds of certified green projects across the nation and internationally provide tangible evidence of what sustainable building design can accomplish in terms of aesthetics, comfort, and energy and resource efficiency. It obviously helps if a business owner or one or more employees are Leadership in Energy and Environmental Design (LEED®) Accredited Professionals.

Roger Woodson, a well-known author and contractor says, "In theory, you don't have to know much about construction to be a builder who subs all the work out to independent contractors, but as the general contractor it is you who will ultimately be responsible for the integrity of the work."

Today more than ever, buildings have a tremendous impact on the environment—both during construction and throughout operation. "Green/ sustainable building" is a loosely defined collection of strategies, such as land use, design, construction, and operation, that reduce environmental impacts. Green building practices facilitate the creation of environmentally sound and resource-efficient, high-performance buildings by employing an integrated team approach to design in which architects, engineers, builders, land planners, and building owners and operators pool their resources to design the structure.

In December 1983 the United Nations founded the World Commission on Environment and Development (WCED) with the mission of addressing growing concerns "about the accelerating deterioration of the human environment and natural resources and the consequences of that deterioration for economic and social development." WCED was followed by the Brundtland Commission

Handbook of Green Building Design, and Construction

in 1987, which produced the "Brundtland Report," with findings that were particularly troubling, including the following:

The "greenhouse effect," one such threat to life support systems, springs directly from increased resource use. The burning of fossil fuels and the cutting and burning of forests release carbon dioxide (CO_2). The accumulation in the atmosphere of CO_2 and certain other gases traps solar radiation near the Earth's surface, causing global warming. This could cause sea level rises over the next 45 years large enough to inundate many low lying coastal cities and river deltas. It could also drastically upset national and international agricultural production and trade systems.

Another threat arises from the depletion of the atmospheric ozone layer by gases released during the production of foam and the use of refrigerants and aerosols. A substantial loss of such ozone could have catastrophic effects on human and livestock health and on some life forms at the base of the marine food chain. The 1986 discovery of a hole in the ozone layer above the Antarctic suggests the possibility of a more rapid depletion than previously suspected.

The report went on to say:

A variety of air pollutants are killing trees and lakes and damaging buildings and cultural treasures, close to and sometimes thousands of miles from points of emission. The acidification of the environment threatens large areas of Europe and North America. Central Europe is currently receiving more than one gram of sulfur on every square meter of ground each year. The loss of forests could bring in its wake disastrous erosion, siltation, floods, and local climatic change. Air pollution damage is also becoming evident in some newly industrialized countries.

The United States was until recently caught up in a substantial building boom that was quite often characterized by inferior design and construction strategies as well as highly inefficient HVAC systems, making buildings the largest contributors to global warming. Several federal and private organizations are making continuous efforts to address these problems; partly because of these efforts, we are now witnessing a surge of interest in green concepts and sustainability to the extent that "green" has now entered the mainstream of the construction industry. Most project owners are aware of the many benefits of incorporating green strategies into their projects and are now increasingly aspiring to achieve LEED certification for their buildings. However, LEED is not the only certification system. Several are currently employed in the United States and generally serve two principal functions: the promotion of high-performance buildings and the facilitation and creation of demand for sustainable construction.

Studies continue to show that green buildings are economically viable, ecologically benign, and sustainable over the long term. With this in mind, the Partnership for Achieving Construction Excellence and the Pentagon Renovation and Construction Program Office recently published the "Field Guide for Sustainable Construction" to assist and educate field workers, supervisors, and managers in making decisions that help the project team meet its sustainable project goals. The salient points outlined in the guide are these:

- *Procurement:* Specific procurement strategies are identified and put in place to ensure that sustainable construction requirements are addressed.
- *Site/environment:* Methods are sought that reduce the environmental impact of construction on the project site and that identify impacts on the surrounding environment.
- *Material Selection:* Select environmentally friendly building materials and products that are nontoxic (and preferably recyclable and renewable) and locally produced, reducing CO_2 emissions and promoting local economies.
- *Waste prevention*: Approaches to reduce and eliminate waste on construction projects are identified.
- *Recycling:* Materials to be recycled and methods to support onsite recycling at each phase of construction.
- *Energy:* Strategies to ensure and improve a building's energy performance, reduce energy consumed during construction, and identify opportunities to use renewable energy sources.
- *Recycled and Renewable Materials:* This means specifying materials that contain recycled content or are reusable to facilitate future reuse of a facility and its systems, equipment, products, and materials. Examples are the use of recycled steel products, high-volume fly ash concrete products and concrete masonry units, as well as wood products certified by the Forest Stewardship Council (FSC).
- *Construction Technologies:* Ascertain which technologies can be used during construction to improve efficiency and reduce waste (especially paper).
- *Health and safety:* Procedures to improve the quality of life for construction workers are identified.
- *Indoor Environmental Quality:* Appropriate methods should be applied to ensure indoor environmental quality (IEQ), such as the use of low-VOC paints and adhesives, as well as Carpet and Rug Institute (CRI™) Green Label Plus carpets and low-emitting certified products. Smoking on premises should be prohibited.

The U.S. Department of Energy (DOE) has developed an Environmental Protection Program, the goals and objectives of which are these:

[T]o implement sound stewardship practices that are protective of the air, water, land, and other natural and cultural resources impacted by DOE operations and by which DOE cost effectively meets or exceeds compliance with applicable environmental; public health; and resource protection laws, regulations, and DOE requirements. This objective must be accomplished by implementing Environmental Management Systems (EMSs) at DOE sites. An EMS is a continuing cycle of planning, implementing, evaluating, and improving processes and actions undertaken to achieve environmental goals.

The following are some of these goals and objectives:

Goal. Protect environment through waste prevention.
Objective: Minimize environmental hazards, protect environmental resources, minimize life-cycle cost and liability of DOE programs, and maximize operational capability by eliminating or minimizing the generation of wastes that would otherwise require storage, treatment, disposal, and long-term monitoring and surveillance.

Goal. Protect environment through reduction of environmental releases.
Objective: Minimize environmental hazards, protect environmental resources, minimize life-cycle cost and liability of DOE programs, and maximize operational capability by eliminating or minimizing the use of toxic chemicals and associated releases of pollutants to the environment that would otherwise require control, treatment, monitoring, and reporting.

Goal. Protect environment through environmentally preferable purchasing.
Objective: Minimize environmental hazards, conserve environmental resources, minimize life-cycle cost and liability of DOE programs, and maximize operational capability through the procurement of recycled-content, bio-based-content, and other environmentally preferable products, thereby minimizing the economic and environmental impacts of managing toxic byproducts and hazardous wastes generated in the conduct of site activities.

Goal. Protect environment through incorporation of environmental stewardship in program planning and operational design.
Objective: Minimize environmental hazards, conserve environmental and energy resources, minimize life-cycle cost and liability of DOE programs, and maximize operational capability by incorporating sustainable environmental stewardship in the commissioning of site operations and facilities.

Goal. Protect environment through post-consumer material recycling.
Objective: Protect environmental resources, minimize life-cycle costs of DOE programs, and maximize operational capability by diverting materials suitable for reuse and recycling from landfills, thereby minimizing the economic and environmental impacts of waste disposal and long-term monitoring and surveillance.

All project team members need to have a clear understanding of LEED certification and the role it can play in improving property owners' competitive edge in an increasingly green market. Certification also gives independent verification that a building has achieved accepted standards in these areas, as outlined in the LEED Green Building Rating System. LEED certification of a project provides recognition of its quality and environmental stewardship. Its rating system is widely accepted and recognized by both public and private sectors, further fueling the demand for green building certification systems.

Since its inception, the LEED Rating System has made significant inroads into the mainstream design and construction industry, and contractors and property developers are realizing that they too can contribute to a project's success in achieving green objectives. This is accomplished first by understanding the LEED process and the specific role it can play in achieving LEED credits and then, through early involvement and participation throughout the different project phases, by incorporating a team approach in an integrated design process. Needless to say, measurable benchmarks are needed to achieve verification and confirm a building's acceptable performance. The American Society of Heating, Refrigerating and Air-Conditioning Engineers (ASHRAE) puts this responsibility of defining design intent requirements squarely on the shoulders of the owner. However, it is not possible to correctly evaluate a building or project unless the criteria on which the project's design and execution were based are made available. A project's plans, specifications, and the like, must therefore be prepared in a manner that can achieve measurable results. If not, a meaningful evaluation to determine if a project has met the required results and original design intent is not possible. Furthermore, before measurable green criteria can be established, it is necessary to first agree on a finite definition of green construction and to specify exactly what is to be accomplished.

The National Association of Home Builders (NAHB®) says that its Model Green Home Building Guidelines were written to facilitate moving environmentally friendly home-building concepts further into the mainstream marketplace; these guidelines represent one of two rating systems that make up NAHBGreen, the National Green Building Program. NAHB states that its guidelines "should be viewed as a dynamic document that will change and evolve as new information becomes available, improvements are made to existing techniques and technologies, and new research tools are developed."

The NAHB point system consists of three levels of green building: Bronze, Silver, and Gold, that are available to builders wishing to use NAHB's guidelines to rate their projects. NAHB stipulates:

At all levels, there are a minimum number of points required for each of the seven guiding principles to assure that all aspects of green building are addressed and that there is a balanced, whole-systems approach. After reaching the thresholds, an additional 100 points must be achieved by implementing any of the remaining line items.

Table 2.1 outlines the points needed to achieve one of the three rating threshold levels for a green building.

While green buildings may appear to be similar to traditional buildings, the conceptual approach in sustainable design substantially differs in that it revolves around a concern for the environment by extending the life span of natural resources and providing human comfort and well-being, security, productivity, and energy efficiency. This approach offers reduced operating costs, including those of energy and water, as well as other, intangible, benefits. For example, according to the Indian Green Building Council (IGBC), which

TABLE 2.1 NAHB Point System

	Bronze	Silver	Gold
Lot design, preparation, and development	8	10	12
Resource efficiency	44	60	77
Energy efficiency	37	62	100
Water efficiency	6	13	19
Indoor environmental quality	32	54	72
Operation, maintenance, and homeowner education	7	7	9
Global impact	3	5	6
Additional points from sections of applicant's choice	100	100	100

administers the LEED India rating system, there are a number of salient attributes of a green building:

- Minimal disturbance to landscapes and site condition
- Use of recycled and environmentally friendly building materials
- Use of nontoxic and recycled/recyclable materials
- Efficient use of water and water recycling
- Use of energy-efficient and eco-friendly equipment
- Use of renewable energy
- Indoor air quality (IAQ) for human safety and comfort
- Effective controls and building management systems

The *Whole Building Design Guide (WBDG)*, a program of the National Institute of Building Sciences (NIBS), also outlines specific objectives and principles of sustainable design:

Objectives
- Avoid resource depletion of energy, water, and raw material
- Prevent environmental degradation caused by facilities and infrastructure throughout their life cycles
- Create built environments that are livable, comfortable, safe, and productive

Principles
- Optimize site potential
- Optimize energy use
- Protect and conserve water
- Use environmentally preferred products
- Enhance indoor environmental quality (IEQ)
- Optimize operations and maintenance procedures

James Woods, executive director of The Building Diagnostics Research Institute notes:

Building performance is a set of facts and not just promises. If the promises are achieved and verified through measurement, beneficial consequences will result and risks will be managed. However, if the promises are not achieved, adverse consequences are likely to lead to increased risks to the occupants and tenants, building owners, designers and contractors; and to the larger interests of national security and climate change.

Alan Bilka, a sustainability design expert with ICC Technical Services, correctly points out:

Over time, more and more "green" materials and methods will appear in the coders and/ or have an effect on current code text. But the implications of green and sustainable building are so wide and far reaching that their effects will most certainly not be limited to one single code or standard. On the contrary, they will affect virtually all codes and will spill beyond the codes. Some green building concepts may become hotly contested political issues in the future, possibly requiring the creation of new legislation and/or entirely new government agencies.

2.2   GREEN BUILDING EVALUATION SYSTEMS

Today there are numerous building rating systems in the United States and around the world. Globally, voluntary systems have played a major role in raising awareness and in popularizing green design; however, most of them have been specifically tailored to suit the building industry of the country where they were developed. For example, in 2006 China's Ministry of Construction introduced a Green Building Evaluation Standard based on a three-star system (i.e., three levels of ratings); this was a first attempt by China to create a local green building standard. The purpose of the Chinese system, which has many striking similarities with the LEED system, is to create voluntary ratings that will encourage green development.

Likewise in 1999, Taiwan's Architecture and Building Research Institute of the Ministry of the Interior formulated a *Green Building Illustration and Assessment Handbook* to promote green building. In the same year, Taiwan introduced a Green Building Evaluation and Labeling System (GBELS) and established a Green Building Committee to evaluate, encourage, and award green building designs (Chinese Architecture and Building Centre, 2007). In India the U.S.-based LEED Rating System is being promoted by the CII Green Business Centre in Hyderabad. With a view to India's agro-climatic conditions—especially the preponderance of non-air-conditioned buildings—it was decided to set up a national rating system, the Green Rating for Integrated Habitat Assessment (GRIHA), applicable to all types of building in India's different climatic zones.

These are but a few of the many evaluation systems in place within the United States and worldwide. Such systems are necessary because buildings have major environmental impacts over their entire life cycle. Today many of our natural resources such as ground cover, forests, water, and energy are being depleted to give way to new building construction. We have found that green buildings only minimally deplete natural resources during their construction and operation. Generally speaking, to be able to make a proper evaluation of how "green" a building is, it is necessary to evaluate and consider the application and incorporation of the following green building design principles and their integration into the design process:

- Sustainable site planning
- Building envelope design should minimize adverse environmental impact
- Building system design that incorporates high-performance/energy-efficient HVAC, lighting, electrical (e.g., ENERGY STAR®), and water-heating systems; assurance that such systems are commissioned
- Integration of renewable energy sources such as solar, wind, and alternative energy to generate energy onsite.
- Water efficiency and waste management
- Use of ecologically sustainable materials and products that have high recycled content, are rapidly renewable, and have minimum off-gassing of harmful chemicals and the like
- Indoor environmental quality to maintain indoor air quality and thermal and visual comfort

2.3   USGBC'S LEED CERTIFICATION AND RATING SYSTEM

Comprehensive documentation about LEED accreditation requirements and reference guides, careers, and e-newsletters can be found on the U.S. Green Building Council (USGBC®) and Green Building Certification Institute (GBCI) websites (www.leedbuilding.org; www.gbci.org). The most appropriate way to contribute to the success of a LEED project is to become familiar with the many requirements and opportunities offered by the program. Success in earning LEED certification for a project starts in the initial planning stage, where stakeholders make a commitment to pursue it. The next step is registering the project and paying an initial flat fee. As part of the newly launched LEED V3, GBCI has assumed responsibility for administrating LEED certification for all commercial and institutional projects registered under a LEED Rating System.

When a project is completed and all the numbers are in, including all supporting documentation, it is submitted for evaluation and certification. Then it is listed on the LEED project list. A summary sheet showing the tally of credits earned becomes available for most certified projects. For assistance in the certification process, an online policy manual gives an overview of program requirements and identifies the policies put in place by the GBCI for the purposes of administering the LEED certification process.

2.3.1 LEED Process Overview

Basically the LEED 2009 Green Building Rating System consists of a set of performance standards used in the certification of commercial, institutional, and other building types in the public and private sectors with the intention of promoting healthy, durable, and environmentally sound practices. A LEED certification is indisputable evidence of independent, third-party verification that a building project has achieved the highest green building and performance measures according to the level of certification achieved. Setting up an integrated project team to include the major project stakeholders such as the developer/owner, architect, engineer, landscape architect, contractor, and asset and property management staff is helpful in jump-starting the process. An integrated, systems-oriented approach to green project design, development, and operations can yield significant synergies while enhancing the overall performance of a building. During the initial project team meetings, the project's goals are clarified and delineated and the LEED certification level sought is established.

Projects must adhere to the LEED Minimum Program Requirements (MPRs) to achieve certification. MPRs describe the eligibility for each system and are intended to "evolve over time in tandem with the LEED Rating Systems." Though there are eight requirements that are standardized for all systems, the thresholds and levels apply differently for each. Nevertheless, LEED projects must comply with all applicable MPRs outlined in the following list. To clarify the minimum program requirements, one of the categories will be used as an example: New Construction and Major Renovations. (The USGBC and GBCI websites should always be checked for the latest updates.)

1. The project must comply with all applicable federal, state, and local building-related environmental laws and regulations where the project is located.
2. The project must consist of a complete, permanent building or space, designed for, constructed on, and operated on already existing land. LEED projects are required to include new, ground-up design and construction or major renovation of at least one complete building. Moreover, construction prerequisites and credits may not be submitted for review until substantial construction is completed.
3. The project must employ a reasonable site boundary:
   (a) The project boundary is to include all contiguous land that is associated with and supports normal building operations for the project.
   (b) The project boundary must normally include only land that is owned by the party that owns the project.
   (c) Projects located on a campus must contain project boundaries so that if all the campus buildings become LEED-certified, 100% of the gross campus land area will be included within a LEED boundary.
   (d) Any given parcel of real property may only be attributed to a single LEED project building.
   (e) Any tampering with a LEED project boundary is completely prohibited.

4. The project must comply with minimum floor area requirements by incorporating a minimum of 1000 square feet (93 square meters) of gross floor area.

5. The project must comply with minimum full-time equivalent (FTE) occupancy rates. One or more FTE must be served, calculated as an annual average to use LEED in its entirety.

6. The project owners must consent to sharing whole-building energy and water usage data with USGBC and/or GBCI for a period of at least five years.

7. The gross floor area of the LEED project must conform to a ratio of minimum building area to site area—the building must not be less than 2% of the gross land area within the LEED project boundary.

8. Registration and certification activity must comply with reasonable timetables and rating system sunset dates, which basically means that if a LEED 2009 project is inactive for four years, GBCI reserves the right to cancel the registration.

2.3.2   How LEED Works

LEED is a point-based system in which building projects earn points for satisfying specific green building criteria. The awarding of points relative to performance is covered under five environmental categories: Sustainable Sites (SS), Water Efficiency (WE), Energy and Atmosphere (EA), Materials and Resources (MR), and Indoor Environmental Quality (IEQ). An additional category, Innovation in Design (ID), addresses sustainable building expertise as well as design measures not covered under the five environmental categories and Regional Priority (RP). Designers can select the points most appropriate to their projects to achieve a LEED rating. A total of 100 base points plus 10 points (6 ID and 4 RP) are possible. The number of points the project earns determines the level of certification it receives: Platinum, Gold, Silver, or Certified. Moreover, LEED 2009 alignment provides a continuous improvement structure that enables USGBC to develop LEED in a more predictable manner.

When USGBC first introduced the LEED Green Building Rating System, Version 1.0, in December 1998, it was considered by all to be a pioneering effort. Since then LEED has inspired and instigated global adoption of sustainable green building practices through the adoption and execution of universally understood and accepted tools and performance criteria. Today LEED has become the leading means for certifying green buildings in the United States. The USGBC recently released a new version, LEED 2009 (formerly known as LEED V3), which was the first major LEED overhaul since Version 2.2 in 2005. LEED 2009 has been significantly transformed by many changes, both major and minor, to the rating system and its priorities.

Many of the changes in LEED 2009 are designed to address criticism levied against earlier versions; they include an entirely new weighting system that refers to the process of redistributing the available points in a manner that a credit's point value more accurately reflects its potential to either mitigate the negative or promote positive environmental impacts of a building. Thus in LEED 2009, credits that most directly address the most significant impacts are given the greatest weight, subject to the system design parameters described previously. This has resulted in a significant change in allocation of points compared with earlier LEED Rating Systems. Generally speaking, the modifications reflect a greater relative emphasis on the reduction of energy consumption and greenhouse-gas emissions associated with building systems, transportation, the embodied energy of water and materials, and, where applicable, solid waste (e.g., for Existing Buildings: Operations & Maintenance).

Additional improvements include an increased opportunity for innovation credits and a new opportunity for achieving bonus points for Regional Priority. A less obvious revision in LEED 2009 is the reduction of maximum possible exemplary performance credits from 4 to 3. The intention here was to return to the original purpose of the credit; that is, encouraging projects to pursue innovation in green building. There are numerous other important modifications and improvements which are discussed next and in the following chapters.

2.3.3   The LEED Points Rating System

LEED is a continually evolving basic point based system that has set the green building standard and has made it the most widely accepted green program in the United States. The various LEED categories differ in their scoring systems based on a set of required "prerequisites" and a variety of "credits" in seven major categories as outlined earlier. In LEED V2.2 for new construction and major renovations for commercial buildings, there were 69 possible points and buildings were able to qualify for four levels of certification. LEED 2009 is a significant improvement on earlier LEED versions and has become much less complicated. The new USGBG LEED Green Building certification levels for all systems are more consistent and are described here:

- *Certified:* 40–49 points
- *Silver:* 50–59 points
- *Gold:* 60–79 points
- *Platinum:* 80 plus points

The number of points available per LEED system has been increased so that all systems have 100 base points as well as 10 possible innovation and regional bonus points, bringing the possible achievable for each category to 110. Figure 2.1 contains pie charts showing the new LEED 2009 for new construction and commercial interiors. Figures 2.2 and 2.3 are examples of buildings that have received various levels of LEED certification.

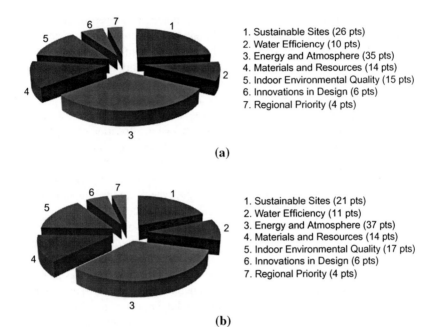

1. Sustainable Sites (26 pts)
2. Water Efficiency (10 pts)
3. Energy and Atmosphere (35 pts)
4. Materials and Resources (14 pts)
5. Indoor Environmental Quality (15 pts)
6. Innovations in Design (6 pts)
7. Regional Priority (4 pts)

(a)

1. Sustainable Sites (21 pts)
2. Water Efficiency (11 pts)
3. Energy and Atmosphere (37 pts)
4. Materials and Resources (14 pts)
5. Indoor Environmental Quality (17 pts)
6. Innovations in Design (6 pts)
7. Regional Priority (4 pts)

(b)

FIGURE 2.1　(a) LEED-NC 2009 and (b) LEED-CI 2009 point distribution system, which incorporates a number of major technical advancements focused on improving energy efficiency, reducing carbon emissions, and addressing additional environmental and human health concerns.

The maximum achievement in earlier versions of LEED-NC, 69 points, in LEED 2009 has been increased to 100, but it remains unclear sometimes how the added 31 points are distributed. Aurora Sharrard, research manager at Green Building Alliance (GBA), says:

The determination of which credits achieve more than 1 point (and how many points they achieve) is actually the most complex part of LEED 2009. LEED has always implicitly weighted buildings' impacts by offering more credits in certain sections. However, in an effort to drive greater (and more focused) reduction of building impact, the USGBC is now applying explicit weightings to all LEED credits. The existing weighting scheme was developed by the National Institute of Standards and Technology (NIST). The USGBC hopes to have its own weighting system for future LEED revisions, but currently LEED credits are proposed to be weighted based on the following categories, which are in order of weighted importance:

- *Greenhouse-gas emissions*
- *Indoor environmental quality fossil fuel depletion*
- *Particulates*
- *Water use*
- *Human health (cancer)*

FIGURE 2.2 Interior of BP America's new Government Affairs Office in Washington, D.C., designed by Fox Architects. The 22,000-square-foot building achieved a LEED Platinum certification. *(Source: Fox Architects.)*

- *Ecotoxicity*
- *Land use*
- *Eutrophication*
- *Smog formation*
- *Human health (noncancer)*
- *Acidification*
- *Ozone depletion*

The new weighting preferences in the LEED 2009 system put much greater emphasis on energy, which is appropriate as this addresses some of the criticism levied against earlier LEED versions. There has also been an increase in the Innovation and Design (ID) credits from 4 to 5. An additional point can be achieved for having a LEED Accredited Professional (LEED AP) on the project team, which brings the total ID points to 6. The introduction of a new category, Regional Priority, also adds another potential 4 bonus points, bringing the total points possible to 110.

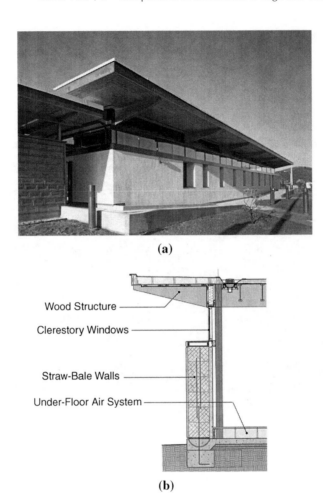

(a)

(b)

FIGURE 2.3   (a) Santa Clarita Transit Maintenance building is one of the first LEED Gold-certified straw-bale buildings in the world. The resource- and energy-efficient facility was designed by HOK and exceeds California Energy Efficiency Standards by more than 40%, securing a new standard for straw-bale construction in high-performance building design. (b) Section view through building's exterior wall. The designers reportedly opted for a solar photovoltaic canopy to shade buses and provide nearly half of the building's annual energy needs. An electronic monitoring system is in place to track thermal comfort, energy efficiency, and moisture levels. *(Source: HOK Architects.)*

2.3.4   The LEED Building Certification Model

Beginning in April 2009, the Green Building Certification Institute assumed responsibility for managing the review and verification process for projects seeking LEED 2009 certification. GBCI, an independent nonprofit organization established in January 2008 with the support of USGBC, provides third-party project certification and professional credentials recognizing excellence in green building performance and practice. The new GBCI building certification infrastructure recently added a network of 10 well-respected certification groups that are accredited to ISO standard 17021. These organizations are recognized for their role in certifying organizations, processes, and products to ISO and other standards. They include

- ABS Quality Evaluations (www.abs-qe.com)
- BSI Management Systems America (www.bsigroup.com)
- Bureau Veritas North America (www.us.bureauveritas.com)
- DNV Certification (www.dnv.us/services/certification)
- Intertek (www.intertek-sc.com)
- KEMA-Registered Quality (www.kema.com)
- Lloyd's Register Quality Assurance (www.lrqausa.com)
- NSF-International Strategic Registrations (www.nsf.org)
- SRI Quality System Registrar (www.sriregistrar.com)
- Underwriters Laboratories-DQS (www.ul.com)

The revised LEED V3 is a significantly improved ISO-compliant certification process that is adaptable and designed to grow with the green building movement. All LEED certification applications will continue to be submitted through LEED-Online. Also, USGBC states that it will continue to administer the development and ongoing improvement of the LEED Rating System and will remain the primary source for LEED and green building education.

It should be noted that all LEED V3 projects use LEED-Online Version 3. In fact, USGBC already has four LEED programs launched or in the process of being launched using the online V3 platform for submittals. These are LEED for Neighborhood Development, LEED Portfolio Program, LEED for Healthcare, and LEED for Retail. Additional information with respect to these programs can be obtained from the GBCI website.

2.3.5   LEED V3: What's New?

With the latest overhaul, a new, greatly improved system has emerged, LEED Online V3, which provides enhanced functionality to improve efficiency and productivity. According to GBCI, the new version is "faster, smarter and a better user experience. It is designed to be scalable and more robust, through improved design, a more intuitive user interface, better communication between project teams and certifying bodies, and upgrades that respond to the changes in the LEED 2009 rating system." On its website, GBCI cites some of the new project management improvement tools incorporated into V3, including the following.

Project organization. The ability to sort, view, and group LEED projects according to a number of project traits, such as location, design or management firm, and so forth.

Team member administration. Increased functionality and flexibility in making credit assignments, adding team roles, and assigning roles to team members. For example, credits are now assigned by team member name rather than by project role.

Status indicators and timeline. Clearer explanation of the review and certification process and highlighting of steps as they are completed in specific projects. The system now displays specific dates related to each phase and step, including target dates on which each review is to be returned to the customer.

Enabling features to support the LEED certification review and submittals process. Enhancements to the functionality of submittal documentation and certification forms:

- *End-to-end process support:* The new system will guide project teams in the certification process, from initial project registration through the various review phases. Furthermore, it will provide assistance to beginners during the registration phase to help them determine the type of LEED Rating System that is best suited to their project.
- *Improved midstream communication:* A mid-review clarification page allows a LEED reviewer to contact the project team through the system when minor clarifications are required to complete the review.
- *Data linkages:* LEED Online V3 automatically fills out fields in all appropriate forms after the user inputs data the first time, which saves time and helps ensure project-wide consistency. Override options are available when required.
- *Automatic data checks:* The new system alerts users when incomplete or required data are missing, allowing them to correct errors before application submission to avoid delays.
- *Progressive, context-based disclosure of relevant content:* After selection of an option, the new system simplifies the completion of forms by only showing data fields that are relevant to the customer's situation and hiding all extraneous content.

2.4 THE GREEN GLOBES RATING SYSTEM

The Green Globes® website (www.greenglobes.com) describes its system as "The Practical Building Rating System" and states,

The Green Globes system is a revolutionary building environmental design and management tool. It delivers an online assessment protocol, rating system and guidance for green building design, operation and management. It is interactive, flexible and affordable, and provides market recognition of a building's environmental attributes through third-party verification.

Green Globes is certainly less complicated than USGBC's LEED Rating System. It employs a straightforward questionnaire-based format, which is written in lay terms and is fairly easy to complete even if an applicant lacks environmental design experience. The questions are typically of a yes/no type and are grouped broadly under seven modules of building environmental performance

FIGURE 2.4   Blakely Hall is a community center and town hall for Issaquah Highlands, a planned community near Seattle. The building consists of 7000 square feet and was built for $1.5M. It is used mainly as a meeting place for numerous clubs and groups. Blakely Hall is the first building in the United States to earn a Green Globes Certification (earned two out of four possible) as well as a LEED.

(management, site, energy, water, resources, emissions, indoor environment). On completion of the questionnaire, a printable report is automatically generated.

Figure 2.4 is a photo of Blakely Hall, the first Green Globes-rated building in the United States. Blakely Hall is a community center and town hall for Issaquah Highlands, a planned community near Seattle, Washington. It earned two GBI Green Globes (out of four possible). The building incorporates a variety of green attributes such as high energy and water efficiency, integration of daylighting, and use of locally sourced materials. The implementation of a construction waste management plan also helped divert more than 97% of waste from landfill. Blakely Hall is an example of a "green" building that has earned various awards including a LEED Silver certification.

The idea and market for green buildings has been growing rapidly, and although there are a number of rating systems available in the United States, the two systems most widely used for commercial structures are LEED and Green Globes (Go Green Plus). LEED focuses largely on assessing new-construction sustainable high-performance buildings, although existing buildings are eligible for rating (Figure 2.5). Green Globes mainly targets owners of existing buildings who want to be more environmentally friendly. In this

FIGURE 2.5 The 80,000-square-foot (7400-square-meter) Integrated Learning Centre at Queen's University in Kingston, Ontario, received a four-leaf rating through the BREEAM/Green Leaf program, which is now accessible online as Green Globes. Designed by B+H Architects of Toronto, the project was completed in 2004. The Ottawa-based firm Green & Gold implemented the BREEAM/ Green Leaf program for the project and helped integrate the building analysis tool into the design process. The lighting, ventilation, and water distribution systems, in particular contributed to the building's high rating. *(Source:* Building Safety Journal, *June 2005; interiorimages.ca.)*

section we analyze and compare Green Globes and LEED. While Green Globes currently has a minute share of the U.S. certification market (about 55 buildings), it is making a determined effort to rectify this situation.

2.4.1 An Overview of the Green Building Initiative and Green Globes

The Green Building Initiative (GBI) is a 501(c)(3) nonprofit educational organization based in Portland, Oregon. Its mission is to accelerate the adoption of sustainable design and construction practices that result in energy-efficient, healthier, and environmentally sustainable buildings by promoting credible and practical green building approaches for residential and commercial construction. Ward Hubbell serves as president of GBI at the discretion of an independent, multistakeholder board of directors that comprises construction professionals, product manufacturers, nonprofit organizations, university officials, and other interested parties.

History and Background

The birth of the Green Globes system lies in the Building Research Establishment's Environmental Assessment Method (BREEAM®), which was exported to Canada from the United Kingdom in 1996 in cooperation with ECD Energy and Environment. Green Globes was initially developed as a rating and assessment system to monitor and assess green buildings in Canada. The Canadian government has been using it for several years under the Green Globes name, and it has been the basis for the Canadian Building Owners and Manufacturer's Association (BOMA Canada) Go Green Plus program. Go Green was adopted by BOMA Canada in 2004 and was chosen by Canada's Department of Public Works and Government Services. Green Globes has also been adopted by the Canada's Continental Automated Buildings Association (CABA) to power a building intelligence tool called Building Intelligence Quotient (BiQ).

The Green Globes environmental assessment and rating system represents more than a decade of research and refinement by a wide range of prominent international organizations and experts. In 1996, with the help of 35 contributors, the Canadian Standards Association first published BREEAM Canada for Existing Buildings. In 1999 ECD Energy and Environment collaborated with Terra Choice, the agency that administers the Government of Canada's Environmental Choice program, to develop a more efficient and streamlined question-based tool that was later introduced as the BREEAM Green Leaf eco-rating program. Later that year the program led to the formation of Green Leaf for Municipal Buildings, associated with the Federation of Canadian Municipalities.

In 2000, BREEAM Green Leaf took another step forward by becoming an online assessment and rating tool under the name Green Globes for Existing Buildings. That same year, BREEAM Green Leaf for the Design of New Buildings was adapted for the Canadian Department of National Defense and Public Works and Government Services Canada. The program underwent a further iteration in 2002 by a panel of experts, including representatives from Arizona State University, the Athena Institute, BOMA, and a number of Canadian federal departments.

In 2002, Green Globes for Existing Buildings went online in the United Kingdom as the Global Environmental Method (GEM), and endeavors were made to incorporate BREEAM Green Leaf for the Design of New Buildings into the online Green Globes for New Buildings. Green Globes for Existing Buildings was adopted and operated by BOMA Canada in 2004 under the name Go Green Comprehensive (now known as Go Green Plus). The Canadian government later announced plans to adopt Go Green Plus for its entire real estate portfolio. All other Green Globes products in Canada are owned and operated by ECD Energy and Environment Canada.

Additionally in 2004, the GBI purchased the rights to promote, develop, and distribute Green Globes for New Construction in the United States. In adapting the system, minor changes were instituted to make it appropriate for the U.S. market (e.g., converting units of measurement and integrating with ENERGY STAR). GBI also committed itself to ensuring that Green Globes continues to reflect best practices, changing opinions, and ongoing advances in research and technology. To that end, in 2005, it became the first green building organization to be accredited as a standards developer by the

American National Standards Institute (ANSI), and Green Globes Rating System is also on track to become the first American National Standard for commercial green buildings. As part of this process, GBI established a technical committee and subcommittees of more than 75 building science experts, including representatives from several federal agencies, states, municipalities, universities, and leading construction firms, in addition to building developers.

In March 2009 GBI and the American Institute of Architects (AIA) signed a Memorandum of Understanding (MOU), which states that the GBI and AIA pledge to work in concert to promote the design and construction of energy-efficient and environmentally responsible buildings. An MOU was also signed between GBI and ASHRAE to collaborate on the adoption of sustainability principles in the built environment.

2.4.2 Defining the Green Globes Rating System

The Green Globes V1 Assessment Protocol covers seven areas, with each area having an assigned number of points to quantify overall building performance. These are listed in Table 2.2. There is a clear emphasis on energy, which takes up more than a third of the total points.

The Process

The scoring for the Green Globe categories is based on a series of questions that are completed via the online questionnaire that is part of the Green Globes Tool. Normally, there are pop-up "tool tips" embedded in the questionnaire to address frequently asked questions and provide clarifications regarding the survey's input data requirements. According to Amy Stodghill, a sustainability

TABLE 2.2 Green Globes' Seven Assessment Categories

Assessment Category	Points	Percentage
Project management	50	5
Site	115	11.5
Energy	360	36
Water	100	10
Resources	100	10
Emissions, effluents, and other impacts	75	7.5
Indoor environment	200	20
Total	**1000**	**100**

consultant who used a free 30-day trial version of the online Environmental Assessment for Existing Commercial Buildings: "It is essentially a 22-page questionnaire/survey covering energy, transportation, water, waste reduction and recycling, site management, air and water emissions, indoor air quality (IAQ), purchasing and communication. It is completed online only and is very user friendly." The time normally required to input data and complete the survey is roughly 2 to 3 hours per building; however, this does not include the time required to research and gather required information.

Stodghill also notes:

Each question is weighted with points (in all totaling up to 1,000). The overall rating is tracked as questions are answered. The overall rating, however, is based on a percentage, not on total points. This way there are no penalties for questions that are not applicable (i.e., Answering 'no' on water efficient irrigation questions will not be counted against you if you do not have any landscaping).

On completion of the questionnaire, the Green Globes system automatically generates a report based on the answers given. The report lists where the building stands in each major category and provides suggestions for raising the applicant's score.

To earn a formal Green Globes rating/certification, a building has to be evaluated by an independent third party recognized, trained, and affiliated with GBI. Both new construction and existing buildings can be formally rated or certified, and within the Green Globes system projects that achieve a score of 35% or more of the maximum 1000 points become eligible for a Green Globes rating of one, two, three, or four globes, as follows:

- *One Globe:* 35–54%
- *Two Globes:* 55–69%
- *Three Globes:* 70–84%
- *Four Globes:* 85–100%

A summary of rating levels and how they relate to environmental achievement is shown in Figure 2.6. However, buildings cannot be promoted as having achieved a Green Globes rating until the information submitted has been assessed by a qualified third party.

According to Green Globes-NC, projects are awarded up to 1000 points based on their performance in seven areas of assessment as follows.

1. Project Management: 50 Points

The Green Globes system emphasizes integrated design, an approach that encourages multidisciplinary collaboration from the earliest stages of a project while considering the interaction between elements related to sustainability. Most decisions that influence a building's performance (such as siting, orientation, form, construction, and building services) are made at the start of the project, and yet it is common, even for experienced designers, to focus on

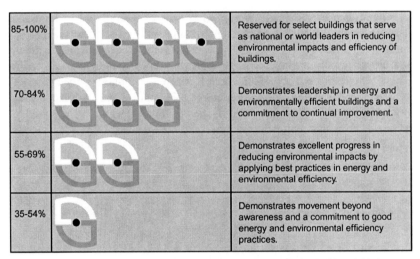

85-100%		Reserved for select buildings that serve as national or world leaders in reducing environmental impacts and efficiency of buildings.
70-84%		Demonstrates leadership in energy and environmentally efficient buildings and a commitment to continual improvement.
55-69%		Demonstrates excellent progress in reducing environmental impacts by applying best practices in energy and environmental efficiency.
35-54%		Demonstrates movement beyond awareness and a commitment to good energy and environmental efficiency practices.

FIGURE 2.6   Globes rating levels in the United States. *(Source: Green Building Initiative.)*

environmental performance late in the process, adding expensive technologies after key decisions have been made. This is costly as well as ineffective.

To ensure that all relevant players are involved, the system tailors questionnaires so that input from team members is captured interactively, even on issues that may at first appear to fall outside their mandate. For example, while site design and landscaping may come under the purview of landscape designers, the questionnaire prompts the electrical engineer's involvement with design issues such as outdoor lighting or security. Thus the Green Globes format promotes design teamwork and prevents a situation where, despite strong individual resources, the combined effort falls short. Also included under project management are environmental purchasing, commissioning, and emergency response.

2.   Site: 115 Points

Building sites are evaluated based on the development area (including site selection, development density, and site remediation), ecological impacts (ecological integrity, biodiversity, air and water quality, microclimate, habitat, and fauna and flora), watershed features (such as site grading, stormwater management, pervious cover, and rainwater capture), and site ecology enhancement.

3.   Energy: 360 Points

To simplify the process of energy performance targeting, Green Globes-NC directs users to the web interface for the ENERGY STAR Target Finder software, which helps to generate a realistic energy consumption target. As a result, an aggressive energy performance goal can be set—with points awarded for design and operations strategies that result in a significant reduction in energy consumption—as compared to actual performance data from real buildings. As previously stated, Green Globes is the only green rating system to use energy data generated through DOE's Commercial Buildings Energy Consumption Survey (CBECS), which is widely considered to be the most accurate and reliable source of energy benchmarking information.

In addition to overall consumption, projects are evaluated based on the objectives of reduced energy demand (through space optimization, microclimatic response to site, daylighting, envelope design, and metering), integration of "right-sized" energy-efficient systems, onsite renewable energy sources, and access to energy-efficient transportation.

4. Water: 100 Points

Projects receive points for overall water efficiency as well as specific water conservation features (such as submetering, efficiency of cooling towers, and irrigation strategies) and onsite treatment (gray water and wastewater).

5. Resources: 100 Points

The resources section covers building materials and solid waste. It includes points for materials with low environmental impact (based on life-cycle assessment); minimal consumption and depletion of resources (with an emphasis on materials that are reused, recycled, bio-based, and, in the case of wood products, certified as having come from sustainable sources); reuse of existing structures; building durability; adaptability and disassembly; and reduction, reuse, and recycling of waste.

6. Emissions, Effluents, and Other Impacts: 75 Points

Points in this section are awarded in six categories, including air emissions, ozone depletion and global warming, protection of waterways and impact on municipal wastewater treatment facilities, minimization of land and water pollution (and the associated risk to occupants' health and to the local environment), integrated pest management, and storage of hazardous materials.

7. Indoor Environment: 200 Points

According to the EPA, indoor air can be up to 10 times more polluted than outdoor air, even in cities where the quality of outdoor air is poor. This has obvious health implications, but the consequences are also economic. A study by Lawrence Berkeley National Laboratory found that improving indoor air at work could save U.S. businesses up to $58 billion in lost sick time each year, with another $200 billion earned in increased worker performance. The Indoor Environment section evaluates the quality of the indoor environment based on the effectiveness of the ventilation system, the source control of indoor pollutants, lighting design and the integration of lighting systems, thermal comfort, and acoustic comfort.

According to GBI, the process for obtaining formal Green Globes rating/certification is quite straightforward; it consists of the following steps:

Step 1: Purchase a subscription to either Green Globes-NC or the Continual Improvement of Existing Buildings (CIEB) program.

Step 2: Log in to Green Globes at the GBI website with a username and password.

Step 3: Select the tool purchased (NC or CIEB) to link to Green Globes.

Step 4: Add a building and enter the basic building information.

Step 5: Use step-through navigation and the building dashboard to complete the survey.

Step 6: Print the report of the building's projected rating and obtain feedback using automatic reports.

Step 7: Order a third-party assessment and Green Globes rating/certification (if the automated report indicates a predicted rating of at least 35% of 1000 points).

Step 8: Schedule and complete a third-party building assessment. Third-party assessment for Green Globes-NC occurs in two comprehensive stages: The first stage is a review of the construction documents developed through the design and delivery process. The second stage is a walk-through of the building post-construction.

Step 9: Receive the Green Globes rating and certification.

2.4.3   Green Globes—An Alternative to LEED?

There has been a great deal of interest in the Green Globes system as it compares to LEED. It is important to bear in mind that there are many similarities between the two, largely because they share common roots and because they share common ideas of green buildings. However, there are several significant differences, highlighted in the following paragraphs.

The origin of the Green Globes Canada system lies in the Building Research Establishment's Environmental Assessment Method (BREEAM), which was developed in the United Kingdom and later published in 1996 by the Canadian Standards Association. One of BREEAM's creators, ECD Consultants, used it as the basis for a Canadian assessment method called BREEAM Green Leaf. At first, BREEAM Green Leaf was created to allow building owners and managers to self-assess the performance of their existing buildings. Green Globes was then developed into a web-based application of Green Leaf by ECD.

The Canadian Green Globes system thus became a web-based green building performance interactive software tool that today competes with the better known, though more complicated and more expensive, USGBC LEED system. Green Globes was introduced to the U.S. market as a potentially viable alternative to LEED. GBI was established to promote the use of the National

Association of Homebuilders' Model Green Home Building Guidelines, and has recently expanded into the nonresidential building market by licensing Green Globes for use in the United States. GBI is supported by various industry groups, including the Wood Promotion Network, that object to some provisions in LEED and that, as trade associations, are prohibited from joining USGBC.

When Green Globes was released in Canada in January 2002, it consisted of a series of questionnaires customized by project phase and the user's role on the design team (e.g., architect, mechanical engineer, landscape architect). A total of eight design phases are now supported. A separate Green Globes module (Green Globes-CIEB) is available for assessing the performance of existing buildings. The questionnaires produce design guidance appropriate to each team member and project phase. Once they have purchased a subscription, Green Globes users can order a Green Globes third-party assessment at any time during or after completion of the questionnaire. After an online self-assessment is completed and payment is made, a GBI representative contacts the project manager or project owner to schedule the third-party assessment and provide contact information. Completion of the pre-assessment checklist, which can be downloaded from the Green Globes Customer Training area, helps prepare for the assessment process.

Formal rating/certification of projects is necessary to provide a mechanism to ensure that new construction project teams or facilities management staff are fully aware of the environmental impact of design and/or operating management decisions. It also offers a visible way to quantify/measure project performance and allows recognition for achievements and hard work. Green Globes is designed to be cost-effective; through its value-added online system and a comprehensive yet streamlined in-person third-party review process, significant savings on consulting fees are made that were normally associated with green certification. There is an annual per-building license fee for use of the online tool as well as a third-party assessment fee. Rates are based on a number of factors, including project size (hectares/acres), number of integrated developments, and location (environmentally sensitive areas). Users can register for the annual license for the online Green Globes tools and choose to purchase a third-party assessment (required for certification). Third-party assessor travel expenses are separately billed (Figure 2.7).

It is estimated that the United States is home to more than 100 million buildings, which adds to the urgency of improving the performance of existing structures as a necessary prerequisite to widespread energy efficiency. According to GBI,

The missing element—until last year when GBI introduced Green Globes-CIEB—was a practical and affordable way to measure and monitor performance on an ongoing basis. Green Globes-CIEB allows users to create a baseline of their building's performance, evaluate interventions, plan for improvements, and monitor success—all within a holistic framework that also addresses physical and human elements such as material use and indoor environment.

FIGURE 2.7 New 356,000-square-foot Newell Rubbermaid Corporate Headquarters in Atlanta, which achieved a two-globe rating using the Green Globes New Construction module. *(Source: Green Building Initiative.)*

Table 2.3 lists some of the costs currently associated with the use of the Green Globes Rating System.

- Green Globes Existing Building Rating/Certification package costs $5270 per building
- New Construction Rating/Certification package costs $7270 per building.

Travel for a GBI assessor to and from the location is invoiced as actual expenses plus 20% after assessment or as a flat fee of $1000 upfront.

The Green Globes building rating system provides a LEED alternative assessment tool for characterizing a building's energy efficiency and environmental performance. It also provides guidance for green building design,

TABLE 2.3 Green Globes Costs

Software Subscriptions Costs	Price
Green Globes CIEB Existing Building one-year subscription	$1000
Green Globes New Construction	
One-project subscription	$500
Three-project subscription	$1500
Ten-project subscription	$2500
Third-Party Assessments/Green Globes Certification	
Green Globes-CIEB assessment/rating	$3500*
Green Globes-NC Stage I assessment	$2000
Green Globes-NC Stage II assessment/rating	$4000*
Green Globes-NC Stage I and II assessment/rating	$6000*

Note: A Green Globes subscription is required for third-party assessment/certification.
*Pricing for buildings of more than 250,000 square feet in size or departing significantly from standard commercial building complexity is custom-quoted prior to performance of assessment services.

operation, and management. Some feel that, compared to LEED, Green Globes' appeal may be enhanced by the flexibility and affordability it provides while simultaneously providing market recognition of a building's environmental attributes through recognized third-party verification. And, from a practical and marketing perspective, it should not be necessary to pursue LEED certification to demonstrate to tenants, customers, clients, and building visitors that a building's owners and managers, when they have Green Globes certification, are taking steps to be more environmentally responsible.

According to Christine Ervin, former president and CEO of USGBC, "Green Globes offers several very appealing features. Interactive feedback on strategies, interactions, and resources can be tailored to twenty different team roles and eight project stages. Numerical assessments are generated at stages for schematic design and construction, designed to coincide with planning and permit approvals."

Green Globes and LEED have many similarities, partly because they both evolved from the same source—BREEAM. For example, Green Globes and LEED are very alike in structure. Both have four levels of achievement, and both share a common set of green building design practices. There are six focus areas for LEED and seven for Green Globes, but the focus areas are in many respects similar.

A University of Minnesota study conducted in 2007 that compared LEED (pre-V3) with Green Globes, found that the systems were very similar. For example, the study found that "nearly 80% of the categories available for points in Green Globes are also addressed in LEED v2.2 and that over 85% of the categories specified in LEED V2.2 are addressed in Green Globes." The study further concluded that LEED was characterized as being more rigorous, rigid, and quantitative whereas Green Globes, while also rigorous, nevertheless maintained greater flexibility. Also, Green Globes focuses primarily on energy efficiency as a goal, and it was found to be easier to work with, less costly, and less time-consuming.

The same study concluded that there was only moderate dissimilarity between the two rating standards, but that LEED has a slightly greater emphasis on materials choices and Green Globes has a greater emphasis on energy saving. Green Globes also more heavily weights energy systems, up to 36% of the total points needed, whereas LEED in its earlier versions limited the energy category to about 25% of the total in the rating system. However, in the new LEED V3 version this has been appropriately addressed.

Furthermore, of the many buildings that have been evaluated with both systems, in all but two instances the systems generated comparable ratings. The two different ratings were only marginally so. It should be noted here that LEED 2009 has addressed many of these issues. In the final analysis, it appears that the primary differences between Green Globes and LEED boil down to cost and ease of use. The University of Minnesota study concluded:

From a process perspective, Green Globes' simpler methodology, employing a user-friendly interactive guide for assessing and integrating green design principles for buildings, continues to be a point of differentiation to LEED's more complex system. While LEED has introduced an online-based system, it remains more extensive and requires expert knowledge in various areas. Green Globes' Web-based self-assessment tool can be completed by any team member with general knowledge of the building's parameters.

GBI currently oversees Green Globes in the United States. It has also become an accredited standards developer under the auspices of the American National Standards Institute and is in the process of establishing Green Globes as an official ANSI standard. The ANSI process has always been consensus-based, involving a balanced committee of varying interests, including users, producers, interested parties, and nongovernmental organizations (NGOs), that conducts a thorough, open, and transparent technical review. Green Globes continues to be monitored by this committee and will continue to be updated through ANSI-approved rules and procedures.

Neither LEED nor Green Globes (or ENERGY STAR for that matter) provides continuous, longitudinal monitoring of energy efficiency or building performance, which means that building measurements and ratings are concluded on a one-off basis and must be reverified later on. This is a significant shortcoming in terms of the practicality of greening existing real estate, since buildings are dynamic and rarely perform in an identical manner week after week. Green Globes-NC is the only environmental rating system that provides early feedback on the process before critical and final decisions are made. This is a proven method for taking advantage of time and cost savings opportunities through integrated design and delivery, while benefiting from a cost-effective and comprehensive third-party assessment program.

Green Globes generates numerical assessment scores in two of the eight project phases: schematic design and construction documentation. These scores can either be used as self-assessments internally or can be verified by third-party certifiers. Projects that have had their scores independently verified can use the Green Globes logo and brand to promote their environmental performance. As mentioned, the Green Globes questionnaire corresponds to a checklist with a total of 1000 points listed in seven categories, as opposed to LEED's 100 points distributed in seven categories (Figure 2.8).

One of the differences between Green Globes and LEED is that the former offers protection against "non-applicable criteria." Thus if a builder marks a criterion as "N/A," he or she will be excused for not gaining points in those areas, which is why the actual number of points available varies by project. For example, if a building code overrides a criterion, the criterion can be marked as "N/A." As another example, if points are available for designing exterior lighting to avoid glare and skyglow, the user marks "N/A" for a project with no exterior lighting; those points are removed from the total number available so as not to penalize the project.

A rating of one or more Green Globes is applied to projects based on the percentage of applicable points they have achieved. In Canada the ratings range from one to five; in the United States the lowest rating has been eliminated and the rest adjusted so that the highest rating is four. Ward Hubbell, executive director of GBI, says that the objective of this adjustment was to have something that applicants are accustomed to—that is, a four-stage system, which is roughly comparable with the four levels of LEED.

It appears that Green Globes is broader than LEED in terms of technical content, including points for criteria such as optimized use of space, acoustical comfort, and an integrated design process. It is difficult to compare the levels

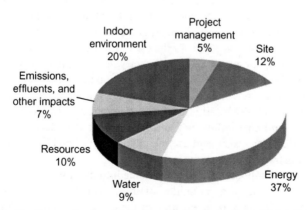

FIGURE 2.8   Pie diagram showing distribution of points in the Green Globes Rating System. *(Source: Building Green LLC.)*

of achievement required to claim points in the two systems because they are organized differently and because the precise requirements in Green Globes are not publicly available. Also, unlike LEED, Green Globes encourages energy reduction but does not require it. LEED calls for a minimum indoor air quality performance, whereas Green Globes does not. A recently published AIA report states that LEED makes it mandatory that builders have "some documentation of the initial building energy and operational performance through fundamental commissioning." This is not a Green Globes requirement.

Recognizing all mainstream forest certification systems is one of Green Globes' main attractions for strong timber-industry lobbying groups supporting GBI in the United States (and where it differs from LEED). Green Globes is more inclusive and opposes favoring Forest Stewardship Council (FSC) over Sustainable Forestry Initiative (SFI) certification; it recognizes timber certified through FSC as well as the American Tree Farm System (ATFS), the Canadian Standards Association (CSA), and SFI. LEED previously referenced only the FSC program; however, independent research has shown that all of these programs are in fact effective. There are more than 390 million acres of certified forest in North America, but less than one-sixth of that acreage is certified by FSC. The consensus is that legislation to encourage green building in states (e.g., Virginia and Arkansas) is likely to include Green Globes in addition to LEED. Furthermore, a number of federal agencies such as the Department of the Interior are reportedly considering Green Globes endorsement.

GBI president Ward Hubbell claims that Green Globes is on a par with LEED with respect to overall achievement levels, and he says, "We did carry out a harmonization exercise with LEED—not credit-by-credit; we compared objectives." The actual development of the Green Globes system in Canada, as well as its subsequent U.S. adaptation, has involved many iterations and participation by a wide range of organizations and individuals. Changes originally made to adapt Green Globes for the U.S. market do not appear to be substantive: converting units of measurement, referencing U.S. rather than Canadian standards and regulations, and incorporation of U.S. programs such as the EPA's Target Finder. Green Globes also awards points for the use of life-cycle assessment methods in product selection, although it does not specify how to apply them.

It is very likely that Green Globes' presence on the American scene has had a beneficial impact on LEED, perhaps prompting it to improve its rating system and release LEED 2009 V3. It is also important to recognize that Green Globes can attract a significant following that for various reasons has been alienated by LEED certifications' costs and complexity. This is good for both the green building industry and the environment.

Green Globes supporters tried to block the introduction of LEED in Canada, but lost a close vote in a committee of the Royal Architectural Institute of Canada that led in 2003 to the creation of the Canada Green Building Council (CaGBC). Alex Zimmerman, president of CaGBC, has levied some criticisms of Green Globes, noting that in Canada Jiri Skopek, president of ECD Energy and Environment Canada, has been the program's primary developer and in the past was its sole certifier. According to Zimmerman, "While there are more certifiers now, it is not clear who they are, how they were chosen, or who they are answerable to." In the United States, GBI responded to this criticism by training a network of independent certifiers to verify Green Globes ratings who have access to reports generated by the Green Globes website as well as other relevant information such as project drawings, energy simulation results, specifications, and commissioning plans.

One of Green Globes' advantages is that it is an economical, practical, and convenient methodology for obtaining comprehensive environmental and sustainability certification for new or existing commercial buildings. It provides a complete, integrated system that has been developed to enable design teams and property managers to focus their resources on the processes of actual environmental improvement of facilities and operations, rather than on costly, cumbersome, and lengthy certification and rating. Other advantages to Green Globes are these:

- A low registration fee is required to have projects evaluated and informally self-assessed.
- Consultants are not necessary for the certification process, reducing costs.
- Certification requirements are generally less cumbersome and less complex than those of other rating/certification systems.
- Online web tools provide a convenient, proven, and effective way to complete the assessment process.
- The entire certification process is fairly rapid, with minimal waiting for final rating/certification.
- The estimated rating is largely known in advance of the decision to pursue certification because the self-assessed score is available to users.
- Upfront commitment to a lengthy and costly rating/certification process is not required.

As of June 2011, 23 states have incorporated Green Globes in green building legislation, regulation, or executive orders, including: Arkansas, Connecticut, Florida, Hawaii, Indiana, Kentucky, Illinois, Maryland, Minnesota, New Jersey, New York, North Carolina, Oklahoma, Oregon, Pennsylvania, South Dakota, South Carolina, and Tennessee, Virginia, and Wisconsin. Acceptance is achieved by passing legislation that is rating system–neutral, meaning that it recognizes Green Globes as an equal option alongside LEED and other credible systems.

2.5 GREEN RATING STANDARDS USED WORLDWIDE

Historically, the first environmental certification system was created in the United Kingdom in 1990 and is known as the Building Research Establishment Environmental Assessment Method (BREEAM). This was followed in the United States by the USGBC's introduction in 1998 of the LEED green building rating system, which was based substantially on the BREEAM rating system. The Green Globes Rating System is an adaptation of the Canadian version of BREEAM and was released in the United States by the GBI in 2005. There are numerous other rating systems in use in countries around the world, each with its pros and cons depending on the type of certification targeted for a specific building. However, LEED, followed by Green Globes, are currently the most widely applied systems for commercial construction in the United States.

Today, USGBC has rating systems for new construction, existing buildings, core and shell, commercial interiors, schools, retail, homes, health care, and neighborhood development, whereas GBI has a rating system for commercial buildings that includes new construction and existing buildings. GBI also partners with NAHB to promote green homes.

Rating and certification systems are required to verify the sustainability and "greenness" of buildings on the market. They basically inform us how eco-friendly and environmentally sound a building is, delineate to what extent green components have been incorporated in it, and identify sustainable principles and practices that have been employed in its construction and maintenance. Moreover, rating or certifying a green building helps remove some of the subjectivity that often surrounds buildings that have not been certified. It also makes a property more marketable by informing tenants and the public about its environmental benefits and disclosing the additional innovation and effort that the owner has invested to achieve a high-performance building. Certification also reveals the level of sustainability achieved.

A holistic approach to design translates into a strategic integration of mechanical, electrical, and materials systems. It often creates substantial efficiencies, the complexities of which are not always apparent. Rating a green building identifies these differences objectively, and quantifies their contribution to energy and resource efficiency. The fact that rating systems typically require independent third-party testing of the various elements means there is less risk that a building's systems will not perform as predicted. Furthermore, formally rating or certifying dramatically reduces the risk of the possibility of falsely marketing a building as green when in fact it is not. The following subsections discuss examples of some of the more widely used rating systems in the United States.

2.5.1   LEED

LEED was developed by the USGBC and continues to be the most widely applied rating system in the United States for commercial buildings. It consists of several rating categories, applicable to different points in a building's life cycle and is discussed in other parts of this chapter. The General Services Administration (GSA), other government departments, and many municipalities, as well as an increasing number of private investors and owners, have instituted policies requiring LEED certification for new construction projects. USGBC holds an annual green building conference, Greenbuild, which helps promote the green building industry.

2.5.2 Green Globes

Green Globes, an interactive, web-based commercial green building assessment protocol as discussed in the previous sections of this chapter. The Green Globes system basically evaluates a building in seven areas and offers immediate feedback on its strengths and weaknesses, automatically generating links to engineering, design, and product sources (www.thegbi.org/greenglobes). Green Globes continues to gain traction. Indeed, its parent company was recently acquired by a highly respected, established company, Jones Lang LaSalle.

2.5.3 ENERGY STAR

ENERGY STAR is a joint EPA/DOE program that since its inception in 1992 has overcome many market barriers and has helped revolutionize the marketplace for cost-effective, energy-efficient products and services. It is now an international standard for energy-efficient consumer products and has been adopted by numerous countries, including the European Union, Canada, Japan, and Australia. ENERGY STAR's only mission is to help businesses and individuals protect the environment through superior energy efficiency; other factors such as indoor air quality, materials, and water conservation are not considered. The program is designed for existing buildings and consists of an Energy Performance Rating System that is free with an online tool. Now encompassing more than 60 product categories for the home and workplace, new homes, and superior energy management within organizations, it basically compares the energy performance of a specific building to that of a national stock of similar buildings. The data entered into the ENERGY STAR Portfolio Manager tool model energy consumption based on a building's size, occupancy, climate, and space type. A minimum of one year's utility information input is required before the property can be assigned a rating (from 1 to 100). In order to apply for and receive the ENERGY STAR label, buildings must achieve a score of 75 or higher.

2.5.4 Other Green Building Standards Worldwide

Many countries throughout the world continue to develop their own standards of energy efficiency for buildings. Only a small number of these systems are currently in use in the United States, but they may still prove influential in the emerging green building industry. Examples include BREEAM in the United Kingdom, Green Star in Australia, and BOMA Go Green Plus in Canada. Following are examples of building environmental assessment tools being used by different countries around the world.

Australia

The Green Building Council of Australia (GBCA) developed a comprehensive, voluntary green building rating system known as Green Star that is accepted as the Australian industry standard. Mandated in three states as a minimum for government office accommodation, the Green Star tools benchmark the potential of buildings based on nine environmental impact categories. There are other standards in use in Australia, such as Energy Efficiency Rating (EER) and National Australian Built Environment Rating System (NABERS). NABERS, a government initiative, is a performance-based rating system for existing buildings that now incorporates the Australian Building Greenhouse Rating (ABGR). It has been renamed NABERS Energy for Offices.

Canada

LEED and Green Globes are the two most widely applied green rating systems in Canada. The Canada Green Building Council, established in December 2002, acquired an exclusive license in 2003 from USGBC to adapt LEED Rating System to Canadian circumstances. The Canadian LEED for Homes rating system was released on March 3, 2009. Canada has also implemented the "R-2000" program, a made-in-Canada home-building technology that promotes construction that goes beyond Canadian building codes to increase energy efficiency and promote sustainability. An optional feature of R-2000 home program is the EnerGuide rating service, available across Canada, which allows home builders and home buyers to measure and rate the performance of their home and confirm that its specifications and standards have been met.

R-2000 is a partnership between the Canadian Home Builders' Association (CHBA) and the Office of Energy Efficiency (OEE) of Natural Resources Canada (NRCan). For close to three decades, the two partners have worked on the components that make up the program. Regional initiatives based on R-2000 include ENERGY STAR for New Homes, Built Green, Novoclimat, GreenHome, Power Smart for New Homes, and GreenHouse. In March 2006, Canada's first green building point of service, Light House Sustainable Building Centre, opened in Vancouver, British Columbia, funded by Canadian government departments and private businesses to help implement green building practices.

China

China reportedly has the largest volume of construction in the world; nearly half of the world's new building construction is now estimated to be Chinese. Yet the green building industry is still in its infancy, and green building demand continues to be driven mainly by multinational companies. There are currently two sets of national building energy standards in China (one for public buildings and another for residential buildings). China has put in place mandatory building energy standards, but these are narrow in scope and currently lack the strong regulatory framework needed to incorporate energy-efficient standards in construction.

Moreover, Ministry of Construction (MoC) enforcement remains problematic, and the central government has established a building inspection program to monitor the implementation of building energy efficiency. Design institutions, developers, and construction companies stand to lose their licenses or certificates under this program if they do not comply with regulations. China also recently launched a new green building standard meant to complement better known labels, such as BREEAM (UK) and LEED, which are presently used only in office buildings for multinationals and upscale apartments.

Multinational companies have taken the lead in promoting green building construction in China by pursuing more stringent LEED certification. This trend was initiated by Plantronics, a California-based electronics company, when it achieved a LEED Gold certification in 2006 for its new manufacturing and design centre in Suzhou. Nokia followed, receiving its first LEED Gold certification globally in 2008 for the Nokia China Campus in Beijing. Other firms, including Siemens and BHP, are seeking or have achieved certification.

In an MOU on building energy efficiency signed by the United States and China, the United States pledged $15 million for a joint U.S.-China clean energy research center. Also, the Chinese MoC recently introduced an "Evaluation Standard for Green Building" (GB/T 50378–2006), which resembles LEED's structure and rating process (which is also being used). Building energy consumption data will be collected by the MoC and will be used to assess building performance; a three-star Green Building certificate will be awarded to qualified buildings. The Green Olympic Building Assessment System (GOBAS), another green building rating system, was initially based on Japan's Comprehensive Assessment System for Building Environment Efficiency (CASBEE). High-performance building projects are being supported by both the government and private business. The World Green Building Council (WorldGBC), a Sustainable Buildings and Climate Initiative (SBCI) member, has assisted the Chinese MoC in establishing the China Green Building Council, which is also supported by USGBC.

France

French President Nicolas Sarkozy instituted the *Le Grenelle Environnement* (Environment Round Table) in 2007 to define France's key points of public policy on ecological and sustainable development issues for the coming five years, and to find ways to redefine its environment policy. The result is a set of recommendations released at the end of October 2007 and put to the French parliament in early 2008. Six working groups were formed composed of representatives of the central government, local governments, employer organizations, trade unions, and NGOs to debate and address various themes such as climate change, energy (the building sector consumes 42.5% and transports 31% of total French energy), biodiversity, natural resources, health and the environment, production and consumption, democracy and governance, competitiveness, development patterns, and environmental employment. Within the framework of Le Grenelle Environnement, performance acceleration is designed to meet the following objectives for tertiary buildings:

- Low-consumption buildings (BBC) by 2010 with minimum requirements concerning the levels of renewable energy and CO_2 absorption materials by 2012
- Passive new buildings (BEPAS) or positive buildings (BEPOS) by 2020
- Labels for refurbishment of existing BBC buildings

All of these developments match with European and international regulations and frameworks.

Additionally, two property companies, AXA Real Estate Investment Management and ING Real Estate, recently set up a rival rating system in the United Kingdom, called Green Rating, which allows owners to compare properties' sustainability across Europe. Green Rating was recently launched in France, Spain, Italy, the Netherlands, and Germany, and is expected to be launched in the United States and Japan in 2010. The advantage of Green Rating is that it is an assessment of the energy efficiency of a building and looks at both the building materials and the waste generated. The audit process has been tested on 50 sites in Europe and is broken into four stages:

- *Data collection:* The site technical manager collects data in six areas— energy use, carbon emissions, water use, waste, proximity to transport links, and employee health.
- *Onsite survey and interview with the site manager or owner:* A Green Rating assessor inspects the building and the equipment.
- *Energy modeling:* A model is created to see how energy is used across areas of the building.
- *Recommendations:* Green Rating advises on improvements to be made. An additional audit could then be carried out a year or two after the first.

Germany

In January 2009 the first German standard for new certificates was developed for sustainable buildings by the *Deutsche Gesellschaft für Nachhaltiges Bauen e.V.* (DGNB—German Sustainable Building Council) and the *Bundesministeriums für Verkehr, Bau und Stadtentwicklung* (BMVBS—Federal Ministry of Transport, Building and Urban Affairs) to be used as a tool for the planning and evaluation of buildings. The DGNB system is clear and easy to understand and covers all relevant topics of sustainable construction. There are six criteria in the evaluation: ecology, economy, social-cultural and functional topics, techniques, processes, and location. Outstanding buildings can achieve awards in the categories Bronze, Silver, and Gold.

The following are a number of German organizations that employ green building techniques:

- The *Solarsiedlung* (Solar Village) in Freiburg, Germany, which features energy-plus houses

- The Vauban development, also in Freiburg
- Houses designed by Baufritz that incorporate passive solar design, heavily insulated walls, triple-glaze doors and windows, nontoxic paints and finishes, summer shading, heat recovery ventilation, and greywater treatment systems
- The new Reichstag building in Berlin, which produces its own energy

India

Green Rating for Integrated Habitat Assessment, a rating system for green buildings developed by the Indian Energy and Resources Institute (TERI), plays a key role in developing green building awareness and strategies in India. It was adopted and endorsed by the Indian government as the National Green Building Rating System. Measures are being taken to spread awareness of GRIHA, and TERI has signed a Memorandum of Understanding with the Union Ministry of New and Renewable Energy to this effect. The aim of GRIHA is to ensure that all types of buildings become green buildings. One of the strengths of GRIHA is that it puts great emphasis on local and traditional construction knowledge and even rates non-air-conditioned buildings as green.

GRIHA uses 32 criteria to evaluate and rate buildings, totaling a maximum of 100 points. A building must score at least 50 to apply for certification. Landscape preservation during construction, soil conservation after construction, and reducing air pollution are some of the qualifying criteria. Buildings will also need to quantify energy consumption in absolute terms, and not percentages alone.

The Confederation of Indian Industry (CII) is also playing an active role in promoting sustainability in the Indian construction sector. CII is the central pillar of the Indian Green Building Council (IGBC), which in turn is licensed by USGBC's LEED. USGBC is currently responsible for certifying LEED New Construction and LEED Core and Shell buildings in India, and certifies all other Indian projects. There are many energy-efficient buildings in India, situated in a variety of climate zones (Figure 2.9).

In June 2007 the Indian Bureau of Energy Efficiency (BEE) launched the Energy Conservation Building Code (ECBC), which specifies energy performance requirements for all new commercial buildings that are to be constructed in India. The ECBC applies energy-efficiency standards for design and construction of any building of a minimum conditioned area of 1000 square meters (10,764 square feet) and a connected demand of 500 KW or 600 KVA of power. On February 25, 2009, the BEE launched a five-star rating scheme for office buildings operated only in the daytime in three climatic zones: composite, hot and dry, and warm and humid.

Israel

Israel approved a green building standard in November 2005 which is awarded to new or renovated residential and office buildings that comply with the requirements and criteria. SI-5281 is a voluntary standard for "Buildings with

FIGURE 2.9   Sohrabji Godrej Green Business Centre in Hyderabad, India. This was the first LEED Platinum-rated green building outside the United States, boasting energy savings of 63%. *(Source: Confederation of Indian Industry.)*

Reduced Environmental Impact." It comprises five sections covering energy, land, water, wastewater, and drainage, and other environmentally related elements. A building that meets the prerequisites in each category and accumulates the minimum number of credit points in every environment-related sphere is eligible for "green building" certification. The standard is based on a point system, so a cumulative score of 55 to 75 points entitles a building to a "green building" label, while a cumulative score of more than 75 points allows it to be certified as an "outstanding green building." Together with complementary standards 5282-1, 5282-2 for energy analysis, and 1738 for sustainable products, this standard provides a system for evaluating environmental sustainability.

The LEED Rating System has also been implemented on several Israeli building projects, and there is a strong industry drive to introduce an Israeli version of LEED in the near future. Because of different climatic conditions and building construction methods in Israel, the LEED Rating System cannot be adopted as is.

Japan

A joint industrial/government/academic project was created in Japan in 2001 with the support of the Housing Bureau of the Ministry of Land, Infrastructure, Transport and Tourism (MLIT). This led to the creation of the Japan Green-Build Council (JaGBC)/Japan Sustainable Building Consortium (JSBC), which in turn created the Comprehensive Assessment System for Building

Environmental Efficiency (CASBEE) system. CASBEE was developed according to the following principals:

- To be structured to award high assessments to superior buildings, thereby enhancing incentives to designers and other stakeholders
- To be applicable to buildings in a wide range of applications
- To be as simple as possible
- To take into consideration issues and problems peculiar to Japan and Asia

CASBEE certification is currently available for New Construction, Existing Building, Renovation, Urban Development, Heat Island, Urban Area + Buildings, Cities, and Detached Home.

The CASBEE system is composed of four assessment tools that correspond to a building's life cycle. Collectively, the four tools, and expanded tools for specific purposes, are known as the "CASBEE Family" and apply to each stage of the design process. Each tool has a separate purpose and target user, with the purpose of accommodating a wide range of uses (offices, schools, apartments, etc.) in the buildings evaluation process. The procedure for obtaining CASBEE certification differs from the LEED procedure in that LEED certification process starts at the beginning of the design process, with review and comments taking place throughout the design and construction of a project. Although CASBEE's latest version for New Construction ranking uses predesign tools, certification consists primarily of site visits once the building is completed.

Malaysia

The primary organization promoting green practices and building techniques in Malaysia is the Standards and Industrial Research Institute of Malaysia (SIRIM). However, the country has now put in place a new rating system, the Green Building Index (GBI), for commercial and residential properties. The GBI was developed by Pertubuhan Akitek Malaysia (PAM) and the Association of Consulting Engineers Malaysia (ACEM). It is a profession-driven initiative to lead the Malaysian property industry toward becoming more environment-friendly (Figure 2.10). The GBI rating system provides an opportunity for developers to design and construct sustainable buildings that can provide increased energy savings, water savings, a healthier indoor environment, better connectivity to public transport, and the adoption of material recycling and greenery for their projects.

The GBI has six key criteria: energy efficiency, indoor environmental quality, sustainable site planning and management, materials and resources, water efficiency, and innovation. Based on the scores achieved, commercial buildings will be rated and then certified as Silver, Gold, or Platinum. Final certification is awarded one year after the building is first occupied. Buildings are required to be reassessed every three years in order to maintain their GBI rating by ensuring that they are well maintained.

FIGURE 2.10   PTM Green Energy Office (GEO) Building is Malaysia's first GBI-certified building and is designed as an administration-cum-research office for Pusat Tenaga Malaysia (Malaysia Energy Centre). The GEO building is built on five acres in Seksyen 9, Bandar Baru Bangi, Selangor, Malaysia. It is among the first Malaysian government office buildings with a design based on green concepts and is environmentally friendly. *(Courtesy of PTM-GreenBuildingIndex.)*

Mexico

The Mexico Green Building Council (CMES) is the principal organization dedicated to promotion of best practices to improve the environmental performance of buildings and foster sustainable building technology and policy. It is an independent nonprofit, NGO that works within the construction industry to promote a broad-based transition toward sustainability. CMES's stated mission is to promote sustainable development through the realization and construction of a superior built environment.

Under a new partnership agreement, ASHRAE and CMES will work together to promote buildings that are healthful, environmentally responsible, comfortable and productive, and profitable. This agreement is part of ASHRAE's new strategy for a global environment, which commits it to working with organizations with shared interests and values.

Mexico City's government is promoting the creation of a standard certification program for green buildings, and its minister of the environment has announced that certified green buildings will be able to obtain up to a 25% discount on property taxes. The Green Building Certification Program will have three levels of certification: lowest level, 21 to 50 points; efficiency level, 51 to 80 points; and excellence level, 81 to 100 points. The higher the level, the greater the property tax discount. At the beginning, the discount will be voluntary, but it is intended to become mandatory in the future.

New Zealand

In July 2005 the New Zealand Green Building Council (NZGBC) was formed as a nonprofit, industry organization dedicated to accelerating the development and adoption of market-based green building practices. In 2006/2007 several major milestones were achieved, including the NZGBC becoming a member of the World Green Building Council, the launching of the Green Star NZ Office Design Tool, and the welcoming of member companies. Green Star is a comprehensive, national, voluntary environmental rating scheme which evaluates the environmental attributes and performance of New Zealand's buildings using a suite of rating tool kits developed to be applicable to each building type and function. It was developed by NZGBC in partnership with the building industry.

South Africa

The Green Building Council (GBC) of South Africa was launched in 2007, with the following stated mission:

To promote, encourage and facilitate green building in the South African property and construction industry through market-based solutions, by:

- *Promoting the practice of green building in the commercial property industry,*
- *Facilitating the implementation of green building practice by acting as a resource centre,*
- *Enabling the objective measurement of green building practices by developing and operating a green building rating system, and*
- *Improving the knowledge and skills base of green building in the industry by enabling and offering training and education.*

The GBC has since developed a suite of Green Star SA rating tools, based on the tools developed by the Green Building Council of Australia, to provide the property industry with an objective measure for green/sustainable buildings and to recognize and reward environmental leadership. Each tool reflects a different market sector (e.g., office, retail, multiunit residential). Green Star SA–Office was the first tool developed and was released in final form (version 1) at the GBC's annual convention and exhibition in November 2008. South Africa is in the process of incorporating an energy standard, SANS 204, that aims to promote energy-saving practices as a basic standard in the South African context. Green Building Media, which was launched in 2007, has also played an instrumental role in green building in South Africa.

United Kingdom

The Association for Environment Conscious Building (AECB) was founded in 1989 and incorporated in January 2005 to increase awareness in the construction industry of the necessity to respect the environment and promote sustainable building in the United Kingdom. The AECB is now under the Energy Performance of Building Directive (EPBD); Europe has made energy certification mandatory since January 4, 2009. A key part of this legislation is that all EU countries must enhance their building regulations and introduce energy certification schemes for buildings. All countries are also required to have inspections of boilers and air conditioners.

A mandatory certificate called the Building Energy Rating system (BER) and an Energy Performance Certificate (EPC) are needed by all buildings that measure more than 1000 square meters (approximately 10,765 square feet) in all European Union nations. These are required as well for houses being bought or sold. Certificates are also required on construction of new homes and for rented homes the first time the property is let after 1 October 2008. The certificate records how energy efficient a property is and provides ratings of A through G. These are similar to the labels now provided with domestic appliances such as refrigerators and washing machines.

The UK Green Building Council (UK-GBC), in March 2009, called for the introduction of a Code for Sustainable Buildings to cover all nondomestic buildings, both new and existing. Although the code is owned by the government, it was developed and is managed and implemented by industry and covers refurbishment as well as new construction. Also, in September 2009 the Welsh Assembly Government planning policy put in place a national standard for sustainability for most new buildings proposed in Wales.

According to its website, the BREEAM assessment process was created in 1990 as a tool to measure the sustainability of new nondomestic buildings in the United Kingdom, with the first two versions covering offices and homes. BREEAM is the leading and widely used environmental assessment method in the United Kingdom for buildings, setting the standard for best practice in sustainable design and providing a means to describe a building's environmental performance. It has been updated regularly in line with U.K. building regulations and it underwent a significant facelift in August 2008, referred to as BREEAM 2008. Credits are awarded in each of the areas listed according to specific performance.

- Management
- Health and well-being
- Energy
- Transport
- Water
- Material and waste
- Land use and ecology
- Pollution

A set of environmental weightings enables the credits to be added together to produce a single overall score and a rating of Pass, Good, Very Good, Excellent, or Outstanding; this is followed by the award of a certificate to the development.

Some of the dramatic changes in BREEAM 2008 were in response to an evolving and changing construction industry and public agenda. They include:

- Introduction of mandatory credits
- Two-stage assessment process introduced (Design and Post-Construction)
- Addition of another rating level (BREEAM Outstanding)
- Modified environmental weightings
- CO_2 emissions benchmarks set to align with the new Environmental Performance Certificate
- Changes to certain specific credits
- Updated Green Guide Ratings, which are available online
- Introduction of BREEAM Healthcare and BREEAM Further Education
- Shell-only assessments

United States

Numerous sustainable design organizations and programs are in place in the United States. The most widely used rating system is the USGBC's LEED, which promotes sustainability in how buildings are designed, built, and operated, and is best known for the development of the LEED Rating System and GreenBuild, USGBC's green building conference and expo which is well known for its promotion of the green building industry and environmental issues.

The USGBC had more than 17,000 member organizations by September 2008 from all sectors of the building industry; its mission is to promote buildings that are environmentally responsible, profitable, and healthy places to live and work. Through its GBCI, USGBC offers industry professionals an opportunity to receive accreditation as green building professionals. In June 2009, LEED underwent a complete overhaul and became a two-tier system known as LEED V3.

The NAHB is a trade association representing home builders, remodelers, and suppliers to the industry; it has formed a voluntary residential green building program called NAHBGreen (www.nahbgreen.org) that incorporates an online scoring tool, national certification, indstry education, and training for local verifiers. The online scoring tool is free to both builders and homeowners. In August 2009 NAHB announced that the number of home builders, remodelers, and other members of the real estate and construction industry who hold the Certified Green Professional (CGP) educational designation now tops 4000.

The Green Building Initiative is a nonprofit network of building industry leaders working to mainstream building approaches that are not only environmentally progressive but practical and affordable for builders to implement. It provides a web-based management/rating tool called Green Globes that includes an assessment protocol, rating system, and guide for integrating environmentally friendly design into both new and existing commercial buildings.

ENERGY STAR is a program established jointly by the EPA and the DOE that focuses on energy-efficient homes and buildings designed to protect the environment while at the same time saving money for homeowners and businesses. The program also rates commercial buildings for energy efficiency and provides qualifications for new homes that must meet a series of energy efficiency guidelines established by the EPA.

Chapter 3

Green Design and the Construction Process

3.1 INTRODUCTION

Building green does not focus only on environmental factors and considerations; it also takes into account how the environment integrates with other factors such as cost, schedule, operations, maintenance, and tenant/employee issues. Research clearly shows that people in contemporary societies such as the United States and Europe generally spend most of their time inside buildings, and we apparently take for granted the shelter, protection, and comfort that our buildings provide. We don't often give much thought to the systems that allows us to enjoy these services unless we are faced with an unfortunate power interruption or some other problem. Moreover, not many people fully comprehend the extent of the environmental consequences that allow us to maintain indoor comfort levels. This may be partly because modern buildings continue to increase in complexity.

A building's functions continue to change, and buildings become increasingly costly to build and maintain, as well as requiring constant adjustment to function effectively over their life cycle. And while sustainable design strategies normally cost no more than conventional building techniques, the real goal of interdependence between strategies, known as holistic design, makes determining the true cost often difficult to assess. Furthermore, we frequently find that the returns on sustainable design are measured by numerous intangibles, such as worker productivity, health, and resource economy. However, for many building owners, developers, and designers it is more likely that determinations on sustainable design strategies will be based on initial construction costs or by a quick return on investment than on the positive returns based on a building's life cycle and the many positive attributes of a green building.

The large numbers of conventionally designed and constructed buildings that exist today are fortifying their negative impact on the environment as well as on occupant health and productivity. Not only that, but these buildings are becoming increasingly expensive to operate and maintain in an acutely competitive market. Owners and developers as well as the construction industry now realize that buildings' contribution to excessive resource consumption, waste generation, and pollution is unacceptable and must be addressed. Reducing negative impacts on our environment, establishing new eco-friendly goals, and adopting guidelines and codes that facilitate the development of green/sustainable buildings, such as outlined by the U.S. Green Building Council (USGBC®), Green Globes®, and similar organizations, must be a priority for this generation.

Of note, the International Green Construction Code (IgCC), which was recently approved after two years of research and development, is another sign of the seriousness with which the federal government regards the negative impact of conventional building construction on the environment. The new law in its final format was officially published in March 2012 and applies to all new and renovated commercial buildings and residential buildings in excess of three stories. This historic code is long overdue and is bound to have a significant impact on future trends, as it sets mandatory baseline standards for all aspects of building design and construction, including site impacts, buildings' energy and water efficiency, building waste, and materials. Local governments and the states have the choice as to whether to adopt the code, but once they do it becomes enforceable.

With this new national building code, the concept of building "green" has ceased to be in the realm of the theoretical and has moved deep into the mainstream of current construction practice; general acceptance by the industry as well as familiarity with green elements and procedures will continue to drive down building costs. The method of building construction and materials that is employed also impacts the development of indoor air quality (IAQ), which can present an array of health challenges. Green buildings can address many of these environmental concerns, which is why it has become an essential component of our society.

Building "green" therefore offers an opportunity to use our resources more effectively while at the same time creating healthier buildings, improving employee productivity, reducing the negative impact on the environment, and achieving significant cost savings over a building's life cycle. Green buildings are referred to sometimes as sustainable buildings; they are structures that are designed, built, renovated, operated, or reused in an ecological and resource-efficient manner.

In today's world of rapidly dwindling fossil fuel, the increasing impact of greenhouse gases on our climate has made sustainable architecture particularly relevant. For this and other reasons, the new national building codes go a long way to addressing the pressing needs to find suitable ways to reduce buildings' energy loads, increase building efficiency, and employ renewable energy resources. Green construction is environmentally friendly because it uses sustainable, location-appropriate building materials and employs building techniques that reduce energy consumption. Indeed, the primary objectives of sustainable design and construction are to avoid resource depletion of essential resources such as energy, water, and raw materials, and to prevent environmental degradation. Sustainability also places a high priority on health issues, which

FIGURE 3.1   The U.S. Green Building Council awarded The Kresge Foundation headquarters, which was built on a three-acre site in Troy, Michigan, a Platinum-level rating, the highest attainable level in the LEED Rating System. This state-of-the-art facility was completed in 2006 and serves as a model of sustainable design and as an educational resource for the local community. The headquarters integrates a nineteenth-century farmhouse and barn—part of the offices for many years—with a new contemporary, two-level, 19,500-square-foot glass and steel building. *(Source: The Kresge Foundation.)*

is partly why green buildings are generally more comfortable and safer to live and work in (Figure 3.1) than conventional buildings.

Familiarity with the IgCC as well as the various "green" certification systems such as Leadership in Energy and Environmental Design (LEED®) and Green Globes is essential for government (federal and state) contractors and certainly is recommended for contractors in the private sector. Even before the introduction of the new green codes, various arms of the federal government required that their public projects meet certain "green" standards whether they be LEED, Green Globes, ENERGY STAR®, and so on, and provided various monetary and tax incentives to do so. For example, the General Services Administration (GSA) requires that all building projects meet the LEED Certified level and target the LEED Silver level. The GSA, however, while strongly encouraging projects to apply for certification, does not require it.

The U.S. Navy, also while requiring appropriate projects to meet LEED certification requirements, does not require actual certification. On the other hand, the U.S. Environmental Protection Agency (EPA) requires all new facility construction and acquisition projects consisting of 20,000 square feet or more to achieve a minimum LEED Gold certification. The U.S. Department of Agriculture also now requires all new or major renovation construction to achieve LEED Silver certification. We have yet to see how the new IgCC will impact organizations such as LEED and Green Globes. Some of the more important reasons for owners and developers to consider green building design and construction are the following:

- Green buildings are generally energy-efficient, which means that they will save on operating costs over time; at a time of sharply rising energy costs, this can be particularly useful.
- Many government agencies provide financial incentives for green building projects, which is a great inducement for building owners and developers to cash in on the "greenness" of their development projects through tax credits, financial incentives, carbon and renewable energy tradable credits, and net metering excess donations. Whether these incentives will be affected by the new codes or not has yet to be determined.
- Not many people realize that in some cases it may actually be cheaper to build green. For example, a building that takes advantage of passive solar energy and includes effective insulation may require a smaller, less expensive HVAC system. Also, purchasing recycled products can often be cheaper than purchasing comparable new products, and incorporating a construction plan that minimizes waste will ultimately save on hauling and landfill charges.
- The market demand for green buildings continues to rise, particularly in high-end residential and prestige corporate projects. A Building Owners and Managers Association (BOMA) Seattle survey, for example, recently found that 61% of real estate leaders believe that green buildings enhance their corporate image and 67% of those also believe that over the next five years tenants will increasingly make the green features of a property an important consideration when choosing space.
- Green buildings are in a much better position to respond to existing and future governmental regulation. Building construction and operations are a major factor in nationwide greenhouse-gas emissions and energy use, and we can expect future government regulations and the new green codes to present a major challenge directed at the building industry.
- Green building helps contribute to conservation, and one of the least expensive ways of stretching a limited resource is to conserve it. If new buildings can be built and operated in a way that conserves energy and materials, these limited resources will go farther and minimize the need for capital-intensive projects to increase them.

Support for green building practices is important because it helps reduce greenhouse-gas emissions, which in turn can help prevent climate change. Greenhouse gases are gases in the atmosphere that are transparent to visible light but which absorb infrared light reflected from the earth, thus trapping heat in the atmosphere. Many naturally occurring gases have this property, including water vapor, carbon dioxide, methane, and nitrous oxide, as do a number of human-made gases such as some aerosol propellants. On the other hand, with conventional methods of construction, owners and developers face a number of challenges such as

- Higher final construction costs
- Prolonged project closeout
- Budget overages
- Missed schedules
- Excessive Change Orders
- Greater potential disputes, arbitration, and litigation
- Inability to solve problems satisfactorily

3.2   GREEN BUILDING PRINCIPLES AND COMPONENTS

While the clear intent of green building is siting, design, construction, and operation to enhance the well-being of a building's occupants, and support for a healthy community and a natural environment with minimal adverse impact on the ecosystem (Figure 3.2), this is not always easy to achieve.

3.2.1   Principals of Green Design

For many years, ASHRAE and the International Code Council (ICC) have worked on the development of codes and standards that can be transformed into the industry standard of care for the design, construction, operation, and maintenance of both commercial and residential buildings in the United States and globally. Prior to the newly launched IgCC, the USGBC had been leading a nationwide green building movement centered on the LEED Green Building Rating System. LEED, which was launched in 2000, has been mandated by many jurisdictions as a de facto building code. The convergence of these efforts in the IgCC may be the most significant development in the building industry over the last decade. We have yet to see how the new IgCC codes will impact organizations, such as LEED and Green Globes, in the years to come.

As ICC chief executive officer Richard P. Weiland says, however, "The emergence of green building codes and standards is an important next step for the green building movement, establishing a much-needed set of baseline regulations for green buildings that is adoptable, usable and enforceable by jurisdictions." Weiland also says, "The IgCC provides a vehicle for jurisdictions to regulate green for the design and performance of new and renovated buildings in a manner that is integrated with existing codes as an overlay, allowing all new buildings to reap the rewards of improved design and construction practices."

In today's competitive world, the practice of sustainable architecture and construction revolves mainly around innovation and creativity. One of the primary attributes of green building is that materials and techniques are employed that do not have a negative impact on the environment. Also, the building's

(a)

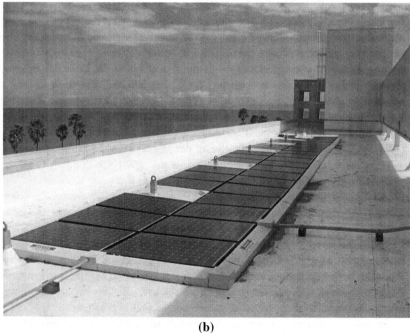

(b)

FIGURE 3.2   (a) Faculty and students on the third-floor terrace of the Donald Bren School of Environmental Science and Management at the University of California at Santa Barbara. Bren Hall is a LEED Gold Pilot project. (b) Rooftop photovoltaic panels.

(c)

FIGURE 3.2—cont'd   (c) Auditorium interior. Bren Hall uses silt fencing, straw-bale catch basins, and scheduled grading activities in accordance with the project's erosion control plan. Building Bren Hall with sustainable materials and methods is estimated to have added only 2% to the building cost, which will easily be offset over time by energy savings. *(Source: Donald Bren School of Environmental Science & Management at www.esm.ucsb.edu/; photos © Kevin Matthews, Artifice Images.)*

inhabitants do not choose materials just because they are more familiar with their use. For example, there are numerous recycled products that can be used in the construction of sustainable structures, such as ceramic floor tiles, which can be made from recycled glass. Also, flooring made from cork oak bark is friendly to the environment since cork harvesting does not harm the trees it is taken from. Bamboo flooring is another suitable alternative to wood that is less expensive and is actually harder than hardwood flooring and more durable.

It is important to address the many traditional building design concerns of economy, utility, durability, and aesthetics. Green design strategies underline additional concerns regarding occupant health, the environment, and resource depletion. To address all these concerns, there are various green design strategies and measures that can be employed, including those outlined next:

- Encourage the use of renewable energy and materials that are sustainably harvested.
- Ensure maximum overall energy efficiency.
- Ensure that water use is efficient, and minimize wastewater and runoff.
- Conserve nonrenewable energy and scarce materials.
- Optimize site selection to conserve green space and minimize transportation impacts.
- Minimize human exposure to hazardous materials.

- Minimize the ecological impact of energy and materials used.
- Encourage use of mass transit, occupant bicycle use, and other alternatives to fossil-fueled vehicles.
- Conserve and restore local air, water, soils, flora, and fauna.
- Minimize adverse impacts of materials by employing green products.
- Orient buildings to take maximum advantage of sunlight and microclimate.

Taking a holistic approach to implementing these strategies puts us in a better position to preserve our environment for future generations by conserving natural resources and protecting air and water quality. It also provides critical benefits by increasing comfort and well-being and helping to maintain healthy air quality. Green building strategies are good for the economy because they reduce maintenance and replacement requirements, lower utility bills, decrease the cost of homeownership, and increase property and resale values. In practical terms, green building is a whole-systems approach to building design and construction that employs features such as

- Using energy-efficient appliances and water-saving devices, fixtures, and technologies
- Building quality, durable structures with good insulation and ventilation
- Taking advantage of the sun and the site to increase a building's capacity for natural heating, cooling, and daylighting
- Recycling and minimizing construction and demolition waste
- Using healthy products and building practices
- Incorporating durable, recycled, salvaged, and sustainably harvested materials
- Landscaping with native, drought-resistant plants and water-efficient practices (Figure 3.3)
- Designing for livable neighborhoods

Integrated Design

It has become almost imperative, to achieve success in green building, to have an integrated design team—designers; a building information modeling (BIM) manager; structural, mechanical, electrical, civil, lighting, plumbing, and landscape engineers; and possibly others, in addition to the contractor—working with the project owner or developer to find the most effective way to meet the owner's goals and objectives. This is aided by adapting the various systems to each other as an integrated whole and recognizing the interconnectivity of the systems and components that cumulatively make up a building and the disciplines involved in its design. Unlike the traditional approach, integrated design correctly assumes that each system affects the functioning of the other systems, which is why these systems must be harmonized if they are to perform together at maximum efficiency. Optimizing the building's performance, and thus reducing the adverse impact on the environment, and minimizing its total cost must be the ultimate objective of sustainability.

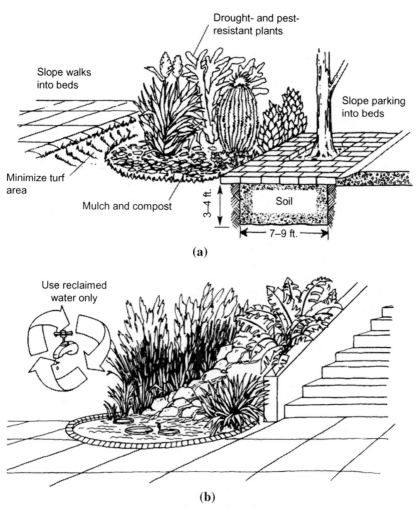

FIGURE 3.3   (a) Sketch showing the use of native- and drought-tolerant plants, which can significantly enhance the environment in addition to providing opportunities for food and decorative gardens. The sketch in (b) shows the use of reclaimed water. *(Source: City of Santa Monica. Green Building Guidelines for Design: Landscape, 2010.)*

It should be apparent from the preceding that the first and most important steps toward sustainability in real estate development is to focus on areas relating to energy efficiency, water efficiency, waste efficiency, and design efficiency, on a per-building and a whole-development basis. The following factors are the main components in achieving green building and are rewarded by the majority of green rating systems, including LEED, Building Research Establishment's Environmental Assessment Method (BREEAM®), and Green Globes.

Site Selection

This is one of the cardinal features of successful green building. It basically emphasizes the reuse and restoration of existing buildings and sites. Site selection is also concerned with rehabilitating contaminated or brownfield sites (determined by a local, state, or federal agency), as well as preserving natural and agricultural resources. Other features of site selection include promotion of biodiversity and maximizing open space by reducing the development footprint, as well as reducing light trespass to minimize light pollution associated with interior light (e.g., existing building and exterior light luminance should not exceed site boundaries). Also included are stormwater management through supporting natural hydrology and reducing water pollution by increasing pervious area and onsite infiltration; reduction of construction waste; reducing the heat island effect; and encouraging use of public or low-environmental-impact transportation options.

The IgCC, however, significantly eliminates development on green fields (undeveloped land), although there are exceptions based primarily on existing infrastructure. It includes clear guidelines for site disturbance, irrigation, erosion control, transportation, heat island mitigation, graywater systems, habitat protection, and site restoration.

Energy Efficiency

In many ways this is the most important issue surrounding green building, and it is also the one element of a project that can most significantly impact reductions in operating costs. Energy efficiency measures may be eligible for federal and state tax credits and other financial incentives as required by the current ASHRAE/IESNA 90.1 Standard. The components of this standard are as follows:

- Building envelope
- Heating, ventilation, and air conditioning
- Water heating, including swimming pools
- Power, including building power-distributed generation systems
- Lighting
- Other electrical equipment

IgCC requirements stipulate that total efficiency must be "51% of the energy allowable in the 2000 International Energy Conservation Code" (IECC), and building envelope performance must exceed that by 10%. IgCC also sets minimum standards for lighting and mechanical systems, and mandates certain levels of submetering and demand-response automation. California approved new green codes ("CalGreen") that took effect in January 2011. David Walls, executive director of the State Building Commission, says that "[t]he new code's mandatory measures will help reduce greenhouse-gas emissions by 3 million metric tons by 2020." As far as California Title 24 standards are concerned, the majority of buildings generally strive to meet this standard. The following strategies contribute to achieving both the IgCC and CalGreen goals.

- Use energy-efficient heat/cooling systems in conjunction with a thermally efficient building shell. Other prudent energy-saving opportunities may exist with heat recovery options and thermal energy storage. High R-value wall and ceiling insulation to be installed; minimal glass to be employed on east and west exposures and light colors for roofing and wall finishes.
- Encourage the incorporation of renewable energy sources such as solar, wind, or other alternative energy into the HVAC system to reduce operational costs and minimize the use of fossil fuels.
- Minimize as much as possible electric loads created by lighting, appliances, and other systems.
- Employ passive design strategies, including building shape and orientation, passive solar design, and the use of natural lighting, to dramatically impact building energy performance.
- Employ modern energy management controls, as improperly programmed controls and outdated technology can mislead a building owner that a building is performing more efficiently than it actually is. Replacing, upgrading, or reprogramming the temperature controls and energy management system will ensure equipment operates at optimum efficiency.
- Develop strategies to provide natural lighting and views where this will improve well-being and productivity. A green building is typically designed to take advantage of the sun's seasonal position to heat its interior in winter and frequently incorporates design features such as light shelves, overhanging eaves, or landscaping to mitigate the sun's heat in summer. Room orientation should generally be designed to improve natural ventilation.
- Install high-efficiency lighting systems with advanced lighting control systems and incorporating motion sensors linked to dimmable lighting controls. Inclusion of task lighting can reduce general overhead light levels.
- Use BIM computer modeling when possible to optimize design of electrical and mechanical systems and the building shell.
- Employ retro-commissioning. Most existing buildings have never been commissioned during construction, and as they age they require regular maintenance. In this respect, retro-commissioning can be extremely useful by resolving problems that occur during the Design or Construction Phases, or by addressing problems that have developed throughout the building's life and thus make a substantial difference in energy usage and savings.

Water Efficiency and Conservation

This establishes maximum consumption of fixtures and appliances and sets specifications for rainwater storage and graywater systems. Of note, the United States annually draws out an estimated 3700 billion gallons more water from its natural water resources than it returns. Many municipalities have legislation in place requiring stormwater and wastewater efficiency measures, while the Energy Policy Act (EPAct) of 1992 already requires water conservation for plumbing fixtures. Implementing water efficiency measures conserves our depleting water resources and preserves water for agricultural uses, in addition to reducing pressure on water-related ecosystems. There are numerous efficiency measures that can be implemented to advance water efficiency and conservation, including

- Employ ultra-low-flush toilets, low-flow showerheads, and other water-conserving fixtures to minimize wastewater.
- Incorporate dual plumbing systems that use recycled water for toilet flushing or a graywater system that recovers rainwater or other nonpotable water for site irrigation.
- Install recirculating systems to be used for centralized hot water distribution, and point-of-use water-heating systems for more distant locations.
- Use a water budget approach that schedules irrigation systems.
- Incorporate self-closing nozzles on hoses and state-of-the-art irrigation controllers.
- Employ micro-irrigation techniques to supply water in nonturf areas; buildings should be metered separately from landscape.

Materials and Resources

Choosing the most appropriate building material is very important because it can have an enormous impact on the natural environment, partly caused by the many processes involved such as extraction, production, and transportation, all of which can negatively impact our ecosystem. But it is also important because some materials may release toxic chemicals that are harmful to building occupants. Green building generally avoids using potentially toxic materials such as treated woods, plastics, and petroleum-based adhesives that can degrade air and water quality and cause health problems. Additionally, building demolition may cause materials to release hazardous or nonbiodegradable material pollutants into the natural environment or into drinking water reserves. Sustainable building materials also reduce landfill waste; IgCC code mandates a minimum of 50% of construction waste to be diverted from landfills and at least 55% of building materials to be salvaged, recycled content, recyclable, bio-based, or indigenous. The code also mandates that buildings must be designed for a minimum of 60 years of life, and must show a service plan that justifies that. The following aspects should be considered when choosing building materials for a project:

- Choose sustainable construction materials and products whenever possible. Their sustainability can be measured by several characteristics such as recycled content, reusability, minimum off-gassing of harmful chemicals, zero or low toxicity, durability, sustainably harvested materials, high recyclability, and local production. Use of such products promotes resource conservation and efficiency, minimizes the adverse impact on the environment, and helps to harmonize the building with its surroundings.

- Employ dimensional planning and other material efficiency strategies to reduce the amount of building materials needed and cut construction costs. For example, the design of rooms to 4-foot multiples minimizes waste by conforming to standard-sized wallboard and plywood sheets.
- If possible, reuse and recycle construction and demolition materials. Using recycled-content products cuts costs and assists in the development of markets for recycled materials that are being diverted from landfills. One example is the use of inert demolition materials as a base course for a parking lot.
- Allocate adequate space to facilitate recycling collection and to incorporate a solid waste management program that reduces waste generation.
- Require waste management plans for managing materials through deconstruction, demolition, and construction.

Employing recycled/reused materials helps to ensure the sustainability of resources. If building projects use only virgin raw materials, these materials will gradually be exhausted. As the availability of raw materials lessens, prices will rise and before long the materials will no longer be obtainable. This trend has already started to impact certain raw materials that are either no longer available or have become very scarce, and can only be obtained as recycled from existing projects. Recycling and reusing materials helps ensure that these materials will be readily available for years to come.

Indoor Environmental Quality and Safety

The adoption of green construction principles can contribute dramatically to a superior interior environment, which in turn can significantly reduce the rate of respiratory disease, allergy, asthma, and sick building syndrome (SBS) symptoms, and enhance tenant comfort and worker performance. Materials such as carpet, cabinetry adhesives, and paint and other wall coverings with zero or low levels of volatile organic compounds (VOCs) release less gas and improve a building's indoor air quality. On the other hand, building materials and cleaning and maintenance products that emit toxic gases, VOCs, and formaldehyde should be avoided as they can have a very negative impact on occupants' health and productivity. Daylighting can also improve indoor environmental quality (IEQ) by boosting the occupant's mood with natural light. Adequate ventilation and a high-efficiency, in-duct filtration system should be provided. Heating and cooling systems that ensure proper ventilation and filtration can have a dramatic and positive impact on indoor air quality. The potential financial benefits of improving indoor environments can be very significant.

To prevent indoor microbial contamination, materials should be chosen that are resistant to microbial growth. Provide effective roof drainage and drainage for the surrounding landscape, as well as proper drainage of air-conditioning coils. Other building systems should be designed to control humidity.

Waste Management Issues

These issues are connected to several areas of green building, from waste reduction measures during construction to waste recycling. In the United States it is estimated that 31.5 million tons of construction waste are produced annually. Furthermore, nearly 40% of solid waste in the United States is produced by construction and demolition.

Commissioning Operation and Maintenance

Green building measures cannot achieve their objectives unless they function as intended according to the specifications and Contract Documents. The incorporation of operating and maintenance factors into the design of a building can contribute to the creation of healthy working environments, higher productivity, and reduced energy and resource costs. Whenever possible, therefore, designers should specify materials and systems that simplify and reduce maintenance and life-cycle costs, use less water and energy, and are cost-effective.

Building commissioning and enhanced commissioning are necessary imperatives that include testing and adjusting mechanical, electrical, and plumbing systems to ensure that all equipment meets the design intent. Commissioning also includes instructing and educating building owners and staff on the operation and maintenance of equipment. As buildings age, their performance generally declines; continued proper performance can only be assured through regular maintenance or through retro-commissioning.

Livable Communities and Neighborhoods

We need to define those structures and strategies that will advance the design of more livable eco-friendly communities and neighborhoods. There are several issues that pertain to community and neighborhood development and that should be addressed, such as the application of ecologically appropriate site development practices, the incorporation of high-performance buildings, and the incorporation of renewable energy.

In addition, the development of new communities and neighborhoods, and the housing in such developments, may involve looking into issues not normally considered in single-structure projects. Such issues may include evaluating the community's location, the proposed structure and density of the community, and the effects of the community on transportation requirements. Other issues that should be considered include setting standards for the community's infrastructure and standards to be applied to specific development projects within the community, as all these factors influence the environmental impacts of the development and the ongoing livability of the community as an integrated whole.

The introduction of the IgCC has clearly impacted the construction industry, which has for some years been part of the mainstream in the United States. Likewise, the escalating costs of energy and building materials, coupled with warnings from the EPA about the toxicity of today's treated and synthetic materials, are prompting architects and engineers to revise their approach to building techniques that employ native resources as construction materials and use nature (daylight, solar, and ventilation) for heating and cooling. Green developments are generally more efficient, last longer, and cost less to operate and maintain than conventional buildings. Moreover, they generally provide greater occupant comfort and higher productivity than conventional developments, which is why most sophisticated buyers and lessors prefer them and are usually willing to pay a premium for green developments.

The Department of Energy (DOE) estimates that buildings in the United States consume annually more than one-third of the nation's energy and contribute approximately 36% of the carbon dioxide (CO_2) emissions released into the atmosphere. This is partly due to the fact that the vast majority of buildings today continue to use mechanical equipment powered by electricity or fossil fuels for heating, cooling, lighting, and maintaining indoor air quality. This means that the fossil fuels used to condition buildings and generate electricity are having an enormous negative impact on the environment; they emit a plethora of hazardous pollutants such as VOCs that cost building occupants and insurance companies millions of dollars annually in healthcare costs. In addition, we have the problem of fossil fuel mining and extraction, which add to the adverse environmental impacts while fomenting price instability, which is causing concern among both investors and building owners. The new IgCC will foster and mandate the creation of buildings that use less energy and both reduce and stabilize costs, as well as have a positive impact the environment.

The DOE early on had the foresight to appreciate the urgent need for buildings that were more energy efficient; in 1998 it took the initiative and decided to collaborate with the commercial construction industry to develop a 20-year plan for research and development on energy-efficient commercial buildings. The primary mission of the DOE's High-Performance Buildings Program is to help create more efficient buildings that save energy and provide a quality, comfortable environment for workers and tenants. The program is targeted mainly to the building community, particularly building owners/developers, architects, and engineers. Today we have the knowledge and technologies required to reduce energy use in our homes and workplaces without having to compromise comfort or aesthetics. The building industry has until recently remained aloof or uninformed and has resisted taking full advantage of these important advances. It is expected that with the new green codes coming into play, future building projects will be designed and operated taking into account the many environmental impacts, to produce healthier and more efficient buildings.

3.2.2   High-Performance and Intelligent Buildings

Many would argue that the terms "high performance," "intelligent," and "building automation" are still being defined or refined, although I suggest that these terms are strongly interrelated. Not surprisingly, high-performance buildings and building automation have become recognizable landmarks in today's contemporary society; they typically consist of programmed, computerized, "intelligent" networks of electronic devices that monitor and control mechanical and lighting systems. The U.S. Energy Independence and Security Act 2007 (PL 110-140) defines a high-performance building as "[a] building that integrates and optimizes on a life-cycle basis all major high performance attributes, including energy [and water] conservation, environment, safety, security, durability, accessibility, cost-benefit, productivity, sustainability, functionality, and operational considerations."

Due partly to rising energy costs, an increasing number of new buildings are incorporating central communications systems to the extent that the "intelligent" building has become an integral part of mainstream America. Even many of today's federal facilities have succeeded in achieving high-performance building that saves energy and reduces environmental impacts. Increasing consumer demand for clean renewable energy and the deregulation of the utilities industry have encouraged and energized growth in green power such as solar, wind, geothermal steam, biomass, and small-scale hydroelectric sources. In addition, President Obama's administration has encouraged small commercial solar power plants to emerge around the country and serve some energy markets within the United States.

The decision to operate a high-performance building requires various proactive management processes for energy and maintenance. It may be prudent and more effective, therefore, when deciding to implement high-performance building projects to initially hold a green design "charrette," or multidisciplinary kick-off meeting, to articulate a clear road map for the project team to follow. A crucial advantage of holding a charrette during the early stage of the design process is that it offers team professionals (with possible assistance from green design experts and facilitators) to brainstorm on design objectives as well as alternative solutions. This goal-setting approach helps identify green strategies for members of the design team and helps facilitate the group's ability to reach a consensus on performance targets and to ensure that these performance targets are achieved.

Designers of sustainable buildings need to pay careful attention to measured performance expectations. Once performance measures are determined, a follow-up is required to establish performance goals and the metrics to be employed for each one. Minimum requirements, or baselines, are typically defined by codes (e.g., the IgCC) and standards, which may differ from one jurisdiction to another. Alternatively, performance baselines can be set to exceed the average performance of a building type, measured against similar buildings recently built or against the performance of the best-documented building of a particular type.

Over the years, several green building rating systems have been developed to set standards for evaluating high performance. To date, the most widely recognized systems for rating building performance in the United States are LEED and Green Globes, which provide various consensus-based criteria to measure performance, along with useful reference to baseline standards and performance criteria. Nevertheless, LEED or Green Globes certification, by itself, does not ensure high performance in terms of energy efficiency, as certification may have been achieved by attaining high marks in other nonenergy-related categories such as Materials and Resources or Sustainable Sites. For this reason, specific energy-related goals must still be set. To some degree, this is being addressed in the United States by the newly adopted national green codes (IgCC) and California's CalGreen, which mandate green specifications.

Integrated design may be considered the cornerstone of the green building process. It is enhanced by the use of computer energy modeling tools such as the Department of Energy's DOE 2.1E, Building Information Modeling (BIM), and other programs. These programs can inform the building team of the impacts of energy use very early in the design process by factoring in relevant information such as climate data, seasonal changes, building massing and orientation, and daylighting. They can also readily prompt investigation and survey of cost-effective design alternatives for the building envelope and mechanical systems by forecasting energy use of various combined alternatives. But before dwelling too much on green design and integrated design, and to understand their full meaning, it may be advisable to first describe the more conventional design process.

The traditional building process is linear and segmented, whereas integrated design is more interactive, more egalitarian, and more consultative. Therefore the traditional design approach usually starts with the architect and the client agreeing on a budget and a design concept, followed by a general massing scheme, typical floor plan, schematic elevations, and usually the general exterior appearance as determined by these design criteria and the intent of the design. Mechanical and electrical engineers are then asked to implement the design and to suggest appropriate systems. Building information modeling programs have only recently been introduced and incorporated into the design process.

Although this will likely change in the near future, the conventional design approach, while greatly oversimplified, remains at this point the main method employed by the majority of general-purpose design consultant firms, which unfortunately tends to keep achievable performance at conventional levels. The introduction of the new green codes will likely encourage a more holistic approach to design and construction, especially since the sequential contributions of the members of the design team in the traditional design process are mainly of a linear structure. Opportunities for optimization are limited during the traditional design process, and optimization in later process stages are usually difficult, if at all viable. This approach has often proven to be inferior and inappropriate, producing high operating costs and often a substandard interior environment. These factors can have a negative impact on a property's ability to attract quality tenants or achieve desirable long-term rentals, and they can lead to reduced asset value for the property.

3.2.3　Building Information Modeling

During the last few decades, we have seen building construction continue to grow in complexity and change under the influence of emerging technologies. To meet these challenges there are a number of new software programs on the horizon that are having a positive impact on the entire design, planning, and construction community. Among them is building information modeling (BIM) software, the latest trend in computer-aided design that is being touted by many industry professionals as a lifesaver for complicated projects because of its ability to correct errors at the design stage and accurately schedule construction. BIM is made up of 3D modeling concepts, information database technology, and interoperable software in a computer application environment that design professionals and contractors can use to design a facility, simulate construction, and accurately estimate project costs. In this regard, Autodesk®, Inc., says:

[B]uilding information modeling (BIM) software facilitates a new way of working collaboratively using a model created from consistent, reliable design information—enabling faster decision-making, better documentation, and the ability to evaluate sustainable building and infrastructure design alternatives using analysis to predict performance before breaking ground.

In fact, some industry professionals forecast that buildings in the not too distant future will be built directly from the electronic models that BIM and similar programs create, and that the design role of architects and engineers will dramatically change (Figure 3.4). Building information modeling is gradually changing the role of drawings in the construction process, improving architectural productivity, and making it easier to consider and evaluate design alternatives.

This modern modeling technology is particularly valuable in sustainable design because it enables project team members to create a virtual model of the structure and all of its systems in 3D, in a format that can be shared with the entire project team, thereby facilitating the integration of the various design teams' work. It allows team members to identify design issues and construction conflicts and resolve them in a virtual environment before the actual commencement of construction, thus directly promoting an integrated team approach. This is discussed in much greater detail in Chapter 5 (Building Information Modeling).

Already BIM technology is being employed by many architectural and engineering consultants, and as its popularity continues to increase it is rapidly becoming pivotal to building design, visualization studies, cost analysis, Contract Documents, 3D simulation, and facilities management. Autodesk Revit®, a BIM software package, continues to make headway in its penetration of the architectural and engineering market, and it is anticipated that within the next few years it will have a significant share of major projects in the United States and possibly other countries.

FIGURE 3.4 Highlands Lodge, Resort, and Spa in Northern California, a joint venture of Q&D Construction and Swinerton Builders, in which Vico, a BIM software package, was used. The five-star hotel and high-end luxury condominium complex has a gross floor area of 406,500 square feet (37,720 square meters) and is situated on a roughly 20-acre site at the Northstar-at-Tahoe Ski Resort in Northern California. *(Source: Vico Software.)*

3.3 HIGH-PERFORMANCE DESIGN STRATEGIES

Although it is difficult to find a definition of a high-performance building that everyone agrees on, perhaps the one characteristic that most endorse is that high-performance buildings reflect design excellence. This may be partly because they are typically designed in a holistic, integrative fashion that allows them to offer benefits such as minimal environmental impact (dramatically reducing greenhouse-gas emissions), savings on energy and natural resources, optimized healthy interiors, and life-cycle cost savings. Yet the real value of high-performance buildings can be easily underestimated when using traditional accounting methods that fail to recognize "external" municipal and regional costs and benefits. Greater accuracy can be achieved when high-performance building cost evaluations adequately address the economic, social, and environmental benefits that typically accompany green buildings.

3.3.1 Green Design Strategies

Improved technology is making it much easier and more cost-effective for designers and engineering professionals to incorporate sustainability into their high-performance design strategies. Likewise, there are many recommended practices that can reduce the environmental and resource impacts of buildings and enhance the health and satisfaction of occupants. The most prominent strategies that come to mind are the following.

Use less to achieve more. The most effective green design solutions are able to address a number of needs with only a few elements. For example, a concrete floor may be simply finished with a colored sealant that reflects daylight for better illumination and eliminates air pollutant emissions from floor coverings. The floor can also be used to store daytime heat and nighttime cold to provide occupant comfort. Thus a carefully designed element serves as structure and finished surface, distributes daylight, and stores heat and cold, saving materials, energy resources, and capital and operating costs.

Incorporate design flexibility and durability. Buildings that are designed with the flexibility to adapt to changing functions over long useful lives reduce life-cycle resource consumption. Durable sustainable structural elements that contain generous service space and are able to accommodate movable partitions can last for many decades, instead of being demolished because they are incapable of adapting to changing building functions. Durable envelope assemblies reduce life-cycle maintenance and energy costs and improve comfort.

To achieve maximum effectiveness, consider combinations of design strategies. Green buildings incorporate increasingly complex systems of interacting and interrelated elements. Intelligent green design must consider the impact of these elements and systems on each other, and on the building as a whole. As an example, the need for mechanical and electrical systems is greatly affected by building form and envelope design. Combining strategies such as daylighting, solar load control, and natural cooling and ventilation can reduce lighting, heating, and cooling loads. Carefully combining these strategies can save resources and money, both in construction and in operation and maintenance.

Take advantage of site conditions. Buildings are usually considered more sustainable when they respond to local microclimate, topography, vegetation, and water resources; they are also usually more comfortable and efficient than conventional designs that rely on technological fixes and ignore their surroundings. As an example, Santa Monica, California, has exemplary solar and wind resources for passive solar heating, natural cooling, ventilation, and daylighting, but has meager local water supplies (some of which have recently been polluted). Taking advantage of such free natural resources and conserving scarce high-priced commodities are appropriate approaches to reducing costs and connecting occupants to their surroundings.

Adopt preventive maintenance rather than repair after the fact. Addressing potential problems from the beginning by applying preventive maintenance is both practical and economically prudent. For example, using low-toxicity building materials and installation practices is more effective than diluting indoor air pollution from toxic sources by employing large quantities of ventilation air.

Another of the attributes of green design is "Smart Growth," which concerns many communities around the country. It relates mainly to controlling sprawl, reusing existing infrastructure, and creating walkable neighborhoods. Locating suitable places to live and work within walking distance or near public transport is an obvious advantage in reducing energy. It is also more logical and resource-efficient to maintain or reuse existing roads and utilities than to build new ones. The preservation of open space, farmland, and undeveloped land strengthens and reinforces the evolution of existing communities and helps maintain their quality of life. It also helps reduce pollution of the environment.

3.3.2   The Integrated Design Process

There are several fundamental differences between the integrated design process (IDP) and the conventional design approach. The IDP approach is basically a collaborative one for designing buildings that emphasizes the development of a holistic or whole-building design in which the owner takes on a more direct and active role in the process and the architect assumes the role of team leader rather than sole decision maker. Additional key players, including the BIM manager; structural, electrical, mechanical, and lighting engineers; and other consultants, become an integral part of the team from the outset and participate in the project's decision-making process—not after completion of the initial design (Figure 3.5). Therefore, from a design perspective the key process difference between green building and conventional design is integration. In the IDP approach, the building is viewed as an interdependent system as opposed to an accumulation of its separate components. The

FIGURE 3.5   Diagram showing the various elements that impact the design of high-performance buildings using the Integrated Design Approach. With this approach, multidisciplinary collaboration is required, including key stakeholders and design professionals, from conception to completion of the project.

objective of looking at all of the component systems together is to ensure that they work in harmony rather than in conflict.

More than at any time in history, the design of buildings today requires the integration of many kinds of information, from different consultants, into a synthetic whole. And to achieve a successful sustainable building project today indeed requires the employment of an integrated design process with clear and precise design objectives, which should be identified early on and held in proper balance during the design process.

The IDP approach to design and construction has become necessary to achieve a successful high-performance building. For example, by working collaboratively as a team the main players (architect, engineers, BIM manager, landscape architect, etc.) can maneuver and direct the ground plane, building shape, section, and planting scheme to provide increased thermal protection and reduce heat loss and heat gain. By reducing heating and cooling loads, the mechanical engineer is able to reduce the size of the mechanical equipment necessary to achieve comfort.

Moreover, the architect and lighting and mechanical engineers can work in unison to design, for example, a more effective interior/exterior element such as a light-shelf, which can serve not only as an architectural feature but also provide needed sunscreening and thus reduce summer cooling loads while at the same time allowing daylight to penetrate deep into the interior. This results in a more efficient environmental performance in addition to ongoing operational savings.

Early in the IDP process, the project owner/client will typically appoint a leader for the project who is proficient in leading a team to design and build the project on the basis of specific requirements in the form of a project brief for space and budgetary capacity. The project brief accompanying this planning activity should describe existing space use, include realistic estimates of both spatial and technical requirements, and contain a space program around which design activity can develop. Depending on a project's size, type, and complexity, there may be a need to bring a construction manager (CM) or a general contractor on board at this point. It has been shown that the best buildings almost always result from active, consistent, organized collaboration among all project players.

On completing the pre-design activities, the architect, designer of record (DOR), and other key consultants, in collaboration with the rest of the team or subconsultants, may produce preliminary graphic proposals for the project or for portions of it via a 3D modeling program (e.g., BIM) or manually. The intention of the preliminary proposals is more to stimulate thought and discussion then to describe any final outcome, although normally the fewer changes initiated before bidding the project, the more cost effective the project will be. It is crucial to involve all relevant consultants and subconsultants early in the process to benefit from their individual insights and to prevent costly changes further along in the process.

Also early in the process, decision-making protocols and complementary design principles must be established to satisfy the goals of the project team's multiple stakeholders while achieving overall project objectives. The final design that emerges will incorporate the interests and requirements of all project team participants, including the owner, while also meeting the overall area requirements and project budget established during the pre-design phase.

By this time a schematic design proposal will be in place, which should include a site location and organization, a 3D model of the project, space allocation, and an outline specification incorporating an initial list of systems and components that form part of the final design. A preliminary cost estimate can also now be made, and depending on the size and complexity of the project, it may be performed by a professional cost estimator or computer program at this point. For smaller projects, this estimate may be part of a preliminary bidding arrangement by one or more of the possible builders. On larger projects, the cost estimate can be linked to the selection process for a builder, assuming other prerequisites are met such as experience and satisfactory references. If a BIM manager is employed, he or she can perform this task.

The schematic Design Phase is followed by the design development phase. This entails going into greater detail for all aspects of the building, including systems, materials, and so on. The collaborative process continues with the architect working hand in hand with the owner and the various contributors and stakeholders. The outcome of this phase is a detailed design on which there is a consensus of all players, who may be asked to sign off. When the project design is developed using an integrated team approach, the end product is usually highly efficient with minimal, if not zero, incremental capital costs and reduced maintenance and long-term operating costs. This avoids having to make costly changes late in the game.

At this point, the development and production of Contract Documents follows, which involves converting the design development information into formats that can be used for pricing, bidding, permitting, and construction. An efficient set of Contract Documents can be achieved by careful scrutiny of and accountability to the initial program requirements as outlined by the design team and the client, in addition to careful coordination and collaboration with the technical consultants on the team. Design, budgetary, and other decisions continue to be made with the appropriate contributions of the various players. Changes in scope during this phase should be avoided as they can significantly impact the project and, once pricing has commenced, can invite confusion, errors, and added costs. Cost estimates may be made at this point, prior to or simultaneous with bidding, to ensure compliance with the budget and to check bids.

Even after the general contractor is selected during the *Construction Phase*, other members of the project team must remain fully involved, as there will remain many outstanding issues to be addressed such as previous decisions that may require clarification, supplier samples and information that must be reviewed for compliance with the Contract Documents, and proposed substitutions that need evaluation. Whenever proposed changes affect the operation of the building, the owner/client must be informed and approval sought. Any changes in user requirements may require modifications to the building's design that will necessitate consultation with the other consultants and subconsultants to assess the implications and ramifications such changes may incur. Changes must be priced and incorporated into the Contract Documents as early as possible.

In the final analysis, the ultimate responsibility for ensuring that the building, up completion, meets the requirements of the Contract Documents lies with the design team. The building's success in meeting program performance requirements can be evaluated through the commissioning and enhanced commissioning processes (preferably employing an independent third party). Here the full range of systems and functions in the building are evaluated, and the design and construction team may be called on to make some required changes and adjustments. Colin Moar, commissioning operations manager for Heery International, says, "To get the best value, hire the commissioning agent to get involved during the concept and schematic design phases." After the building becomes fully operational, a post-occupancy evaluation may be conducted to confirm that the building meets the original and emerging requirements for its use and that it meets the owner's expectations. This is discussed in greater detail in Chapter 15.

3.3.3 Green Building Design and the Delivery Process

The full impact of the IgCC has yet to be determined, but one thing is certain: The process of green building design and construction differs fundamentally from current standard practice. Successful green buildings result from a design process that is strongly committed to the environment and to health issues. Measurable targets challenge the design and construction team, and allow progress to be tracked and managed throughout development and beyond. Employing computer energy simulations offers the ability to assess energy conservation measures early and throughout the design process.

By collaborating early during conceptual design, the expanded design team is able to generate alternative concepts for building form, envelope, and landscaping, and can focus on minimizing peak energy loads, demand, and consumption. Design alternatives are aimed at minimizing the building's construction cost and its life-cycle cost, and their evaluation is on the basis of capital cost as well as reduced life-cycle cost. Assessments include costs and environmental impacts of resource extraction; materials and assembly manufacture; construction; operation and maintenance in use; and eventual reuse, recycling, or disposal.

Computer energy simulation is but one of the tools used to incorporate operational costs into the analysis. It is employed to evaluate a project's effectiveness in energy conservation and its construction costs. Typically, heating and cooling load reductions from better glazing, insulation, efficient lighting, daylighting, and other measures allow smaller and less expensive HVAC equipment and systems, resulting in little or no increase in construction cost compared to conventional designs. Simulations to refine designs and ensure that energy conservation and capital cost goals are met are extremely valuable and demonstrate regulatory compliance. For this reason, simulations are necessary to ensure the project's overall success.

In conventional, non-green buildings, the different specialties associated with project delivery, from design and construction through to building occupancy, are responsive in nature, utilizing restricted approaches to address particular problems. Each specialist typically has wide-ranging knowledge and experience in his or her specific field, and provides solutions to problems that arise solely based on that knowledge and experience. For example, an air-conditioning specialist, if asked to address a problem of an unduly warm room, will suggest increasing the cooling capacity of the HVAC system servicing that room, rather than investigate why the room is unduly warm. The excessive heat gain could, for example, be mitigated by incorporating operable windows or external louvers. The end result, therefore, while often being functional, is nevertheless highly inefficient so that the building ends up comprising different materials and systems with little or no integration between them.

With integrated design, you typically have properly engineered and functioning systems that help ensure the comfort and safety of building occupants. They also empower designers to create environments that are healthy, efficient, and cost effective. Integrated design is a critical factor and a consistent component in the design and construction of green buildings. The summary description outlined in the following sections highlight the benefits of integrated design and the main attributes and characteristics that differentiate the conventional and integrated design processes. Being able to keep the goals and objectives for the project in mind throughout design and construction is certainly one of the unique benefits of integrated design.

3.3.4 Forming the Integrated Multidisciplinary Project Team

It is important that all members of the multidisciplinary team collaborate closely, from the beginning of conceptual design throughout actual design and construction. For sustainable projects, the design team usually has to broaden itself to include certain specialists and other interested parties, such as energy analysts, BIM specialists, materials consultants, cost consultants, and lighting designers; often contractors, operating staff, and prospective tenants are included as well. This enlarged design team provides fresh perspectives reflecting new approaches and provides feedback on performance and cost. The design process becomes a continuous, sustained team effort from conceptual design through commissioning and occupancy.

In most building projects, the *architect* is required to lead the design team and coordinate with subconsultants and other experts. He or she is also required to ensure compliance with the project brief and budget. In some cases, the architect has the authority to hire some or all of the subconsultants; in larger projects, the owner may decide to contract with them directly. The architect usually administers and manages the production of the Contract Documents and oversees the Construction Phase of the project, ensuring compliance with the documents by conducting appropriate inspections and managing submissions approvals and evaluations by the subconsultants. The architect also oversees the evaluation of requests for payment by the builder and other professionals and chairs monthly or biweekly site meetings. Depending on the size and complexity of the project, the owner may hire a BIM manager, whose role and responsibilities will need to be clearly defined.

Involvement at the earliest project phases of *civil, structural, mechanical,* and *electrical engineers* is imperative, as they are an integral part of the project team and essential for achieving a total understanding of the various regulatory and other aspects (e.g., structural, heating, ventilation, and air conditioning) of construction; they may be hired directly by the owner or by the architect. Each consultant produces portions of the Contract Documents that are within his or her specialty, and all participate in assessing their part of the work for compliance.

A *landscape architect* may be hired as an independent consultant depending on the type and size of the project. If one is employed, this should be early in the design process to assess existing natural systems, how they will be impacted by the project, and ways to facilitate accommodation of the project to them. The landscape architect will also organize the arrangement of land for human use involving vehicular and pedestrian ways and the planting of groundcover, plants, and trees. This requires extensive experience in sustainable landscaping, including erosion control, managing stormwater runoff, green roofs, and indigenous plant species.

Other specialized consultants may be required, and as with all contributors to the integrated design process, they should be involved early to incorporate their suggestions and requirements in the design so that their contributions are taken into account to ensure maximum efficiency.

3.4 DESIGN PROCESS FOR HIGH-PERFORMANCE BUILDINGS

Today we are witnessing a rapidly changing world in which building construction practices and advances in architectural modeling technologies have reached a unique crossroad that intersects with changing needs and expectations. And with many successful new building projects taking shape globally, this intersection calls into question the performance level of many of our more typical construction endeavors, forcing us to reevaluate just how far our conventional buildings are falling short of the mark and what needs to be done to meet these new challenges. High-performance outcomes necessitate a far more

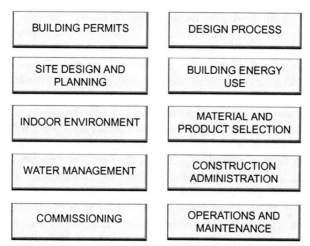

BUILDING PERMITS	DESIGN PROCESS
SITE DESIGN AND PLANNING	BUILDING ENERGY USE
INDOOR ENVIRONMENT	MATERIAL AND PRODUCT SELECTION
WATER MANAGEMENT	CONSTRUCTION ADMINISTRATION
COMMISSIONING	OPERATIONS AND MAINTENANCE

FIGURE 3.6   Main elements of high-performance building design.

integrated design team approach and mark a departure from traditional practices, where emerging designs are handed sequentially from architect to engineer to subconsultant (Figure 3.6). As mentioned earlier, an integrated, holistic approach results in a typically more unified, more team-driven design and construction process that encompasses different experts early on. This process increases the likelihood of creating high-performance buildings that achieve significantly higher targets for energy efficiency and environmental performance than traditionally designed buildings.

The best buildings result from active, consistent, organized collaboration among all players, which is why in the absence of an interactive approach, it would be extremely difficult to achieve a successful high-performance building. The process draws its strength from the knowledge and expertise of all stakeholders (including the owner) across the life cycle of the project in addition to early collaborative involvement in recognizing the need for the building, through planning, design, construction, operation, and maintenance. Building occupancy is part of this process. Also, through a team-driven approach, high-performance buildings are basically utilizing a "front-loading" of expertise. The process typically begins with the consultant and owner leading a green design charrette with all stakeholders (design professionals, operators, and contractors) in a brainstorming session, reflecting a "partnering" that encourages collaboration in achieving high-performance green goals for the new building while breaking down traditional adversarial roles.

By implementing best practices guidelines and an integrated team-driven approach, we maximize the likelihood of achieving superior results in the design and construction of a project. The application of integrated design methods elevates energy and resource efficiency practices into the realm of high performance. This approach differs from the conventional planning and design process of relying on the expertise of various specialists who work in their respective specialties somewhat isolated from each other. The integrated design process encourages designers from all relevant disciplines to be collectively involved in design decision making and to work in harmony to achieve exceptional and creative design solutions that yield multiple benefits at no extra cost.

Design charrettes can be instrumental in complex situations where the interests of the client often conflict, particularly when they are represented by different factions. Charrette team members are expected to discuss and address problems beyond their field of expertise. Although final solutions may not necessarily be produced, important interdependent issues are often studied and clarified. A facility performance evaluation, to confirm that all of the designated high-performance goals have been met and will continue to be met over the life cycle of the project, is also an important consideration. Retro-commissioning is another factor that should be considered to ensure that the building will continue to optimally perform through any potential adjustments and modifications in the future.

It was clearly stated earlier in this chapter that when computer energy simulations are conducted, they should be as early as possible in the design process and continue until the design is complete, to offer a reliable assessment of energy conservation measures and to allow the design team to generate several alternative concepts early on for the building's form, envelope, and landscaping. Computer energy simulation has proven to be an excellent tool to assess both the project's effectiveness in energy conservation and its construction costs. Employing sustainable approaches that reduce heating and cooling loads allows the mechanical consultant to design a more appropriate, more efficient, and less expensive HVAC system, resulting in minimal if any increase in construction cost compared to conventional designs.

Computer simulations have many positive attributes such as allowing us to see how a design can be improved and to ensure that energy conservation and capital cost goals are met, in addition to checking that a design complies with all regulatory requirements. Furthermore, alternative design proposals can be created and readily evaluated on the basis of either capital cost or reduced life-cycle cost. The primary aim of exploring alternative designs is to simultaneously minimize a building's construction cost and its life-cycle cost. But to more accurately assess these costs, a comprehensive approach, involving accurate information on costs and environmental impacts of all aspects of construction, including resource extraction and materials and assembly manufacture, is necessary. Also required is evaluating costs of operation and maintenance from use to final reuse, recycling, and disposal. There are several computer tools available to facilitate life-cycle cost analysis, such as computer energy simulations that can incorporate operational costs into the analysis.

The popularity of high-performance sustainable buildings continues to rise and is emerging as an important market sector both in the United States and globally. At the same time, this increased demand for high-performance buildings has encouraged facility owners, investors, and design professionals to reevaluate their position in the design process. Reassessment of emerging patterns and primary processes in successful high-performance building projects is having a consequential impact on both private and government sectors.

Many government agencies have started to take a serious approach to sustainability, and in January 2006 the Federal Leadership in High Performance and Sustainable Buildings Memorandum of Understanding (MOU) was signed, committing the signatory agencies to federal leadership in the design, construction, and operation of high-performance and sustainable buildings. An important component of this strategy is the implementation of prevalent approaches to meet requirements relating to various sustainable activities such as planning, siting, designing, building, operating, and maintaining high-performance buildings.

The MOU contains a number of guiding principles to be adopted by federal leadership in high-performance, sustainable buildings. These incorporate greater detailed guidance on the principles for optimizing energy performance, conserving water, improving IEQ, integrating design, reducing the impact of materials, and other issues. Since the signing of the MOU, many federal facilities have already succeeded in creating high-performance buildings that save energy and reduce negative impacts on the environment and people throughout the United States.

The Interagency Sustainability Working Group (ISWG), as a subcommittee of the Steering Committee established by Executive Order (E.O.) 13423, initiated the development of guidance to assist agencies in meeting the high-performance and sustainable buildings goals of this order, specifically section 2(f). On December 5, 2008, new guidance on high-performance federal buildings was issued that includes the following:

- Revised guiding principles for new construction
- New guiding principles for existing buildings
- Clarification of reporting guidelines for entering information on the sustainability data element (#25) in the Federal Real Property Profile
- Clarification and explanation of how to calculate the percentage of buildings and square footage that are compliant with the guiding principles for agencies' scorecard input

Whether and how this guidance will be impacted by the new national green codes that were recently issued is not clear.

3.5   GREEN PROJECT DELIVERY SYSTEMS

The most appropriate green project delivery system is determined by the owner during the concept Design Phase. Each delivery system has its characteristic advantages and disadvantages depending on the type and size of the project under consideration. Indeed, selection of the right delivery system is one of the most significant factors impacting a construction project's ability to succeed. But before making a final determination on the delivery system to be employed, the owner needs to have a proper understanding of the attributes and challenges of the various systems. Project delivery is simply a method by which all of the processes, procedures, and components of designing and building a facility are organized and incorporated into an agreement that results in a completed project.

The process begins by fully stating the needs and requirements of the owner in the architectural program from concept design to final Contract Documents. There is a wide range of construction project delivery systems. In this respect, Barbara Jackson, author of *Construction Management Jump Start*, says, "There are basically three project delivery methods: design-bid-build, construction management, and design-build." Jackson continues:

These three project delivery methods differ in five fundamental ways:

- *The number of contracts the owner executes*
- *The relationship and roles of each party to the contract*
- *The point at which the contractor gets involved in the project*
- *The ability to overlap design and construction*
- *Who warrants the sufficiency of the plans and specifications*

Regardless of the project delivery method chosen, three primary players—the owner, the designer (architect and/or engineer), and the contractor—are always involved.

Deciding on the project delivery approach most appropriate for a given project may be the single most pressing question in many owners' minds. To answer this question, the owner must first define and prioritize how to measure the project's success and choose a project delivery that will take the project in that direction. The expectation is that the delivery system chosen will produce the highest-quality and most efficient project at the lowest cost and earliest time. But whichever system is chosen, the owner must maintain realistic expectations and not anticipate perfection, as no project delivery approach is perfect nor can any guarantee a perfect project.

The project delivery approach chosen by the owner will determine the expected tradeoff between the owner's control of the project delivery process and the anticipated risks that come with this decision. Likewise, the owner's choice will govern the amount of involvement, both in time and expertise, required of him or her to make the project delivery successful. This has prompted many owners, especially on large or complex projects, to engage design and construction professionals as independent advisors to assist them in making informed decisions and meet these demands. While these professionals advise, serve, and represent the owner, they should have no other interest in the project other than the owner's protection. Conflict of interest must be avoided at all costs.

3.6 TRADITIONAL GREEN DESIGN–BID–BUILD PROJECT DELIVERY

In most countries around the world, the traditional design–bid–build (DBB) delivery method has been the approach of choice in both public and private construction projects. It remains the project delivery system that is most widely used today and is still required by some states. Moreover, because of its long history, the DBB method is well understood by the majority of owners, contractors, and industry professionals. With DBB, risk is minimized through the owner's control and oversight of both the Design and Construction Phases of the project. The DBB process usually provides the lowest first costs based on submitted tenders, but takes the longest time to execute. However, this method has been somewhat modified and perhaps has become more complex by the inclusion of green features into the equation.

When employing the traditional project delivery system, the owner contracts separately for design and construction to a proposed budget. He or she will typically contract directly with a design professional for complete design of the project, including Contract Documents and professional assistance during the bidding stage. The design professional often provides project oversight and continues to administer the Construction Phase of the project on behalf of the owner. This involves reviewing shop-drawing submittals, monitoring construction progress, and checking payment requests as well as processing contractor requests for information (RFIs) regarding construction documents and addressing Change Order requests. When the plans and specifications (bidding documents) are complete, they are released for bidding and solicitation of tenders to prequalified contractors. Prequalification requires certain information that facilitates this selection. Pertinent information includes proof of experience in similar work, financial capability, a record of exemplary performance by responsible references, and current work in hand to ensure that the contractor is not overloaded.

Allegations of owner favoritism (whether real or perceived) in the selection process can be largely eliminated by allowing all qualified contractors to tender on an equal low-bid basis. The design of the project must be completed prior to contractor bidding and selection. Once the general contractor is selected (normally through a competitive bid process)—in most cases the lowest acceptable bidder—the owner enters into a separate contract with him or her to build the project. This process is generally perceived to be a fair one for contractor selection. Under the DBB project delivery system, the owner retains overall responsibility for project management and all contracts are generally executed directly with him or her. When a lump sum price is agreed to between owner and contractor, the owner can usually rely on the accuracy of the price and is able, with the assistance of the consultant designer, to compare submitted bids to ensure that the best contract price has been obtained. It should be noted that there is no legal agreement between the contractor and the designer of record.

The design–bid–build process has several important advantages. For example, it provides much needed checks and balances between the Design and Construction Phases of the project. It also allows the owner to provide significant input into the process throughout the project's Design Phase. The traditional DBB process also has some disadvantages, the main one being that it is lengthy and time-consuming and the owner often has to address disputes that may arise between the contractor and the design professionals resulting from errors or other unexpected circumstances. With this process, the ultimate estimated cost of construction is unknown until bids are finalized, bearing in mind that the system encourages potential Change Orders that will most likely increase costs. Moreover, there is generally no contractor buy-in to green processes and concepts.

There is also the possibility that construction bids will exceed the project's stated budget (because plans and specifications are completed prior to tendering the project), the consequence of which is either having to abandon the project altogether or having to redesign it to fit within the available budget. Another important consideration with this type of delivery system is that the owner is normally required to make a significant upfront financial commitment in order to have a complete design in hand as part of the Contract Documents before solicitation of tenders.

According to Petina Killiany, associate vice president of PinnacleOne, a leading construction consulting firm, the DBB approach is generally best suited for projects that meet certain requirements such as the following:

- The owner desires the protection of a well-understood design and construction process.
- The owner desires the lowest price on a competitive bid basis for known quantity and quality of the project.
- The owner has the time to invest in a linear, sequential, design–bid–build process.
- The owner needs total design control.

Killiany also maintains that there are certain project success factors that owners sacrifice when using the DBB approach:

First, because there is no input from the contractor during the design phase, their input is lost on what may provide the best value in the trade-off between scope and quality. The construction contract is usually performed on a lump sum basis, any savings are not returned to the owner. Design/bid/build projects normally do not allow for fast track design and construction, and as a result, can take more time than those delivered by other approaches.

It should be noted that if gaps are discovered between the plans and specifications and the owner's requirements, or if errors and omissions are found in the design, it is the owner's responsibility to pay to rectify these mistakes.

3.7   GREEN CONSTRUCTION MANAGEMENT

The *ASHRAE Green Guide* states, "The construction manager method is the process undertaken by public and private owners in which a firm with extensive experience in construction management and general contracting is hired during the design phase of the project to assess project capital costs and constructability issues." In this project delivery system, a "construction manager" is added to the construction team to oversee some of or the whole project independent of the construction work itself. The construction manager's role and responsibilities should be clearly defined—for example, to oversee aspects of the project such as scheduling, cost control, the construction process, safety, the commissioning authority (CxA), and bidding, or to oversee all aspects of the project until final completion.

Joseph Hardesty of Stites & Harbison says,

In many ways, the construction management process is not, by itself, a separate construction delivery system but is a resource the owner can use to assist in the construction project. The added cost of a construction manager must be weighed against the benefits this consultant brings to the project. Often, the architect can fulfill the role provided by a construction manager. However, depending upon the degree of sophistication of the owner's in-house construction staff, and depending upon the complexity of the project, a construction manager can provide an essential element to the construction project.

He adds: "A construction manager is most useful on a large, complex project which requires a good deal of oversight and coordination. A construction manager is also helpful to an owner who does not have a sophisticated in-house construction team. A construction manager can help the owner control costs and avoid delays on complex projects." The two basic types of construction management to consider are the agency construction manager (CM) and the at-risk CM (sometimes called CM/GC).

The *agency CM* offers a fee-based service in which the CM acts as advisor to the owner and is exclusively responsible to the owner, acting on his or her behalf throughout the various stages of the project. The owner will separately commission the general contractor and designer of record. With this method the CM basically acts as an extension of the owner's staff and assumes little risk except for that involved in fulfilling his or her advisory roles and responsibilities. With this method, the general contractor remains responsible for the construction work and still carries out construction management functions relative to internal requirements for project completion. However, the agency CM is not at risk for the budget, the schedule, or the project's performance, nor does the CM contract with subcontractors.

The *at-risk CM* delivery approach does not differ significantly from the traditional DBB method in that the CM replaces the general contractor during the Construction Phase and commits to delivering the project on time and within a guaranteed maximum price (GMP). The CM holds the risk of subletting the construction work to trade subcontractors and guaranteeing completion of the project for a fixed price negotiated at some point either during or on completion of the design process. However, unlike in DBB, during the development and design phases the at-risk CM's role is chiefly to advise the owner on relevant issues.

It is the duty of the owner to weigh the relative advantages and disadvantages of each construction delivery system prior to beginning the project. Petina Killiany lists some of the at-risk CM's advantages over the design−bid−build delivery system:

- *Because construction can often begin before the design is complete, the overall project duration can be shorter;*
- *The owner generally gets better estimates of the ultimate cost of the project during all phases of the project;*
- *The owner benefits from a contractor perspective in making decisions on the trade-offs during the design phase between cost, quality, and construction duration;*
- *Constructability and design reviews by the contractor prior to bidding often result in better designs and lower trade contractor contingencies and bids;*
- *The expertise of the construction manager in pre-qualifying trade contractors helps achieve better performance and workmanship by the trades;*
- *The architect and contractor working together during the design portion can result in a better team effort after the GMP is established.*

However, in some jurisdictions the at-risk CM approach faces the possibility of not being permitted, by statute, to a public owner. Also, because the at-risk CM is not a traditional method of delivery, some owners may not fully understand how to successfully implement this method and, as a result, feel forced to rely on the advice of the CM when he or she should in fact be questioning it. Moreover, the owner should consider the size and complexity of the project, the relative importance of cost or schedule, and the in-house expertise the owner has in place to manage the project before deciding whether this delivery method is appropriate.

It should be noted that when the CM is engaged in an advisory capacity, the service is totally different, for while project owners can't totally avoid risks, it is possible to mitigate them to an acceptable level. Richard Sitnik, a regional director of project management at ARCADIS US, says, "When given appropriate responsibility and the ability to provide effective leadership, project managers (PM)/CMs as advisor promote project success through informed, experience-based decision making, and well-disciplined and regimented project controls." Sitnik also believes that the PM/CM as advisor can provide a wide range of services to the owner throughout the design, bidding, negotiation, and construction phases of the project. The following are some of the more pertinent PM/CM services outlined by Sitnik:

- Perform needs assessments
- Provide direction on alternate project delivery systems
- Assist in the selection of appropriately qualified consultants

- Manage governmental agency approvals
- Identify and manage risks
- Anticipate potential problems before they become costly
- Produce master budgets and schedules
- Establish project controls
- Control costs
- Perform quality controls

The greatest value of PM/CM as advisor occurs when he or she is engaged very early in the design process to initiate the establishment of controls, including over budgets and master schedules. This will contribute not only to greater design efficiency but also to fewer Change Orders in the field and less likelihood of surprises on bid day. The principal role of the PM/CM as advisor is to minimize delays, cost overruns, and failures to meet project objectives. This can be achieved by basically providing the owner with total support and impartial advice and counsel, and guiding the owner to make informed decisions, without compromising the ability to coordinate the multiple agendas and sometimes conflicting interests of the design professionals, contractors, and owners.

On occasion the term "program management" is used. Program management is essentially the same service as PM/CM as advisor, the distinction being that program management is the term applied to large, complex, and multiproject programs. The general benefit to the owner in employing a program manager firm is the expertise and experience it brings to the table, such as assisting the project owner in developing an appropriate overall strategy to manage projects within the program, as individual projects may differ in their requirements and method of construction. When program managers oversee projects that consist of more than one building, allocation of the various roles may differ so that, for example, one building project may comprise an architect, a general contractor, and a PM/CM as advisor, while another may comprise a design-builder or at-risk CM. However, all construction projects are likely to contain some risk, and employing a program manager and/or a PM/CM as advisor minimizes this risk and should be seriously considered for large or complicated projects, particularly in cases where the owner is faced with the risk and responsibility of choosing and implementing a project delivery approach but lacks appropriate in-house technical capability or needs an increase in staff when time frames restrict their use.

3.8   GREEN DESIGN–BUILD PROJECT DELIVERY

There are several definitions of the design–build process. The Design-Build Institute of America (DBIA) describes it this way:

[A]n integrated delivery process that has been embraced by the world's great civilizations. In ancient Mesopotamia, the Code of Hammurabi (1800 BC) fixed absolute accountability upon master builders for both design and construction. In the succeeding

millennia, projects ranging from cathedrals to cable-stayed bridges, from cloisters to corporate headquarters, have been conceived and constructed using the paradigm of design-build.

One of the distinguishing features of the design–build approach is that there is only one contract, meaning that the owner contracts with one entity (the designer/builder) that assumes responsibility for the entire project: its design, supervision, construction, and final delivery. The selection process usually consists of soliciting qualifications and price proposals from various design-builders, usually teams of contractors and designers, before or during the project's conceptual Design Phase. The design–build team is generally led by a contractor (often with a background in engineering or architecture), resulting in the owner issuing a single contract agreement to the contractor, who in turn contracts with a designer for the design. According to Petina Killiany, design–build, when permitted, is generally suited for projects that satisfy the following:

- *The owner is willing to forego control of design and does not seek a highly complex design program/solution.*
- *The owner can provide a complete definitive set of performance specifications and program for design for the design/builder to serve as the basis for the design/ builder's proposal and the owner's contract with the design/builder.*
- *The owner has realistic expectations for the end product and a thorough understanding of the risk of giving up the control of the design.*
- *The owner desires a fast delivery method and is willing to compensate the design-build team for its assumption of risk for design and construction.*

3.8.1   Design–Build Process Basics

Many project owners prefer the design–build project delivery system to DBB because it provides a single point of responsibility for design and construction rather than separate contracts for the Design Phase and then for the construction with two separate entities. This may be the reason it is gaining popularity as the project delivery system (Figure 3.7). Although it has the advantage of removing the owner from contractor and design disputes, it has the disadvantage of eliminating some of the checks and balances that often exist when the Design and Construction Phases are contracted separately. Other disadvantages for the owner include the loss of much of the control of the project that exists under a design–bid–build process, including the loss of the owner/architect advisory relationship, which sometimes results in the project not meeting the owner's expectations.

Nevertheless, interacting with a single entity has obvious advantages for the owner, such as easier coordination and more efficient time management. The design–build contractor or firm will endeavor to streamline the entire design process, construction planning, obtaining permits, and so forth. Also, with the design–build process comes the ability to overlap activities so that certain construction activities on parts of the project can begin even before finalization of the design. There are times when the main contractor may involve other organizations on the project, but in such cases, too, the contractor will be the

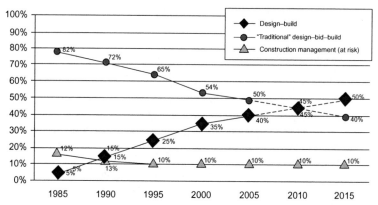

FIGURE 3.7   Graph showing the rising popularity of design–build in nonresidential construction in the United States over the years. *(Source: Design-Build Institute of America; see McClain in Bibliography.)*

one dealing with them and assuming responsibility. This overlapping offers the flexibility to make changes to the design while construction is in progress. With the traditional DBB system, this isn't possible since construction cannot begin prior to finalization of the blueprints and Contract Documents.

There are several important potential advantages and disadvantages for the numerous parties involved in a design–build contract, especially if all parties correctly understand the mechanics of the process as it applies to their project. Kenneth Strong and Charles Juliana of Gordon and Rees, LLP, provide a list of the advantages and disadvantages.

Advantages and Disadvantages of Design/Build

A.   Design/Build Advantages

  1.   *Time Savings:* By combining the selection of a designer and a contractor into one step, the design/build method eliminates the time lost in the DBB process. Further, the design/build contractor is able to start construction before the entire design is completed. For instance, the design/build contractor can start excavation as soon as the foundation and utility relocation design has been prepared. Meanwhile, the design professional can continue design work for the rest of the project during excavation.

  2.   *Cost Savings:* Potential costs savings can be realized with the design-build system because it has high-value engineering capabilities due to the close coordination between the A/E and construction contractor. Construction contractors have direct and real experience with the cost of purchasing and installing materials and, in the design-build system, can share that experience directly with the design professional during the Design Phase of the project. This process has the potential to translate into lower costs, which savings can then be passed on to the owner.

Continued

Advantages and Disadvantages of Design/Build—cont'd

3. *One Point of Contact:* The one point of contact feature for both design and construction is integral to the design/build system. The advantages of this feature are relative—having only one entity to deal with in many instances will outweigh the oversight benefits an owner would otherwise get from contracting separately with a design professional for the project design.

4. *Fewer Change Orders:* A definite advantage of the design/build system is that an owner can expect far fewer change orders on a design/build project. However, if an owner decides it wants a design change during the design/build project, and that change is not covered by the defined scope of the project, that would be considered an extra. Still, in the design/build system, the owner is not liable for any errors the design professional makes because the design professional is part of the design/build team.

5. *Reduced Risk to the Owner:* The shifting of liability for design quality from the owner to the design/build contractor is one of the most significant features of the design/build project delivery system. The advantage to the owner is that it now knows from the outset the cost of that risk. As the design–build contractor is in a better position than the owner to manage and minimize that risk, this is a significant advantage of design/build contracting.

B. Potential Disadvantages to Using the Design/Build Method

1. *Loss of Control of Project Design:* In the design/build system, the shift in responsibility for the design from the owner to the contractor implicitly includes some shift in control. The owner should evaluate the degree to which this loss of control will affect the success of the project. If the owner has specific needs or requirements, it should satisfy itself that it can clearly articulate them in defining the scope of work, or accept the risk that it will have to pay extra to get what it wants via the Change Order process. Change Orders issued to revise scope are not inherently less likely or less expensive in the design/build project delivery method.

2. *Less Project Oversight/Control of Quality:* As has been discussed, one of the advantages of the design/build concept is the cooperation between the design professional and the construction contractor because they both are part of the same team: the design/build contractor. However, this feature can also be a disadvantage, as the architect is no longer the owner's independent consultant and is now working with and for the contractor. For owners who do not have their own design-proficient staff, the loss of the architect's input and judgment may expose them to quality control problems. The owner considering design/build project delivery ignores this issue at its peril. If the owner is one that is used to having the design professional act as its agent, it should make plans to have another entity take that responsibility.

3. *Suitability of Design/Build Teams:* In the DBB methodology, while public agencies are bound by state law to hire the lowest responsive, responsible bidder for construction work, they have more flexibility in selecting designers for their projects. In other words, DBB public owners are allowed to take into account in the selection of a designer more than simply which candidate offered the lowest price. In design/build, the public owner loses the latitude it had in DBB in selecting a design firm. True, the risk for adequacy of the design has been shifted to the design/build contractor, but that is little solace to an owner if the finished project is structurally sound but operationally deficient.

Another potential challenge, or disadvantage, is difficulty in pricing the work. It is often difficult to establish a firm price for a project if the design is incomplete. This is often the situation when the design–build organization is selected. Costly tendering is another issue. Owners are usually expected to pay for the efforts by design–build organizations to formulate their tenders, which normally may include preliminary design work in order to be able to present a cost estimate for the project.

3.8.2   Factors That Impact the Decision to Choose Design–Build

Before deciding whether the design–build methodology is the most appropriate delivery system for a given project, the following factors should be taken into consideration:

- Design–build is an appropriate project delivery system for projects that need to be completed within a tight time frame.
- An important factor that will impact delivery system selection is the type of project to be constructed. An appropriate candidate for design–build is one where the performance and form of the finished project are sufficiently described in a scope document. However, design–build may not be the best method for a project in which the owner's needs are very specific and specialized.
- Several cost-saving benefits in terms of budget can be achieved using the design–build system, in addition to cost savings achieved by shifting more cost control responsibility to the contractor. For example, a construction contractor may want to use certain materials and methods that meet the owner's requirements but were not originally considered by the designer. Any potential cost savings from the contractor's proposed modifications should be passed on to the owner rather than the contractor.

3.8.3   AIA Design–Build Documents

The construction industry has witnessed, during recent decades, the steady increase in popularity of the design–build project delivery system on vertical construction projects. Project owners and contractors, however, have expressed rising concern that the standard AIA forms of agreement for design–build projects do not adequately address their needs. In direct response to these concerns, the American Institute of Architects (AIA) has decided to completely overhaul the design–build forms of agreement, resulting in the introduction of several completely new forms and the retirement of the 1996 series of agreements (A191, A491, and B901).

The new agreements include A141-2004 Agreement between Owner and Design-Builder (replaces A191-1996); A142-2004 Agreement between Design-Builder and Contractor (replaces A491-1996); B142-2004 Agreement between Owner and Consultant, where the owner contemplates using the design–build method of project delivery (no 1996 counterpart); B143-2004 Agreement between Design-Builder and Architect (replaces B901-1996); and

G704-2004 Acknowledgment of Substantial Completion of a Design–Build Project (no 1996 counterpart). In 2008 the AIA also published AIA Document A441™-2008, Standard Form of Agreement between Contractor and Subcontractor for a Design-Build Project, and AIA Document C441-2008, Standard Form of Agreement between Architect and Consultant for a Design-Build Project.

It is true that the design–build delivery system has several advantages for building owners, yet owners have come to realize and understand that with this system they exercise less overall control in guaranteeing that their "intent" is clearly articulated. This has been a great cause for concern, from which has emerged a concept known as "bridging," which is discussed in greater detail in other chapters of this book. Bridging is defined as the owner's means of conveying its intent to the design–build team, and it can take on various forms so that the owner can assume a more expansive role, in which it has much more to say in the design; alternatively, the owner can assume a more limited role and simply set forth its intent in a more conceptual form.

Chapter 4

Green Project Cost Monitoring and Closeout

4.1 INTRODUCTION

The most important objectives of any project are to ensure that it is completed on time, on budget, and according to specifications. In addition to lenders, there are other major players, including owners, contractors, architects/engineers, and material or equipment suppliers, typically involved in the construction of new facilities and major renovations, and that together try to achieve these objectives. The borrower, who is typically the project owner, is often required to present the lender with conceptual designs and specifications, proformas, construction cost estimates, and the like, for the primary purpose of providing enough information for the lending institution to be able to make a loan determination.

After receipt of the various drawings and documents, the lender usually hires a construction consultant to advise and give a professional objective review of the construction loan commitments and payment requests, to fully protect the lending institution. The construction consultant then examines all the documents on behalf of the lender, including conceptual design and specifications and Contract Documents for engineering soundness and compliance with governmental regulations. An assessment of cost comparables is then made for similar projects, in addition to a trade-by-trade breakdown. Once completed, this estimate is compared with the borrower's estimate for general agreement and discrepancies.

Of note is that the lender's interest in each property is subject to rights and restrictions stated and articulated in the loan documents. In the case of new construction, the consultant is usually hired by the lender prior to commencement of the project and basically has the responsibility of administering the project to completion. The consultant therefore assumes that satisfactory access to the property, staff, vendors, and documents will be provided by the borrower. In the event that the borrower fails to cooperate, the lender will apply its leverage to assist the administrator in securing access and all information necessary to monitor the project and to protect the lender's rights. On no account should a lender's representative seek access to any property, staff, vendor, or documents if the borrower refuses such access or restricts the lender's representative from performing its contractual duties as per contract.

4.1.1 Project Evaluation and Analysis

It may be prudent here to clarify what evaluation is. Evaluation is basically a process that achieves the following:

- Supports a specific project, by measuring the extent to which its objectives are achieved, and by highlighting the areas for potential development and improvement
- Identifies achievements
- Facilitates and encourages decisions to be taken, including modifications to objectives and the project methodology
- Makes recommendations for further development of the project

Once it is established that the borrower's estimated costs are in line with typical local costs, the consultant proceeds to prepare a comprehensive review of the project plans and specifications to assure the lender that the design is in compliance with good engineering practice. A detailed written report is prepared and submitted to the lender describing important aspects of the project and including comments on the following:

- Completeness of plans, specifications, and related information and conformance with all applicable building codes and zoning ordinances
- If LEED® or other green certification is being sought, confirmation that documents meet all requirements
- If the International Green Construction Code (IgCC) is applicable, assurance that all documents are in conformance with it
- Design of architectural, structural, HVAC, electrical, plumbing, and fire protection systems, elevators, site improvements, and other relevant information
- Borrower's itemized trades cost breakdown
- Soil borings contents, load tests, engineering reports, and environmental impact studies
- Areas of potential complications that would become a problem to the lender
- Architectural and engineering agreements, material and construction contracts for completeness, function, responsibility, and costs
- Conformity of materials (eco-friendly when possible) specified with the project's overall quality objectives
- Conformity of project scope and design as outlined in the plans and specifications and the project description as set forth in the loan agreement
- Borrower's projected date of construction commencement and date of final completion

It is important to ensure that the preceding are accomplished, for as Paul Eldrenkamp, founding partner of DEAP Energy Group, says, "One of the most striking things about our industry is just how many chances there are to make mistakes. For every opportunity you have to get something right, it seems, there's a thousand chances to get it wrong."

4.2 FRONT-END ANALYSIS

Prior to construction, the lender's representative will often be requested to perform a one-time front-end analysis, which will be summarized in a separate Project Analysis Report (PAR), dated and signed by the consultant(s) performing it. The PAR will include a review of the borrower's plans and specifications to evaluate the completeness of these documents. Moreover, construction lenders frequently request an analysis of the contractor's estimated construction costs to determine if available funds are sufficient to complete the proposed project, and to comment on the project's feasibility.

During this process, all other relevant Contract Documents, such as environmental reports, geotechnical reports, the construction contract, permits, and the like, are also reviewed. If questions or problems arise, the lender normally contacts the borrower, contractor, or their representative, for prompt clarification to accommodate the loan closing time frame. The original front-end analysis report is typically delivered to the lender within roughly three weeks following receipt of the required documents and Notice to Proceed from the lender. It should normally include the following:

- A construction documents review
- A construction costs review
- A pre-closing construction progress inspection
- A pre-closing site inspection
- Notes on attendance at the lender's pre-construction meeting
- Team selection and a pre-project plan

4.2.1 Construction Documents Review

Two complete, half-size, sealed, and signed sets of the plans and specs itemized in the following are forwarded to the consultant. Plans are to be an exact duplicate of those in the set submitted to the building department and the lender. A list of the drawings from the architect's office should accompany the drawings forwarded to the lender's representative, which references each drawing as to the date of preparation and last revision date. If one of the sets is not stamped "Approved" by the building department, a letter should accompany the plans from the architect's office confirming that the documents are an exact duplicate set of those approved. Copies of all revised drawings (with revisions indicated) and specification addenda should be forwarded as issued. One complete set of existing building plans and specs (as-builts if available) should also be forwarded. The documents should include the following:

- Specifications/project manual
- Site plans and offsite plans, if any
- Landscape plans
- Zoning sheets

- Architectural and interior design
- Structural plans and calculations
- HVAC
- Electrical
- Plumbing
- Fire protection
- Parking structure plans and specifications

The exchange of information required by the construction consultant to begin loan monitoring includes the following:

Letter of Agreement. One of the first issues to be resolved is a Letter of Agreement between the construction consultant and the lender spelling out the services to be performed, fee rate, and payment method.

The owner's agreements with contractors. The construction consultant should be familiar with all the parties involved in the project. It should also be noted that there may be other work proceeding in the project that differs from the owner/contractor agreement listed with the lender; these should be taken into account so that the lender may understand potential liabilities on the project. The American Institute of Architects (AIA) Document A101™ standard form of agreement between owner and contractor is widely used where the basis of payment is a stipulated sum (fixed price).

Plans and specifications. The construction consultant should review the architect's or designer's plans and specifications to confirm that the lender's intended understanding of value will be translated to the contractor. At the site, the construction consultant will confirm that materials specified for the project are in fact used and that modifications are appropriately documented for the protection of the parties.

Survey. The instrument survey should be completed at the appropriate time and submitted to the construction consultant for verification of compliance with all zoning requirements.

Title report/deed. The construction consultant should review the title report and take note of any special restrictions or conditions that may be placed on the property and confirm that the project conforms to these restrictions.

Contractor's schedule of costs. Development of an accurately broken-down contractor's schedule of costs into significantly small items avoids overpayment to the contractor and is one of the most important tasks of the construction consultant. These will be the data used to determine the value of draws against completed work to date. It is intended to be used hand in hand with the construction schedule.

Confirmation of utilities. The construction consultant should confirm prior to the release of the financial commitment letter that specified utilities are available to the property or that the contractor has made alternative arrangements.

Building permit. Before building construction is permitted to commence, confirmation that a full building permit was issued should be confirmed.

Sometimes partial building permits are issued and can present a great deal of difficulty for the lender; therefore, it is important to fully investigate the reasons for a partial permit. Changes or conditions requested by the municipality should also be noted.

Release from special entities. Releases from special entities should be confirmed before construction begins. This might include special approvals from design review boards, curb cut permits, and the like.

4.3 REQUISITION FORMAT

Many lenders have developed their own requisition form or format that they prefer the borrower to use. The lender should be consulted to see whether such a form exists. Failing that, the most widely used Application for Payment standard forms are the following:

- AIA Forms: G702 Application and Certificate for Payment
- ConsensusDOCS 291, Application for Payment (GMP): Facilitates the calculation and documentation of progress payments where basis of payment is a guaranteed maximum price
- Engineers Joint Contract Documents Committee (EJCDC) Contract Documents: Another alternative to the AIA contract forms

If the AIA documents are to be used, the requisition should be put together with the line items organized in the AIA G702 format, using as many line items as reasonably possible. Where line items contain more than one trade or work scope, they should be broken down into the individual subcontracts that will be awarded. All subcontractual costs are to be subtotaled prior to adding general conditions, a builder's or developer's fee, and the contingency line items. The AIA contractor form G702™, Application and Certificate for Payment, is a convenient method by which the contractor can apply for payment and the architect can certify payment is due. With respect to AIA Document G702, the AIA requires the contractor to show the following:

- The status of the contract sum to date, including the total dollar amount of the work completed and stored to date
- The amount of retainage (if any)
- The total previous payments (if any)
- A summary of the Change Orders (if any)
- The amount of payment currently being requested

AIA Document G702 serves as both the contractor's application and the architect's certification. Using it can expedite payment and reduce the possibility of error. If the application is correctly completed and acceptable to the architect, the architect's signature certifies/confirms to the owner that a payment in the amount indicated is due to the contractor. Also, this form allows the architect to certify an amount other than the amount applied for, with the architect providing a satisfactory explanation. G703™, Continuation Sheet for G702, breaks the contract sum into portions of the work in accordance with a schedule of values required by the general conditions.

In reviewing the payment requisition, the project architect/administrator (the "administrator") is authorized by the lender only to approve funds commensurate with the value of work in place at the time of the site visit. The administrator will not approve projected or anticipated values of completion. Figure 4.1 is a sample letter addressed to a lender bank confirming the

Name   Date
Title
Name of Bank
Street Address
City, State Zip
(_) _ - _ (tel)
(_) _ - _ (fax)
Email _____

Re: ABC Project No._____
        Loan Disbursement Requirements
        Name of Borrower
        Name of Project

Dear   :

Attached is a copy of ABC's "Monthly Documentation Requirements".

Please review, revise, and fill in the blanks so as to comply with Name of Bank's disbursement procedure requirements/policies and this project's construction loan agreement (CLA). Your attention is specifically directed to Item Nos. 4, 8, 9, 10, 11, 12, 13, and 14.

Please fax an edited copy of this schedule to this office and ABC will then revise the attached schedule and send same to the borrower as per your requirements.

Please do not hesitate to call me should you have any questions at ext. _____.

Sincerely,

ABC GREEN INTERNATIONAL, INC.

Name
Project Manager

___/____
enclosure

SK:\PMO\CHECKLST\MONTHLY LTR

FIGURE 4.1   Sample letter to a lender showing approval of and comment on the monthly document requirements from the project owner (borrower) to facilitate normal disbursement procedure requirements.

MONTHLY DOCUMENTATION REQUIREMENTS
Construction Loan Monitoring Services

In accordance with the Construction Loan Agreement (CLA), the following items are to be provided by the borrower to ABC Green International prior to each monthly advance:

1.	Subcontracts & Purchase Orders	All executed subcontracts and purchase orders greater than $_____ or for which the aggregate cost of services, work, or materials will exceed $_____ .
2.	Payment & Performance Bonds	Copies of executed bonds are to be provided from the GC or CM and all subcontractors as required by the contract and loan agreements prior to releasing payment for same.
3.	Trade Payment Breakdowns ("TPB")	Fully detailed TPBs are to be submitted for every subcontract and purchase order. All TPBs are to be approved by the borrower or the CM, should a CM be used.
4.	Revised Budgets	Any revised direct cost construction budgets or cost-to-complete budgets.
5.	Monthly Job Cost Report	This schedule will typically indicate: the original budget; buyout status to date by showing actual subcontract amounts; scope changes prior to subcontract; approved, pending, and anticipated change orders; total value of work in place; balance to completion, and the total anticipated project cost. This spread sheet report is to be prepared by the borrower if a GC is used or by the CM and then approved by the borrower.
6.	Payment Requisition	A direct cost payment requisition prepared in the format of AIA Document No. G702 with the following modifications: line item amounts are to reflect actual subcontract amounts; work completed is to be on a cost incurred basis - not on a percentage of completion; and stored materials are not to be included in the value of work completed, but tracked on a separate Stored Material Inventory Control Schedule. Retainage is to be withheld as per the CLA requirements. The requisition is to be signed-off by the designer-of-record.

FIGURE 4.1—cont'd

procedure and three-page Monthly Documentation Requirements to allow disbursement of funds.

Prior to the site visit, the requisition columns for related work completed during the period covered, and the columns relating to the total value of work completed to date, should be completely penciled in. The requisition should also include the architect of record's sign-off. To support the value of work completed to date for the various line items, the borrower (project owner) submits the subcontractor's schedule of values, prepared by all of the subcontractors for the consultant's review and approval. These schedules should be previously

MONTHLY PROJECT STATUS REQUIREMENTS - continued

7.	**Partial & Full Waivers of Lien**	Partial waivers are to accompany each subcontractor requisition for subcontracts greater than $_____. Conditional final waivers are to be obtained from each subcontractor and the GC or CM prior to final payment being released.
8.	**Change Orders**	Copies of all Change Orders complete with sign-off by the architect or engineer-of-record should the Change Order result in a deviation from the lender's set of approved contract drawings and/or specifications.
9.	**Change Order Schedule**	A separate schedule identifying each change order, the subcontractor, the add or deduct amount, the date received, and whether it has been approved or if it is pending.
10.	**Stored Material Documentation**	Not all lenders allow funding of stored materials. If funding for stored materials is permitted, review the CLA requirements for the caps on a single stored item and the aggregate dollar amount of materials to be funded at any given time. Consult the CLA to determine if corresponding proportional amounts of general conditions and developer's fee are permitted to be released for stored materials prior to the materials actually being installed. Also, determine whether retainage is to be withheld for stored materials. Typical documentation required by the lender in order to receive funding consists of: • Stored Material Inventory Control Schedule • Subcontracts • Bills of sale • UCC statements • Insurance certificates • Inspection and acceptance certificates
11.	**General Conditions**	General conditions are to be funded on a cost-incurred basis, but the amount disbursed is not to exceed the direct cost percentage of completion. Check to determine whether the CLA permits monthly equal payment funding.
12.	**Retainage**	Check to determine whether the CLA requires retainage on such items as stored materials, general conditions, the CM fee, and the developer's fee.

FIGURE 4.1—cont'd

reviewed and approved by the borrower and his construction manager (CM) or general contractor (GC) to guard against front-loading prior to the consultant's receiving them. Lump-sum-amount subcontractor invoices, while necessary, are on their own not deemed adequate to establish the value of work completed and should be corroborated.

MONTHLY PROJECT STATUS REQUIREMENTS - continued

13.	**Shop Drawing Documentation**	Not all lenders fund for shop drawings; check the CLA. If funding is permitted, have the designer-of-record provide a statement that the shop drawings prepared have been approved. An approved shop drawing invoice complete with back-up documentation for the time expended is also usually required.
14.	**Trade Mobilization Cost**	Not all lenders fund for trade mobilization costs; check the CLA. Some lenders will fund half the up-front mobilization costs leaving the half to be paid upon completion of "de-mobilization" costs. Cost breakdown documentation for the basis of this invoice is usually required.
15.	**Designer-of-Record's Field Observation Reports**	On a weekly basis, all Field Observation Reports are to be forwarded to ABC. Please have ABC placed on the routing list.
16.	**Monthly Compliance Certification**	Certification by the appropriate designers-of-record that the work completed during the payment requisition period was performed in compliance with the lender's set of approved drawings and specifications.
17.	**As-Built Foundation Plan**	Upon completion of the foundation, an as-built foundation survey certified to the bank is to be prepared by a licensed surveyor and accompanied by a statement from the designer-of-record that the foundation was constructed within the required setbacks and to the correct elevations. With respect to a pile foundation, a certified as-driven pile location plan and certified pile log will be required prior to release of funds in addition to an as-built foundation survey.
18.	**Construction Schedule**	All construction schedule revisions.
19.	**Testing Reports**	Copies of all testing reports. Testing results which deviate from the design requirements should be accompanied with a copy of the designer-of-record's response to same. Please have ABC placed on the routing list.
20.	**Work Log**	A trade-by-trade man-day count synopsis for the payment requisition period.
21.	**Photos**	Copies of prints, if monthly construction progress photos are being taken.
22.	**Certificates of Occupancy**	Whether temporary or permanent, as the space or units are accepted by the local governmental authority.

FIGURE 4.1—cont'd

The consultant should be satisfied that the amount requested accurately reflects the value of work in place, and also that the line item has a sufficient balance available to cover completion of the outstanding work. Moreover, unless otherwise directed by the lender, monies and percentages of completion approved are to be based solely on the actual subcontract amount, not the line item's original budgeted amount. Should there be a buyout savings, this "savings" is to be allocated to the requisition's contingency budget. Ideally, the consultant/administrator typically tries to reach agreement on the value of work in place prior to leaving the job site.

Monthly Job Cost Reports The borrower should provide a monthly Job Cost Report/Project Status Report (JCR/PSR), or as required by the lender, that details information relating to the project, such as:

- Actual contract or purchase order costs compared to the original budget's line items
- Total amount of Change Orders (approved and pending)
- Total estimated project cost

The JCR has considerable value as it provides timely information about the status of the project budget and allows the administrator to take whatever action may be necessary to bring the project into compliance with the original budget. It essentially provides, for each item, the quantity and/or percentage completed to date, the cost of the item to date, the estimated cost remaining to complete the item, and the total cost estimate for the item at completion. An estimated cost for the entire project is arrived at by calculating the sum totals of these costs.

4.4   SITE VISITS AND OBSERVATIONS

Normally, construction field observations consist of visits to the site at intervals appropriate to the stage of construction or as otherwise agreed to in writing. This is necessary to monitor the progress and quality of the work and to determine in general if the work is proceeding in accordance with the Contract Documents, and to prepare related reports and communications. Regular site visits for observations and direct communication with the contractor also help facilitate a smooth building process.

4.4.1   Lender's Pre-Construction Meeting

Prior to conducting normal site visits, one of the first and most important steps in the Design Phase is a kick-off or orientation meeting. This is usually scheduled by the project manager at or about the time that the contract with the design professional is executed and approved, depending on the project. The main attendees at this meeting generally include the primary stakeholders and participants with an interest in the project, including design team, cost estimator, commissioning agent, and owner. The agenda generally includes introduction of personnel involved in the project; discussion of administrative procedures; discussion of project scope, budget, and schedule; and a site visit and walkthrough. The design professional/project manager should record attendance and prepare and distribute an agenda and meeting minutes of the design kick-off meeting. The design professional should also prepare a project directory of all participants, including name, title, address, email address, phone, and fax.

Following acceptance of the successful bid and subsequent award of the contract for construction, the owner, along with the contractor and administrator/ consultant, schedules a pre-construction meeting, preferably at the site (and prior to loan closing). The purpose of this meeting is to discuss the specific requirements of the Contract Documents and how they relate to the daily operation of the construction project (Figure 4.2). Also on the agenda should be a discussion of the lines of authority and communication.

The lender, in collaboration with the owner and design team, coordinates and establishes the date, time, and place of the meeting. The purpose of this meeting is to meet collectively with all parties to the construction project (including the owner's representative, lender's representative, architect, primary engineers, contractor's project manager and supervisory staff, as well as major subcontractors and vendors) for discussion of status of the work, construction documents, and contractual relationships, as well as the lender's draw procedures and requirements. Sometimes outside agencies may be invited to attend, such as fire marshals and public utility personnel.

The pre-construction meeting usually covers all of the items listed in the general conditions and supplemental conditions of the contract but in greater detail. It offers an opportunity for the main participants to get to know one another and to discuss certain items in advance so as to alleviate future misunderstandings that might impair the process. Some of the more common issues to be discussed during the pre-construction meeting include

- Introduction of personnel and individual roles and accountabilities
- Names of contacts for the bank and the contractor
- Amount of retainage to be withheld
- Number of draws allowed per month
- Ensuring that a competent superintendent is onsite at all times when work is taking place
- Scheduling/coordination of construction duration, contract dates (start and completion), and hours of operation
- Mobilization and site logistics (site access and security, temporary utilities, temporary facilities)
- Construction coordination issues (RFIs, subcontracts, submittals, shop drawings)
- Schedule issues (Notice to Proceed, work schedule, sequence of work, liquidated damages)
- Payment issues (Application for Payment, schedules of value); submission of draw requests
- Change Orders and additional work
- Completion procedures (substantial completion, final inspection, final punch list, final waivers of lien, final payment)
- Method of payment and advances for materials stored onsite
- Handling of disputes

AGENDA

Germantown Housing Project
Project Kick-off Meeting

ABC Project No. A30115   **Date: August, 7, 2011**

1. **Project Overview**

   a.   What is the development plan?

2. **Project Team**

   a.   Who comprises the project team? What is the project extent of involvement?

   b.   Table of Organization.

3. **Plans & Specifications**

   a.   Are the base building plans and specifications complete?

   b.   Any part of the project being completed on a design/build basis?

4. **Construction Period**

   a.   When is the projected start and completion dates?

   b.   Project duration?

   c.   Schedule updates and who will prepare them?

5. **Project Budget and Contracts**

   a.   Direct cost budget and what are the components? Is there a detailed direct cost estimate and breakdown; when last updated?

   b.   Copies of subcontracts, purchase orders, etc.

   c.   What is timetable on Owner/GC/CM Agreement?

   d.   Any trade payment breakdowns available?

   e.   Any work to be performed with GC's own forces? If so, what are the trades? What percentage of the entire contract amount does such work represent?

   f.   What percentage of the direct cost has been secured by subcontract pricing to date?

FIGURE 4.2   Typical two-page agenda for a pre-construction kick-off meeting. The actual agenda will depend on type of project, project requirements, and circumstances.

6. **Existing Building Condition Survey Reports**

   a.   Envelope/Structural Report

   b.   MEP Systems

7. **Permit Status**

   a.   Are permits issued as a whole or in stages?

   b   Status of documentation review by municipality.

   c.   Schedule of necessary permits.

8. **Bonds**

   a.   Will the GC/CM, if any be securing a 100% performance and payment bond for the project ?

   b.   Will subcontractors be bonded? If so, who?

9. **Architect/Engineer Field Inspection Reports and Punch Lists**

   Will architect/engineer field inspections be conducted, at what frequency, and will reports and punch lists be prepared?

10. **Quality Assurance Program**

   What type of program is in place for quality assurance in addition to the municipality required controlled testing?

11. **Change Orders**

   a.   What is the procedure for Change Order approval?

   b.   A separate schedule is requested identifying each Change Order, the subcontractor, the add or deduct amount or scope change, the date received, and whether it has been approved or is pending review.

FIGURE 4.2—cont'd

4.4.2   Pre-Construction Documents

Following the pre-construction meeting and prior to commencement of construction, the contract administrator or owner and/or architect must ensure that certain documents have been executed between the owner and the contractor, including but not limited to the following:

- Notice of Commencement from owner
- Notice to Proceed from the owner
- Property survey from the owner
- All required permits, licenses, and governmental approvals

- Insurance coverage to be carried by the contractor and all subcontractors
- Bonds—contractor's copies of its performance and payment bonds in addition to proof that subcontractors have furnished surety bonds as required by the Contract Documents

Notice of Commencement

To protect the owner and mandate a Notice to Owner of potential lien claimants, the Notice of Commencement must be prepared and filed before the project begins. This is a legal document prepared by the owner's attorney or financial lending institution and recorded with the Clerk of the County Court. The owner is required to have the notice recorded, and a copy must be posted at the project site. The owner, administrator, and/or architect should also obtain a photocopy of the notice for his or her project files. Financing institutions require the filing of a Notice of Commencement as a provision of the loan agreement.

The Notice of Commencement is a recorded statement considered to be one of the most important documents on a construction project and is the first document filed for the lien process; however, its importance is frequently overlooked by contractors. The Notice of Commencement identifies the name and address of the owner and requires that all persons who provide labor and materials send a Notice to Owner. Recording the Notice of Commencement is necessary and by doing so the owner can require the general contractor to supply releases of lien from all persons who have served a Notice to Owner.

Construction must commence within 90 days from the date the notice was recorded. The notice is effective for one year after it is recorded unless it otherwise provides. Failure to pay attention to the Notice of Commencement can have serious consequences and adversely affect a contractor's ability to recover for the work performed on a project. The three main issues that contractors need to pay attention to regarding their project's Notice of Commencement are

- When does it expire?
- Was the bond attached to it?
- When was it recorded?

By posting the Notice of Commencement at the project site, and on public record, the name of the owner, the contractor, and the surety are provided, so that anyone wishing to file a Notice to Owner may do so. Owners can protect their property from liens by requesting the general contractor to provide proof that all laborers, materialmen, and suppliers have been fully paid. Requiring the general contractor to furnish partial and final releases of lien to the owner prevents those persons from placing liens on the owner's property for nonpayment by the general contractor.

Any work executed at the project site prior to the Notice of Commencement is not covered by lien laws, and while the notice is the owner's responsibility, the administrator, and/or architect should advise the owner of both the need and benefits of such a document. The owner/contractor agreement should stipulate that work is not to commence until the Notice of Commencement has been issued. Two examples of a Notice to Proceed issued by the State of Texas and a homeowner are shown in Figure 4.3.

Texas Department of Housing and Community Affairs
Colonia Self Help Center Program

[Print Form]

Pre-Construction Conference Report and Notice to Proceed

County:	Contract Number:
Homeowner Name(s):	Homeowner Address:

Building Contractor Name:	Building Contractor Address:	Contract Amount:

Statement of Homeowner:
I/We, the undersigned, [hereinafter referred to as Homeowner], participated at _____, on this date, in a pre-construction conference prior to signing a contract for the rehabilitation, reconstruction, or new construction of my/ our property. Homeowner(s) acknowledge that I/we understand the terms of the contract, the explanation of the work to be performed by Building Contractor, the roles of the county and Colonia Self Help Center, and our responsibilities during the construction phase. Homeowner's questions have been adequately answered and I/we are aware that assistance will be provided by the county as requested.

_____   _____   _____   _____
Signature of Homeowner   Date   Signature of Homeowner   Date

Statement of Building Contractor
I, the undersigned, [hereinafter referred to as Building Contractor], hereby certify that I participated in a pre-construction conference with the above-referenced Homeowner and the Colonia Self Help Center's authorized representative at the above-referenced location on this date. Building Contractor understands the procedures to be followed for work write-ups, change orders, work performance, construction, requests for inspections, and requests for payments. Building Contractor understands and hereby certifies that, upon completion, the work performed will meet or exceed all minimum construction standards, specifications, regulations, and codes as required by the Colonia Self Help Center Program and state law. Building Contractor hereby certifies that the work performed will be warranted for a period of one year from date of project completion.

_____   _____
Signature of Building Contractor   Date

Statement of Colonia Self Help Center Representative
I, the undersigned, hereby certify that I participated in a pre-construction conference with the above-referenced Homeowner and Building Contractor at the above-referenced location on this date.

_____   _____
Signature of Colonia SHC Authorized Representative   Date

Notice to Proceed

☐ I/We, the undersigned, hereby authorize the above-referenced Building Contractor to commence work on the property located at _____ within _____ days of the execution of this document. The property will be available to Building Contractor to perform the specified work between _____ a.m. and _____ p.m., seven days a week, unless otherwise specified by Homeowner. If Building Contractor does not commence work within the specified time, the Homeowner may, upon proper notification, consider Building Contractor to be in default.

☐ I elect to withhold authorization to proceed until a later date at which time a separate Notice to Proceed will be issued.

_____   _____   _____   _____
Signature of Homeowner   Date   Signature of Homeowner   Date

Form 14 – Pre-Construction Conference Report and Notice to Proceed   Page 1 of 1
May 1, 2009

(a)   *Continued*

FIGURE 4.3 (a) Typical Notice to Proceed document issued by the Texas Department of Housing and Community Affairs.

SAMPLE NOTICE TO PROCEED

TO: Contractor

FROM: Homeowner Name(s)

SUBJECT: Notice to Proceed with Rehabilitation Construction

Name of Homeowner(s), as owner(s) of the property located at _____Address_____ award the rehabilitation contract to Name of Contractor, on Date. Contractor is hereby notified to commence work set forth in the contract on or before Date.

All work is to be done in accordance with program specifications, conditions provided in the contract, and the work write-up that has my (our) initials on each page and signature on last page.

The project must be fully complete within Number consecutive calendar days after Date. The date of completion of all work is, therefore Date.

If the contractor does not commence work within the specified time, I (We) may upon proper notification, consider the rehabilitation contract to be in default.

_____   _____
Signature of Homeowner   Date

_____   _____
Signature of Homeowner   Date

_____   _____
Signature of Contractor   Date

_____   _____
Signature of Grantee   Date

(b)

FIGURE 4.3—cont'd (b) A homeowner's Notice to Proceed with rehabilitation construction. This is one of the most important documents the contractor should be aware of; failure to obtain one can have serious adverse consequences and significantly impact a contractor's ability to recover for work performed on a project.

Notice to Proceed

The Notice to Proceed is the document that certifies the contractor of the acceptance of its proposal and officially directs it to commence work within a specified time, such as 10 business days. Work to be executed under the owner/contractor agreement generally begins on the date specified, and as articulated in the general conditions of the contract for construction. The Notice to Proceed also triggers the project commencement date by establishing the reference date from which the project duration is measured; often the contract will stipulate that work is to be completed within a stated number of calendar days after the contractor receives its Notice to Proceed.

At this point it is considered good practice for the owner to notify unsuccessful tenderers. The Notice to Proceed implies that the site is free of encumbrances and therefore is available for the contractor's use. However, if there are unresolved issues, then the owner may issue a Letter of Intent stating that it intends to contract with its chosen contractors after resolving any outstanding issues.

The owner and/or architect must recognize that the "Date of Commencement" is the official date for the start of the construction project and is specifically identified in the Notice to Proceed. However, it is often difficult for the contractor to start work on that very date unless it has prior knowledge of the notice and adequate time to mobilize the firm's resources, project team, and equipment.

4.4.3 Site/Project Walk-Through

After awarding the project and when the contractor commences work onsite, the project consultant or administrator representing the lender periodically visits the site and walks the entire project to observe the construction progress. This is in conjunction with and for the purpose of reviewing each monthly construction draw Application for Payment throughout the duration of the construction project, unless notified otherwise by the lender. At the time of the regular monthly site visit, the consultant/administrator conducts a separate walk-through to determine the percentage of completion. This is subsequent to review of the draw Application for Payment. The purpose of the walk-through is to observe and determine, in detail, the quality of workmanship and materials and conformance to the Contract Documents.

The consultant is required to conduct periodic site observation reviews during the construction process. These reviews enable him or her to ascertain whether construction is progressing satisfactorily and is in substantial compliance with plans, specifications, and applicable building codes. The consultant's role includes commenting on the quality of workmanship, materials, stored materials, scheduling, and possible issues. Construction lenders require verification by the consultant that requests for payment of construction funds are accurate and suitable for disbursement. Any issues that need addressing or questions needing answers are discussed with onsite personnel, and if they are significant, they are reported to the lender. They are also promptly submitted to the borrower for explanation or corrections.

The consultant will vary the inspection schedule to meet the lender's needs: for example, from once a week to once a month or as often as needed to verify satisfactory performance and progress at the time of each requisition for payment from the borrower. A written report of each onsite inspection, submitted to the lender within an agreed time frame, should typically include

- A detailed description of the construction progress achieved since the previous inspection of the project

- Observation of quality of work in place and whether construction is proceeding in general accordance with the approved plans and specifications
- A calculated percentage of work in place, overall and by trade
- Comments on whether the work is proceeding according to schedule and an estimated date as to when the project will be completed
- Annotated photographs of the project (i.e., 20–40) showing progress of construction, problem areas, and unacceptable work or conditions
- Unfavorable discrepancies, if any, with recommendations of corrective action

Reports are presented in a form designed to convey accuracy and provide the lender with a feeling of actually "walking through" the site with the consultant.

4.4.4 Photo Documentation

In the "old days" prior to the advent of digital photography, site progress photos were typically $3\frac{1}{2} \times 5$ inches for inserting into standard reports. However, digital cameras have transformed the industry and are almost exclusively used to document construction progress (although video is sometimes used). The objective of photographing a project, whether an existing building or one under construction, is to document representative conditions and use reasonable efforts to document typical conditions present, including material or physical deficiencies, if any. Several formats can be used, but consultants most often use one of two templates depending on the lender's needs. These consist of either two photos per page or six photos per page (Figure 4.4). Captions explaining each photo are helpful to more clearly explain and convey relevant information regarding the project. It is also sometimes helpful to add an arrow pointing to the particular item of interest in the photograph for maximum effect.

Photography is an extremely effective way of recording factual observations. Photographs can provide detail that would be difficult to convey using another medium. Later, notes or captions can be added to for further clarification. If dealing with an existing building or project under construction, the first step is to take photographs from various angles with particular attention to details of work such as defects. This is done prior to writing the report because it will refresh your memory as to what was taking place during the site visit and it will alert you to specific items that may require the attention of both the lender and the contractor. The various photos can also be referenced within the "Field Observations" section of the report.

Photographs should be sorted and placed in a logical manner within a photo template with captions to reflect the various aspects of the project portrayed and included in the report. For most assignments (depending on size, complexity, condition of facility, and at what stage of construction), the number of photographs typically ranges between 20 and 40. The items that should typically be

View showing the elevator lobby on the 5th floor of XYZ Office Park that was previously leased by Doe International. Base building and tenant improvements are substantially complete.

Interior view showing vacant space on the 5th floor XYZ Office Park. Base building is complete.

(a)

FIGURE 4.4 Examples of typical templates using (a) two and (b) six photographs per page. Photographs are almost always required for reports and other documentation.

Interior – View of work in progress on Unit 302 on the 3rd floor of the ABC Apartments. About 230 units are now complete and 12 others are in progress. Unit is scheduled to be completed by January 8, 2012.

Interior – View showing work in progress on the upgrading of the bathroom plumbing in Unit 1104 on the 11th floor. Unit is scheduled to be completed by January 10, 2012.

Interior – View of kitchen of Unit 1120 on the 11th floor of the ABC Apartments. Apartment unit is complete and ready for hand over.

Interior – View of living room in Unit 1120 on the 11th floor of the ABC Apartments, showing status of renovation. Work on unit is complete.

Interior – View of new office workspace on the 1st floor which is scheduled to be completed by third week of January, 2012

Interior – View of newly created Handicap unit (No. 225) on the 2nd floor of the ABC Apartments. View of kitchen area. Unit is temporarily being used as an office.

(b)

FIGURE 4.4—cont'd

photographed depend largely on the type of project but normally include some or all of the following:

- Site (from various angles)
- Exterior/building envelope
- Roof
- Interior
- Structural
- Mechanicals
- Electrical
- Plumbing
- Fire protection/life safety
- Garages/carports
- Elevators (and lobbies)
- Amenities
- ADA facilities
- Detailed photographs should follow
- Stored materials (how will advance delivery of materials to site be treated?)

Once the photographs have been organized in a logical sequence, prepare the photo sheets and number each photo with appropriate captions. On the photo sheets, the various components of the project should be identified. For multistory buildings, each photograph should identify the floor or elevation shown. The object is to convey to the lender the project's progress and any other relevant information. Comments on the photographs should convey a thorough familiarity with the project and highlight information not clearly shown in them.

4.5   LOAN DISBURSEMENTS—DRAW APPLICATION REVIEWS

A lender's construction risk management practices are designed to look out for its interests as well as the borrower's. Both borrower and bank interests lie along the same path toward successful completion of construction. The following are some of the numerous things that could go wrong during the Construction Phase of the loan and many issues that need clarification prior to commencement of the project.

- Ensuring that the loan documents match the approval
- Ensuring that the proposed budget has sufficient funds to complete the project
- Checking that there is adequate equity in the project
- Ensuring that the draw requests balance
- Ensuring that the proposed budget passes the plan and cost review

4.5.1   Value of Work in Place

During the scheduled site meetings, the borrower's payment requisitions are reviewed and evaluated by the consultant/administrator usually on the basis of accurate quantities of work in place and approved. Following onsite inspection, the results are compared with the borrower's requisition for funding for work in place up to the time of the inspection. Any discrepancies should be promptly resolved, preferably prior to submission of the requisition to the lending institution. The main purpose of closely monitoring the flow of construction loan dollars is to ensure that, at any given time during the life of the loan, sufficient funds remain in the undisturbed portion of the loan to complete the project. Any delays in the work should be promptly reported to the borrower and the lender.

For the purpose of calculating the total value of work in place, the contractor must break down the schedule of values into items or quantities of work that can readily be evaluated by the administrator when estimating the work in place. This breakdown is separate from the schedule of values and does not replace it. Its main purpose is to prevent potential disagreements between the contractor and the administrator when evaluating the quantities of work completed. However, the value of work in place should be developed prior to writing the monthly report because developing a number for work in place will exemplify where the emphasis must be when writing the body of the report and the summary. By knowing the total value of work in place and the amount approved for the period covered, the administrator is alerted as to whether the pace of the project is slowing down or speeding up, whether a particular line item is heading for a cost overrun, and whether potential challenges can be expected down the road.

The contractor can then proceed to prepare a certified copy of the Application for Payment in the format outlined in the Contract Documents. The administrator is given a copy of the application and verifies that it is correct as per the site review meeting. The contractor brings to the review meeting all materials required to properly evaluate the application, including stored material invoices, release of liens, and the like. The administrator, owner, and contractor assess the project's current status along with the contractor's Application for Payment, and they agree on the amount due the contractor as outlined in the Contract Documents.

Should the administrator and the contractor not agree on an appropriate amount to be disbursed as per the Application for Payment, the contractor may then prepare and submit an Application for Payment that he or she considers to be appropriate and in line with the work in place. The owner and/or architect, in consultation with the administrator, recommend the amount that is believed should be certified for payment. The certified value of work in place is based essentially on the latest site visit and the latest Application for Payment. In Figure 4.5 we see an example of how the certified current value of work in place is calculated. This calculation is typically included in the Project Status Report sent to the lender and other stakeholders.

4.5.2   Stored Materials Funding

The lender has the final say on whether or not stored materials will be funded and, if so, how. In many cases lenders have indicated their willingness to negotiate this item. However, the lender must exercise great care and ensure that the funding procedures for stored materials comply with Building Loan Agreement

Certified Value of Work-in-Place

Based upon our previous site visit, the GC's Application for Payment No. *(Exhibit "B"),* ABC has determined the current value of work-in-place to be as follows:

Original Direct Cost Budget	$ 156,286,487	
Adjustments as Approved by BA	+ 0	
Adjusted Direct Cost Budget	$ 156,286,487	(1)
Total Value of Work Completed To-Date	$ 43,429,128	
Stored Materials	+ 0	
Subtotal Work Completed and Stored	43,429,128	(2)
Less Retainage	- 2,594,341	
Total Completed Less Retainage	$ 40,834,787	
Less Previous ABC Certification	- 15,376,052	(3)
Current Certification	**$ 25,458,735**	
Cost-to-Complete BLA Adjusted Direct Cost Budget,		
Including Retainage	$ 115,451,700	
Cost-to-Complete Based upon ABC's		
Recommended budget of $165,000,000	$ 124,165,213	

(1) Based upon Borrower's Direct Cost Budget as follows:

Off-Site Work	$	300,000
On-Site Work		23,861,912
Shell/Common Area Construction		59,694,900
Speciality Tenant Construction		6,732,600
Speciality Tenant Allowances		11,290,100
Anchor Tenant Construction		14,640,164
Anchor Tenant Allowances	+	39,766,811
Total Direct Cost Budget	**$**	**156,286,487**

(2) Work Completed and Stored To-Date as Follows:

Off-Site Work	$	0
On-Site Work		25,000,000
Shell/Common Area Construction		13,515,987
Speciality Tenant Construction		66,097
Speciality Tenant Allowances		442,493
Anchor Tenant Construction		737,775
Anchor Tenant Allowances	+	3,666,775
Total Work Completed To-Date	**$**	**43,429,128**

(3) Includes amounts not certified by ABC

FIGURE 4.5   Example of how the current value of work in place is calculated.

(BLA) requirements. Thus, unless specifically authorized by the lender, the consultant/administrator has no contractual authority to approve stored materials. This should be clarified in the general or supplemental conditions. If funding for stored materials is requested, the consultant/administrator notes and reports to the lender how the materials are protected from theft, the elements, and vandalism. He or she should inspect the insurance certificates for these materials to ensure that the lender is named as a co-insured. Likewise, the consultant should inspect the invoice, the bill of sale, a Uniform Commercial Code (UCC) statement, or a contract verifying the cost of the materials in question.

Some of the challenges that stored materials funding faces arise because many lending institutions absolutely prohibit payment for materials until they are physically installed. The lender's main concern is increased exposure in the event of a failure by the general contractor or borrower, since recovery of these materials or their cost has traditionally been difficult, if not impossible. Faced with these problems, lending institutions have often modified their approach to "materials only" payments so that if materials are suitably stored at a bonded warehouse, offsite payments may sometimes be made.

The bottom line is whether payment is to be made for stored materials, and this hinges on the policies of the lender. It is important to state the conditions for stored materials funding in the Contract Documents. The borrower would be well advised to reach agreement on the policy toward such advances early in discussions with the lender, preferably prior to closing the loan. Stored materials should be tracked on a separate stored materials inventory schedule provided by the lender. It is strongly recommended that the consultant/administrator review the BLA regarding the lender's policy of funding for stored materials because funding may be reserved as a lender's business decision. If the lender does decide to proceed with funding, a proportionate amount of the general condition or fee monies may be retained.

4.5.3   Change Orders

Change Orders are in fact changes to the contract (a legal document) and are themselves legal documents. Once a Change Order is executed, it becomes part of the contract and cannot be reversed. The only way to make further modification to a contract is to process another Change Order. The owner is required to provide a schedule and copies of all approved Change Orders as well as a schedule of those pending. Change Orders should not be executed unless they have been previously approved by the owner and there are sufficient funds in the contingency budget to absorb them. After the owner completes the review of the proposal request and approval for a Change Order is obtained, the administrator prepares the order, as outlined in the Contract Documents, utilizing AIA Document G701 or its equivalent (e.g., ConsensusDOCS 202).

The Change Order is produced in three copies and forwarded to the contractor for signature. The contractor's signature on the Change Order request acknowledges that the work will be completed as described there for the stated amount (Figure 4.6). Any additional time, if requested, for the Change Order work is incorporated into the Change Order. Failure by the contractor to request additional time for the Change Order work prohibits the contractor from doing so at a later date. The need for contract changes may be the result of many factors. Among the most common are these:

- Plan deficiency (errors or omissions)
- Modified/amended design criteria

CHANGE
ORDER

AIA DOCUMENT G701

Distribution to:

OWNER	☐
ARCHITECT	☐
CONTRACTOR	☐
FIELD	☐
OTHER	☐

PROJECT:
(name, address)

TO (Contractor):

CHANGE ORDER NUMBER:

INITIATION DATE:

ARCHITECT'S PROJECT NO:

CONTRACT FOR:

CONTRACT DATE:

You are directed to make the following changes in this Contract:

Not valid until signed by both the Owner and Architect.
Signature of the Contractor indicates his agreement herewith, including any adjustment in the Contract Sum or Contract Time.

The original (Contract Sum) (Guaranteed Maximum Cost) was .... $
Net change by previously authorized Change Orders .... $ ....
The (Contract Sum) (Guaranteed Maximum Cost) prior to this Change Order was ... . $
The (Contract Sum) (Guaranteed Maximum Cost) will be (increased) (decreased) (unchanged)
   by this Change Order .... $
The new (Contract Sum) (Guaranteed Maximum Cost) including this Change Order will be .... $
The Contract Time will be (increased) (decreased) (unchanged) by..
The Date of Substantial Completion as of the date of this Change Order therefore is ....

( ) Days

ARCHITECT	CONTRACTOR	OWNER
Address	Address	Address
By	By	By
Date	Date	Date

FIGURE 4.6   Example of a typical Change Order template.

- Specification conflict or ambiguity
- Extra work or unanticipated need
- Contractor-proposed change (material substitution, etc.)
- Settlement of disputes
- Price adjustments for increased or decreased quantities

Once the contractor has signed the Change Order, all three copies are returned to the consultant/administrator or the owner and/or architect for signature and certification. The Change Order is then submitted to the owner for final signature and distribution. The administrator, contractor, and owner and/or architect receive one signed and certified copy for their records. The Change Order is then added to the next monthly Application for Payment. It is recorded in the project log book; one copy is included in the log book and one is transmitted to the central file system.

The Building Loan Agreement should be checked by the administrator for the approval requirements of individual and aggregate Change Order amounts. Once a Change Order is approved, a Notice to Proceed is issued to the contractor. However, before this can happen, the administrator/owner's representative issues a written directive to the contractor asking for a request for proposal (RFP) for the Change Order within 10 days for the subject work. This is then followed with a Notice to Proceed, or a Notice to Proceed immediately, with the work. The Notice to Proceed specifies the manner in which the owner will pay for the work in question. An independent estimate should be done prior to any negotiation with the contractor.

The options available are stipulated in the Contract Documents and usually consist of: (1) an accepted estimate (e.g., using bid prices from recent contracts with similar work and quantities); (2) a time and material estimate; and (3) unit costs (e.g., using the "Means Cost Estimating Guide"). The RFP should also address subcontractor mark-ups, labor rates, and various other requirements regarding Change Order pricing. The contractor should not begin the Change Order work unless he or she receives a written Notice to Proceed. The exact wording and format of the notice varies depending on the nature of the work and the requested pricing method.

Sometimes owners prefer Change Order payments to be separate from the Application for Payment. Likewise, the contractor may request assurance from the owner that adequate funds are available to pay for the Change Order before executing the work. Financing and lending institutions generally stipulate that all Change Orders be processed through their office before being executed. The contractor's performance and payment bond and builder's risk insurance need to be adjusted to reflect substantial changes to the contract sum. The contractor is essentially obligated to execute any Change Order authorized by the owner even if a dispute occurs regarding the actual cost of the Change Order work or its impact on the project schedule. These matters can typically be resolved by exercising provisions included in the Contract Documents.

4.5.4　Lender/Owner Retainage

Retainage is the withholding of certain portions of monies due a contractor for work in place or monies withheld from each progress payment earned by a contractor or subcontractor until a construction project is complete. It acts as an incentive to complete the work and is one of the line items identified in the contractor's Application for Payment. Retainage has been the subject of considerable discussion over recent years, and it is quite apparent that many lending institutions do not have a uniform practice regarding either the amount of retainage or the manner in which it is collected. It is routinely called for in both private and public construction contracts. On public projects, state laws often require retainage and specify the amount and the conditions for releasing it. Otherwise, it is governed by contract.

Most lenders generally prefer using a conservative approach in which a full 10% retainage on all items of construction is withheld and kept throughout the entire construction period. On the other hand, a liberal policy is one in which no retainage is withheld on any item. Between the two extremes, a wide variety of practices are prevalent, including holding a retainage on certain items only or reducing the total amount withheld after a certain point in construction has been reached (usually 50% of project completion). Experienced and knowledgeable borrowers try to negotiate the most liberal agreement possible at the outset of the loan program. It is also common practice for the general contractor to impose retainage on his or her subcontractors as well, although most try where possible to work with their subcontractors to facilitate early payment.

The owner may make payments that reflect adjustments in the retainage amounts as provided in the Contract Documents when the contractor achieves successful execution of the Certificate for Substantial Completion. The reduction in retainage is made with the exception of amounts that have been determined to reflect the costs of remaining work and/or work requiring correction. In these cases, the administrator, in consultation with the owner and/or architect, establishes a value for remaining work and suggests that the owner retain three times the value of the incomplete work. The minimum amount retained for each unacceptable item should reflect the estimated cost to have an alternative contractor brought in to complete or correct the item. This includes any costs for mobilization and/or equipment required to correct or complete any outstanding construction deficiencies.

4.5.5 General Conditions

Many standard general conditions have been developed by numerous trade and professional organizations, but perhaps the most widely used are those that are published by the AIA, AIA Document A201-2007 form of General Conditions, and the ConsensusDOCS 200 form of General Conditions. One of the advantages of using the AIA standard is that most contractors and architects are familiar with it. However, many state organizations and universities have their own standard general conditions depending on the project.

The general conditions set forth the rights and responsibilities of each of the parties such as the owner and contractor in addition to the surety bond provider, the authority and responsibilities of the design professional, and the requirements governing the various parties' business and legal relationships. It is considered to be one of the most essential documents associated with the construction contract. For the contractor, it is imperative to know exactly what is contained in the general conditions and its implications. If the document proves too difficult to read or there is simply not enough time, the project may be put at risk. Some of the general clauses contained in the general conditions that can have a direct affect on the success of a project, if appropriate attention is not paid to them, are outlined in the following paragraphs.

General Provisions

This section[1] includes basic definitions for the contract and roles of the various parties, the work, the drawings and specifications, and other issues such as Change Orders and punch lists. In addition, it clarifies the ownership, use, and overall intent of the Contract Documents.

Owner Responsibilities Among other things, this section outlines the services and information the owner is required to supply depending on the general conditions format used. For example, at the written request of the contractor, prior to commencement of the work and thereafter, the owner must furnish reasonable evidence that financial arrangements are in place to fulfill the owner's obligations under the contract. This evidence is a condition of commencement or continuation of the work. After such evidence has been furnished, the owner may not materially vary such financial arrangements without giving the contractor prior notice. Also, except for permits and fees, which are the responsibility of the contractor under the Contract Documents, the owner will secure and pay for necessary approvals, easements, assessments, and charges required for construction, use, or occupancy of permanent structures, or for permanent changes in existing facilities. Also outlined are the extent of the owner's right to stop the work and the right to carry it out.

Contractor Role and Responsibilities This section lays out the obligations of the contractor under the contract. For example, the contractor warrants all equipment and materials furnished, and work performed, under the contract, against defective materials and workmanship for a specified period (usually 12 months) after acceptance as provided in the contract, unless a longer period is specified, regardless of whether the materials and work were furnished or performed by the contractor or any subcontractors of any tier. This section also includes supervision and construction procedures, materials, labor and workmanship, patents, substitutions, record drawings, shop drawings, product data and samples, taxes, permits, and construction schedules. Additionally, the contractor must, without additional expense to the owner, comply with all applicable laws, ordinances, rules, statutes, and regulations.

---

1. Portions excerpted from AIA A201–2007, with minor edits.

Administration of the Contract This section assigns duties to the architect or as specified for the administration of the contract. Specific clauses dealing with the architect's responsibility for visiting the site and making periodic inspections are included. Also addressed are requests for additional time and how claims and disputes are to be handled. Generally, the owner's (or lender's) representative administers the construction contract. The architect will assist the owner's representative with the administration of the contract as indicated in the Contract Documents. The project administrator will not be responsible for the contractor's failure to perform the work in accordance with the requirements of the Contract Documents.

Subcontracts and Subcontractor Relations This section deals with the general contractor's awarding of subcontracts to specialty contractors for certain portions of the work. The contractor is required here to furnish the owner and the architect, in writing, with the name and trade for each subcontractor, the names of all persons or entities proposed as manufacturers of products, materials, and equipment identified in the Contract Documents, and where applicable the name of the installing contractor. By appropriate agreement, the contractor must require all subcontractors, to the extent of the work to be performed them, to be bound to the contractor by terms of the Contract Documents, and to assume all the obligations and responsibilities, including the responsibility for safety of the subcontractor's work, which the contractor, by these documents, assumes toward the owner and project administrator.

Construction by Owner or Separate Contractors This clause basically states that the owner reserves the right to perform constriction or operations related to the project with his or her own work force and has the right to award separate contracts as stipulated in the Contract Documents. In this respect, "No contractor shall delay another contractor by neglecting to perform his/her work at the appropriate time. Each contractor shall be required to coordinate his/her work with other contractors to afford others reasonable opportunity for execution of their work."

Changes in the Work This section highlights how changes (overhead and profit on Change Orders; time extensions; inclusions) are authorized and processed. "The owner may authorize written Change Orders regarding changes in, or additions to, work to be performed or materials to be furnished pursuant to the contract provisions. The amount of adjustment in the contract price for authorized Change Orders will be agreed on before the Change Orders become effective. Likewise, an order for a minor change in the work may be issued by the architect alone where it does not involve changes to the contract sum.

Time and Schedule Requirements The contractor must acknowledge and agree that time is of the essence on the project. The contract time therefore may only be changed by a Change Order. Contract time is the period of time set forth in the contract for the construction required for substantial and final completion of the entire project or portions of it as defined in the Contract Documents. This part of the contract deals largely with issues relating to project startup, progress, and completion relative to the specific project schedule in addition to issues relating to delay (notice and time impact analysis) and extensions of time. The general conditions should clarify certain issues such as how many days after a delay a contractor has to give notice, and how the notice is to be delivered (verbally, by mail, by registered mail).

Payments and Completion The importance of this section is that it identifies how the contractor will be paid and specifies how applications for progress payments are to be made. The contract sum is stated in the agreement and with the authorized adjustments reflects the total amount payable by the owner to the contractor for performance of the work under the Contract Documents. Before the first Application for Payment, the contractor submits to the architect a schedule of values allocated to various portions of the work, prepared in such form and supported by data to substantiate its accuracy as the architect may require. This schedule, unless objected to by the architect, is the basis for reviewing the contractor's Applications for Payment. The owner's representative may decide not to certify payment and may withhold approval in whole or in part, to the extent reasonably necessary to protect the owner.

Protection of Persons and Property This section of the general conditions addresses safety concerns for both the owner's property and project personnel. According to AIA form 201, "The contractor shall be responsible for initiating, maintaining and supervising all safety precautions and programs in connection with the performance of the Contract." It basically means that the contractor shall conduct operations under this contract in a manner to avoid the risk of bodily harm to persons or risk of damage to any property." Moreover, the contractor is required to comply with applicable safety laws, standards, codes, and regulations in the jurisdiction where the work is being performed.

Insurance and Bonds These issues deal with various insurance and bonding requirements of the parties. For example, the contractor is required to purchase insurance to protect the contractor from claims arising out of or resulting from the contractor's operations under the contract and for which he or she may be legally liable, whether by the contractor or by a subcontractor or by anyone directly or indirectly employed by any of them, or by anyone for whom they may be liable. The owner has the right to require the contractor to furnish bonds covering faithful performance of the contract and payment of obligations arising as stipulated in bidding requirements or specifically required in the Contract Documents, including but not limited to the obligation to correct defects after final payment has been made as required by the Contract Documents on the date of execution of the contract.

Uncovering and Correction of Work This section deals with acceptance of the work in place by the architect or owner's representative and stipulates when the contractor is responsible for uncovering and/or correcting work that is considered unacceptable. The AIA 201 form states:

If a portion of the Work is covered contrary to the architect's request or to requirements specifically expressed in the Contract Documents, it must, if required in writing by the architect, be uncovered for the architect's examination and be replaced at the contractor's expense without change in the Contract Time. The owner may, in its sole discretion, accept work that is not in accordance with the Contract Documents, instead of requiring its removal and correction. In Such case the contract sum is as appropriate and equitable.

Miscellaneous Provisions This section addresses various issues such as successors and assigns, wage rates, tests and inspections, rights and remedies, codes and standards, records, general provisions, and written notice.

Termination or Suspension of the Contract This section deals with the terms under which parties may terminate or suspend the contract. The contractor may terminate or suspend the contract if work is stopped for a period of 30 consecutive days through no act or fault of the contractor or a subcontractor, employees, or any other persons or entities performing portions of the work under direct or indirect contract with the contractor. Similarly, in addition to other rights and remedies granted to the owner under the Contract Documents and by law, the owner may without prejudice terminate the contract under specific conditions, and may also, at any time, terminate the contract in whole or in part for the owner's convenience and without cause.

Lenders often require general conditions to be disbursed in direct proportion to the percentage of completion of the subcontractual costs so that general condition monies will be adequate throughout the project's duration. In some cases, especially in CM and "cost plus a fee" contracts, the borrower is required to pay general conditions on either an equal monthly or a cost incurred basis. The lender should be consulted beforehand to determine its funding policy.

4.5.6 Supplemental Conditions

The supplemental conditions amend or supplement the standard general conditions of the construction contract and other provisions of the Contract Documents. Whereas general conditions can apply to any project of the type being designed and built, supplemental or special conditions usually deal with matters that are project-specific and beyond the standard general conditions' scope, although they generally augment them. These sections may either add to or amend any provisions. The following are examples of project-specific information that may appear in this section:

- Project phasing or special construction schedule requirements
- Safety and security precautions
- Insurance coverage certificates
- Additional bond security
- Cost fluctuation adjustments
- Materials or other services furnished by the owner
- Temporary facilities requirements
- Prevailing wages
- Permits, fees, and notices
- Bonus payment information
- Submittals

4.5.7   Designer of Record/Administrator Sign-Off

Project closeout is the final action to be taken in the construction process. Yet before the designer of record/administrator can sign off and the owner can take possession of the building, certain requirements need to be met. For final completion and before final payment, a number of issues need to be addressed, including the following:

- A Certificate of Occupancy is in place.
- All liens of the GC/CM and trade contractors are released or satisfied and outstanding claims resolved.
- The architect has certified final completion and all "punch list" items are complete.
- The architect has certified that the final inspection has been satisfactorily conducted.
- All warranties, guarantees, and operating manuals have been received.
- Final lien waivers and contractor affidavits have been obtained for all work performed from the architect, the GC/CM, and trade contractors.
- The closeout agreement has been signed, in which all outstanding issues with each trade contractor and with the general contractor have been resolved, and the commencement date for the guarantee period has been confirmed.
- Commissioning has been satisfactorily completed.

4.6   PREPARING THE PROGRESS STATUS REPORT

Regular Project Status Reports (PSRs) help ensure that the lender, owner, and other stakeholders have clear visibility to the true state of a project and that management stays properly informed about project progress, difficulties, and issues by periodically receiving the right kinds of information and updates from the project manager based on site visits and meetings. Frequent communication of project status and issues is a vital part of effective project risk management. The PSRs should let management and stakeholders know whether the project is on schedule to deliver as planned and whether there are issues that need to be addressed.

4.6.1 Draw Applications—Documents Required

The consultant needs to conduct an inventory and review of construction documents, which include but are not limited to the following:

- Plat plan/boundary survey/site plan
- Topography plan
- Environmental site assessment
- Soils investigation report
- Construction plans
- Construction specifications
- Addenda/Change Orders
- Construction contracts(s)/schedule of values
- Architect's contract(s)
- Construction schedule
- Building permits
- Freedom of Information Law (FOIL) documents
- Utility letters
- Estimated variances

4.6.2 Waivers of Lien

A lien is a "hold" against a property that, if unpaid, allows a foreclosure action, forcing the sale of the property. It is recorded with the County Recorder's office by the unpaid contractor, subcontractor, or supplier. Sometimes liens occur when the prime contractor has not paid subcontractors or suppliers. Legally, the property owner (borrower) is ultimately responsible for payment—even if he or she has already paid the prime contractor. The borrower (owner) is required to provide copies of all partial/full waivers of lien from subcontractors, vendors, and the like, to the administrator, normally on a monthly basis, usually with the payment application (Figure 4.7). These partial waivers are to accompany each general contractor and subcontractor requisitions on a monthly basis.

Although most borrowers execute waivers on a monthly basis as the project progresses, some prefer to wait until the project is completed and submit them with the final request for payment. In some cases the borrower may request that, on signing the contract, the subcontractors waive their right to lien the job. Waivers should be properly organized, stating what trade they are for, and included as exhibits in the report. As a project approaches completion, copies should be obtained of all lien waivers. For any that are not forthcoming, the

PARTIAL WAIVER OF LIENS

Project Description: _____
    _____

Period Ending: _____ __, 20_

Work Performed: _____

Work Performed by: _____

Under Contract to: _____

Contract Date: _____

Original Contract Amount: $_____
Change Order Amounts: +_____
Adjusted Contract Amount: $_____

Work Completed to Date: $_____
Less Retainage Not Yet Due: -_____
Net Amount Due to Date: $_____
Less Payments Received to Date: -_____
Total Payment Due: $_____

THE UNDERSIGNED (1) acknowledges receipt of the amount set forth above as payments received to date, (2) to the extent of such payments, waives and releases any claim which it may now or hereafter have upon the land and improvements described above in the project description, (3) that the amount of payments received to the date of this waiver represents the current amount due in accordance with our contact and work completed, and (4) warrants that it has not and will not assign any claims for payment or right to perfect a lien against such land and improvements and warrants that it has the right to execute this waiver and release.

THE UNDERSIGNED further warrants that (1) all workman employed by it or its subcontractors upon this Project have been fully paid to the date hereof, (2) all materialmen from whom the undersigned or its subcontractors have purchased materials used in the Project have been paid for materials delivered on or prior to the date hereof, (3) none of such workman and materialmen has any claim or demand or right of lien against the land and improvements described above, and (4) stipulates that he is an authorized officer with full power to execute this waiver of liens.

THE UNDERSIGNED agrees that _____ and any lender and any title insurer may rely upon this waiver.

                 _____

By   _____

Title   _____

Sworn to me this ____ day
of_____, 200__.

_____
Notary Public

NOTE: Return four (4) signed releases to _____ at _____, _____, __, ____ to the attention of _____ . Additional payments will not be made until the signed releases are returned.

FIGURE 4.7   Example of a partial waiver of lien.

lender should obtain them or they should be deducted from the retainage. The four main types of lien waivers are

- *Conditional lien waiver and release following progress payment:* discharges all claimant rights through a specific date, provided the payments have actually been received and processed, which makes it the safest waiver for claimants.
- *Unconditional lien waiver and release following progress payment:* unconditionally discharges all claimant rights through a specific date with no stipulations.
- *Conditional lien waiver and release following final payment:* releases all claimant rights to file a mechanic's lien with certain provisions if there is evidence that they have been paid to date.
- *Unconditional lien waiver and release following final payment:* provides the safest type of lien waiver for owners; it generally releases all rights of the claimant to place a mechanic's lien on the owner's property unconditionally. However, claimants should issue this type of release only when they are satisfied that their work is complete and that the payment has cleared their bank. Owners should demand this release when the claimant is paid in full.

4.6.3 Testing Reports

The review of testing results should normally be undertaken on a monthly basis. A set of Contract Documents should be submitted to the construction testing agency for its review prior to the commencement of any construction. The testing agency will provide a copy of their current rate schedule for types of test work and a budget estimate for the specific project for the owner's review and to make a determination. After the owner's acceptance and approval, the owner and/or administrator will authorize the testing agency in writing to proceed with required tests. After the administrator reviews the billing with the owner and/or architect, it is given to the owner for payment. The contractor is responsible for coordinating and scheduling required testing activities for the project.

The different types of testing reports received (concrete, mortar, timber, etc.) should be recorded and whether or not they are in compliance with the design specifications. Any test results that fail should be brought to the attention of the owner requesting an explanation before deciding what action is to be taken. A letter may be sent to the designer of record (DOR) noting the failed test and requesting comment, depending on the test's significance and its impact on the project. The owner needs to provide the administrator with copies of all controlled testing reports.

The testing agency normally provides test reports within 48 hours to the administrator, designated engineering consultants, contractor, and owner. It is the administrator's duty to immediately respond, in writing to any test reports indicating that the work fails to conform to the Contract Documents and to ensure that remedial action is taken. Since the contractor is ultimately responsible for the construction means and methods, it becomes his or her responsibility to propose a solution to rectify any construction deficiencies.

4.6.4　Daily Work Log

The day-to-day activities on the job site are generally monitored with daily work logs. The owner should provide a copy of a typical page from the project's daily work log for the day prior to the administrator's site visit and meeting. The contractor's daily log is normally submitted to the administrator on a weekly basis and typically contains the following:

- Daily activities
- Meetings and important decisions
- Unusual events such as stoppages and emergency actions
- Material delivery, equipment onsite, and so forth
- Visitors
- General weather conditions
- Conversation and telephone records
- Problems or potential delays
- Accidents
- Change Orders received (pending or implemented)

4.6.5　Construction Schedule and Schedule of Values

Project schedules are to be provided that consist of monitoring the progress of the contractor and subcontractors relative to established schedules and that update status. After reviewing them, the administrator may discuss how the project stands with respect to the owner's construction schedule and/or the target date for completion. Any factors contributing to delay or progress should be mentioned (good weather, a strike, tight management, etc.). If the initial construction schedule is revised, the owner should submit a copy of the revised schedule to the administrator.

Figure 4.8 shows a construction schedule to monitor how the work is progressing in relation to the contractor's schedule. The Project Construction Schedule shall be prepared in accordance with the Contract Documents in either PERT, BAR, GANTT, or C.P.M. format. It must be updated each period by the contractor and verified by the administrator to ensure that it is always up to date and accurate. The updated schedule forms part of each monthly Application for Payment and also be is included in the administrator's monthly PSR.

A schedule of values lists dollar amounts assigned to each area of work that will be completed on a construction project. It is prepared in accordance with the Contract Documents and equals the total cost of the project. Line items in the schedule are divided into the appropriate specification divisions and broken down to reflect material and labor costs for each item. Unit measurement for

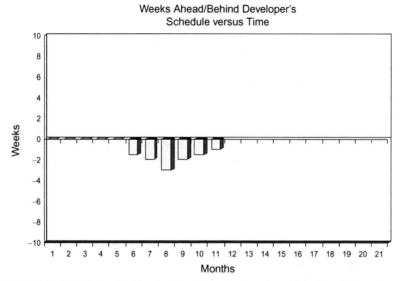

FIGURE 4.8 Typical construction schedule with an actual start date of March 24, 2010, and target completion date of October 24, 2013.

the materials described in the schedule of values are included. These numbers are required to estimate the value of work in place at any specific time. The schedule of values for the project ultimately becomes the source for verifying costs for any additional work and for establishing prices for potential modifications to the Contract Documents (Change Orders). The ConsensusDOCS 293, Schedule of Values form, provides a breakdown of the cost of work elements and can be used with the ConsensusDOCS Application for Payment forms, ConsensusDOCS 291 and 292.

4.6.6 Project Progress Meetings

Periodically and depending on the project, various meeting will take place at the job site throughout the construction process, most of which are well scheduled in advance. Depending on the type and size of the project and the agreement between the lender and the administrator, these meetings are held at regular intervals, usually on either a biweekly or a monthly basis unless an unscheduled special meeting is called to address special issues. The main purpose of these meetings is to discuss and review the project's progress and to provide a forum in which the main participants (administrator, contractor, subcontractor, architect, engineers, and others) can discuss their concerns, including submitted Applications for Payment. They are usually chaired by the project administrator and can be fairly formal with a written agenda. Meetings are normally recorded and the minutes distributed to all the participants within a week of the specified meeting.

4.6.7 Stored Materials Funding and Documentation

It is necessary for all materials stored onsite to be in a secured area. If the contractor is requesting funds for them, a stored materials schedule should be delivered to the project manager (PM), and the Application for Payment and Sworn Statement of General Contractor (AIA Document G702) or equivalent document must list the dollar amounts of all stored materials. All such items must be verifiable by the consulting professional.

This section is included only if there are materials for which funding has been requested by the general contractor that are stored either on- or offsite (Figure 4.9). Where funding is permitted, typical documentation required by the owner or lender should be provided to approve funding. And where the contractor seeks payment for materials that have not been incorporated into the improvements and are not stored onsite, the following backup documents are required to process the request:

- A stored materials statement for the offsite or onsite materials.
- Evidence of insurance on the stored materials (whether at the offsite location or in transit), identifying the type of material, value, and location.
- A bill of sale evidencing the borrower's ownership of the stored materials, including a list of materials, value, and location.
- A letter from the consulting professional or approved third-party inspector stating that materials have been sighted and inventoried and that they are suitably stored and marked for the project in question.
- Evidence that materials stored offsite are in an independent bonded warehouse with prior approval of the owner or lender and at no cost to the owner or lender.
- Inspection and acceptance of any material having architectural finishes by the architect. In no case will payment be made for bulk marble, granite, and so forth, that has not been fabricated and inspected.

Furthermore, the owner or lender may approve funding for stored materials under certain conditions such as in cases where ordering requires a long lead time. Items that are being fabricated and stored with the manufacturer should also be marked and segregated from the manufacturer's other supplies. Prior to request for payment on materials "in process" of being fabricated, the architect or other firm or agency acceptable to the lender should inspect the materials and preferably document the inspection with appropriate photographs.

4.6.8 Subcontracts and Purchase Orders

Subcontracts and purchase orders are typically signed during the general course of construction, and copies should be provided to the administrator. If this is a general contractor project where subcontracts are not known to the owner, then the lender usually requires copies of all major subcontractor trade payment breakdowns prepared on a percentage of completion basis instead of a dollar amount.

MATERIAL REPORT

THIS ORDER IS: PAGE _____ OF _____
COMPLETE   [   ]
                                            P.O.
NOT COMPLETE   [   ]   NO: _____
                                            REQ.
REC. DAMAGED   [   ]   NO.   _____

DATE INSPECTED   _____
ORDERED FROM: Hallick Ironworks
                  27675 Ashgrove Court
                  Sterling, VA

P.O. ITEM NO.	QUANTITY	UNIT	DESCRIPTION	LOCATION	ACCOUNT NO.

CHECKED BY _____

DWG
NO. _____   WAREHOUSE SUPT. _____

FIGURE 4.9   Typical stored materials report.

4.6.9 Payment and Performance Bonds

To assure the owner that the contractor will complete all obligations set out in the contract, a performance bond is often stipulated in the general conditions provided that no action by the owner or the owner's agents prevents or inhibits the contractor from the implementation of the contract requirements. However, it appears that lending institutions have started to lessen their emphasis on "bonded" contracts. A number of lenders have been disappointed to find they are involved in extended litigation after invoking the very bond they looked to for protection. Moreover, bonding companies often will not bond an owner/builder.

Although many title companies previously offered a "completion guarantee" designed to assure interim lenders that their project would be completed, because of extensive losses, they have largely withdrawn such policies. Indeed, the most perilous and defining phase of the delivery of a new project is actual construction. The Construction Phase, because of the infinite number of inherent risks associated with it, rises far above the other phases of the project delivery when claims, failures, problems, and defaults are taken into consideration.

In all cases, the administrator should be aware of both the owner's (and lender's) rights under the contract as well as the obligations necessary to protect them. Should the administrator, while acting as the owner's agent in administering the new project, fail to properly ensure the rights of the owner because of negligence, incorrect documentation, or the untimely issuance of notice, the administrator may certainly be held liable by the owner and will be held responsible for any losses whether moral or financial. The administrator should also be aware of all requirements necessary to ensure the protection of the payment and performance bonds. As a construction expert and owner's representative, the administrator's role in administering the construction contract requires verification of all notices and that all prerequisites are followed by both the owner and the contractor.

Default by the contractor during the Construction Phase of the project is considered the greatest potential risk for both owner and lender. For this reason, it is important for the successful project administrator to exercise both caution and diligence when reviewing the types of payment and performance bonds required by the contract. Normally, the administrator should specify standard AIA bond forms A312 (Performance and Payment Bonds) or an equivalent; these documents are not only respected by the surety industry but have been thoroughly tested and generally upheld through the judicial process. Separate payment bonds and performance bonds with separate bond numbers and power of attorney offer the owner protection equal to double the face value of the construction project. It may be wise to avoid combination payment and performance bonds because of recent legal precedents that have held that the surety was obligated to pay only the face value of the construction project through a combination of claim against both sides of the combined bond.

Payment and performance bonds offer the owner appropriate protection against countless scenarios of contractor default, subcontractor nonpayment, material or supplier nonpayment, and liens, yet the administrator should also recognize the surety and its agents for what they really are—sharp and intelligent businessmen who will carry out all their obligations and responsibilities under the bonds, provided all aspects of the default are in fact the contractor's responsibility. However, the surety may be relieved of its obligations, in whole or in part, by erroneous actions or negligence of the administrator or by the improper actions of the owner, in which case the surety may try to mitigate its losses to the maximum extent possible under the law. Thus by the time the surety becomes involved in a construction project, the relationships between the owner, contractor, and administrator may very well be uncomfortable, to say the least. The surety certainly is not a deep-pocketed, sympathetic benefactor doling out money to those most deserving just because there is a payment and performance bond.

It is necessary for the administrator to read and understand the wording and requirements of each type of bond, for just as every project is unique so is every situation involving the surety. There is no substitute for studying and fully understanding the requirements of the bond to ensure that each and every action required by the administrator is not only fully implemented but fully implemented within the specified time frames delineated in the bond documents. It should also be noted that the statements and opinions expressed here reflect a general attitude of surety and its general position when confronted with differing construction situations. However, the reader should be alerted to the fact that these opinions are not intended to be legal advice but rather are generic in nature.

4.6.10   RFIs and Other Logs

The contractor will sometimes issue requests for information (RFIs) to the administrator or owner for clarification of design information or to present any other questions. The RFI includes a summary log containing the status of each request. It is a formal procedure, and each inquiry and response should be tracked and documented. To avoid potential litigation, it is critically important to get a quick response. RFIs often originate from a subcontractor, vendor, or craftsman who needs certain information to continue working. Most contractors have a standard format to keep the RFIs consistent throughout the contract.

4.6.11   Permits and Approvals

Prior to the commencement of any building construction, the owner is required to obtain a building permit for the project from the local authority. This may be the single most critical aspect of the pre-construction process. Sometimes the permit is obtained prior to the bidding process. If delayed, it can cause considerable hardship and complications for the project. Building permits also contain inspection schedules and zoning ordinances. A copy of the ordinance under which the project was approved as well as copies of all variances, approvals and declaration, and any other zoning information or approvals pertinent to the project should be in the possession of the owner, designer of record, and the administrator. Other permits and approvals need to be obtained such as:

- A print of the zoning sheet highlighting all calculations along with evidence that the zoning computations have been approved by the planning board.
- A comprehensive list of all permits necessary to proceed with the construction of the suggested improvements prepared by the designers of record; also copies of all permits secured to date with the remaining provided as and when they are available.
- A complete list of all agreements between the borrower (owner) and any governmental agreements and between the borrower and any governmental agencies for the construction of the suggested changes and copies of all agreements.
- Copies of all appropriate approval/permits from the landmarks commission, historical preservation group, and other related agencies for historical renovation and restoration work.
- Statements of existing Certificate of Occupancy (CO) and/or CO procedures and any special requirements to obtain a CO on completion of the work; also any Temporary COs issued for partial or phased completion.
- Schedule of building code violations, if any.
- Copies of utility load calculations prepared by the engineers of record and confirmation that existing services are or are not adequate to serve the proposed project in addition to confirmation from the appropriate authority of availability, adequacy, and intent to provide the following utilities for the proposed project: water, gas, electricity, steam, sanitary sewers, and storm sewers.

4.7　FINAL CERTIFICATION AND PROJECT CLOSEOUT

Project closeout and final acceptance can be initiated on receiving notice from the contractor(s) that the work or a specific portion of it is acceptable to the owner and is sufficiently complete, in accordance with Contract Documents, to allow occupancy or utilization for the use for which it is intended and can take place when all contract requirements, warranties, and closeout documents, along with all punch list items, have been resolved. Normally, a "punch list" of unfinished and/or defective work, complete with remedial cost, is prepared for possible escrow purposes. When the owner requests the final loan advance, the administrator collates and examines all permits, approvals, waiver of liens, and other closeout documents specified in the loan agreement. On approval of these documents, the administrator makes another site inspection and verifies and certifies that the work was completed in accordance with the plans, specification, and loan agreement and that there are no outstanding issues on the matter. Documents typically required for project closeout include the following:

- As-built signed/sealed record drawings, which show all changes from the original plans
- As-built record specifications—the contractor's Certificate of Compliance with plans and specifications
- Architect's issuance of Certificate of Substantial Completion and/or Final Acceptance and issuance of final Certificate(s) for Payment after a detailed inspection with the owner's representative is conducted for conformity of the work to the Contract Documents to verify the list submitted by the contractor (s) of items to be completed or corrected
- Architect's certified copy of the final punch list of itemized work stating that each item has been completed or otherwise resolved for acceptance
- Determination of the amounts to be withheld until final completion of outstanding punch list items
- Notification to owner and contractor(s) of deficiencies found in follow-up inspection(s), if any
- Certification from the borrower that all closeout requirements including but not limited to as-built drawings, warranties, operating and maintenance manuals, keys, affidavits, receipts, and releases have been received, reviewed as necessary, and approved for each subcontractor
- Completed commissioning and closeout manual
- Record of approved submittals and samples
- Certification of no asbestos products incorporated in project
- Certificates of use, occupancy, or operation
- Securing and receipt of consent of surety or sureties, if any, for reduction or partial release of retainage or the making of final payment(s) and consents of surety for final payments, if bonds are provided
- Final release of claims and waivers of lien in a form satisfactory to the lender and the title company from all subcontractors, suppliers, and the general contractor, indemnifying the owner against such liens
- Affidavit of payment of debt and claims

4.7.1   As-Built Drawings/Record Drawings

The contractor must furnish as-built record drawings made from the architect/engineer's contract drawings, or subsequent updates thereof, annotated and with actual as-built conditions. Most contracts stipulate that the contractor maintains a set of as-built (sometimes erroneously called record) drawings as the project progresses. These record actual dimensions, locations, and features that may differ from the original Contract Documents and must show all changes in the work relative to the original Contract Documents, as well as additional information of value to the owner's records but not indicated in the original Contract Documents.

When the drawings are in an electronic format, the contractor may be required to correct the drawing files and highlight the modifications.

An example of an as-built item is the location of underground utilities that differs from the original drawings. In such cases, the contractor would be directed (normally in writing) to make the change but the design professional fails to make the necessary modifications because they are the responsibility of the general contractor. It is usually stipulated that the contractor submit a final complete and accurate set of as-built drawings to the owner prior to receipt of final payment. As-built plans represent the existing field conditions at the completion of a project. Accurate project plans are sometimes needed for possible litigation involving construction claims and tort liability suits.

The most qualified individual to note field changes that occurred (called "redline corrections") is usually the resident engineer, architect, or building information modeling (BIM) manager of a completed project. As-built plans are preferably completed using an electronic format such as AutoCad or Micro-Station. Having the drawing saved as a CADD or other digital system file makes it easier to store and update it whenever necessary.

4.7.2   Contractor's Certificate of Compliance

Certificates of Compliance can apply to various issues such as discrimination and affirmative action, subcontractor work, and the like. The general contractor must agree and certify compliance with applicable requirements of the Contract Documents. The Certificate of Compliance is valid for a limited time, say 180 days, and it is the responsibility of the contractor to renew it prior to the expiration date while the project is still in the Construction Phase. A temporary certificate may often be issued for a portion or portions of a building that may safely be utilized prior to final completion.

In this respect, the contractor also agrees to obtain compliance certifications from proposed subcontractors prior to the award of subcontractors exceeding an agreed-on sum according to the Contract Documents. Furthermore, the contractor is required to coordinate the efforts of all subcontractors and obtain any required letters of compliance from the administrator or owner's consultant. The owner is usually required to pay any fee associated with these letters. However, the contractor must reimburse the owner for any costs resulting from failed tests or inspections conducted to obtain a letter of compliance. This reimbursement procedure is spelled out in the Contract Documents, and should be made part of a credit Change Order.

4.7.3   Architect/Administrator's Certificate of Substantial Completion

The Contract Documents define the date of substantial completion, which is considered to be the date on which the administrator, owner, and/or architect certifies that the work, or a designated portion of it, may be beneficially occupied or utilized for its intended use. For this purpose, the AIA G704™ standard form is used for recording the date of substantial completion of the work or a designated portion thereof. This process takes off when the contractor considers the work, or designated portions of it as previously agreed to by the owner, to be substantially complete. The contractor then prepares and submits to the administrator a punch list of items that remain to be completed or corrected.

The AIA G704 form provides for agreement on the time to be allowed for completion or correction and the date when the owner will be able to occupy the project or designated portions it. It also designates responsibility for maintenance, heat, utilities, and insurance. If the administrator concludes that the work is substantially complete, the AIA form is then prepared for acceptance by the contractor and owner. The failure of the contractor to include any items on the list in no way alters his or her responsibility to complete or correct these items per the Contract Documents.

There are variations of the G704–2000, such as the G704-CMa–1992, "Certificate of Substantial Completion, the Construction Manager-Adviser Edition," which serves the same purpose as G704–2000, except that it expands responsibility for certification of substantial completion to include both the architect and the construction manager. There is also the G704DB–2004, a variation of G704–2000 that acknowledges substantial completion of a design–build project. Because of the nature of design–build contracting, in this form the project owner assumes many of the construction contract administration duties performed by the architect in a traditional project. There is no architect to certify substantial completion, so the AIA G704DB–2004 requires the owner to inspect the project to determine whether the work is substantially complete in accordance with the design–build documents and to acknowledge the date when completion occurs. In addition, as an alternative to the AIA documents for a Certificate of Substantial Completion for design–build work, the owner and designer-builder can use ConsensusDOCS 481 and, for a Certificate of Final Completion for design–build work, ConsensusDOCS 482. It is worthwhile to research other ConsensusDOCS, as they are becoming increasingly popular.

Use of the term "beneficial occupancy" generally is an indication that the project or portions thereof are complete to a sufficient degree to allow the owner to utilize the project or portions of it for their intended use. Mechanical systems, life safety systems, telecommunications systems, and any other systems that are required to properly utilize the project or portions thereof, must be complete and in good working order. Items remaining to be completed must be such that their correction does not inconvenience or disrupt the owner's normal operations at the site.

The owner/lender or their representative (administrator) should be consulted to confirm that there are no other evident construction deficiencies that are not on the contractor's punch list. It is important to emphasize that responsibility for preparing the original punch list lies with the contractor. If the administrator is requested to make a substantial completion inspection, and it is obvious that the contractor's punch list is incomplete, the inspection must be discontinued and the contractor so advised. The Certificate for Substantial Completion should not be issued until the administrator can verify that the following conditions are in place:

- Written statement from the contractor that the project, or designated portions thereof, is substantially complete or that construction is sufficiently complete for beneficial occupancy by the owner (with relevant lien waivers).
- Correctly executed Consent of Surety for Reduction in Retainage per the Contract Documents.
- Contractor's punch list with the administrator's supplementary comments added; however, prior to any retainage being released, the administrator must certify substantial completion and administrator (or the designer of record), the general contractor, and the owner must agree on the punch list work to be completed.
- Temporary Certificate of Occupancy (TCO) from appropriate agency with all required permits/approvals.

Normally there is a lender involved in the project, and the Certificate of Substantial Completion is to be prepared by the administrator and certified by the owner and/or architect prior to being submitted to the owner and contractor for their written acceptance of the responsibilities assigned them in the certificate. The Certificate of Substantial Completion also establishes the dates and responsibilities of any transitional arrangements required between the owner and the contractor.

4.7.4 Architect/Administrator's Certified Copy of Final Punch List

The contractor's final punch list is given to the administrator for review of the completed work to determine whether the list is accurate and complete. Items that require correction and/or completion not included in the contractor's punch list, must be supplemented by the administrator. The owner should be informed that the items on the punch list will be rectified and/or completed within the time limit set forth in the Certificate of Substantial Completion. The contractor is also advised that any correction and/or completion of punch list items is to be conducted in a manner so as not to adversely affect or disrupt the owner's occupancy of the facility.

4.7.5 Certificates of Occupancy, Use, and Operation

A Certificate of Occupancy is a document issued by a local government agency or building department certifying that the building in question complies with all applicable building codes, safety codes, health code requirements, and other laws, and basically stating that the building is in a condition suitable for general occupancy. The procedure and requirements for the Certificate of Occupancy vary widely from jurisdiction to jurisdiction and by the type of structure.

In the United States, obtaining a certificate is generally required whenever a new building is constructed, or when a building built for one use is to be used for another (e.g., an industrial building is converted to residential use). Likewise, a certificate is required when the occupancy of a commercial or industrial building changes or ownership of a commercial, industrial, or multiple-family residential building changes. The purpose of this certificate is therefore to document that the use is permitted and that all applicable safety code and health code requirements have been met.

A use and occupancy certificate is required for the space to be used prior to the opening of any business. It is also generally necessary both to be able to occupy the structure for everyday use and to be able to sign a contract to sell the space or close on a mortgage for it. A Certificate of Occupancy is proof that the building complies substantially with the plans and specifications that have been submitted to, and approved by, the local authority. It basically complements a building permit, which is filed by the applicant with the local authority before commencement of construction to signify that the proposed construction will adhere to all relevant ordinances, codes, and laws. Particular attention should be paid to the new International Green Construction Codes and whether they apply.

Often a Temporary Certificate of Occupancy will be applied for. This grants residents and building owners all of the same rights as a Certificate of Occupancy, except that it is valid only for a temporary period of time. In New York City, for example, TCOs usually expire 90 days from the date of issue, although it is not uncommon, and is perfectly legal, for a building owner to reapply for a TCO, following all the steps and inspections required originally, to extend it for another period of time. Temporary Certificates of Occupancy are generally sought and acquired when a building is still under minor construction but a certain section or number of floors are considered to be habitable (e.g., in a high-rise apartment building); after issuance of a TCO, it can legally be occupied or sold.

4.7.6 Final Waivers of Lien

Once the project in hand is finished, the contractor is usually required to complete a final lien waiver (Figure 4.10). The general contractor is also required to obtain conditional final waivers from each subcontractor, vendors, and certain individuals prior to final payment being released. A final lien waiver is basically a document from a contractor, subcontractor, material supplier, equipment lessor, or other party involved in the construction project stating that he or she has received full payment and waives any future lien rights to the property. It should be noted that in the United States, liens cannot be filed against public property. Moreover, some states only use a conditional waiver on progress payment and an unconditional waiver on final payment.

The mechanic's lien process can prove extremely valuable to contractors, subcontractors, material suppliers, and other related parties on a construction

CONTRACTOR/VENDOR FINAL RELEASE AND LIEN WAIVER

The undersigned represents and warrants that it has been paid and has received (or that it will be paid and will receive via proceeds from this pay application) $_____ as full and final settlement under the contract/agreement dated _____ (including any amendments or modifications thereto) (the "**Contract**") between the undersigned and _____ ("**Contractor/Vendor**") for the _____ Project owned by _____ ("**Owner**") (PO Number: _____). In consideration for this final payment, and other good and valuable consideration, receipt of which is acknowledged, the undersigned makes the following representations and warranties:

1. The undersigned and Owner have fully settled all terms and conditions of the Contract (including any amendments or modifications thereto), as well as any other written or oral commitments, agreements, and/or understandings in connection with the Project.

2. The undersigned has been paid in full (or it will be paid in full via proceeds from this pay application) for the labor, services, and materials in connection with the Contract, including all work performed or any materials provided by its subcontractors, vendors, suppliers, materialmen, laborers, or other persons or entities.

3. The undersigned has paid in full (or it will pay in full via proceeds from this pay application) all its subcontractors, vendors, suppliers, materialmen, laborers, and other person or entity providing services, labor, or materials to the Project; there are no outstanding claims, demands, or rights to liens against the undersigned, the Project, or the Owner in connection with the Contract on the part of any person or entity; and no claims, demands, or liens have been filed against the undersigned, the Project, or the Owner relating to the Contract.

4. The undersigned releases and discharges Owner from all claims, demands, or causes of action (including all lien claims and rights) that the undersigned has, or might have, under any present or future law, against Owner in connection with the Contract. The undersigned hereby specifically waives and releases any lien or claim or right to lien in connection with the Contract against Owner, Owner's property, and the Project, and also specifically waives, to the extent allowed by law, all liens, claims, or rights of lien in connection with the Contract by the undersigned's subcontractors, materialmen, laborers, and all other persons or entities furnishing services, labor, or materials in connection with the Contract.

5. The undersigned shall indemnify, defend, and hold harmless Owner from any action, proceeding, arbitration, claim, demand, lien, or right to lien relating to the Contract, and shall pay any costs, expenses, and/or attorneys' fees incurred by Owner in connection therewith.

The undersigned makes the foregoing representations and warranties with full knowledge that Owner shall be entitled to rely upon the truth and accuracy thereof.

DATED:

(Contractor/Vendor company name)
By:
Title:

STATE OF
COUNTY OF

I, a Notary Public for the above County and State, certify that _____ personally came before me this day and acknowledged that he/she is _____ [title] of _____ [company name], and that he/she, as _____ [title], being authorized to do so, executed the foregoing on behalf of _____ [company name]. Witness my hand and official seal this ____ day of _____, 20___.

Notary Public

My Commission Expires:

NOTICE: THIS DOCUMENT WAIVES RIGHTS UNCONDITIONALLY AND STATES THAT YOU HAVE BEEN PAID FOR GIVING UP THOSE RIGHTS. THIS DOCUMENT IS ENFORCEABLE AGAINST YOU IF YOU SIGN IT, EVEN IF YOU HAVE NOT BEEN PAID.

FIGURE 4.10   Two examples of final waiver of lien formats.

FINAL WAIVER OF LIENS

Project Description: Contract Date:
                                                             Work Performed:
                                                             Work Performed by:
                                                             Under Contract to:

Listed below is the final information regarding the above contract:

      Contract Price $_____
      Net Extras/Deductions   +_____
      Adjusted Contract Price $_____
      Amount Previously Paid   -_____
      Balance Due-Final Payment $_____

The undersigned, being duly sworn, deposes, certifies and says that:

(i) He (She) is an officer of, and is duly authorized to make this affidavit, waiver and release on behalf of _____ ("Contractor").

(ii) Contractor has received in full all payments (plus applicable retention) due through the date of this instrument for all labor, services, equipment and materials (sometimes referred to as the "work") furnished to _____ ("Owner") on the job of above project.

(iii) Contractor has paid in full or otherwise satisfied all of its obligations for labor, materials, equipment and services and all other indebtedness associated with the performance of Contractor's work on the Project, including without limitation payment in full to, or other satisfaction of, all persons and entities (the "Subcontractors") which have furnished labor, services, equipment or materials to Contractor.

(iv) In consideration of the payments received, and upon receipt of the applicable retention, Contractor forever waives, releases and relinquishes any and all claims and rights to a mechanic's lien, stop notice, bond right, equitable claim or right to any fund, and right to a labor and material bond or other bond on the Project and all other rights and claims that Contractor has on the Project.

(v) Contractor guarantees to Owner that the work furnished by Contractor (including work furnished by the Subcontractors) on the Project is and, after receipt of the applicable retention, shall be lien free, that the Subcontractors have no right to any mechanic's lien, stop notice, bond right, equitable claim or right to any fund, any right to a labor and material bond or other bond on the Project or other rights and claims with respect to the Project, and Contractor agrees to indemnify _____ and _____ against any claim or lien asserted through or under Contractor with respect to the Project, including without limitation any claim or lien asserted by any person who has furnished labor, materials, equipment or services to Contractor.

(vi) The undersigned further guarantees that all portions of the work furnished and installed by them are in accordance with the contract and that the terms of the contract with respect to these guarantees will hold for the period specified in said contract.

_____

_____

_____

Sworn to me this _____ day of
                                        By _____
_____, 200__.
                                        Title _____
_____
Notary Public

NOTE: Return four (4) signed releases to _____ at _____, _____ to the attention of _____.
       Payment will not be made until the signed releases are returned.

FIGURE 4.10—cont'd

project in enforcing their claims if carried out according to the laws of the various states or the federal government. These parties are entitled to be paid for their material or labor contributions to the improvement of real property. Most lien waiver forms can be obtained online or from local office supply stores or professional organizations such as the AIA.

4.7.7 Miscellaneous Issues

Commissioning and Warranties

Establish commissioning procedures and dates for the commencement of all warranties. Commissioning and warranty review services consist of the following:

- Monitoring compliance by the GC/CM with commissioning of operating systems, and the like; the GC/CM must obtain from trade contractors and give the owner all required warranty documents and operating manuals, "as-built" (record) drawings, and so forth.
- Consultation with and recommendation to the administrator and owner for the duration of warranties in connection with inadequate performance of materials, systems, and equipment under warranty.
- Inspection(s) prior to expiration of the warranty period(s) to evaluate adequacy of performance of materials, systems, and equipment.
- Documenting defects and/or deficiencies and assisting owner in providing instruction to contractor(s) for rectifying noted defects and deficiencies.

Architect's Supplemental Instructions

The project architect/administrator may issue additional instructions or authorize minor changes in the work not involving an adjustment in cost or requiring an extension of time by issuing a document called the Architect's Supplemental Instructions (ASI). These are often documented by AIA form G710, Architect's Supplemental Instructions; this form is intended to assist the project architect and administrator in performing its obligations as interpreter of the Contract Documents in accordance with the owner/architect agreement and the general conditions. The administrator prepares and issues an ASI for all additional work that is not included in the Contract Documents and that does not modify the contract sum or extend the contract time. Any changes are to be effected by a written order signed by the architect/administrator or project manager directing the contractor to execute the work promptly.

All ASIs need to be recorded in the project log book. The administrator may prepare them for the owner and/or architect's signature. The administrator is generally encouraged whenever possible to try to resolve small incidental issues through the ASI, which are to be forwarded to the contractor for signature as an acknowledgment that the work described will not modify the contract sum or contract time.

Time Extensions

In many projects, the contractor may feel, during the course of construction, a need to put forward a request for an extension in the contract time. This can be for one of several legitimate reasons that are totally beyond his or her control, such as:

- Inclement weather
- Owner-requested changes or additions to the original scope of the work (e.g., Change Order or Construction Change Directive)
- Delays caused by slow responses to RFIs
- Late material shipments from suppliers
- Slow processing of submittals or shop drawings
- Labor strikes

Inclement weather is one of the most common reasons for time extension requests. The contractor should make allowances for a normal amount of severe weather in the Project Construction Schedule. Time extensions are granted only for abnormally severe weather, defined as weather that is both detrimental to construction activities and more frequent than usually experienced during that time of year. It is important to note that adverse weather during certain phases of construction (e.g., pouring of concrete floors or foundations) can affect the construction schedule more adversely than good weather can benefit it during other phases. The contractor is typically required by the construction documents to notify the administrator, owner, and/or architect of any potential claim for additional time due to delay within 20 days of the start of the delay. When reviewing claims for time extensions, the contractor's daily log should be reviewed to verify that the bad weather occurred during the specified time and that lost time for that period was actually what the contractor experienced.

Time extensions requested for delinquent or late material deliveries should be verified and the contractor should be asked to furnish verification of the original date on which the material order was to be delivered. In many cases, the contractor or the subcontractor failed to place the order in sufficient time to ensure that delivery would meet the schedule. Time extensions may consist of simple requests for additional time or the requests may be more complex and include related cost reimbursement. Except for exceptional circumstances, it may be prudent to retain all time extension requests until the project's completion.

Shop Drawing Submittal and Review Procedure

The purpose of the shop drawing submittal process is to ensure that the provided products, materials, equipment, and so forth, are in compliance with the Contract Documents. This is why review and approval of shop drawing submittals are required prior to fabrication, installation, and/or use of a submitted product.

Furthermore, the importance of this process cannot be overemphasized because a delay in it can cause a delay in the overall completion of the project. There are many items associated with construction that cannot be ordered out of a product brochure or off the shelf.

In many cases items need to be fabricated in a shop or manufactured specifically for the project. To confirm the owner's intent, "shop drawings" are required, which are essentially the supplier's or fabricator's version of information shown on the drawings in the Contract Documents. The review/approval of shop drawings is a careful and methodical process. Shop drawings are typically submitted by subcontractors or vendors and contain greater detail and configurations of the item in question sufficient to fabricate and erect it. Shop drawing submittals also include design drawings, detailed design calculations, fabrication drawings, installation drawings, erection drawings, lists, graphs, operating instructions, catalog sheets, data sheets, samples, schedules, and similar items. Once completed, the drawings are submitted to the general contractor, who sends them off to the project administrator (or architect) for final approval. After approval, they are returned to the general contractor and subcontractor to begin fabrication.

Many items in the construction process, including steel rebar bends, steel beams, trusses, architectural woodwork, and ornamental metalwork, require shop drawings. In the case of structural steel, for example, shop drawings may include welding details and connections that are not typically part of the structural engineer's drawings. The term *submittal* often refers to the totality of shop drawings, product data, and samples; all of these documents are submitted to the owner or owner's representative for approval prior to fabrication and manufacture of the items they represent. Once approved, submittals may become part of the Contract Documents and should be incorporated into a submittal log. To be included in the Contract Documents, the shop drawings must have been in existence at the time of the signing of the construction contract and be incorporated by reference into the contract. This includes drawings that are added later as contract modifications and that are signed by the owner and the contractor, such as Change Orders and Construction Change Directives.

According to Arthur F. O'Leary, author of *A Guide to Successful Construction* (Part 1—Learning to Live with This 'Necessary Evil'):

To the construction industry, shop drawings seem to be a necessary evil. Contractors find them expensive to produce and architects find them unappealing to review. Both find them time-consuming and costly to administer. We seemingly cannot construct buildings without them; but they have become a perennial source of annoyance and confusion and more importantly, a significant source of professional liability claims against architects. Undiscovered mistakes in shop drawings will often lead to unexpected or undesired construction results as well as exorbitant economic claims against architects, engineers, and contractors. Some shop drawing anomalies have resulted in costly construction defects, tragic personal injuries, and catastrophic loss of life.

O'Leary goes on to say:

The principal reason architects and engineers need to review the shop drawings is to ascertain that the contractor understands the architectural and engineering design concepts and to correct any misapprehensions before they are carried out in the shop or field. They review shop drawings of any particular trade or component to determine if the contract drawings and specifications have been properly understood and interpreted by the producers and suppliers.

The shop drawings should prove to the architect's satisfaction that the work of the contract would be fulfilled. If the shop drawings indicate that the work depicted will not comply with the intent of the contract drawings and specifications, the architect has an opportunity to notify the contractor before the costs of fabrication, purchase, or installation have been incurred.

The administrator should employ the following procedures for the processing of all shop drawings and related product data for the project:

- Shop drawings must be submitted in the format required by the Contract Documents. Those received in any other format may be returned to the contractor with a "Not Reviewed" note attached to the submittal.
- Shop drawings are to be initially reviewed by the contractor and stamped accordingly. Those that do not bear the contractor's approval stamp may be returned to the contractor to be resubmitted as required. Additionally, shop drawings containing excessive errors and/or that clearly indicate that the contractor's review was inadequate may be returned.
- Submittals are to be date-stamped on receipt and with a standard office review stamp directly below the date.
- The Standard AIA G-712™, Shop Drawing Review form, is to be used for submittals. A separate sheet must be used for each specification division. Log submittals are to be by CSI number and numbered chronologically.
- Shop drawing submittals are to be as required by the Contract Documents. All transmittals to the administrator and consultants will be recorded in the shop drawing log in the same manner as those submittals reviewed by the owner and/or architect. The shop drawing number should be written in the upper corner of the transmittal for ease in tracking.
- The architect's shop drawing review will be conducted using a printed copy of the submittal. All correct items and deficient items needing correction are to be marked, preferably using different colored markers. All questions during the review process need to be noted.
- The corrected drawings with appropriate review comments, date stamp, and approval stamp with necessary action indicated are to be transmitted back to the contractor. The return submittal is recorded in the shop drawing log with a submittal number, date, and status of transmitted item recorded. If the submittal is rejected, this should be noted.
- Marked-up copies of the submittals should be retained for reference. It is suggested that the copies be marked accordingly (e.g., "Mark-up," "Final," "Rejected").

Once approved, the shop drawings should be distributed to the relevant parties, including the administrator, owner, general contractor, and architect of record. Often steps have to be taken in the shop drawing review process to avoid any unnecessary work or assumption of responsibility by the administrator. These include the following:

- Avoid accepting responsibility for such things as verifying field dimensions and confirming compatibility with other submitted items, as this work is clearly described in the Contract Documents as the responsibility of the general contractor.
- Normally, shop drawings are processed within 10 working days. The contractor should be advised in writing if the review for a particular submittal is anticipated to take longer than this.
- Shop drawings and project samples should be kept in a file cabinet at the administrator's desk and not in the central file system.
- When the shop drawings for specialized equipment are furnished by the fabricator, it would be prudent to take the following precautions:
  - The administrator's responsibilities should be carefully reviewed as they relate to the shop drawing.
  - Statements in the owner/architect agreement that may relieve the architect of responsibility of design and construction work by others should not be solely relied on.
  - It is advisable to consider bringing in a specialist (engineer) to design and approve the equipment if it presents unusual risks—not a fabricator. The specialist review should be requested prior to approving the equipment.
  - Consider having design and construction details checked by a qualified specialist if they are outside the administrator's expertise.

Freedom of Information Letters

The Freedom of Information Law (FOIL) allows the general public access to records maintained by the government. FOIL requests are often used as a way to check each subject property for building and code compliance, as well as to locate its CO. The research log provides space for keeping track of the numerous agencies and officials that are invariably contacted for assistance in locating the municipal departments that record and maintain such information. It is necessary to submit FOIL requests as soon as possible since most agencies, under the Freedom of Information Act, are usually given anywhere from 7 to 10 business days to respond. The information found for FOIL requests is typically listed by agency, state, municipality, and then building or fire department. Responses to these requests are to be used as exhibits in the Project Status Report. Note that some states, such as Virginia, stipulate that they are not required to provide such information, even under Freedom of Information statutes, if the person or entity making the request is located outside of the state. In such instances, the administrator should be notified immediately.

4.8   QUALITY CONTROL AND QUALITY ASSURANCE

Although quality control and quality assurance are important concepts, many project managers and design professionals lack a deep understanding of their meanings and the differences between them. In fact, both these terms are often used interchangeably to refer to ways of ensuring the quality of a service or product. However, the terms are different in both meaning and purpose. The ISO 9000 standard defines quality control as "the operational techniques and activities that are used to fulfill requirements for quality," whereas it defines quality assurance as "all those planned and systematic activities implemented to provide adequate confidence that an entity will fulfill requirements for quality." That is, quality control refers to quality-related activities associated with the creation of project deliverables. It is used to verify that deliverables are of acceptable quality and that they are complete and correct. Quality assurance, on the other hand, refers to the process used to create the deliverables and can be performed by a manager, client, or third-party reviewer.

Quality assurance is based on a process approach. Quality monitoring and its assurance ensure that processes and systems are developed and adhered to in a manner that the deliverables are of superior (or at least acceptable) quality. This is intended to produce defect-free goods or services with basically minimum or no rework required. Quality control, however, is a product-based approach. It checks whether the deliverables satisfy specific quality requirements as well as the specifications of the customer. Should the results prove negative, suitable corrective action is taken by quality control personnel to rectify the situation.

Another major difference between quality control and quality assurance is that assurance of quality is generally done before starting a project whereas quality control generally begins once the product has been manufactured. During the monitoring process, the requirements of the customer are defined; based on those requirements, the processes and systems are established and documented. After the product is manufactured, the quality-control process typically begins. Based on client requirements and standards developed during the quality guarantee process, quality-control personnel check whether the manufactured product satisfies those requirements or not. Assurance of quality is therefore a proactive or preventive process to avoid defects, whereas quality control is a corrective process to identify the defects to correct them.

The majority of activities falling under the scope of quality assurance are conducted by managers, clients, and third-party auditors. Such activities may include process documentation, developing checklists, establishing standards, and conducting internal and external audits. Designers, engineers, inspectors, and supervisors on the shop floor or project site perform quality control activities. These activities are varied and include performing and receiving inspection, final inspection, and other activities.

Quality control and quality assurance are, to a great extent, interdependent. The quality-assurance department relies predominantly on feedback provided by the quality-control department. For example, should there be a recurrent problem regarding the quality of a product, the quality-control department provides necessary feedback to quality monitoring and assurance personnel that there is a problem in the process or system that is causing product quality issues. After determining the principal cause of the problem, the quality assurance department instigates changes to the process to rectify the situation and to ensure that there are no quality issues to worry about in the future.

Similarly, the quality control department follows the guidelines and standards established by the quality assurance department to check and ensure that deliverables meet quality requirements. For this reason, both departments are fundamental to maintaining the high quality of deliverables. Moreover, although quality control and quality assurance are different processes, the strong interdependence between them can sometimes lead to confusion among design professionals and contractors.

Chapter 5

Building Information Modeling

5.1 BRIEF HISTORY AND OVERVIEW

What is BIM? Building information modeling (BIM) is one of the more promising developments in the architecture, engineering, and construction fields. It is changing the way contractors and engineers do business, but its application is still relatively new and there is much to learn. One way to learn is from observing how other businesses are using BIM and their trials and tribulations along the way. BIM was introduced over a decade ago mainly to distinguish the information-rich architectural 3D modeling from the traditional 2D drawing. It is being acclaimed by its advocates as a lifesaver for complicated projects because of its ability to correct errors early in the design stage and accurately schedule construction.

Although over recent years, the term "building information modeling" or "BIM" has gained widespread popularity, it has failed to gain a consistent definition. According to Patrick Suermann, PE, a National Building Information Model Standard (NBIMS) testing team leader, "BIM is the virtual representation of the physical and functional characteristics of a facility from inception onward. As such, it serves as a shared information repository for collaboration throughout a facility's life cycle." The National Institute of Building Sciences (NIBS) sees it as "a digital representation of physical and functional characteristics of a facility...and a shared knowledge resource for information about a facility forming a reliable basis for decisions during its life cycle, defined as existing from earliest conception to demolition." But generally speaking, BIM technology allows an accurate virtual model of a building to be constructed digitally. Completed computer-generated models contain accurate and well-defined geometry and pertinent data required to facilitate the construction, fabrication, and procurement activities necessary to realize the final building.

BIM consists mainly of 3D modeling concepts in addition to information database technology and interoperable software in a desktop computer environment that architects, engineers, and contractors can use to design a facility and simulate construction. This technology allows members of the project team to generate a virtual model of the structure and all of its systems in 3D and to be able to share that information with each other. Likewise, the drawings, specifications, and construction details are fundamental to the model, which includes attributes such as building geometry, spatial relationships, quantity character-istics of building components, and geographic information. These allow the project team to quickly identify design and construction issues and resolve them in a virtual environment well before the Construction Phase in the real world.

BIM is therefore primarily a process by which you generate and manage building data during a project's life cycle. It typically uses three-dimensional, real-time, dynamic building-modeling software to manage and increase produc-tivity in building design and construction. The process produces the building information model, which encompasses all relevant data relating to building geometry, spatial relationships, geographic information, and quantities and properties of building components. Construction technology for the BIM process is continuing to improve with the passing of time as contractors, archi-tects, engineers, and others continue to find new ways to improve the BIM process. One of the many significant advantages of using modern BIM design tools, as Chuck Eastman, director of Digital Building Laboratory, states, is:

[They now] define objects parametrically. That is, the objects are defined as parameters and relations to other objects, so that if a related object changes, this one will also. Parametric objects automatically re-build themselves according to the rules embedded in them. The rules may be simple, requiring a window to be wholly within a wall, and moving the window with the wall, or complex defining size ranges, and detailing, such as the physical connection between a steel beam and column.

But before one can give a precise definition of BIM, one must resolve the ambiguity over whether it is or is not fundamentally different from CAD or CADD. In the author's opinion, BIM is not CAD, nor is it intended to be. CAD is a replacement for pen and paper, a documentation tool, and CAD files are basic data consisting of elements that are lines, arcs, and circles—and some-times surfaces and solids—that are purely graphical representations of building components. Moreover, early definitions asserting that BIM is basically a 3D model of a facility are incorrect and do not reflect the truth, nor do they ade-quately communicate the capabilities and potential of digital, object-based, interoperable building information modeling processes and tools and modern communications techniques.

BIM programs today are design applications in which the documentation flows from and is a derivative of the process, from schematic design to construc-tion to facility management. Furthermore, with BIM technology, an accurate virtual model of a building can be constructed digitally, and when completed, the computer-generated model will contain all the relevant data and accurate geometry needed to support the construction, fabrication, and procurement activities required to execute the project. Ken Stowe, of AEC Division at Autodesk[®], reaffirms this and comments:

The construction industry is in the early stages of an historic transformation: from a 2D environment to a model-based environment. The benefits are many and are enjoyed by various members of the project team. Some firms are leading in planning and directing

*the whole team in BIM participation, implementing best practices, and making a point
of measuring those benefits. The savings can be in the millions of dollars. The project
durations are being reduced by weeks or months.*

It is sometimes difficult to determine who first coined the term "BIM."
Some claim Charles M. Eastman at Georgia Tech coined the term, the theory
being based on a view that the term is basically the same as "building product
model," which Eastman has used extensively in his publications since the late
1970s. Others believe it was first coined by architect and Autodesk building
industry strategist Phil Bernstein, FAIA, who reportedly used the actual term
"building information modeling," which was later accepted by Bentley
Systems and others. (See Figure 5.1.) It is claimed that Graphisoft® produced
the original BIM—in the original terminology "virtual building"—software,
known as ArchiCAD. But many firms and organizations made contributions
to BIM's continuing development.

For example, Skidmore, Owings & Merrill (SOM) is one such pioneering
firm that made significant contributions to the development and use of BIM.
Early on, SOM created a multipurpose, database-driven, modeling system
known as AES, or architecture engineering system, and single-handedly

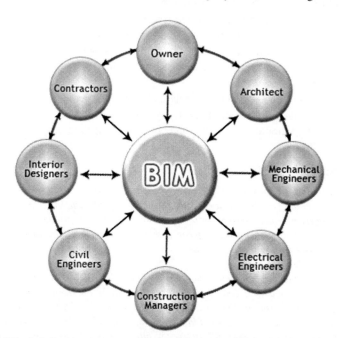

FIGURE 5.1 Relationship of BIM to the various stakeholders and project team members. BIM
technology continues to manifest itself as the most feasible and reliable option in the building construction industry. It can minimize errors and omissions made by the project team by allowing the
use of conflict detection technology, where the computer informs team members whenever parts of
the building are in conflict. *(Source: ADVENSER Engineering Services Private Ltd.)*

pioneered its development. AES is regarded by some as the precursor to today's BIM tools. As noted at http://som.com/content.cfm/brief_history_4:

In the future, SOM envisions BIM as a vehicle for real-time performative design simulation and environmental analysis, enabled through new visual and tactile feedback systems. This will allow architects to focus on building performance that can truly be validated—obtaining and interpreting data as one simultaneously designs—and will encompass new modes of collaboration. SOM envisions the architect/engineer in a pivotal role in this new virtual design and construction collaborative environment: as the conceiver of ideas and the manager of knowledge.

Dana (Deke) Smith, FAIA, executive director of the buildingSMART alliance[™] who has been involved with the development of building information modeling since its inception, says: "One of the basic principles and metrics for BIM implementation is the ability to enter data one time and then use it many times throughout the life of the project." Smith identifies the following 10 principles of BIM:

1. Coordinate and plan with all parties before you start.
2. Ensure all parties have a life-cycle view—involve them early and often.
3. Build the model then build to the model.
4. Detailed data can be summarized (the reverse is not possible).
5. Enter data one time, then improve and refine over life.
6. Build data sustainment into business processes—keep data alive.
7. Use information assurance and metadata to build trust—know data sources and users.
8. Contract for data—good contracts make good projects.
9. Ensure that data are externally accessible yet protected.
10. Use international standards and cloud storage to ensure long-term accessibility.

Smith believes the following:

We are still all too often slaves to the stovepipes that have been our industry's tradition, where information is collected for a specific instance and then not reused by others. There are currently many reasons for this: perceived intellectual property concerns, perceived liability issues, organizations pushing their own agenda, proprietary approaches, and simply not knowing that someone already entered the information because of poor ability to collaborate.

One group taking this challenge head-on is buildingSMART International. buildingSMART International is a coalition of more than 50 countries worldwide who are focused on implementing an open-standard, BIM approach to interoperability of information for building construction and facility maintenance. The North American chapter of this group is the buildingSMART alliance. While it is our belief that the final goal will be an international, standards-based, information exchange, the primary goal of interoperability remains at the foundation of this effort, using whatever format is universally easiest to use at the time.

Today, we have several organizations with initiatives under way to develop a national BIM standard. In 2007, the first version of this standard (NBIMS Version 1) was passed, but it has failed to take hold in the architecture, engineering, and construction (AEC) community mostly because of its reliance on the IFC (Industry Foundation Class) file format for 3D modeling. After several years, the National Institute of Building Sciences' buildingSMART alliance developed version 2 of the National BIM Standard–United States, which is a significant improvement on version 1. The United Kingdom has also come out with its own AEC (UK) BIM standards.

Multiple federal agencies have implemented BIM initiatives, from the GSA and the Army Corps of Engineers to the U.S. Coast Guard and Sandia National Laboratories. Finith Jernigan, FAIA, president of Design Atlantic, says, "To prosper in today's fast changing and unpredictable markets, you need new ways of doing business more effectively." And although BIM is not a technology, it does require appropriate technology to be effectively implemented.

5.2 BASIC BENEFITS, CHALLENGES, AND RISKS OF USING BIM

According to Sam Neider, director and co-founder of Proactive Controls Group in Pittsburgh,

BIM allows the reduction of risk through better information throughout the process. So when you look at a project, not only are you gaining efficiencies via clash detection, coordination, scheduling, etc., you are also reducing owner's risk of exposure for schedule and budget overruns, for claims, etc. Looking at the current economy, owners that would put down big dollars to do a project are no longer doing so. So what will help convince them to do so? You need to convince them there is a much better risk scenario out there and that is what BIM (and IPD) is delivering.

The rapid embracing of BIM is fundamentally changing the way AEC project teams work together to communicate, resolve problems, and build efficient projects faster and at less cost. In today's highly competitive construction market, it is no longer sufficient to execute a project in the real world of concrete, girders, sheet metal, pipe, and racks. In many cases, requests for proposals (RFPs) on most large projects now require contractors and subcontractors to execute the project first in the virtual world using BIM, and understandably so.

5.2.1 Benefits of Using BIM

Effective use of BIM can have a dramatic impact on a project through improved design, enhanced constructability, and quicker project completion, saving time and money both for the owner and for the project team. BIM is also emerging as the solution to reduce waste and inefficiency in building design and construction, although some organizations are taking a wait-and-see approach, seeking clear evidence for a return on the investment BIM would entail. The most significant benefits of BIM are these:

- Lower net costs and risks for owners, designers, and engineers.
- Development of a schematic model prior to the generation of a detailed building model, allowing the designer to make a more accurate assessment of the proposed scheme and evaluate whether it meets the functional and sustainable requirements set out by the owner; this helps increase project performance and overall quality (Figure 5.2).
- Improved productivity due to easy retrieval of information.
- Improved coordination of construction documents.
- Coordination of construction, which reduces construction time and eliminates Change Orders.
- Reduced contractor and subcontractors' costs and risks.
- Accurate and consistent 2D drawings generated at any stage of the design, which reduces the amount of time needed to produce construction drawings for the different design disciplines while minimizing the number of potential errors in the construction drawings process.
- Increased speed of project delivery.

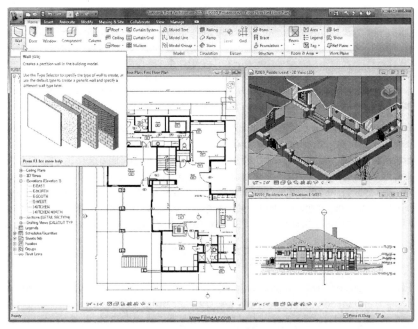

FIGURE 5.2 Use of Autodesk BIM solutions to achieve better design results. According to information from Autodesk, "BIM helps improve the way work gets done by providing more insight and greater predictability."

- Embedding and linking of vital information such as vendors for specific materials, location of details, and quantities required for estimation and tendering.
- Visualization by the project team and owner of the design at any stage of the process with the understanding that it will be dimensionally consistent in every view, thereby improving monitoring efficiency and reducing operating costs.
- Savings for realtors, appraisers, and bankers.
- Coordination and collaboration by multiple design disciplines, shortening the design period while helping to reduce potential design errors and omissions; also greater insight and early detection of possible design problems, allowing for better performance prediction.
- Safer buildings for first responders.

Ken Stowe says, "There are ten measurable ways for project teams to benefit from a comprehensive BIM solution. They fall into two categories, each with 5 ways to leverage BIM." The categories are shown in Table 5.1.

TABLE 5.1 Ways to Benefit from Comprehensive BIM Solutions

Better Planning, Cost Forecasting, and Control	Lean Project Teamwork and Communications
1. Model-to-cost integration means that more design options can be quickly and accurately priced for capital cost and compared to building performance gains.	6. The ability to affordably simulate building performance leads to better decisions for structure, comfort, lighting, energy performance, resource conservation, and materials performance.
2. 3D visualization invites richer participation, resulting in fewer RFIs and Change Orders.	7. Coordinated documents in a lean and automated system, dramatically reducing wasted effort and rework, and fostering confidence in the specialty trades.
3. 3D coordination for the subcontractors means clash-free geometry in the field, reducing rework.	
4. 4D construction simulation and communications heighten the power of planning for safety and field efficiency.	8. Rich digital teamwork leads to early builder and owner guidance for more constructible designs and efficiencies during maintenance.
5. 3D geometry fosters confidence in pre-fabrication, enabling higher quality, lower labor costs, and accelerated schedules.	9. Confidence in the geometry enables just-in-time deliveries, leading to safer and better-orchestrated field work.
	10. Stability in the design configuration means labor productivity improvements, leading directly to high-performing construction.

For a more detailed analysis and a list of the benefits of BIM, it is worth visiting BIM Wiki at http://communities.bentley.com/products/building/building_analysis_design/w/building_analysis_and_design_wiki/bim-benefits.aspx. It should be noted that each member of the project team will have a different concept of what is BIM's most beneficial aspect. To assist us, BIM Wiki breaks these benefits into specific groups and then goes into greater detail, for each group, as follows:

1. Benefits at planning	1.1. Benefits to the planner/designer
	1.2. Benefits to the cost engineer
	1.3. Benefits to the owner
2. Benefits at design	2.1. Benefits to the architectural designer
	2.2. Benefits to the electrical designer
	2.3. Benefits to the mechanical designer
	2.4. Benefits to the plumbing designer
	2.5. Benefits to the landscape designer
	2.6. Benefits to the structural designer
	2.7. Benefits to the telecom designer
	2.8. Benefits to the civil engineering designer
	2.9. Benefits to the cost engineer
	2.10. Benefits to the specifications writer
	2.11. Benefits to the owner
3. Benefits at construction	3.1. Benefits to the construction manager
	3.2. Benefits to the construction contractor
	3.3. Benefits to the owner
4. Benefits at operations	4.1. Benefits to the occupant
	4.2. Benefits to the owner
5. Benefits at maintenance	5.1. Benefits to the occupant
	5.2. Benefits to the owner

BIM Wiki then renders additional detail; for example, for the plumbing designer the benefits are seen as:

- Fixture schedules can be synchronized or linked to the architect's schedule with a mere key stroke if desired.
- Fixture schedules, plans, riser diagrams, sections, and details can be automatically synchronized.
- The designer and his/her collaborators can visualize fixture layout and piping in 3D throughout the design process.
- Collisions and interferences can be determined immediately and automatically by software and integrated designs. No more RFIs to process because BIM contains the information they need.

- Riser diagrams can be developed once and then automatically synchronized with the plans. All engineering data (such as drainage fixture units) can be automatically and continuously followed in plan and analyzed in a variety of views and filters.
- Revisions to the plan, including architecture, can be checked in much less time compared to CAD or drafting methods.
- The designer can add, delete, and modify fixtures and outlets easily with automatic update to the engineering data and the model.

The National Institute of Standards and Technology (NIST) issued a report in August 2004 entitled "Cost Analysis of Inadequate Interoperability in the U.S. Capital Facilities Industry" (NIST GCR 04-867). The report concludes that, as a conservative estimate, $15.8 billion is lost annually by this industry caused by inadequate interoperability due to "the highly fragmented nature of the industry, the industry's continued paper-based business practices, a lack of standardization, and inconsistent technology adoption among stakeholders."

According to Chris Rippingham, BIM engineer at San Francisco-based DPR Construction, "Our thinking is that if we can sit at the table with the other great minds in the project—the architects, MEP and structural engineers, and our key subcontractors—as early as possible, then we can all deliver the most efficient building." He goes on to say, "We definitely try to collaborate as much as possible even in situations where the contract doesn't obligate us to do that, but with our experience in integrated delivery that's our normal way of working."

Integration and transition of models have led to increased cost effectiveness in employing BIM, and although the transition process may be long and expensive, the ultimate benefits of BIM are certainly worth the investment. When BIM is properly used, it coordinates the mechanical, electrical, and plumbing (MEP) trades, expands pre-fabrication opportunities, increases productivity, eliminates most rework, and reduces labor costs, at the same time improving the consistency of the final work product. This is echoed by the EMCOR Group, headquartered in Norwalk, Connecticut, which considers it well worth the contractor's time, effort, and financial investment to make the transition. EMCOR also emphasizes that successful implementation and use of BIM require significant investments in technology, staff, and training. With its subsidiaries, such as Dynalectric, EMCOR has considerable firsthand experience with this transition, having more than 200 trained BIM professionals. And with increasing use of BIM among trade contractors, combined with improved delivery methods, the building industry continues to move closer to realization of major BIM benefits—including substantial cost and time savings.

BIM technology also gives apartment building owners, architects, engineers, contractors, and fabricators affordable access to a full range of interactive tools for refitting of buildings for enhanced energy conservation, thereby helping to "green" these buildings. Monster Commercial, a commercial real estate information service, states that "Utilizing sophisticated parametric change technology, BIM software enables energy savings assessments for every conceivable aspect of a project—from floor plan designs to high-tech thermal imaging analysis."

5.2.2   Risks Associated with Applying BIM to Sustainable Projects

As mentioned earlier, the advent of BIM technology is expected to have a tremendous impact on the construction industry while promising to improve the overall design systems for construction projects. It is also rapidly becoming the dominant system for project design delivery. And while the use of BIM benefits far outweigh its negatives and have the potential to reduce risk, it is nevertheless imperative to have a total understanding of the impact of these challenges and risks on a project. Moreover, we find that BIM begins to blur the allocations of responsibility developed over generations, to the extent that it has become necessary for all members of the project team (architects, engineers, contractors, owners, and developers) to adopt legal safeguards. It may take years to fully comprehend the scope of BIM's legal ramifications and the sources of potential risk that are generated by using it on a green project.

Mike Bordenaro, co-founder of BIM Education Co-op, says, "The biggest hindrance to achieving these benefits is the lack of universal acceptance, support and implementation of Open Standards as represented by the work of the buildingSMART alliance." The Alliance for Construction Excellence and other organizations list some of the challenges that a project would face by employing BIM and, as a team, determine how to address these challenges in the rapidly changing technology of today. Contract provisions dealing with potential risks that are unique to BIM should not be overlooked by the parties and should be addressed in design–build/BIM contracts. Perhaps the most anticipated potential liabilities/risk challenges that confront a BIM project are these:

- *Information management:* The function of a BIM model includes the projection of values, cost savings, and/or efficiencies, but what happens if, when the building is completed, these projections fail to materialize?
- *Ownership of the model at various stages of the project:* Does the owner have sole ownership of the model after construction?
- *Copyright:* How should intellectual property rights be addressed? It is important to license and secure all intellectual property used within the model and to ensure that proprietary or copyrighted information integrated in the BIM model is not at risk of copyright or patent infringement. Also, it should be pre-determined who owns the data in the BIM model and who determines their use?
- *Control concerns:* To create a BIM model requires coordination. Which party is responsible for this, and who has access to the models and to what degree? Does the party managing the modeling process take on any additional liability exposure? There is often a concern that the model may change without all stakeholders' input.

- Concern that design and construction fees do not support the BIM process (or training for BIM).
- The BIM process will be a complete paradigm shift for the design and construction industry.
- Legal/insurance language and procedures will need future clarification.
- What is the level of detail from the design team and at what level of accuracy and reliance is the information provided by the team to the BIM model?
- BIM modeling allows the design–build team to improve on the project's potential aesthetics, but what happens when those projections do not meet expectations?
- What are the project-specific standards for file sharing, and is there a recognized protocol for the preservation of different replications of the model for historical purposes as well as for possible conflict?
- Two-dimensional documents are generated from the design model for permits and distributed to the general contractor and subcontractors along with the 3D design model.
- Shop drawings and submittals may only be eliminated for subcontractors and vendor/suppliers participating in the development of the model. There will be others who have no need to access the model (toilet accessories, components, etc.).
- What happens when a BIM model is based on or includes faulty information provided by specialty equipment vendors?

It is apparent from this list that with the advent of BIM technologies, methodologies, and processes, it is extremely important to identify who or what holds key project details—the architect or the general contractor (GC), the model or the drawings—because with that ownership comes great reward or risk. Identifying these legal risks should assist the parties in addressing the unique challenges that are associated with the use of BIM, particularly since there are few standard-form contracts currently on the market that adequately address them. It should be noted that contracts play a pivotal role in defining deliverables, interactions with project stakeholders, and risk obligations for commercial building projects.

On June 30, 2008, ConsensusDOCS 301, BIM Addendum, was released, which is described as a product of industry consensus on current best practices in the use of BIM techniques and technology. ConsensusDOCS contracts were developed by a coalition of 35 leading industry associations representing owners, contractors, subcontractors, designers, and sureties. Additionally, 301 is the first contract form that is specifically applicable to projects using BIM and remains the only industry standard document to adequately address the legal uncertainties associated with it. But while the BIM Addendum reflects a good starting point, it is nevertheless designed to be used with other traditional standard form contracts and should not be used as a standalone contract. And as with any addendum, the parties should carefully review underlying agreements and address any inconsistencies that may exist between the addendum and the base Contract Documents.

Project stakeholders should be made aware that there is apparently no case law that currently addresses BIM legal risks, as there is no case law that tests the adequacy of the BIM Addendum. However, for project participants to be in a better position to avoid potential conflict, they must examine the unique potential BIM risks for their project and address them upfront.

5.3 INTEGRATED PROJECT DELIVERY—SHARING INTELLIGENT DATA FOR SUSTAINABLE SOLUTIONS

The type of communication that will take place should be discussed among the various parties involved in a project, as well as how data sharing should be approached. Advantage should be taken of the three-dimensional information, such as that provided by building information modeling, that allows all members of the building team to visualize the many components of a project and to determine how they will work together. BIM and other 3D tools convey the idea and intent of the designer to the entire building team and lay the groundwork for integrated project delivery (IPD).

Integrated project delivery is a recently adopted innovative approach to the design and construction of buildings. Charles Thomsen, a fellow of both the American Institute of Architects and the Construction Management Association of America and a leading expert in this field, defines it thus: "Integrated project delivery is an approach to agreements and processes for design and construction, conceived to accommodate the intense intellectual collaboration that twenty-first century complex buildings require." At the core of IPD is the capability to have all data that affects a project stored in one unified database. Thomsen says IPD projects can generally be characterized by eight common themes:

- A legal relationship
- A management committee
- An incentive pool
- A no-blame working environment
- Design assistance
- Collaboration software
- Lean construction
- Integrated leadership

The main objective of IPD is to promote maximum collaboration, open sharing of project goals and risks, and maximization of the knowledge and insights of all participants (Figure 5.3). The end result of this process is an increase in value to the owner and a reduction in waste and inefficiencies as well as increased productivity throughout the various phases of design, fabrication, and construction. When IPD teams are first formed, they are typically faced with the challenge of determining how they will organize themselves to collaborate effectively, what processes they should follow, in what sequence. Previously, sequential

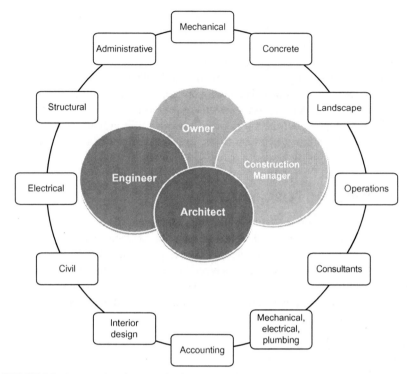

FIGURE 5.3 Integrated project organization.

exchange of paper-based documentation was the main information-sharing method for project participants in the vast majority of design and construction projects. By challenging traditional methods of delivery, IPD has paved the way for improved, faster, and less costly building projects, streamlined information and material supply chains, and provided more efficient processes throughout the building design and construction industry.

While the sequential exchange method has typically worked well in most idealized project situations, particularly where there is adequate time to fully develop the design intent, evaluate it, and then modify it as often as necessary prior to fabrication and installation, in complex project situations owners and their project design and construction teams find themselves having to work in parallel and with incomplete information.

The fundamental essence of IPD revolves around the concept of having all stakeholders involved on a project working together as early as possible—preferably during schematic design—to accumulate, combine, and focus their expertise toward the development of a project prior to anything being designed. Furthermore, as an inducement, shared risk and shared reward contracts are established upfront with the clear understanding that all parties must work together for the good of the project. An integrated project delivery guide, developed jointly by the American Institute of Architects (AIA) Documents Committee and the AIA California Council, is a tool to assist owners, architects, engineers, contractors, and other key stakeholders in moving toward unified models and improved design, construction, and operations processes.

From the preceding, it is clear that BIM and IPD have changed the way we hire and mentor in addition to rethinking our approach to handling data requirements. Here the underlying principle that needs to be remembered is the realization that adopting BIM and IPD workflows means improved communication both internally within the organization and externally with clients and trade partners.

5.4　BUILDING FORM WITH BIM

Building information modeling transcends traditional 2D drawings to include the additional dimensions of height and time. And because BIM allows active material input in the model (including data required for LEED® credits), any proposed changes can quickly be analyzed to confirm that they do not adversely impact the project's objectives (Figure 5.4). Likewise, by providing a long-term repository for data representing design intent, LEED and green documentation

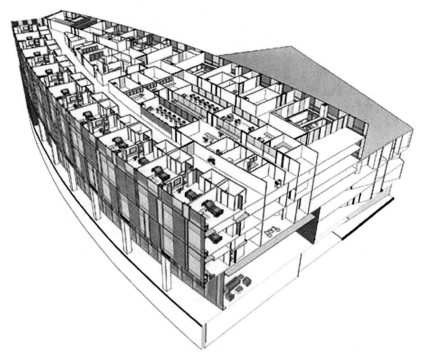

FIGURE 5.4　Use of BIM to "green an apartment building." *(Source: Monster Commercial, www.monstercommercial.com/tag/building-information-modeling/.)*

are much less likely to be deleted or overlooked in the BIM model. Moreover, BIM allows architects, engineers, and contractors to weigh the costs and benefits of the majority of building components and their interrelationships. According to Michael Laurie, P.Eng., "BIM software enables energy savings assessments for every conceivable aspect of a project—from floor plan designs to high-tech thermal imaging analysis." He goes on to say that with BIM "every drawing, view, enhancement, and modification is preserved within one single digital file that can be emailed or wirelessly accessed for accurate tracking, scheduling, and communication."

In addition, the affirmation of design options by the BIM model facilitates the reduction of design contingencies and allows the construction schedule to be consolidated. The provision of more accurate scheduling and improved cost take-offs reduces risk and construction contingencies, thus allowing the owner the potential to incorporate even more green features that may not have initially fitted the project's cost model. BIM also allows architects, engineers, and contractors the luxury of importing/exporting data from a wide variety of related software with ease, thereby simplifying design coordination and reducing the potential for loss of green goals in the transfer between software and individual users.

The employment of BIM from the outset can help reduce green design fees partly because the data embedded in the model reduce multiple inputting processes and any design modifications are immediately reflected. In the past such design variations might typically have resulted in a Change Order, whereas now they may be checked quickly and with minimal effort to determine their impact on the project. There are clear signals that the industry is irreversibly headed in these new directions. It is worthwhile noting that up to 20 LEED credits can be validated and documented by using BIM.

5.4.1 Customizing BIM

There is a saying that necessity is the mother of invention, and this would definitely apply to BIM in the contracting industry. Much of BIM's development has been focused largely on serving the needs of architectural and engineering firms. One of its key advantages is its flexibility in allowing project participants to customize existing elements or create new ones, which can then be incorporated into the design. Indeed, customization can usually provide great benefits in the use of BIM software. However, it has been shown that a number of firms that use BIM technology are not reaping the greatest benefits because of difficulties such as the lack of communication between the various participants in design and construction.

To address such issues, it is strongly recommended bringing on board a new professional specializing in the application of BIM technology, standards, and modeling who would undertake the coordination needed in BIM contexts. Such a *BIM manager* (the precise title of such a specialist is immaterial) would be an integral part of the project team and would be a small investment compared to his or her potential benefits. Many AEC-sector companies are already employing BIM managers, although their precise function and responsibilities frequently require defining. This is particularly important because to this day software vendors have yet to build a BIM product that comes out of the box with content that totally meets the requirements of the various industry contractors and fabricators.

Over the course of several years, Dynalectric, an electrical subcontractor and an EMCOR company, has become an industry leader in BIM applications. And through the implementation of virtual design and construction (VDC), it has been able to develop and execute electrical services and systems. To achieve this, a layer standard had to be determined and keyed (e.g., a layer for conduit, a layer for hangers), which allows the system to be designed to automatically place each component type on the appropriate layer. These latest efforts include embedding calculations into BIM that are aimed at improving pre-fabrication and making the project team more efficient, allowing it to achieve schedule and cost savings while making the overall project more efficient. As a result of this in-house customization (doing everything from schedules and take-offs to automatic engineering calculations), when a BIM engineer models an electrical system on a project, it is very accurate indeed.

5.4.2   Making the Transition

As previously discussed, successful implementation and utilization of BIM requires significant investments in technology, staff, and training. There is no "magic wand" solution, which is why in adopting the process one's eyes should be wide open. Furthermore, after taking the decision to transition to BIM, careful consideration should be given to the purchase of BIM software. Once the goals and priorities are analyzed and determined, one can proceed, bearing in mind that any transition from CAD to the new BIM technology will require additional investment in, among other things, more powerful hardware, software, and servers, as well as high-speed telecommunications to support the process. It will then require additional investment to customize the application software.

The transition to BIM will also require the hiring of a specialist BIM manager—someone with a solid foundation in BIM technology and computer skills who understands the process of collaboration in addition to possessing the technical and intellectual capabilities to integrate this knowledge into the BIM model. The building shown in Figure 5.5 was almost unanimously selected by the jury as the winner in the category Design/Delivery Process Innovation Using BIM, and it has a LEED Gold certification. M. A. Mortenson Company was committed to using BIM in all aspects of the Hall project, and created the role of design coordinator to manage interdisciplinary model creation, coordination, and interoperability.

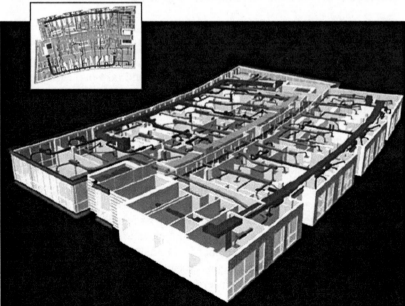

FIGURE 5.5   The full 3D model of a building (*top*) and the mechanical systems layout model (*bottom*) that allowed the Benjamin D. Hall Interdisciplinary Research Building to be designed with an extra floor. *(Source: www.mortenson.com/Resources/Images/11321.pdf.)*

BIM is frequently associated with Industry Foundation Classes (IFCs) and aecXML, which are data structures for representing the information it uses. IFCs are described as object-based file formats with a data model developed in BIM by buildingSMART of the International Alliance for Interoperability (IAI) to facilitate interoperability between software platforms in the building industry. However, aecXML is a specific XML mark-up language that typically employs IFC to create a vendor-neutral means to access data generated by BIM. It was developed for use mainly in the AEC and facility management industries, in conjunction with BIM software. There are other data structures on the market, but most of these are proprietary.

In addition to BIM technology being used for new construction, there are continuing attempts at creating BIM software that can be applied to older, pre-existing facilities. For this they typically reference key metrics such as the Facility Condition Index (FCI). The FCI is essentially used in facilities management to provide a benchmark to compare the relative condition of a group of facilities. However, to validate the accuracy of these models, they need to be monitored over time because whenever an attempt is made to model an existing building, numerous assumptions must be made regarding data relating to materials, design standards, building codes, construction methods, and the like. This makes it far more difficult and complex to modify an information model of an existing building from the design stage.

5.5   BUILDING SYSTEMS WITH BIM

A BIM modeler can easily extract intelligent property data from a BIM model for the purposes of engineering, pre-fabrication, and take-off. However, this isn't possible in off-the-shelf software; Dynalectric had to customize its BIM system to enable its staff to perform these functions. In addition, Dynalectric has in place a 4000+ object library that contains custom objects intelligently linked together to both populate the model and derive information from it, including scheduling, seismic calculations, and so forth. The benefits of customized BIM software are many, including annotating drawings and creating schedules. For example, merely by employing pertinent intelligent property data to a conduit and parts, the BIM modeler or staff can easily get a take-off of the conduit to determine the number of linear feet or quantity of hangers. In this way, it is possible to annotate conduit elevations by simply pulling up live data from the actual model components in the drawing.

Moreover, customization has in this case enabled the BIM modeler and the project team to easily perform engineering calculations. Likewise, a set of routines can be easily formulated that works with the intelligent property data to calculate strut loads and is capable of re-calculating automatically as the user changes the length of the object. This system in turn enables the BIM modeler and the project team to perform live engineering calculations, including "what-if" scenarios, quickly and accurately.

5.5.1 Virtual Best Practices

Over the last decade we have witnessed BIM technologies emerge from the research and development (R&D) arena and enter the mainstream of green construction. And it must be apparent from this that BIM has become much more than an electronic drawing tool, now allowing project participants and stakeholders full collaboration. Generally, the BIM manager (or the general contractor) is responsible for designing and implementing the BIM execution plan. This will vary from project to project, but includes determining what is to be modeled and at what level of detail, as well as facilitating mechanical, electrical, and plumbing (MEP) coordination.

However, with today's BIM technology each trade is given architectural and structural models from the owner. The trades then commence routing their systems. The use of design BIM systems facilitates detection of internal conflicts, and the employment of model-viewing systems such as Autodesk Navisworks® offers the ability to detect and highlight conflicts between the models and other information imported into the viewer. Each trade contractor would post regularly, say on a weekly basis, its systems to an FTP or other shared website. The BIM manager (or GC) could then assemble all of the models into a composite using, say, NavisWorks, which would enable project team members to integrate and share data and drawings from the various software programs. The composite BIM can be viewed, manipulated, modified, and analyzed for conflicts among the trades, and can then be used to hammer out modifications to resolve the conflicts. This process continues throughout the project—floor by floor and quadrant by quadrant—until all conflicts are finally resolved and "fit" in the virtual building. This integration of key systems into the model BIM can greatly reduce conflict issues.

The continuous development of BIM will mean greater and improved opportunities for owners, architects, engineers, constructors, and manufacturers, allowing them to model reality in the built environment with greater accuracy and reliability, particularly with the use of plug-ins, network (BIM) servers brought online, enhanced third-party interface and analysis tools able to function with major BIM software packages, and local hardware with increased capacity to handle, store, and manipulate larger files on a cost-effective basis. Combining of the most advanced virtual-reality modeling technology in computer science and applying the most current business practices in the construction industry, the BIM technology world is progressing so rapidly that much of the published data may already be out of date.

5.5.2 AIA Document E202

Because BIM is a relatively new technology, there were some legal challenges and other issues that necessitated clarification. To help clear up these legal issues, in 2008 the AIA released Document E202™, which lays out standard procedures and responsibilities for BIM models. Most important, however, it serves as a standard contract for projects using BIM. This document also establishes certain rules and regulations, such as who owns the model, how it is used, and the party responsible for each model element. Because of the unique nature of individual projects, Document E202 cannot give a blanket declaration for each one; rather, it lays out a legally binding framework of rules and then allows for adaptation (AIA 2008, p. 1).

AIA Document E202 has been a huge boon to BIM-based contracts. People all across the building industries recognize AIA and have embraced its efforts in simplifying the complex legal environment around BIM. Because BIM is in many respects still new, many of those who deal with construction law simply do not know how to work with it. Document E202 created a standard BIM-based contract that addresses many of the legal issues and challenges faced when using BIM.

5.6 THE FUTURE OF BIM AND ITS USE WORLDWIDE

Building information modeling is now used widely all over the world in countries such as the United States, the United Kingdom, France, Germany, Finland, Denmark, Australia, Malaysia, and Singapore. Moreover, internationally it is increasingly gaining the attention of the building industry and organizations involved in architecture, engineering, and construction (AEC) in addition to owners and operators of building projects and other structures. BIM standards efforts in the United States, Europe, and elsewhere in the global arena assume that this digital information is shareable, is interoperable among different stakeholders' information systems, is based on accepted open standards, and is definable in contract language. Discussing the future of integrated BIM, Dennis Neeley, AIA, product director of Reed Construction Data, believes the following:

Owners need to start immediately setting standards for their BIM projects. They need to provide the objects that their designers will use, or they need to get the manufacturers that they work with to provide the objects. They need to be consistent across all projects. Standardization on space designs, assemblies and objects and the data attached and associated is critical. The Reed SmartBIM site (www.SmartBIM.com) shows the concept, the Spaces section shows how an owner could created complete models of each of their spaces populated with the equipment and furnishing needed, including services like power, communications, etc. During construction, the BIM project must be updated with changes and substitutions. These steps will insure the downstream value and use of the BIM projects. These BIM projects can be integrated into companies GIS systems. The sooner owners get integrated BIM projects (BIM, IPD, and FM) the sooner they will see unbelievable savings.

It should be noted that the application of global BIM standards will necessitate the reflection of "business views" of information exchanged between AEC and owner/operator interests. These standards will build on those already in use.

5.6.1 United Kingdom

Many firms in the United Kingdom continue to resist using building information modeling, partly because of the seismic change in culture that it would necessitate. However, this has started to change now that the British government plans to make BIM compulsory on all public projects, in the belief that this new technology will facilitate improved ways of working that will reduce cost and add long-term value to the development and management of public-sector buildings. To help establish BIM on public projects, a task group has been formed, headed by the chairman of the UK Government BIM Group Mark Bew, to draw a road map and to phase in BIM over a period of five years. Recognizing that BIM isn't a mature technology in the United Kingdom, the level at which it will be mandated on projects will reflect the ability of the industry to adapt. However, if this plan is to succeed, the U.K. construction industry will need to go through a steep learning curve. Still, one example of the successful employment of BIM is Heathrow Airport Terminal 5, where its use reduced project costs by £210 million.

We are now witnessing a surprising upsurge in support for new working methods in the United Kingdom as industry challenges intensify. This is spurring the U.K. AEC industry to make a significant move toward incorporating BIM in its projects. One of the main drivers for the AEC industry to move toward using BIM has been the need to accelerate productivity. Pete Baxter, Autodesk vice president of the Middle East and Africa division, says: "BIM methods of working have been shown to create major efficiencies by eliminating inaccuracies, waste and clashes—and at the same time maintaining transparency and accountability. It's no surprise that new working challenges have tipped the balance in BIM's favor."

In the United Kingdom, the Construction Project Information Committee (CPIC) has the responsibility for providing best-practice guidance on the content, form, and preparation of construction production information, and for making sure this best practice is disseminated throughout the construction industry. CPIC, which is made up of representatives from many major U.K. institutions, has proposed a definition of BIM for adoption throughout the U.K. construction industry and has, moreover, invited all industry parties to debate the subject to facilitate reaching an agreed starting point. One of the main stumbling blocks to adopting a good working method that can significantly improve the quality and sustainability of deliveries from the design and construction team to the owner is the lack of a clear definition of the term "BIM" as well as the proliferation of interpretations that currently exist.

5.6.2 France

In France, a number of organizations are pushing for a more integrated adoption of BIM standards to improve software interoperability and cooperation among players in the building industry. Such organizations include the *Fédération*

Française du Bâtiment (FFB) and the French arm of buildingSMART International, which is supporting IFCs. Software-editing companies, on the other hand, such as Vizelia, were early adopters of IFCs and are now reaping the benefits of the full potential of BIM in the newly emerging green building business.

According to McGraw-Hill Construction's "SmartMarket Report" (2010), "France has the highest adoption rate of BIM among construction professionals surveyed at 38%, although it is only slightly higher than rates in the U.K. and Germany." The report goes on to say, "A very high percentage of French adopters (72%) use BIM on 30% or more projects." It concludes that French users see the most value from BIM through reduced conflicts during construction (76%) and improved collective understanding of design intent (71%).

5.6.3   Germany

The "SmartMarket Report" says, "German adopters as a group use BIM 47% of the time on 30% or more of their projects." In Germany architects have the highest adoption rate among industry professionals with (77%) followed by engineers (53%) and finally contractors (10%).

The U.S. Army Corps of Engineers' implementation of BIM systems has also arrived in Europe, The Corps now requires BIM to be used for many projects in its various mission areas and across its divisions, and that number continues to steadily increase. Jim Noble, Engineering Branch chief for the Europe District, says, "Many German firms in private industry are on board with BIM. The challenge for us is that many architects the Bauämter uses do not have much experience with BIM." It is expected to take some time to fully incorporate BIM into the Europe District construction process largely because of the legal issues as outlined in signed agreements between the U.S. and German governments that stipulate how BIM projects are to be accomplished in Germany. "Our job now is to get together with our partners, agree on some parameters, starting points and interpretations and move forward. We're doing just that," according to Noble.

5.6.4   Finland

A more extensive implementation of BIM in Finland has been achieved than in neighboring Scandinavian countries. Moreover, Finland now requires BIM to be used on all public-sector projects. In recent surveys it was shown that architects are the main users of BIM in their projects (approximately 93%) and engineers use is roughly 60%. It should also be noted that there is a great commitment in Finland's public sector toward wider BIM implementation. Evidence of this is the BIM guidelines that have been drafted as a result of the R&D ProIT project conducted with industry-wide support. These guidelines are in the Finnish language and cover general fundamentals of product modeling in construction projects, architectural design projects, and structural design projects, as well as product modeling in building services design projects. Although these guidelines describe product modeling in detail, they do not provide adequate data exchange specifications, thus providing potential for further guideline development.

BIM advocates in Finland's private sector are also quite active, and a number of corporations, such as Skanska Oy and Tekes, are earnestly conducting R&D in BIM. Likewise, research organizations and universities as well as the Association of Finnish Contractors and the state client (Senate Properties) are active in promoting implementation of BIM in the industry.

5.6.5 Norway

Graphisoft Norway and Solibri, Inc., have partnered in response to an increasing need for BIM quality assurance and model checking in Norway. For this reason, BIM is being promoted and used by various Norwegian public organizations and contractors including the civil state client Statsbygg and the Norwegian Homebuilders Association. Norway also recently produced BIM guidelines based on the experiences from the Statsbygg's *Høgskolen I Bodø* (HIBO) project (prepared in coordination with the NBIMS standard in the United States).

The private sector has also been active in promoting building information modeling. For example, Selvaag Bluethink is developing BIM and ICT solutions based on BIM. Norway's SINTEF is the leading organization conducting research in BIM and is a part of Erabuild, a network of national R&D programs focusing on sustainable tools to improve construction and operation of buildings. Moreover, Norway is recognized as being among the first few countries to develop an International Framework for Dictionaries (IFD) standard for the construction industry.

5.6.6 Denmark

In 2007 Denmark (like Finland) mandated that BIM be used on all public-sector projects. The overall use of BIM continues to increase there. According to a survey conducted in 2006 (cited in Kiviniemi et al., 2008), roughly 50% of architects and 40% of engineers were using BIM for some parts of their projects. One of the leading Danish user-driven organizations is bips which has a strong influence in the implementation of IT in the Danish construction industry. Moreover, the mandatory demands for BIM from Danish state clients have moved its use to a higher level.

In the public sector, Denmark boasts at least three public agencies that have initiated the implementation of BIM. These are the Palaces and Properties Agency, the Danish University and Property Agency, and the Defense Construction Service. Although government projects do not represent a large portion of the total property area, their impact on the market created by the IFC requirements is significant. There are other government agencies, such as

Gentofte Municipality and KLP Ejendommehave, that have also adopted the requirements from Denmark's Digital Construction project–a government initiative.

The Danish government has aggressively outlined its requirements for using BIM in government projects. These requirements are called *Byggherrekravene* (Det Digitale Byggeri, 2007). Starting in January 2007, all architects, designers, and contractors participating in government construction projects are required to adopt a number of new digital routines, approaches, and tools. Under the Digital Construction program initiated by the Danish Enterprise and Construction Authority, a package of guidelines relating to 3D was developed. These guidelines are concerned with both setting up and fulfilling requirements in file- and database-based CAD/BIM applications.

Bips, the Danish membership organization that promotes IT, collaboration, and productivity in construction, is also developing BIM guidelines for the private sector and has adopted the results from the Digital Construction project. It is also promoting new working methods for the Danish construction industry. The Danish Enterprise and Construction Authority is supporting BIM R&D in Denmark as well as in other Danish organizations and universities, such as Aalborg University, which is focused on IFC model servers and 3D models; and Aarhus School of Architecture, which is focusing on product configuration, design intent, and IFC model server; and the Technical University of Denmark, which is focused on interoperability.

5.6.7   Hong Kong

The construction industry in Asia Pacific is undergoing fundamental changes and is rapidly automating and streamlining its processes to stay abreast with the international business ecosystem. In this respect, Hong Kong is considered to be one of the most advanced countries in the region to adopt building technologies, and it has played an important role in setting up standards in the industry. By deploying cutting-edge technologies, Hong Kong has demonstrated its effectiveness and efficiency in completing world-class building projects.

5.6.8   China

China is currently experiencing the world's largest construction boom, and building information modeling is providing a competitive advantage to Chinese architects and engineers. Moreover, the 2008 Beijing Olympics and the 2010 World Expo in Shanghai have prompted the investment of billions of dollars in new construction in those cities, which already have some of the highest commercial and industrial rents in Asia.

The recent building frenzy taking place in China, along with burgeoning pollution problems, has induced many Chinese architects to take a keen interest in sustainable design. It goes without saying that using a building information

FIGURE 5.6   This auditorium complex in central China is 15,000 square meters (161,460 sq. ft.).

model facilitates the complex design evaluations and analyses that support key aspects of sustainable design, and it enables the project team to balance China's construction growth with environmental concerns. Moreover, China's rapid building growth presents enormous challenges as well as enormous opportunities. By embracing BIM, the Chinese building industry is able to take advantage of the productivity benefits that surround a digital building methodology, thus giving designers a distinctive competitive edge in the midst of a whopping construction boom.

Wuhan Architectural Design Institute (WADI) is one of China's main multi-discipline architectural design firms, with 625 employees, including 242 architects. In 2004, WADI selected Autodesk Revit® to help it transition directly from its existing 2D-drafting solution (AutoCAD) to BIM. In preparation for bidding and using the Revit architecture, one WADI architect was reportedly able in just four days to produce all of the schematic design and presentation documentation for the building shown in Figure 5.6.

5.6.9   Singapore

Construction and Real Estate Network (CORENET) is Singapore's principal organization for the development and implementation of BIM for government projects. CORENET, a major IT initiative, was launched in 1995 by Singapore's Ministry of National Development. It provides various information services including e-information systems such as eNPQS and e-Catalog to its clients, as well as e-submission and integrated plan checking. IT standards are being adopted in Singapore's construction industry based on the guidelines of the International Alliance for Interoperability (buildingSMART). Singapore also now requires the adoption of BIM for various kinds of approvals such as building plan approvals and fire safety certifications, and BIM integrated plan-checking guidelines are now operational.

Although BIM is a relatively recently developed technology in an industry that is often slow to change, most early adopters feel confident that it will quickly grow to play an ever more crucial role in building design and construction. Building information modeling has been defined as an integrated process that allows architects, engineers, builders, owners, and other stakeholders to explore a project's key physical and functional characteristics digitally—prior to construction. BIM is the future, and it is certainly here now.

Chapter 6

Green Building Materials and Products

6.1   INTRODUCTION

Basically any material that is used for the purpose of construction can be considered building material. And when designing a green building, it is important to carefully consider the choice of materials to be used to achieve your sustainability goals. The U.S. Green Building Council (USGBC®) believes that building materials choices are fundamental to achieving success in sustainable design. This is due to many factors, including the extensive network of extraction, processing, and transportation steps required. Additionally, the numerous activities required to create building materials may have a negative impact on the environment by polluting the air and water, destroying natural habitats, and depleting natural resources. Also, incorporating green products into a project should not imply sacrifice in performance or aesthetics, nor does it necessarily entail higher project cost.

For thousands of years man has used naturally occurring materials, such as clay, sand, reed, wood, and stone, to construct his habitat. But humans have developed and supplemented these with many man-made products—some more and some less synthetic. Building materials manufacture has long been an established industry in many countries around the world, from China to Iraq, and their use is typically compartmented into specific specialty trades (e.g., carpentry, roofing, flooring, and plumbing).

The concept of repair and reuse of a building instead of tearing it down and building a new structure is a highly effective strategy for minimizing environmental impact. And refurbishment and rehabilitation of existing building components helps minimize any potential negative impact on the environment and saves natural resources, including the raw materials, energy, and water resources required for new construction. It also helps reduce pollution that might take place as a byproduct of manufacturing, extraction, and transportation of raw materials, in addition to minimizing the creation of solid waste that often ends up in landfills.

North Carolina has a Building Reuse and Restoration Grants Program that provides grants for the restoration and upfitting of vacant buildings in rural

Handbook of Green Building Design, and Construction

communities or in economically distressed urban areas, as well as for the expansion and renovation of buildings currently occupied by certain types of businesses. Many rating systems, including the USGBC's Leadership in Energy and Environmental Design (LEED®) Rating System also recognize the importance of building reuse. Reusing a building can contribute to earning points under LEED "Materials Resource Credits on Building Reuse."

6.1.1 Definition of Green Building Materials and Products

Green building means different things to different people—it is a multifaceted concept that lends itself to numerous interpretations. Likewise, green building materials defy easy definition; they are eco-friendly and are composed of renewable, rather than nonrenewable, resources. In general, building materials are called "green" because they have minimal or no negative impact on the environment and in some cases may even have a positive impact. It should be noted that there is no perfect green material, but in practice there is an upsurge in the number of materials on the market that reduce or eliminate negative impacts on people and the environment. And especially today, with building and construction activities worldwide consuming billions of tons of raw materials each year, it has become increasingly imperative to employ green building materials and products and so help in the conservation of dwindling nonrenewable resources internationally.

Many green products are made from recycled materials, which helps the environment and puts waste to good use in addition to reducing the energy required to make them. Building materials are also considered green when they are made from renewable resources that are sustainably harvested. An example of this is flooring that is made from sustainably grown and harvested lumber or bamboo. Durability is another characteristic of sustainability, and some building materials are considered green because they are very durable. One example is a durable form of siding that outlasts less durable products, resulting in substantial savings in energy and materials over the lifetime of a property. Additional benefits can be achieved when a durable product is made from environmentally friendly materials, such as recycled waste.

6.1.2 Natural versus Synthetic Materials

Building materials generally fall into one of two classifications, natural and synthetic. Natural building materials are those that have not gone through any process or that have been minimally processed by industry, such as timber or glass. Synthetic materials, on the other hand, are made in industrial settings and have gone through processing after considerable human manipulations; they include plastics and petroleum-based paints. Both have their advantages and disadvantages.

Clay, mud, stone, and fibrous plants (e.g., bamboo and reeds) are among the most basic natural building materials. They are being used together by people all over the world to create shelters and other structures to suit their local habitat. In general, stone and brush are used as basic structural components in these buildings, while mud is generally used to fill in the space between, acting as a form of mortar and insulation.

In many cases, it has been clearly demonstrated that natural materials cannot cope with or meet the required specifications of the industrial challenges faced by the construction industry today. Plastic is a case in point and a good example of a typical synthetic material. The term "plastic" covers a range of synthetic or semisynthetic organic condensation or polymerization products that can be molded or extruded into objects, films, or fibers. The term is derived from the material's malleable nature when in a semiliquid state. Plastics vary immensely in heat tolerance and are hard-wearing and highly adaptable, can be molded and cast in a variety of forms, and can mimic and perform the task of most other building materials. Plastics continue to be viewed as a potential replacement for other natural building materials. Combined with this adaptability, their general uniformity of composition and lightness facilitate their use in almost all industrial applications.

A material's "greenness" is generally based on certain criteria, such as its durability, and whether it is renewable and resource-efficient in its manufacture, installation, use, and disposal. Other considerations are whether the material is eco-friendly and supports the health and well-being of occupants, construction personnel, and the public; whether it is appropriate for the application; and what the environmental and economic trade-offs among alternatives are.

Considerable research remains to be conducted to enable satisfactory evaluation of alternatives and selection of the best material for a project. Appropriate material selection should consider a number of factors, including the material's impact throughout its life cycle (from raw material extraction to use and then to reuse, recycling, or disposal). The areas of impact to consider at each stage in the life cycle of a material include:

- Energy required for extraction, manufacturing, and transport
- Natural resource depletion; air and water pollution; hazardous and solid waste disposal
- Energy performance in useful life and durability
- Impact on indoor air quality; exposure of occupant, manufacturer, or installer to harmful/toxic substances; moisture and mold resistance; cleaning and maintenance methods

Properties that typical green building materials and products may share include:

- Readily recyclable or reusable when no longer required
- Sustainably harvested from rapidly renewable resources such as genuine linoleum flooring, bamboo flooring, wool carpets, strawboard, and cotton-ball insulation (made from denim scrap); using rapid renewables helps reduce the use and depletion of finite raw materials

- Durability
- Wood or wood-based materials that meet the Forest Stewardship Council's (FSC) principles and criteria for wood building components
- Post-consumer recycled content
- Reuse, refurbishing, remanufacturing, or recycling potential
- Manufactured from a waste material such as straw or fly ash or a waste-reducing process
- Minimally packaged and/or wrapped with recyclable packaging
- Locally extracted and processed, which means less energy used in extraction, processing, and transport to the job site to help regional economies
- Water efficiency
- Manufactured with a water-efficient process
- Energy-efficient in use
- Generates renewable energy

6.1.3   Storage and Collection of Recyclables

Storage and Collection of Recyclables is a Materials & Resources prerequisite in most of the LEED Rating System categories. The intent of this prerequisite is essentially to facilitate the reduction of waste that is hauled to and disposed of in landfills by encouraging storage and collection of recyclables. For this reason, LEED stipulates that an area be provided that is dedicated to recycling inside the building so that occupants can have the option to recycle their paper, cardboard, glass, plastic, and metals. This recycling means that there is a reduction in the need for virgin materials as well as a significant reduction in the amount of waste otherwise going to landfills.

It should be noted that LEED has not set any specific standards or requirements for this area, but the USGBC guidelines (Table 6.1) state that the area should be easily accessible, serve the entire building(s), and be dedicated to

TABLE 6.1 Recycling Storage Area Guidelines Based on Overall Building Square Footage

Commercial Building (square footage)	Minimum Area (square footage)
0 to 5000	82
5001 to 15,000	125
15,001 to 50,000	175
50,001 to 100,000	225
100,001 to 200,000	275
200,001 or greater	500

Source: USGBC.

the storage and collection of nonhazardous materials for recycling. The average waste per employee is estimated to be about three pounds per day! It is therefore important that building occupants have the option to maintain good recycling programs throughout the life span of the building.

6.2 LOW-EMITTING MATERIALS

Today, most major manufacturers and suppliers offer a low-emitting option for any of the materials recommended for credits in the Indoor Environmental Quality (IEQ) section of the LEED building design and construction reference guides. "Green" appears to be the latest trend, and everyone's trying to hop on the bandwagon. People are increasingly seeking out "healthy" buildings to live and work in. LEED addresses low-emitting materials within the Indoor Environmental Quality section of this book. Among the most straightforward credits to earn of the LEED IEQc4 credits is IEQc4.5, particularly if you choose to meet it by using GREENGUARD Indoor Air Quality–certified furniture.

With respect to the New Construction (NC) Rating System, credits MR4.1-4 relate to low-emitting materials such as adhesives and sealants, paints and coatings, flooring systems, and composite wood and agrifiber products. Many of the environmental impacts associated with building materials have already taken place by the time the materials are installed. Pollutants are emitted during extraction from the ground or harvesting from forests and during manufacture. Energy has therefore been invested throughout production. Certain materials, such as those containing ozone-depleting HCFCs and volatile organic compounds (VOCs), may continue emitting pollutants during their life cycle. Some materials will also have negative environmental impacts associated with their disposal.

Important considerations to bear in mind when selecting building materials and products include the following:

- Avoid materials and products that generate substantial amounts of pollution (VOCs, HCFCs, etc.) during manufacture or use.
- Specify salvaged building materials or those produced from waste or that contain post-consumer recycled content.
- Avoid materials made from toxic or hazardous constituents (benzene, arsenic, etc.).
- Specify materials with low embodied energy (the energy used in resource extraction, manufacturing, and shipping).
- Help the regional economy and the environment by using materials and products manufactured regionally.
- Encourage environmentally responsible forestry by using wood or wood-based material that meets the FSC's principles and criteria for wood building components, and avoid materials that unduly deplete limited natural resources such as old-growth timber.

Of note, there are various resource-efficient products available at no extra charge, while others may cost more. Likewise, if the installation differs from standard practice, it may raise labor cost if an installer is unfamiliar with a product.

6.2.1   Adhesives, Finishes, and Sealants

The New Construction credit template for IEQc4.1, Low-Emitting Adhesives and Sealants, for example, requires the project manager or specifier to list all indoor adhesives, sealants, and sealant primer products to be used on the project, and to input the following information:

- Name of product manufacturer
- Product name and model
- Product VOC content (g/L)
- Source of VOC data
- South Coast Air Quality Management District (SCAQMD) Rule #1168 as of 2007, Allowable VOC Content (g/L), as indicated in Figure 6.1. Aerosol adhesives not covered by Rule #1168 must meet Green Seal Standard GS-36 requirements in effect on October 19, 2000. All indoor air contaminants that are odorous, potentially irritating, and/or harmful to the comfort and well-being of installers and occupants should be avoided or minimized.

An important characteristic of sealants is that they can increase the resistance of materials to water or other chemical exposure, while caulks and other adhesives can assist in controlling vibration and strengthen assemblies by spreading loads beyond the immediate vicinity of fasteners. These properties enhance durability of surfaces and structures, but they do so at a cost because very often they are shown to be hazardous in manufacture and application. Moreover, construction adhesive formulas often contain in excess of 30% volatile petroleum-derived solvents, such as hexane, to maintain liquidity until application. This has caused workers to become exposed to toxic solvents; also, as the materials continue to off-gas during curing, occupants may be potentially exposed to emissions for extended periods.

Industry tests indicate that water-based adhesives work as well as or better than solvent-based adhesives, and can pass all relevant American Society of Testing Materials (ASTM) and American Plywood Association (APA) performance tests. Also, water-based adhesives can be purchased at comparable costs from numerous manufacturers and, when purchased in bulk, larger containers can often be returned to vendors for refill. Most stains and sealants also emit potentially toxic VOCs into indoor air. One way of managing this problem is by employing materials that do not require additional sealing, such as stone, ceramic and glass tile, and clay plasters. The toxicity and the air and water pollution generated by the manufacture of chlorinated hydrocarbons such as methylene chloride strongly reinforce the need for responsible, effective alternatives, such as plant-based, nontoxic, or low-toxicity sealant formulations.

Architectural Applications (SCAQMD 1168)	VOC Limit (g/L less water)	Specialty Applications (SCAQMD 1168)	VOC Limit (g/L less water)
Ceramic tile	65	Welding: ABS (avoid)	325
Contact	80	Welding: CPVC (avoid)	490
Drywall and panel	50	Welding: plastic cement	250
Metal to metal	30	Welding: PVC (avoid)	510
Multipurpose construction	70	Plastic primer (avoid)	650
Rubber floor	60	Special-purpose contact	250
Wood: structural member	140		
Wood: flooring	100	**Sealants and Primers (SCAQMD 1168)**	
Wood: all other	30	Architectural porous primers (avoid)	775
All other adhesives	50	Sealants and nonporous primers	250
Carpet pad	50	Other primers (avoid)	750
Structural glazing	100		
		Aerosol Adhesives (GS-36)	**VOC Limit**
Substrate-Specific Applications		General-purpose mist spray	65%
Fiberglass	80	General-purpose web spray	55%
Metal to metal	30	Special-purpose aerosol adhesives	70%

VOC weight limit is based on grams/liter of VOC minus water. Percentage is by total weight

FIGURE 6.1   Adhesives, sealants, and sealant primers: SCAQMD Rule #1168. VOC limits are listed and correspond to an effective date of July 1, 2005, and a rule amendment date of January 7, 2005. VOC weight limit is based on grams/liter of VOC minus water. Percentage is by total weight. *(Source: USGBC.)*

   LEED requires that all adhesives and sealants used on building interiors (defined as inside the weatherproofing system and applied onsite) need to comply with the reference standards shown in Figure 6.1. Environmentally preferable cleaning methods and products can lessen indoor air pollution and solid/liquid waste generation. Safe cleansers are readily available and are competitively priced and eco-friendly. The improper use and disposal of some common cleaning and maintenance products can contribute to indoor-air contamination, toxic waste, and water pollution.

   The active ingredients in cleaners are surfactants, for which biodegradability is a key factor. Even low surfactant concentrations in runoff have been shown to pose risks to the environment. Petroleum-derived surfactants generally break down more slowly than vegetable-oil–derived fatty acids; some materials are even resistant to municipal sewage treatment. The harmful effects of toxins can be minimized by implementing the following:

● Storing hazardous materials outside the building envelope.
● Selecting materials with a durable finish that do not require frequent stripping, waxing, or oiling (such as linoleum, cork, or colored concrete).

- Selecting products that have approved third-party or government agency certification: Green Seal, Scientific Certification Systems (SCS), Environmental Protection Agency (EPA) Environmentally Preferable Purchasing Program, General Services Agency, California Integrated Waste Management Board (CIWMB) Recycled Content Product Directory.
- Whenever possible, selecting biodegradable, nontoxic cleansers.
- Avoiding cleansers, waxes, and oils that are labeled as toxic, poisonous, harmful or fatal if swallowed, corrosive, flammable, explosive, volatile, requiring "adequate ventilation" or safety equipment, or causing cancer or reproductive harm.
- Placing mats at all building entrances to minimize stripper use; refinishing only areas where the finish surface is wearing; cleaning regularly by dust-mopping and/or vacuuming frequently and wet-mopping with a liquid cleaner.

6.2.2 Paints and Coatings

Paints generally consist of a mixture of solid pigment suspended in a liquid medium and applied as a thin (often opaque) coating to a surface for protection and/or decoration. Primers are the first coat in a paint system (i.e., basecoats), whose main function is to increase the adhesion between the substrate and the total paint system (i.e., subsequent coats of paint or varnish). Sealers are also basecoats and are applied to a surface with the main function of helping reduce the absorption of subsequent coats of paint or varnish and preventing bleeding through the finish coat by sealing in aggressive chemicals (e.g., alkalinity). Paint was first used on a large scale thousands of years ago by the Egyptians and Babylonians in their buildings and temples.

It should be noted that the LEED Guidelines for Paints, Coatings, and Primers may vary from one LEED program to another, although USGBC has made a determined effort to make this credit more consistent through the introduction of LEED 2009. Also, LEED considers paints and coatings as only those used on the interior of the building, since exterior paints will not affect a building's "indoor air quality." Generally speaking, the USGBC requires that paints and coatings applied onsite and used on the interior of the building (defined as inside the weatherproofing system) must comply with the following referenced standards:

- Architectural paints, coatings, and primers applied to interior walls and ceilings: Do not exceed the VOC content limits established in Green Seal Standard GS-11, Paints, Second Edition, May 12, 2008 (see Table 6.2).
- Anticorrosive and antirust paints applied to interior ferrous metal substrates: Do not exceed the VOC content limit of 250 g/L established in Green Seal Standard GC-03, Anti-Corrosive Paints, Second Edition, January 7, 1997.
- Clear wood finishes, floor coatings, stains, and shellacs applied to interior elements: Do not exceed the VOC content limits established in SCAQMD Rule #1113, Architectural Coatings, January 1, 2004. Table 6.2 shows the allowable VOC levels stipulated by SCAQMD.

TABLE 6.2 Allowable VOC Levels in Paints and Finishes

Type	Limit (g/L)
Paints	
Flat	50
Nonflat	50
Primers, sealers, and undercoats	100
Quick-dry enamels	50
Finishes	
Clear wood	Varnish, 350 Lacquer, 550
Floor coatings	100
Sealers	Waterproofing, 250 Sanding, 275 All others, 200
Shellacs	Clear, 730 Pigmented, 550
Stains	250

Note: Grams/liter less water and exempt compounds, according to SCAQMD.

Santa Cruz County, California, officials often point out that paint can have significant environmental and health implications in its manufacture, application, and disposal. Most paint, even water-based "latex," is derived from petroleum. Its manufacture requires substantial energy and water and creates air pollution and solid/liquid waste. Volatile organic compounds are typically the pollutants of greatest concern in paints. Those from the solvents found in most paints (including latex) are released into the atmosphere during manufacture and application and for weeks or months after. VOCs emitted from paint and other building materials are associated with eye, lung, and skin irritation, headaches, nausea, respiratory problems, and liver and kidney damage. Manufacturers continue to employ the latest technology to reduce the VOCs found in these paints while maintaining costs at a reasonable level.

Although exposure to solvents emitted by finish products can be significant, renewable alternatives (e.g., milk paint) address many of these concerns. In this case there may be a premium to pay, and some products may not be suitable for exterior applications. Paint manufacturers have started to produce reformulated low- and zero-VOC latex paints with excellent performance in both indoor and outdoor applications; they can be purchased for the same price as older high-VOC products or less. Paints that meet GS-11 standards meet stringent performance criteria, are low in VOCs and aromatic solvents, and do not contain heavy metals, formaldehyde, or chlorinated solvents.

There are other alternative paints on the market such as silicate paints, which are solvent-free and may be used on concrete, stone, and stucco. Silicate paints have many advantages; they are odorless, nontoxic, vapor-permeable, naturally resistant to fungi and algae, noncombustible, colorfast, light-reflective, and even acid rain-resistant. Though silicate paints are more expensive, their extraordinary durability provides some compensation, plus the fact that they cannot spall or flake off and will only crack if the substrate cracks.

6.2.3   Flooring Systems

There are many kinds of flooring systems available in today's marketplace, from nonmagnetic access flooring systems to carpet, wood block, and resinous flooring. Each system is designed to meet different requirements and to satisfy the rigorous demands of high-traffic commercial, residential, and institutional applications as well as the aesthetic requirements for high-visibility public and private facilities.

Carpet

Carpet is a very controversial material and, in most cases, is not considered to be green. As with any product, significant environmental impacts can occur throughout a carpet's life cycle (i.e., its manufacture, use, and disposal). By considering a variety of life-cycle attributes, from the materials used to manufacture and install carpet to recycling and disposal issues, purchasers can make informed decisions, including the potential health concerns it presents. Various volatile organic compounds can be emitted from carpet materials, although VOC emissions from new carpet usually fall to very low levels within 48 to 72 hours after installation if good ventilation is provided.

Most carpet products are synthetic, usually derivatives of nonrenewable petroleum products; their manufacture requires substantial energy and water and creates harmful air and solid/liquid waste. Today, however, many carpets are being manufactured with recycled content (e.g., plastic bottles), and a growing number of carpet manufacturers are refurbishing and recycling used carpets into new ones. At the end of their useful life, most carpet tends to end up in landfills, and the EPA states that "Over four billion pounds of carpet enter the solid waste stream in the United States every year, accounting for more than 1% by weight and about 2% by volume of all municipal solid waste (MSW).

Carpet remains the most popular floor covering in the United States and is installed on nearly 70% of our floors. Synthetic carpeting is the most common line of products, constructed from petroleum-based materials that have been linked to health concerns. Carpets and their backings, and the adhesives used with them, have been shown to off-gas many unhealthy VOCs, all of which pollute indoor and outdoor air. Nylon is the most popular fiber used as the face fiber in commercial carpet. Polypropylene (olefin), polyethylene terephthalate (PET) polyester, and recycled PET are also employed in carpet face fiber.

In general, carpet made from PET and polypropylene face fiber is not as durable as carpet made from nylon face fiber. Redesigned carpets, new adhesives, and natural fibers are now available that emit low or zero amounts of VOCs. For improved air quality, selected carpets and adhesives should meet a third-party standard, such as the Carpet and Rug Institute (CRI) Green Label Plus or the State of California's Indoor Air Emission Standard 1350.

From an environmental standpoint, natural fibers are an eco-friendly and preferable carpeting option because they are renewable and biodegradable. A traditional material is wool, and in many parts of the world sheep are still raised specifically for carpet fibers. Wool carpets are more durable than synthetics; resist dust mites, moisture, and fire; and can be more comfortable underfoot. They often use jute backing, upping the sustainable nature of these carpet products. Other natural fiber options include sisal, seagrass, abaca, coir, and wool floor coverings. One of the disadvantages of carpets is that they tend to harbor more dust, allergens, and contaminants than many other materials (Figure 6.2). Durable flooring, such as a concrete-finish floor, linoleum, cork, or reclaimed hardwoods, is generally preferable in helping to improve indoor air quality.

The LEED intent of low-emitting carpet systems is to reduce the quantity of indoor air contaminants that are odorous and potentially irritating and/or harmful to the comfort and well-being of installers and occupants. Additionally, for LEED credits, all carpet installed within a building's interior must meet or exceed testing and product requirements of the Carpet and Rug Institute (CRI) Green Label Plus program for VOC emission limits. Carpet pads installed within a building's interior must also meet or exceed CRI's limits. Adhesives and sealants used in carpet installation must comply with SCAQMD, Rule #1168.

The EPA has developed five guiding principles to help federal government purchasers incorporate environmental considerations into purchasing decisions, which provide a framework purchasers can use to make environmentally preferable purchases. They include environmental factors as well as traditional considerations of price and performance as part of the normal purchasing process.

- Emphasize pollution prevention early in the purchasing process.
- Examine multiple environmental attributes throughout a product's or service's life cycle.
- Compare relative environmental impacts when selecting products and services.
- Collect and base purchasing decisions on accurate and meaningful information about environmental performance.

It should be noted that in LEED 2009 NC, C&S, and CI, the title of EQ 4.3 has been changed from "Carpet" to "FlooringSystems," and this credit has been substantially expanded. From 2009 onward, LEED has stipulated that all hard-surface flooring must be certified as compliant with the FloorScore® standard by an independent third party. FloorScore is a program developed by the Resilient Floor Covering Institute (RFCI) together with Scientific Certification Systems to test and certify flooring products and flooring adhesive products for compliance with indoor air quality (IAQ) emissions targets. Flooring products covered by FloorScore include vinyl, linoleum, laminate, wood,

FIGURE 6.2   FLOR carpet squares laid in a flexible and practical "tile" format. The tiles are made from renewable and recyclable materials and are available in a range of colors, textures, and patterns. More than 2 million tons of carpets are landfilled in the United States each year. *(Courtesy: FLOR®, Inc.)*

ceramic, rubber, wall base, and associated sundries. A FloorScore certification means healthier, cleaner air and therefore healthier living and working conditions.

Polyvinyl Chloride/Vinyl

Polyvinyl chloride (PVC), also referred to as "vinyl," is one of the most widely used synthetic materials in building and construction because of its durability, versatility, and cost. In addition to flooring, PVC is common in pipes, vinyl siding, vinyl flooring, wire insulation, conduit, window frames, packaging, wall covering, roofing, and many other products. PVC is generally transparent with a bluish tint. It is attacked by many organic solvents but has a very good resistance to oils and a low permeability to gases. In its rigid form PVC is available in sheets that can readily be welded to produce tanks, trays, and troughs. It is not recommended for use above 158 °F (70 °C) although it can be taken to 176 °F (80 °C) for short periods. PVC is important because it accounts for nearly 50% of total plastic use in construction and because it is increasingly recognized as problematic.

Vinyl is commonplace today, with about 14 billion pounds being produced annually in North America. It is inexpensive, and not all of its alternatives have yet worked out all their negative issues. Moreover, as the USGBC suggested in its long-awaited report on PVC, all materials have potential pitfalls, from indoor air quality to disposal. PVC is difficult to recycle for many reasons including its high chlorine content, which makes recycling complicated and expensive because it cannot be mixed with other plastics.

It is said that vinyl composition tile (VCT) accounts for more square footage than any other category of resilient flooring. Today, millions of square feet of VCT have been installed around the world, in commercial buildings, retail stores, supermarkets, hospitals, and schools. It has been extensively used because of its benefits: good strength relative to its weight, durability, water resistance, and adaptability. Vinyl tends to be inexpensive, in part because production typically requires roughly half the energy required to produce other plastics. Products made from vinyl can be resistant to biodegradation and weather and are effective insulators. The physical properties of vinyl can be tailored for a wide variety of applications. Many firms are increasingly concerned about the difficulty in recycling VCT and the negative environmental impacts this creates, and are therefore struggling to find appropriate alternatives.

Vinyl/PVC's main problem is its "toxic life cycle," which begins and ends with hazards, most stemming from chlorine, its primary component. Chlorine makes PVC more fire-resistant than other plastics. The production of PVC requires hazardous chemicals such as vinyl chloride (a simple chemical made of chlorine, carbon, and hydrogen), which causes cancer, and very hazardous chemicals are byproducts of that same production, including dioxin and poly-chlorinated biphenyls (PCBs). Lead, cadmium, and other heavy metals are sometimes added to vinyl as stabilizers; phthalate plasticizers, which give PVC its flexibility, pose potential reproductive risks. Also, some consumer products such as phthalates can over time leach out or off-gas harmful chemicals, exposing building occupants to materials linked to reproductive system damage and cancer in laboratory animals. Manufacturing vinyl or burning it in incinerators produces toxic byproducts, including dioxins, which are among the most toxic chemicals known to man. Research has shown that the health effects of dioxin, even in minute quantities, include cancer and birth defects.

Polyvinyl chloride is one of the most environmentally hazardous consumer materials produced. It is a strong thermoplastic material that is made from vinyl chloride monomer (VCM) and ethylene dichloride (EDC), both of which are carcinogens and acutely toxic. The production of PVC causes the release of these toxic carcinogens into the environment, and there is no way to confidently quantify these hazards and upset condition impacts for a life-cycle assessment (LCA) or risk analysis. Clean air regulation and liability concerns have been effective in reducing total VCM releases since 1980, while PVC use has roughly tripled. Most PVC products are believed to be basically harmless when properly used. However, some of the additives and softeners can leach out of certain vinyl products. And although PVC resin is inert in normal use, older PVC products are frequently contaminated with traces of VCM (many of the older landfills have been releasing toxic fluids for decades), which can leach into the surrounding environment and contaminate drinking water.

There are many possible substitutes on the market that may cost more or require different maintenance, but several can outlast plastics with proper care. Moreover, for many applications, particularly indoors where occupants can be directly exposed to off-gassing plasticizers, substitution of vinyl is clearly prudent for maintaining the health and well-being of occupants. Here are some potential examples of possible material alternatives:

- Flooring made from cork, natural linoleum, tile, finished concrete, or earth
- Stucco, lime plaster, reclaimed wood, fiber-cement, and FSC-certified wood siding
- Natural wall coverings instead of vinyl wallpaper
- Windows framed with fiberglass, FSC-certified wood, or possibly wood-based composites utilizing formaldehyde-free binders
- Glass shower doors instead of vinyl curtains

Tile

Tile production is a process that dates back to ancient Babylonian and Egyptian times. Tile typically starts out in the earth, where the raw materials are quarried and refined. Once the raw materials are quarried, prepared, and properly mixed, the tiles can be formed. Tiles are primarily made from fired clay (porcelain and other ceramics), glass, stone, or cement; they provide a useful option for flooring, countertops, and wall applications whose principal environmental requirement is durability. Tile is very durable, even in high-traffic areas, eliminating the waste and expense of replacements.

Tile production, however, is energy-intensive, although tile made from recycled glass requires less energy than tile made from virgin materials. Among tile's positive attributes is that it does not burn, does not retain liquids, and does not absorb fumes, odors, or smoke, and, when installed with low- or zero-VOC mortar, can contribute to a building's good indoor air quality (IAQ). But such performance can only be achieved if the tile has the appropriate surface hardness for the location. Tile hardness is measured on the Porcelain Enamel Institute (PEI) scale of 0 to 5, with 0 being the least hard, indicating that a tile should not be used as flooring, and 5, signifying a surface designed for very heavy foot traffic and abrasion. Floor tiles can easily last as long as the building they are installed in if properly maintained.

The environmental impacts of mining, producing, and delivering a unit of tile require important considerations, although ceramic tile production today has a lower environmental impact compared with other materials, thanks to technological and production innovations by the ceramics industry. Most of the tiles used in the United States today are imported (roughly 75%). The remaining amount represents about 650 million square feet of ceramic tile produced by U.S. factories each year, together with the billions of square feet manufactured globally. This requires mining millions of tons of clay and other minerals and substantial energy to fire material into hardened tile.

Once the raw materials are processed, a number of steps are put in motion to obtain the finished product. These steps include batching, mixing and grinding, spray-drying, forming, drying, glazing, and firing. In modern facilities, many of these steps are now accomplished using automated equipment. Stone, while requiring relatively little energy to process, nevertheless requires significant energy to quarry and ship. Selecting tile produced regionally may dramatically reduce the energy use and pollution of transport and thus facilitate achieving a LEED credit.

In the United States, more than 95% of the tile industry's product consists of glazed or unglazed floor tile and wall tile, including quarry tile and ceramic mosaic tile (Figure 6.3). Due to the industry's focus on decorative tiles, it

FIGURE 6.3   Interior showing installation of Crystal Micro double-loading polished tile. *(Source: Foshan Yeshengyuan Ceramics Co. Ltd.)*

has become completely dependent on the economic health of the construction and refurbishing industries. The only real difference between the production process for ceramic glazed tile and ordinary ceramic tile is that the glazed tile includes a step known as glazing. There are many ways to glaze ceramic tile; basically, it requires a liquid made from colored dyes and a glass derivative known as flirt that is applied to the tile, either using a high-pressure spray or by direct pouring. This gives a glazed look to the ceramic tile. Tile in low-traffic areas, particularly roofing, may use lower-impact water-based glazes. Glazed tiles have the advantage of being practically stain-proof, even though they can be more slippery. Unglazed tiles, on the other hand, are generally more slip-resistant but may require a sealant.

In addition, the integral color and generally greater thickness of unglazed tiles tends to make them more durable than glazed tiles. Factory-sealed tiles can help minimize or eliminate a source of indoor VOC emissions. Glass floor tile can also offer a nonskid surface appropriate for Americans with Disabilities Act (ADA) compliance. When installing stone tile, especially for countertop applications, a nontoxic sealer should be used for the grout and tile surface. Also, it is important that the final product meet certain specifications regarding physical and chemical properties. These properties are determined by standard tests established by the American Society of Testing and Materials (ASTM). ASTM tests measure properties such as abrasion resistance, chemical resistance, mechanical strength, dimensional stability, water absorption, frost resistance, and linear coefficient of thermal expansion.

6.2.4 Earthen Building Materials

Earthen building materials were used from Neolithic times—even before the invention of writing. But the techniques and methods for earth construction are numerous and vary with culture, climate, and resources. The primary types of earthen building materials include adobe bricks made from clay, sand, and straw; rammed earth compressed with fibers for stabilization; cobs made of clay, sand, and straw that are stacked and shaped while wet; and wattle-and-daub and earth plasters and finishes. Provided they are obtained locally, earthen building materials can reduce or eliminate many of the environmental problems posed by conventional building materials since they are plentiful, nontoxic, biodegradable, and reusable. Well-built earthen buildings are known to be durable and long-lasting and require little maintenance.

For thousands of years, people throughout the world have built comfortable homes and communities with earthen materials that provide excellent shelter. Though the domestic popularity of earthen materials waned during the twentieth century, a revival has emerged since the 1970s. By contrast, modern "stick-frame" construction, which requires specialized skills and tools, has been standard practice in the United States only since the end of World War II and remains today uncommon in many parts of the world. Main considerations in regard to earthen construction include

- Earthen construction is generally labor-intensive, although minimal skill is required.
- Earthen walls are thick and may take up a high percentage of floor area on a small site, making its use inappropriate.
- Multistory and cob structures require post-and-beam designs.

It may be more difficult to obtain necessary permits in certain jurisdictions, although code recognition and structural testing are available in most states.

Unit production costs differ in relation to local conditions, including availability of soil and its suitability for stabilization. If the work is done primarily by building professionals, the square-foot cost of earthen construction is comparable to conventional building methods. Advantages and benefits of earthen materials include:

- Abundance of the raw material—earth
- Durable and require low maintenance
- Eco-friendly with minimal environmental impact, provided materials come from local sources
- High thermal insulating properties; thermal mass helps keep indoor temperatures stable, particularly in mild to warm climates
- High sound insulation
- No waste generated during construction
- Biodegradable or reusable
- Construction is inexpensive and simple, with high workability and flexibility, requiring few special skills or tools
- Pleasing aesthetics when well designed
- Highly resistant to fire
- Not susceptible to insects or rodents
- Inert, containing no toxic substances; therefore, require no toxic treatments and do not off-gas hazardous fumes and so are good for chemically sensitive individuals

Earthen flooring, also called adobe flooring, varies in its construction, but is generally durable, inexpensive, ecologically sustainable, and a uniquely aesthetic complement to a home or office. Since "dirt" is plentiful and indigenous, earthen flooring can save money and virtually eliminates the waste, pollution, and energy necessary to manufacture a floor. The use of earth floors in the United States is still most often confined to outbuildings and sheds, but, if properly installed, they can also be used in interior spaces (Figure 6.4). For interior use, earth floors must be properly insulated, moisture-sealed, and protected from capillary action of water by sealing with a watertight membrane underlayment. Often, an earthen floor may be constructed of two or three layers. A typical earthen floor might include 70% sand and 30% clay, with lots of chopped straw for much-needed tensile strength.

FIGURE 6.4 Method for installing an earth floor. *(Source: Brian "Ziggy" Liloia, "Building My Cob House" blog.).*

Prior to proceeding with construction, the removal of any vegetation under the floor area is needed, followed by ramming of the area. The ground must be dry before installation of the floor. After the surface is moisture-proofed, a foundation of stone, gravel, or sand is installed, 8 to 10 inches (20−25 centimeters) deep. An insulating layer such as a straw-clay mixture is then installed. The key to a good earthen floor is the proper mixture of dirt, clay, and straw. Stabilizers such as starch paste, casein, glue, or Portland cement are sometimes added to obtain a harder floor. Earthen floors are first troweled to a smooth finish and then usually sealed with an oxidizing oil such as linseed or hemp, which hardens them. Sweep or vacuum any loose debris and dust; light mopping or sponging is also possible. Time should be given for the moisture to dry before applying the oil.

Considerations and attributes of earthen floor installation include the following:

- Earthen material is generally inexpensive when found locally.
- Construction waste is eliminated; any excess earth can be reincorporated into the landscape.
- Earthen materials are easy to process and require little or no transport, and therefore produce minimal to zero pollution. Even when produced by a machine, a finished earthen slab is estimated to have 90% lower embodied energy than finished concrete.

- Earthen floors are durable with proper maintenance and repairable, and when properly sealed they can be swept or moist-mopped; a stabilized earthen flooring is not dusty.
- Earthen floors are labor-intensive to install. This is not a problem in developing countries where labor is very cheap.
- In high-traffic areas such as entries or workspaces, flagstones or other protective materials may be required.
- Earthen floors are more durable than vinyl because they are repairable, but more vulnerable to scratching and gouging than hard tile or cement.
- Few U.S. contractors are experienced in installing earthen flooring.

6.2.5 Windows

Windows are vital elements in construction because they provide ventilation, natural light, views, and a connection to the outside world in addition to significantly improving the health, comfort, and productivity of a building's occupants. Drafty, inefficient, poorly insulated, or simply poorly chosen windows can present a major source of unwanted heat loss, discomfort, and condensation, and thus compromise the energy efficiency of a building's envelope. Window manufacture, whether of aluminum, plastic, steel, or wood, will require energy and will likely generate air pollution. Energy efficiency is one of the main considerations in reducing the environmental impacts of a window, followed by waste generated in manufacturing and general durability. However, these negative aspects have largely been addressed by modern technology. Figure 6.5 shows the various components of a window. According to Gregg D. Ander, FAIA:

In recent years, windows have undergone a technological revolution. High-performance, energy-efficient window and glazing systems are now available that can dramatically cut energy consumption and pollution sources: they have lower heat loss, less air leakage, and

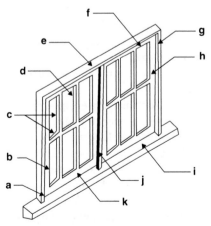

Window Elements

a. Jamb
b. Stile
c. Muntin
d. Pane
e. Head
f . Top rail
g. Jamb
h. Stile
i . Sill
j . Mullion
k. Bottom rail

FIGURE 6.5 Window showing individual elements.

warmer window surfaces that improve comfort and minimize condensation. These high-performance windows feature double or triple glazing, specialized transparent coatings, insulating gas sandwiched between panes, and improved frames. All of these features reduce heat transfer, thereby cutting the energy lost through windows.

Windows come with a variety of glazing options. Each option offers a different thermal resistance or R-value. R-value is a thermal resistance measure that is used in the building industry. A high R-value indicates that the window has a greater resistance to heat flow and a higher insulating value than a window with a low R-value. It should be noted that R-values are approximate and vary with temperature, type of coating, type of glass, and distance between glazings. R-value is the inverse of U-factor ($R = 1/U$) and is expressed in units of hr.sq. ft°F/Btu. U-value gives you the rate of heat flow due to conduction, convection, and radiation through a window resulting from a temperature difference between interior and exterior. The higher the U-factor, the more heat is transferred (lost) through the window. Here are typical examples of R-values from lowest to greatest resistance:

- Single glazing and acrylic single glazing are similar: $R = 1.0$.
- Single glazing with a storm window and double glazing are similar: $R = 2.0$.
- Double glazing with a low-E (low-emissivity) coating and triple glazing are similar: $R = 3.0$.
- Triple glazing with a low-E coating: $R = 4.0$.

It is interesting to note that for a conventional insulated stud wall $R = 14.0$.

According to Ecology Action of Santa Cruz (2004), residential window frames are typically made from aluminum, wood, vinyl, or fiberglass, or combinations of wood and aluminum or vinyl (i.e., "clad"). Older, single-pane windows rarely perform as well as new windows and should preferably be reused only in unheated structures such as greenhouses or barns. Each material has a different cost, insulating ability, and durability, as shown here:

- Wood is a natural material that requires continuous maintenance (stain or paint) for durability. Likewise, the wood source should be certified by an accredited organization such as the FSC.
- Fiberglass is energy-intensive to manufacture, but is strong and durable and has excellent insulation value.
- Aluminum and steel are poor insulators and very energy-intensive to manufacture. Also, over time aluminum will oxidize, leaving a dull pitted appearance. When using metal-framed windows, look for recycled content and frames insulated with "thermal breaks" to limit the loss of heat to the exterior.
- Vinyl is a product of the plastics industry, offers good insulation, but is highly toxic in its manufacture and if burned. Vinyl windows usually cannot be painted, but do offer lifetime free maintenance. High-efficiency windows typically utilize dual or triple panes with low-E coatings and gas fill (typically argon) between panes to help control heat gain and loss. Factory-applied

low-E coatings on internal glass surfaces are more durable and effective than films. High-quality, efficient windows are widely available from local retailers. To make an informed choice, consider only windows that have National Fenestration Rating Council (NFRC) ratings. The EPA ENERGY STAR® label for windows can be a useful summary of these factors.

Factors that can impact the full frame R-value of a window are

- Type of glazing material used (e.g., glass, transparent coatings, suspended film, treated glass, single, double, or triple glazing).
- Number of air chambers created by multiple layers of suspended film or glass panes.
- Type of gas (e.g., argon), if any, used to fill the air space(s).
- Air and water "tightness" of the window, which can significantly affect a window's performance.
- Thermal resistance and attributes of the frame and spacer materials, which affect a window's performance.

The National Fenestration Rating Council (NFRC) is a nonprofit, public/private organization created by the window, door, and skylight industry, and is composed of manufacturers, suppliers, researchers, architects and designers, code officials, utilities, and government agencies. The NFRC has developed a window energy rating system based on whole-product performance and provides performance ratings in five categories:

- The U-factor is a measure of a window's ability to keep heat inside or outside a building. U-factor values generally range from 0.25 to 1.25 and are measured in $Btu/h./ft.^2/°F$. The lower the U-factor, the better the window insulates. Seek values of 0.4 or lower.
- The solar heat gain coefficient (SHGC) summarizes a window's capability to block heat caused by sunlight. SHGC is measured on a scale of 0 to 1; values typically range from 0.25 to 0.80. The lower the SHGC, the less solar heat the window transmits. SHGC values of less than 0.4 are preferable.
- Visible light transmittance (VLT) measures the amount of light the window lets through. Desired VLT varies with taste and application. VLT is measured on a scale of 0 to 1; values generally range from 0.20 to 0.80. The higher the VLT, the more light you can see.
- Condensation resistance measures how well the window resists water build-up. It is scored on a scale from 0 to 100. The higher the condensation resistance, the better. Condensation can contribute to mold growth, although new, high-quality windows (with a low U-factor) are generally better equipped to resist condensation than older windows.
- Air leakage (AL) measures the rate at which air passes through joints in the window. AL is measured in cubic feet of air passing through one square foot of window area per minute. The lower the AL value, the less air leakage. Most industry standards and building codes require an AL of 0.3 $cf·m/ft^2$.

6.2.6 Miscellaneous Building Elements

Gypsum Wallboard (Drywall)

Gypsum board is the most common indoor building material in the United States. In the United States and Canada it is manufactured to comply with ASTM Specification C 1396, which was designed to replace several existing ASTM specifications, leaving one reference standard for all gypsum board products. This standard is to be applied whether the core consists of natural ore or synthetic gypsum.

Gypsum wallboard, also known as drywall or plasterboard, is a plaster-based wall finish that is available in a variety of standard sizes; 4 feet wide by 8 feet high is the most common. Thicknesses vary in 1/8-inch increments from 1/4 to 3/4 inch. Gypsum wallboard, which is also known by its proprietary names drywall and Sheetrock®, is ubiquitous in construction. It is a benign substance (basically paper-covered calcium sulfate), but it has significant environmental impacts because it is used on a vast scale; domestic construction uses an estimated 30 billion square feet per year. Advantages of gypsum board include its low cost, ease of installation and finishing, fire resistance, nontoxicity, sound attenuation, and availability. Disadvantages include difficulty in curved-surface application and low durability when subject to damage from impact or abrasion.

Gypsum board manufacturers are increasingly relying on "synthetic" gypsum as an effective alternative to natural gypsum. It is estimated that roughly 45% of the gypsum used by U.S. manufacturers in 2010 was of the synthetic variety. Synthetic gypsum and natural gypsum have similar general chemical compositions ($CaSO_4.2\ H_2O$). The vast majority of the synthetic gypsum used by the industry is a byproduct of the process used to remove pollutants from the exhaust created by the burning of fossil fuels for power generation. If synthetic gypsum were not used to manufacture gypsum panel products, it would end up in landfills.

Though synthetic gypsum board use is growing in popularity, and reclaimed gypsum board can easily be recycled into new gypsum panels that conform to the same quality standards as natural and synthetic gypsum, doing this may not be practical because gypsum is an inexpensive material that can require significant labor to separate and prepare for recycling. Gypsum-board face paper is nearly 100% recycled from newsprint, cardboard, and other post-consumer waste streams, but most recycled gypsum in wallboard products is post-industrial, made from gypsum board manufacture.

Ecology Action, a nonprofit environmental consultancy, states that the main environmental impacts of gypsum include habitat disruption from mining, energy use and associated emissions in processing and shipment, and solid waste from disposal. Some of these impacts can be significantly reduced by the use of "synthetic" or recycled gypsum board. Synthetic gypsum, which is now used in about 30% of drywall, is a byproduct of coal-fired power plants. It is sometimes confused with fly ash, another coal combustion product with which it has very little in common. In excess of 80% of coal fly ash sold in the United States is used in gypsum board.

New technologies have helped in the development of several new gypsum board products that are coming on the market and that are more environmentally friendly and superior in many ways to the traditional gypsum board. One such example is the new eco-friendly EcoRock drywall, which has significantly changed and improved the drywall product, from its basic material elements to its production processing methods. EcoRock is a fully recyclable and highly attractive alternative. It is manufactured from 80% post-industrial recycling, exploits material from steel and cement plant waste, and can be safely discarded in landfills. It is naturally cured and dried, which means that 80% less energy is required than in traditional methods used in the manufacturing process. Moreover, it contains no gypsum, thus eliminating the need for high intensive energy consumption during production, and improves air quality by eliminating airborne mercury. EcoRock creates 60% less dust, is resistant to termites, and is 50% more resistant to mildew and mold.

Siding

Siding is the external covering or cladding applied to the outermost surface of an exterior wall, with the main function of providing protection from adverse effects of extreme weather, moisture and excess water, and the heat and ultraviolet radiation of the sun. There are many types of siding materials that one can choose from, such as plastic (vinyl), fiber-cement, wood, composite, metal, and masonry, to name but a few. Selecting siding that is reclaimed, recyclable, or incorporating recycled material will reduce waste and pollution. However, the environmental impact of a siding product will vary considerably according to the material it is made from.

Siding may be formed of horizontal boards or vertical boards (also known as weatherboarding in some countries), shingles, or sheet materials. In all these cases, avoiding wind and rain infiltration through the joints is a challenge that is met by overlapping, by covering or sealing the joint, or by creating an interlocking joint (e.g., a tongue-and-groove or rabbet). Creating rigid joints between the siding elements is not practical because materials will contract or expand according to the changing temperature and humidity. Moreover, siding may be attached directly to the building structure (studs in the case of wood construction) or to an intermediate layer of horizontal planks called sheathing. There are many types of siding, as described in the following paragraphs.

Vinyl siding is made from a PVC plastic and is very widely used. Unlike wood or cedar, it won't rot or flake. Vinyl siding has grown in popularity because it requires little maintenance and is generally less costly to purchase and install than most other siding materials. The main drawbacks of vinyl are that it cracks, fades, and grows dingy over time. It is also controversial because of environmental concerns.

Earth or lime plasters may last a long time and don't require much maintenance. Cement or lime is commonly added to improve hardening and durability, but the relatively low (or zero) overall cement content of natural plasters means that the material requires relatively small amounts of pollution and energy use to prepare and install. Deep eaves or overhangs are often needed to protect the siding from extended moisture exposure and are critical to extending the natural plaster's useful life.

Fiber-cement siding has proven to be extremely durable, and many products are backed by 50-year or lifetime warranties. It is made from cement, sand, and cellulose fiber. In addition to its improved eco-friendliness, fiber-cement siding also comes with a lower price tag than other materials. Moreover, it is fire-resistant and pest-resistant and emits no pollutants in use. However, it does possess a high embodied energy due to its cement content.

Quality cement stucco is another alternative material that can be extremely durable, which helps minimize long-term waste, but it is also energy-intensive to manufacture. Cement substitutes such as fly ash or rice-hull ash can mitigate the environmental cost of stuccos.

Metal siding comes in a variety of metals, styles, and colors. It is very durable and recyclable, and typically contains significant postconsumer recycled content. It is generally energy-intensive to manufacture, although recycled steel and aluminum require far less energy than virgin ore. Metal siding is most often associated with modern, industrial, and retro buildings. Utilitarian buildings often use corrugated galvanized steel sheet siding or cladding, and corrugated aluminum cladding is common where a more durable finish is required.

Composite siding such as hardboard is made of newspaper or wood fiber mixed with recycled plastic or binding agents. It is highly durable and generally resists moisture and decay. It also often has significant recycled content and is not prone to warping or cracking like wood. Composites generally do not require frequent repainting, and some need not be painted at all, saving waste and valuable resources.

Wood siding is one of the oldest types of siding and is popular in old and historic homes. Its main disadvantage is that it requires more maintenance (polishing and painting) than many of the other siding options. If it is not well maintained, wood can easily be the least durable option, generating significant waste. Wood siding is also more vulnerable to warping, splitting, and damage by insects and termites. Among its positive attributes is that it is renewable and requires relatively little energy to harvest and process. Wood siding can also be made of unpainted weather-resistant woods such as redwood.

As with most green products, selecting the most appropriate siding is often a matter of weighing trade-offs in longevity, biodegradability, insulation, maintenance, and, sometimes, cost. Final selection will also depend largely on your objectives and require you to prioritize the attributes that best meet them. Such considerations should include:

- Selecting the most durable siding product that is appropriate. Siding failures that allow water into the wall cavity can lead to expensive repairs, the waste of damaged components, and the environmental costs of replacement materials. Fire resistance is a feature that helps reduce the financial and environmental impact of rebuilding, particularly in high-risk areas.

- For existing buildings, consideration should be given to refinishing existing siding to minimize waste, pollution, and energy use.
- Preference should be given to selecting materials that are biodegradable, have recycled content, and/or are recyclable.
- Reclaimed or remilled wood siding should be used to reduce demand for virgin wood and waste. Painted wood should be tested for lead contamination prior to use.
- New wood siding should display an FSC-certified label.
- Although vinyl is durable, it is not considered a green building material. Attributed disadvantages include pollution generated in manufacturing, air emissions, human health hazards of manufacturing and installation, the release of dioxin and other toxic persistent organic pollutants in the event of fire, and the difficulty in recycling.

6.2.7 Roofing

A roof system's primary function is to protect against and manage the weather elements, particularly precipitation, thereby protecting the interior and structural components of a building from deterioration. There are two basic types of roof construction, sloped or pitched and flat. Sloped roofs are generally covered with individual pieces of overlapped shingling material to prevent water penetration. Flat roofs are basically watertight membranes that should have just enough of a slope to allow water to run off. Flat roofs are more popular in hot, arid climates, particularly in developing countries where they are used in the evenings for sleeping. In the United States and Europe, however, flat or low-slope roofs are typically selected when the roof is expected to accommodate rooftop equipment.

The most critical characteristics of roofing materials are moisture/water resistance, dependability, and durability. The development of new materials and processes is needed to minimize creation of new health or environmental safety problems. The transporting and processing of the materials are coming under increasingly strict regulation to protect the health of workers involved in production and distribution. Roofing material extraction, manufacture, transport, and disposal pollute air and water, deplete resources, and damage natural habitats. Roofing materials comprise an estimated 12 to 15% of construction and demolition waste.

For a roof to be environmentally sustainable it must be durable and long-lasting and ideally also contain recycled or low-impact materials. Roofs that are environmentally friendly can provide several advantages such as aesthetically pleasing design, reduced life-cycle costs, and environmental benefits such as reduced landfill waste, energy use, and impacts from harvesting or mining of virgin materials. It takes roughly the same materials, energy, and labor to manufacture and install a 50-year-warranted roof as to install a 30-year roof, yet the first option is "greener" because disposal and the roof's life-cycle is extended, thereby providing a better investment. Moreover, a properly installed

50 + -rated roof can reduce roofing waste by 80 to 90% over its lifetime, relative to a roof with a warranty of only 20 years.

Normally, mild climates are better suited for passive temperature controls that reduce winter heating and reduce or eliminate the need for mechanical cooling. The need for air conditioning is generally less in mild climates because operable windows and skylights are often employed that can easily provide ventilation and cooling, particularly in smaller buildings. But even larger commercial buildings can be cooled effectively in mild climates without the use of air conditioning provided care is taken in the initial design and the design of the roofing structure minimizes heat gain. And although cool roofing does not renew resources, it is often a highly cost-effective way to conserve them. Likewise, electricity from solar photovoltaic panels reduces demand for fossil fuels and is therefore environmentally friendly.

Choosing Roofing Materials

Choosing the right roofing materials requires taking into consideration numerous factors, including climate and weather conditions in the project area, the roof's life-span expectancy, budget, aesthetic preferences, and various sustainability factors. Some primary considerations that should impact the type of roof to be chosen include:

- Ability to resist heat flow into the interior, whether through insulation, radiant barriers (highly reflective material that inhibits heat transfer by thermal radiation), or both.
- Capacity to reflect sunlight and re-emit surface heat. Cool roofs can reduce cooling loads and urban heat island effects while extending roof life.
- Ability to reduce ambient roof air temperatures through evaporation and shading, as in the case of vegetated green roofs.
- Recyclability and/or capability of being reusable at the end of its useful life to minimize waste, pollution, and resource use. Roofs with high post-consumer recycled content (up to 30%) are preferable.
- Nonhalogenated fire-retardant roofing membranes (i.e., materials that do not contain bromine or chlorine) that meet fire code requirements. Burning of polyvinyl chloride (PVC) and thermoplastic polyolefin or olefin (TPO) produces strong acids and persistent toxic organic pollution, such as dioxin.

For local fire code compliance, the roof may require protective ballast such as concrete tile. Existing PVC and thermoplastic polyolefin (TPO) roofing membranes, as well as underlying polystyrene insulation, are increasing being recycled as federal construction-specification requirements generate increased demand in the industry. The following are some of the roofing options for residential and commercial applications:

- Clay or cement tiles are very durable and made from abundant materials, but they are heavy and expensive. Ensure that the structure can take additional weight.

- Recycled plastic, rubber, or wood composite shingles are generally durable, lightweight, and sometimes recyclable, but they are not biodegradable.
- Composition shingles are very popular and typically have a lifetime of 15 to 30 years, although some manufacturers offer up to 50-year warranties. The composition shingle has a fiberglass mat core that gives it flexibility and provides some fire resistance. The exterior of the composition shingle has a weather-resistant asphalt coating embedded with crushed rock. Composition shingle roofs can be recycled, but are susceptible to algae growth unless an anti-algae coating is applied.
- Fiber-cement is durable and fire- and insect-proof, but is heavy and not renewable or biodegradable; however, it may be used as inert fill at demolition.
- Metal (e.g., aluminum, steel) is a durable, fire- and insect-proof, and recyclable material. It typically contains recycled content, but manufacture is generally energy-intensive and causes pollution and habitat destruction.
- Built-up roofs are said to have a track record exceeding 150 years, but their durability depends largely on the structure, installation, flashing, and membrane chosen. Also, most membranes are not made from renewable resources, although some may contain recycled content. Built-up roofs are extremely reliable, but their market share has declined significantly over recent decades mainly due to the increasing cost of tar. Of note, high-VOC products emit air pollution during installation.
- Vegetated green roofs are most commonly installed on low-pitched roofs (at least one inch of rise for every foot of run to facilitate drainage). They help reduce the negative aspects of conventional roofing, while adding green space to the property.
- Wood shakes have a rustic appearance and are biodegradable, but they are also flammable and not very durable. Wood shakes are not typically considered to be a "green" option for areas that are fire-prone.

New technologies are emerging that encourage the promotion of green building and green roofs, because green roofs offer many economic, social, and environmental benefits. Today we are witnessing increasing efforts to create usable space on existing rooftops and/or new roofs to allow additional living space. Green or "living" roofing, which involves the use of vegetation as the weathering surface, has proved very successful because it helps reduce extremes in rooftop temperature that shorten the life of a roof (leading to increased construction and demolition waste), conserves energy, and extends the useful life of the roof. An important factor in creating these spaces is the need to use lightweight and recycled materials and to help with managing stormwater runoff because traditional drainage systems using pipe and stone are not plausible. Green roof systems are a natural and cost-effective way of providing additional clean air through the transference of CO_2 and oxygen between the vegetation and the atmosphere, thereby helping to control urban air pollution.

The main reasons that green roofing has proven to be effective is that the large surface area of soil and plants helps to reradiate heat; it also provides shade and insulation for the waterproof roof membrane, while the plants' transpiration provides cooling. The net result of this is a 25- to 80-degree decrease in peak roof temperature and a much reduced cooling energy demand (up to 75%). Cooler roof temperatures will also reduce the urban heat island effect and thus help reduce the cooling load for surrounding buildings.

In addition to their environmental benefits, green roofs provide aesthetic and cost benefits (as well as benefits to local property values). The soil and vegetation in many common designs can retain up to 75% of a one-inch rainfall and will filter the remainder. This onsite stormwater management helps reduce demand on stormwater infrastructure, saving resources and money for the entire community. To help address this issue, New York City property owners are assessed a tax based on the volume of stormwater runoff, whereas they are offered a tax reduction if the building has a green roof.

Green roofs can often facilitate urban wildlife microhabitat. Although not a replacement for wild land, a vegetated roof can accommodate birds, beneficial insects, and native plants far better than tar and gravel. Contemporary green roof designs generally contain a mixture of hard and soft landscaping (Figure 6.6). It is very important, therefore, that the selected drainage/retention

FIGURE 6.6 Example of a 70,000-square-foot vegetated roof on the LDS Assembly Hall building, Salt Lake City. *(Source: American Hydrotech, Inc.)*

layer is capable of supporting any type of landscape—from roadways and paths to soil and trees—so as to permit excess water to drain unobstructed underneath. (Figure 6.7).

Green roof systems generally contain certain essential layers and components to be viable. The main layers generally consist of vegetation, soil, filter and waterproofing membrane, root barrier, drainage and irrigation system, roof membrane, and sometimes an insulation layer. American Wick Drain (AWD), a leader in roof garden and prefabricated drainage systems, notes that the main components, materials, and locations to take into consideration when contemplating the installation of a green or vegetated roof are:

Structural support. Roof and roof garden systems are required to have an underlying structural system in place to support additional weights resulting from use of normal building materials such as concrete, wood, and so forth. The structural engineer needs to consider the load of the roof-garden system in the initial design phase.

Roofing membrane. The design engineer has a number of roofing membrane options available at his/her disposal. The final membrane choice may be decided by several factors such as loads imposed by the rooftop garden, by available membrane protection elements in the rooftop-garden system, by root penetration properties of the membrane, and by membrane drainage and aeration requirements.

Membrane protection. The roofing membrane may require protection from installation damage, long-term water exposure, UV exposure, drainage-medium loads, chemical properties, or growing-medium loads.

Root barrier. A root barrier filter fabric may be required between the growing medium and the lower components to allow excess water in the growing medium to drain while preventing small particles in the growing medium from moving into the drain core. It may also be needed to eliminate or mitigate potential root penetration into the roofing membrane, drainage medium, or water-storage medium. The optimum location is above the membrane, drainage medium, and water-storage medium so that all three components are protected. Alternate locations may require multiple roof-guard elements to avoid long-term root penetration. While some systems have the root barrier placed under the drainage layer, Amergreen, Inc., for example, which specializes in green roofs, states that it designs the root barrier "at the most effective location for system performance—above the drainage layer and in direct contact with the soil medium." Some roofs do not incorporate a specific insulation layer unless a higher insulation is required. Insulation (usually based on a rigid synthetic board foam) may be installed above the structural support, depending on the thermal design of the structure. Insulation may also be installed above or below the roofing membrane, whichever proves more effective.

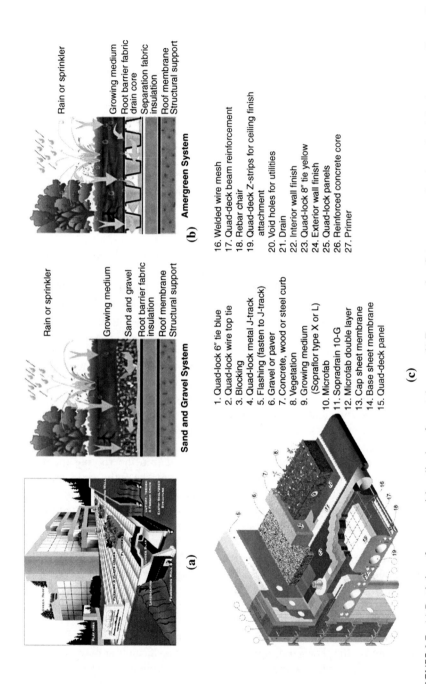

Sand and Gravel System

Rain or sprinkler

Growing medium
Sand and gravel
Root barrier fabric
insulation
Roof membrane
Structural support

(b) Amergreen System

Rain or sprinkler

Growing medium
Root barrier fabric
drain core
Separation fabric
insulation
Roof membrane
Structural support

1. Quad-lock 6″ tie blue
2. Quad-lock wire top tie
3. Blocking
4. Quad-lock metal J-track
5. Flashing (fasten to J-track)
6. Gravel or paver
7. Concrete, wood or steel curb
8. Vegetation
9. Growing medium
   (Sopraflor type X or L)
10. Microfab
11. Sopradrain 10-G
12. Microfab double layer
13. Cap sheet membrane
14. Base sheet membrane
15. Quad-deck panel

16. Welded wire mesh
17. Quad-deck beam reinforcement
18. Rebar chair
19. Quad-deck Z-strips for ceiling finish
    attachment
20. Void holes for utilities
21. Drain
22. Interior wall finish
23. Quad-lock 8″ tie yellow
24. Exterior wall finish
25. Quad-lock panels
26. Reinforced concrete core
27. Primer

(a)

(c)

FIGURE 6.7 (a) Rendering of a commercial application of green elements. (b) Two types of roofing system: sand and gravel and the Amergreen, Inc. roofing system. (*Source: American Wick Drain Corporation.*) (c) Drawing of a Quad-lock roofing system detail (*Source: Quad-lock Insulating Concrete Forms.*)

Drainage medium. An appropriate drainage medium is required to ensure the proper range of water content in the growing medium. An excess of water can have an adverse effect and cause root rot. Insufficient water can result in poor vegetation growth. There are various drainage-medium options ranging from gravel to materials designed specifically for this purpose. Plastic materials usually combine the drainage function with water storage and aeration in addition to protecting the membrane from roots and potentially damaging materials in the growing medium.

Aeration. A vital element to promote optimal vegetation growth. The aeration medium usually serves as a drainage medium as well. The open channels incorporated in the prefabricated drain core are designed to provide necessary air to the plant roots.

Growing medium. There are a number of natural and manufactured materials that may be used as a growing medium. Soil is often mixed with other materials to reduce weight, to provide better structure for roots, and to provide essential nutrients, oxygen, and water.

Water storage. While the growing medium will store a certain quantity of water, additional water storage may be required to provide more efficient growth of vegetation. Most prefabricated plastic roof-garden products have water-storage capability. The plastic cones of the prefabricated core provide positive water storage reservoirs. Likewise, sand and gravel may hold a certain amount of water.

Vegetation. This provides the upper and most visible layer of the roof-garden system. There is a wide variety of typical landscaping and garden plants that may be employed for a rooftop garden. An experienced landscape architect should preferably be involved in plant selection to ensure that the vegetation chosen is appropriate both for the geographic region and for a rooftop-garden environment. Moreover, careful consideration should be given to the type of vegetation chosen as this will inevitably affect the selection of other components of the roof-garden system such as root protection, water and aeration needs, and drainage requirements—all of which are affected by plant selection.

Extensive and Intensive Green Roofs

There are basically two primary categories of green roofs—extensive and intensive—although a green roof is frequently designed to contain features of both, and is then referred to as a semi-extensive or semi-intensive green roof (sometimes spelled *greenroof*). The roof's function or the objective of the roof space usually determines the final design (e.g., whether it is intended as an ecological cover or whether it is intended for human recreation, vegetable gardening).

Extensive green roofs (sometimes called eco-roofs or low-profile roofs) typically contain a layer of soil medium that is relatively thin (two to six inches) and lightweight (10–50 pounds per square foot for the entire system when fully saturated with water). They are lightweight, relatively easy to install, durable, very low maintenance, and cost effective. Extensive green roofs are usually built when the main goal is for an ecological roof cover with limited human access. Ideal plants for this type of roof are low growing, horizontally spreading root ground covers, having a general maximum plant height of 16 to 24 inches. Alpine-type plants are typically successful because they have the necessary attributes for green roofs, including high drought, wind, frost, and heat tolerance. Appropriate plants include sedums and other succulents, flowering herbs, and certain types of grass.

Intensive green roofs (sometimes called high-profile) often look like traditional roof gardens because they are designed to accommodate trees and gardens. Soil can be as deep as needed to accommodate the desired tree or plant species, but deeper, denser soil dramatically increases dead load, requiring a stronger and more expensive structure, greater maintenance, and either terracing or a relatively flat roof. The engineered soil media for an intensive green roof usually contains about 45 to 50% organic material to 50 to 55% mineral material and, when fully saturated with water, weighs between about 80 to 120 pounds per square foot or more. The inclusion of architectural and decorative elements such as waterfalls, ponds, gazebos, and so forth, is possible, and these green roofs provide recreation spaces and encourage interaction between nature and a building's occupants. It should be noted that there are green-roof options available for almost any building type or location. Among the many benefits attributed to green roofs are

- They provide greater insulation and more moderate rooftop temperatures, which reduce cooling and heating requirements, saving energy and money. Research by the National Research Council of Canada found that extensive green roofs reduce the daily energy demand for air conditioning in the summer by over 75% (Liu, 2003).
- They are considered a best practice because they facilitate filtration and detention of stormwater, reducing pollution and the cost of new and expanded infrastructure as paved areas increase. Green roofs can reduce the amount of stormwater runoff volume to sewer systems by 50 to 90% and peak stormwater runoff flow by 75 to 90%, resulting in decreased stress on sewer systems at peak flow periods. By comparison, a typical roof will retain 10% of stormwater runoff.
- They naturally absorb dust and filter harmful particulates and airborne pollutants. They also have superior noise attenuation, especially for low-frequency sounds. An extensive green roof, for example, can reduce sound from outside by 40 decibels, while an intensive one can reduce sound by 46 to 50 decibels (Peck et al., 1999).
- They are an effective fire retardant. Green roofs generally have a much lower burning heat load (the heat generated when a substance burns) than do conventional roofs.

- They reduce ambient air temperatures, lowering urban heat island effects and helping to enhance the microclimate of surrounding areas, because one of vegetation's natural functions is to cool the air by the release of water through plants into the atmosphere.
- They extend the life of roof membranes, by decreasing exposure to large temperature fluctuations that can cause microtearing, and protect the roof from ultraviolet radiation, extreme temperatures, and mechanical damage. Plant species, soil depth, and root-resistant layers are carefully matched to ensure that the roof membrane is not damaged by the roots themselves.
- Lightweight extensive systems can be designed with dead loads comparable to standard low-slope roofing ballast. Structural reinforcement may not be necessary, and cost can be comparable to conventional high-quality roofing options.
- They can increase a building's marketability, resulting in higher rents and increased resale value. A recent study by UMass Boston estimates that green roofs can increase a property's value by an average of 15%. Green roofs can also facilitate employee recruiting and decrease employee and tenant turnover.
- They can transform rooftop eyesores into attractive assets.

6.2.8   Wood

There are thousands of species of wood, which can probably best be separated into two broad categories: hardwoods and softwoods.

Wood Types

The distinction between hardwood and softwood is botanical and does not refer to the strength or hardness of the wood. "Hardwood" is a term generally applied to trees that lose their leaves in winter, whereas "softwood" generally describes evergreens such as pine and redwood. It should be noted that hardwoods are not necessarily hard and softwoods are not necessarily soft (e.g., balsa is a hardwood and white cypress is very strong softwood). The different types of wood have a multitude of uses, and in many cases are interchangeable.

Pressure-treated lumber, such as CCA pressure-treated wood, has been popular over the decades partially due to its resistance to rotting, insects, and microbial agents. Existing CCA-treated lumber, however, poses a challenge because arsenic is acutely toxic and carcinogenic and has been shown to leach into surrounding soils; this has prompted it being largely phased out in a cooperative effort between manufacturers and the Environmental Protection Agency (effective December 31, 2003). CCA is a chemical wood preservative containing chromium, copper, and arsenic. It has been classified by the EPA as a restricted-use product—that is, for use only by certified pesticide applicators.

Reuse of CCA pressure-treated wood would help conserve forest resources and keep a potentially useful resource out of landfills. Still, permitting its reuse would allow CCA to continue to leach arsenic into soils. CCA-treated wood should not be composted or disposed of in green-waste or wood-waste bins. It is now mandated to be disposed of in a lined landfill or as class I hazardous waste. The burning of CCA-treated wood is highly toxic. Newer alternative wood treatments that are less toxic, such as copper azole (CA) and alkaline copper quaternary (ACQ), are more corrosive than CCA. To address this, and in order to minimize rust and prevent staining, manufacturer-recommended fasteners should be employed. Consider the following points when working with wood products:

- Reuse wood in good condition.
- Repair and/or refinish existing decks, railings, or fencing with nontoxic materials.
- Build with durable materials such as plastic lumber. The composition of plastic lumber varies widely, from 100% post-consumer recycled content to 100% virgin plastic resin.
- The majority of plastic lumber products currently on the market are made from polyethylene, although some manufacturers are also employing polystyrene (PS) and PVC. In addition, there are some plastic lumber products that rely on a commingled mix of different types of plastics, collected mainly from municipal recycling programs.
- If the structural elements will be in contact with soil and water, consider the following:
  - Use heartwood (the dense inner part of a tree trunk), yielding the hardest timber from decay-resistant species such as redwood or cedar that has been FSC-certified as harvested from a responsibly managed forest.
  - When the use of pressure-treated lumber is required, the two water-resistant preservatives currently employed are CA and ACQ, which are significantly less toxic than CCA.
  - Avoid remaining stocks of CCA.
- For fencing, consider eco-friendly alternatives such as a living fence of bushes, shrubs, or live bamboo in urban settings or fencing made of a rapidly renewable material such as cut bamboo.

As wood is a renewable material, it requires less energy than the majority of other materials to process into finished products, as opposed to the significant negative environmental impacts caused by the logging, manufacture, transport, and disposal of wood products. Standard logging practices are known to cause erosion, pollute streams and waterways with sediments, damage sensitive ecosystems, reduce biodiversity, and lead to a loss of soil carbon. These impacts can be reduced by the minimization of wood use through the substitution of suitable alternatives (e.g., reusing salvaged wood, selecting wood from responsibly managed forests, controlling waste, and minimizing redundant components).

Where salvaged or reclaimed wood is unavailable or not applicable (e.g., for structural applications), specify products that are certified by an approved and accredited organization such as the FSC.

Engineered lumber (also known as composite wood or man-made wood) consists of a range of derivative wood products that are manufactured by pressing or laminating together the strands, particles, fibers, or veneers of wood with a binding agent to produce a range of different types of building products such as structural framing lumber and trim material. Engineered wood is normally straighter, more stable, and structurally consistent than dimensional lumber. Its superior strength and durability allow it to displace the use of large (and increasingly unavailable) mature timber. In joist and rafter applications, the reconstituted products have proven to be particularly useful because of their ability to span long distances with less sagging than similarly sized conventional lumber. Engineered lumber is also less susceptible to humidity-induced warping than equivalent solid woods, although the majority of particle-based and fiber-based boards require treatment with an appropriate sealant or paint to prevent possible water penetration.

There are numerous applications for engineered products. Generally, such products are engineered to meet precise application-specific design specifications and are tested to meet national or international standards. Employing engineered lumber instead of large-dimension rafters, joists, trusses, and posts can save money and reduce total wood use by up to 35%. Engineered lumber also allows for wider spacing of members, which in turn allows for increasing the insulated portion of walls. Other advantages include its ability to form large panels from fibers taken from small-diameter trees, small pieces of wood, and wood that has defects; these panels can be used in many engineered-wood products, especially particle-based and fiber-based boards. Engineered-wood products are used in a number of ways, usually in applications similar to those for solid wood products, but many builders prefer engineered products because they are economical and typically longer, stronger, straighter, more durable, and lighter than comparable solid lumber.

Engineered-wood products also have several disadvantages in comparison to dimensional lumber. For example, they require more primary energy for their manufacture than solid lumber; they are less fire resistant and have adhesives that can potentially release toxins into the environment; and they are prone to moisture damage. An expressed concern with some resins is the release of formaldehyde in the finished product, often seen with urea-formaldehyde-bonded products. Working with engineered wood products can therefore potentially expose workers to toxic constituents that could cause harm. The applications of engineered-wood products are varied and include being used as columns, beams, joists, girders, rafters, studs, and bracing. Although engineered wood is generally more expensive than dimensional lumber, the cost can be offset to some degree by labor savings and improved quality of the product.

Wood adhesives have been important in helping use timber resources efficiently. The main function of an adhesive is to bond wood components such as veneer, particles, strands, and fibers. Moreover, an adhesive must provide the necessary strength immediately following manufacture as well as after long-term use. Prior to the introduction of synthetic adhesives in the 1930s, the adhesives used for bonding wood were generally made from natural polymers found in plants and animals.

Today, natural adhesives continue to be used (but to a much lesser extent) in some nonstructural products; they do not provide the necessary strength and durability required for many of today's engineered wood products. To address the needs of contemporary engineered-wood products, polymer scientists have developed various synthetic adhesives that are designed to perform a variety of functions in product applications. These human-made polymers resemble natural resins in physical characteristics, but they can be tailored to meet specific woodworking requirements. The choice of an adhesive is determined by several factors; these include cost, structural performance, moisture resistance, fire performance, adhesive curing, and so forth. Generally, there are two primary categories of adhesives employed in engineered wood; the first category is for structural products, and the second is for interior nonstructural products. This first group of resins includes:

- Phenolic (also called phenol formaldehyde or PF), which has a yellow/ brown or dark reddish color and is available as liquid and powder or in film form. It is commonly used for exterior exposure products and to produce softwood plywood for severe service conditions.
- Resorcinol, which is purple in color, waterproof/boil-proof, and typically used for structural wood beams.
- Melamine-formaldehyde resin (MF), which is white in color, heat- and water-resistant, and preferred for exposed surfaces in more costly designs. Its use is limited to a few special applications such as marine plywood, where the need for a light-colored water-resistant adhesive justifies its cost.
- Methylene diphenyl diisocyanate (MDI), which is an aromatic diisocyanate that, although not benign, is the least hazardous of the commonly available isocyanates. It is generally expensive, generally waterproof, and does not contain formaldehyde.
- Polyurethane (PU) resin, which, like MDI, is also generally expensive, waterproof, and does not contain formaldehyde. It is used largely in the sphere of coatings and adhesives because of its high reactivity, high flexibility in formulation and application technologies, adhesion and mechanical properties, and resistance to adverse weather.

Typical examples of structural products include OSB, plywood, end-jointed lumber, glued laminated timber (glulam), I-joists, and structural composite lumber.

The second group of adhesives includes those formulated from materials of natural origin such as animal, vegetable, casein, and blood glues. They lack the temperature capabilities or environmental durability evidenced by structural adhesives. Nevertheless, they generally provide an instantaneous bond due to their pressure-sensitive or hot-melt characteristic. Nonstructural adhesives are generally used for interior applications, although many new products are seeing exterior applications as well. These include urea formaldehyde resin (UF), which is widely used for the manufacture of interior-grade plywood and also for the manufacture of particleboard. UF is considered to be the most widely used thermosetting resin for wood; although not waterproof, it is nevertheless popular because it is inexpensive. The low resistance to heat and moisture of natural adhesives makes their use appropriate for indoor, nonstructural wood products.

Sheathing is the structural covering of plywood or oriented-strand board (OSB) that is applied to studs and roof/floor joists to provide shear strength and serve as a base for finish flooring or a building's weatherproof exterior. Sheathing is considered to be the second most wood-intensive element that is used in wood-frame construction. Exterior gypsum sheathing comes in various sizes, including a 1/2-inch thick, 2-foot-wide product with a tongue and groove edge, and a 1/2-inch and 5/8-inch thick, 4-foot wide square-edge product. Five-eighths-inch exterior gypsum sheathing has a Type X core, for use in fire-rated assemblies. Engineered-wood sheathing materials do have some environmental trade-offs because the wood fibers are typically bound with formaldehyde-based resins. Interior-grade plywood typically contains UF, which is less chemically stable than the PF found in water-resistant exterior-grade plywood and OSB. This advantage makes exterior-grade plywood preferable for indoor applications as it emits less toxic and suspected carcinogenic compounds.

There are numerous alternatives to these wood-intensive conventional and engineered materials. For example, fiberboard products rated for structural applications (e.g., Homasote®, a 100%-recycled nailable structural board) are considered alternatives to plywood and OSB. Structural-grade fiber-cement composite siding combines sheathing and cladding, providing shear strength and protection from the elements while reducing labor costs for installation. It is promoted as being eco-friendly because it requires fewer trees, won't burn, won't rot, extends the life of a paint job, and usually is warranted for 50 years.

This product is relatively new but has become increasingly popular since its introduction, in part because it can be manufactured to have the realistic appearance of wood, stucco, or masonry. Its main drawback, however, is that it is heavy to lift and its installation requires specialized cutting tools. Water-resistant exterior-grade gypsum sheathing is one of the options that can be employed as an underlayment for various exterior siding materials such as wood, stucco, metal or vinyl siding, and masonry veneer to reduce wood requirements. The panel is manufactured with a wax-treated, water-resistant core faced with water-repellent paper on the face and back surfaces as well as on the long edges. Structural insulated-panel construction provides interior and exterior sheathing as well as insulation in precut, factory-made panels. And by designing for disassembly, sheathing materials can be readily reused or recycled. Also, designs that combine bracing with nonstructural sheathing can provide necessary strength while increasing insulation and reducing wood requirements.

Medium-density fiberboard (MDF) is an engineered composite wood product that is nonstructural and somewhat similar to particleboard. MDF is one of the most rapidly growing composite board products to enter the market in recent years. It typically consists of low-value wood byproducts such as sawdust combined with a synthetic resin, such as UF, or other suitable bonding system, joined together under heat and pressure. Additives are often introduced during the manufacturing process to impart additional characteristics. MDF panels are therefore manufactured with a variety of physical properties and dimensions that allow the end product to be designed with the characteristics and density needed. MDF is widely employed in the manufacture of furniture, kitchen cabinets, laminate flooring, paneling, door parts, shelving, millwork, and moldings. Generally, it can provide an excellent substitute for solid wood in many applications, except when the stiffness of solid wood is needed.

MDF can be dangerous to use if the correct safety precautions are not taken. It contains UF, which may be released from the material through cutting and sanding, and which may cause irritation to the eyes and lungs. It is necessary to have proper ventilation when using it, and facemasks are needed when cutting or sanding with machinery because the dust produced can be very dangerous. However, although MDF is highly toxic to manufacture, it does not emit VOCs in use. It will accept a wide variety of sealers, primers, and coatings to produce a hard, durable tool surface, but it is not suitable for high-temperature applications. MDF-type panels can also be made using waste wood fiber from demolition wood and waste paper.

The properties of MDF will likely vary from country to country, and region to region, based on where it is produced; thus the properties of MDF board produced in China will likely differ from MDF board produced in Romania or Indonesia. MDF also comes in several densities depending on the intended application. Its surface is generally flat, smooth, uniform, dense, and free of knots and grain patterns. The homogenous density profile of MDF allows intricate and precise machining and finishing techniques for superior finished products. Trim waste is significantly reduced when using MDF compared to other substrates. Stability and strength are important attributes of MDF, which can be machined into complex and delicate patterns that require precise tolerances. Moreover, its smooth surfaces make MDF an excellent base for veneers and laminate applications.

Some of the current environmental pollution problems are created by burning and dumping of agricultural residues, and together with concern for the conservation of future forest resources, they have generated considerable interest in finding suitable outlets to utilize the large amounts of crop residues being produced annually. And because agricultural residues are abundant and renewable annually, it has become evident that they are excellent alternative sources to replace wood and wood fiber. Increasing constraints on residue burning have also been a prime motivator for their introduction. This new environmentally friendly technology for turning agricultural residues into eco-friendly boards entails compressing the agricultural residue materials with nonformaldehyde glues; the panels provide an excellent alternative to plywood sheets 3/8 inch and thicker and can be used in much the same way as medium-density fiberboard. They can also replace OSB and MDF for interior walls and partitions.

Agricultural-residue (ag-res) boards are made from waste wheat straw, rice straw, jute, coconut coirs, bagasse, cotton stalks, casuarina leaves, banana stem, and even sunflower seed husks. Ag-res boards are aesthetically pleasing, often stronger than MDF, and just as functional. Under heat and pressure, microscopic "hooks" on the straws link together, reducing or eliminating the need for binders. The use of new soy adhesives promises improved performance and economics for the ag-composites industry. They are also expected to be safer to handle and to reduce VOC emissions.

Homasote is a brand name that has become synonymous with the product generically known as cellulose-based fiber wallboard. It is a panel product made of 100% post-consumer recycled newspaper fiber and has actually been in production longer than plywood and OSB (about 100 years). It has many potential fiberboard applications specific to sound control in floors and walls, tackable wallboard, fire protection for roof decks, concrete expansion joints, low-emission indoor air quality, and thermal insulation. Homasote is weather-resistant, structural, and extremely durable, and it has two to three times the strength of typical light-density wood fiberboard. Furthermore, it is nontoxic, wax-emulsified for moisture and mold resistance, and integrally protected against termites and fungi, ensuring a healthy environment, conserving natural resources, and reducing solid waste in landfills.

Framing

Advanced framing, or optimum-value engineering (OVE), as it is sometimes referred to, consists of a variety of framing techniques designed to reduce the amount of lumber used and waste generated in the construction of wood-framed structures. This helps reduce material cost and use of natural resources while at the same time increasing energy efficiency by providing more space for insulation. It also helps reduce the processes of extraction and manufacture, as well as transport and lumber disposal, which deplete resources, damage natural habitats, and pollute air and water.

Another problematic issue here is that dimensional lumber supplies depend on larger trees that need decades to mature. Although OVE framing techniques are accepted by code, for one reason or another, they have received limited market penetration and acceptance by builders, framers, and consumers (less than 1% of the residential building market), and this is despite the long-term experience and significant resources that continue to support its use. Provisions for several key OVE framing practices can be found in model U.S. building codes (ICC, 2012).

Modern OVE advanced framing techniques include the spacing of studs at 24-inch on center (o.c.); 2-foot modular designs that reduce cutoff waste from standard-sized building materials (Figure 6.8(a)); in-line framing that reduces the need for double top plates; building corners with two studs; and insulated headers over exterior building openings (or no headers for non-load-bearing walls).

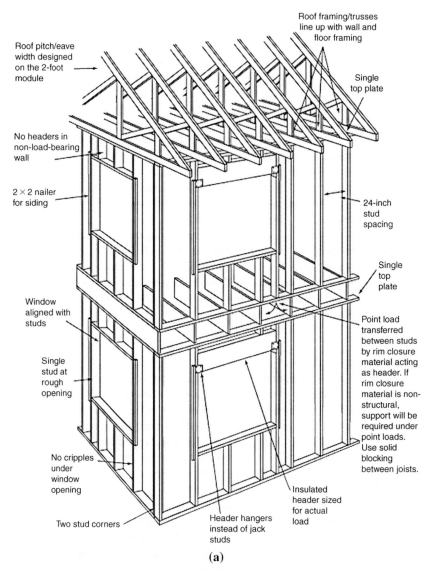

Roof framing/trusses line up with wall and floor framing

Roof pitch/eave width designed on the 2-foot module

Single top plate

No headers in non-load-bearing wall

2 × 2 nailer for siding

24-inch stud spacing

Single top plate

Window aligned with studs

Point load transferred between studs by rim closure material acting as header. If rim closure material is non-structural, support will be required under point loads. Use solid blocking between joists.

Single stud at rough opening

No cripples under window opening

Two stud corners

Header hangers instead of jack studs

Insulated header sized for actual load

(a)

FIGURE 6.8 (a) Isometric drawing illustrating advanced framing techniques used in residential construction. *(Source: Adapted from Building Science Corporation.)*

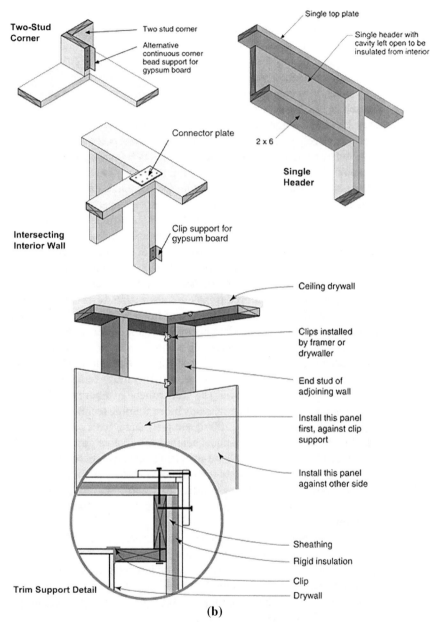

(b)

FIGURE 6.8—cont'd (b) Isometric drawings illustrating OVE framing details used in residential construction. *(Source: Adapted from Building Science Corporation.)*

The advantage of spacing studs at 24-inch o.c. rather than 16-inch o.c. is that it reduces the amount of framing lumber required to construct a home and replaces framing members with insulation. This allows the wall to achieve a higher overall insulating value and to cost less to construct than a conventionally framed wall while still meeting structural requirements.

Still, to achieve maximum success, advanced framing techniques should be considered at the earliest phases of the design process. For this reason, preliminary building design and planning decisions can significantly impact the ability to effectively implement OVE practices or offset potential benefits that can be achieved by using them. Likewise, including OVE framing details on construction plans can greatly facilitate proper implementation (Figure 6.8(b)). Another key aspect of appropriately selecting OVE framing techniques is the careful consideration of all factors that can impact the final result, including trade-offs that may affect detailing and installation of nonstructural components (e.g., flooring, trim, and siding) to ensure that the lumber savings are justifiable.

Homeowners and builders alike can benefit from advanced framing techniques by providing structurally sound homes that require less material and have lower labor costs than a conventionally framed house. Moreover, additional construction cost savings can be achieved from less construction waste to be disposed of, which also helps the environment. The use of fewer studs in OVE improves energy efficiency and enhances insulation values because fewer studs help maximize the insulated wall area through replacing lumber with increased insulation material and reduction of thermal bridging (conduction of heat through framing).

Conventional framing is often found to be structurally redundant, using wood unnecessarily for convenience. The Department of Energy's Office of Building Technology points out that with advanced framing techniques savings in material costs of some $500 per 1200 square feet can be achieved in addition to 3 to 5% of labor costs and 5% in annual heating and cooling costs. While it is true that advanced framing is more wood-efficient than conventional framing, it is also true that some alternative structural technologies such as insulated structural systems, straw-bale or earthen construction, and high-recycled-content steel framing with thermal breaks place fewer demands on our forest resources than OVE framing techniques.

OVE framing techniques have without a doubt been proven effective, yet some techniques are not allowed under certain circumstances (i.e., areas prone to high winds or with seismic potential) or in some jurisdictions. Local building officials should therefore be consulted early in the design phase to verify or to obtain acceptance of these techniques. According to the Partnership for Advancing Technology in Housing (PATH):

The OVE techniques that can be practical to implement (if visibly marked on the construction plans) and have noticeable material savings, thermal benefits, and contribute to the quality of the framing job include:

- *Right sizing headers according to the IRC 2006*
- *Three-stud insulated corners*

- *Ladder-blocking for intersecting walls*
- *24" o.c. floor joist framing (using L/480 deflection limits).*

The use of 2-foot modules helps to make the best use of common sheet-goods' sizes and reduce waste and labor. PATH goes on to say:

Future design and construction guidelines for OVE framing must address OVE decision-making factors and potential trade-offs with respect to serviceability considerations, energy-efficiency impacts, installation practices, and manufacturer requirements for various assemblies, components, and finishes. In addition, OVE guidance in the area of wall bracing would fill a needed void created by increased complexity of wall bracing provisions in modern building codes. Current guides tend to stress material savings without in-depth consideration of some of the cost and performance trade-offs involved. The goal should be to optimize, not maximize, the use of OVE framing in the context of all costs and objectives associated with a building project.

Additional OVE framing practices, such as the use of single headers or band-joist headers, should be incorporated into prescriptive building codes to facilitate their use without incurring the added cost of professional engineering for each application.

Structural insulated panels (SIPs) are high-performance building panels used in floors, walls, and roofs mainly for residential and light commercial buildings. The panels usually consist of two sheets of rigid structural facing board such as oriented-strand board or plywood that is applied to both sides of a core of rigid foam plastic such as expanded polystyrene (EPS) that is four or more inches thick. Alternative skin material can be used for specific purposes. The result of this simple sandwich is a strong structural building system for building walls, roofs, and floors that is significantly more energy efficient and cost effective, and that yields improved R-values compared to traditional framing. SIPs are manufactured under factory-controlled conditions and can be custom-designed for each project. In addition to SIPs' excellent insulation properties, they offer airtight assembly, noise attenuation, and superior structural strength. Though SIP panels may initially cost more per square foot than conventional construction, total construction costs are often minimized due to reduced labor and faster completion.

The superior insulation afforded by the SIP system, as well as cost savings due to the reduction in construction waste, is especially significant when compared to conventional stick or steel stud systems. SIP panels can be delivered precut to the precise dimensions required, and each panel contains the structure, insulation, and moisture barrier of the wall system. OSB is the most common sheathing and facing material in SIP, reducing wood use by as much as 35% and reducing pressure on mature forests by allowing the use of smaller farm-grown trees for structural applications. OSB differs from traditional plywood in that it has no gaps, laps, knots, or voids. In most applications, OSB sheathing is also dimensionally stronger and stiffer than comparable dimensional plywood boards, and is a low-emitting material. However, one of an SIP panel's adverse characteristics and for which it has been often criticized is its environmental footprint. Usually the resin adhesive that the OSB strands are bonded with isn't eco-friendly and contains formaldehyde, which can be toxic, allergenic, and carcinogenic. Recent technological developments appear to be addressing this issue and making SIPs more environmentally friendly.

SIPs are generally chosen for their versatility, strength, cost effectiveness, and energy efficiency, and are engineered and custom-manufactured to give the designer greater control over the project, which includes materials and costs. Another advantage of insulated structural systems is that they integrate a building's structure and insulation into a single component. These characteristics also make them suitable for a wide range of residential and commercial applications. SIP wall assemblies are custom-made according to specifications and drawings and therefore tend to be well sealed, enhancing energy efficiency. As with any tightly sealed structure, moisture control and well-designed ventilation are critical. SIP construction systems can make a significant contribution to good indoor air quality; the plastic insulating foams used in their manufacture (expanded polystyrene or polyurethane/polyisocyanurate) are chemically stable.

SIP cores can be made of several materials, of which expanded polystyrene (EPS) is the most common. EPS requires less energy to manufacture than some of the other options and is more recyclable than polyurethane or polyisocyanurate. Many products now offer a one-hour fire rating when installed with 5/8-inch or thicker gypsum sheathing. Another advantage of EPS foam is that it is expanded with pentane, which does not contribute to ozone depletion or global warming; additionally, it is often recaptured at the factory for reuse, adding to its value.

Although polyurethane and polyisocyanurate have insulation properties per inch of foam superior to those of EPS and offer greater resistance to thermal breakdown, they are unlikely to be recycled. Moreover, polyurethane and polyisocyanurate use HCFC blowing agents, which contribute to global warming and ozone depletion (although to a lesser degree than CFCs). Research is currently under way to develop more suitable eco-friendly alternatives for use in SIPs: new resins derived from soy.

The beauty of using straw-core SIPs is that they are made from waste agricultural straw, are renewable and recyclable, and their pressed-straw core does not require a binding agent. The drawback of using straw-core SIPs is that they offer less insulation per inch of thickness and are significantly heavier than other options; energy used in shipping is a significant consideration when using straw-core SIPs. Preplanning is one of the keys to successful SIP construction, and there are a number of factors to consider when building with SIPs:

- SIP designs should be to standardized panel dimensions. Also, to minimize waste, SIP panels should be ordered precut to meet project requirements when delivered to the job site, including window and door openings. This will save 20 to 30% on framing labor and approximately 30% on waste costs.
- Plumbing and electrical runs need to be predetermined so that the manufacturer can accommodate these needs.

- A tighter house means smaller HVAC systems (up to 40% savings), so when sizing the heating system, consider the thermal performance of SIPs to save money upfront and energy over time. Oversized heating and cooling systems are inefficient.
- With improved indoor air quality, smaller or no air purification systems are required in many climates
- Roofs using SIP systems often do not require ventilation, making them appropriate for low-slope roofs. If local jurisdiction mandates ventilated roofs, consider SIPs with integrated air channels or upgrading from composition roofing.
- Check to see whether the SIP supplier or manufacturer is willing to take back any offcuts for recycling.

Insulated concrete forms (ICFs) are forms or molds that have built-in insulation for accepting reinforced concrete; they are rapidly becoming a mainstream preferred building product. The forms consist of large interlocking modular units that are dry-stacked (without mortar) and filled with concrete. The forms lock together somewhat like LEGO® bricks and serve to create a form for the structural walls of a building. Concrete is pumped into the cavity every several feet to form the structural element of the walls. ICFs usually employ reinforcing steel (rebar) before concrete placement to give the resulting walls flexural strength similar to that of bridges and concrete high-rise buildings. They also employ an insulating material as permanent formwork that becomes a part of the finished wall.

After the concrete has cured or firmed up, the ICFs are left in place permanently to increase thermal and acoustic insulation and render greater fire protection. ICFs can also accommodate electrical and plumbing installations (Figure 6.9). The end result leaves you with a high-performing wall that is structurally sound, insulated, and strapped, has a vapor barrier, and is ready to accept final exterior and interior finishes. ICFs can generally be considered "green" because they are durable, produce little or no waste during construction, and significantly improve the thermal performance of concrete walls. Also, there are no CFCs, HCFCs, formaldehydes, or wood to rot and mold.

There are essentially three main types of ICF systems on the market. Each type addresses significantly different construction issues and different completed results to the owner. The comparisons are typically based on a variety of test standards, criteria, and calculations. The different features of the three types are shown here:

- The solid monolithic concrete forms or flat system forms consist of an even thickness of concrete throughout the walls, like a conventionally poured wall.
- The waffle-grid system creates a waffle pattern where the concrete is thicker at some points than others.
- The screen grid or post-and-beam system forms consist of detached horizontal and vertical columns of concrete.

(a)

(b)

FIGURE 6.9 Application of insulating concrete formwork with the Quad-Lock® system. This can be considered a "green" material as it is durable, produces little or no waste during construction, and greatly improves the thermal performance of concrete walls and floors. Also, Quad-Lock panels are molded of fire-retardant Expanded Polystyrene (EPS), which is a foamed insulation that has a zero ozone-depletion rating. *(Source: Quad-Lock Building Systems.)*

An important characteristic of standard concrete is that it is a dense material with a high heat capacity that can be utilized as thermal mass, thereby reducing the energy required to maintain comfortable interior temperatures. One of concrete's negative attributes is that it is not a good insulator and standard formwork is therefore waste-intensive. Additionally, toxic materials are frequently required to separate the formwork from the hardened product. ICF is able to address these weaknesses by reducing solid waste, air and water pollution, and (potentially) construction cost. ICF wall systems have proven to be thermally superior, enhancing their usefulness for passive heating and cooling; comfort is also enhanced and energy costs are reduced. Any possible higher initial costs that are incurred can be offset or minimized by the downsizing of the heating/cooling system.

A variety of materials can be employed in the manufacturer of ICF systems, including lightweight foamed-concrete panels, rigid foams such as expanded polystyrene and polyurethane, and composites that combine concrete with mineral wool, wood waste, paper pulp, or expanded polystyrene beads. Likewise, there are several ICF systems currently on the market that substitute straw bales or fiber-cement for polystyrene, such as Baleblock™ and Faswall®. Rigid foams used in ICFs are generally less green because they do not have significant recycled content and are less likely to be recyclable at the end of their life. However, this does not preclude them from being reused in fill or other composite concrete products to meet market demands.

A distinct attribute of ICFs is that they offer the structural and fire resistance benefits of reinforced concrete; structural failures due to fire are therefore not commonplace. By adding flame-retardant additives, polystyrene ICFs tend to melt rather than burn, and interior ICF walls tend to contain fires far better than wood frame walls, improving overall fire safety. As is the case in most heated structures, a key design consideration for ICF walls is moisture control. Solid concrete walls sandwiched in polystyrene blocks tend to be very well sealed to enhance energy efficiency, but they consequently also tend to seal water vapor within the structure. Potential mold growth and impaired indoor air quality are serious health concerns that require attention. A simple approach to resolving this is by incorporating mechanical ventilation. Certain systems, such as straw bale and RASTRA (i.e., 85% recycled polystyrene), tend to be more vapor-permeable, reducing this concern. Several ICF products such as RASTRA are more eco-friendly because they are made from recycled postconsumer polystyrene (foam) waste products.

6.2.9 Concrete

Concrete is a composite building material made up of three basic components: water, aggregate (rock, crushed stone, sand, or gravel), and a binder or paste such as cement. The cement hydrates after mixing and hardens into a stone-like durable material with which we are all familiar. Concrete has a low tensile strength and is generally strengthened by the addition of steel reinforcing bars; this is commonly referred to as reinforced concrete.

Over the centuries, concrete has proven to be strong, durable, yet inexpensive, and it is widely used as a structural building material in the United States and throughout the world. As a result of the increasing scale of concrete demand, the impacts of its manufacture, use, and demolition are widespread. Habitats are disturbed from materials extraction; significant energy is used to extract, produce, and ship cement; and toxic air and water emissions result from cement manufacturing. It is estimated that approximately one ton of carbon dioxide is released for each ton of cement produced, resulting in 7 to 8% of man-made CO_2 emissions. And although concrete is generally only 9 to 13% cement, it nevertheless accounts for 92% of concrete's embodied energy. Cement dust contains free silicon-dioxide crystals, the trace element chromium, and lime, all of which can have negative impacts on worker health and the environment. Mixing concrete requires large amounts of water and generates alkaline wastewater and runoff that can contaminate vegetation and waterways.

Admixtures are often added to a concrete mix so as to achieve certain specific goals. Here are some of the main admixtures that are used and what they are designed to achieve:

- *Accelerating admixtures* are added to concrete to reduce its setting time and to accelerate achieving early strength. The amount of reduction in setting time will vary according to the amount of accelerator used. Although calcium chloride is a low-cost accelerator, specifications will often require a nonchloride accelerator to prevent the corrosion of reinforcing steel.
- *Retarding admixtures* are frequently required in hot weather conditions to facilitate delaying the setting time. They are also used to delay set in more difficult jobs or for special finishing operations such as exposing aggregate. Retarders also often act as a water reducer.
- *Fly ash* is a residue from coal combustion. It can replace 15 to 30% of the cement in the concrete mix. Fly ash is quite popular as a cement substitute and its use improves concrete performance, giving greater compressive strength, decreased porosity, greater durability, improved workability, and more resistance to chemical attack, although the curing time is increased. Using fly ash also creates significant benefits for the environment.
- *Water-reducing admixtures* reduce the amount of water needed in the concrete mix. The water/cement ratio will be lower while the concrete's strength will be greater. Most low-range water reducers reduce the water needed in the mix by 5 to 10%.
- *Air-entraining admixtures* should be used whenever concrete is exposed to freezing, thawing, and de-icing salts. They entrain microscopic air bubbles in the concrete, so that when the hardened concrete freezes, the frozen water inside expands into these air bubbles instead of damaging the concrete.

The incorporation of local and/or recycled aggregate (e.g., ground concrete from demolition) is an excellent way to reduce the impacts of solid waste, transit emissions, and habitat disturbance.

In nonstructural applications, concrete use may be reduced by trapping air in the finished product or through the use of low-density aggregates. Trapped air displaces concrete while enhancing insulation value and reducing weight and material costs without compromising durability and fire resistance. Similar insulation and weight-reducing benefits are provided by other low-density aggregates such as vermiculite, perlite, pumice, shale, polystyrene beads, and mineral fiber. Cast-in-place or precast concrete and concrete-masonry unit (CMU) considerations include the following:

- Recycle demolished concrete onsite for use as aggregate or fill material for new projects, or recycle at local landfills.
- Whenever possible, redeploy portions of existing structures, such as slabs or walls that are in satisfactory condition.
- Employ precast systems to minimize waste of forming material and to reduce the impact of wash water on soils.
- Incorporate the maximum amount of fly ash, blast-furnace slag, silica fume, and/or rice-husk slag appropriate to the project, to reduce cement use by 15 to 100%.
- Employ alternative material substitutes for concrete such as insulating concrete forms (ICF) to reduce waste, enhance thermal performance, and shorten construction schedules. Likewise, use cellular, foamed, autoclaved-aerated (AAC), and other lightweight concretes to add insulation value while reducing weight and concrete required. Also use earthen and rapidly renewable materials, such as rammed earth, cob, or straw bale, to help reduce the need for insulation and finish materials in both residential and commercial projects.
- Use nontoxic form-release agents when possible.
- Minimize waste by carefully planning concrete material quantities.
- Consider fabric-based form systems for footings to achieve faster installation and greater wood savings.
- Reduce wood waste and material costs by employing steel or aluminum concrete forms, which unlike many wood forms can be reused many times over.
- Use permeable or porous/pervious surfaces to allow water to percolate into the soil to filter out pollutants and recharge the water table.

Urban and suburban sites typically contain large areas of impermeable surfaces, which cause a number of problems. It is estimated that up to 75% of urban surface area is covered by impermeable/impervious pavement, which is a solid surface that doesn't allow water to penetrate but forces it to run off, inhibiting groundwater recharge, contributing to erosion and flooding, conveying pollution to local waters, and increasing the complexity and expense of stormwater treatment. Also due to the heat-absorbing quality of asphalt and other paving materials, sites with high ratios of impermeable surfaces increase ambient air temperatures and require more energy for cooling, thereby creating a heat island effect.

One of the main characteristics of permeable surface/paving (also known as porous or pervious surface), on the other hand, is that it contains voids that allow water to percolate into the soil to filter out pollutants and recharge the water table. Pervious paving may incorporate recycled aggregate and fly ash, which help reduce waste and embodied energy. Pervious paving is suitable for use in parking and access areas, as it has a compressive strength of up to 4000 psi. It also mitigates problems with tree roots; percolation areas encourage roots to grow deeper. Enhanced heat exchange with the underlying soil can decrease summer ambient-air temperature by 2 to 4 °F.

Concrete poured-in-place applications require onsite formwork, which acts as a mold to give shape to walls, slabs, and other project elements as they cure to a satisfactory strength and which is removed afterward (Figure 6.10). Plywood and milled lumber are the most common form materials, contributing to construction waste and the impacts of timber harvesting and processing. Wooden formwork can be made from salvaged wood and typically can be disassembled and reused several times.

Form release agents are materials that prevent the adhesion of freshly placed concrete to the forming surface (which is usually plywood, overlaid plywood, steel, or aluminum). Such materials prevent concrete from bonding to the form, which can mar the surface when forms are dismantled. There are two principal categories of release agents available; these are barrier (nonreactive or passive) and reactive (chemically active). Barrier release agents prevent adhesion by creating a physical film or barrier between the forming surface and the fresh concrete. Reactive or chemically active release agents, on the other hand, work by the process of a chemical reaction with the calcium (lime) that is available in fresh concrete. A soapy film is created that prevents adhesion. Also, because this is a chemically reactive process, there is generally little to no residue on the forming surface or concrete, thus providing for a cleaner process.

Traditional form releasers such as diesel fuel, motor oil, and home heating oil are carcinogenic and are now prohibited by a variety of state and federal regulations, including the Clean Air Act, because they expose construction personnel (and potentially occupants as well) to VOCs. Low- and zero-VOC water-based form-release compounds that incorporate soy or other biologically derived oils dramatically reduce health risks to construction staff and occupants and often make it easier to apply finishes or sealants when necessary. Many soy-based options are generally less expensive than their petroleum-based counterparts.

The design of concrete formwork necessitates that all factors that will adversely affect the formwork's pressure be taken into consideration. These factors include the rate of placement, concrete mix, and temperature. The rate of placement should generally be lower in the winter than in the summer. It doesn't matter how many cubic yards are actually placed per hour or how large the project is.

FIGURE 6.10   Workers setting concrete formwork for the walls of a high-level waste facility pit. *(Source: Bechtel Corporation.)*

What does matter is the rate of placement per height and time (height of wall poured per hour). Moreover, the forms should also be of sufficient strength and stability to enable them to carry all live and dead loads that may be encountered before, during, and after placing the concrete. Most exterior APA panels can be used for concrete formwork because they are manufactured with waterproof glue.

6.3 BUILDING AND MATERIAL REUSE

The term *building reuse* generally means leaving the main portion of the building structure and shell in place while performing a "gut rehab," as it is known in the trade.

6.3.1 Building Reuse

The intent of this LEED credit is to extend the life cycle of existing building stock, reusing building materials and products, retain cultural resources, reduce waste, protect virgin resources, and reduce environmental impacts of new buildings as they relate to materials manufacturing and transport.

This is particularly important for green building because repairing a building rather than tearing it down saves natural resources as well as significantly reducing materials ending up in the landfill. Reuse is also important because of the embodied energy that is in the production, manufacture, transportation, and construction of new materials. It discourages the production of new products and minimizes the negative impact of embodied energy through reduction in raw material extraction. A key factor in building reuse is the durability of the original structure.

In some states, including North Carolina, grants are provided to renovate vacant buildings in rural counties or in economically distressed urban areas. Note that disaster-recovered materials such as trees uprooted by tornadoes or hurricanes are not eligible for LEED credit.

Maintaining Structural Elements: Existing Walls, Floors, and Roof

LEED requirements for new construction are to maintain a minimum of 50, 75, or 95% (for up to three points) of the existing building structure (based on surface area), including structural floor and roof decking as well as the envelope (exterior skin and framing but excluding window assemblies and nonstructural roofing material). It is possible to achieve a credit by maintaining a minimum of 50% (by area) of interior nonstructural elements of an existing building, such as interior walls, doors, refurbished wood floors, and ceiling systems, in the new building. Hazardous materials that are remediated as a part of the project scope

TABLE 6.3 Minimum Building Structure Reuse

Building Reuse (%)	Points
55	1
75	2
95	3

are to be excluded from the calculation of the percentage maintained. The credit will not apply if the project includes an addition to an existing building where the square footage of the addition is more than twice the square footage of the existing building.

Table 6.3 shows the minimum building structure reuse required for achieving LEED credits for new construction (always check the USGBC website for the latest updates for any certification category). However, for core and shell you are required to maintain a minimum of 25, 33, 42, 50, or 75% of existing walls, floors, and roof for up to five credits. Schools must maintain 55 or 75% of existing walls, floors, and roof for up to two credits.

Potential Technologies and Strategies

Consider the use of salvaged, refurbished, or reused materials from previously occupied buildings, including structure, envelope, and elements. Hazardous materials that pose contamination risk to building occupants and that are remediated as a part of the project scope are excluded from the calculation of the percentage maintained. Upgrade components that would improve energy and water efficiency, such as windows, mechanical systems, and plumbing fixtures. However, mechanical, electrical, plumbing or specialty items and components should be excluded for this credit. Furniture may be included only if it is included in the other MR credits.

Interior Nonstructural Elements

Retain 50% of interior nonstructural elements for new construction and schools and 40 and 60% for commercial interiors. The intent here, according to LEED, is to extend the life cycle of existing building stock, conserve resources, retain cultural resources, reduce waste, and reduce environmental impacts of new buildings as they relate to materials' manufacturing and transport.

LEED Requirements

Maintain at least 50% (by area) of existing interior nonshell and nonstructural elements (interior walls, doors, floor coverings, and ceiling systems) of the completed building (including additions). If the project includes an addition to an existing building, this credit is not applicable if the square footage of the addition is more than two times the square footage of the existing building.

In terms of potential technologies and strategies, LEED requires that consideration be given to the reuse of existing buildings, including structure, envelope, and interior nonstructural elements. Hazardous elements that pose contamination risk to building occupants are to be removed, and components that would improve energy and water efficiency, such as mechanical systems and plumbing fixtures, should be upgraded. For the LEED credit, the extent of building reuse needs to be quantified, and the owner/developer must provide a report prepared by a qualified person outlining the extent to which major building elements from a previous building were incorporated into the existing building. The report should include pre- and post-construction details highlighting and quantifying the reused elements such as foundations, structural elements, and façades. Windows, doors, and similar assemblies may be excluded.

6.3.2   Materials Reuse

Materials reuse should be 5 and 10% for new construction, schools, and commercial interiors (30% for furniture and furnishing), and 5% for core and shell. The intent is to reuse building materials and products to protect and reduce demand for virgin resources and to reduce waste, thereby reducing impacts associated with the extraction and processing of virgin resources.

Requirements of LEED

Use salvaged, refurbished, or reused materials such that the sum of these materials constitutes at least 5, 10, or 30% (for commercial interiors, furniture, and furnishings) of the total value of materials on the project based on cost. Mechanical, electrical, and plumbing components and specialty items such as elevators and equipment are not to be included in this calculation. Include only materials permanently installed in the project. Furniture may be included, providing it is included consistently in MR credits 3 through 7. Most credits in the Materials & Resources category are calculated using a percentage of total building materials.

LEED Potential Technologies and Strategies

Include the identification of opportunities to incorporate salvaged materials into the building design, and research potential material suppliers. Salvaged materials such as beams and posts, flooring, paneling, doors and frames, cabinetry and furniture, brick, and decorative items should be considered. The difference between reuse and recycling is that reuse is essentially the salvage and reinstallation of materials in their original form, whereas recycling is the collection and remanufacture of materials into a new material or product, typically different from the original. Biodegradable material breaks down organically and may be returned to the earth with none of the damage associated with the generation of typical waste materials.

Construction and demolition are estimated to be responsible for about 30% of the U.S. solid-waste stream. Real-world case studies by the Alameda County Waste Management Authority, for example, have concluded that more than 85% of that material, from flooring to roofing to packaging, is reusable or recyclable. For this reason, reusing materials slated for the landfill has become an extremely eco-friendly way to build so as to avoid negative elements such as the extraction, manufacture, transport, and disposal of virgin building materials that pollute air and water, deplete resources, and damage natural habitats.

Salvaging materials from renovation projects and specifying salvaged materials can reduce the costs of material while adding character to projects and maximizing environmental benefits, such as reduced landfill waste, reduced embodied energy, and reduced impacts from harvesting/mining of virgin materials (logging old-growth or tropical hardwood trees, mining metals, etc.). On the other hand, some materials require remediation or should not be reused at all. For example, materials contaminated by hazardous substances such as asbestos, arsenic, and lead paint must be treated and/or disposed of properly. Avoiding materials that will cause future problems is critical to long-term waste reduction as well as the health of communities and the environment.

Factors that impact the selection of reusable building materials include the following:

- Existing building shells, when appropriate, should be reused to yield the greatest overall reduction in project impacts.
- Materials from remodeling or renovation should be reused onsite.
- Products containing hazardous materials such as asbestos, lead, or arsenic should be disposed of properly or remediated prior to reuse.
- Building materials composed of one substance (e.g., steel, concrete, wood, etc.), or that are readily disassembled are generally easiest to reuse or recycle.
- For remodels and redevelopment, adequate time should be allowed in the construction schedule for deconstruction and recycling.
- Inefficient fixtures, components, and appliances (e.g., toilets using more than 1.6 gallons per flush, single-pane windows, and refrigerators or other appliances over five years old) should be replaced.
- Salvaged materials can vary in availability, quality, and uniformity. Ensure that materials are readily available to meet project needs before specifying them.
- Materials should be carefully evaluated to ensure that they offer the best choice for the application. They need to be durable and preferably readily disassembled for reuse, recycling, or biodegrading at the end of the useful life of the building.
- Materials composed of many ingredients, such as vinyl siding, OSB, or particleboard, are generally not recyclable or biodegradable.

6.4   CONSTRUCTION WASTE MANAGEMENT

The overall intent of the LEED's Construction Waste Management credit is to avoid materials going to landfills during construction by diverting construction waste, demolition debris, and land-clearing debris from landfill disposal and incinerators; redirecting recyclable recovered resources back to the manufacturing process; and redirecting reusable materials to appropriate sites.

6.4.1   LEED Requirements

Recycle and/or salvage at least 50 or 75% of debris—that is, nonhazardous construction, demolition, and packaging (95% for extra credit). Develop and implement a construction waste management plan that at a minimum identifies and quantifies the materials generated during construction that are to be salvaged, recycled, refurbished, or diverted from disposal, and note whether such materials will be sorted onsite or commingled. Typical items include recycled cardboard, metal, brick, acoustical tile, concrete, plastics, clean wood, glass, gypsum board, carpet, and insulation, as well as doors and windows, ductwork, clean dimensional wood, paperboard, paneling, cabinetry, plastic used in packing, and the like. Mechanical, electrical, and plumbing (MEP) systems may now be included, although this is not clear from the LEED reference books (Figure 6.11). Excavated soil, rocks, vegetation, hazardous materials, and land-clearing debris do not contribute to this credit. Calculations can be done by weight or by volume but must be consistent throughout.

Documentation is required for each credit a project attempts to achieve using the LEED system to prove the activity was completed. LEED letter templates are to be used to certify that requirements are met for each prerequisite and credit. Additional documentation may also be required. The contractor is generally responsible for completing the required LEED documentation for these two credits since the responsibility for construction waste management lies with the contractor.

The LEED letter template is to be signed by the architect, owner, or other responsible party, tabulating total waste material, the quantities diverted, and the means by which they were diverted, and declaring that the credit

FIGURE 6.11   Truck from DRC Emergency Services unloading construction and demolition debris at Birmingham's New Georgia Landfill, which is being used for brush, tree, construction, and demolition debris from the April 27, 2011, tornadoes. *(Source:* Birmingham News/*Joe Songer.)*

requirements have been met. As a portion of the credits in each application may be audited, the contractor should be prepared with backup documentation. Most LEED projects require a waste management plan with regular submittals tracking progress. The plan should indicate how the required recycling rate is to be achieved, including materials to be recycled or salvaged, cost estimates comparing recycling to disposal fees, materials-handling requirements, and how the plan will be communicated to the crew and subcontractors. All subcontractors are required to adhere to the plan in their contracts. Considerations relating to construction waste reduction should include the following:

- The smaller the project, the less material used, reducing both solid waste and operating costs.
- Design assemblies should match the standard dimensions of the materials to be used.
- Consider disassembly design so that materials can be readily reused or recycled.
- Track recycling through the construction process (the general contractor should keep records such as receipts of recyclable and waste diversion pickups).
- Designate a site in the construction area for the separation process.
- Employ clips and stops to support drywall or wood paneling at top plates, end walls, and corners. Clips can provide the potential for two-stud corners, reducing wood use, easing electrical and plumbing rough-in, and improving thermal performance.
- Materials attached with removable fasteners are generally quicker, cheaper, and more feasible to deconstruct than materials installed with adhesives, although adhesives distribute loads over larger areas than fasteners used alone.
- When possible, make use of existing foundations and structures in good condition to reduce waste, material requirements, and possibly labor costs.
- Design for flexibility and changing use of spaces.
- Specify materials such as structural insulated panels, panelized wood framing, and precast concrete that can be delivered precut for rapid, almost waste-free installation.
- For wood construction, consider 24-inch o.c. framing with insulated headers, trusses for roofs and floors, finger-jointed studs, and engineered-wood framing and sheathing materials.
- Whenever practical, specify materials with high recycled content.

According to the U.S. Environmental Protection Agency:

Commercial construction typically generates between 2 and 2.5 pounds of solid waste per square foot, the majority of which can be recycled. Salvaging and recycling C&D waste reduces demand for virgin resources and the associated environmental impacts. Effective construction waste management, including appropriate handling of non-recyclables, can reduce contamination from and extend the life of existing landfills. Whenever feasible, reducing initial waste generation is environmentally preferable to reuse or recycling.

The agency goes on to say:

The Construction Waste Management Plan should ideally recognize project waste as an integral part of overall materials management. The premise that waste management is a part of materials management, and the recognition that one project's wastes are materials available for another project, facilitates efficient and effective waste management.

It is also important for waste management requirements to be taken into account early in the design process and to be a topic of discussion at both pre-construction and ongoing regular job meetings to ensure that contractors and appropriate subcontractors are fully informed of the implications of these requirements on their work prior to and throughout construction. Furthermore, the EPA states:

Plan implementation of the waste management should be coordinated with or part of the standard quality assurance program and waste management requirements should be addressed regularly throughout the project. If possible, adherence to the plan would be facilitated by tying completion of recycling documentation to one of the payments for each trade contractor.

6.5   RECYCLED MATERIALS

Material recycling can be defined in several ways depending largely on the different processes that recycling is involved in. It can be as simple as reusing a given product beyond its intended use, such as passing old clothes on to charities, the poor, or relatives to avoid throwing them out so that someone else can make good use of them. However, recycling is more commonly associated with the practice of recovering old goods from the waste stream and reincorporating them into the manufacturing process, thus allowing them to be turned into new products. The recycling of waste materials into new products helps prevent the waste of potentially useful materials and reduces the potential consumption of fresh virgin materials. Additionally, it lessens energy usage, reduces air pollution (by incineration), and reduces the need for "conventional" waste disposal. Recyclable materials are a key component of modern waste reduction and can take many forms, including different kinds of plastic, paper, glass, metal, textiles, and electronics.

According to the *Environmental Building News* (BuildingGreen.com):

Recycled content refers to the portion of materials used in a product that have been diverted from the solid waste stream. If those materials are diverted during the manufacturing process, they are referred to as pre-consumer recycled content (sometimes referred to as post-industrial). If they are diverted after consumer use, they are post-consumer.

Post-consumer content is generally viewed as offering greater environmental benefit than pre-consumer content. Although pre-consumer waste is much more vast, it is also more likely to be diverted from the waste stream. Post-consumer waste is more likely to fill limited space in municipal landfills and is typically mixed, making recovery more difficult.

To claim that it is using pre-consumer recycled content, a company must be able to substantiate that the material it is using would have become garbage, had it not purchased it from another company's waste stream, for example. If a manufacturer routinely collects scraps and feeds them back into its own process, that material does not qualify as recycled.

Recycled content is the most widely cited attribute of green building products. The LEED intent for Materials & Resources (MR) Credit 4.1 is to protect virgin resources by increasing demand for building products with recycled content. These are the LEED requirements:

Use materials with recycled content such that the sum of post-consumer recycled content plus one-half of the pre-consumer content constitutes at least 10% (based on cost) of the total value of the materials in the project. The recycled content value of a material assembly shall be determined by weight. The recycled fraction of the assembly is then multiplied by the cost of assembly to determine the recycled content value.

Mechanical, electrical and plumbing components and specialty items such as elevators shall not be included in this calculation. Only include materials permanently installed in the project. Furniture may be included, providing it is included consistently in MR Credits 3–7.

Recycled content is to be defined in accordance with the International Organization for Standardization (ISO) document "ISO 14021—Environmental Labels and Declarations—Self-Declared Environmental Claims (Type II Environmental Labeling)."

Many federal, state, and local government agencies around the nation have established "buy recycled" programs aimed at increasing markets for recycled materials. These include the California Department of Resources, Recycling and Recovery (CalRecycle), San Mateo County; the Iowa Program, Montgomery County; and others. Such programs support the Department of General Services (DGS) and other state agencies, as well as local governments, in establishing policies and practices for purchasing recycled-content products (RCP), in addition to supporting activities that promote waste reduction and management. Likewise, a principal goal of these programs is supporting all recycling activities to reduce solid-waste disposal, and many communities in the United States now offer regular curbside collection or drop-off sites for certain recyclable materials. Materials collection in itself, however, is insufficient for making the recycling process work. Successful recycling also requires that manufacturers produce viable products from the recovered materials and, in turn, that there is a market ready to purchase products made of recycled materials.

Recyclability is a characteristic of materials that maintains useful physical or chemical properties after serving their original purpose and therefore allows them to be reused or remanufactured into additional products through a recognized process. In fact, many national and international companies constantly seek an environmental-marketing edge by advertising the recycled content of their products, which is often undocumented or uncertified and can be misleading. Such claims come under the jurisdiction of the Federal Trade Commission (FTC), which first published definitions for common environmental terms in its Green Guides in 1992. LEED Rating System offers credit for recycled-content materials, referencing definitions from ISO 14021. However, these definitions leave a lot of gray areas, which many manufacturers often interpret in their own favor. Third-party certification of recycled content is useful in maintaining a high standard and offering the ability to verify any claims made regarding sustainability.

Waste is not a luxury we can afford, although we have to bear it. Yet the extraction, manufacture/transport, and disposal of building materials continue to clog our landfills, pollute our air and water, deplete our resources, and damage our natural habitats. The California Integrated Waste Management Board (CIWMB) notes that construction and demolition (C&D) waste comprised 22% of California's solid-waste stream in 2004. Probably more than 85% of that material, from flooring to roofing, and much of that percentage can be salvaged for reuse or recycling. In addition to C&D waste, we must also consider the material in our recycling bins, our used bottles, paper, cans, and cardboard, which can provide suitable raw materials for recycled-content products. But keeping a material out of the landfill is only the first step to putting "waste" back into productive use. The "waste" has to be reprocessed into a new, quality product, and that product must be capable of being sold to an entity that recognizes its benefits. The reprocessing of our "waste" as the raw material for new products increases demand for recycling and it encourages manufacturers to employ more recycled material, continuously strengthening this cycle.

The benefits of using recycled-content materials are many and include reduced pollution, reduced solid waste, reduced energy and water use, reduced greenhouse-gas emissions, and healthier indoor air quality. Here is a partial list of materials that are readily recyclable and that generally may cost less to recycle than to dispose of as garbage:

- Acoustical ceiling tiles
- Asphalt
- Asphalt shingles
- Cardboard
- Carpet and carpet pads made of plastic bottles or sometimes used carpet (up to half of all polyester carpet made in the United States contains recycled plastic)
- Concrete containing ground-up concrete as aggregate, fly ash (a cementitious waste product from coal-burning power plants), asphalt, brick, and other cementitious materials
- Countertops made with everything from recycled glass to sunflower-seed shells

- Drywall made with recycled gypsum and Homasote wallboard made from recycled paper
- Fluorescent lights and ballasts
- Insulation, such as cotton made from denim, newspaper processed into cellulose, or fiberglass with some recycled-glass content
- Land-clearing debris (vegetation, stumpage, dirt)
- Metals (pipes, rebar flashing, steel, aluminum, copper, brass, and stainless steel)
- Paint (use a hazardous waste outlet)
- Plastic film (sheeting, shrink wrap, packaging)
- Plastic and wood-plastic composite lumber from plastic and wood chips (ideal for outdoor decking and railings)
- Tile containing recycled glass
- Window glass
- Wood (includes engineered products; nails are acceptable)

To achieve maximum benefit when selecting a recycled-content building material, the following points should be taken into consideration:

- Choose materials that contain the highest recycled content possible. For example, a recycled product that is 70% recycled is preferable to one that is only 10% recycled and 90% virgin material.
- Choose materials with high post-consumer recycled content. Some "recycled" content is waste from manufacturing processes. Reducing manufacturing waste is important, but recycling postconsumer material is necessary to close the loop.
- Choose materials that are appropriate for the application at hand.
- Salvaging (reusing) whole materials is preferable to recycling, and all but eliminates waste, energy, water use, and pollution.
- When possible, choose materials that are both recycled and recyclable or biodegradable at the end of their useful life. Ideally, a material may be continuously recycled back into the same product.

Reclaimed wood has many applications, including but not limited to flooring, siding, furniture, and in some cases as structural members. Reusing wood from an existing building onsite should be carefully considered; where appropriate, look to salvage yards and onsite deconstruction sales for a portion of a project's material needs. It is important to note that salvaging or reusing wood can reduce solid waste, save forest resources, and save money. Moreover, reclaimed wood is often available in dimensions, species, and old-growth quality that are no longer available today. Table 6.4 lists examples of reusable (RU), recyclable (RC), and biodegradable (B) building materials.

TABLE 6.4 RU, RC, and B Building Materials

Reusable	Recyclable	Biodegradable
Bricks	Asphalt	Earthen materials
Doors and windows	Bricks	Gypsum wallboard
Earthen materials	Concrete, ground and used as aggregate	Linoleum flooring
Gypsum wallboard		Straw bales
Lighting fixtures	Metal: steel, aluminum, iron, copper	Wood and dimensional lumber, such as beams, trusses, studs, and plywood
Metal: steel, aluminum, iron, copper	Wood and dimensional lumber, such as beams, trusses, studs, and plywood	Wool carpet
Plumbing		
Unique and antique products that may no longer be available		
Wood and dimensional lumber, such as beams, trusses, studs, and plywood		

Deconstruction consists of the systematic disassembly of a building, with the purpose of recovering valuable materials for reuse in construction, renovation, or manufacturing into new wood products, thereby preserving the useful value of its component materials. Deconstruction is preferable to demolishing; the combination of various tax breaks, new tools, and increasing local expertise is making it easier to keep materials out of the landfill. It has grown by leaps and bounds in recent years, due mainly to new for-profit and nonprofit entities throughout the United States. Although deconstruction takes longer and may initially cost more than demolition, it is nevertheless likely to reduce the overall project cost. Waste reduction has the benefits of minimizing energy use, conserving resources, and easing pressure on landfill capacity.

6.6   REGIONAL MATERIALS

Regional materials are those that are extracted, harvested, and manufactured within a 500-mile radius of the project site. The main LEED intent here is to reduce material transport by increasing demand for building materials and products that are extracted and manufactured within the region where the project is located, thereby supporting both use of indigenous resources and the regional economy, as well as reducing the negative environmental impacts associated with transportation.

6.6.1   LEED Requirements

Use a minimum of 10 or 20% (based on cost) of total building materials and products that are extracted, harvested, recovered, or manufactured regionally within a radius of 500 miles of the site (Figure 6.12). To calculate, either the default 45-percent rule or the actual materials cost may be used. All mechanical, electrical, plumbing, and specialty items, such as elevator equipment, need to be excluded. If only a fraction of the product/material is extracted, harvested, recovered, or manufactured within 500 miles of the site, then only that percentage (based on weight) may contribute to the regional value. Furniture may be included only if it is included throughout MR Credits 3 through 7. Of note, "manufacturing" refers to the final assembly of components into the building product that is furnished and installed by the contractor. Thus, if the hardware comes from Los Angeles, California, the lumber from Vancouver, British Columbia, and the joist assembled in Fairfax, Virginia, the location of the final assembly destination is considered to be Fairfax. One or two points can be earned for using materials that are both harvested and manufactured within a 500-mile radius of the site. In LEED for Commercial Interiors, and in older versions of LEED for New Construction (LEED-NC) and LEED for Core & Shell (LEED-C&S), one point is given for merely manufacturing within that radius and a second point for harvesting as well.

By simply tracking the materials that are typically produced and supplied within 500 miles of the project site, it is possible to achieve the 20%-credit threshold without impacting cost. In some cases, however, the 20% threshold can only be achieved by targeting certain materials (e.g., specific types of stone or brick) or limiting the number of manufacturers whose products are to be

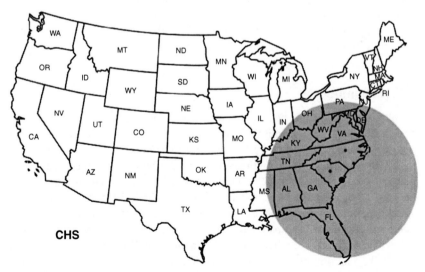

FIGURE 6.12 Map program that is capable of drawing any required radius for any chosen location. *(Source: Free Map Tools, www.freemaptools.com.)*

considered in the project bids. In such instances, there is a possibility of incurring additional costs.

To verify that the extraction/harvest/recovery site is located within a 500-mile radius of the project site, project teams are required to attest to the actual mileage between the site and the manufacturer and, likewise, attest to the distance between the project site and the extraction site for each indigenous material in the submittal template. Alternatively, a statement on the manufacturer's letterhead indicating that the point of manufacture is within 500 miles of the LEED project site will be accepted as part of the documentation and credit submittals. The benefit of using indigenous materials is that it reduces transportation distances and the associated environmental impacts.

6.7   RAPIDLY RENEWABLE MATERIALS

Rapidly renewable materials are numerous and include bamboo, cork, insulation, linoleum, straw bale, wheat board, wool, and the like. These are considered sustainable because they are natural, non-petroleum-based building materials (petroleum-based materials are nonrenewable) that can be grown and harvested within 10 years. LEED states that the intent of using rapidly renewable materials is to "reduce the use and depletion of finite raw materials and long-cycle renewable materials by replacing them with rapidly renewable materials." LEED MR Credit 6.0 states that to be eligible for credits, rapidly renewable materials must be equal to no less than 2.5% of the cost of a building project in terms of value. It has been found that the use of rapidly renewable resources can often save land as well as other resources that usually go into conventional materials. Moreover, because of their shorter harvesting cycles, rapidly renewable materials provide environmental benefits and are able to sustain a community for a longer period than more finite sources can.

6.7.1   Bamboo

Bamboo is considered to be one of many rapidly renewable resources for LEED certification under MR Credit 6. Furthermore, some bamboo lines are also made with no formaldehyde, thereby contributing to EQ 4.4—Low Emitting Materials. Partly because of its rapid regeneration, bamboo has emerged as an alternative to other types of wood commonly used in the United States and abroad. In the past it was used as a basic material for making household objects and small structures, but continuous research and engineering efforts have enabled its true value to be realized. The bamboo plant meets all the criteria of being rapidly renewable because when harvested sustainably it will regrow from the same root stalk, maturing in just a few years, whereas trees like oak and maple take far longer to grow and are incapable of regrowing from the same plant; also, they can be harvested only once.

Technically, bamboo is not a wood; it is a giant grass that comes in 1500 varieties and produces hard, strong, dimensionally stable wood. It can be found in many tropical regions of Asia, Africa, and South America. It has been used both as a building material and for furniture construction for thousands of years (Figure 6.13(a)). In addition to being considered a fast-growing woody plant, it is one of the most versatile and sustainable building materials available. Bamboo can reach maturity in months in a wide range of climates, is exceedingly strong for its weight, and can be used both structurally and as a finish material. Likewise, it can be clear-cut and will grow right back.

(a)

(b)

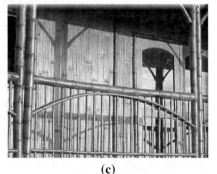

(c)

FIGURE 6.13   (a) Bamboo plants in Kyoto, Japan. (b) Kitchen cabinets made of bamboo. (c) Use of bamboo as exterior siding. *(Sources: (a) Wikipedia, photo by Paul Viaar, 2004; (b) and (c) Bamboo Technologies.)*

Bamboo's rapid regeneration, strength, and durability make it an environmentally superior alternative to conventional hardwood flooring. While some hardwood trees may require decades to reach maturity, bamboo can usually be harvested on a four- or five-year cycle, and the mature forest will continue to send up new shoots for decades. Pine forests are known to have the most rapid growth among tree species, but bamboo grass species used in flooring can grow much faster—more than three feet per day—and produce nearly twice as much harvestable fiber per year. At the same time, bamboo can yield a product that is 13% harder than rock maple and with a durability comparable to red oak. Nevertheless, one has to be careful with bamboo, as an inferior manufacturing process can sometimes neutralize the hardness benefits of the material itself. Indeed, a poor-quality topcoat will scratch no matter how durable the flooring. Bamboo canes have a natural beauty when exposed, which is why it is often used for paneling, furnishings, and cabinetry (Figure 6.13(b, c)).

Many developing countries around the world, especially those with more tropical climates and where bamboo often grows into larger-diameter canes, have a long vernacular tradition for its use in structures. However, the general use of bamboo in the United States is limited by a lack of architects and/or engineers trained in bamboo design, joining systems, harvesting, and treatment and strength of relevant species, as well as a lack of capable carpenters with the skills to economically and efficiently build bamboo structures. One of the issues confronting the sustainability of bamboo is the transportation it requires to get to North America. The majority of bamboo currently used in hardwood flooring comes from Hunan Province in China, which means it has to be shipped across the ocean and then put on trucks or trains to get to its final destination.

When using bamboo, particular care should be given to joinery details, since its strength comes from its integral structure, and it cannot be joined with many of the traditional methods used with wood. In this respect, the ancient ways of building with bamboo can be especially informative. Also, exposing bamboo to heavy moisture can eventually destroy it by fungus, which is what happens when it is exposed to adverse weather. Even with direct exposure, it is possible to prevent this by injecting the bamboo with chemical formulae of varying levels of environmental acceptability.

Tests have shown bamboo to be an extremely strong fiber having twice the compressive strength of concrete and roughly the same strength-to-weight ratio of steel in tension. Some bamboo fibers also have greater shear strength than structural woods, and take much longer to come to ultimate failure. However, bamboo has the ability to bend without breaking, which makes it unsuitable for floor structures due to our low tolerance for deflection and our unwillingness to accept a floor that has a "bouncy" feel. An appropriate substitute for the standard oak floor is a 3/4-inch-thick bamboo-finished floor because it installs the same way, is harder, and expands less. Likewise, vertically laminated flooring and plywood products that consist of layers of bamboo compressed with a binder can create a durable, resilient finish material. And when well maintained, bamboo floors can last decades. Like most interior-grade hardwood plywood, bamboo flooring is typically made with a urea-formaldehyde binder, which can emit tiny amounts of formaldehyde. To counter this and minimize indoor air pollution caused by use of UF, it is important to choose high-quality products, particularly from manufacturers that provide independent air quality testing data.

6.7.2 Cork

Ninety-nine percent of the world's cork grows in the sunny Mediterranean. Cork is a natural, sustainable product harvested from the bark of the cork oak, *Quercus suber*. It can be first harvested when it is 25 years old, when the virgin bark is carefully cut from the tree. It should be noted that from the three layers of the cork bark harvested, only the middle layer is used to make cork products. Moreover, the harvesting of cork does not require a single tree to be cut down, and only a percentage of the bark is removed from each tree, thus allowing it to maintain its protection while regenerating. Following this, the tree can be regularly "stripped" of its cork every 9 years for roughly 200 years without any harm coming to it. This helps to encourage long-term management of this renewable resource. It is estimated that an 80-year-old cork tree can produce nearly 500 pounds of cork. And unlike synthetic vinyl flooring, cork provides a resilient building alternative with a life span of 50+ years compared to the 10- to 20-year life span of synthetic flooring. It also has a negligible impact on energy performance at its point of use.

Cork applications are becoming increasingly popular due to its unusual characteristics: a combination of beauty, durability (can last for decades), insulation, and renewability. (See Figure 6.14.) Likewise, modern cork floors are durable and fire-resistant, and provide thermal and acoustic insulation. In addition to being soft on the feet, cork is antimicrobial and inherently resistant to mold and mildew, has low off-gassing from natural oils, does not produce chemicals during the

(a)   (b)

FIGURE 6.14 (a) Residential interior using cork flooring. (b) Cork pattern detail. *(Source: Globus Cork.)*

manufacturing process (dioxin specifically) and is completely biodegradable. Cork adapts well to weight and will recover from large amounts of pressure, which makes it appropriate for use in kitchens and laundry rooms. Cork floors are usually covered with an acrylic finish but may alternatively be covered with polyurethane for bathroom or kitchen applications.

Cork floors are sometimes considered as a natural alternative to carpet because cork provides the majority of the benefits of carpet without its liabilities. Carpet can attract and hold indoor pollutants in its fibers, whereas cork is easier to thoroughly clean, sheds no dust or fibers, and is naturally antistatic. In addition to its hypoallergenic properties, cork offers thermal and acoustic insulation. But the benefits of using cork go beyond human health; they include less landfill waste (it can be recycled back into the manufacturing process to minimize further waste), many products are locally obtainable, it has exemplary aesthetics, and it involves reduced ecological impacts of harvesting/mining the raw materials. However, there is significant regulation for cork harvesting (unlike bamboo), and to minimize potential damage to trees and eco-systems, countries that harvest cork monitor the frequency at which the resource can be harvested.

The extraction, manufacture, transport, and disposal of synthetic flooring materials, on the other hand, pollutes air and water, depletes resources, damages natural habitats, and can have negative health impacts.

6.7.3   Insulation

The majority of insulation used in buildings is for thermal purposes, but the term also applies to acoustic insulation, fire insulation, and impact insulation (e.g., for vibrations caused by industrial applications). Insulation therefore helps to protect a building's occupants from heat, cold, and noise; in addition, it reduces pollution while conserving the energy needed to heat and cool a building. Insulation materials will often be chosen for their ability to perform several of these functions at once, and well-insulated building envelopes are primary considerations in comfort and sustainability. Environmentally preferable insulation options can offer additional benefits, such as reduced waste and pollution in manufacture and installation, as well as more efficient resource use, better recyclability, improved R-value, and reduced or eliminated health risks for installers and occupants. The comfort and energy efficiency of a home or office depend on the R-value of the entire wall, roof, or floor (i.e., whole-wall R-value), not just the R-value of the insulation.

Fiberglass, which is usually the material of choice for insulating ceilings and walls, consists of extremely fine glass fibers. Its popularity is often based on economics, even though its use may present potential health risks. It is advisable that all fiberglass insulation used be formaldehyde-free with a minimum of 50% total recycled content (minimum 25% post-consumer). Some products are manufactured with heavier, intertwined glass fibers to reduce the amount of fibers becoming airborne and also to mitigate the fraction of fibers that can enter the lungs. One of the issues with fiberglass fibers is that they are friable and can easily become airborne, particularly during installation. These fibers can be inhaled, and some health experts claim that this particulate matter is carcinogenic.

Fiberglass insulation is similar to other glass products in that it is made primarily from silica heated to high temperatures, requiring significant energy and releasing formaldehyde. Short-term effects that may be experienced during installation or other contact include irritation to eyes, nose, throat, lungs, and skin. Longer-term effects are controversial, but OSHA now requires fiberglass insulation to carry a cancer warning label. Binders in most fiberglass batts contain toxic formaldehyde that continues to slowly emit for months or years after installation, potentially contaminating indoor air.

There are various environmentally preferable insulation options, including recycled cotton, which insulates as well as fiberglass and offers superior noise reduction. Cotton insulation typically comes in batt form and is easy to work with. It is also soft, not irritating during installation, and poses no health risk to the installer (unlike fiberglass) or occupants, nor does it use a formaldehyde-based binder. To maximize its energy performance, the insulation should be fitted to completely fill the wall and ceiling cavities without being compressed by pipes or wires. Cotton insulation can retain up to 15% moisture, which may not create problems in wall assemblies that are dry or are able dry out between cycles of water loading. However, care should be taken to avoid repeated wetting and drying, which can cause the borate treatment to seep out and encourage the growth of mold. Also, cotton insulation costs roughly twice as much as fiberglass, bearing in mind that insulation material costs are generally a very small percentage of the total cost of construction. In addition, the excellent sound and safety qualities of cotton insulation give it an edge over fiberglass.

BioBased Insulation® is essentially polyurethane spray-in-place foam that is manufactured with annually renewable soybean oil, which allows it to be classified as a rapidly renewable product. Such innovative products have excellent thermal and sound insulating properties that can help to provide a healthy, comfortable, energy efficient, and durable residential or commercial building. Depending on the size, cost, and amount of the project insulated, BioBased Insulation can contribute a significant portion of LEED's required 2.5% of the total value of all building materials and products used in the project.

Likewise, cellulose (recycled newspaper) insulation can be an acceptable alternative because it generates a nontoxic, fire-retardant insulation product. It also acts as an effective protective shield to reduce the transmission of heat or sound and is suitable for insulation of timber frame walls, attics, and lofts. It poses no health risk and offers superior R-value per inch. Both cellulose and cotton are treated with borate, which is not toxic to humans and makes both materials more resistant to fire and insects than fiberglass. Some cellulose insulation products are manufactured entirely from recycled newspaper that might otherwise end up in a landfill.

Sprayed polyurethane foam is sometimes used in large- to mid-scale applications, and is sprayed onto concrete slabs, into wall cavities of an unfinished wall, or through holes drilled in sheathing or drywall into the wall cavity of a finished wall to provide insulation, a vapor barrier, and additional shear strength. Although the cost of sprayed polyurethane foam can be high compared to traditional insulation, it is offset by many advantages, including increasing structural stability (unlike loose fill), and blocking airflow through expansion and sealing off leaks, gaps, and penetrations. Sprayed cementitious foams, such as Air Krete®, have similar properties. Air Krete is environmentally friendly, nontoxic, nonhazardous, fireproof, nonexpansive, insect-resistant, mold-proof, zero-VOC emitting, and insoluble in water.

6.7.4   Linoleum

Linoleum is a highly durable, environment friendly, resilient material that is used mainly for flooring. It is made from a mixture of natural materials such as solidified linseed oil (linoxyn), recycled wood flour, ground cork dust, pine rosin, and mineral pigments that are mounted onto a jute-fiber backing. Linoleum is also naturally antibacterial and biodegradable. The manufacture of linoleum requires mixing oxidized linseed oil (or a combination of oxidized linseed oil and tall oil) and rosin with the other raw materials to form linoleum granules. The granules are then pressed onto a jute backing, creating linoleum sheets. These sheets are then hung and allowed to cure in special drying rooms to achieve the required flexibility and resilience. Maximum waste reduction is achieved by recycling back any linoleum remnants into the production process. Linoleum manufacturing should be conducted in accordance with ISO 14001 standards. LEED credits may be given for purchasing local materials.

Although flexible vinyl flooring (often incorrectly referred to as linoleum) largely displaced linoleum in the marketplace in the 1960s, linoleum has made a dramatic reappearance as a flooring choice for those who are environmentally conscientious. In addition, being made of organic materials and purportedly nonallergenic in nature, high-quality linoleum continues to be in use in many places (especially in nonallergenic homes, hospitals, and health care facilities). The two materials are quite different. First the cost of linoleum is higher, but it offers certain performance advantages that are superior to vinyl, like lasting longer, being inherently antistatic, and antibacterial. Moreover, linoleum is all natural, requires less energy, and creates less waste in its manufacture, and it can be chipped and composted at the end of its useful life. Maintenance of linoleum is likewise less labor-intensive and less expensive as it does not require sealing, waxing, or polishing as frequently as vinyl. On the other hand, flexible vinyl flooring remains a more prolific generator of solid waste because it is manufactured from toxic materials and will typically last about 10 years; it is neither biodegradable nor generally recyclable. Linoleum also emits far fewer VOCs when installed with a low-VOC adhesive and does not exude the phthalate plasticizers that are an increasing concern for human health.

Vinyl tile is still favored over many other kinds of flooring materials in various commercial and institutional applications where high-traffic is anticipated, because of its characteristically low cost, durability, and ease of maintenance. The durability of hard vinyl composition tile (VCT) may be comparable to linoleum, but recycling has until recently been impractical, which is why VCT tile usually ends up in a landfill. However, new technology has allowed VCT to contain increasingly high percentages of recycled content, reducing energy consumption and waste generation. But vinyl products can nevertheless be harmful because their manufacture consumes petroleum and involves the generation of hazardous wastes and air pollution. Important attributes of linoleum include:

- It is made from all natural nontoxic materials, and these natural raw materials are available in abundance.
- It does not contain formaldehyde, asbestos, or plasticizers.
- It is very durable, often lasting for 25 to 40 years; this helps reduce waste associated with the frequent replacement of flexible vinyl flooring.
- It is 100% biodegradable at the end of its useful life.
- It is resilient, quiet, and comfortable.
- It can be 100% recycled. As a common alternative to incineration, linoleum can be safely added to landfill refuse sites, where natural decomposition takes place.
- It is easy to clean and maintain using gentle detergent with a minimal amount of water. However, linoleum floors can be kept in satisfactory condition for long periods without the need for major maintenance.
- While its resistance to temporary water exposure makes it suitable for use in kitchens, its sensitivity to standing water is a concern for use in bathrooms.
- Its natural bactericidal and antistatic properties helps control dust and dirt and the subsequent growth of household mites and/or bacteria.
- It contains virtually no trace of toxic material and therefore very low VOC emissions (no off-gassing) when installed with appropriate adhesives, and is thus naturally beneficial to air quality.
- Square-foot cost may be comparable to high-quality flexible vinyl flooring, although flexible vinyl is commonly replaced within 10 years (as opposed to 25 to 40 years for linoleum) and is toxic to manufacture; it is neither biodegradable nor recyclable.
- It is the same color all the way through, which permits gouges and scratches to be buffed out, reducing long-term costs and waste.

Linoleum is also considered to be a rapidly renewable resource, which has environmental advantages over finite raw material and long-cycle renewable resource extraction.

6.7.5　Straw-Bale Construction

Straw and reeds have been used as building materials in the Middle East for thousands of years. Today, straw-bale building consists of stacking rows of compressed blocks (bales) of straw (often in running bond), on a raised footing or foundation, with a moisture barrier between. This can be implemented as fill for a wall cavity (non-load-bearing) or as a structural component of a wall in which the bales may actually provide the support for openings and roof (load-bearing). The most common non-load-bearing approach is using a post-and-beam framework that supports the basic structure of the building, with the bales of straw being employed as infill (serving mainly as insulation and plaster substrate). This method is also the main one that is permitted in many jurisdictions, although many localities now have specific codes for straw-bale construction.

Until recently, "field bales," bales that were created on farms with baling machines, were used, but lately higher-density "recompressed" bales have come into use that are increasing the loads that may be supported. Whereas field bales may be capable of supporting roughly 600 pounds per linear foot of wall, the high-density bales are designed to support up to 4000 pounds per linear foot. This is particularly important in northern regions, where there is a potential for snow loading that can exceed the strength of the straw-bale walls.

In wet climates, it is necessary to apply a vapor-permeable finish that precludes the use of cement-based stucco commonly used on load-bearing bale walls, since the interior and exterior sides of a bale wall are usually covered by stucco, plaster, clay, or other treatment. This type of construction can offer structural properties superior to the sum of its parts. Both load- and non-load-bearing straw-bale designs divert agricultural waste from the landfill for use as a building material with many eco-friendly qualities.

Straw-bale construction is increasing in popularity in many parts of the country, partly because it lends itself well to an owner−builder project, so that today there are thousands of straw-bale homes in the United States. However, in designing load-bearing straw-bale buildings, architects and engineers must take into consideration the possible settling of the straw bales as the weight of the roof and other elements compress them. It is also important to ensure that the straw is kept dry, or it will eventually rot. For this reason it is generally best to allow a straw-bale wall to remain breathable; any incorrectly applied moisture barrier may invite condensation to collect and potentially undermine the structure. Additionally, the skin on the straw-bale walls should be treated to resist infestation of rodents and insects.

Straw-bale houses can typically save about 15% of the wood that would normally be needed in a conventionally framed house. However, it should be noted that straw bales do not hold nails as well as wood, and thus nailing surfaces need to be provided. Also, because of the specialized work that goes into plastering both sides of straw-bale walls (to provide thermal mass), and the extra expense needed to protect them from moisture, the cost of finishing a straw-bale house sometimes exceeds that of standard construction. Moreover, because straw-bale walls are thick, they may constitute a high percentage of the floor area, and this may be problematic on a small site. Nevertheless, the final product often provides excellent value because of the superior insulation and wall depth achieved. Advantages of employing straw-bale construction methods include:

- It provides extemporary thermal insulation, thereby enhancing occupant comfort. It is estimated that a well-built straw-bale home can save you nearly 75% on heating and cooling costs. Applying interior plaster to straw-bale houses increases the "thermal mass" of the home, which helps to stabilize interior temperature fluctuations.
- It provides superior acoustic insulation, which is particularly helpful for homeowners seeking to block out exterior noise emanating from traffic or airplanes in urban environments.
- It is environmentally friendly and does not require toxic treatment, thereby helping chemically sensitive individuals.
- Straw bales are inexpensive (or free), and owners, builders, and volunteers are able to contribute significantly to labor.
- It offers much greater fire resistance (roughly three times that of conventional construction).
- Typically, a traditional "stick frame" home of 2×6 construction will often have an insulating value of R-14, whereas with a properly insulated roof, straw-bale construction can increase this to R-35 to R-50.
- It reduces construction waste, which has a positive impact on the environment. The main building material is a waste product, and any excess straw can be used onsite in compost or as soil-protecting ground cover.
- Straw bales are biodegradable or reusable at the end of their useful life.
- It has potential for major reductions in wood and cement use, particularly in load-bearing straw-bale designs. In a load-bearing assembly, the wood in the walls can be completely eliminated, except for around the windows.
- Conventional foundations and roofs can be employed with straw-bale buildings.
- It is highly resistant to vermin (including termites).
- It provides great potential for aesthetic flexibility from conventional linearity to organic undulation.
- Using straw as insulation means that unhealthy insulation materials cease to be required. For example, fiberglass insulation generally has formaldehyde in it, a known carcinogen. Bale walls also eliminate the use of plywood (which often contains unhealthy glues) in the walls.

As can be deduced from the preceding, straw-bale construction provides a notable alternative building material that helps to reduce or eliminate many of the problems that plague the environment (Figure 6.15).

FIGURE 6.15   Interior of a residence built of straw bales. *(Source: StrawBale Innovations, LLC.)*

6.7.6   Wheat Board

One of the more popular renewable materials is wheat board, which is a fiber-composite byproduct of wheat straw. This material is environmentally friendly, has no formaldehyde, and can be used to create, among other things, quality furniture and cabinets. Wheat board is a durable material that is produced in 4×8 sheets of various thicknesses that possess superior properties compared to wood-based composite panels. It can be painted or laminated with a wide variety of surface treatments, sealed, stained, or varnished. It can also be shaped in a wide variety of designs, and is used in millwork, cabinetry, and finished-product applications for a renewable, nontoxic alternative to commercial MDF. Wheat board can also provide a sustainable alternative to traditional wood flooring and gypsum walls.

In the past, wheat board was burned or added to landfills. Today, there are a number of manufacturers that produce wheat board in the United States, Canada, and Europe. It is a viable substitute for wood and benefits the environment by reducing deforestation and lessening both air pollution and landfill use. In addition to its versatility, wheat board can also help a building project earn crucial LEED MR credits for rapidly renewable materials (Credit 6—Rapidly Renewable Resources), recycled content (Credit 4.1/ 4.2—Recycled Content), and indoor air quality (Credit 4.1—Low-Emitting Adhesives and Sealants).

6.8   GREEN OFFICE EQUIPMENT

Today the nation is transitioning to a more service-focused economy, a large portion of which is small businesses, in which the amount of energy-consuming office equipment will also increase. This is why for some time now enlightened designers, building owners, and leading-edge organizations have been searching for ways to reduce their environmental footprint by becoming greener, both at home and at the workplace. Likewise, manufacturers of office equipment are responding to consumer desire for more environmentally sound products. One of the main driving forces behind this trend has been the ENERGY STAR program, which was begun by the EPA in 1992.

Modern research shows that office equipment accounts for approximately 16% of an office's energy use, and buildings generally contribute about 40% of U.S. CO_2 emissions; therefore, even minor changes at the office can potentially have a significant impact. And although the energy component generally represents about 30% of operating expenses in a typical office building, making it the single largest and most manageable operating expense in the provision of office space, many facility operators unlike their informed counterparts, often fail to focus on the energy-consuming office equipment or appliances when they consider energy consumption. Instead, they unwittingly focus on a building's operations such as lighting and air conditioning and its shell components.

People often don't realize that the cost of energy to operate office equipment or an appliance over time can be substantially more than the original cost of the equipment itself. Moreover, choosing the most energy-efficient models available can have a positive impact on the environment while at the same time saving money. Still, more often than not, building owners and tenants decide to choose the cheapest and least expensive equipment or appliance on the market, which means that the owner/tenant ends up spending the minimum amount upfront on an appliance or piece of equipment, only to find that they are paying through the nose for years to come in recurring monthly energy costs. It is also important that all nonessential electrical equipment (TVs, copiers, computers, VCRs, etc.) be switched off at night or when not in use.

It is no secret that much of the energy and water that appliances and office equipment consume promptly translates into additional fuel being burned at power plants, which in turn contributes to air pollution, and waste of our limited natural resources. The good news is that continuous technological advances have helped create efficient new appliances and office equipment that can use only one-half to one-third as much energy as those purchased only a decade ago. The Lawrence Berkeley National Laboratory reports that with the replacement of older appliances and equipment such as refrigerators, dishwashers, clothes washers, thermostats, heating equipment, and incandescent lighting with ENERGY STAR equipment, enough energy and water can be saved to provide an average after-tax return on investment of over 16%. This is a substantially better return than the stock market. Moreover, in today's era of escalating energy costs and climate change, energy efficiency has become the most important aspect of greening a business.

In the United States, most retailers carry efficient, durable appliances and office equipment. Appliances and equipment with the ENERGY STAR label are typically preferred, although they are not necessarily always the most efficient of all available models. Nevertheless, ENERGY STAR products do usually perform significantly better than federal minimum efficiency standards. ENERGY STAR is a label from the EPA that identifies various energy-saving products in more than 60 product categories, including appliances, lighting, office equipment, and consumer electronics. According to a 2002 EPA report, ENERGY STAR-labeled office buildings generally generate utility bills that are 40% less than the average office building. In addition, there are often rebates and/or incentives for the purchase of energy- and water-saving appliances. Inefficient office equipment not only draws power, but also emits heat that can contribute to higher cooling bills.

Factors that should be considered when selecting office equipment and appliances:

- Always buy appliances and equipment that use the least energy and preferably are ENERGY STAR-qualified. The ENERGY STAR label indicates the most efficient light bulbs, computers, printers, copiers, refrigerators, televisions, windows, thermostats, ceiling fans, and other appliances and equipment.
- Install appliances and equipment that use the least water: low-flow showers, faucets, toilets, urinals, hose valves, and so forth (zero-water urinals, 1-gpf toilets, and 0.5-gmp faucets are all readily available).
- Choose durable appliances and equipment that meet long-term needs.
- For space and water heating select natural-gas appliances; gas is often more cost effective and can reduce overall energy, but, like other fossil fuels, it is not a renewable resource.
- Incorporate sealed-combustion and direct-vent furnaces and water heaters to increase IAQ.
- Consider using occupancy sensors in offices to minimize unnecessary lighting, as well as "smart" energy-efficient power strips that combine an occupancy sensor with a surge protector; smart power strips will shut down devices (such as monitors, task lights, space heaters, and printers) that can be safely turned off when space is unoccupied. The latest versions can be found with remote-control shutoffs, main shutoffs that can power down peripheral equipment, and motion-detecting shutoffs.
- Install low-cost energy monitors that can provide an accurate display of the cost and energy use of individual equipment. Research shows that this step alone can lead to energy savings of up to 40%.

By using green office equipment, there is also the potential to achieve LEED credit points. For example, for existing buildings the requirements for electronics can be found in the MR 2.1 section under "Sustainable Purchasing: Durable Goods, Electric," which states:

One point is awarded to projects that achieve sustainable purchases of at least 40% of total purchases of electric-powered equipment (by cost) over the performance period. Examples of electric-powered equipment include, but are not limited to, office equipment (computers, monitors, copiers, printers, scanners, fax machines), appliances (refrigerators, dishwashers, water coolers), external power adapters, and televisions and other audiovisual equipment.

As for residential construction, ENERGY STAR offers home buyers many of the features they desire in a new home, in addition to the energy-efficient improvements that deliver better performance, greater comfort, and lower utility bills. To earn the ENERGY STAR label, a home is required to meet strict guidelines for energy efficiency as set by the EPA. Such homes are typically 20 to 30% more efficient than conventional homes.

6.9   FORESTRY CERTIFICATION AND CERTIFIED WOOD

Forest certification was launched more than a decade ago to help protect forests from destructive logging practices. It involves the green labeling of companies and wood products that meet specific standards of "sustainable" or "responsible" forestry, while at the same time providing a means for independent organizations to develop standards of good forest management and for independent auditors to issue certificates to forest operations that comply with those standards. The intention is therefore to reflect a seal of approval and a means of notifying consumers that a wood or paper product comes from forests managed in accordance with strict environmental and social standards. For example, a person shopping for wall paneling or furniture would seek a certified forest product to ensure that the wood was harvested in a sustainable manner from a healthy forest and not procured from a tropical rainforest or the ancestral homelands of a forest-dependent indigenous people.

The primary purpose of forest certification is that it provides market recognition for forest producers that meet a set of agreed-on environmental and social standards. The importance of this certification is that it verifies that the forests are well managed, as defined by a particular standard, and ensures that certain wood and paper products come from responsibly managed forests. This is particularly important today since current forestry practices have a multitude of negative effects on the environment such as soil erosion and loss of wildlife habitat. The intent of the LEED Certified Wood credit is to encourage such environmentally responsible forest management programs. The Green Globes® Rating System, unlike the LEED Rating System, does not have an FSC-only policy. Moreover, there has been much controversy relating to certification in the LEED system and LEED has encountered considerable criticism.

In response to the USGBC's recently released new rating system, LEED 2012, the American Forest Foundation (AFF) states that while recognizing the significance of the proposed changes, it believes that the story for wood remains unchanged and the increased recognition of American Tree Farm System wood remains unchanged. AFF believes that "While wood is an energy efficient, renewable, carbon sequestering material, LEED has done very little to promote the environmental benefits of wood or to encourage builders to choose wood products—essentially blocking wood from the growing green building market."

LEED has also been at odds with the National Lumber and Building Material Dealers Association (NLBMDA) regarding certification. Earlier rewrites of the certified wood policy in the LEED Rating System unfortunately failed to get a two-thirds majority vote from USGBC members to become policy. Without a two-thirds majority, the policy failed to pass under LEED rules, and the certified wood credits will remain unchanged. Melissa Harden, a public affairs manager at AFF, says,

Builders and architects can collect few credits under LEED for using wood, and the credits that are related to wood products are even more restrictive. For example, under the current LEED system, the forest certification credit does not recognize the two largest forest certification standards in North America, the American Tree Farm System® and the Sustainable Forestry Initiative; only wood certified by the Forest Stewardship Council is recognized. LEED 2012 is even more restrictive in only recognizing "FSC Pure" certified wood products. Very few North American wood products are certified as FSC Pure.

Harden goes on:

While LEED 2012 does offer some new changes with seeming potential for increased recognition of wood products, these new changes come with additional questions. For example, LEED 2012 would allow materials, like wood, to achieve recognition through performance-based Life Cycle Assessment (LCA) and Environmental Product Declarations (EPDs).

Although LEED 2012 recognizes products that are "bio-based," it relies on the Department of Agriculture's definition and database of BioPreferred products. The BioPreferred® program currently does not include products with a mature market and therefore includes the majority of paper and wood products.

The criteria for LEED Certified Wood Credits continue to change, and it is best to be constantly updated by checking with the USGBC website. Previously, LEED awarded credit to projects that used wood certified to the standards of the FSC for at least half of their wood-based materials. The USGBC has now purportedly broadened the credit to recognize any forest certification program that meets its criteria. The change is partly in response to criticism that LEED favors one program, FSC, over others—in particular, the Sustainable Forestry Initiative (SFI), a rival to the FSC that is portrayed by some environmentalists as being less rigorous. Nevertheless, the revision brings the credit into line with a trend in LEED toward transparent criteria to determine which third-party certification programs to recognize.

At present the FSC remains the only certification program granted a "certified wood credit" in the LEED system, although LEED claims that it will recognize other programs that are found to be compliant with this benchmark. On the other hand, wood certification programs that are not found to be in alignment with the benchmark would have a clear and transparent understanding of what modifications are needed to receive recognition under LEED. A number of other programs, including SFI and the Canadian Standards Association (CSA), may face some difficulty with parts of the LEED benchmark system in obtaining certification.

In recent years, many progressive companies have decided, whenever possible, to buy wood and paper from FSC-certified suppliers, although they still use products from other sources to meet their needs. According to the USGBC, homebuilders can currently earn up to 6.5 points out of a possible 136 for using wood that is FSC-certified wood. Specifically, they can earn a half point each for using FSC-certified wood in the following:

- Exterior wall framing
- Exterior wall siding
- Flooring
- Floor framing
- Interior wall framing
- Decking
- Cabinets
- Counters
- Doors
- Trim
- Window framing
- Roof framing
- Sheathing

Scientific Certification Systems (SCS) has emerged as a global leader in certifying forest management operations and wood product manufacturers. It first developed its Forest Conservation Program in 1991, and in 1996 the FSC accredited it as a certification body, enabling it to evaluate forests according to FSC principles and criteria. Through a well-developed network of regional representatives and contractors, SCS provides timely and cost-effective certification services globally. Implementation requirements to achieve the LEED credit include:

- A minimum of 50% wood-based materials and products used that are certified in accordance with an accredited certifier (e.g., the FSC) for wood building components such as framing, flooring and subflooring, wood doors, and so forth.
- The basis is cost of sustainable wood products as determined by the ratio of total cost of wood material purchased for the project, unlike previous credits that were based on total materials purchased.
- MEP and elevator equipment is excluded.
- Furniture may be included in calculations.
- Contractor does not require the certification number but the supplier does.

To receive the new pilot credit, LEED projects must validate that, based on value, at least 10% of nonstructural products meet any of the following requirements:

- Product environmental claims are be verified by a third party.
- Products are to be certified to third-party multi-attribute performance standards.
- The product manufacturer must complete a life-cycle assessment report or a third-party verified Environmental Product Declaration for the product.

Many thought that the certified wood controversy had ended last December with USGBC's decision to keep the FSC wood as the one wood that gets a point for being sustainably harvested. However, a new and very confusing controversial credit has recently been introduced that appears to get a foot in the door for other certification systems; it was recently noted in a press release from UL Environment, which does third-party certification. The company explains that the credit is designed to promote use of certified products.

The recently released pilot credit will be used as a "testing ground" or "trial run" before its formal adoption into the LEED Rating System. This new pilot credit is important to the sustainable building community because certified products provide a mechanism for market transformation, and the credit awards points for improving performance, transparency, and evaluation of the environmental impact of products and materials. The president of UL Environment, Steve Wenc, echoes this: "This move toward increased performance, transparency, authenticity and third-party verification of manufacturers' claims will help transform the market." We have yet to see if it does.

6.10 LIFE-CYCLE ASSESSMENT AND COST ANALYSIS OF BUILDING MATERIALS AND PRODUCTS

Life-cycle cost analysis (LCCA) is discussed in greater detail in Chapter 10.

6.10.1 Life-Cycle Assessment

Life-cycle assessment (LCA, also known as life-cycle analysis) is the analysis method most directly related to sustainability. It is a technique for assessing environmental impacts associated with all stages of a product's life from cradle to grave, which begins with the extraction of raw materials from the earth to create the product through materials processing, manufacture, distribution, use, repair and maintenance, and disposal or recycling. It ends at the point when all materials are returned to the earth. LCA is a tool that takes into account the entire life cycle of a product. It assesses the many stages of a product's life from the point of view that they are interdependent, meaning that one operation leads to the next.

An important characteristic of LCA is that it enables estimation of the cumulative environmental impacts emanating from all stages in the product life cycle, often including impacts not considered in more traditional analyses (e.g., raw material extraction, material transportation, ultimate product disposal). Inclusion of these impacts throughout the product's life cycle provides for a more accurate assessment of a material's true environmental impact and allows LCA to offer a more complete view of the environmental features of the product or process and a fundamentally improved picture of the true environmental compromises in product and process selection. It also takes into account any resale or salvage value recovered during or at the end of the time period examined. This type of analysis allows for comprehensive and multidimensional product comparisons.

6.10.2   Life-Cycle Cost Analysis

According to the National Institute of Standards and Technology (NIST), life-cycle cost (LCC) is defined as "the total discounted dollar cost of owning, operating, maintaining, and disposing of a building or a building system" over a period of time. Life-cycle costing is therefore an economic evaluation methodology for assessing the total cost of acquiring, owning, and operating a facility over a period of time, and it takes into consideration relevant costs of alternative building designs, systems, components, materials, or practices in addition to the multiple impacts on the environment (both positive and negative) that building materials and certain products have. LCCA can prove especially useful when comparing project alternatives that fulfill similar performance requirements but differ with respect to initial and operating costs, so that one can be selected for maximum net savings. One of the advantages of LCCA is that it can be applied to both large and small buildings as well as to isolated building systems. However, LCCA is not helpful for budget allocation.

Whenever possible, anticipate and determine from the outset the potential health and safety issues that may emerge during the construction, occupation, maintenance, alteration, and disposal of the facility. Many astute building owners apply the principles of LCCA in decisions they make regarding construction or improvements to a facility. While initial cost has always been an important factor in the decision-making process, it is only one of several that impact the final decision.

Related to LCA, life-cycle costing measures the opportunity cost of one investment versus another and provides data as to which might provide a better return on investment (ROI). It provides a systematic evaluation of the financial ramifications of a material, a design decision, or a whole building and, unlike a simple payback analysis, examines both the initial cost of a project and the expected operations, maintenance, financing, useful life, and any salvage value that the project may have at the end of its life. LCC's tools can also help calculate future factors such as payback period, cash flow, present value, internal rate of return, and other financial criteria. It can then calculate a present value of the future investment using a discount rate percentage that is specific to the investor's requirements. This can go a long way toward comprehending how a modest upfront cost for environmentally preferable materials or design features can provide a very sound investment over the life of a building.

Lowest life-cycle cost (LLCC) is considered the most straightforward and easy-to-interpret system of economic evaluation. There are other commonly used measures, such as internal rate of return, savings-to-investment ratio, net savings (or net benefits), and payback period. If the same criterion and length of study period are used, then these systems are all compatible with the lowest LCC measure of evaluation. Quantity surveyors, architects, cost engineers, and others might choose any or several of these techniques to evaluate a project. The approach to making cost-effective choices for building projects can be quite similar whether it is called cost estimating, value engineering, or economic analysis. Open-book accounting, when shared across the whole project team, helps everyone to see and appreciate the project's actual costs. The LCCA should typically be performed early in the design process while there remains an opportunity to refine or modify the design in a way that would reduce LCC.

Sieglinde Fuller of NIST says, "The first and most challenging task of an LCCA, or any economic evaluation method for that matter, is to determine the economic effects of alternative designs of buildings and building systems and to quantify these effects and express them in dollar amounts." Fuller also believes that when viewed over a 30-year period, initial building costs have been shown to generally account for approximately just 2% of the total, while operations and maintenance costs equal 6%, and personnel costs reflect the lion's share at 92% (Romm, 1994).

A variety of building-related costs are associated with acquiring, operating, maintaining, and disposing of a building or building system. These costs typically fall into one of several categories (see Chapter 10), including:

- First costs: purchase, acquisition, construction
- Operation, maintenance, and repair costs
- Fuel costs
- Replacement costs
- Residual values: resale, salvage, or disposal
- Finance charges: loan interest payments
- Nonmonetary benefits or costs

It is not necessary to include costs that are minor and insignificant or costs within each category that are irrelevant to making a valid investment decision. For relevancy, costs should vary from one alternative to another. Significance is achieved when the costs are large enough to make an appreciable difference in the LCC of a project alternative. For calculation purposes, costs are entered as base-year amounts in today's dollars. The LCCA method is then applied, which accelerates these amounts to their future year of occurrence and then discounts them back to the base date to convert them to current dollar values.

Of note, detailed construction cost estimates are not required for preliminary economic analyses of alternative building designs or systems. Detailed estimates are usually unavailable until the design is fairly advanced and the possible contingency for cost-reducing modifications has been missed. LCCA can be repeated throughout the different stages of the design process whenever more detailed cost information becomes available. To start with, construction costs may be estimated by referencing historical data from similar facilities, or they can be estimated using government or private-sector cost-estimating guides and databases.

Detailed cost estimates rely mainly on cost databases such as the *RS Means Building Construction Cost Data*. They are usually prepared at the submittal stages of design and are based on quantity take-off calculations. There are also several well-known testing organizations such as ASTM International and NIST, as well as various trade organizations, that have reference data available for materials and products they represent or have tested. To avoid or minimize cost overruns, the discerning owner/developer needs to have:

- A complete sustainable design that meets planning and statutory requirements and that will not later necessitate modification
- A project brief that is comprehensive, unambiguous, and consistent
- Green goals that are appropriate and unlikely to be subject to modification during the course of the project
- A coordinated sustainable design that from the beginning takes into account factors such as maintenance, health and safety, indoor air quality, and the like
- An uncomplicated payment mechanism that incentivizes the parties to achieve common and agreed objectives
- Clear leadership with a qualified project team and appropriate management controls in place
- Project estimates using BIM technology or similar to provide realistic and comprehensive cost estimates
- An appropriate, clear, and unambiguous risk allocation and contingency that is clear and unambiguous

6.11   THIRD-PARTY CERTIFICATION

The LEED Third-Party Certification Program is an internationally recognized green building certification system and benchmark for the design, construction, and operation of high-performance green buildings. It certifies that a building or community was designed and built employing strategies and methods intended to improve performance in metrics such as energy savings, water efficiency, and improved indoor environmental quality. According to Alice Soulek, vice president of LEED Development:

Third-party certification is the hallmark of the LEED™ program," and "Moving the administration of LEED™ certification under GBCI will continue to support market transformation by delivering auditable third-party certification. Importantly, it also allows USGBC to stick to the knitting of advancing the technical and scientific basis of LEED™.

Moving administration of the LEED-certification process to the Green Building Certification Institute (GBCI), a nonprofit organization established in 2007 with the support of USGBC, was a wise decision, and it is having far-reaching positive ramifications for the USGBC and its influential LEED Rating System. Working with selected certification bodies, GBCI is now in a better position to deliver a substantially improved, ISO-compliant certification process that will continue to grow with the green building movement.

USGBC has decided to outsource LEED certification to independent, accredited certifiers overseen by GBCI. In that respect, LEED V3 has announced the names of the certification bodies for its updated Rating System. The companies are well known and respected for their role in certifying organizations, processes, and products to ISO and other standards. A list of the members can be found in Chapter 2, Section 2.3.4. This development in the certification process was undertaken as an integrated part of a major update to the technical Rating System that was put in place as LEED 2009. The update also includes a comprehensive technology upgrade to LEED-Online aimed at improving the user experience and expanding its portfolio management capabilities.

To acquire LEED credits, third-party testing and certification are required so as to provide an independent analysis of manufacturers' environmental performance claims based on established standards. This provides building owners and operators with the tools to have an immediate and measurable impact on their buildings' performance. Sustainable building strategies should be considered early in the development cycle. An integrated project team will include the major stakeholders of the project, such as the developer/owner, asset and property-management staff, BIM manager, architect, engineer, contractor, and landscape architect.

Making choices based on third-party analysis is often easier than LCA, but it is vitally important to determine the independence, credibility, and testing protocols of the third-party certifiers. Michelle Moore, senior vice president of policy and public affairs at USGBC, says, "We believe in third-party certification," and continues:

The USGBC provides independent third-party verification to ensure that a building meets these high performance standards. As part of this process, USGBC requires technically rigorous documentation that includes information such as project drawings and renderings, product manufacturer specifications, energy calculations, and actual utility bills. This process is facilitated through a comprehensive online system that guides project teams through the certification process. All certification submittals are audited by third-party reviewers.

Moore also believes that the separation of LEED from the certification process will bring it into alignment with norms established by the ISO for certification programs.

In conclusion, for healthy IAQ, green building materials and methods should typically have zero or low emissions of toxic or irritating chemicals and be moisture- and mold-resistant. Green materials and products are typically manufactured with a low-pollution process from nontoxic components, have low maintenance requirements, and do not require the use of toxic cleansers. This may explain why most green materials do not emit VOCs, particularly indoors, and are free of toxic materials such as chlorine, lead, mercury, and arsenic. While individual products do not carry LEED points, they can nevertheless contribute to them. Green building strategies include the monitoring of indoor pollutants and poor ventilation by radon and carbon monoxide detectors. The use of ozone-depleting gases such as halons and HCFCs are to be avoided.

Chapter 7

Indoor Environmental Quality

7.1 INTRODUCTION

One of the major concerns we face today in homes, schools, and workplaces is poor indoor environmental quality (IEQ), which can lead to poor health, learning difficulties, and productivity problems. This is particularly worrying since the majority of us spend most of our time indoors (especially in the United States); it is not surprising therefore that we should expect our indoor environment to be healthy and free from a plethora of hazardous pollutants. Indoor pollution is found to exist under many diverse conditions, from dust and bacterial build-up in ductwork to secondhand smoke and the off-gassing of paint solvents, all of which are potential health hazards.

Studies by the American College of Allergies show that roughly 50% of all illness is aggravated or caused by polluted indoor air (Figure 7.1). Moreover, cases of building-related illness (BRI) and sick building syndrome (SBS) continue to rise. In fact, recent studies point to the presence of more than 900 possible contaminants, from thousands of different sources, in a given indoor environment. It is not surprising therefore that indoor air pollution is now generally recognized as having a greater potential impact on public health than most types of outdoor air pollution, causing numerous health problems from respiratory distress to cancer.

Furthermore, a building interior's air quality is one of the most pivotal factors in maintaining its occupants' safety, productivity, and well-being. This heightened public awareness has led to a sudden surge of building occupants demanding compensation for their illnesses. Tenants are suing not only building owners but also architects, engineers, and others involved in the building's construction. Building owners are shifting the blame by making claims against the consultant, the contractor, and others involved in the facility's construction. But while architects and engineers to date have not been a major target of publicity or litigation arising out of indoor air quality (IAQ) issues, nevertheless, the potential scope and cost of some of the incidents have led to everyone associated with a project being blamed when the inside air of a building appears to be the cause of its occupants becoming sick. This is causing great concern among design professionals because it can ultimately result in a loss of reputation, as well as time and financial losses.

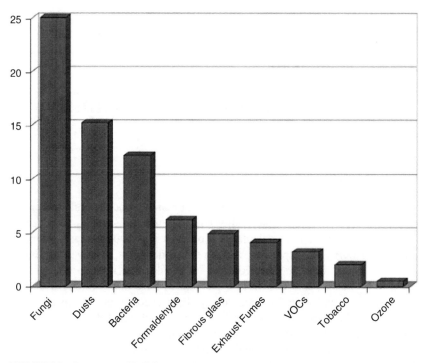

FIGURE 7.1   Percentage of buildings that have an inappropriate concentration of contaminants. *(Source: HBI Database.)*

Because of intense competition to maintain high occupancy rates, forward-thinking owners and managers of offices and public buildings find themselves under increasing pressure to meet or exceed the demands of the marketplace in attracting and retaining tenants. Furthermore, technological breakthroughs are bringing down the cost of facility-monitoring systems and making them more affordable for a wider range of building types. By reducing the cost of facility monitoring, many financial and maintenance obstacles are removed, making permanent monitoring systems an appropriate consideration for a broader range of facilities managers. Schools, health care facilities, and general office buildings can benefit from measuring many of their environmental conditions and using that information to respond to occupant complaints, optimize facility performance, and keep energy costs in check.

In addition, feedback from the indoor environment can be used to establish baselines for building performance and to document improvements to indoor air quality. Facility-monitoring systems can be valuable instruments for improving indoor air quality, identifying energy savings opportunities, and validating facility performance. Automating the process of recording and analyzing relevant data and providing facility managers adequate access to this information can improve their ability to meet the challenge of maintaining healthy, productive indoor environments.

7.1.1 Causes of Indoor Pollution

Poor indoor air quality is usually the result of sources that release gases or particles into the air. Inadequate ventilation is generally considered the single most common cause of pollutant build-up (Figure 7.2(a)) because it can increase indoor pollution levels by not bringing enough outdoor air in to dilute emissions from indoor sources and by not removing indoor air pollutants to the outside. High temperature and humidity levels can also increase concentrations of some pollutants. Another common cause of pollutant build-up is inefficient filtration (Figure 7.2(b)). But despite fundamental improvements in air filter technology, too many buildings continue to persist in relying on inefficient filters, or continue to be negligent in the maintenance of acceptable filters.

There are several factors that can trigger an investigation into indoor air quality contamination, including the presence of biological growth (mold), unusual odors, adverse health concerns of occupants, and a variety of other symptoms or observations, such as respiratory problems, headaches, nausea, irritation of eyes, nose, or throat, fatigue, and so on. Any information that is extracted from continuous monitoring can help minimize the total investigative time and expense needed to respond to occupant complaints; the information can also be used proactively in the optimization of building performance. Indoor environmental quality (IEQ) and energy efficiency may be classified into three basic categories: (1) comfort and ventilation, (2) air cleanliness, and (3) building pollutants.

Within these basic categories, facility-wide monitoring systems are available that can provide independent measurement of a range of parameters, such as temperature, humidity, total volatile organic compounds (TVOCs), carbon dioxide (CO_2), carbon monoxide (CO), and airborne particulates. Unfortunately, until recently there has been insignificant federal legislation controlling indoor air quality. This has changed with the adoption of the new International Green Construction Code (IgCC) in the United States. Also, several engineering societies such as the American Society of Heating, Refrigerating, and Air-Conditioning Engineers (ASHRAE) have established guidelines that have

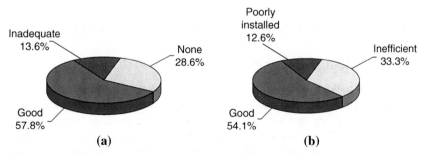

FIGURE 7.2 (a) Inadequate ventilation is the single most common cause of pollutant build-up. (b) Inefficient filtration is the second most important factor in indoor pollution. *(Source: HBI Database.)*

been generally accepted by designers as minimum air design requirements for commercial buildings. ASHRAE has established these two procedures for determining minimum acceptable ventilation rates:

Ventilation rate procedure. This stipulates a minimum ventilation rate based on space functions within a specified building type and is based on respiration rates resulting from occupants' activities.

Indoor air quality procedure. This requires the monitoring of certain indoor air contaminants below specified values. Air-sampling techniques require the use of a device to impinge organisms from a specific volume of air and place them in a sterile agar growth medium. The sample is then incubated for a specified period of time (say seven days). The colonies are counted and the results recorded. When testing the air of a potentially contaminated area, it is best practice to have comparative samples of air from both the contaminated area and outside the potentially contaminated building.

7.1.2   Sick Building Syndrome

Sometimes building occupants complain of symptoms that do not appear to fit the pattern of any specific illness and are difficult to trace to any specific source. This has been labeled sick building syndrome (SBS) and is a fairly recent phenomenon. It is a term used to describe situations in which building occupants experience acute health and discomfort effects that appear to be attributed to time spent in a building, but often no specific illness or cause can be identified. Complaints may be localized in a particular room or zone, or may be widespread throughout the building. Factors that may impact SBS include noise, poor lighting, thermal discomfort, and psychological stress.

The Environmental Protection Agency's (EPA) Indoor Air Quality website contains pertinent information regarding strategies for identifying the causes of SBS as well as finding possible solutions to the problem. According to industry IAQ standards, SBS is diagnosed if significantly more than 20% of a building's occupants complain of adverse health effects such as headaches, eye irritation, fatigue, and dizziness, and so on, over a period of two weeks or more, but without a clinically diagnosable disease being identified, and if the SBS symptoms disappear or are diminished when the complainant leaves the building.

7.1.3   Building-Related Illness

In contrast, building-related illness is the general term used to describe symptoms for a medically diagnosable illness that is caused by or related to occupancy of a building and which can be attributed directly to airborne building contaminants. The causes of BRIs can be determined and are typically related to allergic reactions and infections. It has been known for some time that indoor environments strongly affect human health. The EPA, for example, has estimated that pollutant concentration levels (e.g., volatile organic compounds) inside a building may be two to five times higher than outside levels. A 1997 study by W. J. Fisk and A. H. Rosenfeld (Fisk and Rosenfeld, 1997) estimates that the cost to the nation's workforce of upper respiratory diseases in 1995 reached $35 billion in lost work, which doesn't include an estimated additional $29 billion in health care costs. The report suggests that by just having healthier and more efficient indoor environments, these costs could be reduced by 10 to 30%. To fully profit from the fiscal, physical, and psychological benefits of healthy buildings, projects need to incorporate a comprehensive, integrated design and development process that seeks to:

- Ensure adequate ventilation
- Provide maximum access to natural daylight and views to the outdoors
- Eliminate or control sources of indoor air contamination
- Prevent water leaks and unwanted moisture accumulation
- Improve the psychological and social aspects of space

The current marketplace shows that many building products contain chemicals that evaporate or "off-gas" for significant periods of time after installation. When substantial quantities of these products are used inside a building, or products are used that have particularly strong emissions, they pollute the indoor air and can be hazardous. Some products readily trap dust and odors and release them over time. Building materials, particularly when damp, can also support growth of mold and bacteria, which can cause allergic reactions, respiratory problems, and persistent odors, (i.e., SBS symptoms). There are currently several environmental rating methods for buildings, but it is not always clear whether these methods assess the most relevant environmental aspects or whether other considerations lie behind the specific methods chosen. General concern for occupant health continues to increase with increased awareness, and this has translated into public demand for more exacting performance requirements for materials selection and installation, improved ventilation practices, and better commissioning and monitoring protocols.

The Insurance Information Institute (III) reports a dramatic increase in IAQ-related lawsuits in the United States and that there are currently thousands of IAQ-related cases pending. This follows several lawsuits with large damage awards in recent years brought by building occupants suffering from health problems linked to chemicals off-gassed from building materials, which are setting legal precedents across the country. The flood of IAQ-related lawsuits has prompted insurance companies to reexamine their policies and their clients' design and building methods. An effective way to reduce health risks and thus minimize potential liability is to follow a rigorous selection procedure for construction materials aimed at minimizing harmful effects to occupants.

7.2 FACTORS THAT AFFECT INDOOR ENVIRONMENTAL QUALITY

A report on IEQ released in July 2005 states that there are a growing number of people suffering a range of debilitating physical reactions from exposures to everyday materials and chemicals found in building products, floor coverings, cleaning products, and fragrances, among others. In addition, there are those who have developed an acute sensitivity to various types of chemicals, a condition known as multiple chemical sensitivity (MCS).

7.2.1 Indoor Air Quality

The quality of the indoor environment can have a profound impact on the health and productivity of employees and tenants, and studies consistently reinforce the correlation between improved IEQ and occupants' health and well-being. The adverse affects to building occupants caused by poor air quality and lighting levels, the growth of molds and bacteria, and off-gassing of chemicals from building materials can be significant. One of the chief characteristics of sustainable design is to support the well-being of building occupants by reducing indoor air pollution. This can best be achieved through the selection of materials with low off-gassing potential, appropriate ventilation strategies, providing adequate access to daylight and views, and providing optimum comfort through control of lighting, temperature levels, and humidity.

Inorganic Contaminants

Inorganic substances such as asbestos, radon, and lead are among the leading indoor contaminants, exposure to which can create significant health risks.

Asbestos

This is a generic term given to a variety of naturally occurring, hydrated fibrous silicate minerals that possess unique physical and chemical properties that distinguish them from other silicate minerals. Such properties include thermal, electric, and acoustic insulation; chemical and thermal stability; and high-tensile strength, all of which have contributed to asbestos' wide use in the construction industry (Table 7.1). High concentrations of airborne asbestos can occur during demolition and after asbestos-containing materials are disturbed by cutting, sanding, and other activities. Asbestos-containing materials are also found in concealed areas such as wall cavities, below ground level, and other hidden spaces. In many older establishments, asbestos-based insulation was used on heating pies and on the boiler. An adequate asbestos survey requires the inspector to perform destructive testing (i.e., opening walls, etc.) to inspect areas likely to contain suspect materials.

The Environmental Protection Agency (EPA) and the U.S. Consumer Product Safety Commission (CPSC) have banned several asbestos products. But

TABLE 7.1 Partial List of Materials That May Contain Asbestos

• Acoustical ceiling texture	• Gray roofing paint
• Asphalt flooring	• High-temperature gaskets
• Base flashing	• HVAC duct insulation
• Blown-in insulation	• Incandescent light fixture backing
• Boiler/tank insulation	• Joint compound/wallboard
• Breaching insulation	• Laboratory hoods/table tops
• Brick mortar	• Laboratory fume hood
• Built-up roofing	• Mudded pipe elbow insulation
• Caulking/putties	• Nicolet (white) roofing paper
• Ceiling tiles/panels/mastic	• Packing materials
• Cement board	• Paper fire box in walls
• Cement pipes	• Pipe insulation/fittings
• Cement roofing shingles	• Plaster/wall joints
• Chalkboards	• Poured flooring
• Construction mastics	• Rolled roofing
• Duct tape/paper	• Roofing shingles
• Ductwork flexible connections	• Sink insulation
• Electrical cloth	• Spray-applied insulation
• Electrical panel partitions	• Stucco
• Electrical wiring insulation	• Subflooring slip sheet
• Elevator brake shoes	• Textured paints/coatings
• Fire blankets	• Vapor barrier
• Fire curtains/hose	• Vermiculite
• Fire doors	• Vinyl floor tile/mastic
• Fireproofing	• Vinyl sheet flooring/mastic
• Furnace insulation	• Vinyl wall coverings

Note: This is intended as a general guide to show which types of materials may contain asbestos and is not all-inclusive.

asbestos-containing material became more of a high-profile public concern after federal legislation known as the Asbestos Hazard Emergency Response Act (AHERA) was enacted in 1987. Today, asbestos can still be found in older homes, in pipe and furnace insulation materials, asbestos shingles, millboard, textured paints and other coating materials, and floor tiles. Asbestos is considered the most widely recognized environmentally regulated material (ERM) during building evaluations.

Health Risks The risk of airborne asbestos fibers is generally low when the material is in good condition. However, when the material becomes damaged or if it is located in a high-activity area (family room, workshop, laundry, etc.) the risk increases. Increased levels of exposure to airborne asbestos fibers cause disease. When these fibers get into the air, they may be inhaled and remain and accumulate in lung tissue, where they can cause substantial health problems, including lung cancer, mesothelioma (a cancer of the chest and abdominal linings), and asbestosis (irreversible lung scarring that can be fatal). Symptoms of these diseases do not show up until many years after exposure. Studies indicate that people with asbestos-related diseases were usually exposed to elevated concentrations on the job, although some developed disease from exposure to clothing and equipment brought home from job sites. While the process is slow, and years may pass before health problems are evidenced, the results and thus the risk are well established.

Radon

Radon (Rn) is a natural odorless, tasteless, radioactive gas that is emitted from the soil as a carcinogenic byproduct of the radioactive decay of radium-226, which is found in uranium ores. The byproduct can, however, cling to dust particles that, when inhaled, settle in bronchial airways. Generally, radon is drawn into a building environment by the presence of air pressure differentials. The ground beneath a building is typically under higher pressure than the basement or foundation. Air and gas move from high-pressure areas to low-pressure areas. The gas can enter the building through cracks in walls and floors, as well as penetrations associated with plumbing, electrical openings, sump wells, and so on, in building spaces coming in close contact with uranium-rich soil. Vent fans and exhaust fans also put a room under negative pressure and increase the draw of soil gas, which can increase the level of radon within a building.

Radon exposure becomes a concern when it gets trapped in buildings and indoor levels of concentrations build up, which is why adequate ventilation is necessary to prevent the gas from accumulating in buildings to dangerous levels, as this can pose a serious health hazard. Where radon is suspected, a survey should be conducted to measure its concentrations in the air and determine whether any actions will be required to reduce the contamination. Radon levels will vary from region to region, season to season, and one building to another. They are typically at their highest during the coolest part of the day when pressure differentials are at their greatest.

High concentrations of radon in the air are often an indication of possible radon contamination of the water supply (if a private water supply is present). In this case, a water test for radon is the prudent first step. Should high concentrations of radon be found in the water, an evaluation of ventilation rates in the structure as well as air quality tests are highly recommended. Generally speaking, high radon concentrations are more likely to exist where there are large rock masses, such as in mountainous regions. The EPA recommends that buildings should be tested every few years to assess the safety of radon levels.

Mitigation Everything being equal, elevated radon levels should not necessarily deter investors from purchasing a property, as the problem can usually be easily be resolved—even in existing buildings—without having to incur great expense. However, lowering high radon levels requires technical knowledge and special skills, which means that a trained radon reduction contractor who understands how to fix radon problems should be used. The EPA has published several brochures and instructional aides regarding radon-resistant construction. This is perhaps the most cost-effective way to handle a radon problem, as it is easier to build the system into the building than retrofit later. Also, EPA studies suggest that elevated radon levels are more likely to exist in energy-efficient buildings than otherwise. If your building has a radon system built in, the EPA recommends periodic testing to ensure that it is working properly and that the radon level in your building has not changed.

Health Risks The principal health hazard associated with exposure to elevated levels of radon is lung cancer. Research suggests that while swallowing water with high radon levels may also pose risks, these are believed to be much lower than those from breathing air containing radon. However, the real threat is not so much from the radon gas itself but from the products that it produces, such as lead, bismuth, and polonium, when it decays. The risk is greatest for people with diminished lung capacity, asthma sufferers, smokers, and so on.

Testing Methods There are many methods that can be used for radon testing. For short-term testing, consultants typically use electret ionization chambers, which generally last about a week. The chambers work by incorporating a small charged Teflon® plate screwed into the bottom section of a small plastic chamber. When the radon gas enters the chamber, it begins to decay and creates charged ions that deplete the charge on the plate. By registering the voltage prior to deployment and then reading the voltage on recovery, a mathematical formula is used to calculate the radon concentration levels within the building.

Lead

For several decades, lead has been recognized as a harmful environmental pollutant, and in late 1991 the Secretary of the Department of Health and Human Services described it as the *"number one environmental threat to the health of children in the United States."* There are many ways in which humans may be exposed to lead: as mineral particles in ambient air, drinking water, food, contaminated soil, deteriorating paint, and dust. Lead is a heavy metal and does not break down in the environment; it continues to be used in many materials and products to this day. Lead is a natural element, and most of it in use today is inorganic, entering the body when an individual breathes (inhales) or swallows (ingestion) lead particles or dust once it has settled. Lead dust or particles cannot penetrate the skin unless the skin is broken. Organic lead, however, such as the type used in gasoline, can penetrate the skin.

Lead Levels in the Indoor Environment   Because of its widespread use and the nature of its individual uses, lead has for some time been known to be a common contaminant of interior environments. For centuries, lead compounds such as white lead and lead chromate have been used as white pigments in commercial paints. In addition to their pigment properties, these compounds were valued because of their durability and weather resistance, which made their use more viable particularly in exterior white paints.

Old lead-based paint is considered the most significant source of lead exposure in the United States today. In fact, the majority of homes and buildings built before 1960 contain heavily leaded paint. Even as recently as 1978, there were homes and buildings that used lead paint. This paint may have been used on window frames, walls, the building's exterior or other surfaces. Because of potentially serious health hazards and negative publicity, lead content was gradually reduced until it was eliminated altogether in 1978 (in the United States). In commercial buildings, lead was used primarily as a paint preservative.

Lead piping has sometimes been used in older buildings, and while not legally required to be replaced, it can create a health hazard because it is frequently found to be deteriorating and leaching into the building's drinking water. In some buildings, lead solder has also been used in copper pipe installation, but in most jurisdictions this procedure is now banned because of water contamination resulting from its deterioration. The potential for water contamination can often be removed by chemical treatment of the water. Where this cannot be accomplished, the piping may have to be replaced.

Health Risks   Lead is a highly toxic substance that affects a variety of target organs and systems in the body, including the brain and the central nervous system and the renal, reproductive, and cardiovascular systems. High levels of lead exposure can cause convulsions, coma, and even death. However, the nervous system appears to be the main target for lead exposure. Effects of lead poisoning depend largely on dose exposure. Contact with lead-contaminated dust is the primary means by which most children are exposed to harmful levels. Pregnant women, infants, and children are more vulnerable to lead exposure than adults because lead is more easily absorbed into growing bodies, and the tissues of small children are more sensitive to its damaging effects. In adults, high lead levels have many adverse effects, including kidney damage, digestive problems, high blood pressure, headaches, diminishing memory and concentration, mood changes, nerve disorders, sleep disturbances, and muscle or joint pain.

Testing Methods Various testing methods and procedures can be used to identify the presence of lead paint. In the field, the most widely applied is an X-ray fluorescent (XRF) lead-in-paint analyzer. The XRF analyzer is normally held up to the surface being tested for several seconds. It then emits radiation, which is absorbed and then fluoresces (is emitted) back to the analyzer. The XRF unit breaks down the signals to determine if lead is present and, if so, in what concentration. It can normally read through up to 20 layers of paint, but it is expensive and should only be used by trained professionals.

Combustion-Generated Contaminates

There are many combustion byproducts such as fine particulate matter, carbon monoxide (CO), and nitrogen oxides. Tobacco smoke is another source. Combustion (burning) byproducts are essentially gases and tiny particles created by the incomplete burning of fuels. These fuels (such as natural gas, propane, kerosene, fuel oil, coal, coke, charcoal, wood, and gasoline, and materials such as tobacco, candles, and incense), when burned, produce a wide variety of air contaminants. If fuels and materials used in the combustion process are free of contaminates and combustion is complete, emissions are limited to carbon dioxide (CO_2), water vapor (H_2O), and high-temperature reaction products formed from atmospheric nitrogen (NO_x) and oxygen (O_2). Sources of combustion-generated pollutants in indoor environments are many and include wood heaters and woodstoves, furnaces, gas ranges, fireplaces, and car exhaust (in an attached garage). Other combustion-generated contaminates include respirable particles (RSP), aldehydes such as formaldehyde (HCHO) and acetaldehyde, as well as a number of volatile organic compounds (VOCs); fuels and materials containing sulfur produce sulfur dioxide (SO_2). Particulate-phase emissions may include tar and nicotine from tobacco, creosote from wood, inorganic carbon, and polycyclic aromatic hydrocarbons (PAHs).

Carbon Dioxide

Carbon dioxide (CO_2) is a colorless, odorless, heavy, incombustible gas that is found in the atmosphere and formed during respiration. It is typically obtained from the burning of gasoline, oil, kerosene, natural gas, wood, coal, and coke. It is also obtained from carbohydrates by fermentation, by reaction of acid with limestone or other carbonates, and naturally from springs. CO_2 is absorbed from the air by plants in a process called photosynthesis. Although it is not normally a safety problem, a high CO_2 level can indicate poor ventilation, which in turn can lead to a build-up of particles and more harmful gases such as carbon monoxide that can negatively impact people's health and safety. CO_2 is used extensively in industry as dry ice or carbon dioxide snow, in carbonated beverages, in fire extinguishers, and so on.

Carbon Monoxide

Carbon monoxide (CO) is an odorless, colorless, lighter-than-air, nonirritating gas that interferes with the delivery of oxygen throughout the body. It is the leading cause of poisoning deaths in the United States and occurs when there is incomplete combustion of carbon-containing material such as coal, wood, natural gas, kerosene, gasoline, charcoal, fuel oil, fabrics, and plastics. The fact that CO cannot be seen, smelled, or tasted makes it especially dangerous because you are not aware when you are being poisoned. Moreover, doctors frequently misdiagnose CO poisoning.

Testing Methods The only reliable method currently used to test for the presence of carbon monoxide is an electronic carbon monoxide alarm. In the home, detectors should be placed in areas where the family spends most of its time, such as the family room, bedroom, or kitchen, but placed far enough away from obvious and predictable sources of CO, such as a gas stove, to avoid false alarms.

Nitrogen Dioxide

Nitrogen dioxide (NO_2) is a colorless, odorless gas that irritates the mucous membranes in the eye, nose, and throat and causes shortness of breath with exposure to high concentrations. It is also a major concern as an air pollutant because it contributes to the formation of photochemical smog, which can have significant impacts on human health.

Organic Contaminants: Aldehydes, VOCs/SVOCs, and Pesticides

Modern industrialized societies have developed such a massive array of organic pollutants that it is becoming increasingly difficult to generalize in a meaningful way as to sources, uses, or impacts. The main organic compounds include VOCs, the very volatile organic compounds (VVOCs), semivolatile organic compounds (SVOCs), and particulate organic materials (POMs). POMs may comprise components of airborne or surface dusts. Organic compounds often pose serious indoor contamination problems and include the aldehydes, VOCs/SVOCs, which include a large number of volatile as well as less volatile compounds, and pesticides and biocides, which are largely SVOCs. Organic compounds that are known to be contaminates of indoor environments include a large variety of aliphatic hydrocarbons, aromatic hydrocarbons, oxygenated hydrocarbons (such as aldehydes, ketones, alcohols, ethers, esters, and acids), and halogenated hydrocarbons (primarily those containing chlorine and fluorine). Volatile organic compound concentration levels are generally higher in indoor environments than in outdoor air.

In recent years, we have witnessed a steady increase in the number of identified VOCs. They are characterized by a wide range of physical and chemical attributes, the most important of which are their water solubility and whether

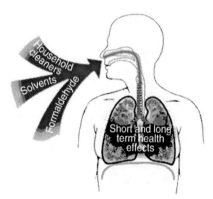

FIGURE 7.3   Inhalation of volatile organic compounds. *(Source: AirAdvice, Inc., Portland, OR.)*

they are neutral, basic, or acidic. VOCs are released into the indoor environment by extensive sources. They pose many health hazards such as some being potent narcotics that cause a depression in the central nervous system; others can cause eye, nose, and throat irritation, headaches, loss of coordination, nausea, and damage to the liver, kidneys, and central nervous system (Figure 7.3). A number of these chemicals are suspected or known to cause cancer in humans.

Formaldehyde

Formaldehyde (HCHO) is a colorless, pungent-smelling gas. It is one of the more common VOCs found indoors and is an important chemical used widely by industry to manufacture building materials and numerous household products. It is also a byproduct of combustion and certain other natural processes and thus may be present in substantial concentrations both indoors and outdoors. On condensing, it forms a liquid with a high vapor pressure, and due to its high reactivity, it rapidly polymerizes with itself to form paraformaldehyde. Formaldehyde, by itself or in combination with other chemicals, serves a number of purposes in manufactured products, including as a component of glues and adhesives and as a preservative in some paints and coating products.

Some of the attributes of urea-formaldehyde (UF) resin include high-tensile strength, flexural modulus, heat distortion temperature, low water absorption, mold shrinkage, high surface hardness, and volume resistance. UF copolymeric resins are present in many building materials such as wood adhesives, which are used in the manufacture of pressed-wood products including particleboard, medium-density fiberboard (MDF), plywood, finish coatings (acid-cured), and textile treatments, as well as in the production of UF foam insulation (UFFI). However, most people are unaware that formaldehyde is given off by materials other than UFFI. Certain types of pressed-wood products (composition board such as MDF, paneling, etc.), carpeting, and other material can be formaldehyde sources. Many of these products use a UF-based resin as an adhesive. Some continue to give off formaldehyde much longer than UFFI. Like the majority of VOCs, formaldehyde levels decrease substantially with time and/or with increased ventilation.

Health Risks For some people, formaldehyde can be a respiratory irritant, and continuous exposure to it can be dangerous. More specifically, chronic, low-level, continuous, and even intermittent exposure to formaldehyde can cause chemical hypersensitivity and provide an accelerating factor in the development of chronic bronchitis and pulmonary emphysema.

Polychlorinated Biphenyls

Polychlorinated biphenyls (PCBs) are oils used primarily as a coolant in electrical transformers. Although production and sale of PCBs were banned by the EPA in 1979, a large number of PCB-filled transformers remain in use. It has also been estimated that some 2 million mineral oil transformers still contain some percentage of PCB. PCBs may also be found in light ballasts and elevator hydraulic fluids. They are a suspected carcinogen, but if properly sealed or contained, they do not pose a hazard.

Hydrocarbons

Hydrocarbons are a class of organic chemical compounds consisting only of the elements hydrogen (H) and carbon (C); they are colorless, flammable, toxic liquids. Hydrocarbons are cardinal to our modern way of life and its quality and are one of the Earth's most important energy resources. They are the principal constituents of petroleum and natural gas, and are also derived from coal. The bulk of the world's hydrocarbons are used for fuels and lubricants, as well as for electrical power generation and heating.

The symptoms associated with exposure to aliphatic hydrocarbons may include watery eyes, nausea, vomiting, dizziness, weakness, and central nervous system effects such as depression, convulsions, and, in extreme cases, coma. Other symptoms may include pulmonary and gastrointestinal irritation, pulmonary edema, bronchial pneumonia, anorexia, anemia, nervousness, pain in the limbs, and numbness. Benzene is found in most hydrocarbons and is considered to be one of the more serious contaminants; it is known to cause leukemia. Air-quality tests may be necessary as well as tests for contaminants in the soil around the foundation.

Pesticides

Pesticides are chemical poisons designed to control, destroy, or repel plants and animals such as insects (insecticides), weeds (herbicides), rodents (rodenticides), and mold or fungus (fungicides). They include active ingredients (those intended to kill the target) and inert ingredients, which are often not "inert" at all. Pesticides are generally toxic and can be absorbed through the skin, swallowed, or inhaled, and as such they are unique contaminants of indoor environments. Studies show that approximately 16 million Americans are sensitive to pesticides because their immune systems have been damaged as a result of prior pesticide exposure. In addition, pesticides have been linked to a wide range of serious and often fatal conditions: cancer, leukemia, miscarriages, genetic damage, decreased fertility, liver damage, thyroid disorders, diabetes, neuropathy, stillbirths, decreased sperm counts, asthma, and other autoimmune disorders (e.g., lupus).

Pesticides are carefully regulated by the federal government, in cooperation with the states, to ensure that they do not pose unreasonable risks to human health or the environment. There are currently more than 1055 active ingredients registered as pesticides, which are formulated into thousands of different pesticide products available in the marketplace, including some of the most widely used over the past 60 years that are persistent and have become globally distributed.

Biological Contaminants

Mold and mildew, viruses, bacteria, and exposures to mite, insect, and animal allergens are biological pollutants arising from various sources such as microbiological contamination, (e.g., fungi, bacteria, viruses), mites, pollens, and the remains and dropping of pests such as cockroaches. Of particular concern are biological contaminants that cause immunological sensitization manifested as chronic allergic rhinitis, asthma, and hypersensitivity pneumonitis. Pollutants of biological origin can also significantly impact indoor air quality and cause infectious disease through airborne transmission.

One of the major contributors to poor IAQ, mold growth, and unhealthy buildings is the presence of moisture, but by controlling the relative humidity level, the growth of some sources of biologicals can be minimized. Standing water, water-damaged materials, rainwater leaks, and wet surfaces also serve as a breeding ground for molds, mildews, bacteria, and insects, as well as contaminated central heating, ventilating, and air-conditioning (HVAC) systems, which can distribute these contaminants through the building.

A method often used for deterring rainwater intrusion into walls is the rain screen approach, which incorporates cladding, an air cavity, a drainage plane, and an airtight support wall to offer multiple moisture-shedding pathways. The concept of the rain screen is simple; it separates the plane in a wall where the rainwater is shed and where the air infiltration is halted. In terms of construction, this means that there is an outer plane that sheds rainwater but allows air to freely circulate, and an inner plane that is relatively airtight.

Mold and Mildew

Mold and mildew are forms of musty-smelling fungi that thrive in moist environments. Their function in nature is primarily to break down and decompose organic materials such as leaves, wood, and plants. They grow, penetrate, and infect the air we breathe. There are thousands of species of molds, which include pathogens, saprotrophs, aquatic species, and thermophiles. Molds are part of the natural environment growing on dead organic matter and are present everywhere in nature; their presence is only visible to the unaided eye where mold colonies grow (Table 7.2).

Different mold species vary enormously in their tolerance to temperature and humidity extremes. The key to controlling indoor mold growth is to control moisture content and the temperatures of all surfaces, including interstitial surfaces within walls. Mold generally needs a temperature range between 40 and 100 °F to grow, and maintaining relative humidity levels between 30% and 60% helps control mold and many of these known biological contaminants. Winter humidification and summer dehumidification controls/modules can supplement central HVAC systems when climate excesses require additional conditioning measures.

Exposure to fungus in indoor air settings has emerged as a health problem of great concern in both residential environments and workplaces. Fungi are

TABLE 7.2 List of Typical Molds Found in Damp Buildings

Fungal species	Substrate	Possible metabolites	Potential health effects
Alternaria alternata	moist windowsills, walls	allergens	asthma, allergy
Aspergillus versicolor	damp wood, wallpaper glue	mycotoxins, VOCs	unknown
Aspergillus fumigatus	house dust, potting soil	allergens	asthma, rhinitis, hypersensitivity pneumonitis
		many mycotoxins	toxic pneumonitis infection
Cladosporium herbarum	moist windowsills, wood	allergens	asthma, allergy
Penicillium chrysogenum	damp wallpaper, behind paint	mycotoxins	unknown
		VOCs	unknown
Penicillium expansum	damp wallpaper	mycotoxins	nephrotoxicity?
Stachybotrys chartarum (atra)	heavily wetted carpet, gypsum board	mycotoxins	dermatitis, mucosal irritation, immunosuppression

Source: California Department of Health Services, Environmental Health Investigations Branch, "Mold and Indoor Air Quality," 2003.

primitive plants that lack chlorophyll and therefore feed on organic matter which they digest externally and absorb, or they must live as parasites. True fungi include yeast, mold, mildew, rust, smut, and mushrooms. When mold spores land on a damp spot indoors, they can grow and start digesting whatever they are growing on.

Four vital elements are needed for mold to grow: viable spores, a nutrient source (organic matter such as wood products, carpet, and drywall), moisture, and warmth. The mere presence of humid air in itself is not necessarily conducive to mold growth, except where air has a relative humidity (RH) level at or above 80% and is in contact with a surface. Mold spores are carried by air currents and can reach all surfaces and cavities of buildings. When the surfaces and/or cavities are warm and contain the right nutrients and amounts of moisture, the mold spores grow into colonies and gradually destroy the things they grow on. Likewise, by removing any of the four essential growth elements, the growth process is inhibited or nonexistent.

To execute a mold remediation project, the first step requires determining the growth's root cause. The next step is to evaluate the extent of the growth, which is usually done through visual examination. Since old mold growth may not always be visible, investigators may need to use instruments such as moisture meters, thermal imaging equipment, or borescope cameras to identify moisture in building materials or "hidden" mold growth within wall cavities, HVAC ducts, and so on. Toxic molds and fungi are a significant source of airborne VOCs that create IAQ problems, as can be seen in Figure 7.4. Toxicity can arise from inhalation or skin contact with toxigenic molds. Some molds produce toxic liquid or gaseous compounds, known as mycotoxins, in addition to infectious airborne mold spores that often cause serious health problems to residents and workers.

Bacteria and Viruses

Many millions of people around the globe suffer daily from viral infections of varying degrees of severity and at immense cost to the economy, including the costs of medical treatment, lost income due to inability to work, and decreased productivity. In fact, viruses have been identified as the most common cause of infectious diseases acquired in indoor environments, particularly those causing respiratory and gastrointestinal infections. The most common viruses causing respiratory infections include influenza viruses, rhinoviruses, corona viruses, respiratory syncytial viruses (RSVs), and parainfluenza viruses (PIVs); those responsible for gastrointestinal infections include rotavirus, astrovirus, and Norwalk-like viruses (NLVs). Some of these infections, such as the common cold, are widely spread but are not severe, while infections like influenza are relatively more serious.

Bacteria and viruses are minute in size and readily become airborne and remain suspended in air for hours, which makes them a cause of considerable concern due to because their ability to transmit infectious diseases. While there

FIGURE 7.4   (a) Mold on a ceiling growing out of control; this can be found in damp buildings. (b) An extreme case of toxic mold growth in the process of being treated. (c) Mold remediation expert examining mold infestation prior to writing remediation estimate. *(Sources: (a) courtesy MOLD-KILL.com; (b) courtesy Applied Forensic Engineering; (c) courtesy Mario Alvarez, GEC Environmental. All used with permission.)*

are many ways for the infection to spread, the most significant, from an epidemiological point of view, is airborne transport. Microorganisms can become airborne when droplets are given off during speech, coughing, sneezing, vomiting, or atomization of feces during sewage removal. Q fever is another emerging infectious disease among U.S. soldiers serving in Iraq.

Liquid and solid airborne particles (aerosols) in indoor air originate from many indoor and outdoor sources. These particles may differ in size, shape, and chemical and biological composition. Size signifies the most important characteristic affecting particle fate during transport, and it is also significant in affecting their biological properties. Bacterial aerosols have also been found to be a means to transmit a number of major diseases, as shown in Table 7.3.

According to Professor Lidia Morawska of Queensland University of Technology in Australia, the degree of hazard created by biological contaminants, including viruses, in indoor environments is controlled by a number of factors:

- The type of virus and potential health effects it causes
- Mode of exit from the body

TABLE 7.3 Infectious Diseases Associated with Bacterial Aerosols

Disease	Causal Organism
Tuberculosis	*Mycobacterium tuberculosis*
Pneumonia	*Mycoplasma pneumoniae*
Diphtheria	*Corynebacterium diphtheriae*
Anthrax	*Bacillus anthracis*
Legionnaires' disease	*Legionella pneumophila*
Meningococcal meningitis	*Neisseria meningitides*
Respiratory infections	*Pseudomonas aeruginosa*
Wound infections	*Staphylococcus aureus*

- Concentration levels
- Size distribution of the aerosol containing the virus
- Physical characteristics of the environment (temperature, humidity, oxygenation, UV light, suspension medium, etc.)
- Air circulation pattern
- Operation of the heating, ventilation, and air-conditioning system

The physical characteristics of the indoor environment, as well as the design and operation of building ventilation systems, are of paramount importance. Ducts, coils, and recesses of building ventilation systems often provide fertile breeding grounds for viruses and bacteria that have been proven to cause a wide range of ailments from influenza to tuberculosis. Likewise, a number of viral diseases may be transmitted in aerosols derived from infected individuals. Many infectious viral diseases and associated causal viruses transmitted through air are listed in Table 7.4.

TABLE 7.4 Infectious Diseases Associated with Viral Aerosols

Disease	Virus/Bacteria Type
Influenza	*Orthomyxovirus*
Cold	*Coronavirus*
Measles	*Paramyxovirus*
Rubella	*Togavirus*
Chicken pox	*Herpes virus*
Respiratory infection	*Adenovirus*

Rodent, Insect, and Animal Allergens

According to the Illinois Department of Public Health, a typical large city in the United States, such as Chicago, can receive more than 10,000 complaints each year about rodent problems and performs tens of thousands of rodent control inspections and baiting services. Effective measures need be taken to prevent rodents, insects, and pests from entering the home or office. Cockroaches, rats, termites, and other pests have plagued commercial facilities for far longer than computer viruses. According to the National Pest Management Association, pests can cause serious threats to human health, including such diseases as rabies, salmonellosis, dysentery, and staph. In addition to presenting a serious health concern, they distract from a facility's appearance and value.

Rats

Large communities of rats exist today within and beneath cities, traveling unnoticed from building to building along sewers and utility lines. Each rat colony has its own territory, which can span an entire city block and harbor more than 100 rats. As they explore their territories, rats and mice discover new food sources and escape routes. A rat's territory or "home range" is generally within a 50- to 150-foot radius of the nest, while mice usually have a much smaller range, living within a 10- to 30-foot radius of the nest (Figure 7.5). In places where all their needs (food, water, shelter) are met, rodents have smaller territories.

Insects

Today, more than 900,000 species of insects have been identified, and additional species are being identified every day. Some of these are known sources of inhalant allergens that may cause chronic allergic rhinitis and/or asthma. They include cockroaches, crickets, beetles, moths, locusts, midges, termites,

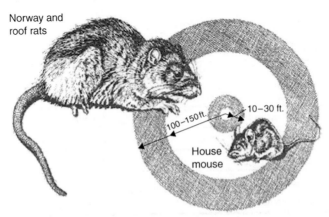

Norway and roof rats

100–150 ft.

10–30 ft.

House mouse

FIGURE 7.5 Typical home range of rodents. *(Source: Illinois Department of Public Health.)*

and flies. Insect body parts are especially potent allergens for some people. Cockroach allergens are potent and are commonly implicated as contributors to SBS in urban housing and facilities with poor sanitation. Most of the allergens from cockroaches come from the insect's discarded skin (Figure 7.6). As the skin disintegrates over time, it becomes airborne and is inhaled. Cockroaches have been reported to spread at least 33 kinds of bacteria, six kinds of parasitic worms, and at least seven other kinds of human pathogens.

Mites Mites are microscopic insects that thrive on the constant supply of shed human skin cells (commonly called dander) that accumulate on carpeting, drapes, furniture coverings, and bedding (Figure 7.7). The proteins in the combination of dust mite droppings and skin shedding are what cause allergic reactions in humans. Dust mites are perhaps the most common cause of

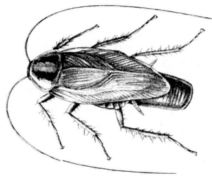

FIGURE 7.6 American cockroach.

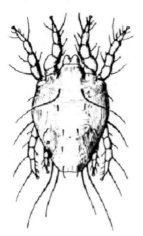

FIGURE 7.7 Dust mite. *(Source: Environment, Health and Safety Online, Atlanta, GA.)*

perennial allergic rhinitis. They are the source of one of the most powerful biological allergens and flourish in damp, warm environments. It is estimated that
up to 15% of people are allergic to dust mites, which, due to their very small size
(250–300 microns in length) and translucent bodies, are not visible to the naked
eye. To be able to make an accurate identification, one needs at least a 10X
magnification.

Dust mites have eight hairy legs, no eyes or antennae, a mouthpart group in
front of the body (which resembles a head), and a tough, translucent shell. They
have multiple developmental stages, commencing with an egg, then a larva, and
then by several nymph stages, and finally the adult. They also prefer warm,
moist surroundings like the inside of a mattress, particularly when someone
is lying on it, but they may also accumulate in draperies, carpet, and other areas
where dust collects. The favorite food of mites appears to be dander (both
human and animal). Humans generally shed about 1/5 ounce of dander a week.
Dust mite populations are usually highest in humid regions and lowest in areas
of high-altitude and/or dry climates.

Ants There are in excess of 20 varieties of ants invading homes and offices
throughout the United States, particularly during the warm months of the year
(Figure 7.8). Worldwide, there are more than 12,000 species, but of these only
a limited number actually cause problems. Destructive ant species include
fire and carpenter ants. Fire ants are vicious, unrelenting predators and have
a powerful, painful sting. Between 1991 and 2001, it was determined that
Hymenoptera stings accounted for 533 deaths attributed to severe allergic reactions to fire ant stings.

Termites Termites can pose a major threat to structures, which is why it is
important to address any termite infestation as soon as possible. A qualified
termite control company or inspector should look for the many tell-tale signs
termites usually provide, such as small holes in wood, straw-shaped mud tubes,
crumbling drywall, shed wings, and sagging doors or floors (Figure 7.9).

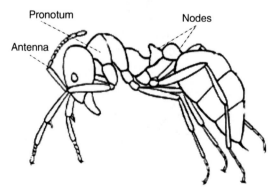

FIGURE 7.8   Primary features of the ant.

FIGURE 7.9   Termite.

Animal Allergens

Allergens are produced by many mammalian and avian species and can be inhaled by humans and cause immunological sensitization as well as symptoms of chronic allergic rhinitis and asthma. These allergens are normally associated with dander, hair, saliva, and urine of dogs, cats, rodents, and birds, although pollens, ragweed, and a variety of other allergens can find their way indoors. Ragweed is known to cause what is commonly referred to as "hay fever," or what allergist/immunologists refer to as *allergic rhinitis*. In the United States, seasonal allergic rhinitis, which is caused by breathing in allergens such as pollen, affects more than 35 million people.

7.2.2   General Steps to Reducing Pollutant Exposure

Pollutant source removal or modification is the best approach whenever sources are identified and control is possible. These may include:

- Routine maintenance of HVAC systems
- Applying smoking restrictions in the home and the office
- Venting contaminant source emissions to the outdoors
- Proper storage and use of paints, pesticides, and other pollutant sources in well-ventilated areas and their use during periods of nonoccupancy
- Allowing time for building materials in new or remodeled areas to off-gas pollutants before occupancy

Most of the mechanical ventilation systems in large buildings are designed and operated not only to heat and cool the air but also to draw in and circulate outdoor air.

One cost-effective method to reduce indoor pollutant levels is to increase ventilation rates and air distribution. At a minimum, HVAC systems should be designed to meet ventilation standards in local building codes. In practice, however, many systems are not operated or adequately maintained to ensure that these design ventilation rates are followed. Often IAQ can be improved by operating the HVAC system to at least its design standard, and to ASHRAE Standard 62-2001 if possible. When confronted with strong pollutant sources, local exhaust ventilation may be required to exhaust contaminated air directly from the building. The use of local exhaust ventilation is particularly advised to remove pollutants that accumulate in specific areas such as restrooms, copy rooms, and printing facilities. Air cleaners can also be a useful adjunct to source control and ventilation, although they are somewhat limited in their application. Air cleaners are discussed in Section 7.3.5.

Indoor air pollution is currently ranked among the top four environmental risks in America by the EPA, which may explain why for many forward-thinking real estate property managers it is becoming a standard of doing business to have their buildings routinely inspected as part of a proactive IAQ-monitoring program.

Investigating Indoor Air Quality

The procedure for investigating IAQ may be characterized as a cycle of information gathering, hypothesis formation, and hypothesis testing. It typically begins with a walk-through inspection of the problem area to gather information relating to the four basic factors that influence IAQ:

- The building's occupants
- The building's HVAC system
- Possible pollutant pathways
- Possible sources of contamination

7.2.3   Thermal Comfort

Defining thermal comfort is somewhat elusive other than to say that it is a state of well-being and involves temperature, humidity, and air movement among other things. An often heard complaint facility managers get from building occupants is that their office space is either too cold or too hot. Studies show that people of different cultures generally have different comfort zones; even people belonging to the same family may feel comfortable under different conditions, and keeping everyone comfortable at the same time is not an easy matter. Regarding levels of thermal satisfaction, the Center for the Built Environment states:

Current comfort standards specify a "comfort zone" representing the optimal range and combinations of thermal factors (air temperature, radiant temperature, air velocity, humidity) and personal factors (clothing and activity level) with which at least 80 percent of the building occupants are expected to express satisfaction.

This is the goal outlined by ASHRAE in the industry's gold standard of comfort: Standard 55, Thermal Environmental Conditions for Human Occupancy. This standard also specifies the thermal conditions deemed likely to be comfortable to occupants.

As previously mentioned, employee health and productivity are greatly influenced by the quality of the air in the indoor environment. When temperature extremes—too cold or too hot—become the norm indoors, all building occupants suffer. In spaces that are either very hot or very cold, individuals must expend physiological energy to cope with the surroundings; this is energy that could be better utilized to focus on work and learning, particularly since research has shown that people simply don't perform as well, and attendance declines, in very hot and very cold workplaces. Poor air quality and lighting levels, off-gassing of chemicals from building materials, and the growth of molds and bacteria can all adversely affect building occupants. Sustainable design supports the well-being of building occupants and their desire to achieve optimum comfort by reducing indoor air pollution. This can be achieved by a number of strategies such as selection of materials with low off-gassing potential, providing access to daylight and views, appropriate ventilation strategies, and controlling lighting, humidity, and temperature levels.

7.2.4   Noise Pollution

Much research has been conducted on noise pollution, which is considered to be a form of energy pollution in which distracting, irritating, or damaging sounds are freely audible. Noise and vibration from sources including HVAC systems, vacuums, pumps, and helicopters can often trigger severe symptoms, including seizures, in susceptible individuals. In the United States, regulation of noise pollution was stripped from the EPA and passed to the individual states in the early 1980s. Although two noise-control bills it passed remain in effect, the EPA can no longer form relevant legislation. Needless to say, a noisy workplace is not conducive to getting work done. What is not so apparent is that constant noise can lead to voice disorders for paraprofessionals in the office, where many employees spend time on the telephone or routinely use their voices at work. And although good engineering design can mitigate noise pollution levels to some extent, it is frequently not to acceptable levels, particularly if a significant number of individual sources combine to create a cumulative impact. Nevertheless, humans, whether tenants or building occupants, have a basic right to live in environments that are relatively free from the intrusion of noise pollution, even though this may not always be possible.

Defining Noise

The City of Berkeley's Planning and Development Department defines sound as pressure variations in air or water that can be perceived by human hearing and the objectionable nature of sound as caused by its pitch or loudness. In addition to the concepts of pitch and loudness, there are several methods to measure noise. The most common is the decibel (dB). On the decibel scale, zero represents the lowest sound level that a healthy, unimpaired human ear can detect. Sound levels in decibels are calculated on a logarithmic basis. Thus, an increase of 10 dBs represents a tenfold increase in acoustic energy, a 20-dB increase is 100 times more intense (10×10), and so on. The human ear likewise responds logarithmically, and each 10-dB increase in sound level is perceived as approximately a doubling of loudness.

Impact of Noise

Sound is of great value; it warns us of potential danger and gives us the advantage of speech and the ability to express joy or sorrow. But sometimes it can be undesirable. For example, sound may interfere with and disrupt useful activities. Sometimes, too, certain types of music (e.g., pop, opera) may become noise at certain times (e.g., after midnight), in certain places (e.g., a museum), or to certain people (e.g., the elderly). There is therefore a value judgment among people as to when sound becomes unwanted noise, which is why it is difficult to offer a clear definition of "good" or "bad" noise levels in any attempt to generalize its potential impact on people.

Health Effects

Wikipedia points out that elevated noise in the workplace or home can "cause hearing impairment, hypertension, ischemic heart disease, annoyance, sleep disturbance, and decreased school performance. Changes in the immune system and birth defects have been attributed to noise exposure, but evidence is limited." Hearing loss is one of the potential disabilities that can occur from chronic exposure to excessive noise, but it may also occur in certain circumstances such as after an explosion. Noise exposure has also been known to induce dilated pupils, elevated blood pressure, tinnitus, hypertension, vasoconstriction, and other cardiovascular impacts.

The Occupational Safety and Health Administration (OSHA) has a noise exposure standard that is set at just below the noise threshold where hearing loss may occur from long-term exposure. The impact of noise on physical stress reactions can be readily observed when people are exposed to noise levels of 85 dB or higher. The safe maximum level is set at 90 dB averaged over 8 hours. If the noise is above 90 dB, the safe exposure dose becomes correspondingly shorter. Adverse stress reaction to excessive noise can be broken down into two stages. The first stage is where noise is above 65 dB, making it difficult to have a normal conversation without raising one's voice. The second is the link between noise and socioeconomic conditions that may further lead to undesirable stress-related behavior and increased workplace accident rates, or in many cases may stimulate aggression and other antisocial behaviors when people are exposed to chronically excessive noise.

Major Sources of Noise

In the United States, cities can only adopt noise exposure standards for noise levels emanating from trucks, trains, or planes and then not permit land uses to be developed in areas with excessive noise for an intended use. Cities also play a role in enforcing state vehicle code requirements regarding muffler operation and may set speed limits or weight restrictions on streets that impact noise generation. However, a city's actions are typically proactive with regard to nontransportation sources and reactive for sources outside the city's control.

Noise abatement and reduction of excessive noise exposure can be accomplished by reducing the noise level at the source, increasing the distance between the source and the receiver, and placing an appropriate obstruction (e.g., a wall) between the noise source and the receiver.

A noise wall may sometimes be the only practical solution since vehicular noise is exempt from local control and relocation of sensitive land uses away from freeways or major roads is not practical. Yet with noise walls we have both a positive side, reducing the noise exposure to affected persons, and a negative side, effectively blocking the line of sight between source and receiver. A properly sited wall can reduce noise levels by almost 10 dB, which for most people translates to being about half as loud as before. Unfortunately, the social, economic, and aesthetic costs of noise walls are high. While they screen the traffic from receivers, they may also block beautiful views of trees, parks, and water, and may give drivers a claustrophobic feeling of being surrounded by massive walls.

7.2.5   Daylighting and the Daylight Factor

For millions of years the sun has been our principal source of light and heat, and we have become almost totally dependent on it for our health and survival. The sustainable design movement is now returning to nature because of its increasing concern with global warming, carbon emissions, and sustainable design, and has started to take positive steps to increase the use of managed admission of natural light in both residential and nonresidential buildings. Daylighting has come to play a pivotal role in programs, such as LEED® and the IgCC, and now has increased recognition in California's Title 24 Energy Code. According to Craig DiLouie of the Lighting Controls Association, daylighting is defined as "the use of daylight as a primary source of illumination to support human activity in a space." Direct sunlight is a most powerful source and has the greatest impact on our lives; it not only provides visible light but also provides ultraviolet and infrared (heat) radiation (Figure 7.10).

Assessing the daylight quality in a room traditionally consisted of a manual average daylight factor calculation, or was based on a computerized version of the manual method. The term "average daylight factor" is sometimes construed to be the average daylight factor on all surfaces, whereas the output from most computer-based calculations reflect the average daylight factor derived from a series of points on the working plane. However, in light of recent technological advances, daylight design is rapidly moving forward and is now able to provide the kind of information that accommodates all of the requirements of the daylight consultant, the architect, and the end user. The ideal package should integrate natural lighting and electrical lighting calculations, and take into account an evaluation of the thermal impact on window design. Table 7.5 lists recommended illumination levels for various locations and functions.

Daylight availability in a room is normally expressed by a measure commonly called the daylight factor (DF). However, according to Chris Croly, a building

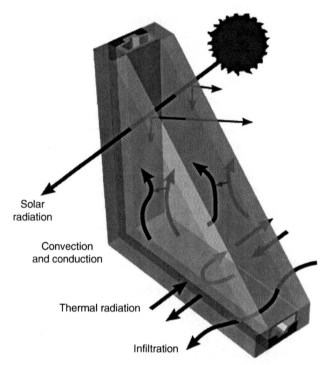

FIGURE 7.10 Three major types of energy flow that occur through windows are nonsolar heat losses and gains in the form of conduction, convection, and radiation; solar heat gains in the form of radiation; and airflow, both intentional as ventilation and unintentional as infiltration. *(Source: DOE.)*

services engineering associate with BDP, Dublin, and Martin Lupton, formerly a director with BDP:

The calculation of daylight factor using traditional methods becomes particularly difficult when trying to assess the effects of transfer glazing, external overhangs, or light shelves. Modern radiosity or ray tracing calculations are now readily available and are easy to use but still generally offer results in the form of daylight factor or lux levels corresponding to a particular static external condition.

Croly and Lupton define DF as "the ratio of the internal illuminance to that on a horizontal external surface located in an area with an unobstructed view of a hemisphere of the sky." In other words, it is the ratio of outside luminance over inside luminance, expressed in percent. The higher the DF, the more natural light is available in the room. The impact of direct sunlight on both illuminances must be considered separately and is not included. The DF can be expressed as:

$$DF = 100 * E_{in}/E_{ext}$$

TABLE 7.5 Recommended Illumination Levels

Area	Footcandles
Building surrounds	1
Parking area	5
Exterior entrance	5
Exterior shipping area	20
Exterior loading platforms	20
Office corridors and stairways	20
Elevators and escalators	20
Reception rooms	30
Reading or writing areas	70
General office work areas	100
Accounting bookkeeping areas	150
Detailed drafting areas	200

where E_{in} represents the inside illuminance at a fixed point and E_{ext} represents the outside horizontal illuminance under an overcast or uniform sky, as defined by the *Commission Internationale de l'Eclairage* (CIE).

Daylighting Strategies

Research over the years clearly demonstrates that effective daylighting saves energy and improves the quality of the visual environment; it also reduces operating costs and enhances occupant satisfaction. Thus, while daylight can reduce the amount of electric light needed to adequately illuminate a workspace and therefore reduce potential energy costs, allowing too much light or solar radiation into a space can have a negative effect, resulting in heat gain and offsetting any savings achieved by reduced lighting loads. Some architectural/design firms known for their sustainable design inclinations (e.g., HOK and Gensler) design the majority of their buildings to be internally load-dominated, meaning that the buildings need to be cooled for most of the year. Strategies for improved daylighting include use of miniature optical light shelves, light-directing louvers, light-directing glazing, clerestories, roof monitors and skylights, light tubes, and heliostats. It is important to appreciate that whatever tools are applied in daylighting design, to be successful they will involve the integration of several key disciplines, including architectural, mechanical, electrical, and lighting. As with sustainable design in general, an integrated design approach is needed where team members are brought into the design process early to ensure that daylighting concepts and strategies are satisfactorily implemented throughout the project (Figure 7.11).

Advanced daylighting strategies and systems can significantly enhance the quality of light in an indoor environment as well improve energy efficiency by minimizing lighting, heating, and cooling loads and thereby reduce a building's electricity consumption. By providing a direct link to the outdoors, daylighting helps create a visually stimulating and productive environment for building occupants, at the same time significantly reducing total building energy costs.

Note that when light hits a surface, part of it is reflected back. This reflection normally takes the form of diffused (nondirectional) light and is dependent on the object's reflection. The reflection of the outside ground is typically on the order of 0.2 or 20%. This means that in addition to the direct sunlight and skylight components, there also exists an indirect component that can make a significant contribution to the lighting inside a building, especially since the light reflected off the ground hits the ceilings, thus adding to its brightness.

Daylighting and Visual Comfort

Designing for natural light has for many years been a challenge that designers often face because of fluctuations in light levels, colors, and direction of the light source. This has led architects and engineers to make some unwise design decisions that in the 1960s and 1970s culminated in hermetically sealed office blocks that were fully air-conditioned and artificially lit. This in turn led to a

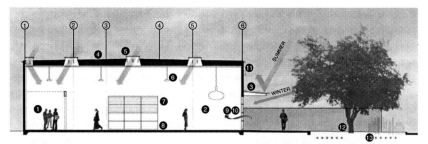

1. Private office
2. Open office
3. Sunshade with building integrated photovoltaics
4. Roof with building integrated photovoltaics
5. Skylight
6. Energy-efficient and occupancy sensor controlled light fixtures
7. Electrochromic glass
8. Radiant heat floor
9. Natural ventilation
10. High-performance glass
11. Reduction of outdoor light pollution
12. Water-efficient landscaping
13. Ground-source heat pump

FIGURE 7.11   Schematic of an integrated approach to the design of a building's various systems in the new IDeAs headquarters, San José, CA. *(Courtesy Integrated Design Associates, Inc.)*

sharp increase in complaints and symptoms attributed to BRIs and SBS. Gregg Ander, FAIA, says that "In large measure, the art and science of proper daylighting design is not so much how to provide enough daylight to an occupied space, but how to do so without any undesirable side effects." Designers should adopt practical design strategies for sustainable daylighting that increase visual comfort by applying three primary techniques. These are basic lighting approaches that reflect the strategies of sustainability and thus support the larger ecological goal.

- *Architectonic:* What has made daylighting design so difficult until recently is the lack of specific design tools. Today most large architectural practices have a diverse team of consultants and design tools that enable them to undertake complex daylighting analysis, whereas the typical school or small office does not have this capability or the budget for it.
- *Human Factors:* Because of the impact on people and their experience, designers need to achieve the best lighting levels possible while avoiding glare and high-contrast ratios. These can usually be avoided by not allowing direct sunlight to enter a workspace through the use of shading devices, for example.
- *Environmental:* Use of the natural forces that impact design and resource and energy conservation.

The Heschong Mahone Group (HMG), a California architectural consulting firm, conducted a study that concluded that students who received their lessons in classrooms with more natural light scored as much as 25% higher on standardized tests than students in the same school district whose classrooms had less natural light. This appears to confirm what many educators have suspected, that children's capacity to learn is greater under natural illumination from skylights or windows than from artificial lighting. The logical explanation given is that "daylighting" enhances learning by boosting the eyesight, mood, and/or health of students and teachers.

Another investigation by HMG looked at the relationship of natural light to retail sales. The study analyzed the sales of 108 stores that were part of a large retail chain. The stores were all one story and virtually identical in layout, except that two-thirds had skylights while the others did not. The study specifically focused on skylighting as a means to isolate daylight as an illumination source and avoid the other qualities associated with daylighting from windows. When they compared sales figures for the various stores, they discovered a statistically compelling connection between skylighting and retail sales performance, and found that stores with skylight systems had increased sales by 40%—even though the design and operation of all the stores were remarkably uniform, except for the presence of skylights in some. The study showed that, all other things being equal, an average non-skylit store in the chain would likely increase its sales by an average of about 40% just by adding skylights.

Technology is moving at a rapid pace, and architects are now increasingly specifying high-performance glass with spectrally selective coatings that allow only visible light to pass through and keep out the infrared wavelength. This eliminates most infrared and ultraviolet radiation while allowing the majority of the visible light spectrum through the glass. But even with high-performance glass, much of the light can be converted to heat. Glass with high visible-light transmittance still allows light energy into a building, and when this light energy hits a solid surface, it is absorbed and reradiated into the space as heat.

Combining daylighting with efficient electric lighting strategies can provide substantial energy savings. The building's planning module can often give indications on how best to organize the lighting. In any case, the lighting system must correlate to the various systems in place, including structural, curtain wall, ceiling, and furniture. Likewise, initial lighting costs may rise when designing for sustainability and implementing energy-efficient strategies. These energy-saving designs may require items such as dimmable ballasts, photocells, and occupancy sensors, all of which are not typically covered in most traditional project budgets (Figure 7.12). However, these items are normally included if an integrated design approach is employed and if daylight strategies are appropriately employed at an early phase of the project's design.

Daylight has many positive attributes, the main one being enhancing the psychological value of space. Likewise, the introduction of daylight into a building reduces the need for electric lighting during the day while helping to link indoor spaces with the outdoor spaces for building occupants. However, natural light also has its negatives, such as glare, overheating, variability, and privacy issues. It is left to the designer, therefore, to find ways to increase the positive aspects of using natural light while reducing the negative aspects. Addressing glare requires keeping sunlight out of the field of view of building occupants and protecting them from disturbing reflections. Addressing over-heating means adding appropriate exterior shading, filtering incoming solar radiation, or even using passive controls such as thermal mass. Furthermore, addressing variability and privacy issues requires creative ways to block or alter light patterns and compensate with alternative sources of light.

In recent years the implementation of daylighting strategies at an early stage of a building's design has become vital for the success of the building's lighting strategy. This is because, previously, simple tools for predicting the performance of advanced daylighting strategies were not available to the designer. The data output from daylighting studies can be extremely useful for fine-tuning and finalizing the building's orientation, massing, space planning, and interior finishes. Innovative daylighting systems are designed to re-direct sunlight or skylight to areas where it is most needed, and avoid glare.

These systems use optical devices that initiate reflection and refraction and/or use the total internal reflection of sunlight and skylight. With today's advancing technology, they can be programmed to actively track the sun's movement or passively control the direction of sunlight, skylight, and other shading

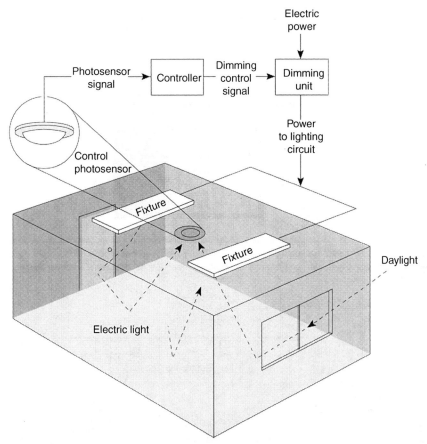

FIGURE 7.12   Schematic of a room utilizing a photoelectric dimming system. The ceiling-mounted photosensor reads both electric light and daylight in the space and adjusts the electric lighting as required to maintain the design level of total lighting. *(Source: Ernest Orlando, Lawrence Berkeley National Laboratory.)*

systems (Figure 7.13). Some owners are being driven by the financial and competitive pressure of powerful market forces to seek architectural solutions such as highly glazed, transparent façades. While these trends may offer clear potential benefits, they also expose owners to real risks and costs associated with their use. The following are significant potential risks associated with highly glazed façades:

- Increased sun penetration and excessive brightness levels that exceed good practice, which may cause or heighten visual discomfort
- Adequate tools may not always be available to reliably predict thermal and optical performance of components and systems, and to assess environmental quality

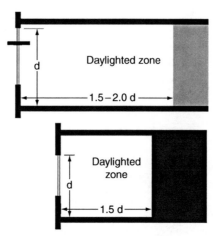

FIGURE 7.13 A rule of thumb for daylight penetration with typical depth and ceiling height is 1.5 times head height for standard windows, or 1.5 to 2.0 times head height with a light-shelf, for south-facing windows under direct sunlight. *(Source: Ernest Orlando, Lawrence Berkeley National Laboratory.)*

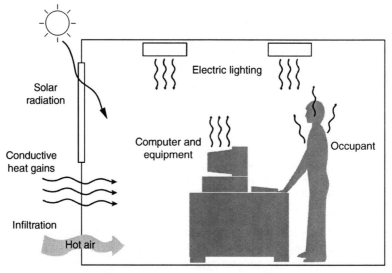

FIGURE 7.14 Graphic of various cooling-load sources. *(Source: Ernest Orlando, Lawrence Berkeley National Laboratory.)*

- Buildings using transparent glazing that generally use greater cooling loads and cooling energy, which has the potential for thermal discomfort (Figure 7.14)
- Increased cost of automated shading systems and purchasing lighting controls using dimming ballasts and difficulty in commissioning systems after installation

- Technical difficulty and the high cost of reliably integrating dimmable lighting and shading controls with each other and with building automation systems to ensure effective operation over time
- Uncertainty of occupant behavior with the use of automated, distributed controls in open-landscaped office space and the potential for conflict between different needs and preferences

Large glazed spaces in work areas (as distinct from corridors, lobbies, etc.) require much better sun and glare control to reap potential benefits and minimize possible risks. Appropriate solutions must be delivered by systems that can rapidly respond to exterior climate and interior needs. One of the challenges facing manufacturers is how to provide such needed increased functionality at lower cost and lower risk to owners. Because of various advantages and disadvantages, lighting consultants often recommend the use of switching for spaces where nonstationary tasks are performed, such as corridors, and continuous dimming for spaces where stationary tasks are performed, such as offices. It has been shown that daylight harvesting using continuous dimming equipment automatically controlled by a photosensor can generate 30 to 40% savings in lighting energy consumption, thereby significantly reducing operating costs.

Shades and Shade Controls

The greatest benefit of harvesting daylight can be achieved by implementing a shading strategy that is tailored to the building. In hot climates, exterior shading devices have been found to work well to both reduce heat gain and diffuse natural light prior to entering the work space (Figure 7.15). Examples of such devices include light shelves, overhangs, vertical louvers, horizontal louvers, and dynamic tracking or reflecting systems. Thus, for example, exterior shading of the glass can eliminate up to 80% of the solar heat gain. Shades and shade control strategies are based on the perception that occupants of commercial buildings typically prefer natural light to electric light, and the goals of a shade system normally include maximizing the use of natural light within a glare-free environment while avoiding direct solar radiation on occupants by intercepting sunlight penetration. Such a strategy may also include facilitating occupant connectivity with the outdoors through increased glazing and external views.

7.2.6   Views

Research over the years has shown that windows providing daylight and ample views can dramatically affect building occupants' mental alertness, productivity, and psychological well-being. David Hobstetter, a principal of Kaplan-McLaughlin-Diaz, a San Francisco-based architectural practice, affirms this: "Dozens of research studies have confirmed the benefits of natural daylight and views of greenspace in improving a person's productivity, reducing absenteeism and improving health and well-being." Though some educators opine

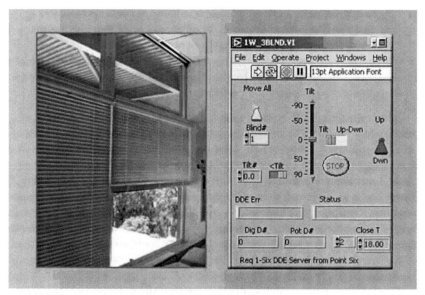

FIGURE 7.15   Venetian-blind system at a Lawrence Berkeley National Laboratory office building equipped with a "virtual instrument" panel for the Integrated Building Environmental Communications System (IBECS) control of blind settings. *(Courtesy High-Performance Commercial Building Systems.)*

that views out of windows may be unnecessarily distracting to students, the California Energy Commission's (CEC) 2003 study of the Fresno school district found that a varied view out of a window that included vegetation or human activity and objects in the far distance supported better learning.

Such findings confirm results of earlier research, such as a 1984 hospital study that concluded that post-operative patients with a view of vegetation took far fewer painkillers and experienced faster recovery times than patients looking at plain concrete walls. Another revealing study noted that computer programmers with views spent 15% more time on their primary task, while workers without views spent 15% more time talking on the phone or to one another.

Building occupants generally relish contact with the outside world, even if only through a windowpane, and landscapes, not surprisingly, are preferred to cityscapes. In many countries around the world, views, whether high-rise or otherwise, are normally considered perks. Moreover, some researchers contend that the view from a window may be even more important than the daylight it admits. The CEC's study of workers in the Sacramento Municipal Utility District's call center found that better views were consistently associated with better performance: "Workers with good views were found to process calls 7 to 12% faster than colleagues without views. Workers with better views also reported better health conditions and feelings of well-being, while their counterparts reported higher fatigue."

Researchers have also concluded that views of nature improve attention spans after extended mental activity has drained a person's ability to concentrate. Among the main building types that can most benefit from the application of daylighting are educational buildings such as schools, administrative buildings such as offices, maintenance facilities, and storage facilities such as warehouses.

7.3   VENTILATION AND FILTRATION

In ancient times, buildings, whether a Babylonian palace, an Egyptian temple, or a Roman castle, were ventilated naturally using either "badgeer/malqafs" (wind shafts/towers) or some other innovative method (Figure 7.16) since mechanical systems did not exist at the time. Andy Walker of the National Renewable Energy Laboratory says:

Wind towers, often topped with fabric sails that direct wind into the building, are a common feature in historic Arabic architecture, and are known as "malqafs." The incoming air is often routed past a fountain to achieve evaporative cooling as well as ventilation. At night, the process is reversed and the wind tower acts as a chimney to vent room air.

It is not surprising that with today's increased awareness of the cost and environmental impacts of energy use, natural ventilation has once again come to the fore and become an increasingly attractive method for reducing energy use and cost, and for providing acceptable indoor environmental quality. Natural ventilation systems use the natural forces of wind and buoyancy, that is, pressure differences, to move fresh air through buildings. These pressure differences can be a result of wind, temperature differences, or differences in humidity. The amount and type of ventilation achieved will depend to a large extent on the size and placement of openings in the building. In today's polluted environment, inadequate ventilation is one of the main culprits causing increased indoor pollutant levels by not bringing in enough outdoor air to dilute emissions from indoor sources and by not removing these indoor air pollutants to the exterior.

7.3.1   Ventilation and Ductwork

Appropriate ventilation is vital for the health and comfort of building occupants. It is specifically needed to reduce and remove pollutants emitted from various internal and external sources. Good design combined with optimum air tightness is a prerequisite to healthy air quality, occupant comfort, and energy efficiency. Sufficient air supply and movement can be tested and analyzed to determine the efficiency of an HVAC system. Regular maintenance of ductwork is pivotal to achieving both a better indoor environment and system stability. Ductwork can be evaluated, cleaned, and sealed to prevent airflow and potential quality issues. All ductwork should be analyzed by a professional trained and certified by the National Air Duct Cleaning Association (NADCA).

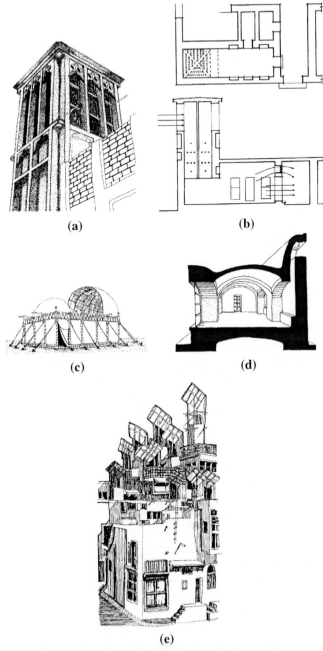

(a)   **(b)**

(c)   **(d)**

(e)

FIGURE 7.16   Various types of wind catchers ("badgeer"/"malqaf") used in traditional and ancient architecture. (a) Multidirectional traditional Dubai wind catcher. (b) Plan and section of a Dubai wind catcher. (c) Ancient Assyrian wind catcher. (d) Section through a traditional wind scoop. (e) Traditional Pakistani wind catchers.

7.3.2 Air Filtration

It is surprising that, to date, we lack federal standards for air filter performance. Air-cleaning filters are designed to remove pollutants from indoor air, to improve the indoor environment, and to produce cleaner air. Proper filtration removes dirt, dust, and debris from the air you breathe, and it reduces pollen and other allergens that can cause asthmatic attacks and allergic reactions. Filters awarded the high-efficiency particulate air (HEPA) accolade have to satisfy certain standards of efficiency such as those set by the Department of Energy (DOE). To qualify as HEPA by government standards, an air filter must remove 99.97% of all particles larger than 0.3 micrometer from the air that passes through. A filter that qualifies as HEPA is superior to those that do not qualify, but is also subject to interior classifications. And although air-cleaning devices may help to control the levels of airborne allergens, particles, and in some cases gaseous pollutants in a facility, they may not decrease adverse health effects from indoor air pollutants.

The marketplace is currently flooded with various types of air filters, including mechanical filters, electronic filters, hybrid filters, and gas-phase and ozone generators—some of which are designed to be installed inside the ductwork of a facility's central heating, ventilation, and air-conditioning system to clean the air in the whole facility. Other types include portable room air cleaners designed to clean the air in a single room or specific areas and not intended for complete facility filtration. Filters are often rated according to the minimum efficiency reporting value (MERV), which is a filter-rating system devised by ASHRAE to standardize and simplify filter efficiency ratings for the public. In the MERV Rating System, the higher the rating, the higher the efficiency of the air filter; thus a MERV 12 filter removes smaller particles from the air than a MERV 8 filter. Before making a determination on the most appropriate filter for a project or home, adequate research is required (see ANSI/AHRI Standard 680 (I-P)-2009).

7.3.3 Air Purification

Pollutants are released more or less continuously from many sources, such as building materials, furnishings, and household products such as air fresheners. Some sources, related to activities carried out in the home or workplace, release pollutants intermittently. These include smoking; the use of unvented or malfunctioning stoves, furnaces, or space heaters; the use of solvents in cleaning and hobby activities; the use of paint strippers in redecorating activities; and the use of cleaning products and pesticides. High pollutant concentrations can remain in the air for extended periods after some of these activities cease if appropriate action is not taken. While air filtration removes particulates, air purification is required to remove the things a filter fails to, such as odors and gases. Chemicals in paints, carpets, and other building materials (i.e., VOCs) are harmful to building occupants and should be removed through air purification. There is also an increasing concern regarding the presence of biological infectious agents, and one way to address this is through air purification on a regular basis.

7.3.4 Amount of Ventilation

There are various ways for outdoor air to enter and leave a building: infiltration, natural ventilation, and mechanical ventilation. Outdoor air can infiltrate a building through openings, joints, and cracks in walls, floors, and ceilings and around windows and doors. Natural ventilation involves air moving through opened windows and doors. Air movement associated with infiltration and natural ventilation is a consequence of air temperature differences between the indoor and outdoor air, and of wind. When insufficient outdoor air enters a home, pollutants can accumulate to elevated levels to a degree that they can pose health and comfort problems. In the event that natural ventilation is insufficient to achieve good air quality, there are a number of mechanical ventilation devices. These include outdoor-vented fans that will intermittently remove air from a room, such as bathroom and kitchen, and air-handling systems that utilize fans and ducts to continuously remove indoor air and distribute filtered and conditioned outdoor air to strategic points throughout the building. The rate at which outdoor air replaces indoor air is known as the air exchange rate. Insufficient air infiltration, natural ventilation, or mechanical ventilation lower the air exchange rate and cause rising pollutant levels.

Residents or occupants are occasionally in a position to take appropriate action to improve the indoor air quality of a space by removing the source, altering an activity, unblocking an air supply vent, or opening a window to temporarily increase ventilation. In other cases, however, the building owner or manager is the only one in a position to remedy the problem. Building management should be prevailed on to follow the EPA's IAQ Building Education and Assessment Model (I-BEAM). I-BEAM expands and updates the EPA's existing *Building Air Quality: A Guide for Building Owners and Facility Managers* (known as "BAQ"). BAQ is considered to be complete state-of-the-art guidance for managing IAQ in commercial buildings. Building management should also be encouraged to follow the EPA and National Institute for Occupational Safety and Health (NIOSH) guidelines; the document can be downloaded in PDF format.

7.3.5 Ventilation Improvements

Increasing the amount of outdoor air coming indoors is another approach to lowering the concentrations of indoor air pollutants. Many heating and cooling systems, including forced-air heating systems, do not mechanically bring fresh air into the house. This can often be addressed by opening windows and doors or running a window air conditioner with the vent control open. In residences, local bathroom or kitchen fans that exhaust outdoors can be used to remove contaminants directly from the room where the fan is located and to increase the outdoor air ventilation rate. Good ventilation is especially important when undertaking short-term activities that can generate high levels of pollutants such as painting, paint stripping, or heating with kerosene heaters. Such activities should preferably be executed outdoors whenever possible. The following design recommendations can help achieve better ventilation in buildings:

- Buildings that are naturally ventilated should preferably be narrow, as wide buildings pose greater difficulty in distributing fresh air to all areas using natural ventilation.
- Use of mechanical cooling is recommended in hot, humid climates.
- Occupants should be able to operate window openings.
- Use of fan-assisted cooling strategies should be given consideration.
- Decide whether an open- or closed-building ventilation approach potentially offers the best results. (A closed-building approach is more appropriate in hot, dry climates where there is a large diurnal temperature range from day to night. An open-building approach is more effective in warm and humid areas, where the temperature difference between day and night is relatively small.)
- Maximize wind-induced ventilation by siting buildings so that summer wind obstructions are minimal.
- When possible, provide ventilation to the attic space as this greatly reduces heat transfer to conditioned rooms below. Ventilated attics have been found to be approximately 30°F cooler than unventilated attics.

Air Cleaners

Air cleaners come in a variety of types and sizes ranging from relatively inexpensive table-top models to larger, more sophisticated and expensive systems. Some air cleaners are highly effective at particle removal, while others, including most table-top models, are much less so. It should be noted that air cleaners are generally not designed to remove gaseous pollutants. The effectiveness of an air cleaner is expressed as a percentage efficiency rate that depends on how well it collects pollutants from indoor air and how much air it draws through the cleaning or filtering element. The latter is expressed in cubic feet per minute. It needs noting that even an efficient collector that has a low air circulation rate will not be effective; neither will an air cleaner with a high air circulation rate but a less efficient collector. All that being said, the long-term performance of any air cleaner depends to a large extent on maintaining it in accordance with the manufacturer's directions.

Another critical factor in determining the effectiveness of an air cleaner is the level and strength of the pollutant source. Table-top air cleaners, in particular, may not be capable of adequately reducing amounts of pollutants from strong nearby sources. Persons who are sensitive to particular pollutant types may find that air cleaners are useful mostly when used in conjunction with collaborative efforts to remove the source of pollution.

Ventilation Systems

Most large commercial buildings have mechanical ventilation systems that are typically designed and operated to heat and cool the air, and to draw in and circulate outdoor air. However, ventilation systems themselves can be a source of indoor pollution and contribute to indoor air problems if they are poorly designed, operated, or maintained. They can sometimes spread harmful biological contaminants that have steadily multiplied in cooling towers, humidifiers, dehumidifiers, air conditioners, or inside surfaces of ventilation ducts. For example, problems arise when, in an effort to save energy, ventilation systems are incorrectly programmed and bring in inadequate amounts of outdoor air. Other examples of inadequate ventilation occur when the air supply and return vents within a space are blocked or placed in a manner that prevents the outdoor air from reaching the breathing zone of building occupants. Improper location of outdoor air intake vents can also bring in contaminated air, particularly exhaust from automobiles and trucks, fumes from dumpsters, boiler emissions, or air vented from kitchens and bathrooms.

For naturally ventilated spaces, refer to California's Collaborative for High-Performance Schools (CHPS) *Best Practices Manual*, Appendix C, "A Field Based Thermal Comfort Standard for Naturally Ventilated Buildings." For mechanically ventilated spaces, designers should refer to ASHRAE Standard 55-1992, Addenda 1995 "Thermal Environmental Conditions for Human Occupancy."

7.4 BUILDING MATERIALS AND FINISHES: EMITTANCE LEVELS

Although several studies have been conducted over the years to investigate the impact of pollution emitted by building materials on indoor air quality and relating the results to ventilation requirements, there have been few systematic experiments in which building materials are initially ranked according to their pollution strength, and then their impact on indoor air quality in real rooms analyzed. Such studies would allow us to quantify the extent to which using low-polluting building materials would reduce the energy used for ventilation of buildings without compromising indoor air quality.

7.4.1 Toxic Building Materials

The Healthy Building Network (HBN) identifies the primary building materials that are considered toxic and that have unacceptable high-VOC emittance levels. Examples are given in the following subsections.

Polyvinyl Chloride

HBN singled out polyvinyl chloride (PVC), or vinyl, for elimination because of its uniquely wide and potent range of chemical emissions throughout its life cycle, including many of the target chemicals listed next. It is virtually the only material that requires phthalate plasticizers, frequently includes heavy metals, and emits large amounts of VOCs. In addition, during manufacture PVC produces a large quantity of highly toxic chemicals, including potent carcinogen dioxins, vinyl chloride, ethylene dichloride, PCBs, and others. Moreover, when burned it releases hydrochloric acid and more dioxins. It is therefore prudent to avoid products made with PVC.

Volatile Organic Compounds

As discussed earlier, VOCs consist of thousands of different chemicals, such as formaldehyde and benzene, which evaporate readily into the air. Depending on the level of exposure, they can cause dizziness; headaches; eye, nose, and throat irritation; or asthma; some cases they can also cause cancer; induce longer-term damage to the liver, kidneys, and nervous system; and stimulate higher sensitivity to other chemicals. When dealing with wet products such as paints, adhesives, and other coatings, ensure that the products contain no or low VOCs. Look for the Green Seal when using certified paints or paints with less than 20 g/L VOCs; the seal was initiated by a nonprofit organization that has been actively identifying and promoting sustainability in the marketplace since 1989. For adhesives and coatings, make sure they are South Coast Air Quality Management District (SCAQMD)-compliant.

To minimize VOCs in flooring and carpet, wall coverings, ceiling tiles, and furniture, it is advisable to use only CA 01350-compliant products. A number of programs currently use the CA 01350 testing protocol to measure the actual levels of individual VOCs emitted from the material and to compare it to allowable levels set by the state of California. These include CHPS, the Carpet and Rug Institute's (CRI) Green Label Plus, Scientific Certification Systems' Indoor Advantage™, the Resilient Floor Covering Institute's FloorScore®, and GREENGUARD's Children & Schools Certification Program. Try to avoid flooring that requires waxing and stripping, a process that releases more VOCs than the original material. Care should be taken to ensure that there is no added formaldehyde present in all composite-wood products and insulation.

Phthalates

Di(2-ethylhexyl) phthalate (DEHP) and other phthalates have attracted considerable adverse publicity for their use in PVC medical products and in toys; concerns have been raised about their impact on the development of young children. Phthalates are, however, also used widely in flexible PVC building materials and have been linked to bronchial irritation and asthma. It is important therefore to avoid using products with phthalates (including PVCs).

Heavy Metals

Even though heavy metals are known to be health hazards, they continue to be used for stabilizers or other additives in building materials. Lead, mercury, and organotins are all known potent neurotoxins, which are particularly damaging to the brains of fetuses and growing children. Cadmium is a carcinogen and can cause a variety of kidney, lung, and other damage. Look for products that do not contain heavy metals that are health hazards.

Halogenated Flame Retardants

The use of halogenated flame retardants (HFRs) in many fabrics, foams, and plastics is known to have saved many lives over the years. However, HFRs, including polybrominated diphenylethers and other brominated flame retardants, have been found to disrupt thyroid and estrogen hormones, which can cause developmental effects such as permanent changes to the brain and the reproductive system. The use of products that have HFRs should be avoided.

Perfluorocarbons

Numerous treatments for fabric and some building materials have been based on perfluorocarbons (PFCs) that, like HFRs, are characteristically highly persistent and bioaccumulative and hence concentrate at alarming levels in humans. Perfluorooctanoic acid (PFOA), a major component of treatment products (e.g., Scotchgard™, STAINMASTER®, Teflon®, and GORE-TEX®) has been linked to a range of developmental and other adverse health effects. Thus, avoid all products that are treated with a PFC-based material.

7.4.2   Resources for Locating Healthy Building Materials

Because of the great diversity of materials that are used in the construction and manufacturing industries, it is difficult to produce a single building materials list or certification that covers all relevant health and environmental issues. For example, the Healthy Building Network's PVC-Free Alternatives Database (www.healthybuilding.net/pvc/alternatives.html) is a Construction Specifications Insititue-prepared listing of alternatives for a wide range of building materials. The following list should be useful.

> **Green Seal-certified products.** Paints and coatings that meet the Green Seal VOC content standards are listed on the Green Seal website; these materials do not contain certain excluded chemicals and meet typical LEED performance requirements.
> **EcoLogo™.** This is an environmental standard and certification mark founded in 1988 by the Government of Canada and now recognized internationally. The EcoLogo program provides customers—public, corporate, and consumer—with assurance that the products and services bearing the logo meet stringent standards of environmental leadership. EcoLogo is a Type I eco-label, as defined by the ISO and has been successfully audited by the Global EcoLabelling Network as meeting ISO 14024 standards for eco-labeling. Products EcoLogo certifies include carpet, adhesives, and paint.
> **Collaborative for High-Performance Schools low-emitting materials.** CHPS (www.chps.net/dev/Drupal/node/445) maintains a database that lists

products that have been certified by the manufacturer and an independent laboratory to meet the CHPS low-emitting materials criteria, Section 01350, for use in typical classrooms. Certified materials include adhesives, sealants, concrete sealers, acoustical ceilings, wall panels, wood flooring, composite-wood boards, resilient flooring, and carpet. Note that this list also includes paints, but CA 01350 is not yet a replacement for low-VOC screening.

Green Label Plus. This basically confirms that the Carpet and Rug Institute (a trade association) assures customers that approved carpet products meet stringent requirements for low chemical emissions; it also certifies that these carpets and adhesives meet CA 01350 VOC requirements. Any architect, interior designer, government specifier, or facility administrator who is committed to using green building products just needs to look for the Green Label Plus logo, as this signifies that the carpet product has been tested and certified by an independent laboratory and meets stringent criteria for low emissions.

FloorScore. Scientific Certification Systems (www.scscertified.com) certifies for the Resilient Floor Covering Institute (a trade association) that resilient flooring meets CA 01350 VOC requirements.

GREENGUARD® for Children & Schools. Air Quality Sciences certifies that furniture and indoor finishes meet the lower of CA 01350 VOC or 1/100 of threshold limit values (TLVs). The GREENGUARD Environmental Institute (GEI) has formulated performance-based standards to define goods with low chemical and particle emissions for use indoors; these primarily include building materials, interior furnishings, furniture, cleaning and maintenance products, electronic equipment, and personal care products. The standard establishes certification procedures including test methods; allowable emissions levels; product sample collection and handling, testing type and frequency; and program application processes and acceptance. GEI now certifies products across multiple industries.

Brominated flame retardants/halogenated flame retardants and per-fluorochemicals. Although no listings are yet screened for these emerging problem chemicals, all halogen-based flame retardants are likely to be problematic. Flame retardants are added to plastics, particularly fabrics and foams. PBDEs are the most widely used.

7.5 INDOOR ENVIRONMENTAL QUALITY BEST PRACTICES

An overview of the new IgCCs clearly affirms that indoor environmental quality is a critical component of sustainable buildings. Ventilation, thermal comfort, air quality, and access to daylight and views are all factors that play a pivotal role in determining indoor environmental quality.

The Architectural and Transportation Barriers Compliance Board (Access Board), an independent federal agency devoted to accessibility for people with disabilities, contracted with the National Institute of Building Sciences (NIBS) to establish an Indoor Environmental Quality Project as a first step in implementing an action plan. The NIBS issued a project report on indoor environmental quality in July 2005, which revealed that a growing number of people in the United States suffer a range of debilitating physical reactions caused by exposure to everyday materials and chemicals found in building products, floor coverings, cleaning products and fragrances, and others. This condition is known as multiple chemical sensitivity (MCS). The range and severity of these reactions are varied. In addition, the Access Board received numerous complaints from people who report adverse reactions from exposures to electrical devices and frequencies, a condition referred to as electro-magnetic sensitivity (EMS).

The Access Board responded to these concerns by sponsoring a study on ways to tackle the problem of indoor environmental quality for persons with MCS and EMS as well as for the general population. In conducting this study for the board, the NIBS brought together a number of interested parties to explore the relevant issues and to develop an appropriate action plan. The report includes, among other things, recommendations on improving indoor environmental quality that address building products, materials, ventilation, and maintenance issues.

The following are some of the steps and best practices that can be applied to ensure good IEQ[1]:

- Conduct a faculty-wide IEQ survey/inspection of the facility, noting odors, unsanitary conditions, visible mold growth, staining, presence of moisture in inappropriate places, poorly maintained filters, personal air cleaners, hazardous chemicals, uneven temperatures, and blocked vents.
- Determine operating schedule and design parameters for HVAC systems and ensure that adequate fresh air is provided to prevent the development of indoor air quality problems and to contribute to the comfort and well-being of building occupants. Maintain complete and up-to-date ventilation system records.
- Ensure that appropriate preventive maintenance (PM) is performed on HVAC systems, including but not limited to outside air intakes, the inside of air-handling units, distribution dampers, air filters, heating and cooling coils, fan motors and belts, air distribution ducts and variable air volume (VAV) boxes, air humidification and controls, and cooling towers.
- Manage and review processes with potentially significant pollution sources, such as renovation and remodeling, painting, shipping and receiving, pest control, and smoking. Ensure that adequate controls are instituted on all renovation and construction projects and evaluate control impacts on IEQ.
- Control environmental tobacco smoke by prohibiting smoking within buildings or near building entrances. Designate outdoor smoking areas at least 25 feet from openings serving occupied spaces and air intakes.

---

1. Portions of this list excerpted from "Indoor Enviornmental Quality," Harvard University Campus Services.

- Control moisture inside buildings to inhibit mold growth, particularly in basements. Dehumidify when necessary and respond promptly to floods, leaks, and spills. Use of porous materials in basements should be monitored and restricted whenever possible.
- When mold growth is evidenced, immediate action should be taken to remediate.
- Choose low-emitting materials with minimal or no volatile organic compounds. This particularly applies to paints, sealants, adhesives, carpet and flooring, furniture, and composite-wood products and insulation.
- Monitor CO_2 and install carbon dioxide and airflow sensors to provide occupants with adequate fresh air when required.
- To maintain occupants' thermal comfort, include adjustable features such as thermostats or operable windows.
- Window size, location, and glass type should be selected to provide adequate daylight levels in each space.
- Window sizes and positions in walls should be designed to take advantage of outward views and have high visible transmittance rates ($>50\%$) to ensure maximum outward visibility.
- Incorporate design strategies that maximize daylight and views for building occupants' visual comfort.
- Educate cleaning staff regarding the use of appropriate methods and products, cleaning schedules, materials storage and use, and trash disposal.
- A process for complaint procedures should be established and IEQ complaints promptly responded to.
- Discuss with occupants how they can participate in maintaining acceptable indoor environmental quality.
- Permanent entryway systems such as grilles or grates should be installed to prevent occupant-borne contaminants from entering the building.
- A Construction IAQ Management Plan should be in place so that during construction materials are protected from moisture damage, and control particulates through the use of air filters.

There are also a number of suggested IEQ-related recommendations that tenants should follow to ensure that all building occupants maintain a healthy indoor environment, including:

- The use of air handlers during construction must be accompanied by the use of filtration media with a MERV of 8 at each return grill as determined by ASHRAE 52.2-1999.
- Replace all filtration media immediately prior to occupancy; when possible, conduct a minimum 2-week flush-out with new filtration media with 100% outside air after construction is completed and prior to occupancy of the affected space.
- Contractors must notify the property manager 48 hours prior to commencement of any work that may cause objectionable noise or odors.

- Protect stored onsite materials and installed absorptive materials from moisture damage.
- All applied adhesives must meet or exceed the limits of the SCAQMD Rule #1168. Also, sealants used as fillers must meet or exceed Bay Area Air Quality Management District Regulation 8, Rule 51.
- Ensure that all paints and coatings meet or exceed the VOC and chemical component limits of Green Seal requirements.
- Ensure that carpet systems meet or exceed standards of the CRI's Green Label Indoor Air Quality Test Program.
- Composite-wood and agrifiber products should not contain any added UF resins.
- Contractors should provide protection and barricades where needed to ensure personnel safety, and should comply with OSHA at a minimum.

Finally, it should be remembered that the air quality of a building is one of the most important factors in maintaining employee productivity and health. Toward this end, IEQ monitoring helps minimize tenant complaints of BRI and SBS, and a cohesive proactive IEQ-monitoring program can be a powerful tool that can be used to achieve this goal. Several national and international organizations, including the EPA, OSHA, ASHRAE, ASTM, USGBC®, and others are currently in discussions concerning formulating new standards and updating and improving existing national indoor air quality standards. The following are some relevant standards, codes, and guidelines

- ASHRAE Standard 62-1999—Ventilation for Acceptable Indoor Air Quality
- ASHRAE Standard 129-1997—Measuring Air Change Effectiveness
- International Green Construction Code
- SCAQMD Rule #1168
- Bay Area Air Quality Management District Regulation 8, Rule 51
- Canada's Environmental Choice/EcoLogo, Eco Concepts, Inc.
- "Best Sustainable Indoor Air Quality Practices in Commercial Buildings," GreenBuilding, Inc.
- "Guidelines for Reducing Occupant Exposure to Volatile Organic Compounds from Office Building Construction Materials," California Department of Health Services
- CRI Green Label Indoor Air Quality Test Program

Chapter 8

Water Efficiency and Sanitary Waste

8.1 INTRODUCTION

The concept of conserving our natural resources has in recent years become part of society's green culture. However, with respect to water efficiency, Randhir Sahni, AIA, president of Llewelyn-Davies Sahni, says:

The United States is notoriously water inefficient. For example, there are many development sites that have water and sewer service, but no buildings because the market evaporated or use changed. So what happens? The municipal utility district (MUD) has to put bacteria or animal manure in the sewer plant in order to operate and maintain the facilities.

One of the more prominent issues facing us today is water conservation and using water more efficiently. In this respect, the U.S. Green Building Council's (USGBC®) Leadership in Energy and Environmental Design (LEED®) certification program has raised the bar for water conservation and now requires all projects to reduce water use by at least 20% as a prerequisite to LEED certification. Earlier versions of LEED awarded a point for a 20% reduction.

This prerequisite was first introduced in LEED 2009 and is significantly more demanding; plus, it does not apply to earlier LEED versions. The baseline is determined by assuming that all fixtures meet national codes, as laid out on a fixture-by-fixture basis in the credit requirements. As for LEED for Existing Buildings, the Operations & Maintenance threshold depends on when the facility was originally constructed or last renovated. Of note, those following the same Water Efficiency (WE) requirements as LEED for New Construction include: LEED for Commercial Interiors, LEED for Core & Shell, LEED for Schools, LEED for Retail, and LEED for Healthcare.

Recent EPA estimates lace the amount of freshwater—that is, water needed for drinking, industry, and sanitation—at about 2.5% of the world's total. Roughly one-third of this is readily accessible to humans via lakes, streams, and rivers. Demand for freshwater continues to rise, and if current trends continue, experts project that demand for it will double within the next three decades. Since 1950, the U.S. population has increased by almost 90%. In that same time span, public demand for water has increased by 209%. Americans now use an average of 100 gallons of water per person each day. This increased demand has put tremendous stress on water supplies and distribution systems, threatening both human health and the environment.

Reacting to this potential crisis, the South Nevada Water Authority has put into place a Water Efficient Technologies program that offers financial incentives for capital expenditures when businesses retrofit existing equipment with more water-efficient technologies. Likewise, the EPA has launched WaterSense®, a water-oriented counterpart to the ENERGY STAR® program that promotes water efficiency and aims to boost the market for water-efficient products, programs, and practices.

In addition, local codes do not always keep pace with some of the new green codes—for example, the International Green Construction Code (IgCC) and emerging technologies that are not code-compliant but are nevertheless available in the marketplace. These include, graywater systems, rainwater collection systems, high-efficiency irrigation systems, recirculating shower systems, regulations controlling hot-water delivery, recirculation of hot water, insulation of hot-water piping, demand-type tankless water heaters, water softeners, and drinking water treatment systems. All of these are being implemented through EPA's WaterSense. The EPA estimates that toilets account for roughly 30% of the water used in residences, and Americans annually waste 900 billion gallons by the use of old, inefficient toilets. By replacing an older toilet with a WaterSense-labeled model, a family of four could reduce total indoor water use by about 16% and, depending on local water and sewer costs, save more than $90 annually.

Moreover, water conservation translates into energy conservation and savings. If just one in every ten homes in the United States installed WaterSense faucets or aerators in their bathrooms, this could result in an aggregate saving of about 6 billion gallons of water and more than $50 million in the energy costs to supply, heat, and treat that water. The EPA also estimates that if the average home were retrofitted with water-efficient fixtures, there would be a savings of 30,000 gallons of water per year. If only one out of every ten homes upgraded to water-efficient fixtures (including ENERGY STAR-labeled clothes washers), the resultant savings could reach more than 300 billion gallons and nearly $2 billion annually. This could have a significant positive economic impact on small plumbing contractors and small businesses throughout the various sectors. In fact, the recent increased demand and focus on water efficiency can provide a powerful catalyst to helping the emerging water and energy conservation market to revitalize these industries across the country at a time when most small-business owners are suffering because of tough economic times.

8.2   WASTEWATER STRATEGY AND WATER REUSE/RECYCLING

According to the U.S. Department of Energy (DOE) estimates, commercial buildings consume approximately 88% of the potable water in the United States. This offers facilities' managers a unique opportunity to make a huge impact on overall U.S. water consumption. Benchmarking a facility's water use and implementing measures to improve overall efficiency go a long way to achieving less water use. Likewise, in spite of the limited emphasis by LEED on water efficiency, water-efficient design should be one of the main goals of any project, particularly since our nation's growing population is placing considerable stress on available water supplies. And even though the U.S. population has nearly doubled in the last five or six decades, public demand for water has more than tripled! This increased demand is adding to the stress on water supplies and distribution systems, and depleting reservoirs and groundwater can put our water supplies, human health, and environment at serious risk. According to the EPA, lower water levels can contribute to higher concentrations of natural or human pollutants. Using water more efficiently helps maintain supplies at safe levels, protecting human health and the environment.

The EPA estimates that an American family of four uses about 400 gallons of water per day. About 30% of this is used outdoors for various purposes, including landscaping, cleaning sidewalks and driveways, washing cars, and maintaining swimming pools. Nationally, landscape irrigation accounts for almost one-third of all residential water use. That amounts to more than 7 billion gallons per day. Water Efficiency is one of the principal categories of the LEED Rating System, and the number of WE credits available depends on the type of certification sought (New Construction, Commercial Interiors, Schools, and so on). However, meeting LEED's Water Efficiency Credit 3—Water Use Reduction is no longer a sure thing, even for commercial office buildings. Moreover, recent feedback from the Green Building Certification Institute (GBCI) states that municipally treated process water is no longer acceptable for alternative compliance paths for WE Prerequisite 1 and WE Credit 3 (LEED V3), and municipally supplied graywater may not be used to gain water savings in this prerequisite.

For New Construction (NC) certification, a total of 10 possible points (5 points were allotted in previous versions) can be achieved for WE LEED V3 certification (WE credits for LEED-Homes: maximum 15 points possible). The main WE categories and topics to know for LEED⁻NC are

- WE Credit 1: Water-Efficient Landscaping (4 points)
  - Reduce by 50%
  - No potable use or no irrigation
- WE Credit 1.1: Reduce by 50% (2 points)
- WE Credit 1.2: No Potable Water Use or No Irrigation (2 points in addition to WE Credit 1.1)
- WE Credit 2: Innovative Wastewater Technologies (2 points)
- WE Credit 3: Water Use Reduction (2–4 points)
  - 20%
  - 30%

Landscaping irrigation is the main source of outdoor water consumption, accounting for about 30% of the 26 billion gallons of daily water consumption.

The intent of water-efficient landscaping in the LEED Rating System is to reduce (by at least 50%) or eliminate potable water consumption and natural surface or consumption of subsurface water resources available on or near the project site and used for landscape irrigation. The following are best practice strategies

- Use the most appropriate plant material for the project climate.
- Use native or adapted plants to reduce or eliminate irrigation.
- Use high-efficiency equipment when irrigation is required.
- Use climate-based controllers.

On occasion, landscape design strategies alone are unable to achieve a project's irrigation efficiency goals, in which case attempts should be made to meet efficiency demands through optimization of the irrigation system design. For example, use of high-efficiency drip, micro, and subsurface systems can reduce the amount of water required to irrigate a given landscape. The USGBC reports that drip systems alone can reduce water use by 30 to 50%. Climate-based controls, such as moisture sensors with rain shutoffs and weather-based evapotranspiration controllers, can further reduce demands by allowing naturally occurring rainfall to meet a portion of the landscaping's irrigation needs.

To earn a LEED WEc3 credit, a reduction is needed in the use of potable water for irrigation by 50 to 100% compared with a baseline irrigation system typical for the region. Because landscape irrigation can account for nearly 40% of the average office building's potable water consumption, reducing or eliminating its use for landscaping can save both water and money. For LEED certification, one point is awarded for a 50% reduction in water consumption for irrigation from a calculated mid-summer baseline case, and a total of two points is awarded for a 100% reduction. While LEED V3 has made it increasingly difficult to obtain WE points, this should have a positive impact on architects and plumbing engineers by continually challenging them to develop creative solutions that reduce building potable water consumption.

To facilitate greening the water supply, it is necessary to tap alternate water sources. LEED recognizes two alternate sources: rainwater collection and wastewater recovery. Rainwater collection involves collecting and holding rainfall in onsite cisterns, underground tanks, or ponds during rainfall. This water can then be used during dry periods by the irrigation system. Wastewater recovery can be achieved either onsite or at the municipal scale. Onsite systems capture graywater (which does not contain human or food-processing waste) from the building and apply it to irrigation. Reductions are attributed to any combination of the following approaches:

- Use a high-efficiency micro-irrigation system, such as drip, micromisters, and subsurface irrigation systems.
- Replace potable (drinking) water with captured rainwater, recycled wastewater (graywater), or treated water.
- Use water treated and conveyed by a public agency that is specifically for nonpotable purposes.

- Factor in plant species, density, and microclimate, and install landscaping that does not require permanent irrigation systems.
- Apply xeriscape principles to all new development whenever possible. Xeriscaping is the use of low-water, drought-resistant plants and plants that are accustomed to local rainwater patterns.

Additionally, groundwater seepage that is collected and pumped away from the immediate vicinity of foundations and building slabs is eligible for use in landscape irrigation to meet the intent of this credit. It must be demonstrated, however, that doing so does not impact the site stormwater management systems. When a landscaping design incorporates rainwater collection or wastewater recovery in particular, it is essential to assemble a team of experts and establish project roles at an early stage in the process. Rainwater collection and wastewater treatment systems stretch over multiple project disciplines, making it particularly important to clearly articulate responsibilities. Having an experienced landscape architect on board is pivotal for a water-efficient landscape and irrigation system design. Early planning is highly recommended to take advantage of the available LEED points for water-efficient landscaping credits.

Several of the LEED credits deal with graywater and black water. Graywater has several definitions; it is typically considered to be untreated wastewater that has not come into contact with toilet waste, shower water, or water from sinks (other than the kitchen), bathtubs, wash basins, and clothes washers. Graywater use is generally for both indoors and outdoors. When used outdoors, it is usually filtered and then used for watering landscape. Indoor graywater, on the other hand, consists of recycled water and is used mainly for flushing toilets. Graywater has other applications, including construction activities, concrete mixing, and cooling for power plants. The Uniform Plumbing Code (UPC) defines graywater as untreated household wastewater that has not come in contact with toilet waste, whereas the International Plumbing Code (IPC) defines it as wastewater discharged from lavatories, bathtubs, showers, clothes washers, and laundry sinks; some jurisdictions allow the inclusion of kitchen sinks in this list. Black water lacks a specific definition that is accepted nationwide, but is generally considered to constitute toilet, urinal, and kitchen sink water (in most jurisdictions). However, depending on the jurisdiction, implementing graywater systems that reuse wastewater from showers and sinks for purposes like flushing toilets or irrigation may encounter code compliance restrictions.

8.2.1 Reclaimed-Water versus Graywater Systems

The recycling of water and putting it back to use is commonly thought of in two different water usage strategies: reclaimed water and graywater, and it is important to distinguish between the two, although some mistakenly use the terms *reclaimed water* and *graywater* interchangeably.

Simply put, reclaimed water is wastewater effluent/sewage that has been treated according to high standards at municipal treatment facilities and meets reclaimed water effluent criteria. Its treatment takes place offsite, and it is delivered to a facility. Reclaimed water is most commonly used for nonpotable purposes, such as landscaping, agriculture, dust control, soil compaction, and processes such as concrete production and cooling for power plants. The use of reclaimed water is increasing in popularity, especially in states like California, where openness to innovative, environmentally friendly concepts prevails, especially in the face of a very real and critical water crisis. For example, Orange County, California, has recently started delivering purified wastewater, providing one of the first "toilet-to-tap" systems to be employed in the nation.

On the other hand, graywater is the product of domestic water use such as showers, washing machines, and sinks, and does not normally include wastewater from kitchen sinks, photo lab sinks, dishwashers, or laundry water from soiled diapers. These sources are typically considered to be black water producers because they contain serious contaminants and therefore cannot be reused. Moreover, graywater use is a point-of-source strategy—that is, graywater collected from a building is reused in the same building.

8.2.2　Innovative Wastewater Technologies

The intent of LEED's Innovative Wastewater Technologies credit is to reduce wastewater generation and potable water demand, and increase the recharge of local aquifers. To achieve this credit requires a 50% reduction in potable water used for building sewage (black water) conveyance that is the product of flush fixtures. You can reduce potable water demand by using water-conserving fixtures, reusing nondrinking water for flushing, or reusing water treated onsite to tertiary standards (with the treated water infiltrated or used onsite). Tertiary treatment is the final stage of treatment before water can be discharged back into the environment. If it is used, the water must be treated by biological systems, constructed wetlands, or a high-efficiency filtration system. Of note, the City of San José's Water Efficient Technologies (WET) program is now in place that offers financial incentives to commercial and multifamily property owners who install water-efficient devices and implement new water-saving technologies. Examples of effective water-efficient technologies strategies include:

- Ultra-high-efficiency toilets and efficient retrofits
- Efficient showerheads and retrofits
- Water-free and high-efficiency urinals
- Other ultra-low-water consumption products
- Conversion of a sports field from grass to an artificial surface
- Retrofitting of standard cooling towers with qualifying, high-efficiency drift elimination technologies

Strategies for meeting one of LEED's WE compliance requirements, reducing potable water use for sewage conveyance, falls into two categories that can be implemented either independently or in concert. As shown earlier, by simply meeting demand efficiently, the use of ultra-high-efficiency plumbing fixtures can reduce the water required for sewage conveyance in excess of the 50% requirement. To use a typical example, composting toilets (not normally used in commercial facilities) and water-free urinals use no water. These two technologies alone can eliminate a facility's use of potable water for sewage conveyance, qualifying both for this credit's point and potentially for a LEED Innovation in Design point for exemplary performance. Should the selected plumbing fixtures alone prove to be inadequate to reach the 50% reduction threshold, or if ultra-high-efficiency plumbing fixtures are not selected, the water necessary for toilet and urinal flushing can be reduced by a minimum of 50%, or eliminated entirely, by applying strategies such as rainwater collection or wastewater treatment.

An excellent example of how this credit can be achieved is provided by the Southface Eco Office in Atlanta, Georgia (Figure 8.1). It provides a showcase of state-of-the-art energy, water, and waste-reducing features. Targeting LEED Platinum certification, this facility was able to completely eliminate the use

FIGURE 8.1   The new Southface Eco Office in Atlanta, a 10,000-square-foot facility seeking a LEED Platinum rating that was designed as a model for environmentally responsible commercial construction achievable utilizing existing off-the-shelf materials and technology.

of potable water for sewage conveyance using a variety of complementary strategies. For example, foam flush composting toilets and waterless urinals are used in the staff restrooms. Composting toilets require only six ounces of water per use, which significantly reduces the volume of water required for sewage conveyance. Water requirements in the facility's public restrooms are also reduced through the employment of a combination of dual-flush toilets, ultra-high-efficiency toilets, and water-free urinals. The remaining reduced volume of water required for sewage conveyance is supplied by rainwater collected from a roof-mounted solar array and stored in a rooftop cistern, in addition to a supplemental inground storage tank.

Early involvement of an experienced and knowledgeable team of local code officials is critical for the successful design and implementation of nonpotable water supply systems. Furthermore, dual-plumbing lines for nonpotable water within the building are fairly easy to plan for during the design phase but much more difficult to retrofit after construction is complete and the building is occupied.

8.2.3   Water Use Reduction

The intent of the Water Use Reduction credit, according to LEED, is to "maximize water efficiency within buildings to reduce the burden on municipal water supply and wastewater systems." One point was previously awarded for reducing water use by 20%. In LEED V3 it is a prerequisite and becomes two points for reducing annual potable water use by 30%. Of note, in LEED V3 a 35% saving awards three points whereas four points can be achieved for a 40% saving. The fixtures governed by this credit include water closets, urinals, lavatory faucets, showers, and kitchen sinks. Water-using fixtures and equipment such as dishwashers, clothes washers, and mechanical equipment (nonregulated uses), which are not addressed by this credit, may qualify for the LEED Innovation in Design point.

The use of proven, cost-effective technologies can facilitate achieving the required percentage reduction necessary to earn points for the Water Use Reduction credit. The use of low-flow lavatory faucets with automatic controls (0.5 gallons per minute, 12 seconds per use) is normally sufficient to achieve a 20% reduction, qualifying for the prerequisite. An additional 14% reduction can be achieved by the use of water-free urinals that, when combined with low-flow faucets, should exceed the 30% reduction threshold, thereby earning points. Here, too, the first step in the optimization process, reducing demands, does not apply. It's not possible to design away occupants' needs to use the restroom, wash their hands, or take a shower. Strategies for water use reduction therefore fall into the same two categories identified for Innovative Wastewater Technologies—either meeting demands efficiently or fulfilling them in alternate, more environmentally appropriate means. The two credits complement one another and water savings related to the Innovative Wastewater Technologies credit also contribute to the Water Use Reduction credit. John Starr, AIA, and Jim Nicolow, AIA, of Lord, Aeck & Sargent state that among the LEED Water Efficiency credits:

Water Use Reduction can often be achieved without the early planning and design integration required by the other two credits. Most alternative plumbing fixtures use conventional plumbing supply and waste lines, allowing these fixtures to be substituted for less-efficient standard fixtures at any point in the design process, and even well into the construction process.

8.2.4 System Approaches

Local municipalities and individual facilities continue to struggle to meet water needs in the face of dwindling water supplies. This has led to the emergence of a variety of reclaimed water and graywater system approaches. These systems range in their size and complexity. Toward the high end are the multibuilding installations that draw wastewater from municipal sources, followed by the middle tier, which includes buildings that have installed storage tanks capable of collecting thousands of gallons of water from rainfall, sinks, and steam-condensate that is then treated and funneled to water reuse sources. There are also the more affordable under-counter systems that are simpler and on a smaller, yet significant, scale; these carry out on-the-spot treatment of water that flows down sink drains and is then pumped directly into toilet tanks. More complex systems should be built into new construction rather than be retrofitted at a later date, whereas on-the-spot collection systems can be implemented at any time. It is important when specifying sustainable systems and technologies to remain within budget by setting goals and performing research upfront to determine the added value and payoff of the systems being used.

Graywater Demonstrations

The volume of graywater produced in a particular building depends largely on the type of facility. For example, a typical office building may not yield as much graywater as a college dorm or multiuse retail and condominium building; the benefits are all about economies of scale and deriving value from the system, no matter how large or how small it may be. Let us consider the amount of potable water that a typical four-person household can save. On average, each person uses 80 to 100 gallons of water per day, with toilet flushing being the largest contributor to this use. The combined use of kitchen and bathroom sinks is only 15% of the water that comes into a home, which is significant considering that 100% of the water that comes into the home has been treated and made potable for drinking.

Thus, with the largest single source of freshwater use in the home capable of using graywater instead of potable water, the household can make real gains on re-using water that is totally appropriate for toilet flushing. For households and small commercial facilities, the best solution may be to use a graywater system that incorporates a reservoir, which is installed under the sink and attached to the toilet. This system is designed so that the toilet draws first from the collected water in the reservoir. The system remains connected to the fresh-water pipes so that, should flushing deplete the water stored in the reservoir, the toilet can then secondarily draw from outside water. Because toilets are the largest consumers of water in households, such systems are able to save up to 5000 gallons per year.

Differing graywater policies and regulations between states are significantly impacting the extent to which facilities and homeowners can deploy graywater systems. The state of Arizona for example, has graywater guidelines to educate residents on methods to build simple, efficient, and safe graywater irrigation systems. For those who follow these guidelines, their system falls under a general permit and automatically becomes "legal," which means that the residents don't have to apply or pay for any permits or inspections. California, on the other hand, has a graywater policy that is restrictive, which usually makes it difficult and unaffordable to install a permitted system. Many states have no graywater policy and don't issue permits at all, while others issue experimental permits for systems on a case-by-case basis.

The term *recycling* is usually reserved for waste such as aluminum cans, glass bottles, and newspapers. Water can also be recycled and, indeed, through the natural water cycle the earth has recycled and reused for millions of years. *Water recycling*, though, generally refers to using technology to speed up these natural processes. The recycling of water by any means provides substantial benefits, including reduction of stress on potable water resources, reduction of nutrient loading to waterways, reduction of strain on failing septic tanks or treatment plants, less energy and chemical use, and a cost lower than that for potable water.

Long-Term Savings

It is worth noting that as little as a decade ago, purchasing environmentally friendly building components that met LEED compliance standards may have added more than 10% to total building costs, whereas today plumbers, engineers, and other specifiers are now discovering that they can adopt higher sustainability standards without necessarily incurring extra costs. And where they do have to spend extra, the long-term payoff more than compensates when you factor in long-term operating costs, including water and wastewater utility bills, plus the energy it takes to heat water for faucets, showerheads, and the like.

According to Flex Your Power©, California's energy efficiency marketing and outreach campaign, utilities account for about 30% of an office building's expenses. A 30% reduction in energy consumption can lower operating costs by $25,000 a year for every 50,000 square feet of office space. This has raised public awareness and created greater notice of how companies and facilities expend water and energy; both users and communities are holding building owners accountable for their use of precious local resources. Engineers need to stay abreast and monitor water and energy efficiency options in restrooms and elsewhere in their facilities to minimize operating costs and help ensure that buildings meet LEED standards as well as the new IgCC standards.

Construction Waste Management

Commercial construction typically generates between 2 and 2.5 pounds of solid waste per square foot, the majority of which is recyclable. Salvaging and recycling construction and demolition (C&D) waste can substantially reduce demand for virgin resources and its associated environmental impacts. Additionally, effective construction waste management, including appropriate handling of nonrecyclables, can reduce contamination from and extend the life of existing landfills. Whenever feasible, therefore, reducing initial waste generation is environmentally preferable to reuse or recycling.

From the outset, a project's construction waste management plan should recognize project waste as an integral part of overall materials management. The premise being that waste management is a part of materials management, and the recognition that one project's wastes are materials available for another project, facilitates efficient and effective waste management. Moreover, waste management requirements should be included as a topic of discussion, both during the pre-construction phase and at ongoing regular job meetings, to ensure that contractors and appropriate subcontractors are fully aware of the implications of these requirements on their work prior to and throughout construction. Furthermore, waste management should be coordinated with or part of a standard quality assurance program, and its requirements should be addressed regularly throughout the project. All topical applications of processed clean wood waste and ground gypsum board as a soil amendment must be implemented in accordance with local and state regulations.

8.3   WATER FIXTURES AND WATER USE REDUCTION STRATEGIES

Today's mainstream market is flooded with thousands of plumbing fixtures and fittings that can help save water, energy, and money. These include but are not limited to aerators, metering and electronic faucets, and pre-rinse spray valves. But when selecting energy-efficient equipment, it is vital to select quality products that meet conservation requirements without compromising performance. A product should deliver the consistent flow required while maintaining the water and energy savings the industry demands. And with restroom fixtures accounting for most of a typical commercial building's water consumption, the best opportunities for increasing efficiency can be found there. Fortunately, there is increased public awareness combined with an increasing number of available higher-efficiency plumbing fixtures.

One of the best ways to increase water efficiency in buildings is through plumbing fixture replacement and implementation of new technologies, particularly since significant water efficiency improvements over conventional practice are now readily achievable. Replacing older, high-flow water closets and flush valves with models that meet current UPC and IPC requirements is important. While current codes require a lower flow rate for new fixtures, existing buildings often have older, high-flow flush valves. Despite the tremendous water savings available by updating fixtures, facility managers often avoid the upgrade due to concerns about clogging. Solid waste removal must be 350 grams or greater. Fixtures pass or fail based on whether they completely clear all test media in a single flush in at least four of five attempts. Toilets that pass qualify for the EPA WaterSense label. It should be noted that when the Energy Policy Act (EPAct) of 1992 was first enacted, many facility managers at the time experienced problems with the low-flow fixtures clogging due to fixture design issues. These have long since been addressed (Table 8.1).

The value of selecting water-efficient fixtures is not only in reducing sewer and water bills; efficient water use also reduces the need for expensive water supply and wastewater treatment facilities, and helps maintain healthy aquatic and riparian environments. Moreover, it reduces the energy needed to pump, treat, and heat water. Water is employed in a product's manufacture, during a product's use, and in cleaning, which means that water efficiency and pollution prevention can occur during several product life-cycle stages. Mark Sanders, product manager for Sloan Valve Company's AQUS® Water Reuse System says that graywater and reclaimed water strategies make good use of water resources, especially when implemented in conjunction with efficient plumbing systems.

The maximum volume of water discharged, using both original-equipment tank trim and aftermarket closure seals, must be tested according to the protocol detailed on the WaterSense website. There are two primary approaches to measuring water volume: gallons per flush (gpf) for toilets and urinals and

TABLE 8.1 Comparison of Plumbing Fixture Water-Flow Rates

Plumbing Fixture	Before 1992	EPA 1992	Current Plumbing Codes
Toilets	4 to 7 gpf	1.6 gpf	1.6 gpf
Urinals	3.5 to 5 gpf	1.0 gpf	1.0 gpf
Faucets*	5 to 7 gpm	2.5 gpm	0.5 gpm
Showerheads*	4.5 to 8 gpm	2.5 gpm	2.5 gpm

*At 80 psi flowing water pressure.
Source: Domestic Water Conservation Technologies, Federal Energy Management Program, U.S Department of Energy, Office of Energy Efficiency and Renewable Energy, National Renewable Energy Laboratory, October 2002.

gallons per minute (gpm) for flow-type fixtures such as lavatories, sinks, and showers. Metered faucets with controlled flow rates for preset time periods are measured in gallons per cubic yard. The maximum volume of water that may be discharged by the toilet, when field adjustment of the tank trim is set at its maximum water use setting, should not exceed 1.68 gpf for single-flush fixtures; for dual-flush fixtures, it should not exceed 1.40 gpf in reduced-flush mode and 2.00 gpf in full-flush mode.

For LEED purposes, baseline calculations should be computed by determining the number and gender of users. As a default, LEED lets you assume that females use toilets three times per day and males once per day in addition to using the urinal two times per day. Both males and females use bathroom faucets three times each day and the kitchen sink once for 15 seconds. The following section discusses the various types of water-efficient fixtures on the market.

8.3.1 Toilet and Urinal Types

By using water more efficiently, we can help preserve water supplies for future generations, save money, and protect the environment.

High-Efficiency Toilets

The signing into law of the National Energy Policy Act of 1992, requiring that toilets sold in the United States after January 1994 use no more than 1.6 gallons (6 liters) per flush, was a significant step in water conservation. This mandate to conserve has encouraged manufacturers to produce a new generation of high-efficiency toilets (HETs) that use technologies such as pressure assist, gravity flush, and dual flush to remove waste using as little water as possible. Of these new technologies, the dual-flush method has the advantage of intuitive flushing, where the operator can decide electively that less water is required and so use 1 gallon (3 liters) or less per flush instead of the 1.6-gallon maximum.

Currently, two basic types of toilet fixtures dominate the marketplace: ultra-low flush toilets (ULFTs), aka "low-flow" or "ultra-low-flow," and high-efficiency toilets (HETs). ULFTs are defined by a flush volume between 1.28 gpf and 1.6 gpf. HETs are defined by a flush volume 20% below the 1.6-gpf maximum or less, equating to a maximum of 1.28 gpf. Dual-flush fixtures are included in the HET category. This 20% reduction threshold serves as a metric for water authorities and municipalities designing more aggressive toilet replacement programs and, in some cases, establishing an additional performance tier for financial incentives such as rebate and voucher programs. It is also a part of the water efficiency element of many green building programs that exist throughout the United States. Unfortunately, this standard currently applies only to tank-type toilets. Flushometer valve toilets have yet to be studied in the same way as tank types, and testing needs to be performed on the flushometer valve with the various bowls on the market so that the pair can then be rated.

Even though toilets purchased for new construction and retrofits are required to meet the new standards, millions of older, inefficient toilets remain in use. As water and sewer costs keep rising, low-flow toilets are becoming increasingly attractive to the American consumer, and local and state governments use rebates and tax incentives to encourage households to convert to these new technologies. The advantages of low-flow toilets in conserving water and thus reducing demand on local water treatment facilities are obvious. According to the EPA, the elimination of inefficient old-style toilets would save the nation about 2 billion gallons of water a day. With a growing population and an antiquated water treatment infrastructure, water conservation will continue to be a major concern to the public.

Dual-Flush Toilets

Dual-flush toilets can help make bathrooms more environmentally friendly. They handle solid and liquid waste differently from standard American toilets, giving the user a choice of flushes. A dual-flush toilet has an interactive design that helps conserve water, which has become popular especially in countries where water is in short supply and in areas where water supply and treatment facilities are older or overtaxed. The EPA estimates that by the year 2013, 36 states will possibly experience water shortages as a result of increased water usage and inefficient water management from aging regional infrastructures. However, using less water to flush liquid waste, while logical, may face cultural biases in the United States that make accepting such an innovative approach to personal waste removal difficult to accept. Interest in low-flow and dual-flush toilets is on the rise in the United States, partly due to increased government regulation and the rising cost of water, and the introduction of incentives in many states for making changes in the way we use the commode.

The method by which water is used to remove waste from the bowl impacts the amount of water needed to get the job done. Standard toilets use a siphoning action, which basically employs a tube to discharge waste. A high volume of water enters the toilet bowl when the toilet is flushed, fills the siphon tube, and pulls the waste and water down the drain. When air enters the tube, the siphoning action stops. Dual-flush toilets employ a larger trapway (the hole at the bottom of the bowl) and a wash-down flushing design that pushes waste down the drain. Because no siphoning action is involved, the system requires less water per flush, and the larger trapway diameter facilitates the exit of waste from the bowl.

Combined with the savings from using only half-flushes for liquid waste, a dual-flush toilet can save up to 68% more water than a conventional low-flow toilet. The larger-diameter trapway is the main reason a dual-flush toilet doesn't clog as often as a conventional toilet while requiring less water to flush efficiently and saving more water than a low-flow toilet when flushing liquid waste. However, it should be noted that a dual-flush unit is typically slightly more expensive than comparable low-flow toilet designs. Also, dual-flush toilets typically retain only a small amount of water in the bowl, and flushing doesn't always remove all the waste. Even in full-flush mode, occasional streaking will occur.

Composting Toilets

Composting toilets are dry toilets that use a predominantly aerobic processing system that treats excreta, typically with no water or small volumes of flush water, via composting or managed aerobic decomposition. Because they require little or no water to function effectively, they are particularly suitable (although not exclusively) for use as an alternative to flush toilets in locations where main water and sewerage connections or waste treatment facilities are unavailable, or in locations where water consumption needs to be minimized to the greatest extent possible (Figure 8.2). A composting toilet can save more than 6600 gallons (24,984 liters) of water per person a year.

It is estimated that the average American uses 74 gallons (280 liters) of water per day, one-third of which splashes down a flushing toilet. Older toilets may use up to 7 gallons (26.5 liters) per flush, whereas federal law stipulates that only 1.6 gallon (6.1 liter) be used for low-flow models in new homes. Not using water to flush a toilet also cuts out all the energy expended down the line, from the septic system to the treatment plant. That could be beneficial to our waterways.

To function properly, self-contained composting toilets require appropriate ventilation that can keep the smell out of the bathroom while providing enough oxygen for the compost to break down. Some toilets achieve this by employing fans and a heater powered by electricity (some models do not require electricity). The composter also has to be kept at a minimum temperature of 65 °F (18.3 °C), so for those living off the grid, a heater could potentially require more electricity than is used in the rest of the house. The heater doesn't have to run all the time, however, and one model may only operate at a maximum of 540 watts for about 6 hours a day. As self-contained models are relatively small, the power needed for fans is fairly minimal. A fan may need about 80 to 150 watts, which is roughly the same amount of power used by a light bulb. The use of solar panels to power the fans and heater are possible alternatives. Composting toilets use the natural processes of decomposition and evaporation to recycle human waste. The waste that enters the toilets is over 90% water, which is evaporated and carried back to the atmosphere through the vent system. The small amount of remaining solid material is converted to useful fertilizer by natural decomposition.

High-Efficiency Urinals

High-efficiency urinals (HEUs) use 0.5 gpf or less—at least half the amount of water used to flush the average urinal (i.e., as opposed to the baseline value of 1.0 gpf). The California Urban Water Conservation Council (CUWCC), in cooperation with water authorities and local agencies, defined them as fixtures

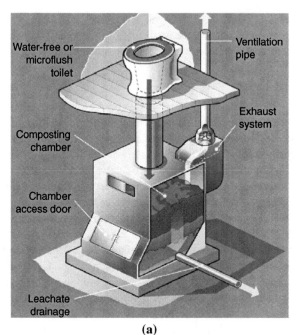

(a)

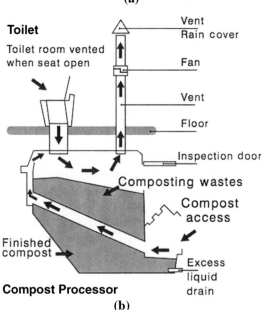

(b)

FIGURE 8.2   (a) Composting toilet operation. (b) Sectional diagram of a composting toilet. There are plans to install a composting toilet in a South Bronx community garden. *(Sources: (a) HowStuffWorks. com; (b) New York Daily News.)*

that have an average flush volume lower than the mandated 1.0 gpf. There are also zero water-consumption urinals. Based on data from studies of actual usage, HEUs save 20,000 gallons of water per year with an estimated 20-year life. They therefore not only help the environment but are making a significant difference in water usage and water bills. In addition to HEUs, there are ultra-low-water urinals that use only 1 pint (0.125 gallons) of water to flush. These combine the vitreous china fixture with either a manual or sensor-operated flush valve, providing effective, low-maintenance flushing in public restrooms while reducing water consumption by up to 88%.

Water-Free Urinals

With water-free urinals, we see a significant improvement over traditional urinals in both maintenance and hygiene, in addition to savings on water as well as sewage and water supply line costs. Water-free urinals and HEUs are part of the next generation of water-efficient plumbing products and contribute to LEED credits for water use reduction. These fixtures employ a special trap with a lightweight biodegradable oil that lets urine and water pass through but prevents odor from escaping (Figure 8.3). Also, there are no valves to fail and no flooding. Periodic upkeep is required to clean the fixture and maintain the liquid seal device. Installation is easy whether in new or retrofit applications. The initial cost of a water-free urinal is often less than that of conventional no-touch fixtures, lowering your initial investment. The urinal can be used to accumulate water-efficient LEED credits, including innovation points. In lieu of no-water urinals, graywater or rainwater harvesting can be implemented.

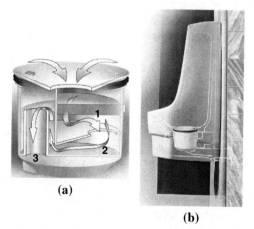

(a)

(b)

FIGURE 8.3   Water-free urinal: (a) The cartridge acts as a funnel directing flow through the liquid sealant (1), preventing any odors from escaping (2); the cartridge then collects uric sediment (3). (b) The remaining liquid, which is noncorrosive and free of hard water, is allowed to pass freely down the drainage pipe. *(Courtesy Sloan Valve Company.)*

8.3.2 Faucets and Showerheads

EPAct 1992 lays out specific requirements for faucet flow rates. For example, residential lavatory faucets must be regulated by an aerator to 2.2 gpm or less; kitchen faucets, to 2.5 gpm or less. LEED V3 has adjusted the baseline rate for public faucets from 2.2 gpm down to 0.5 gpm. The updated baseline standard applies to V3 editions of LEED-NC, LEED-CS, LEED-Schools, and LEED-CI. Version 3 distinguishes between "public" and "private" commercial restrooms and residential bathrooms. Commercial faucet requirements vary according to fixture type: Handle-operated models are regulated by aerator to 0.5 gpm, while self-closing and sensor-operated models are limited to less than 0.25 gallons per cycle. However, today we find that the technology available greatly exceeds EPAct regulations. While kitchen faucets may require about a 2.5-gpm flow rate to fill a pot in a timely fashion, studies have shown that residential lavatory faucets are satisfactory for the user even when reduced to a 0.5-gpm flow rate, and conservation-minded specifiers have started to recommend aerators that deliver this.

Electronic Faucets

Use of electronic faucets is an easy way to save energy, and although they are more costly than a traditional faucet, they pay for themselves in water and energy savings in a short period of time. The electronic faucet has a sensor feature that prevents it from being left on and prevents excess dripping. According to EnergyStar.gov, "hot water leaking at a rate of one drip per second from a single faucet can waste up to 1661 gallons of water over the course of a year." Electronic faucets typically come equipped with several standard features, including the choice of electric plug-in (AC) and battery (DC) power options. Of note, recent research has shown that electronic faucets are more susceptible to contamination with bacteria (especially *Legionella*) than manual faucets, and may pose a potential risk for health care–associated infections. This has led to their cancellation in several newly constructed hospitals.

Metering Faucets and Aerators

Metering faucets and aerators are less expensive than electronic faucets, yet they can deliver similar energy-saving results. Metered faucets, which are more common in commercial washrooms, are generally mechanically operated fixtures that deliver water (at no more than 0.25-gal/cycle) and then self-close. The manual push feature prevents the faucets from being left on after use and prevents unnecessary waste while scrubbing hands. The typical metering faucet's cycle time can be adjusted to deliver the desired amount of water per minute. Many of these devices are designed to allow the user to adjust the temperature before operation. However, in the majority of commercial washrooms, sensors are increasingly becoming the standard. But engineers appear to have reached the limit of water efficiency for sensor models: 0.08 gal/cycle. Nevertheless, it is not user demand or engineering limitations that have determined this to be the limit; it is due to the fact that other environmental considerations come into play.

Aerators add air into the water stream to increase the feeling of flow and are a very common faucet accessory. They are capable of controlling the flow to less than 1.5 gpm, and provide a simple and inexpensive low-flow/low-energy solution. Aerators come in a variety of models to provide the exact flow that complies with local plumbing codes.

Flow-Optimized Showerheads

A 10-minute shower can use between 25 and 50 gallons of water because a typical high-flow showerhead uses 6 to 10 gpm. Flow-optimized single- and three-function showerheads have a flow rate of 1.75 gpm, making them some of the first water-efficient showerheads to offer up to a 30% water savings from the industry-standard 2.5-gpm showerheads without sacrificing performance. This can also contribute toward maximizing LEED points.

The flow rate of 2.5 gpm is both the EPAct requirement and the LEED baseline. Attempts to reduce the flow rate still further are mostly met with very unhappy users. Some users even remove the flow restrictors from their fixtures, producing rates of 4 to 6 gpm, which is clearly not green by any standard. Likewise, flow rates below 2.5 gpm risk failure of certain types of thermostatic mixing valves, leading to scalding of the user. Before specifying valves and showerheads, it may be prudent to consult the manufacturer of the valve; this information may help alleviate this problem altogether.

8.3.3 Baseline Water Consumption Calculations

To achieve the WE LEED credit, you must first determine the baseline model for water usage in the building. The primary factors in this calculation are the types of fixtures in the building, the number of occupants, and the specified fixture's flow or flush rate. When evaluating a building's water efficiency, the USGBC offers a helpful method that allows you to benchmark annual water use and compare that use to current standards.

Establishing FTE Occupants' Water Use

First you must establish water use based on past annual use records or on estimates of building occupancy. This should be followed by estimating a theoretical water use baseline based on the types of fixtures in the building and the number of building occupants. To determine the number of occupants in the building, the number of full-time equivalent (FTE) occupants must be known (acquired from the LEED administrator). The FTE is typically broken down 50/50 for males and females except in cases where the type of building is meant primarily for one gender—for example, a women's gym. In cases that do not adhere to a strict 50/50 split for male and female occupants, an explanation of the design case rationale is recommended. This can be included in the narrative section of the LEED online template for this credit. The FTE should include the transient (visitor) building occupants whom the building is designed for, in addition to the primary occupants. Projects that contain both FTE and transient occupants require separate calculations for each type of occupancy.

In Table 8.2, we have an example used by the USGBC to illustrate the calculation process. It represents potable water calculation for sewage conveyance for a two-story office building with a capacity of 300 occupants. The calculations are based on a typical 8-hour workday and a 50/50 male/female ratio. Male occupants are assumed to use water closets once and urinals twice in a typical workday (default), and females are assumed to use water closets three times each day (default). The reduction amount is the difference between the design case and the baseline case.

In Table 8.3, we show the baseline case being used in line with the EPAct 1992 fixture flow rates. When undertaking these calculations, the number of occupants, number of workdays, and frequency data should remain the same. Furthermore, volumes of graywater or rainwater harvesting should not be included. The baseline case in Table 8.3 estimates the amount of potable water per year used for sewage conveyance to be 327,600 gallons. This means that a

TABLE 8.2 Design Case for Water Use Calculation

Fixture Type	Daily Uses	Flow Rate (gpf)	Occupants	Sewage Generation (gal)
Low-flow water closet (male)	0	1.1	150	0
Low-flow water closet (female)	3	1.1	150	495
Composting toilet (male)	1	0.0	150	0
Composting toilet (female)	0	0.0	150	0
Water-free urinal (male)	2	0.0	150	0
			Total daily volume (gal)	495
			Annual workdays	260
			Annual volume (gal)	128,700
		Rainwater or graywater volume (gal)		(36,000)
			Total annual volume (gal)	**92,700**

Source: USGBC.

TABLE 8.3 Water Use According to EPAct 1992

Fixture Type	Daily Uses	Flow Rate (gpf)	Occupants	Sewage Generation (gal)
Water closet (male)	1	1.6	150	240
Water closet (female)	3	1.6	150	720
Urinal (male)	2	1.0	150	300
			Total daily volume (gal)	1260
			Annual workdays	260
			Total annual volume (gal)	**327,600**

Source: USGBC.

reduction of 72% has been achieved in potable water volumes for this purpose. Using this strategy can earn one point in LEED's Rating System.

Of note, the baseline calculation is based on the assumption that 100% of the building's indoor plumbing fixtures comply with the requirement of the 2006 UPC or the 2006 IPC fixture and fitting performance requirements. Once the baseline has been established for the building, actual use can be compared and measures can be implemented to reduce water use and increase overall water efficiency. Although this baseline methodology is specific to LEED, it can be used in buildings that are not seeking LEED certification.

For calculation purposes on LEED projects, the precise number of fixtures is not important unless there are multiple types of the same fixture specified throughout the building. For example, if there are public restrooms with different water closets on the second floor, their use is to be accounted for as a percentage of the FTE in the LEED credit template calculations. By applying EPAct 1992's fixture and flow rates to FTE building occupants, the baseline quantity use can be established. To determine the estimated use by building occupants, FTE calculations for the project must be used consistently throughout the baseline and design case calculations.

8.4 RETENTION AND DETENTION PONDS, BIOSWALES, AND OTHER SYSTEMS

Stormwater runoff is generally generated when precipitation from rain and snowmelt events flows over land or impervious surfaces and is unable to percolate into the ground. As the runoff flows over the land or impervious surfaces (paved streets, parking lots, and building rooftops), it accumulates debris, chemicals, sediment, or other pollutants that can adversely affect water quality if it is discharged untreated. The most appropriate method to control stormwater discharges is the application of best management practices (BMPs). Because stormwater discharges are normally considered point sources, they require coverage under a National Pollutant Discharge Elimination System (NPDES) permit. Using rainwater collection systems such as cisterns, underground tanks, and ponds can substantially reduce or eliminate the amount of potable water used for irrigation. Rainwater can be collected from roofs, plazas, and paved areas. Prior to its use in irrigation, it should be filtered by a combination of graded screens and paper filters.

A retention pond essentially consists of a body of water that is used to collect stormwater runoff for the purpose of controlling its release. Such ponds do not have outlets, streams, creek ditches, and the like, and after the water collects, it is released through an atmospheric phenomenon such as evaporation or infiltration. Moreover, retention ponds differ from detention ponds in that a detention pond has an outlet such as a pipe to discharge the water to a stream. A detention pond is similar to a retention pond in that it is used to collect stormwater runoff for the purpose of controlling its release. However, the pipe that a detention pond contains is sized to control the release rate of the runoff. Although neighborhood ponds serve several purposes, none of these include swimming or wading.

It is important that the pond is of sufficient depth (at least 8–10 feet) to prevent stagnation and algae growth, and to handle the amount of stormwater runoff that is expected to enter it. A typical problem that is encountered with ponds is the build-up of bacteria such as *E. coli*. Because of the limited water flow and the tendency of wildlife such as geese to gather, these ponds can become breeding grounds for dangerous bacteria. With proper design and maintenance, they can be very attractive, but doing this may require extra planning and more land. Also without maintenance, these ponds can turn into major liabilities.

Local requirements for rainwater harvest and wastewater treatment vary from location to location and jurisdiction to jurisdiction, which is why early involvement and input from local code officials is important. The owner should assemble an experienced team, including the architect, the landscape architect, civil and plumbing engineers, and the rainwater system designer, early in the design process if realistic, cost-effective efficiency goals are to be achieved. Developers often try to do away with retention ponds and replace them with pervious concrete pavement. While perhaps more expensive than typical concrete pavement, the cost of pervious pavement can be partially or fully offset by reducing or eliminating the need for drainage systems and retention ponds and their associated maintenance costs. In addition to the cost savings, elimination of retention ponds can also help meet LEED's goal of reducing site disturbance and therefore help earn additional LEED points.

Whether it is practical to incorporate detention ponds or not will depend largely on site development, not on large-scale land development. The costs of building an extended detention basin versus a traditional detention pond should be studied, bearing in mind that the same storage volume will have to be provided. Then, after adding the cost for pumps, controls, additional storage, and thousands of linear feet of pipe, the decision needs to be made whether all that extra cost outweighs the cost of losing, say, 12 to 15% of your land for a traditional detention pond. Historically, this water was conveniently forced into city storm drains or into retention ponds, becoming someone else's problem. Water from rainstorms and snow-melt needs to be carefully managed in order to conserve it for times of need, to better clean it before it starts its journey back to local aquifers, and to lessen the burden of excessive water runoff on municipal drainage systems.

A system of interlocking, porous pavers resting on a multilayer bed of crushed stones and gravel of different sizes can be used for a parking area. This allows water to diffuse through the surface of the parking lot, slowing the rush of water into the ground and permitting the surrounding landscaping to absorb the water while it is diverted toward bioswales surrounding the property. Bioswales are gently sloped areas of the property designed to collect silt and other rainwater runoff while slowing down the speed with which water collects (Figure 8.4). The swales are designed so that water is diverted in a manner that does not encourage erosion of the ground and soil. The planting of native vegetation in the bioswale can facilitate water absorption, and lengthy root systems can help prevent soil erosion while needing minimum maintenance. The

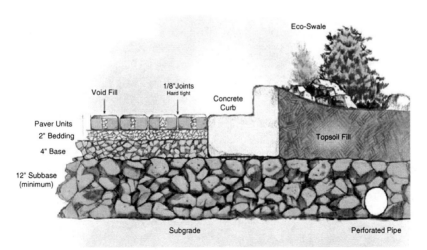

FIGURE 8.4 Typical bioswale (Eco-Swale®) or bioaquifer storm system. *(Source: Other World Computing.)*

advantage of native plants is that they are hearty and can manage well during periods of dry, hot weather, yet make use of and manage the flow of water from unexpected storms. This ability to combine nature with a well-planned surface system can make for an attractive design in addition to being an extremely efficient source of water management and filtering.

Chapter 9

Impact of Energy and Atmosphere

9.1 INTRODUCTION

Even though the United States currently relies on outdated, inefficient power systems that fail to balance energy supply and demand, more and more organizations are now turning to green building practices, mainly because green buildings can reduce the environmental impact of construction and improve the health and wellness of occupants. Green building clearly represents the future of the American construction industry, and it is already creating a wealth of new opportunities and challenges. Moreover, green construction practices can open doors to local and state financial incentives.

The need to jump on the green bandwagon is becoming even more pressing when we realize that in the United States alone, there are nearly 5 million commercial buildings. Furthermore, commercial and residential buildings together account for roughly one-third of all energy consumed, as well as for two-thirds of the total electricity used in the country. This is why it is so important for companies to incorporate green construction and green systems. Still, we have yet to see how implementation of the new 2012 International Green Construction Code (IgCC) will impact green building systems.

Rob Watson, founding father of LEED and an international pioneer in the modern green building movement, says:

Buildings are literally the worst thing that humans do to the planet. Nothing consumes more energy; nothing consumes more materials; nothing consumes more drinking water, and human beings spend up to 90% of their time indoors so if they are getting sick from their environment, in fact, they are getting sick from their indoor environment not from their outdoor environment.

Watson also believes that it is necessary for us to change our paradigms and way of living. This is why many property owners and operators in the process of planning for new construction or retrofitting of existing buildings have come to realize how critically important it is to rethink our approaches to sustainability and green building and to make our buildings more cost-effective and healthy places to live and work in. We can now achieve this through the use of green strategies, such as integrated design processes, that enable us to create high-performance buildings where all systems and components work together to produce enhanced overall functionality and environmental performance while meeting the needs of owners and tenants. Moreover, through integrated design we can now create net zero energy buildings (NZEB): buildings that, on an annual basis, draw from outside sources equal or less energy than produced onsite from renewable sources.

There is also the need to adhere to codes designed to protect the environment, conserve energy, and preserve natural resources. The recently launched IgCC by the International Code Council (ICC) is an example and addresses many of these issues. The intent of the IgCC is to significantly reduce energy usage and greenhouse gases through mandatory green building design and performance in new and existing commercial buildings. Other primary objectives of the IgCC include preserving natural and material resources both in site development and in land use, improving indoor air quality, supporting the use of energy-efficient appliances and renewable energy systems, and employing water resource conservation measures.

It is important to understand the many aspects that impact the design and construction of green buildings that are both healthy and cost effective. First, however, it is prudent to conduct an overview of Leadership in Energy and Environmental Design (LEED®) V3 Rating System, particularly the Energy and Atmosphere category, and briefly discuss some of the changes and new requirements for acquiring LEED credits in this category. It is important to note that many of the questions in the LEED exam tend to focus on the energy credits, especially strategies to optimize energy performance. It is therefore especially prudent to pay particular attention to this category. For the latest updates relating to the LEED 2009 certification and test requirements, visit the GBCI and USGBC® websites: www.gbci.org and www.usgbc.org.

The LEED V3 requirements have changed significantly from their predecessors, with increasing emphasis on sustainable sites, water efficiency, and energy and atmosphere. In terms of possible credits and points, Energy and Atmosphere must be considered the most important of the seven categories in the new LEED 2009 system. For certification purposes, Energy and Atmosphere can now earn up to 35 points out of 100 + 10 (Table 9.1). It should be stressed, however, that no individual product or system in itself can be LEED-certified; it can only help contribute to the completion of LEED requirements.

The significance of the dramatic changes to the LEED 2009 scoring system cannot be overstated, particularly in how they relate to energy modeling. Energy and Atmosphere Prerequisite 2 (Minimum Energy Performance) and Credit 1 (Optimize Energy Performance) have changed significantly. Thus, the threshold for the prerequisite has changed from 14 to 10% and the points awarded in the Optimize Energy Performance credit have increased from a 1- to 10-point scale, to a possible 1- to 19-point scale, awarding basically 9 extra points for the same percentage improvement over the baseline building. But what is perhaps even more interesting is that the baseline itself has changed. LEED project

TABLE 9.1 Point Allocation for Categories in LEED 2009

Credit	NC	CI	EB	C&S	Schools
Sustainable Sites	26	21	26	28	24
Water Efficiency	10	11	14	10	11
Energy and Atmosphere	35	37	35	37	33
Materials/Resources	14	14	10	13	13
Indoor Environmental Quality	15	17	15	12	19
Innovation in Design	6	6	6	6	6
Regionalization	4	4	4	4	4
Total Points	**110**	**110**	**110**	**110**	**110**

teams are mandated to use the ASHRAE® standard referenced in the applicable Reference Guide and are permitted to use addenda within the most recent supplement to that standard. The new LEED V3 is largely governed by the 2007 update of ASHRAE 90.1 (as opposed to the previous version, ASHRAE 90.1-2004). The main modifications relating to LEED requirements include mandatory compliance with Appendix G of ASHRAE 90.1-2007.

Building construction values in the United States have, over the years, become stricter. For example, in climate zone 3A minimum compliance for roof insulation has increased from R-15 to R-20 and wall insulation has increased from R-13 to R-16.8. Although glass compliance has remained unchanged, different U-values have been introduced based on the type of glass, with the former assembly values of U-0.57 and SHGC-0.25 remaining consistent. Thus, using highly efficient glass remains an appropriate method for earning percentage points against the baseline. The significance of this change will dramatically increase in building types where skin loads represent a large percentage of the peak HVAC load (i.e., office buildings) but be less significant in spaces where persons and ventilation loads dominate the sizing of HVAC equipment such as in schools and assembly areas.

In the United States, many state and local governments adopt commercial energy codes to establish minimum energy efficiency standards for the design and construction of buildings, and the majority of these energy codes are based on ASHRAE 90.1 or the International Energy Conseervation Code® (IECC). It should be noted that several organizations have also produced standards for energy-efficient buildings, but ASHRAE and USGBC are perhaps the best known. ASHRAE Standard 90.1-2007, Energy Standard for Buildings Except Low-Rise Residential Buildings, was established by the USGBC as the commercial building reference standard for LEED 2009, which launched on

April 27, 2009. The latest ASHRAE version is the last in a long succession from the original, Standard 90-1975, which is becoming increasingly more stringent.

To qualify for LEED certification, HVAC control systems and lighting control systems earn very few points on their own, perhaps three or so. However, with the addition of the necessary sensors and building controls, the number of achievable points grows considerably—to as much as 25 points and even more with the use of fully integrated building systems. This is because, with integrated systems, the building can earn multiple LEED points as well as cost savings by taking such measures as having a zone's occupancy sensor control both the HVAC and the lighting systems. A comprehensive building operation plan needs to be developed that addresses the heating, cooling, humidity control, lighting, and safety systems. Additionally, there is a need to develop a building automation control plan. Energy conservation measures (ECMs) are recommended to ensure that the building will have the highest percentage of energy savings below the baseline building for the lowest upfront capital costs. Some of these recommended measures may involve minor upfront capital costs.

For Energy and Atmosphere Credit 3 for Commercial Interiors: Measurement and Verification, points can be earned in one of two ways. For projects less than 75% of the total building area, either install submetering equipment to measure and record energy use in the tenant space (2 points) or negotiate a lease whereby the tenant pays the energy costs, which are not included in the base rent (3 points). For projects that constitute 75% or more of the total building area, install continuous metering equipment for one of several end uses such as lighting systems and controls or boiler efficiencies (5 points). Some of the other topics that require close attention in preparing for the LEED exam in the Energy and Atmosphere category are discussed next.

LEED EA Prerequisite 1: Fundamental Commissioning of Building Systems. Basically, the commissioning plan involves verification that the facility's energy related systems are all installed and calibrated, and are performing according to the Owner's Project Requirements (OPR) and basis of design (BOD). The building is to comply with the mandatory and prescriptive requirements of ASHRAE 90.1-2007 in order to establish the minimum level of energy efficiency for the building type. The plans and data produced as a result of the building commissioning lay the groundwork for later energy efficiency savings. This prerequisite is discussed in detail in Chapter 11. It is extremely important and should be completely understood. Several questions on this subject frequently turn up on the exams.

LEED EA Prerequisite 2: Minimum Energy Performance. The intent of this prerequisite is to establish the minimum level of energy efficiency for the project. It is important to remember to comply with the mandatory and prescriptive provisions of ASHRAE 90.1-2007 or a state code, whichever is more stringent.

LEED EA Prerequisite 3: Fundamental Refrigerant Management. The intent is to reduce ozone depletion. This can be achieved by zero use of CFC-based refrigerants in new HVAC&R systems. For existing construction, a comprehensive CFC phase-out conversion prior to project completion is required if reusing existing HVAC equipment (EPA's Montreal Protocol, 1995).

LEED EA Credit 1: Optimize Energy Performance. The intent is to increase levels of energy performance in comparison to prerequisite standards. Option 1: Whole Building Energy Simulation using an approved energy modeling program (1 to 19 points for NC and Schools and 3 to 21 points for SC). Option 2: Prescriptive Compliance Path: Comply with the *ASHRAE Advanced Energy Design Guide* (1 point) appropriate to the project scope; the facility must be 20,000 square feet or less and must be office occupancy or retail occupancy. Option 3: Prescriptive Compliance Path: Advanced Buildings "Core Performance" Guide (1 to 3 points). The facility must be less than 100,000 square feet.

> **Exemplary Performance.** For project teams pursuing Option 1, new construction must exceed the ASHRAE 90.1-2007, Appendix G, baseline performance rating by 50% (previously 45.5% for NC); for existing buildings it must exceed this rating by 46% (previously 38.5% for NC) to be considered under the Innovation in Design category.

LEED EA Credit 2: Onsite Renewable Energy. This credit awards 1 to 7 points for NC and schools and 1 to 4 for CS. The intent is to encourage increased use of renewable energy self-supply and reduce impacts associated with fossil fuel energy use. For the minimum renewable energy percentage for each point threshold see the Reference Guide.

> **Exemplary Performance: Onsite Renewable Energy.** For NC and Schools, projects can earn credit for exemplary performance by showing that onsite renewable energy accounts for at least 15% of annual building energy costs. For the CS category, onsite renewable energy must account for at least 5% of annual building energy costs to earn an exemplary performance credit.

LEED EA Credit 3: Enhanced Commissioning. This credit (2 points) is discussed in Chapter 11. The basic intent is to begin the commissioning process early in the design and implement additional activities after systems performance verification.

> **Exemplary Performance: Enhanced Commissioning.** For NC, CS, and Schools, projects that conduct comprehensive envelope commissioning may be considered for an innovation credit. These projects will need to demonstrate the standards and protocol by which the envelope was commissioned.

LEED EA Credit 4: Enhanced Refrigerant Management. This credit awards 2 points for NC and CS and 1 point for Schools. The intent is to reduce ozone depletion while complying with the Montreal Protocol. Option 1: Do not use refrigerants. Option 2: Select refrigerants and HVAC&R that minimize or eliminate the emission of compounds that contribute to ozone depletion and global warming. Meet or exceed requirements set by the maximum threshold for combined contributions to ozone depletion and global warming potential.

LEED EA Credit 5: Measurement and Verification. This credit awards 3 points for NC and 2 point for Schools. The intent is to provide for ongoing measurement and accountability of building energy consumption. Option 1: Develop and implement a measurement and verification (M&V) plan consistent with Option D: Calibrated simulation as specified in the International Performance Measurement and Verification Protocol (IPMVP) Volume III: Concepts and Options for Determining Energy Savings in New Construction, April 2003. The M&V period must last not less than one year of post-construction occupancy. Option 2: Develop and implement an M&V plan consistent with Option B IPMVP Volume III: Concepts and Options for Determining Energy Savings in New Construction, April 2003. The M&V period must last not less than one year of post-construction occupancy.

LEED EA Credit 6: Green Power. This credit awards 2 points for NC, CS, and Schools). The intent is to encourage and develop the use of grid-source, renewable energy technologies. Option 1: Use the annual electricity consumption results of EA Credit 1: Optimization Energy Performance to determine baseline electricity use. Option 2: Determine the baseline electricity consumption by using the DOE Commercial buildings Energy Consumption Survey database. Renewable energy certificates (RECs) provide the renewable attributes associated with green power. RECs can be provided at a much more competitive cost than from local utilities. A facility is not required to switch its current utility to procure offsite renewable energy for this credit.

> **Exemplary Performance: Green Power.** For NC, CS, and Schools, projects that purchase 100% of their electricity from renewable sources may be considered for an innovation design credit.

Finally, fire protection systems should never be compromised as they serve the purpose of life safety. However, just like other building systems, they should be designed, sourced, installed, and maintained in a manner that is environmental friendly and that reduces their impacts on the environment as discussed in Section 9.8.

9.2   THE BUILDING ENVELOPE

The building envelope consists of all exterior elements of a building: roof, exterior walls, foundations, windows, doors, and floors. These elements together form a barrier that separates the interior of the building from the outdoor environment. There are four basic functions of the building envelope. These include adding structural support, controlling moisture and humidity, regulating temperature, and controlling air pressure changes. The building envelope is a complicated but integral entity of a building, and yet, surprisingly, it is frequently the most neglected. By serving these different functions, the envelope also affects ventilation and energy use within the building. It generally influences the heat exchange between the building and its environment, and regulates the penetration of solar energy and humidity exchange.

Moreover, due to the varied and sometimes competing functions associated with the building envelope, an integrated, synergistic approach is the most appropriate because it considers all phases of the facility life cycle. Furthermore, to function efficiently, the envelope's main components must be properly designed, constructed, and maintained, and all of its elements should interact in a systematic manner to affect the inflow and outflow of heat, air, moisture, and sound. This is particularly important since it is estimated that energy losses through the building's exterior walls, floors, roof, windows, and doors account for 10 to 25% of the energy used by most buildings, depending on outdoor conditions and construction of the building elements. The better the building envelope's overall performance, the better the health, comfort, and productivity of the occupants, and the lower the utility and maintenance bills. The two chief parameters that govern energy losses are the difference in temperature between the indoor and outdoor environment and the envelope's ability to resist heat transfer due to conduction, convection, infiltration, and solar radiation absorption.

There are numerous envelope systems currently in use, each consisting of multiple components and complex technologies. These components need to be properly detailed and maintained for an envelope to operate at maximum efficiency. By addressing such issues of energy efficiency, moisture infiltration, aesthetics, and occupant comfort, the building envelope elements and component systems can enhance design opportunities and minimize potential risk. Using a "sustainable" approach to building design and construction supports an increased commitment to environmental stewardship and conservation, resulting in an optimal balance of cost, environmental, societal, and human benefits while achieving the goals and objectives of the proposed facility.

The condition of the building envelope is vitally important since failures can result in serious safety and health problems, as well as potential structural damage. For these reasons, it is prudent to retain a licensed engineer or architect with sufficient experience in building envelope issues and who is capable of conducting a proper investigation as well as prepare drawings and specifications that may be needed for repair. Typically, applying a holistic approach to the investigation—viewing the entire building envelope (foundations, roofs, walls, etc.) and the building's structural and mechanical systems—is often the best way to identify the real cause of the problem. Also, when a problem is encountered, a hands-on, close-up inspection (not just binocular inspection) is imperative to achieving a correct diagnosis. It is important not to forget that what works for one building may not work for another.

The National Institute of Building Sciences (NIBS)—under contract from the Army Corps of Engineers, the Naval Facilities Engineering Command (NAVFAC), the Air Force, the General Services Administration (GSA), the Department of Energy (DOE), and the Federal Emergency Management Agency—has developed comprehensive federal guidelines for exterior envelope design and construction for institutional/office buildings. The NIBS Design Guide is intended to provide comprehensive guidance on the design and construction of high-quality, long-lasting enclosures for offices and other public buildings. But although intended to significantly improve the performance of building envelopes within the public sector, it is expected that the guide will also provide a great resource for architects and building owners in the private sector as well.

The building envelope's prime functions are to provide shelter, security, solar and thermal control, moisture control, indoor air quality control, access to daylight and views to the outside, fire resistance, acoustics, cost effectiveness, and aesthetics. And having a well-insulated envelope is crucial to creating an environmentally sound building. The reduction of heat transfer through the envelope will help minimize energy used to maintain the interior climate, while at the same time help reduce both utility bills and the environmental costs of fossil fuel use.

9.2.1   Exterior Wall Systems

One approach to characterizing building wall systems is based on the function they serve. Based on this, they can be divided into three broad categories: veneer, structural/load-bearing, and non-load-bearing. They can also be divided into various categories based on materials used as shown next.

Masonry

Masonry has been used in building construction from prehistoric times to the present day. Today, masonry wall systems are employed to form durable cladding systems with various aesthetic effects. In addition to being used for exterior cladding, masonry walls can serve as a portion of the structural framing for the building. They are also known for their increased fire resistance. Masonry walls are usually constructed onsite, where the units are laid in mortar to various heights; the strength of the assembly is normally achieved during curing of the mortar. Masonry can also form structural elements (typically bearing walls, columns, or pilasters) and/or the finished cladding system.

Stone

Most thin stone wall systems used for exterior building envelopes often take the form of stone panels ranging in thickness from three-quarter inch to two inches. These panels are typically fabricated from granite, while marble, limestone, travertine, and sandstone are also used but to a lesser extent. Overall panel dimensions can vary significantly depending on the panel's design and its function, and also depending on the strength of the stone used in the panel and the architectural affect desired (Figure 9.1).

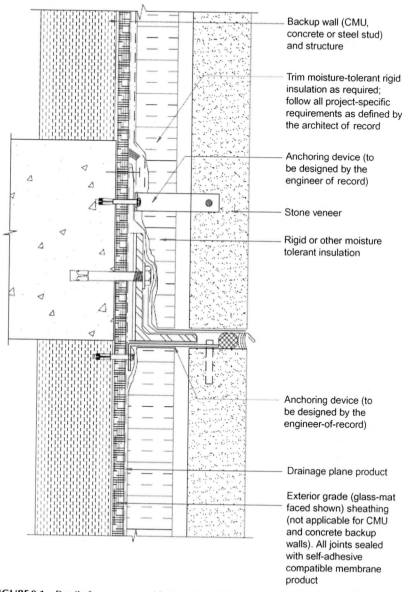

Backup wall (CMU, concrete or steel stud) and structure

Trim moisture-tolerant rigid insulation as required; follow all project-specific requirements as defined by the architect of record

Anchoring device (to be designed by the engineer of record)

Stone veneer

Rigid or other moisture tolerant insulation

Anchoring device (to be designed by the engineer-of-record)

Drainage plane product

Exterior grade (glass-mat faced shown) sheathing (not applicable for CMU and concrete backup walls). All joints sealed with self-adhesive compatible membrane product

FIGURE 9.1   Detail of stone veneer with through-wall flashing; not intended for use with an actual project. *(Courtesy Whole Building Design Guide)*

Concrete

For all the seeming permanence of concrete, it has come under attack from both natural and man-made forces since the time it was first formed and poured. The relative rate of degradation resulting from these assaults depends on a wide variety of factors, of which only some are controllable. Concrete is a man-made, composite construction material, composed of cement and other cementitious materials such as fly ash and slag cement, aggregate (generally a coarse mix of gravel or crushed rocks such as limestone or granite, plus a fine aggregate such as sand), chemical admixtures, and water. It is used more than any other man-made material in the world for everything from high-rise buildings to bridges to roads and everything in-between. Concrete is discussed in greater detail in Chapter 6 "Green Building Materials and Products."

Exterior Insulation and Finish System

The exterior insulation and finish system (EIFS) is an exterior wall covering that the *Whole Building Design Guide (WBDG)* describes as "an exterior wall cladding that utilizes rigid insulation boards on the exterior of the wall sheathing with a plaster appearance exterior skin." The concern that has arisen with EIFS is that if not properly installed or maintained, moisture can penetrate through openings in the cladding and become trapped. Over the years, variations of the EIFS have been developed. In the case of a wood-framed structure, the trapped water is absorbed by the wood, and wood rot, decay, fungus, and insect infestations become problems, none of which are externally visible. Note that all EIFSs are proprietary systems and their components should not be modified beyond the limits stated in the manufacturer's literature. According to the *WBDG*:

EIFS is available in two basic types: a barrier wall system or a wall drainage system. Barrier EIFS wall systems rely primarily on the base coat portion of the exterior skin to resist water penetration. Therefore, all other components of the exterior wall must either be barrier type systems or be properly sealed and flashed to prevent water from migrating behind the EIFS and into the underlying walls or interiors. Wall drainage EIFS systems are similar to cavity walls; they are installed over a weather barrier behind the insulation that acts as a secondary drainage plane. The weather barrier must be properly flashed and coordinated with all other portions of the exterior wall to prevent water from migrating into the underlying walls or interiors.

Curtain Wall

A curtain wall is an outer covering of a building in which the outer walls are nonstructural but are attached to the building structure and used to keep out the weather. Because the curtain wall is nonstructural it can be made of a light-weight material (e.g., aluminum-framed walls containing infills of glass, metal panels, or thin stone), reducing construction costs. The use of glass as the curtain wall has the advantage of allowing natural light to penetrate deeper within the building. Of note, the curtain wall façade does not carry any dead-load weight from the building other than its own. In addition, a curtain wall is designed to resist air and water infiltration, resist sway induced by wind and seismic forces acting on the building, and its own dead-load weight forces.

Curtain walls generally fall into two basic categories that are based on their method of fabrication and installation. These are *unitized* or *modular systems* (Figure 9.2a,b) and *stick systems* (Figure 9.3a,b). In the stick system, the curtain wall frame (mullions) and glazing panels are installed and connected together piece by piece. In the unitized system, the curtain wall is composed of large units that are assembled and glazed in the factory, shipped to the site, and erected on the building.

Siding, Cladding, and/or Weatherboard

Siding (also known as cladding and weatherboard) is an exterior finish material that is installed over the wall framing or building support structure. Siding's main functions is to improve the exterior appearance of the building and also to help keep external elements out. Although siding itself may not be water- or wind-resistant, if combined with other waterproofing elements (e.g., building wrap and insulation), siding systems can help complete the exterior walls. Siding comes in many forms and materials, which are generally chosen based on the building's design and aesthetics (for further details see Chapter 6 "Green Building Materials and Products."

Windows, Exterior Doors, and Skylights

Windows, exterior doors, and skylights are basically holes in the walls and roof that admit light and people. Doors and windows are important elements of the exterior closure system because they provide visual and physical access between the exterior and interior environments and can affect lighting costs by taking maximum advantage of natural daylight. The design and installation of these penetrations are the areas of most frequent water and air infiltration into buildings, which is why energy-efficient doors, windows, and skylights are helpful in reducing heat flow between indoors and outdoors. For this reason, special weatherproofing is often installed around the perimeter of the door (Figure 9.4). It is estimated that having optimum window and door placement with energy-efficient glazing can reduce energy consumption by 10 to 40%.

9.2.2 Weatherproofing

Leakage problems in construction have become the number-one cause of lawsuits, which is perhaps why in today's competitive construction market, building owners and investors seek a structure that is aesthetically pleasing and remains leak-free for as many years as possible. Moisture intrusion into buildings causes billions of dollars in property damage annually. Furthermore, moisture is the number-one cause of mold and mildew growth and should be

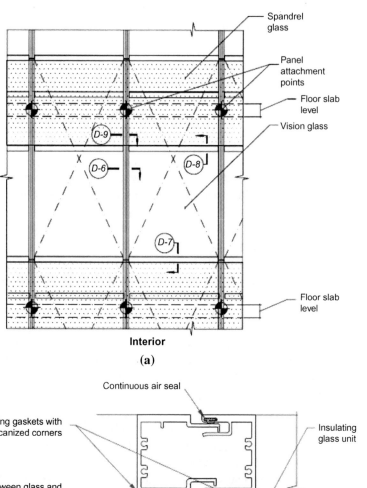

Spandrel glass

Panel attachment points

Floor slab level

Vision glass

D-9

D-6   X   X   D-8

D-7

Floor slab level

Interior

(a)

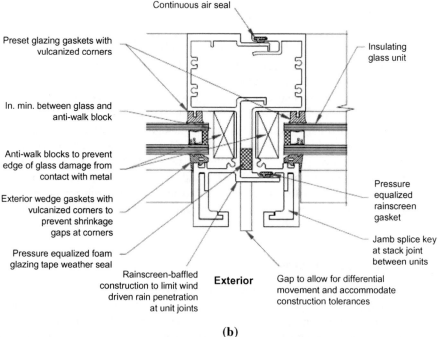

Continuous air seal

Preset glazing gaskets with vulcanized corners

Insulating glass unit

In. min. between glass and anti-walk block

Anti-walk blocks to prevent edge of glass damage from contact with metal

Exterior wedge gaskets with vulcanized corners to prevent shrinkage gaps at corners

Pressure equalized rainscreen gasket

Jamb splice key at stack joint between units

Pressure equalized foam glazing tape weather seal

Rainscreen-baffled construction to limit wind driven rain penetration at unit joints

Exterior

Gap to allow for differential movement and accommodate construction tolerances

(b)

FIGURE 9.2   (a) Typical elevation—unitized curtain wall system. (b) Vision glass jamb detail—unitized-curtain wall system. Diagrams are not intended for use with an actual construction project. *(Courtesy Whole Building Design Guide)*

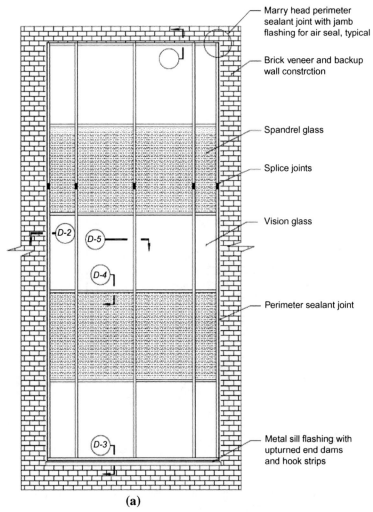

Marry head perimeter sealant joint with jamb flashing for air seal, typical

Brick veneer and backup wall constrction

Spandrel glass

Splice joints

Vision glass

Perimeter sealant joint

Metal sill flashing with upturned end dams and hook strips

(a)

FIGURE 9.3 (a) Typical elevation—stick-built curtain wall system (not for use in actual construction).

removed before it contaminates the entire building and its occupants (Figure 9.5). One of the critical elements in maintaining a weatherproof envelope is the performance of the building joints. All buildings require joints and how you seal them is an important factor in determining the overall performance and durability of the envelope. But moisture can intrude into a building in a number of ways, whether from improperly designed and/or constructed vapor barriers in walls, roofs, and floors or through leaks in walls, floors, roofs, windows, and doors caused by improper design, construction, and maintenance.

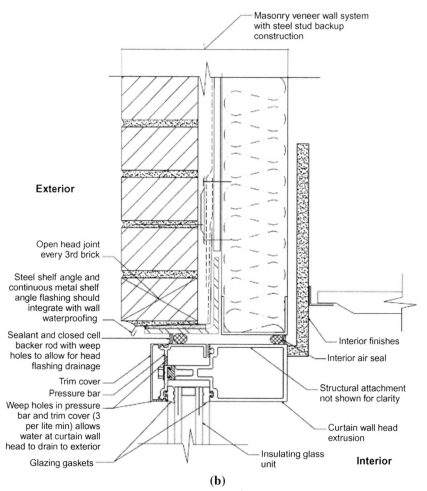

FIGURE 9.3—cont'd   (b) Curtain wall head detail—stick-built system (not for use in actual construction). *(Courtesy Whole Building Design Guide)*

9.3   INTELLIGENT ENERGY MANAGEMENT SYSTEMS

Over the past decade, we have seen energy demand growing faster than supply, which means that for any level of sustainability to be achieved, there is a clear need for more efficient generation, delivery, and consumption of energy—that is, intelligent energy management systems. Indeed, not only can intelligent energy management technologies provide immediate solutions, but correctly implemented, they can cut energy use, spending, and emissions, as well as provide a solid foundation to build tomorrow's smarter energy infrastructure.

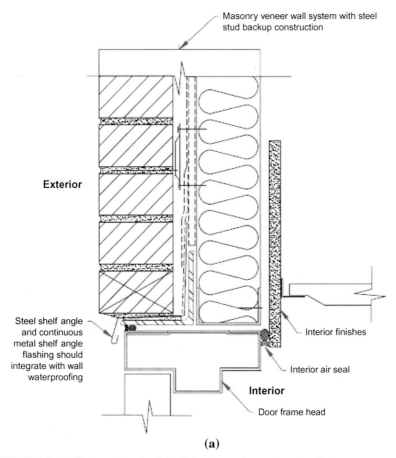

(a)

FIGURE 9.4   (a) Section of door head detail (not for use in actual construction).

Building automation basically consists of a programmed, computerized "intelligent" network of electronic devices that monitor and control a building's mechanical and lighting systems. As more and more buildings are incorporating central communications systems, the computer-integrated building has not only become a reality but has become an integral part of mainstream America and one of the landmarks of today's society. The intent is to create an intelligent sustainable building that can reduce energy and maintenance costs. In addition, increasing consumer demand for clean renewable energy and the deregulation of the utilities industry have spurred growth in green power—solar, wind, geo-thermal steam, biomass, and small-scale hydroelectric. Small commercial solar power plants are beginning to emerge and have started to serve some energy markets.

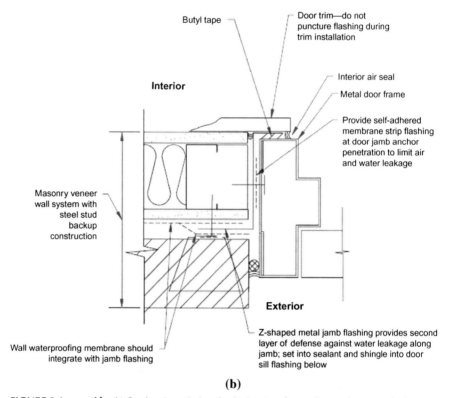

Butyl tape

Door trim—do not puncture flashing during trim installation

Interior

Interior air seal

Metal door frame

Provide self-adhered membrane strip flashing at door jamb anchor penetration to limit air and water leakage

Masonry veneer wall system with steel stud backup construction

Exterior

Z-shaped metal jamb flashing provides second layer of defense against water leakage along jamb; set into sealant and shingle into door sill flashing below

Wall waterproofing membrane should integrate with jamb flashing

(b)

FIGURE 9.4—cont'd   (b) Section through door jamb plan (not for use in actual construction).

FIGURE 9.5   Effects of water penetration contributing to mold growth.

9.3.1   Building Automation and Intelligent Buildings

The concept of intelligent buildings was introduced in the United States in the early 1980s. With this approach, buildings employ high-technology electronics extensively to achieve desired results, which basically consist of integrating four primary groups (energy efficiency, life safety, telecommunications, and workplace automation) into a single computerized system. There are various definitions of intelligent buildings and sustainability; the one proposed by the Intelligent Building Institute is as follows:

[A]n intelligent building is one that provides a productive and cost-effective environment through optimization of its four basic elements—structure, systems, services, and management—and the interrelationships between them. Intelligent buildings help business owners, property managers and occupants to realize their goals in the areas of cost, comfort, convenience, safety, long-term flexibility and marketability.

Regarding sustainability, the *ASHRAE GreenGuide* defines it as "Providing for the needs of the present without detracting from the ability to fulfill the needs of the future." Thomas Hartman, professional engineer and building automation expert, believes there are three cardinal elements of an intelligent building (Hartman, 2007):

1. *The occupants:* An intelligent building is one that provides easy access, keeps people comfortable, environmentally satisfied, secure, and provides services to keep the occupants productive for their purpose in the building.
2. *Structure and systems:* An intelligent building is one that at a bare minimum significantly reduces environmental disruption, degradation, or depletion associated with the building while ensuring a long-term useful functional capacity for the building.
3. *Advanced technologies:* An intelligent building is one that because of its climate and/or use is challenged to meet elements 1 and 2, and succeeds in meeting those challenges through the use of appropriate advanced technologies.

Many in the engineering field today understand an intelligent building to be one that incorporates computer programs to coordinate many building subsystems to regulate interior temperatures, HVAC, and power. The goal is usually to reduce the operating cost of the building while maintaining the desired environment for occupants (Figure 9.6). Often people fail to realize that this is really about the use of advanced technologies to dramatically improve the comfort, environment, and performance of a building's occupants while minimizing the external environmental impact of its structure and systems. The key phrase here is "comfort of its occupants," which is what it is all about. In the final analysis, intelligent buildings help property owners and developers as well as tenants to achieve their objectives in the areas of comfort, cost, safety, long-term flexibility, marketability, and increased productivity.

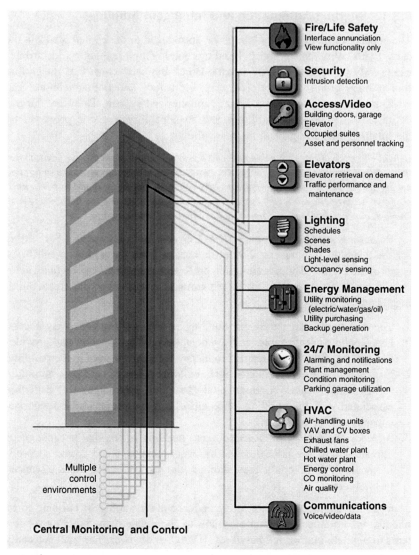

FIGURE 9.6 An intelligent building is one that can merge building management requirements with IT systems to achieve optimized system performance as well as simplified general facility operations. In this illustration, a single IP network is compared with a multiple proprietary network. *(Source: BPG Properties.)*

Although there are numerous commercial off-the-shelf building automation systems (BASs) now on the market, and the majority of facility and building managers recognize the potential value of such systems as a powerful energy-saving tool, if it weren't for the initial costs involved there would be no hesitation in employing them. As an example, one basic BAS that is readily available saves energy by widening temperature ranges and reducing lighting in unoccupied spaces, and reduces costs for electricity by shedding loads when electricity is higher priced. But Bill Lydon, chief editor of *InTech*, suggests a degree of caution: "It is easy to label a product with the term *commercial off-the-shelf* (COTS), and lately it is being done more often as a way to lead buyers to believe it should be the only criteria to select a product. Automation and control professionals should consciously make decisions based on their operational goals."

In this respect, Kristin Kamm, a senior research associate at E Source, notes that the most common strategies that BASs employ to cut energy use include:

- *Scheduling: Scheduling turns equipment on or off depending on time of day, day of the week, day type, or other variables such as outdoor air conditions.*
- *Lockouts: Lockouts ensure that equipment doesn't turn on unless it's necessary. For example, a chiller and its associated pumps can be locked out according to calendar date, when the outdoor air falls below a certain temperature, or when building cooling requirements are below a minimum.*
- *Resets: When equipment operates at greater capacity than necessary to meet building loads, it wastes energy. A BAS can ensure that equipment operates at the minimum needed capacity by automatically resetting operating parameters to match current weather conditions. For example, as the outdoor air temperature decreases, the chilled-water temperature can be reset to a higher value.*
- *Diagnostics: Building operators who use a BAS to monitor information such as temperatures, flows, pressures, and actuator positions may use that data to determine whether equipment is operating incorrectly or inefficiently, and to troubleshoot problems. Some systems also use the data to automatically provide maintenance bulletins.*

As a building owner or operator, you need to have the ability to monitor your building(s) whenever you wish to do so, 24 hours a day. With Internet-based systems, it is now possible to monitor your building from any location that has an Internet connection. Building automation systems can be custom-designed to fit your needs—whether for energy management, building automation, or environmental control. Some intelligent buildings have the capability to detect and report faults in mechanical and electrical systems, especially critical systems. Many also have the ability to track individual occupants to adapt building systems to an individual's wants and needs (e.g., setting a room's temperature and lighting levels automatically when a resident enters), as well as anticipate forecasted weather, utility costs, or electrical demand. There are other nonenergy uses for automation in a building such as scheduling preventive maintenance, monitoring security, and monitoring rent or consumables charges based on actual usage. Typical elements and components frequently used in building automation include the following:

- *Controller*. One of main advantages of today's controllers is that they allow users to network and gain real-time access to information from multiple resource segments in a building's network, creating an "intelligent building." These controllers come in a variety of sizes and capabilities to control common building devices. Usually the primary and secondary buses are chosen based on what is provided by the controllers.

- *Occupancy sensors*. These devices were originally designed for use with security systems. They have been refined and enhanced to control lighting and HVAC in both commercial and residential spaces. There are different types of occupancy sensors (e.g., infrared, ultrasonic, and dual-tech) designed to meet a wide range of applications. Occupancy is usually based on time-of-day schedules, but override is possible through different means. Some buildings can sense occupancy in their internal spaces by an override switch or sensor. Sensors can be either ceiling- or wall-mounted, depending on the type and application (Figure 9.7). Acclimate™ incorporates both thermal and photoelectric technologies that interact to maximize detection. It is a fire sensor with an onboard microprocessor with advanced software that makes adjustments to reduce false alarms.

- *Lighting*. With today's building automation systems, lighting can be turned on and off depending on the time of day or by occupancy sensors and timers. There are many different control systems, including time-based and optimizer parameter–based where a level of illuminance or particular use of lighting is required. One typical example is to turn the lights in a space on for a half-hour after the last motion sensed. A photocell placed outside a building can sense darkness and the time of day, and modulate lights in outer offices and the parking lot. This is discussed in greater detail in Section 9.5.3.

FIGURE 9.7　Multicriteria intelligent sensor. *(Source: System Sensor.)*

- *Air-handling unit (AHU).* Less temperature change is required with most air handlers because they typically mix the return and outside air. Analog or digital temperature sensors may be placed in the space or room, the return and supply air ducts, and sometimes the external air. Actuators are placed on the hot- and chilled-water valves, the outside air, and return-air dampers. This in turn can save money by using less chilled or heated water (not all AHUs use chilled/hot-water circuits). Some external air needs to be introduced to keep the building's air quality healthy. The supply fan (and return if applicable) is started and stopped based on time of day, temperature, building pressure, or a combination of all three.

- *Constant-volume air-handling unit (CAV).* This is a less efficient type of air handler because the fans lack variable-speed controls. Instead, CAVs open and close dampers and water-supply valves to maintain temperatures in the building's spaces. CAVs heat or cool spaces by opening or closing chilled- or hot-water valves that feed their internal heat exchangers. Generally, one CAV serves several spaces, but larger buildings may incorporate many.

- *Variable-volume air-handling unit (VAV).* The VAV is a more efficient unit than the CAV; this system brings in outside air and returns air to the AHU, where the temperature and humidity of the incoming air can be controlled. VAVs supply pressurized air to VAV boxes, usually one box per room or area. A VAV air handler can change the pressure to the boxes by changing the speed of a fan or blower with a variable-frequency drive or (less efficiently) by moving inlet guide vanes to a fixed-speed fan. The amount of air is determined by the needs of the spaces served by the boxes. Some VAV boxes also have hot-water valves and an internal heat exchanger. The valves for hot and cold water are opened or closed based on the heat demand for the spaces it is supplying. A minimum and maximum CFM must be set on VAV boxes to ensure adequate ventilation and proper air balance. The main AHU components are the supply fan, heating coil, cooling coil, filter, and humidity control.

- *VAV hybrid systems.* In many large systems or in systems that have undergone renovation, it is not unusual to find a combination of constant-volume and variable-volume zones in a single air handler. The hybrid system is basically a variation of VAV and CAV. Health care, laboratory, and process applications are its most common users. In this system the interior zones operate as in a VAV system, but the outer zones differ in that the heating is supplied by a heating fan in a central location, usually with a heating coil fed by the building boiler. The heated air is ducted to the exterior dual-duct mixing boxes and dampers controlled by the zone thermostat calling for either cooled or heated air as necessary.

- *Central plant.* The main function of a central plant is to supply the AHUs with water. It may supply chilled-water, hot-water, and condenser-water systems, as well as transformers and an auxiliary power unit for emergency power.

If well managed, these can often help each other. For example, some plants generate electric power at periods of peak demand using a gas turbine and then use the turbine's hot exhaust to heat water or power an absorptive chiller.

- *Chilled-water system.* This is normally used to cool a building's air and equipment. Chilled-water systems usually incorporate chiller(s) and pumps. Analog temperature sensors are used to measure the chilled-water supply and return lines. The chiller(s) are sequenced on and off to ensure that the water supply is chilled. The efficiency of these systems is their ability to integrate and directly interface with the complete building automation system via various communications protocols.

- *Condenser-water system.* Cooled condenser water is supplied to the chillers through cooling tower(s) and pumps. To ensure that the condenser-water supply to the chillers is constant, speed drives are commonly employed on the cooling tower fans to control temperature. Proper cooling tower temperature ensures proper refrigerant head pressure in the chiller. Analog temperature sensors measure the condenser-water supply and return lines. The cooling tower set point depends on the refrigerant being used.

- *Hot-water system.* The hot-water system supplies heat to the building's AHUs or VAV boxes. It has a boiler(s) and pumps. Analog temperature sensors are placed in the hot-water supply and return lines. Some type of mixing valve is typically incorporated to control the heating water loop temperature. The boiler(s) and pumps are sequenced on and off to maintain a constant supply.

- *Alarms and security.* The vast majority of intelligent building automation systems today incorporate some form of alarm capabilities. If an alarm is tripped, it can be programmed to notify someone. Notification can be implemented via a computer, pager, cellular phone, or audible alarm. Security systems can also be interlocked to a building automation system. If occupancy sensors are present, they can also be used as burglar alarms. This is discussed in greater detail in Section 9.8.

There are a large number of proprietary protocols and industry standards on the market, including ASHRAE, Building Automation and Control Network (BACnet™), Digital Addressable Lighting Interface (DALI), Dialogic® Distributed Signal Interface (DSI), Philips Dynalite's DyNet, ENERGY STAR®, OpenRemote's KNX, LonTalk®, and Texas Instrument's ZigBee. The latest details of these systems can be found by researching the Internet as they are outside our scope.

9.4   MECHANICAL SYSTEMS: AIR CONDITIONING, HEATING, AND VENTILATION

Most tenants living in American cities take for granted that the buildings they live and work in will have appropriate mechanical heating, ventilation, and air-conditioning (HVAC) systems in place. It is understood that these systems

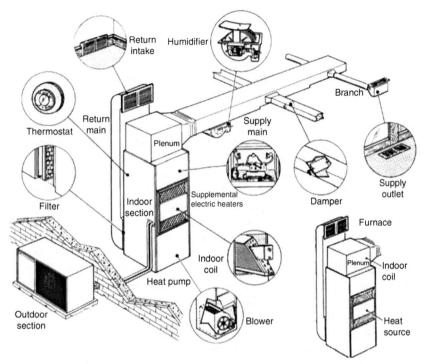

FIGURE 9.8   Basics of HVAC systems. *(Source: Southface Energy Institute.)*

are designed to provide air at comfortable temperature and humidity levels and to be free of harmful concentrations of air pollutants (Figure 9.8). Moreover, the technological advances and continuous development of air-conditioning systems have brought about fundamental changes in the way we design projects because they have allowed investors to build larger, higher, and more efficient buildings than previously possible. But even though buildings today are being designed with increasingly sophisticated energy management and control systems (EMCS) for monitoring and controlling the conditions of interior space, we nevertheless frequently discover that a building's HVAC equipment routinely fails to satisfy the performance expectations of its designers and owners; these failures often going unnoticed for extended periods of time.

  The introduction of new technologies and developments in computers and electronics equipment have made it possible to create HVAC systems that are smarter, smaller, and more efficient. These advancements have reshaped how the systems are installed, how they are maintained, and how they operate. Among the more important developments in HVAC equipment design in recent years is VAV, which basically involves a technique for controlling the capacity of an HVAC system. This means that with these systems, persons who have conditioned air circulating in, on, or around them can control the temperature in their personal space. For example, if two individuals are on the same system and one seeks to increase the temperature, the system can heat that person's space and cool the other's. Another advantage is that VAV can change the volume of air delivered to the space and also damper off a space that is not used or occupied, thereby increasing efficiency. The fan capacity control, especially with modern electronic variable-speed drives, reduces the energy consumed by fans, which can be a substantial part of the total cooling energy requirements of a building. Also, to attain partial load-cooling capacity, dehumidification with VAV systems is greater than it is with constant-volume systems that modulate the discharge air temperature.

Recent estimates indicate that buildings in the United States annually consume about 42% of America's energy and 68% of its electricity, of which HVAC systems consume a significant percentage. Energy sources that provide power to an HVAC system are usually gas, solid fuels, oil, or electricity, and the conducting medium is usually water, steam, or gas. The heating and cooling source equipment comprises components that use the energy source to heat or cool the conducting medium. The heating and cooling units (e.g., air conditioners and AHUs) are instrumental in modifying air temperatures in the interior spaces. The assessment of the HVAC system is one of the main components of a general baseline evaluation.

Researchers and others have known for decades that physical comfort is critical to work effectiveness, satisfaction, and physical and mental well-being, and occupants may even be driven to distraction trying to adjust the comfort in their spaces. As discussed in Chapter 7, Indoor Environmental Quality, we know that uncomfortable conditions in the workplace such as noise, inadequate lighting, uncomfortable temperature, high humidity, ergonomics, and other physiological stressors invariably restrict occupants' ability to function to their full capacity, leading in many cases to lower job satisfaction and increases in building-related illness (BRI) symptoms. And since humans generally spend most of their time indoors, health, well-being, and comfort inside buildings are crucial because they help us breathe easier and focus our attention better.

9.4.1 Refrigerants: Hydrochlorofluorocarbons, and Chlorofluorocarbons

Refrigerants are fluids (chemical compounds) used as the heat carrier; they change from gas to liquid and then back to gas in the refrigeration cycle. The two refrigerant families most often used in air-conditioning systems are hydrochlorofluorocarbons (HCFCs) and chlorofluorocarbons (CFCs). Refrigerants are used primarily in refrigerators/freezers, air-conditioning, and fire suppression systems. Mike Opitz, formally certification manager for LEED for Existing Buildings, says:

Chemical refrigerants are the heart of a large majority of building HVAC and refrigeration equipment. These manufactured fluids provide enormous benefits to society, but in recent decades have been found to have harmful consequences when released to the atmosphere: all refrigerants in common use until the 1990s caused significant damage to the protective ozone layer in the earth's upper atmosphere, and most also enhanced the greenhouse effect, leading to accelerated global warming.

For example, in EA Credit 4 (Enhanced Refrigerant Management) for NC, Schools, and CS, points can be earned either by not using refrigerants (Option 1) or by selecting environmentally friendly refrigerants and heating, ventilation, air conditioning, and refrigeration (HVAC&R) systems that minimize or eliminate the emission of compounds that contribute to ozone depletion and global warming (Option 2).

On September 21, 2007, these and other issues were addressed when parties to the Montreal Protocol, including the United States, overwhelmingly agreed to accelerate the phase-out of HCFCs to protect the ozone and combat climate change, with adjustments beginning in 2010 to production and consumption allowances for developed and developing countries. Production of CFCs ceased in 1995, and the phase-out of HCFCs will have a significant impact on proposed real estate purchases that still utilize this equipment. The 2007 Montreal Protocol definitely energized the green building movement and equipment manufacturers to make changes in the types of refrigerants used in certain equipment because of general environmental concerns and the desire for more suitable environmental alternatives. It is no longer a question of whether facilities managers will upgrade their HVAC and other equipment, but when and how. This means that owners and administrators will need to take the long-term view when making decisions that can impact their capital investments.

In accordance with the Montreal Protocol, the U.S. Environmental Protection Agency (EPA) is now obligated to phase out HCFC refrigerants used in heat pump and air-conditioning systems because of their impact on ozone depletion. CFC refrigerants manufacture has been banned in the United States since 1995. To date, the main alternatives are hydrofluorocarbons (HFCs) and HFC blends, although there are several potential non-HFC alternatives as well. DuPont has produced a complete family of easy-to-use, non-ozone-depleting HFC retrofit refrigerants for CFC and HCFC equipment. But while HFCs may be suitable as short- to medium-term replacements, they may not be suitable for long-term use due to their high global warming potential (GWP) and their impact on the environment.

In some categories of LEED Rating System, Minimum Energy Performance and Fundamental Refrigerant Management are included as a prerequisites, the intent being to establish a minimum level of energy efficiency for the building system and reduce ozone depletion potential (ODP), GWP, and to support early compliance with the EPA and the Montreal Protocol. The LEED requirements mandate zero use of CFC-based refrigerants in new HVAC&R systems. When reusing existing base-building HVAC, a comprehensive CFC phase-out conversion must be conducted prior to project completion. Also, various categories of LEED award one credit point for using non-ozone-depleting HFC refrigerants.

9.4.2   Types of HVAC Systems

There are basically two approaches to conditioning a room or building. The first is using a radiant system; the second is using a forced-air system. Radiant systems usually involve running hot or chilled water through pipes that loop around the structure and radiate into the conditioned space via a floor surface or radiator pipe. Forced-air systems use a fan to push air through a duct system, where it is conditioned by a coil on a furnace or air handler before being returned to the space. While there are a wide variety of HVAC systems in use today, no system is right for every application. In order to service specific needs, there are a number of different HVAC systems available (e.g., single zone/multiple zone, constant volume/variable air volume). The most common classification of HVAC systems is by the carrying media used to heat or cool a building. The two main transfer mediums for this purpose are air and water, which take them to emitters. On smaller projects, electricity is often used for heating, although some systems now use a combination of transfer media. HVAC systems range in complexity from standalone units that serve individual rooms or zones to large centrally controlled systems serving multiple zones in a building or complex.

9.4.3   Heating Systems

The three basic components of heating systems used today to regulate temperature for commercial and residential buildings are the fuel source, the energy conversion plant, and the energy distribution system. Heating systems can be either central or local. The most commonly used setup is the central-heating system, where the heat is concentrated in a single central location from where it is then circulated for various heating processes and applications. Some of the more common heating systems currently in use are described next.

Electric Heating

Electric heating is a process in which electrical energy is converted to heat. Common applications include space heating, water heating, and industrial processes. An electric heater is a device that transforms electrical energy into heat. Electric heaters contain electric resistors, which act as heating elements. The use of electricity for heating is becoming increasingly popular in both residences and public buildings. Although it generally costs more than energy obtained from combustion of a fuel, the convenience, cleanliness, and reduced space needs of electric heat often justify its use. Heat can be provided from electric coils or strips used in varying patterns such as convectors in or on the walls, under windows, or as baseboard radiation in part or all of a room. Heating elements or wires can even be incorporated in ceilings or floors to radiate low-temperature heat into a space. Also, by the incorporation of a heat pump, the overall cost of electric heating can be reduced significantly.

Electric Baseboard Heating

Electric baseboard systems are a fairly common heat source for low-cost installation and quiet operation, heating the room (particularly bedrooms) by a process call electric resistance. These types of heaters are zonal, controlled by
thermostats located within each room. Inside baseboard heaters are electric
cables, which warm the air that passes through them. Electric baseboard heaters
are typically installed along the lower part of outside walls to provide perimeter
heating. Room air heated by the resistance element rises and is replaced by
cooler room air, establishing a continuous convective flow of warm air while
in operation. By superior design and proper placement under a window area,
the electric baseboard case causes air to flow naturally.

Central Heating

Central heating is most often used in cold climates to heat a building or group of
buildings, although most modern commercial buildings, including offices,
high-rise residential, hotels, and shopping malls are today provided with some
form of central heat. There are many different types of central-heating systems
on the market, most of which comprise a central boiler (which is actually a heat
generator because the water is not "boiled"; rather, it peaks at 82–90°C) or furnace to heat water, pipes to distribute the heated water, and heat exchangers or
radiators to conduct this heat to the air. In large systems, steam or hot water is
usually employed to distribute the heat. However, there is no such thing as a
standard central-heating system, and each project requires the system to be
tailored to meet its own requirements; with advanced controls, a correctly
programmed central-heating system, when optimized, is able to constantly
monitor and automatically adjust the system basically on its own. The term
"district heating" is generally applied to systems in which a large number of
buildings are supplied with steam from central boiler rooms operated by a
public utility.

The main difference between central heating and local heating is that with
central heating heat generation occurs in one place such as a furnace room in a
house or a mechanical room in a large building. The most common method to
generate heat involves the combustion of fossil fuel in a furnace or boiler. The
resultant heat is distributed typically by forced air through ductwork, by water
circulating through pipes, or by steam fed through pipes. Increasingly, buildings
utilize solar-powered heat sources, in which case the distribution system
normally uses water circulation. With most modern systems, a pump is used
to circulate the water and ensure an equal supply of heat to all the radiators.
The heated water is often fed through another heat exchanger inside a storage
cylinder to provide hot running water. Forced-air systems send air through
ductwork; the air can be reused for air conditioning and can be filtered or put
through air cleaners. The heating elements (radiators or vents) are ideally located in the coldest part of the room, typically next to the windows. An important

characteristic of a central-heating system is that it provides warmth to the whole interior of a building or portions of a building as required.

Furnace Heating

A furnace consists of a heating system component that is designed to heat air for distribution to various building spaces. Small-capacity furnaces that rely on natural convection for heat distribution can be classified as local and can usually effectively condition a single space only. However, furnaces equipped with fans to circulate air over greater distances or to several rooms are used for residential and small commercial heating systems and are found in most central-heating systems in use today. Furnaces typically use natural gas, fuel oil, propane, and electricity for the heat source as well as onsite energy collection (solar energy) and heat transfer (heat pumps). Natural-gas furnaces are available in condensing and noncondensing models. The cooling can be packaged within the system, or a cooling coil can be added. When direct-expansion systems with coils are used, the condenser can be part of the package or remote. The efficiency of new furnaces is measured by annual fuel utilization efficiency (AFUE), a measure of seasonal performance.

Today's furnaces are designed to be between 78 and 96% AFUE. Traditional "power combustion" furnaces are 80 to 82% AFUE. Above 90% AFUE, a furnace is "condensing," which generally means it recaptures some of the heat wasted in traditional systems by condensing escaping water vapor. Furnaces, both large and small, are usually automatically responsive to remote thermostats that control their operation. Oil-fired or gas-fired furnaces only need the control of burners to regulate heat. Furnaces that use solid fuels, however, require the admission of additional fuel to the system and the removal of ashes from the stoker or grates.

Radiant Heating

Radiant heating is increasing in popularity because it is clean, quiet, efficient, dependable, and invisible. It is provided in part by radiation in all forms of direct heating, but the term is usually applied to systems in which floors, walls, or ceilings are used as the radiating units. Steam or hot-water pipes are placed in the walls or floors during the construction process, and radiant heating systems circulate warm water through continuous loops of tubing. The tubing system transfers the heat into the floor and upward into virtually any surface, including carpeting, hardwood, parquet, quarry and ceramic tile, vinyl, and concrete. If electricity is used for heating, the panels containing the heating elements are mounted on a wall, baseboard, or ceiling. Radiant heating provides uniform heat and is both efficient and relatively inexpensive to operate. Efficiency is high because radiant heat raises the inside-surface temperature, thereby providing comfort at a lower room-air temperature than other systems can provide.

Warm-Air Heating

Gravity and forced-air are the two basic types of warm-air heating systems currently on the market. The gravity system operates by air convection and is based on the principle that when air is heated it expands, becomes lighter, and rises. Cooler air is dense and therefore falls. The difference in air temperature creates the convection or motivation for air movement. The return of a gravity system must be unrestricted, and even a filter is considered too restrictive. This is necessary to develop positive convection and better distribution. The furnace consists of a burner compartment (firebox) and a heat exchanger. The heat exchanger is the medium used to transfer heat from the flame to the air, which moves via ducts to the various rooms.

Besides being the medium of heat transfer, the heat exchanger keeps the burned fuels separate from the air. Often the furnace is arranged so warm air passes over a water pan in the furnace for humidification before circulating through the building. As the air is heated, it passes through the ducts to individual grilles or registers in each room (which may be opened or closed to control the temperature) on the upper floors. The chief problem in this type of system lies in obtaining adequate air circulation, that is, the system may not heat a facility adequately if the warm-air ducts are insufficiently large in diameter and not slanted upward from the furnace, or properly insulated to prevent heat losses.

As in gravity warm-air heating systems, the heat exchanger is the medium of heat transfer and separates the burned fuel from the air that moves through the building. Forced-circulation systems typically have a fan or blower placed in the furnace casing which blows air through an evaporator coil, which cools the air (Figure 9.9). This cool air is routed throughout the intended space by means of a series of air ducts, thus ensuring the circulation of a large amount of air even under unfavorable conditions. The ability to use the same equipment to provide air conditioning throughout the year has given added impetus to the use of forced-circulation warm-air systems in residential installations. In addition, when combined with cooling, humidifying, and dehumidifying units, forced-circulation systems may be effectively used for heating and cooling in various types of buildings.

Hot-Water Heating

Hot-water systems typically have a central boiler, in which water is heated to a temperature of from 140 to 180 °F (60–83 °C), and then circulated by means of pipes to some type of coil units, such as radiators, located in the various rooms. Circulation of the hot water can be accomplished by pressure and gravity, but forced circulation using a pump is more efficient because it provides flexibility and control. In the rooms, the emitters give out the heat from their surfaces by radiation and convection. The cooled water is then returned to the boiler. There are combination systems that use ducts for supplying air from the central AHU and water to heat the air before it is transferred into the conditioned space.

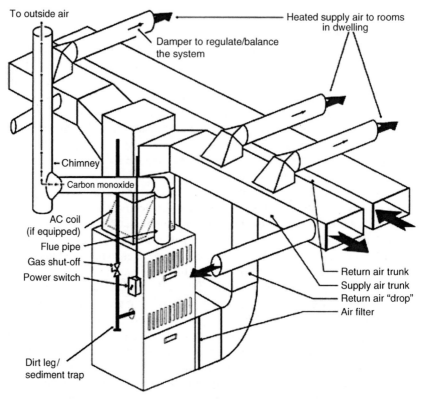

To outside air

Heated supply air to rooms in dwelling

Damper to regulate/balance the system

Chimney

Carbon monoxide

AC coil (if equipped)

Flue pipe

Gas shut-off

Power switch

Return air trunk

Supply air trunk

Return air "drop"

Air filter

Dirt leg/ sediment trap

FIGURE 9.9 Forced warm-air system operation. *(Source: warmair.com.inc.)*

Combination boiler heating systems are most commonly used in central-heating systems. Running on mains pressure water eliminates both the need for tanks in the loft and the need for a hot-water cylinder as the water is instantly heated when needed. Hot-water circulating systems generally provide convenience and save water, but have proven not be cost effective and expend large amounts of energy.

Generally speaking, hot-water systems use either a one-pipe or a two-pipe system to circulate the heated water. The one-pipe system uses less pipe than the two-pipe arrangement, which is why it is less expensive to install. However, it is also less efficient because larger radiators or longer baseboards are required at the end of the loop because this part gets less heat. The operation of a one-pipe system is fairly simple: Water enters each radiator from the supply side of the main pipe, circulates through the radiator, and flows back into the same pipe. In boiler hydronic systems there are different ways to arrange the piping depending on the budget at installation time and the efficiency level required. As with many hydronic loop systems, the two-pipe direct return needs balancing valves, and in both systems an expansion tank is required to compensate for variations in the volume of water in the system.

Modern layouts generally use a two-pipe layout, in which radiators are all supplied with hot water at the same temperature from a single supply pipe, and the water from these then flows back to the furnace to be reheated through a common return pipe. Although the two-pipe system requires more pipe work, it is more efficient and easier to control than the one-pipe system. Another advantage of the two-pipe direct-return and reverse-return loop over the one-pipe loop is that it can be zoned. Zoning offers additional control over where and when heat is required, which in turn can reduce heating costs. The two-pipe system is thus more efficient and easier to control than the one-pipe system. In both systems an expansion tank is required to compensate for variations in the volume of water. Closed expansion tanks contain about 50% air, which compresses and expands to compensate for volume changes in the water. Another system that is sometimes used is the sealed hot-water system. This is basically a closed system that does not need water tanks because the hot water is supplied directly from the mains.

Steam Heating

Steam systems closely resemble hot-water systems except that steam rather than hot water is circulated through the pipes to the radiators. Steam is often used to carry heat from a boiler to consumers as heat exchangers, process equipment, and so forth. Sometimes steam is also used for heating purposes in buildings. Steam heating systems closely resemble their hydronic counterpart except that steam rather than hot water is circulated through the pipes to the radiators and no circulating pumps are required. The steam condenses in the radiators and/or baseboards, giving up its latent heat. Both one-pipe and two-pipe arrangements are employed for circulating the steam and for returning to the boiler the water formed by condensation. Three main types of steam systems used are: air-vent systems, vapor systems, and vacuum, or mechanical-pump systems.

Each heating unit in a one-pipe gravity-flow system has a single pipe connection through which it simultaneously receives steam and releases condensate. All heating units and the end of the supply main are sufficiently above the boiler water line so that condensate is able to flow back to the boiler by gravity. In a two-pipe system, steam supply to the heating units and condensate return from the heating units are through separate pipes. Air accumulation in piping and heating units discharges from the system through the open vent on the condensate pump receiver. Piping and heating units must be installed with proper pitch to provide gravity flow of all condensate to the pump receiver.

Vacuum systems resemble vapor systems in that each radiator is equipped with an inlet valve and a steam trap; however, they have a vacuum pump in the return piping. A partial vacuum is maintained with the pump in the system so that the steam, air, and condensate circulate more readily. The air is expelled into the atmosphere when the condensate and air return to a central point, from which the condensate is pumped back into the boiler.

Heat Pumps

A heat pump is a device that extracts available heat from one location and transfers it to another, unlike a furnace, which creates heat. Heat pumps are actually air conditioners that run in reverse to bring heat from outdoors into the interior. This works by heating up a piped refrigerant in the outdoor air, then pumping the heat generated by the warmed refrigerant inside to warm the indoor air. This type of system works best in moderate climates and becomes less efficient in very cold winter temperatures, when electrical heat is needed for auxiliary demand. Heat pumps are most efficient when the outside temperature is in the 50 °F range. This is because as the outdoor temperature begins to drop, the heat loss of a space becomes greater, requiring the heat pump to operate for longer stretches of time for it to be able to maintain a constant indoor temperature. Also, the fact that most heat pumps use atmospheric air as their heat source presents a problem in areas where winter temperatures frequently drop below freezing, making it difficult to raise the temperature and pressure of the refrigerant. As with furnaces, heat pumps are usually controlled by thermostats. A typical residential application of a water pump system is illustrated in Figure 9.10.

Reverse-Cycle Chillers

Reverse-cycle chillers (RCCs), a recently introduced heat pump variant, are a type of heating system that uses hot water as a heat source. When the cycle is reversed, RCCs have the ability to cool the room. This means the system serves a dual purpose of providing both heating and cooling. RCCs basically heat or cool an insulated tank of water and then distribute the heating or cooling either through fans and ducts or radiant floor systems. The need for auxiliary electric

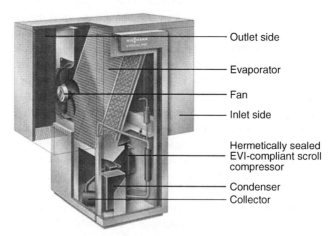

Outlet side

Evaporator

Fan

Inlet side

Hermetically sealed
EVI-compliant scroll
compressor

Condenser

Collector

FIGURE 9.10 Vitocal 350A air-water heat pump. The heating output of this heat pump is 10.6 to 18.5 kilowatts. *(Source: Viessmann Werke.)*

heating coils and defrosting cycles to prevent icing of the refrigerant is eliminated, making these systems more suitable for cold climates than traditional heat pumps. Newer models also now offer solar-powered hot-water heating for the unit; however, these systems still require an exterior condenser unit similar to traditional HVAC systems.

Geothermal Heat Pumps

Geothermal heat pumps (GHPs)—sometimes referred to as geoexchange, earth-coupled, ground-source, or water-source heat pumps—have been in use for several decades. This is a technology that is gaining wide acceptance for both residential and commercial buildings. Studies show that approximately 70% of the energy used in a geothermal heat pump system is renewable energy from the ground and this system is more than 45% more energy-efficient than standard options. According to ENERGY STAR, "geothermal heat pumps (GHPs) are among the most efficient and comfortable heating and cooling technologies currently available, because they use the earth's natural heat to provide heating, cooling, and often, water heating."

A GHP uses the relatively constant temperature of the ground or water several feet below the earth's surface as its source of heating and cooling. The earth's constant temperature is what makes geothermal heat pumps one of the most efficient (relative to air-source heat pumps), comfortable, and quiet heating and cooling technologies available today. GHPs also last longer, need little maintenance, and do not depend on the temperature of the outside air. Some models of geothermal systems are available with two-speed compressors and variable fans for more comfort and energy savings.

There are four basic classifications of ground-loop systems. Three of these—horizontal, vertical, and pond/lake—are closed-loop systems. The fourth is the open-loop option; these systems use well or surface water as the heat exchange fluid that circulates directly through the GHP. Once it has circulated through the system, the water returns to the ground through the well, a recharge well, or surface discharge. Which one of these is best depends on climate, soil conditions, available land, and local installation costs at the site. All of these approaches can be used for residential and commercial building applications. The first three classifications are described in Table 9.2.

Hybrid Systems

Another technology option is the hybrid system, which uses several different geothermal resources, or a combination of a geothermal resource with outdoor air (i.e., a cooling tower). Hybrid approaches are particularly effective where cooling needs are significantly greater than heating needs.

Geothermal heat pumps are appropriate for retrofit or new facilities, where both heating and cooling are desired, and business owners around the United States are now installing them to heat and cool their buildings. These systems

TABLE 9.2 Ground-Loop System Classifications

System	Description
Horizontal	This type of installation is generally the most cost effective for residential installations, particularly for new construction, where sufficient land is available
Vertical	These systems are often used by large commercial buildings and schools because the land area required for horizontal loops would be prohibitive. Vertical loops are also used where the soil is too shallow for trenching, and they minimize the disturbance to existing landscaping.
Pond/lake	This may be the lowest cost option if the site has an adequate body of water. A supply line pipe is run underground from the building to the water and coiled into circles at least eight feet under the surface to prevent freezing.

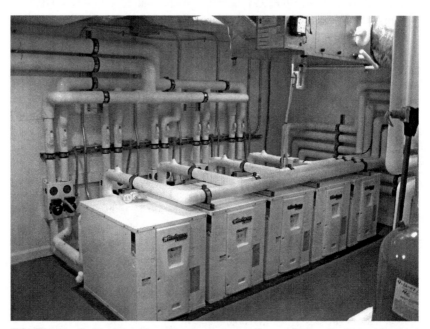

FIGURE 9.11　Residential geothermal heat pump system. *(Source: Climate Heating and Cooling.)*

can also be located indoors because there is no need to exchange heat with the outdoor air. Although this technology may be more expensive to install than traditional HVAC systems, it greatly reduces gas or electric bills through reduced energy, operation, and maintenance costs, thus allowing for relatively short payback periods (Figure 9.11). Conventional ductwork is generally used to distribute heated or cooled air from the geothermal heat pump throughout the building.

It is important to check when selecting ground-source heat pumps that the models chosen qualify for the ENERGY STAR label or meet the recommended levels of coefficient of performance (COP) and energy efficiency ratio (EER). Efficiency is measured by the amount of heat a system can produce or remove using a given amount of electricity. A common measurement of this performance is the seasonal energy efficiency ratio (SEER). The Federal Appliance Standards, which took effect on January 23, 2006, require new standards for central air conditioners of a minimum of 13 SEER. Most manufacturers now offer SEER 10, 11, 12, and 13 models, and some offer SEER 14. This translates into five separate efficiency options, with model numbers usually keyed to the SEER numbers so they are easy to recognize. The most efficient models, however, generally involve two-speed compressor systems and variable fans and an increased heat exchange area, and thus cost significantly more. But even though the installation price of a geothermal system can be several times that of an air-source system of the same heating and cooling capacity, the additional cost can be recouped in energy savings in 5 to 10 years. System life is estimated at 25 years for the inside components and 50-plus years for the ground loop. There are approximately 50,000 geothermal heat pumps installed in the United States annually.

No doubt, with the ongoing development of new technologies and innovations, heat pump performance will continue to improve. Thus, for example, the introduction of two-speed compressors allows heat pumps to operate close to the heating or cooling capacity that is needed at any particular moment. This saves large amounts of electrical energy and reduces compressor wear. Also, some heat pumps are equipped with *variable-speed* or *dual-speed* motors on their indoor fans, outdoor fans, or both. The variable-speed controls for these fans attempt to keep the air moving at a comfortable velocity, minimizing cool drafts and maximizing electrical savings. Another advance in heat pump technology is a device called a *scroll compressor*, which compresses the air or refrigerant by forcing it into increasingly smaller areas. The scroll compressor uses two interleaved scrolls to pump, compress, or pressurize fluids such as liquids and gases. Some reports estimate that heat pumps with scroll compressors provide 10 to 15 °F (5.6–8.3 °C) warmer air when in heating mode, compared to existing heat pumps with piston compressors.

Solar Thermal Collectors

A solar thermal collector is a device designed specifically to collect heat by absorbing sunlight and may be used to heat air or water for building heating. A solar collector operates on a very simple basis. The radiation from the sun heats a liquid that goes to a hot-water tank. The liquid heats the water and flows back to the solar collector. Water-heating collectors may replace or supplement a boiler in a water-based heating system. Air-heating collectors may replace or supplement a furnace. Solar collectors are considered to be one of the renewable energy technologies with the best economics. They have an estimated lifetime of 25 to 30 years or more and require very little maintenance except control of antifreeze and pressure. Solar water-heating collectors may also provide heated water that can be used for space cooling in conjunction with an absorption refrigeration system.

Solar systems are basically either active or passive. The terms *passive* and *active* refer to whether the systems rely on pumps or only on thermodynamics to circulate water. As solar energy in an active solar system is typically collected at a location remote from the spaces requiring heat, solar collectors are normally associated with central systems. Solar water-heating collectors may also provide heated water that can be used for space cooling in conjunction with an absorption refrigeration system. These systems are generally used in hotels and homes in sunny climates such as those found in southern Europe. Since the sun provides free energy, a saving of up to 70% of the energy that would otherwise be used for heating the water is possible. Besides the economical reward, there is a significant environmental advantage. By using solar collectors for water heating an average family can save up to one ton of CO_2 per year. Figure 9.12 illustrates an active indirect solar system.

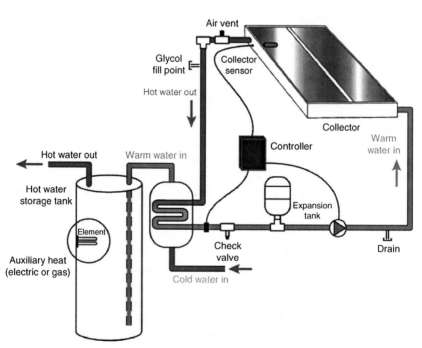

FIGURE 9.12   An active indirect solar system is preferred in climates with extended periods of below-freezing temperatures. *(Courtesy Southface Energy Institute.)*

9.4.4   Ventilation

It shouldn't surprise anyone that proper ventilation is a prerequisite to healthy indoor living. It can be accomplished passively through natural ventilation, or actively by forced ventilation through mechanical distribution systems powered by fans. The subject is discussed in greater detail in Chapter 7 (Indoor Environmental Quality).

9.4.5   Air-Conditioning Systems

ASHRAE's definition of an air-conditioning system is one that accomplishes four specific objectives simultaneously: air temperature control, air humidity control, air circulation control, and air quality control. Air-conditioning systems are typically designed to provide heating, cooling, ventilation, and humidity control for a building or facility. It is much easier to incorporate it into new modern offices and public buildings under construction than to retrofit existing buildings because of the bulky air ducts that require installation. Moreover, systems must be carefully maintained to prevent the growth of pathogenic bacteria in the ducts and ensure efficient operation. Air-conditioned buildings (especially high-rise buildings) often have sealed windows because open windows would disrupt the attempts of the control system to maintain constant air quality.

Currently there are many different types of air-conditioning systems on the market, depending on individual needs and requirements. Some use direct-expansion coils for cooling such as window units, package units, split air conditioners, packaged terminal air conditioner, like those used in hotels, and mini-split ductless air-conditioners. Other types use chilled water, and these are typically commercial models for large commercial buildings. Whichever type is used, the coils in the system are required to be brought to a temperature colder than the air.

When operating properly, air conditioners use direct-expansion coils or chilled-water coils to remove the heat from the air as the air is blown across the coils. The evaporator coil in an air-conditioning system is responsible for absorbing heat. The evaporator and condenser both comprise tubing, usually copper, surrounded by aluminum fins. As air (or water in a chiller) passes over the evaporator coils, a heat exchange process takes place between the air and the refrigerant. The refrigerant absorbs the heat and evaporates in the indoor evaporator coils, draws the heat out of the air, and cools the facility. Finally, the hot refrigerant gas is pumped (compressed) into the outdoor condenser unit, where it returns to liquid form. The main types of air-conditioning systems currently available include those described next.

Vapor Compression Refrigeration

The vapor compression refrigeration system is the most widely used refrigeration system today for air conditioning of large public buildings, private residences, hotels, hospitals, theaters, and restaurants. This approach involves the operation of a vapor compression refrigeration cycle to induce heat to travel in a direction contrary to gross environmental temperature differences. During the overheated period, the outside air temperature is expected not to be just above the balance point temperature but also above the indoor air temperature. Under such conditions, heat flow will be from higher to lower temperature (from outside to inside). To maintain thermal comfort during overly warm periods requires that heat be removed from a building, not added to it. Through a series of artificially maintained temperature and pressure conditions in a heat transfer fluid (refrigerant), a refrigeration system can induce heat to flow from inside a cooler building to an outside that is warmer. These systems all have four essential components:

- *A gas compressor:* This is a mechanical device that increases the pressure of a gas by reducing its volume (similar to a pump).
- *A condenser:* This is a device or unit used to condense a substance from its gaseous to its liquid state, typically by cooling it.
- *An expansion or throttle valve* (also called a valve).
- *An evaporator.*

Exterior Wall or Window Air-Conditioning Units

Window and through-the-wall electric air-conditioning units are often used in single-zone applications that do not have central air conditioning installed such as small buildings and trailers. They are also used in retrofit situations in conjunction with an existing system. Basically, these are small ductless units with casings extending through the wall; they are generally noisy and designed to cool small areas, though some larger units may be able to cool larger spaces. The primary advantage of a wall unit over a window unit is that it does not occupy window space. Note that whenever a window unit is installed, part of the window becomes unusable (reducing incoming daylight).

Removing the cover of an unplugged window unit shows that it comprises of a number of components, including a compressor, an expansion valve, a hot coil (on the outside), a chilled coil (on the inside), two fans, and a control unit (Figure 9.13). The wall unit air conditioner works by removing hot air from the room into the unit; the hot air that enters is brought over the air-conditioning condenser and cooled. The cooled air is then pushed back into the room. Many of the newer units incorporate significant innovations (e.g., electronic touchpad controls, energy-saver settings, and digital temperature readouts). A timer is another improvement now available on new air-conditioning (AC) units as it allows you to set them to cycle on and off at certain times of the day (i.e., they might stay off all day and turn on an hour before you are scheduled to come home from work).

Central Air-Conditioning Systems

In most modern buildings, air conditioning has become one of those amenities frequently taken for granted, and often, particularly in relatively warm climates, we find it has become more the rule than the exception. In addition

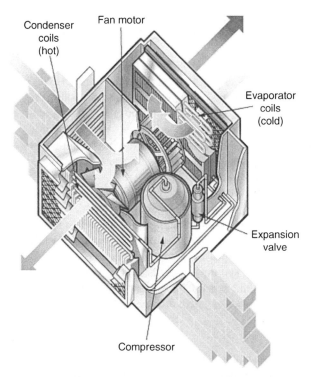

Condenser coils (hot)

Fan motor

Evaporator coils (cold)

Expansion valve

Compressor

FIGURE 9.13 The components of a basic window AC unit, as shown, are the compressor, an expansion valve, a hot coil (on the outside), a chilled coil (on the inside), two fans, and a control unit. *(Source: HowStuffWorks.com.)*

to cooling, AC systems are designed to dehumidify and filter air, making it more comfortable and cleaner. Central air-conditioning systems consist of three main components: the outdoor unit (condenser and compressor), the indoor unit (blower coil or evaporator), and the indoor thermostat to regulate the temperature. The success of a central air system is dependent on these three components appropriately functioning together.

Likewise, the design of an air-conditioning system depends on, among other things, the type of structure in which the system is to be placed, the amount of space to be cooled, the function of that space, and the number of occupants using it. For example, a room or building with large windows exposed to the sun, or an indoor office space with many heat-producing lights and fixtures, requires a system with a larger cooling capacity than a space with minimal windows in which cool fluorescent lighting is used. Also, a space in which the occupants are allowed to smoke will require greater air circulation than a space of equal capacity in which smoking is prohibited. Air-conditioned buildings often have sealed windows, because open windows would disrupt the HVAC system's ability to maintain constant indoor air conditions. This must be taken into consideration in any HVAC system design.

With new technology and the "intelligent building" craze, many central HVAC systems today are being designed to serve more than one thermal zone, with its main components being located outside of the zone or zones being served—usually at a convenient central location in, on, or near the building. Central air-conditioning systems are extensively installed in offices, public buildings, theaters, stores, restaurants, and other building types. Although they provide fully controlled heating, cooling, and ventilation, they need to be installed during construction.

In recent years, these systems have increasingly become automated by computer technology for energy conservation. In older buildings, indoor spaces may be equipped with a refrigerating unit, blowers, air ducts, and a plenum chamber in which air from the interior of the building is mixed with air from the exterior. Such installations are used for cooling and dehumidifying during the summer months, and the regular heating system is used during the winter. Figure 9.14 shows the general components of a small central air-conditioning system that can be used in small commercial buildings or residences.

Split Systems

A split air-conditioner consists of two main parts: the outdoor unit and the indoor unit. It generally implies that the condenser and compressor are housed in an outdoor cabinet and the refrigerant-metering device and the

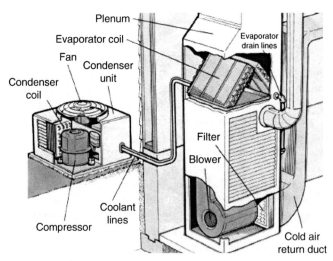

FIGURE 9.14 Domestic central air conditioners are made up of two basic components: the condenser unit, located outside the house on a concrete slab, and the evaporator coil above the furnace. These components in turn comprise several elements. *(Source: HowStuffWorks.com.)*

evaporator are housed in an indoor cabinet. With many split air-conditioners, the indoor cabinet also houses a furnace or a part of the heat pump. The cabinet or main supply duct of this furnace or heat pump also houses the air conditioner's evaporator coil. For reverse-cycle applications, the heat exchanger can be exposed to outside air, becoming the evaporator, and the inside heat exchanger can become the condenser (Figure 9.15).

Split systems may have a variety of configurations. The four basic components of the vapor compression refrigeration cycle—compressor, condenser, refrigerant-metering device, and evaporator—can be grouped in several ways. The grouping of components is based on practical considerations (e.g., available space, ease of installation, and noise reduction in occupied spaces). Generally, split systems are more expensive to purchase but potentially less expensive to install, and a ductless system is practical for homes that don't already have ductwork. The big advantage of a ductless mini-split system is that you can adjust temperature levels for individual rooms or areas.

Where the condensing unit is quite large (e.g., in department stores, businesses, malls, and warehouses), it is normally located on the roof. Alternatively, there may be many smaller units on the roof, each attached to a small inside air handler that cools a specific zone in the building. The split-system approach

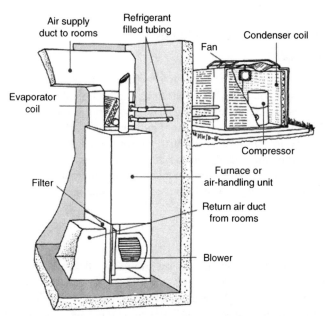

FIGURE 9.15 A drawing illustrating how a split system works. With a typical "split system," the condenser and the compressor are located in an outdoor unit; the evaporator is mounted in the air-handling unit, which is often a forced-air furnace. With a "package system," all components are located in a single outdoor unit that may be located on the ground or on the roof of the structure. *(After: HomeTips.com.)*

may not always be suitable for larger buildings, and particularly multistory buildings, because problems start to appear such as that the distance between the condenser and the air handler exceeds pipe distance limitations or that the amount of ductwork and the length of the ducts cease to be viable.

Packaged Systems

The main difference between a package unit and a split-system unit is that a split system uses indoor and outdoor components to provide a complete comfort system whereas a package, or self-contained, unit requires no external coils, air handlers, or heating units. Packaged air-conditioning models are used for medium-sized halls and multiple rooms on the same floor, and are usually for applications where air conditioning of more than 5 tons is required. Packaged units commonly use electricity to cool and gas to heat. They typically have all components in a single outdoor unit located either on the ground or on the roof. A packaged rooftop unit is a self-contained AHU, typically used in low-rise buildings and mounted directly onto roof curbs, discharging conditioned air into the building's air duct distribution system. AHUs come in many capacities, from just over 1 ton to several hundred tons that contain multiple compressors and are designed for single- or multiple-zone application.

9.4.6 Basic HVAC System Types

All-Air Systems

All-air systems represent the majority of systems currently in operation. They basically transfer cooled or heated air from a central plant via ducting, distributing air through a series of grilles or diffusers to the room or rooms being served. The overall energy used to cool buildings with all-air systems includes that necessary to power the fans that transport cool air through the ducts. Because the fans are usually placed in the air stream, fan movement heats the conditioned air (Figure 9.16(a)), adding to the thermal cooling peak load. For optimum efficiency, constrictions and sharp changes of direction should be avoided.

All-Water Systems

Heating systems represent the largest group of all-water systems. Robert McDowall, author of *Fundamentals of HVAC Systems*, says, "When the ventilation is provided through natural ventilation, by opening windows, or other means, there is no need to duct ventilation air to the zones from a central plant. This allows all processes other than ventilation to be provided by local equipment supplied with hot and chilled water from a central plant." McDowall also notes that "both the air-and-water and all-water systems rely on a central supply of hot water for heating and chilled water for cooling."

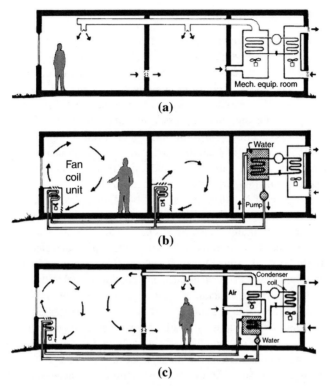

FIGURE 9.16 The three basic HVAC systems, shown here, are (a) all-air, (b) all-water, and (c) air-water. *(Source: Adapted from Norbert Lechner in Mechanical Equipment for Heating and Cooling, December 2002.)*

The conditioning effect in these systems is distributed from a central plant to conditioned spaces via heated or cooled water. Water is an effective heat transfer medium, so distribution pipes are often of relatively small volume (compared to air ducts). On the other hand, water cannot be directly dumped into a space through a diffuser, but requires a more sophisticated delivery device. All-water heating-only systems employ a variety of delivery devices such as baseboard radiators, convectors, unit heaters, and radiant floors. All-water cooling-only systems are rare, with valance units being the most common delivery device for such systems (Figure 9.16(b)). When full air-conditioning is contemplated, the most appropriate delivery device may be the fan-coil unit. All-water systems are generally the most expensive to install and own, and are classed as the least energy-efficient in terms of transfer of energy.

Air-Water Systems

Air-water systems are another category of central HVAC systems that distribute conditioning effects by means of heated or chilled water and heated or cooled air (Figure 9.16(c)).

9.4.7 HVAC System Requirements

HVAC professionals should be aware that building spaces such as cavities between walls can support platforms for air handlers, and that plenums defined or constructed with materials other than sealed sheet metal, duct board, or flexible duct must not be used for conveying conditioned air, including return air and supply air. Care should be taken to ensure that ducts installed in cavities and support platforms are not compressed in a manner that would cause reductions in the cross-sectional area of the ducts. Connections between metal ducts and the inner core of flexible ducts must be mechanically fastened, and openings must be sealed with mastic, tape, or other duct closure systems that meet the codes and standards of local jurisdictions.

Moreover, in most jurisdictions national building codes stipulate that access be provided to certain components of mechanical and electrical systems. This is usually required for maintenance and repair, and includes such elements as valves, fire dampers, heating coils, mechanical equipment, and electrical junction boxes. Commercial construction usually takes advantage of ceiling plenums to run horizontal ducts, while vertical ducts are contained within their own chases. Depending on the type of structure and the depth of the plenum, large ducts may occupy much of this depth, leaving little if any space for recessed light fixtures. Where the plenum is used as a return-air space, most local and national building codes prohibit the use of combustible materials (e.g., wood or exposed wire) within the space in commercial building projects.

Certain occasions in commercial construction may necessitate the use of access flooring (typically in computer rooms), which consists of a false floor of individual panels raised by pedestals above the structural floor. This is designed to provide sufficient space to run electrical and communication wiring as well as HVAC ductwork. Sometimes small pipes are designed to run in a wall system, whereas larger pipes may need deeper walls or even chase walls to accommodate them. Fan systems that exhaust air from the building to the outside must be provided with back draft or automatic dampers. Gravity ventilating systems must have an automatic or readily accessible manually operated damper in all openings to the exterior, except combustion inlet and outlet air openings and elevator shaft vents.

9.4.8 Common HVAC Deficiencies

The majority of HVAC deficiencies have been found to be maintenance-related. And when maintenance of the equipment is deferred or performed by unqualified personnel, the system will likely increasingly experience problems. However, when properly maintained, a building's HVAC system can enjoy a substantial life span. HVAC deficiencies fall into two main categories: issues that are fairly simple to address (e.g., filter or belt replacement) and complex issues requiring the attention of specialized personnel such as pump or boiler replacement. One deficiency often encountered is inadequacy of the system for the size of the facility. Most designers of mechanical systems now use computer analysis software to determine heating and cooling loads, but a general rule of thumb in an assessment is to compare the actual tonnage of the unit to the standard design tonnage using the following formulas:

$$\text{BTU of unit} \div 12{,}000 = \text{actual tonnage of unit}$$
$$\text{Square footage of building} + 350 = \text{design tonnage}$$

Deficiencies and other issues that need to be checked for conformance when conducting HVAC evaluations of existing buildings include the following:

- Evidence of abnormal component vibrations or excessive noise
- Evidence of unsafe equipment conditions, including instability or absence of safety equipment (guards, grills, or signage)
- Presence of drafts in the room or space being cooled
- Location of the thermostats
- Whether the building has an exhaust system and whether the toilets are vented independently or mixed with the common area venting system
- Insufficient air movement to reach all parts of the room or space being cooled or, alternatively, the presence of drafts in the room or space being cooled
- Return air at the return registers not at least 10 to 15°F warmer than the supply air
- Evidence of leaking caused by inadequate seals
- Whether there is a fresh air make-up system in the building
- Evidence of fan alignment deficiencies, deterioration, corrosion, or scaling

9.4.9 HVAC Components and Systems

A key function of an HVAC system is to provide building occupants with healthy and comfortable interior thermal conditions (Figure 9.17). Such systems generally include a number of active mechanical and electrical systems employed to provide thermal control. Control of the thermal environment is one of the key objectives of virtually all occupied buildings. Numerous systems and components are used in combination to provide fresh air as well as temperature and humidity control.

HVAC system components can generally be grouped into the following three functional categories: source components, distribution components, and delivery components.

Source components generally provide or remove heat or moisture. There are four basic types of heat source employed in buildings: Onsite combustion (fuel such as natural gas or coal); electric resistance (converting electricity to heat); a solar collector on the roof connected to the furnace; and a

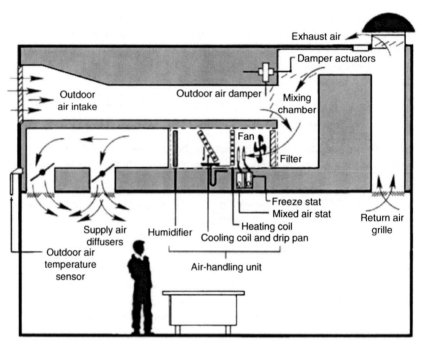

FIGURE 9.17 Typical HVAC system components. These deliver conditioned air to a building or space to maintain thermal comfort and IAQ. *(Source: Terry Brennan, Camroden Associates.)*

heat pump in the furnace. Choosing a heat source for a given building depends on several factors such as source availability, fuel costs, required system capacity, and equipment costs.

Distribution components are used to convey a heating or cooling medium from a source location to portions of a building that require conditioning. Central systems produce a heating and/or cooling effect in a single location, which is then transmitted to the various spaces that require conditioning. Three transmission media are commonly used in central systems: air, water, and steam. Hot air can be used as a heating medium, cold air as a cooling medium. Hot water and steam can be used as heating media, while cold water is a common cooling medium. A central system always requires distribution components to convey the heating or cooling effect from the source to the conditioned locations.

Delivery components basically serve as an interface between the distribution system and the occupied spaces. The heating or cooling effect produced at a source has to be properly delivered to each space to promote comfort and well-being. In air-based systems, the dumping of heated or cooled air into each space does not provide the control over air distribution required of an air-conditioning system. Likewise, with water-based systems, the heated or cooled media (water or steam) cannot just be dumped into a space. Some means of transferring the conditioning effect from the media to the space is required. "Delivery devices" are designed to provide the interface between occupied building spaces and distribution components.

Normally, when compact systems only serve a single zone of a building, they frequently incorporate all three functions in one piece of equipment, whereas systems that are intended to condition multiple spaces (central systems) usually employ distinctly different equipment elements for each function. Furthermore, for each commercial property type, some systems perform better than others. However, from a lender's perspective, performance is judged by how well the needs of tenants and owners are met regarding comfort, operating costs, aesthetics, flexibility, and reliability.

Ductwork

The main objective of duct design is to provide an efficient distribution network of conditioned air to the various spaces within a building or complex. To achieve this, ducts must be designed to facilitate airflow and minimize friction, turbulence, and heat loss and gain. Optimal air distribution systems have correctly sized ducts with minimal runs, smooth interior surfaces, and minimum direction and size changes. Ducts that are badly designed and installed can result in poor air distribution, poor indoor air quality (IAQ), occupant discomfort, additional heat losses or gains, increased noise levels, and increased energy consumption. Duct system design requirements and construction can be impacted by the design of the building envelope. A duct system's overall performance can also be impacted by the materials used. Fiberglass insulation products are currently used in the majority of duct systems installed in the United States, and serve as key components of well-designed, well-operated, and well-maintained HVAC systems that provide both thermal and acoustical benefits for the life of the building. Other materials commonly used for low-pressure duct construction include sheet metal (galvanized steel), black carbon steel, aluminum, stainless steel, fiberglass-reinforced plastic, polyvinyl steel, and concrete.

HVAC professionals strongly advise a preventive maintenance inspection program that will help identify system breaches that are typically more prevalent at duct intersections and flexible connections. Supply and return air may both utilize ductwork, which may be located in the ceiling cavity or below the floor slab depending on system configuration. In single-duct systems, both cool air and hot air use the same duct, and in double-duct systems, separate ducts are used for cooling and heating. It should be noted that air flowing through a duct system will encounter friction losses through contact with the duct walls and in passing through the various devices (e.g., dampers, diffusers, filters, and coils). To overcome these losses, a fan can provide the energy input required to overcome friction losses and can circulate air through the system.

Whether a central HVAC system requires the use of several fans for air supply and return and for exhaust air is determined by the size of the ductwork to be used. The use of fire dampers will be required whenever ducts penetrate a fire wall. Modern fire dampers contain a fusible link that melts and separates when a particular temperature is reached, causing them to slam shut in the event of a fire. Another major factor that impacts the fabrication and design of duct systems and that has to be considered is acoustics. Unless the ducting system is properly designed and constructed, it can act like large speaker tubes transmitting unwanted noise throughout the building. Splitters and turning vanes are often used to reduce the noise generated within the ducts.

Grilles, Registers, and Diffusers

Grilles, registers, and diffusers are generally employed in conjunction with ductwork and assist in controlling the return, collection, and supply of conditioned air in HVAC systems. A *grille* is basically a decorative cover for return-air inlets; it does not have an attached damper and in most cases has no moving parts. However, a grille can be used for both supply air and return air. The same is not true for a register or a diffuser. Grilles are also used to block sightlines to prevent persons from seeing directly into return-air openings. A *diffuser* is an airflow device designed primarily to discharge supply air into a space, mix it with the room air, and minimize unwelcomed drafts. *Registers* are adjustable grille-like devices that cover the opening of a duct in a heating or cooling system, providing an outlet for heated or cooled air to be released into a room. They are similar to diffusers except that they are designed and used for floor or sidewall air supply applications or sometimes as return-air inlets. Grilles, registers, and diffusers introduce and blend fresh air with the air of another location. When fitted together, they take in fresh air, the registers blow it out, and the diffusers scatter it in the required space.

Thermostats

Thermostats are devices whose principal function is to control the operation of HVAC systems, turning on heating or cooling as required. New technologies have had an enormous impact on thermostat design. According to APS, a subsidiary of Pinnacle West Capital Corp.,

Modern programmable thermostats provide the basic function of maintaining comfortable indoor temperatures, but they include other valuable features as well. First, they can be programmed to automatically raise or lower the temperature of your facility according to schedules that you define. Manufacturers claim that you can save three to four percent for each degree you lower your thermostat in the winter and raise it in the summer. Most programmable thermostats allow you to input weekday and weekend schedules. The most sophisticated thermostats will control humidity, outdoor air ventilation, and inform you when the air-conditioning filters need to be changed. Some modern thermostats can also include a communications link and demand management features that can be used to reduce air conditioning system energy use during periods of peak electrical demand or high electricity costs. Programmable thermostats can also be combined with HVAC zone control systems to provide optimal comfort and efficiency throughout the facility. When combined with programmable thermostats, manufacturers claim that zoning systems can save up to 30% on heating and cooling costs while providing superior comfort.

Boilers and heating systems use thermostats to prevent overheating and to control the temperature of the circulating water. Thermostats are fitted to hot-water cylinders, boilers, and radiators in rooms. Thermostat location should be coordinated with light switches, dimmers, and other visible control devices. They are typically placed 48 to 60 inches above the floor and away from exterior walls and heat sources.

Control of an HVAC system is pivotal to its successful operation. HVAC systems can have single-zone or multizone capabilities. In a single-zone system, the entire building is considered one area, whereas in a multizone system, the building is divided into various zones, allowing specific control of each. In fact, many of today's newest HVAC systems are designed to incorporate individually controlled temperature zones to improve occupant comfort and provide the ability to manage the heating or cooling of individual rooms or spaces by use. Additionally, they allow us to adjust individual room temperatures for individual preferences and to close off airflow to areas that are rarely used. A zoned HVAC system is typically provided with a series of dampers, which can save on the installation cost of multiple-unit systems. But whether using one HVAC unit using zone dampers or using multiple HVAC units, zoning can save on utility and maintenance costs. More important, perhaps, the design of the duct system for today's zoning is an important factor in a comfortable and efficient zoning system.

Likewise, zoning and high-efficiency equipment can substantially increase the overall energy performance of a home or office while maintaining rising energy costs at a manageable level. Zoning systems can automatically direct the flow of the conditioned air to zones needing it and at the same time automatically switch over and provide the opposite mode to other zones. This eliminates the need for constant balance and outlet adjustments based on continuously changing indoor conditions. But zoning represents just one of several actions that are designed to improve HVAC performance and give building occupants personal climate control throughout their environment.

Boilers

Boilers are heating-system components designed to generate steam or hot water for distribution to various building spaces (Figure 9.18). As water cannot be used to directly heat a space, boilers are only used in central systems where hot water is circulated to delivery devices (e.g., baseboard radiators, unit heaters, convectors, or AHUs). Once the delivery device is heated with hot water, the water is returned to the boiler to be reheated and the water circulation loop continues. Generally speaking, hot-water boilers are more efficient than

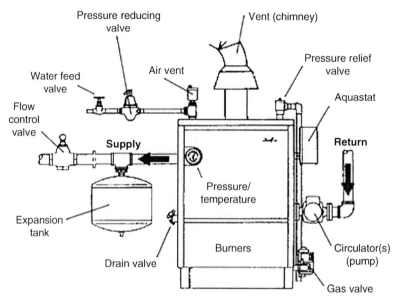

FIGURE 9.18 Gas-fired hot-water boiler. *(Courtesy Home-Cost.com.)*

steam boilers for several reasons. For example, there is less heat loss throughout the hot-water piping and the shell of the boiler because a hot-water boiler operates at a lower temperature than a steam boiler. This means there is less heat loss throughout the entire boiler and piping system. Also, because a hot-water boiler operates at a lower temperature, it requires less fuel or energy to convert into heat. An onsite solar energy collection system may serve in lieu of a boiler. Heat transfer systems (heat pumps) likewise may serve as a substitute for a boiler. Constructed of cast iron or steel, and occasionally copper, boilers can be fired by various fuel sources including natural or propane gas, electricity, coal, oil, steam or hot water, and wood.

Chilled-Water Systems

Chilled-water cooling systems remove heat from one element (water) and move it into another element, which is either ambient air or water. They are a key component of air-conditioning systems for large buildings, although they typically use more energy than any other piece of equipment. A chilled-water system is similar to an air-conditioning system in that it is compressor-based, but a chiller cools liquid while an air-conditioning system cools air. Other components are a reservoir, recirculating pump, evaporator, condenser, and temperature controller. Chillers vary in terms of condenser cooling method, cooling specifications, and process pump specifications. The cooling fluid used is usually a mix of ethylene glycol and water.

Chillers can be *air-cooled, water-cooled*, or *evaporatively cooled*. Water-cooled chillers incorporate cooling towers, which improve their thermodynamic effectiveness as compared to air-cooled chillers. This is due to heat rejection at or near the air's wet-bulb temperature rather than at the higher, sometimes much higher, dry-bulb temperature. Evaporatively cooled chillers are more efficient than air-cooled chillers but less efficient than water-cooled chillers. Air-cooled chillers are usually located outside and consist of condenser coils cooled by fan-driven air. Water-cooled chillers are typically located inside a building, and the heat from these chillers is carried by recirculating water to outdoor cooling towers. Evaporatively cooled chillers are basically water-cooled chillers in a box. These packaged units cool the air by humidifying it and then evaporating the moisture.

Chilled-water systems are mainly employed in modern commercial and industrial cooling applications, although there are some residential and light commercial HVAC chilled-water systems in use. One of the reasons behind the popularity of chilled-water systems is that they use water as a refrigerant. Water is much less expensive than refrigerant, which makes them cost effective especially in commercial HVAC air-conditioning applications. Thus, instead of running refrigerant lines over a large area of the building, water pipes are run throughout the building and to evaporator coils in air handlers for HVAC air-conditioning systems. The chilled water is pumped through these pipes from a chiller, where the evaporator coil absorbs heat and returns it to the chiller to reject the heat. Maintaining chilled-water systems well and operating them smartly can yield significant energy savings.

There are two main types of chillers commonly used today: the *compression chiller* and the *absorption chiller*. Absorption chillers use a heat source (e.g., natural gas or district steam) to create a refrigeration cycle that does not use mechanical compression. During the compression cycle, the refrigerant passes through four major components within the chiller: the evaporator, the compressor, the condenser, and a flow-metering device such as an expansion valve. The evaporator is the low-temperature (cooling) side of the system, and the condenser is the high-temperature (heat rejection) side of the system. Compression chillers, depending on size and load, use different types of compressors for the compression process. Mechanical compression chillers, for example, are classified by compressor type: reciprocating, rotary screw, centrifugal, and frictionless centrifugal. Modern dual-compressor centrifugal chillers offer many advantages over conventional chillers. From a performance point of view, the chiller is most efficient at 50% capacity. At this point, only one compressor is operating and the evaporator and condenser are twice the size normally used for the compressor. One advantage that a dual-compressor chiller offers over a VFD chiller is that it does not require significant condenser-water temperature relief to provide the savings.

Factors that can impact the choice of a water-cooled chiller or an air-cooled unit include whether a cooling tower is available or not. The water chiller option is often preferred over the air-cooled unit because it costs less, has a higher cooling capacity per horsepower, and consumes less energy per horsepower. Compared to water, air is a poor conductor of heat, making the air-cooled chiller much larger and less efficient, which is why it is less frequently used unless it is not possible to construct a water-cooling tower.

Cooling Towers

The main function of a cooling tower is removing heat from the water discharged from the condenser so that the water can be discharged to the environment or recirculated and reused. Cooling towers are only used in conjunction with water-cooled chillers and vary in size from small rooftop units to very large hyperboloid structures. Cooling towers are also characterized by the means by which air is moved. *Mechanical draft cooling* towers are the most widely used in buildings, and rely on power-driven fans to draw or force the air through the tower. They are normally located outside the building.

The two most common types of mechanical draft towers in the HVAC industry are *induced draft* and *forced draft*. Induced-draft towers have a large propeller fan at the top of the tower (discharge end) to draw air upward through the tower while warm condenser water spills down. This type requires much smaller fan motors for the same capacity than forced-draft towers (Figures 9.19). Forced-draft towers utilize a fan at the bottom or side of the structure. Air is forced through the water spill area and discharged out the top of the structure. After the water has been cooled in the cooling tower, it is pumped to a heat exchanger or condenser in the refrigeration unit, where it picks up heat again and is returned to the tower.

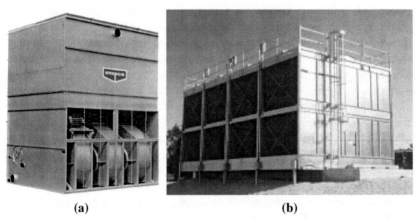

(a)   **(b)**

FIGURE 9.19 (a) Forced-draft towers have fans on the air inlet to push air either counter-flow or cross-flow to the movement of the water. (b) Induced-draft towers have a large propeller fan at the top of the tower to draw air counter-flow to the water. Induced-draft towers are considered to be less susceptible to recirculation, which can result in reduced performance. *(Source: McQuay International.)*

Condensers

Condensers are components that form an essential part of air-conditioning systems to cool and condense the refrigerant gas that becomes hot during the evaporation stage of the cooling process. There are two common condenser types: air-cooled and cooling tower. Cooling is accomplished through the use of air, water, or both.

Air Filters

The main purpose of using air filters is to remove particles and contaminants from the air. They are a critical component of the air-conditioning system; without them, these systems would become dirty and the interior environment would be filled with pollutants and become unhealthy (Figure 9.20). It is critically important to change filters periodically to minimize pollution and improve IAQ. Although no individual product or system in itself can be LEED-certified, proper employment of air filtration systems provides tangible ways to improve IAQ and energy efficiency and can contribute to the completion of LEED prerequisites and credits. This is why the right filter media strategy is important and can help buildings become "greener" and meet LEED and other green building rating system criteria.

FIGURE 9.20   Clogged filter that has been removed from an AC unit.

The LEED 2009 Reference Guide (depending on the LEED certification targeted) should be consulted for more detailed information relating to achievable credits. For example, for EQ Credit 3.1: Construction IAQ Management Plan: during Construction, the guide reads as follows: "If permanently installed air handlers are used during construction, filtration media with a Minimum Efficiency Reporting Value (MERV) of 8 shall be used at each return-air grille, as determined by ASHRAE 52.2-1999." MERV values vary from 1 to 16, and the higher the MERV value is, the more efficient the filter will be in trapping airborne particles.

After construction is completed and prior to occupancy (after all interior finishes are completed), new MERV 13 filters should be installed followed by a two-week building flush-out by supplying a total air volume of 14,000 cubic feet of outdoor air per square foot of floor area. Also for NC EQ Credit 5: Indoor Chemical & Pollutant Source Control, the guide states, "In mechanically ventilated buildings, provide regularly occupied areas of the building with new air filtration media prior to occupancy that provides a MERV of 13 or better. Filtration should be applied to process both return and outside air that is to be delivered as supply air."

It is necessary to understand ASHRAE 52.1 and 52.2 in order to be able to identify what to look for when selecting the right filter to meet IAQ and energy efficiency requirements to help achieve green building standards. On this point, Dave Matela of Kimberly-Clark Filtration Products, says,

One of the biggest determining factors is filtration efficiency, which defines how well the filter will remove contaminants from air passing through the HVAC system. Initial and sustained efficiency are the primary performance indicators for HVAC filters. Initial efficiency refers to the filter's efficiency out of the box or immediately after installation. Sustained efficiency refers to efficiency levels maintained throughout the service life of the filter. Some filters have lower initial efficiency and do not achieve high efficiency until a "dirt cake" builds up on the filter. Other filters offer both high initial, as well as sustained, efficiency, meaning they achieve an ideal performance level early and maintain that level.

Of the various types of air-conditioning filters on the market, the most common types are: conventional fiberglass disposable filters (1 inch and 2 inch), pleated fiberglass disposable filters (1 inch and 2 inch), electrostatic filters, electronic filters, and carbon filters. Most air-conditioning filters are sized 1-1/2 to 2 square feet for each ton of capacity for a home or commercial property. Applying the MERV rating is a good way to help evaluate the effectiveness of a filter. Another consideration is airflow through the HVAC system. Leaving a dirty air filter in place or using a filter that is too restrictive may result in low airflow and possibly cause the system to malfunction. Of note, there are various types of filters with a MERV of 13, each having different design requirements and pressure drop. It would be prudent, therefore, for building owners and designers to consult with a certified air filter specialist (CAFS) to obtain the best information on the optimum filters and pre-filters to obtain LEED certification. Filters should be selected for their ability to protect general indoor air quality as well as HVAC system components.

9.5 ELECTRICAL POWER AND LIGHTING SYSTEMS

The production and transmission of electricity is relatively efficient and inexpensive, although, unlike other forms of energy, electricity is not easily stored and therefore must typically be used as it is being produced. Electrical systems can provide a facility with accessible energy for heating, cooling, lighting, and equipment (telecommunication devices, personal computers, networks, copiers, printers, etc.) and appliance operation (e.g., refrigerators and dishwashers). Electricity has witnessed dramatic developments in the last few decades, comprising the fastest-growing energy load within a building. More than ever, facilities today need electrical systems to provide the power with which most vital building systems operate. These systems control the energy required in the building and distribute it to the location utilizing it. Most frequently, distribution line voltage carried at utility poles is delivered at 2400/4160 volts. Transformers step down this voltage to predefined levels for use within buildings. In an electric power distribution grid, the most common form of electric service is through the use of overhead wires known as a *service drop*, which is an electrical line running from a utility pole to a customer's building or other premises; it is the point from where customers receive their power from the electric utilities.

In residential installations in North America and in countries that use North American systems, a service drop comprises two 120-volt lines and a neutral line. When these lines are insulated and twisted together, they are referred to as a *triplex* cable. In order for these lines to enter a customer's premises, they must usually first pass through an electric meter and then the main service panel, which will usually contain a "main" fuse or circuit breaker. This circuit breaker controls all of the electrical current entering the building at once, and controls a number of smaller fuses/breakers, which protect individual branch circuits. There is always a main shutoff switch to turn off all power; when circuit breakers are used, this is provided by the main circuit breaker. The neutral line from the pole is connected to an earth ground near the service panel, often a conductive rod driven into the earth. For residential applications, the service drop provides the building with two separate 120-volt lines of opposite phase, so 240 volts can be obtained by connecting a circuit between the two 120-volt conductors, while 120-volt circuits are connected between either of the two 120-volt lines and the neutral line. Circuits of 240 volts are used for high-power devices and major appliances, such as air conditioners, clothes dryers, ovens, and boilers, while 120-volt circuits are used for lighting and ordinary small appliances. As these are only "nominal" numbers, it means that the actual voltage may vary.

In many places around the world, including Europe, a three-phase 416Y/ 230 system is used. The service drop consists of three 240-volt wires, or phases, and a neutral wire which is grounded. Each phase wire provides 240 volts to loads connected between it and the neutral. Each of the phase wires carries a 50-Hz alternating current, which is 120 degrees out of phase with the other two. The higher voltages, combined with the economical three-phase transmission scheme, allow a service drop to be longer than in the North American system, and allow a single drop to service several customers. Commercial and industrial service drops are usually much larger and more complex, and so a three-phase system is used. In the United States, common services consist of 120Y/208 (three 120-volt circuits that are 120 degrees out of phase, with 208 volts line to line), 240-volt three-phase, and 480-volt three-phase. In Canada, 575-volt three-phase is common, and 380 to 415-volt or 690-volt three-phase is found in many other countries. Generally, higher voltages are used for heavy industrial loads and lower voltages for commercial applications.

The difference between commercial and residential electrical installations can be quite significant, particularly with large installations. While the electrical needs of a commercial building can be simple, consisting of a few lights for some small structures, they are often quite complex, with transformers and heavy industrial equipment. When electrical or lighting system deficiencies become evident and need attention, they are usually measurable and include power surges, tripped circuit breakers, noisy ballasts, and other more obvious conditions (e.g., inoperative electrical receptacles or lighting fixtures) that are frequently discovered or observed during a review of the system. As illustrated in Figures 9.21 and 9.22, there are a number of typical deficiencies found in both the electrical and the lighting systems.

In commercial buildings, the major load placed on a given electrical system usually comes from the lighting requirements; therefore, the distribution and management of electrical and lighting loads must always be monitored on a regular basis. Lighting management should also be periodically checked because building space uses change and users relocate within the building. It is also highly advisable for the lighting system to be integrated with the electrical system in the facility. Lighting systems are designed to ensure adequate visibility for both the interior and exterior of a facility and comprise an energy source and distribution elements, which normally consist of wiring and light-emitting equipment.

There are several different electrical codes in various jurisdictions throughout the United States. Some of the larger cities (e.g., New York and Los Angeles) have created and adopted their own electrical codes. The National Electrical Code (NEC) and the National Fire Protection Code (NFPC), published by the National Fire Protection Association (NFPA), cover almost all electrical system components. The NEC is commonly adopted in whole or in part by municipalities. Inspection of the electrical and lighting system should

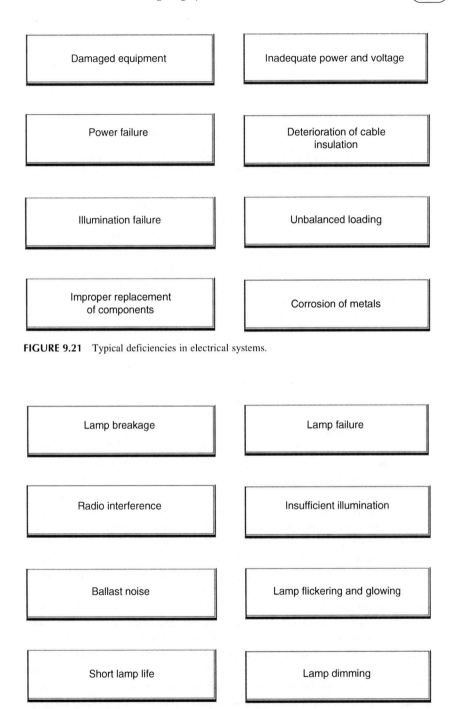

FIGURE 9.21 Typical deficiencies in electrical systems.

FIGURE 9.22 Typical deficiencies in lighting systems.

include a determination of general compliance with these codes at the facility. With very large facilities employing complex electrical equipment, it may be necessary to operate it under engineering supervision or, alternatively, to have a full-time facilities manager.

9.5.1   Understanding Amps, Volts, and Watts

The electrical service in most countries around the world is brought into a building at either 240 volts or 120 volts. These numbers are called "nominal," meaning that the actual voltage may vary. Most modern buildings receive 240-volt service, a total achieved by the provision of two individual 120-volt incoming power lines. Older buildings and electrical services often delivered only 120 volts. While knowing the available voltage level is important, this information alone is insufficient as it does not indicate the amount of electrical power available inside a building. To calculate this, we need to know the service voltage at a building in addition to the service amperage. However, before proceeding it would be prudent to have a basic understanding of some common electrical terms that apply to electrical systems.

Amperage

This is a unit of electrical current. The amperage, or amps (A or amp), provided by an electrical service is the flow rate of "electrical current" that is available. Appliances will typically have an amp rating or, if only a wattage is quoted, amps = wattage/voltage. Practically speaking, the voltage level provided by an electrical service, combined with the ampacity rating of the service panel, determines the electrical load or capacity. Branch circuit wire sizes and fusing or circuit breakers are used to typically set the limit on the total electrical load or the number of electrical devices that can be run at once on a given circuit. Thus, for example, if you have a 100-amp current flow rate in place, you may be able to run approximately ten 10-amp electric heaters simultaneously. If you have only 60 amps available, you won't be able to run more than 6 such heaters without risking overheating the wiring, tripping a circuit breaker, blowing a fuse, or causing a fire, which is why it is important to avoid overloading the system.

To be able to determine the amount of electrical service a facility receives, the service ampacity and voltage must be known. The safe and proper service amperage available at a property is set by the smallest of the service conductors, the main disconnect fuse or switch, or the rated capacity of the electric panel itself. The main fuse/circuit breaker (CB) is the only component that actively limits amperage at a property by shutting off loads drawing more than the main fuse rating. The main breakers or fuses are allowed to have lower overcurrent protection than the capability of the service equipment (panel) and conductors (entrance cable).

Voltage

A volt (V) can be defined in several ways; often it is defined as the potential difference across a conductor when a current of one ampere dissipates one watt of power. Practically speaking, a volt is a measure of the strength of an electrical source at a given current or amperage level. If we bring 100 amps into a building at 240 volts, we have twice as much power available as if we bring in 100 amps at 120 volts. However, if we exceed the current rating of a wire, it will get hot, risking a fire. This is why fuse devices are employed to limit the current flow on electrical conductors to a safe level and thereby prevent overheating and potential fires. Moreover, as previously mentioned, a "240 V" circuit is a nominal rating which implies that the actual voltage level will vary. In many countries the actual voltage level varies around the nominal delivered "voltage rating," and, in fact, depending on the quality of electrical power delivered on a particular service, voltage will also vary continuously around its actual rating. Most electrical power systems are prone to slight variations in voltage due to demand or other factors. Generally, this difference is inconsequential, as most appliances are built to tolerate current a certain percentage above or below the rated voltage. However, severe variations in current can damage electrical equipment, which is why installing a voltage stabilizer is always advisable where sensitive electronic equipment is used.

Wattage

In electricity, a watt is a unit of measure of electric power and is equal to current (in amperes) multiplied by voltage (in volts). Most people use a very simple mathematical formula to determine how many watts an electrical circuit can carry or how many watts an electrical device will require: watts = volts × amps. In buildings, the unit of electricity consumption measure is the watt hour, which is usually in thousands, called kilowatt hours (kWh). In larger buildings, not only is the total consumption rate measured but the peak demand is measured as well.

9.5.2   Electrical Components

Even though electric service is vital to all residential and commercial buildings, it is nevertheless often one of the last components to be installed during the construction process.

Service Connections

Planning the design, construction, and timing of installation of electric service on a construction project should be contemplated from the very early stages. The service connection equipment basically provides a connection between the power company service and the facility and also measures the amount of electricity a facility uses. From here a meter either feeds a disconnect switch or a main breaker or fuse panel. The connection can be located either overhead or underground. The service connection should be checked for type (i.e., voltage, amperage) and general condition, and whether the total power adequately serves the facility's requirements. The equipment should be clean and free from overgrown planting and debris. A detailed discussion of installation requirements of electric service connections is outside our scope.

Switchgear and Switchboards

The function of switching equipment is to control the power supply in the facility and all the services arriving on the site (service drop). This consists of the wires from the main line, a transformer, a meter, and a disconnect switch. The main service switch is the system disconnect for the entire electrical service, and is generally used in combination with metering, the disconnect switch, and protective and regulating equipment to protect and control motors, generators, transformers, and transmission and distribution lines. To avoid excessive voltage drop and flicker, the distance from the transformer to the meter should not exceed 150 feet. In commercial construction, the panel and disconnect should preferably be located outside the building but may be located inside the building if accessible from an exterior door.

A switchboard is made up of one or more panels with various switches and indicators that are used to route electricity and operate circuits. The main switchboard controls and protects the main feeder lines of the system. Switchgear and switchboards should be readily accessible, in good condition, and have protective panels and doors. They should also be checked for evidence of overloading or burn marks. Switchboard covers should not normally be removed.

Switchgears are typically concentrated at points where electrical systems make significant changes in power, current, or routing such as electrical supply substations and control centers. Switchgear assemblies range in size from smaller, ground-mounted units to large walk-in installations and can be classified as outdoor or indoor. Commercial and industrial assemblies are usually indoors, while utilities and cogeneration facilities are more likely to have outdoor gear. Manufactured for a variety of functions and power levels, all switchgear conforms to standards set by the Institute of Electrical and Electronic Engineers (IEEE), the National Electrical Manufacturers Association (NEMA), or the American National Standards Institute (ANSI).

Meters

There are basically two methods of measuring electric consumption in a building. In residential applications, only the total electric consumption is measured. In larger facilities, both total consumption and peak rate demand are measured. This is because large peaks require the utility company to build more power-generating capacity to meet them. Commercial services of up to 200 amps single-phase may have service panels similar to those found in residences.

Larger services may require standalone switchboards with one or more meters. In a multiple-occupancy building, there may be separate meters for each tenant or common metering.

Panelboards

Electrical panelboards and their cabinets house an assembly of circuit breakers and control and protect the branch circuits. From the panelboards, the power generation can be monitored and the power generated can be distributed. In addition to controlling and protecting the branch circuits, panelboards are designed to provide a central distributing point for the branch circuits for a building, a floor, or part of a floor. Each breaker serves a single circuit, and the overload protection is based on the size and current-carrying capacity of the wiring in that circuit. A building may have a number of panelboards and a main panel, with a disconnect switch for the entire building. The following are three lighting panel types: (1) plug-in circuit breakers (1-pole), (2) bolt-on circuit breakers (1-pole, 2-pole, 3-pole), and (3) fusible switch.

To estimate the electric service panel ampacity: Evidence of a tag (normally paper) or embossed rating on fuse pull outs on the panel itself, often includes the amperage rating of the panel. This information is usually present in newer panels on a panel side or on the panel cover. Actual dimensions of an electric panel are not a reliable determinant of ampacity. For example, many larger panels can be fitted with a variety of bus-bar and main switch assemblies of varying ampacity.

Aluminum Wiring

During the 1970s, aluminum (instead of copper) wiring became quite popular and was used extensively. Since that time, however, aluminum-wired connections have been implicated in a number of house fires, and most jurisdictions no longer permit their use in new installations. Aluminum-wired connections were found to have a very high probability of overheating compared with copper-wired connections and were therefore a potential fire hazard. Over the years, a large number of connection burnouts have occurred in aluminum-wired homes, and according to the U.S. Consumer Product Safety Commission, many fires have resulted, some involving injury and death. The main problem with aluminum wiring is a phenomenon known as "cold creep." When aluminum wiring warms up, it expands, and when it cools down, it contracts. However, unlike copper, when aluminum goes through a number of warm/cool cycles, it begins to lose some of its tightness. To add to the problem, aluminum oxidizes, or corrodes, when in contact with certain types of metal, so the resistance of the connection increases. This causes it to heat up and corrode/oxidize still more until eventually the wire may start getting very hot and melt the insulation or fixture it's attached to, and possibly even cause a fire without ever tripping the circuit breaker.

Although aluminum wire "alloys" were introduced in the early1970s, they did not adequately address most of the connection failure problems. Aluminum wiring is still permitted and used for certain applications, including residential service entrance wiring and single-purpose higher-amperage circuits such as for 240-volt air conditioning or electric ranges. Although the fire risk from single-purpose circuits is much less than for branch circuits, field reports indicate that these connections remain a potential fire hazard.

A simple method of identifying aluminum wiring is to examine the wire sheathing for the word *aluminum*. If you can't find it embossed in the wire sheathing, then look for silver-colored wire instead of the copper-colored wire used in modern wiring. Without opening any electrical panels or other devices, it is possible to still look for printed or embossed letters on the plastic wire jacket where wiring is visible at the electric panel. Some aluminum wire has the word *aluminum* or a specific brand name, such as "Kaiser," "Alcan," "aluminum," or "AL/2," plainly marked on the plastic wire jacket. Some white colored plastic wire jackets are inked in red; others have embossed letters without ink and are hard to read. Shining a light along the wire may make it easier to identify. Of note, the fact that no aluminum wiring is evident in the panel does not necessarily mean that none is present. Aluminum may have been used for parts of circuits or for some but not other circuits in the building.

Service Outlets and Receptacles

Service outlets include convenience receptacles, motors, lights, and appliances. Receptacles are commonly known as *outlets* or sometimes erroneously as *wall plugs*—a plug is what actually goes into the outlet. It is preferable for outlets to be three-prong, where the third prong is grounded. For large spaces or areas, all the outlets should not be on the same circuit so that when a fuse or circuit breaker trips due to an overload, the space will not be plunged into complete darkness—without power. Important specifications for electrical receptacles include number of poles and grounding method. Today's electrical receptacles have a variety of features; some include surge protection against mild- to moderate-spikes or peaks in the electrical supply, while others have a locking mechanism or a power light.

Grounding

The grounding of a service to earth is basically a safety precaution and is necessary mainly to protect against lightning strikes or other high-voltage line strikes. Grounding in a commercial building might be to a rod inside a switchboard, to a steel cold-water pipe in the plumbing system, or to the steel frame of a building. Other methods of grounding are also used depending on the equipment or system to be grounded. Grounding also drains any static charges away as quickly as they are produced. Ground wires typically are covered with green insulation but sometimes may be without cover.

Motors, Switches, and Controls

Motors are devices that convert any form of energy into mechanical energy, especially an internal-combustion engine or an arrangement of coils and magnets that convert electric current into mechanical power. Basically, there are four types of motors in general use.

- *DC motor:* This is a rotating electric machine designed to operate from a direct voltage source. It is used for small-scale applications and for elevators, where continuous and smooth acceleration to a high speed is important.
- *Stepper/switched-reluctance (SR) motors:* These are brushless, synchronous electric motors that can divide a full rotation into a large number of steps. The motor's position can be controlled precisely, without any feedback mechanism. Stepper motors are basically similar to SR motors; in fact, the latter are very large stepping motors with a reduced pole count, and generally are closed-loop commutated. The main advantage of stepper motors is that they can achieve accurate position control without the requirement for position feedback. Stepper motors operate differently from normal DC motors, which rotate when voltage is applied to their terminals.
- *AC induction motor:* This is the most common and simple industrial motor; it can be either three-phase AC or single-phase AC. The three-phase AC induction motor is a rotating electric machine designed to operate from a three-phase source of alternating voltage and usually applies to larger equipment. These motors are characterized by extreme reliability and remain constant in rpm, unless heavily overloaded. The single-phase motor is a rotating machine that has both main and auxiliary windings and a squirrel-cage rotor.
- *Universal motor:* This is a rotating electric machine similar to a DC motor but designed to operate either from direct current or single-phase alternating current and vary in speed based on the load. The universal motor is usually found in mixers, hand drills, and similar appliances.

Motors should always be protected against overload by *thermal relays*, which shut off the power when any part of the motor or housing overheats.

Switches and controls are devices that direct the flow of power service to the electrical equipment. Safety switches are installed in locations where service cut-off is available in case of emergencies. They include toggle switches, dials, and levers.

Emergency Power

For certain facilities it is an absolute necessity to have standby power with which to ensure continued electrical service when a shutdown of the standard power service takes place. Emergency power is required for life support systems, fire and life safety circuits, elevators, and exit and emergency lighting. Facilities that require full operation during emergencies or disasters such as

FIGURE 9.23   Amtrak generator for emergency backup power.

hospitals and shelters, always have backup power. Computer facilities, to ensure continued storage and survival of data, also commonly have emergency power. For major equipment, a diesel engine generator with an automatic starting switch and an automatic transfer switch is often provided (Figure 9.23) while for lighting, battery units are installed. The typical AC power frequency in the United States is 60 cycles per second or 60 Hertz, whereas in Europe, 50 Hertz is the standard.

Transformers

Transformers are devices that convert an alternating current (AC) circuit of a certain voltage to a higher or lower value, without change of frequency, by electromagnetic induction. They are used to step up voltage (and so are called "step-up" transformers) to transmit power over long distances without excessive losses, and subsequently step down voltage (called "step-down" transformers) to more usable levels. While a transformer changes the voltage of an AC in a circuit to a higher or lower value, it has practically no effect on the total power in the circuit.

Transformers come in two distinct types: wet or dry. There are also subcategories of each main type. Lower-voltage types are dry and typically noise generating, with minimal requirements for insulation and avenues for ventilation of heat generated by voltage changes. For wet or liquid-filled transformers,

the cooling medium can be conventional mineral oil. Some wet-type transformers use less flammable liquids such as high fire point hydrocarbons and
silicones. Wet types are typically more efficient than dry types and usually
have a longer life expectancy. There are some drawbacks, however. For example, fire prevention is more important with liquid-type units because the liquid
cooling medium used may catch fire (although dry-type transformers are also
susceptible) or even explode. Wet-type transformers typically contain a type
of fire-resistive fluid or mineral oil such as PCBs, and, depending on the application, may require a containment trough for protection against possible leaks
of the fluid, which is why they are preferred predominantly when placed
outdoors.

For lower-voltage, indoor-installed distribution transformers of 600 volts
and below, the dry-type transformer is preferred even though it has minimal
requirements for insulation and avenues for ventilation of heat generated by
voltage changes. Dry-type transformers come in enclosures that have louvers
or are sealed. The location of transformers should be carefully considered,
and there should be clear access to surrounding exterior transformers and
adequate ventilation and access for interior transformers, which should be
inside a fireproof vault. Onsite transformers in parking lots may require
bollards or other protection (Figure 9.24). Transformers should be analyzed
for PCBs and their registration number noted. In addition, transformers tend
to make a certain amount of noise (hum), which should addressed if it causes
irritation.

FIGURE 9.24   Outdoor transformer protected by bollards.

9.5.3   Lighting Systems

The main function of good lighting, whether natural or artificial, is to provide visibility and allow us to see so that we can perform our tasks, thus making a space useable. Different artificial sources produce different kinds of light and vary significantly in their efficiency, which is the calculated lumen output per watt input. Another primary objective must be to minimize energy use while achieving visibility, quality, and aesthetic objectives. The quality and quantity of lighting affects the ambience, security, and function of a facility as well as the performance of its employees. Regrettably, U.S. lighting design does not readily translate overseas—not when different regions have their own voltage requirements, product standards, construction methods, and conceptions about what light is meant to achieve.

Interior Lighting

Because lighting typically accounts for a significant percentage of annual commercial business and residential electric bills, it is important to understand the relationship between a source of light, the surfaces that reflect light, and how we see light, and the need to have a common comprehensive lighting language. It goes without saying that without a light source we cannot see, and without surfaces to reflect light there is nothing to see. Advances in lighting technology can significantly reduce the amount of money that is spent for lighting a facility. More important, interior lighting should meet minimum illumination levels (Table 9.3). It is important to determine the amount of light required for the activity that will take place in a space. Typically, the levels needed for visibility and perception increase with high-accuracy activities, as the size of details decreases, as contrast between details and their backgrounds is reduced, and as task reflectance is reduced. However, interior lighting must not exceed allowed power limits. Interior lighting includes all permanently installed general and task lighting shown on the plans, but does not include specialized lighting for medical, dental, or research purposes or display lighting for exhibits in galleries, monuments, and museums. For this reason, there are many types of interior lighting systems that address these needs and enable us to make full use of a facility around the clock. The most common categories of interior lighting systems are as follows.

Fluorescent Lamps

This type of fixture has long been preferable to incandescent lighting in terms of energy efficiency. Fluorescent lighting is far more efficient than incandescent lighting and has an average life of 10 to 20 times longer (fluorescent lamps last up to 20,000 hours of use), and it uses roughly one-third as much electricity as incandescent lighting with comparable output. Compact fluorescent lamps (CFLs) are similar in operation to standard fluorescent lamps, but are manufactured to produce colors similar to incandescent lamps. New developments

TABLE 9.3 Recommended Illumination Levels for Various Functions

Area	Foot-candles
Building surrounds	1
Parking area	5
Exterior entrance	5
Exterior shipping area	20
Exterior loading platforms	20
Office corridors and stairways	20
Elevators and escalators	20
Reception rooms	30
Reading or writing areas	70
General office work areas	100
Accounting/bookkeeping areas	150
Detailed drafting areas	200

with fluorescent technology, including the high-efficacy T-5, T-8, and T-10 lamps, have pushed the energy efficiency envelope even further. Recently, attention has been paid to the mercury content of fluorescents and the consequences of mercury releases into the environment. As with all resource use and pollution issues, reduction is the best way to limit the problem. Even with low-mercury lamps, however, recycling of old lamps remains a high priority.

A fluorescent fixture typically consists of the lamp and associated ballast, which controls the voltage and the current to the lamp. Replacing standard incandescent light bulbs with CFLs will reportedly slash electrical consumption in homes and offices where incandescent lighting is widely used. By reducing the amount of electricity used, corresponding emissions of associated carbon dioxide, sulfur dioxide, and nitrous oxide are reduced. CFL technology continues to develop and evolve, and is now capable of replacing most of the light fixtures that were originally designed for incandescent light bulbs.

Incandescent Lamps

Incandescent lamps have relatively short lives (typically 1000 to 2000 hours of use) and are the least efficient of common light sources. In fact, only about 15% of the energy they use comes out as light and the rest becomes heat. Nevertheless, they remain popular because they produce a pleasant color that is similar to natural sunlight and they are the least expensive to buy. Incandescent lamps come in various shapes and sizes with different characteristics. The most common incandescent outdoor lighting options are metal halide and high-pressure sodium. Environmental issues include lamp efficacy (lumens per watt), luminaire efficiency, controllability of the light source, potential for PV power, and control of light pollution. To control light pollution, full-cut-off luminaires should be specified. It also makes very good sense to use, whenever possible, environmentally friendly, commercial outdoor lighting systems. For example, ENERGY STAR lights consume only about 20% of the energy consumed by traditional lighting products, thus providing substantial savings in money, energy, and greenhouse-gas emissions.

Tungsten halogen lamps are a type of incandescent lamp that has become increasingly popular in recent years. They produce a whiter, more intense light than standard incandescent lamps and are typically used for decorative, display, or accent lighting. They are about twice as efficient as regular incandescent lamps and typically last two to four times longer.

High-Intensity Discharge

This category of high-output light source consists of a lamp within a lamp that runs at a very high voltage. There are basically four types of high-intensity discharge (HID) lamps: high-pressure sodium (HPS), mercury vapor, metal halide gas, and low-pressure sodium. HID lights require ballasts for proper lamp operation (similar to fluorescent lights). The efficiency of HID sources varies widely, from mercury vapor, with a low efficiency (almost as low as incandescent), to low-pressure sodium, which is an extremely efficient light source. Color rendering also varies widely from the bluish cast of mercury vapor lamps to the distinctly yellow light of low-pressure sodium.

Fiber Optics

This is an up-and-coming technology, providing an alternative that is superior to conventional interior and exterior lighting systems. The technology possesses enormous information-carrying capacity, is low cost, and is immune to many of the disturbances that often afflict electrical wires and wireless communication systems. Fiber-optics technology is based on the use of hair-thin, transparent fibers to transmit light or infrared signals. The fibers are flexible and consist of a core of optically transparent glass or plastic surrounded by a glass or plastic cladding that reflects the light signals back into the core. Light signals can be modulated to carry almost any other sort of signal, including sounds, electrical signals, and computer data, and a single fiber can carry hundreds of such signals simultaneously, literally at the speed of light. The superiority of optical fibers for carrying information from one location to another is leading to their rapidly replacing many older technologies. A typical fiber-optic lighting system can be broken down into two basic components: a light source, which generates the light, and the fiber optics, which deliver the light.

Although fiber optic lighting offers unique flexibility compared to conventional lighting, it does have its limitations. Areas of high ambient light should be avoided as they tend to "wash out" the color. However, fiber optics can often be installed in areas not accessible to conventional lighting. Good ventilation is very necessary for all illuminators. Light-colored reflective surfaces are preferable for end light or sidelight applications. Dark surfaces absorb light and should only be used to provide contrast. Typical applications include cove lighting, walkway lighting, and entertainment illumination (Figure 9.25). One cannot overemphasize the crucial role that optical fibers played, and continue to play, in making possible the extraordinary growth in world-wide communications that has occurred over the last two or three decades, and that is vital in enabling the proliferating use of the Internet and the creation of the "Information Age." In fact fiber-optics systems were even used aboard the NASA space shuttle *Endeavor* during its February 2000 mission.

It is a well-known fact that all electric light sources experience lumen depreciation; thus, the useful life of a lighting installation becomes progressively less during its operation due to dirt accumulation on the surface and aging of the equipment. The rate of reduction is influenced by equipment choice and

FIGURE 9.25   Glass block walls with optical fiber interior. *(Source: Lumenyte International Corp.)*

environmental and operating conditions. In lighting scheme design, we must take account of this deficiency by the use of a maintenance factor and plan suitable maintenance schedules to limit the decay.

Exterior Lighting

It is generally necessary to have adequate outdoor lighting around buildings, and there are many innovative, energy-efficient lighting solutions for outdoor applications. The adequacy of outdoor lighting is an important factor in maintaining good security in parking lots and other outdoor areas. The inadequacy of exterior lighting has been the basis of many lawsuits alleging that the facility owner was negligent in providing a proper level of security. In older buildings, it is likely that the outdoor security lighting is inadequate largely because, for many years, lighting in parking lots and other outdoor areas was a low priority for lighting designers. Exterior lighting should be carefully designed, and sufficient thought should be given to its placement, intensity, timing, duration, and color. It should also meet the requirements of the Illuminating Engineering Society of North America (IES or IESNA).

Outdoor lighting used to illuminate statues, signs, flags, or other objects mounted on a pole, pedestal, or platform, and spotlighting or floodlighting used for architectural or landscape purposes, must use full cut-off or directionally shielded lighting fixtures that are aimed and controlled so that the directed light is substantially confined to the object intended to be illuminated. Facility evaluations are often required to identify inadequate exterior lighting conditions. Full cut-off lighting fixtures are required for all outdoor walkway, parking lot, canopy, and building/wall-mounted lighting, as well as all lighting fixtures located within those portions of open-sided parking structures that are above ground. An open-sided parking structure is one that contains exterior walls that are not fully enclosed between the floor and ceiling.

Many cities and towns around the country have enacted ordinances concerning "light pollution." These ordinances often set limits on the amount and type of light that can be used for outdoor parking. It is important to consult your local jurisdiction before making any changes in the lighting system. Moreover, to meet code requirements, automatic controls are typically required for all exterior lights. The control may be a directional photocell, an astronomical time switch, or a building automation system with astronomical time switch capabilities. It should automatically turn off exterior lighting when daylight is available. Lights in parking garages, tunnels, and other large covered areas that are required to be on during daylight hours are exempt from this requirement. Incandescent and high-intensity discharge is the most common type of exterior lighting. Illumination levels should be adequate and in good condition. The growth of trees and other types of landscaping is another challenge that may have to be addressed because it can have a significant impact on outdoor lighting. Often, a well-designed lighting system becomes ineffective due to tree growth to a point where large portions of the light is blocked out. This can be addressed by arranging to have trees and landscaping regularly trimmed so that the lighting system is not adversely affected.

Emergency Lighting

NFPA 101 2006 stipulates that emergency illumination (when required) must be provided for a minimum period of 1.5 hours to compensate for the possible failure of normal lighting. NFPA also requires emergency lighting to be arranged to provide initial illumination of not less than an average of 1 foot-candle and a minimum at any point of 0.1 foot-candle measured along the path of egress at floor level. In all cases, an emergency lighting system must be designed to provide illumination automatically in the event of any interruption of normal lighting (NFPA 101 2006 7.9.2.3). Emergency lighting and LED signs typically use relatively small amounts of energy and have a long life expectancy. Although LED fixtures may cost more than incandescent fixtures, reduced energy costs and labor savings will often quickly make up the difference.

For most facilities, an emergency lighting system is necessary in the event of a power failure or other emergency because it enables the occupants to exit safely. Emergency lighting can consist of individual battery units placed in all corridors and areas that may require sufficient lighting for building users exiting and in interior and some exterior exitways. These batteries are continuously recharged while the power is on, and take over when power is lost. Alternatively, the lighting can be powered by a central battery unit. Fluorescent lamps will require some method of power conversion as batteries are typically 12 volt.

9.5.4 Harmonics

Loads connected to electricity supply systems may be broadly categorized as either linear or nonlinear. There was a time when almost all electrical loads were linear—those that weren't made up such a small portion of the total that they had little effect on electrical system operation. That all changed, however, with the arrival of the solid-state electronic revolution. Today, we are immersed in an environment rich in nonlinear loads, including a variety of solid state devices, such as desktop computers, uninterruptible power supply (UPS) equipment, inverters, induction motors, variable-speed drives, and electronic fluorescent lighting ballasts. Operation of these devices represents a double-edged sword. While they provide greater efficiency, they can also cause serious consequences to power distribution systems, by creating high levels of harmonic distortion. In reality, total harmonic distortion is hardly perceptible to the human ear; however, even though the voltage distortion caused by increasing penetration of nonlinear loads is often accommodated without serious consequences, power quality is compromised in some cases unless steps are taken to address this phenomenon.

Although most of the loads connected to the electricity supply system draw power that is a linear (or near linear) function of the voltage and current supplied to it, these linear loads do not normally cause disturbance or adversely impact other users of the supply system. Some types of loads, however, cause a distortion of the supply voltage/current waveform due to their nonlinear impedance. Harmonic distortion can surface in electric supply systems through the presence of nonlinear loads of sufficient size and quantity. The severity of the problem depends on local and regional supply characteristics, the size of the loads, their quantity, and how they interact with each other. Utility companies are clearly concerned about emerging problems caused by increasing concentrations of nonlinear loads resulting from the growing proliferation of electronic equipment, particularly computers and their AC-to-DC power supply converters and electronic controllers. By taking a closer look at linear and nonlinear loads, we can get a better understanding of the hows and whys of this distortion.

Harmonic Reduction

Reducing harmonic voltage and current distortion from nonlinear distribution loads, adjustable-frequency drives (AFDs) can be achieved through several basic approaches. However, in the presence of excessive harmonic distortion, it is highly recommended to bring in a specialized consultant to correct the issue. Some of the methods used by harmonics specialists to reduce harmonic distortion may include a DC choke, line reactors, 12-pulse converters, 12-pulse distribution, harmonic trap filters, broadband filters, and active filters. Whatever approach is used, it must meet the guidelines of the IEEE. Furthermore, control of lower-order power-line harmonic emissions from nonlinear loads is rapidly becoming one of the most severe and complex electrical challenges facing the electrical industry, and is one that requires close cooperation between utilities, equipment manufacturers, premises owners, and end users if it is to be addressed.

9.6   SOLAR ENERGY SYSTEMS

Solar power, which at its simplest is the raw energy created by the sun's rays, can be either active or passive. Solar technologies use the sun's energy and light to provide heat, light, hot water, electricity, and even cooling for homes, businesses, and industry. In fact, many believe that passive and active solar energy is the energy source of the future, for nowhere on the planet can we find any other energy source as powerful without sacrificing our natural resources and environment.

9.6.1   Solar Options

There are many solar options today that can replace much of today's regular energy needs, saving money and benefiting the environment by cutting down on the use of fossil fuels. New technologies continue to be developed at a rapid pace that will help us harness this enormous natural asset.

Active Solar Energy

Solar electric systems are environmentally friendly because they do not generate emissions of greenhouse gases or other pollutants, and so do not have an adverse impact on global climate. They have proven to be reliable and are pollution-free. They make use of our most important renewable source of energy, the sun. Also, photovoltaic systems for homes and businesses are becoming more accessible and affordable.

Solar Photovoltaics

Photovoltaic (PV) materials convert sunlight into useful, clean electricity. By adding PV to your home or office, you can generate renewable energy, reduce your own environmental impact, enjoy protection from rising utility costs, and reduce greenhouse-gas emissions. Electricity is only one of many uses for solar energy. The sun, of course, is essential to your garden and it can heat water very cost effectively, but the most fundamental use of solar energy is in overall building design. Good design uses solar radiation to passively and/or actively heat your building, and to help keep it cool. Solar energy is also increasingly being used for street lighting (Figure 9.26). Building integrated photovoltaic systems (BIPV) offer additional design options, allowing electricity to be generated by windows, shades and awnings, roofing shingles, and PV-laminated metal roofing, for example. BIPV options can be used in retrofits or new construction.

As previously stated, solar energy is a renewable resource that is environmentally friendly, and unlike fossil fuels it is available in abundance, free, and immune to rising energy prices. The many ways that solar energy can be used include providing heat, lighting, mechanical power, and electricity. It helps minimize the impact of pollution from energy generation, which is considered to be the single largest contributor to global warming. Renewable energy can clean the air, stave off global warming, and help eliminate our nation's dependence on fossil fuels. The recent upsurge in consumer demand for clean, renewable energy and the deregulation of the utilities industry have spurred growth in green power: solar, wind, geothermal steam, biomass, and small-scale hydroelectric power sources. This energy demand is further being served by the emergence of small commercial solar power plants around the country.

For decades, solar technologies in the United States and around the world have used the sun's energy and light to provide heat, light, hot water, electricity, and even cooling, for homes, businesses, and industry. The types of renewable technologies available for a particular facility depend largely on the application and what sort of energy is required, as well as a building's design and access to the renewable energy source. Building facilities can use renewable energy for space heating, water heating, air conditioning, lighting, and refrigeration. Commercial facilities include assembly and meeting spaces, educational facilities,

FIGURE 9.26 Solar-powered LED street lighting with automatic on/off lasts for four to five nights after being fully charged. *(Source: Hankey Asia.)*

supermarkets, restaurants, hospitals and clinics, hotels and motels, stores and service businesses, offices, factories, and warehouses.

In the LEED Rating System, onsite renewable energy credits are not always easily achieved, particularly in urban locations. Essentially, you need to generate 2.5 to 7% of the building's electricity from wind, water, or solar energy, which, due to the many site constraints in a city environment, leaves us with little more than solar energy to focus on.

9.6.2 Solar System Basics

Solar electric systems, also known as PV systems, convert sunlight into electricity. When interconnected solar cells convert sunlight directly into electricity, they form a solar panel or "module," and several modules connected together electrically form an array. Most people picture a solar electric system as simply the solar array, but a complete system consists of several other components. The working of a solar collector is very simple (Figure 9.27). The energy in sunlight takes the form of electromagnetic radiation from infrared (long) to ultraviolet

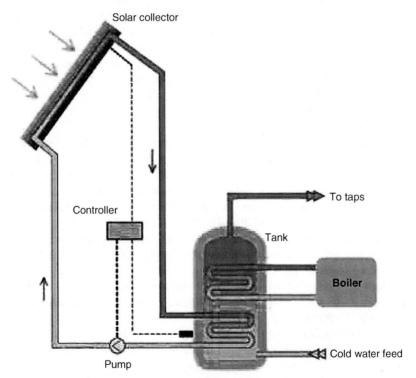

FIGURE 9.27 Solar thermal collector operation.

(short) wavelengths. This radiation heats a liquid that goes to a hot-water tank. The liquid heats the water and flows back to the solar collector.

The solar energy that strikes Earth's surface at any particular time largely depends on weather conditions, as well as location and orientation of the surface, but overall it averages approximately 1000 watts per 10 square feet (equivalent to 1 square meter) under clear skies with the surface directly perpendicular to the sun's rays. This solar thermal heat is able to provide hot water for an entire family during the summer. The collector size needed per person is just over 16 square feet (1.5 square meters). An average family of four people therefore needs a collector of about 65 square feet (6 square meters). The most common component equipment generally used in on-grid and off-grid solar-electric systems is described next, although systems vary and not all equipment is necessary for every system type. Indeed, understanding the basic components of PV systems and how they function is not particularly difficult.

Solar Electric Panels

Photovoltaic is the technical word for solar panels that create electricity. Photovoltaic material most commonly uses highly purified silicon to convert sunlight directly into electricity. When sunlight strikes the material, electrons are dislodged, creating an electrical current that can be captured and harnessed. The photovoltaic material can consist of several individual solar cells or a single thin layer, either of which makes up a larger solar panel. Panels are usually mounted on either a stationary rack or a tracking rack that follows the movement of the sun (Figure 9.28), and as they have no moving parts, solar electric panels operate silently. The life expectancy of a typical system is 40 to 50 years. Panels are generally warranted for 20 to 25 years.

PV has over recent years been making significant inroads as supplementary power for utility customers already served by the electric grid. In fact, grid-connected solar systems now comprise a larger market share than off-grid applications. However, compared to most conventional fuel options, photovoltaics remain a very small percentage of the energy make-up both within the United States and globally. Still, with increasing concerns about global warming, more and more individuals, companies, and communities are choosing PV for a variety of reasons, including environmental, economic, emergency backup, and fuel and risk diversification. The economics of a photovoltaic system for a home or business is not just the solar resource but rather a combination of the solar resource, electricity prices, and local/national tax and other incentives.

One of the critical aspects of solar design is siting. For example, solar panels like full sun, facing within 30 degrees of south and tilting within 30 degrees of the site's latitude. A 1-kilowatt system requires about 80 square feet of solar electric panels. Stationary racks can be roof or pole-mounted. Tracking racks are typically pole-mounted.

FIGURE 9.28   This photo, taken by Airman 1st Class Nadine Barclay of the Nellis Solar Power Plant, shows the largest photovoltaic power plant in North America. The 70,000 solar panels sit on 140 acres of unused base land in Nevada, forming part of a solar PV array that will generate in excess of 25 million kilowatt-hours of electricity annually and supply more than 25% of the power the base uses. *(Source: Wikimedia Commons; photo by Nadine Y. Barclay, October 2007)*

Inverter. An *inverter* converts the system's DC electricity produced by the PV modules into usable 120-volt AC electricity, which is the most common type for powering lights, appliances, and other needs. Grid-tied inverters are utilized to synchronize the electricity they produce with the grid's utility grade AC electricity, allowing the system to feed solar-made electricity to the utility grid. Inverters are typically warranted for 5 to 10 years.

Array mounting rack. Mounting racks provide a secure platform on which to anchor the PV panels, ensuring that they are fixed in place and correctly oriented. Panels can be mounted on a rooftop, on top of a steel pole set in concrete, or at ground level. A photovoltaic array is the complete power-generating unit, comprising one or more solar PV modules (solar panels) that convert sunlight into clean solar electricity. The solar modules need to be mounted facing the sun and avoiding shade for best results. Solar panels generate DC power, which can be converted to AC power with an inverter.

Wiring. Selecting the correct size and type of wire will enhance the performance and reliability of the PV system. The size of the wire must be sufficiently large to carry the maximum current expected without undue voltage losses.

Battery bank. This is used to store solar-produced electricity for evening or emergency backup power. Batteries may be required in locations that have limited access to power lines, as in some remote or rural areas. If batteries are part of the system, a *charge controller* may be required to protect the batteries from being overcharged or drawn down too low. Depending on the current and voltages for certain applications, the batteries are wired in series and/or parallel.

Charge controller. The main function of a charge controller is to protect the battery bank from overcharging. This is achieved by monitoring the battery bank, and when the bank is fully charged, the controller interrupts the flow of electricity from the PV panels. Modern charge controllers usually incorporate maximum power point tracking (MPPT), which optimizes the PV array's output, thereby increasing the energy it produces.

System meter. This is used to measure and display several different aspects of a solar electric system's performance and status, tracking how full the battery bank is; how much electricity the solar panels are producing or have produced; and how much electricity is in use.

Array DC disconnect. The DC disconnect is used to safely interrupt the flow of electricity from the PV array. It is an essential component when system maintenance or troubleshooting is required. The disconnect enclosure houses an electrical switch rated for use in DC circuits and, if required, may also integrate either circuit breakers or fuses.

Main DC disconnect. Battery-based systems require *disconnect switches* to allow the power from a solar electric system to be turned off for safety purposes during maintenance or emergencies. It also protects the inverter-to-battery wiring against electrical fires. A disconnect typically consists of a large, DC-rated breaker mounted in a sheet metal enclosure.

AC breaker panel. This is the point at which all a property's electrical wiring meets with the provider of the electricity, whether that is the grid or a solar electric system. The AC breaker panel typically consists of a wall-mounted panel or box that is normally installed in a utility room, basement, garage, or on the exterior of the building. It contains a number of labeled circuit breakers that route electricity to the various spaces throughout a structure. These breakers allow electricity to be disconnected for servicing, and also protect the building's wiring against electrical fires.

Kilowatt-hour meter. Homes and businesses with a grid-tied solar electric system will often have AC electricity both coming from and going to the electric utility grid. A bidirectional kWh meter is able to simultaneously keep track of how much electricity flows in each direction, which tells you how much electricity is being used and how much the solar electric system is producing.

Backup generator. Off-grid solar electric systems can be sized to provide electricity during cloudy periods when the sun doesn't shine. But sizing a system to cover a worst-case scenario, like several cloudy weeks during the winter, can result in an unduly large system that will rarely be used to its full capacity. Engine generators can be fueled with biodiesel, petroleum diesel, gasoline, or propane, depending on the design. These generators produce AC electricity that a battery charger (either standalone or incorporated into an inverter) converts to DC energy, which is stored in batteries.

9.6.3 Solar Energy Systems

Solar electric systems are attracting increasing attention because they are environmentally friendly and do not generate emissions of greenhouse gases or other pollutants, thereby reducing global climate impacts. Solar panels reflect a visible demonstration of concern for the environment, community education, and proactive forward thinking. The three most widely used types of solar electric systems are grid-tied, grid-intertied with battery backup, and off-grid (standalone). Each has distinct applications and component needs.

Grid-tied solar systems (alternating current), also known as on-grid, or grid-intertied, generate electricity for your home or business and route the excess power into the electric utility grid. This type of solar electric system does not require storage equipment (i.e., batteries) because it generates solar electricity and routes it to the electric utility grid, offsetting a home's or business' electrical consumption and, in some instances, even turning the electric meter backwards. Living with a grid-connected solar electric system doesn't really differ from living with grid power, except that some or all of the electricity used comes from the sun. The crucial issue relative to the photovoltaic panels is the technical aspects of tying into the electricity grid. Applications of this type require the use of grid-tied inverters that meet the requirements of the utilities.

It is important that the systems do not emit "noise," which can interfere with the reception of equipment (e.g., televisions); switch off in the case of a grid failure; and retain acceptable levels of harmonic distortion (i.e., quality of voltage and current output waveforms). This type of system tends to be an optimum configuration from an economic viewpoint because all the electricity is utilized by the owner during the day and any surplus is exported to the grid. Meanwhile, the cost of storage to meet nighttime needs is avoided because the owner simply draws on the grid in the usual way. Also, with access to the grid, the system does not need to be sized to meet peak loads. This arrangement is termed *net metering* or *net billing*. The specific terms of net metering laws and regulations vary from state to state and utility to utility, which is why for specific guidelines the local electricity provider or state regulatory agency should be consulted.

The standalone, grid-tied solar system with battery backup (alternating current) is the same as the grid-tied system except that battery storage (battery bank or generator backup) is added to enable power to be generated even when the electricity grid fails. Incorporating batteries into the system requires more components, is more expensive, and lowers the system's overall efficiency. But for homes and businesses that regularly experience utility outages or have critical electrical loads, having a backup energy source is invaluable. The additional cost to the customer can be quantified against the value of knowing that the power supply will not be interrupted.

The standalone off-grid solar electric system without energy storage (direct current) is a configuration (i.e., without any energy storage device) that consists of a PV system whose output is dependent on the intensity of the sun. In this system, the electricity generated is used immediately and therefore the application must be capable of working on both DC and variable power output. Standalone off-grid electric systems are most common in remote locations where there is no utility grid service. They operate independently from the grid to provide all the electricity required by a household or small business. The choice to live off-grid may be because of the prohibitive cost of bringing utility lines to remote locations, the appeal of an independent lifestyle, or the general reliability a solar electric system provides. However, those who choose to live off-grid often need to make adjustments to when and how they use electricity to allow them to live and work with the limitations of the system's capabilities.

To meet the greatest power needs in an off-grid location, the PV system may need to be configured with a small diesel generator. This increases its capability as it no longer has to be sized to cope with the worst sunlight conditions available during the year. The diesel generator can also provide backup power, but its use is minimized during the rest of the year to keep fuel and maintenance costs to a minimum.

For any module with a defined peak power, the actual amount of electricity in kilowatt hours that it generates will depend primarily on the amount of sunlight it receives. The electrical power output of a PV module is the current that it generates (dependent on its surface area) multiplied by the voltage at which it operates. The larger the module, or the solar array (the number of modules connected together), the more power is generated. A linear current booster (LCB) can be added to convert excess voltage into amperage to keep a pump running in low-light conditions. An LCB can boost pump output by 40% or more. For safety considerations, PV arrays are normally earthed.

With respect to energy production, each kilowatt produced by unshaded stationary solar electric panels generates about 1200 kilowatt-hours of electricity per year. A 1-kilowatt, dual-axis tracking system will generate about 1600 kilowatt hours per year. Power is generated during peak daylight hours. Solar power exhibits a very good peak coincidence with commercial building electrical loads. Dual-axis tracking systems, where the panels follow the sun, require periodic maintenance as other systems do.

Passive Solar Energy

Passive solar heating and cooling represent an important strategy for displacing traditional energy sources in buildings and are an effective method of heating and cooling through utilization of sunlight. The sun's energy arrives on earth in the primary forms of heat and light. To be successful, building designs must carefully balance their energy requirements with the building's site and window orientation. The term "passive" indicates that no additional mechanical equipment is used other than the normal building elements. Solar gains are generally introduced through windows, and minimum use is made of pumps or fans to distribute heat or effect cooling. Passive cooling minimizes the effects of solar radiation through shading or generating airflows with convection ventilation.

Correct building orientation, thermal mass, and insulation are specified in conjunction with careful placement of windows and shading. The thermal mass absorbs heat during the day and radiates it back into the space at night. To do this, passive solar techniques make use of building elements (e.g., walls, windows, floors, and roofs) in addition to exterior building elements and landscaping, to control heat generated by solar radiation. Solar heating designs collect and store thermal energy from direct sunlight in a manner that provides energy-efficient space and stable year-round temperatures, yet they are quiet and comfortable.

Daylighting design is another solar concept that optimizes the use of natural daylight and contributes greatly to energy efficiency. The quantity and quality of light around us help determine how well we see, work, and play. Light impacts our health, safety, comfort, morale, and productivity. Whether at home or in the office, it is possible to save energy and still maintain good light quantity and quality. However, there are many benefits to using passive solar techniques, including simplicity, price, and the design elegance of fulfilling one's needs with materials at hand. Some of the advantages of passive solar designs include the following:

- At little or no cost, it can easily be designed into new construction and can in some cases be retrofitted into existing buildings.
- Pays dividends over the life of the building through reduced or eliminated heating and cooling costs.
- Indoor air quality is improved through elimination of forced-air systems.
- Sites with good southern exposure are most suitable.
- Retrofitting is rarely as effective as initial design.

LEED offers credits in its Indoor Environmental Quality section, Daylight and Views. The *intent* of the credits appears to be to reduce electric lighting, increase productivity, and provide building occupants with a connection between indoor and outdoor spaces by incorporation of daylight and views into regularly occupied spaces.

The LEED requirements for these credits are to achieve daylight (through computer simulations) in a minimum of 75 or 90% of regularly occupied spaces, and to achieve a daylight illuminance level of a minimum of 25 foot-candles and a maximum of 500 foot-candles in a clear sky condition on September 21 at 9:00 am and 3:00 pm. A combination of side lighting and/or top lighting may be used to achieve the total daylighting zone required, which is at least 75% of all the regularly occupied spaces. Sunlight redirection and/or glare control devices should be provided to ensure daylight effectiveness, avoid high-contrast situations, and avoid impeding of visual tasks. Exceptions for areas where tasks would be hindered by daylight will be considered on their merits. It should be stressed that the USGBC Reference Guide or its website should be consulted for the latest updated requirements, including possible exemplary performance credits.

9.7 FEDERAL TAX CREDITS FOR ENERGY EFFICIENCY

There are many tax credits and incentives available for *consumers*.[1] For example, one is 10% of cost up to $500 or a specific amount from $50 to $300. This expired on December 31, 2011. It had to be for an existing home and your principal residence. New construction and rentals did not qualify. This was applicable for

- Windows and doors
- Insulation
- Roofs (metal and asphalt)
- HVAC
- Water heaters (nonsolar)
- Biomass stoves

Tax credits are available at 30% of the cost, with no upper limit, through 2016 for existing homes and new construction. Second homes qualify but rentals do not. This is applicable for (1) geothermal heat pumps, (2) solar energy systems, and (3) small wind turbines.

Tax credits are available for 30% of the cost, up to $500 per 0.5 kW of power capacity, through 2016 for existing homes and new construction. The home must be your principal residence. Rentals and second homes do not qualify. This is applicable for fuel cells and microturbine systems.

A tax deduction for *commercial buildings* of up to $1.80 per square foot is available to owners or designers of new or existing commercial buildings that save at least 50% of the heating and cooling energy of a building that meets ASHRAE Standard 90.1-2001. Partial deductions of up to $0.60 per square foot can be taken for measures affecting any one of three building systems: the building envelope, lighting, or heating and cooling. These tax deductions are available for systems placed in service from January 1, 2006 through December 31, 2013. In addition, there are many other tax credits for efficient cars, home builders, home improvements, and so forth.

---

1. From ENERGY STAR—U.S. Environmental Protection Agency and U.S. Department of Energy.

9.8   FIRE SUPPRESSION AND PROTECTION SYSTEMS

According to the Fire Suppression Systems Association (FSSA) 43% of businesses closed by fire never reopen and another 29% fail within three years. These are stark statistics. Fire suppression systems are used in conjunction with smoke detectors and fire alarm systems to improve and increase public safety. Suppression systems are governed by the codes in the NFPA 13 Handbook and include: sprinkler systems (wet, dry, pre-action, and deluge), gaseous agents, and wet/dry chemical agents. When planning for fire protection, an integrated approach is needed in which system designers analyze the building's components as a total package. To achieve an optimum symbiosis between these components, an experienced system designer, such as a fire protection engineer, should be involved in the very early stages of the planning and design process. With the increasing number of high-rise, high-performance buildings being built both nationwide and globally, the planning for fire protection has taken on a real urgency. We should start seeking out sustainable environmentally friendly fire suppression approaches to reduce environmental impacts during design and testing and also to help a project earn LEED credits.

Fire protection systems play an increasingly pivotal role in overall building design and construction and should never be compromised, because they serve the purpose of life safety. Indeed, it is frequently argued that the life safety system is the most important system to be evaluated in a facility, particularly when it comes to high-rise structures. Furthermore, like any other building system, green concepts and specifications can be applied to their design, installation, and maintenance in a manner that reduces their harmful impacts on the environment. Moreover, there have been significant advances recently in fire detection technology and fire suppression systems, in addition to an ongoing development of international and national codes and standards, all of which have made possible the "greening" of fire-safety systems and are taking on increasing importance for building owners and property developers.

For optimum efficiency, the various components of modern fire protection systems should work in cohesion to detect, contain, control, and/or extinguish a fire in its early stages—and to survive during the fire. And the installation of environmentally friendly fire protection technology can help earn credits under LEED for new or retrofitted buildings. A facility's type, size, and function will generally determine the complexity of the life safety system used.

In some smaller structures, the system may comprise only smoke detectors and fire extinguishers. In larger, more complex buildings, a complete fire suppression system, such as fire sprinklers, is installed throughout the facility. An important aspect in the assessment of any life safety system includes verification that periodic maintenance, inspection, and testing of the main components of the system are being conducted. Figure 9.29 illustrates several types of life safety systems normally employed to address fire safety requirements. Each of these gives rise to their own issues that need to be taken into account in facility surveys.

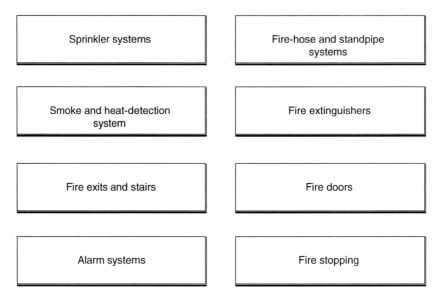

FIGURE 9.29 Typical fire suppression system components.

The extent of a life safety system survey and the expertise required to perform such an evaluation varies greatly from facility to facility depending on its size and complexity. Fire detection and prevention technologies have become increasingly sophisticated, intelligent, and powerful in recent years. Frank Monikowski and Terry Victor, of SimplexGrinnell, highlight some of the advances and emerging technologies that can be found in today's life safety systems in the following list:

- *Control-mode sprinklers*—standard manufactured sprinklers that limit fire spread and stunt high heat release rather than extinguish a fire; they also "pre-wet" adjacent combustibles.
- *Suppression sprinklers*—operate quickly for high-challenge fires, and are expected to extinguish a fire by releasing a high density of water directly to the base of the fire.
- *Quick-response sprinklers*—provide quicker response and are now required for all light-hazard installations.
- *Residential sprinklers*—designed specifically to increase the survivability of an individual who is in the room where a fire originates.
- *Extended-coverage sprinklers*—designed to reduce the number of sprinklers needed to protect a given area. These come in quick-response, residential, and standard-response types, and are also available for both light- and ordinary-hazard occupancies.

- *Special sprinklers, such as early suppression fast response (ESFR)*— designed for high-challenge rack storage and high-pile storage fires. In most cases, these sprinklers can eliminate the expense and resources needed to install in-rack sprinkler heads.
- *Low-pressure sprinklers*—provide needed water coverage in multistory buildings where pressure may be reduced. These low-pressure sprinklers bring a number of benefits: reduced pipe size, reduction or elimination of a fire pump, and overall cost savings.
- *Low-profile, decorator, and concealed sprinklers*—designed to be more aesthetically pleasing.
- *Sprinkler system valves*—smaller, lighter, and easier to install and maintain and, therefore less costly.
- *Fluid delivery time computer program*—simulates water flowing through a dry system in order to accurately predict critical "water-to-fire" delivery time for dry-pipe systems.
- *Cost-efficient CPVC piping*—for light-hazard and residential sprinkler systems.
- *Advanced coatings on steel pipes*—designed to resist or reduce micro-biologically influenced corrosion (MIC) and enhance sprinkler system life.
- *Corrosion monitoring devices*—alert users of potential problems.
- *More efficient coordination in evaluating building sprinkler system need*— including site surveys, accurate measurements, and the use of CAD and hydraulics software to ensure that fire sprinkler system designs respond to the specific risks and the physical layout of the premises.

According to the NFPA, more than 43 million people in the United States have a disability, which is partly why the NFPA recently developed and issued a new *Emergency Evacuation Planning Guide for People with Disabilities*. This document provides general information to assist designers in identifying the needs of people with disabilities related to emergency evacuation planning. It covers five general categories of disabilities: *mobility impairments, visual impairments, hearing impairments, speech impairments, and cognitive impairments*. The four elements of evacuation information needed by occupants are *notification, way finding, use of way,* and *assistance*.

9.8.1 LEED Contributions

Fire sprinklers have been stopping fire growth and minimizing greenhouse and toxic gas production for over a century. Yet they are not given any credit in the USGBC's LEED certification program. In fact, fire suppression systems are only indirectly referenced in LEED certification documents. For example, with LEED for New Construction (LEED-NC) V3 Energy and Atmosphere (EA) Credit 4, Enhanced Refrigerant Management, and LEED for Existing Buildings: Operations & Maintenance (EBOM) V3 EA Credit 5, Refrigerant

Management, the intent is reducing ozone depletion, supporting compliance with the Montreal Protocol, and minimizing direct contributions to global warming.

It appears that credits can be earned with the installation/operation of fire suppression systems that do not contain ozone-depleting substances such as chlorofluorocarbons (CFCs), hydrochlorofluorocarbons (HCFCs), and halons. Likewise, LEED credits in the Innovation in Design category can be obtained for fire suppression systems. The *LEED Reference Guide* in the relevant category should be consulted, but generally, to earn those points, it is necessary to document and substantiate the innovation and design processes used. Dominick G. Kasmauskas, who is with the National Fire Sprinkler Association, says that "The fire sprinkler industry plans to work with the USGBC to develop a credit for fire sprinklers in future editions of LEED based on the environmental benefits of sprinkler systems."

9.8.2 Sprinkler Systems

Since the dawn of history, people have used water as an extinguishing agent, and today it is still the preferred choice for modern fire protection in the form of sprinklers and other methods. Sprinklers are the most common, widely specified, and most effective fire suppression system in commercial facilities—particularly in occupied spaces. The various types of sprinkler systems are outlined next. In situations where sprinklers are not feasible because of special considerations (e.g., water from sprinklers would damage sensitive equipment or inventory) alternative fire suppression systems might be decided on, such as gaseous/chemical suppression. In the final analysis, the type of sprinkler system used depends largely on a building's function.

Automatic Sprinkler Systems

Optimized sprinkler system designs offer an effective means of addressing environmental impact and sustainability. They are the most common, widely specified, and most effective fire suppression system in commercial facilities, particularly in occupied spaces. Furthermore, automatic sprinkler systems are now not only required in new high-rise office buildings, but in many American cities they are mandated by code that existing high-rises be retrofitted with them. This is because sprinklers are the most widely specified and most effective fire suppression system for commercial facilities, particularly in occupied spaces. There are several types of sprinkler systems that are commonly used; these include wet- and dry-pipe, pre-action, deluge, and fire cycle. Of these, wet-pipe and dry-pipe are the most common. In a wet-pipe system, the sprinklers are connected to a water supply, enabling immediate discharge of water at sprinkler heads triggered by the heat of the fire. In a dry-pipe system, the sprinklers are under air pressure that, when the pressure is eased by the opening of the sprinkler heads, fills the system with water.

Careful attention should be given to proper connections for flow and flow testing when designing automatic sprinkler systems. Likewise, flexible connections and arm-overs may be employed to provide a means for facilitating the relocation of sprinklers with minimal need for additional materials if the system designer incorporates appropriate flow restrictions due to friction losses. As mentioned earlier, sprinklers are not feasible due to special considerations sometimes, so alternative fire suppression systems, such as gaseous/chemical suppression, may be considered. But in the final analysis, the type of sprinkler system decided on depends mainly on a building's function. Of note, the majority of today's fire sprinklers incorporate the latest advances in design and engineering technologies, thereby providing a very high level of life safety and property protection. The features and benefits now available are making fire sprinkler systems more efficient, reliable, and cost effective. And as the benefits of sprinkler systems become better understood and more obvious, and the cost more affordable, their installation in residential structures is becoming more common. However, these sprinkler systems typically fall under a residential classification and not a commercial one.

The main difference between commercial and residential sprinkler systems is that a commercial system is designed to protect the structure and the occupants from a fire, whereas most residential systems are primarily designed to suppress a fire in a manner that allows for the safe escape of the building occupants. While these systems will often also protect the structure from major fire damage, this consideration nevertheless remains of secondary importance. In residential structures, sprinklers are typically omitted from closets, bathrooms, balconies, and attics because a fire in these areas would not normally impact an occupant's escape route. When a system is operating as intended, it is highly reliable, but like any other mechanical system, sprinkler systems require periodic maintenance and inspection in order to sustain proper operation. In the rare event a sprinkler system fails to control a fire, the root cause of failure has often been found to be the lack of proper maintenance. Figure 9.30 is an example of a fire sprinkler control valve assembly including pressure switches and valve monitors.

Wet-Pipe Systems

Wet-pipe sprinkler systems are the most common and have the highest reliability. Wet systems are typically used in buildings where there is no risk of freezing. The systems are simple, with the only operating component being the automatic sprinkler. A water supply provides pressure to the piping, and all of the piping is filled with water adjacent to the sprinklers. The water is held back by the automatic sprinklers (Figure 9.31) until activated. When one or more of the automatic sprinklers is exposed to sufficient heat, the heat-sensitive element releases, allowing water to flow from that sprinkler. Each sprinkler

FIGURE 9.30   Fire sprinkler control valve assembly including pressure switches and valve monitors. *(Source: Wikipedia.)*

operates individually. Sprinklers are manufactured to react to a specific range of temperatures, and only sprinklers subjected to a temperature at or above their specific temperature rating will operate. Figure 9.32 is a drawing of a typical wet-pipe sprinkler system. The principal disadvantage of these systems is that they are not suited for subfreezing environments.

Dry-Pipe Systems

This is the second most common sprinkler system type currently in use after the wet-pipe system. A dry-pipe sprinkler system is one in which pipes are filled with pressurized air or nitrogen rather than water. This air holds a remote valve, known as a dry-pipe valve, in a closed position. The dry-pipe valve is located in a heated space and prevents water from entering the pipe until a fire causes one or more sprinklers to be activated. Once this happens, the air escapes and the dry-pipe valve releases. To prevent the larger water supply pressure from forcing water into the piping system, the design of the dry-pipe valve intentionally includes a larger valve clapper area exposed to the

FIGURE 9.31 Ceiling-mounted sprinkler head. *(Source: Sujay Fire & Safety Equipment.)*

specified air pressure, as compared to the water pressure. Water then enters the pipe, flowing through open sprinklers onto the fire. However, regulations (e.g., NFPA 13 2007 ed. Sections 7-2 and A7-2) typically stipulate that these systems can only be used in spaces in which the ambient temperature may be cold enough to freeze the water in a wet-pipe system, thus rendering it inoperable. For this reason, we often find dry-pipe systems used in unheated buildings and refrigerated coolers.

Activation of the system takes place when one or more of the automatic sprinklers are exposed to sufficient heat, allowing the maintenance air to vent from that sprinkler. Each sprinkler operates individually. As the air pressure in the piping drops, the pressure differential across the dry-pipe valve changes, allowing water to enter the piping system. Delays can be experienced in dry-pipe systems, since the air pressure must drop before the water can enter the pipes and suppress the fire. Dry-pipe systems are therefore not as effective as wet-pipe systems in fire control during the initial stages of the fire, although to aid in faster activation, dry-pipe valves may employ quick opening devices connected to them.

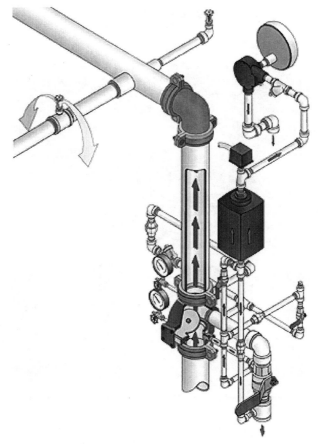

FIGURE 9.32   Typical wet-pipe sprinkler system.

Deluge and Pre-Action Systems

Deluge and pre-action systems are less common and are used mainly in environments that require special sprinkler protection. They are activated by fire detection systems. Deluge and pre-action systems represent only a small percentage of market share of the total sprinkler systems currently in operation. Deluge systems are similar to pre-action systems except that the sprinkler heads are open and the pipe is not pressurized with air. This means that in a deluge system the heat-sensing operating element is removed during installation so that all sprinklers connected to the water-piping system remain open by the operation of a smoke or heat detection system. These detection systems are normally installed in the same area as the sprinklers so that when the detection system is activated water readily discharges through all of the sprinkler heads.

These systems are typically used in high-hazard areas because they provide a simultaneous application of water over the entire hazard, and where rapid fire spread is a major concern such as power plants, aircraft hangars, and chemical storage facilities. Water is not present in the piping until the system operates. Because the sprinkler orifices are open, the piping is at ambient air pressure. To prevent the water supply pressure from forcing water into the piping, a *deluge valve* is used in the water supply connection, which is a mechanically latched nonresetting valve that stays open once tripped. Because the heat-sensing elements in the automatic sprinklers have been removed (resulting in open sprinklers), the deluge valve must be opened as signaled by a specialized fire alarm system. The type of fire alarm activation device used is based largely on the hazard (e.g., smoke detectors or heat detectors). The activation/initiation device signals the fire alarm panel, which in turn signals the deluge valve to open. Activation can also be achieved manually, depending on the system goals. Manual activation is usually via an electric or pneumatic fire alarm pull station, which signals the fire alarm panel, which in turn signals the deluge valve to open, allowing water to enter the piping system. Water flow effectively takes place from all sprinklers simultaneously.

Pre-action sprinkler systems are specialized systems that combine a fire detection system with a sprinkler system. They are typically used to ensure reliable protection against false alarms in locations where accidental activation is undesired, such as in museums containing rare art works or computer suites. These sprinkler systems employ the basic concept of a dry-pipe system in that water is not normally contained in the pipes. They differ from the dry-pipe system, however, in that water is held from entering the piping by an electrically operated valve, known as a pre-action valve. Valve operation is controlled by a signal from the fire detection system, not by a fall in pressure after a sprinkler has opened. If there is a fault in the fire detection system, a pre-action system is switched over to operate as a normal dry system. Pre-action systems are hybrids of wet, dry, and deluge systems, depending on the exact system goal.

Pre-action systems can be either single interlock or double interlock. The operation of single interlock systems is similar to a dry system except that these systems require a "preceding" and supervised event (typically the activation of a heat or smoke detector) to activate water introduction into the system's piping due to opening of the pre-action valve (i.e., a mechanically latched valve). The operation of double interlock systems is similar to a deluge system except that automatic sprinklers are used. Following detection of the fire by the fire alarm system, it basically converts from a dry system into a wet system.

Water Mist Systems

Water mist systems consist of environmentally friendly components that normally force water and pressurized gas together through stainless steel tubes that are much narrower in diameter than pipes used in traditional sprinkler systems.

The water mist system produces a fine mist with a large surface area that absorbs heat efficiently through vaporization. It is totally safe for humans because it uses water as the extinguishing medium. With these systems, fires are suppressed using three main mechanisms:

- As the water droplets contact the fire, they convert to steam. This process absorbs energy from the surface of the burning material.
- As the water turns into steam, it expands greatly. This removes heat and lowers the temperature of the fire and the air surrounding it.
- The water and the steam act to block the radiant heat and prevent the oxygen from reaching the fire (thus starving it of oxygen) so the fire smothers.

Water mist systems can be useful for suppressing fires in gas turbine enclosures and machinery spaces and are FM-approved (i.e., Factory Mutual) for such applications. They are ideally suited for cultural heritage buildings where large amounts of water can potentially cause unacceptable damage to irreplaceable items and in retrofits where space is often limited. Water mist systems are also often used to protect passenger cruise ships where the system's excellent performance and low total system weight have made them very popular.

Foam-Water Systems

A foam-water sprinkler system is a special application that discharges a mixture of water and low-expansion foam concentrate, resulting in a foam spray from the sprinkler. These systems are generally more economical than a water-only system, when evaluated for the same risk, and provide for actual extinguishment of the fire and a lower water demand. The conversion assists in the reduction of property loss and loss of life and, in many cases, reduction of insurance rates. But while foam concentrates and expanded foams are generally considered to be safe with regard to exposure to humans, they can, unless specifically indicated, adversely impact the environment if allowed to flow freely into watershed areas. The base properties of typical foaming agents include nitrates, phosphorous, and organic carbon. It should be noted that the use of halons in fire suppression systems was phased out in the early 1990s to comply with the Montreal Protocol because they were determined to cause significant damage to the ozone layer. Moreover, halons have a long life in the atmosphere and a high global warming potential.

One of the characteristics of a foam-water system is that almost any sprinkler system—wet, dry, deluge, or pre-action—can be readily adapted to include the injection of aqueous film-forming foam (AFFF) foam concentrate to combat high-risk situations. These systems are typically used with special-hazard occupancies associated with high-challenge fires, such as flammable liquids and airport hangars. Added components to the sprinkler system riser include bladder tanks to hold the foam concentrate, concentrate control valves to isolate the sprinkler system from the concentrate until activation, and proportioners for mixing the appropriate amount of foam concentrate with the system supply water. The main standard that delineates the minimum requirements for the design, installation, and maintenance of foam-water sprinkler and spray systems is NFPA 16: Standard, Foam-Water Sprinkler and Foam-Water Spray Systems.

The checklist that follows is provided by the New York Property Insurance Underwriting Association to help identify general problems that may arise in typical sprinkler systems. The list is intended to identify what is required to be done and to ensure that the sprinkler system is properly maintained.

- Are sprinkler heads free of paint, dust, and grease?
- Are the sprinkler heads obstructed by stored material? There should be no less than 18 inches of clearance at each head. Obstructions will diminish the operation of the head.
- Are the sprinkler pipes used to support lighting or other objects?
- Are there extra sprinkler heads and wrenches located at the control area for maintenance purposes?
- Is the OS&Y valve chained in an open position to avoid disabling of the system?
- Are the sprinkler heads directed properly for their location?
- Is there a sprinkler contractor that supervises and inspects the system as required by NFPA and ISO? Is a service log maintained?
- Are the sprinkler alarms activated to protect your property in the event of accidental discharge or fire?
- Has the occupancy classification of the material in the building changed since its installation so that the sprinkler system is now ineffective?
- Is the heat supply in the premises adequate for the operation of a wet-pipe system?

Fire Hose and Standpipe Systems

Michael O'Brian, president of Code Savvy Consultants, says:

Standpipes are a critical tool that requires preplanning on first responding apparatus in order to be used effectively. The initial approval process for these systems is critical and the fire prevention bureau can assist responding crews by ensuring proper installation and maintenance of these systems. Standpipe systems vary in design, use, and location. These factors vary based on the adopted code; the use, size, and type of building they are installed in. Typically, model codes refer to NFPA 14, Standpipe and Hose Systems, for the design, installation, and maintenance of these systems.

Standpipe systems consist of piping, valves, outlets, and related equipment designed to provide water at specified pressures and installed exclusively for the fire department or trained occupant to use for fighting fires. These systems are used in conjunction with sprinklers or hoses and basically consist of a water pipe riser running vertically through the building, although sometimes a building is provided with only piping for the standpipe system.

Standpipe systems can be wet or dry. Dry systems are normally empty and are not connected to a water source. A Siamese fitting is located at the bottom end of the pipe, allowing the fire department to pump water into the system. In a wet-type system, the pipe is filled with water and attached to a tank or pump. This type also contains Siamese fittings for the fire department's use. O'Brian says:

Many buildings are required to have an Automatic Class I standpipe system with a design pressure of 100 psi. Based on friction loss, municipal water supply, and pressure loss for the height of the standpipe a fire pump may need to be designed into the system. Due to the pressure requirements standpipes are limited to a maximum height of 275 feet. Those buildings over 275 feet in height will require the standpipe systems to be split in different pressure zones.

Model fire and building codes stipulate, among other things, the requirements for the installation of standpipe systems. The specific type of system is based on the occupancy classification and building height. The three main classifications of standpipe systems are as follows:

- *Class I* signifies that the standpipe is equipped with a 2.5-inch fire hose connection for fire department use and those trained in handling heavy fire streams. These connections must match the hose thread utilized by the fire department and are typically found in stairwells of buildings.
- *Class II* signifies that the standpipe is directly connected to a water supply and serves a 1.5-inch fire hose connection that provides a means for the control or extinguishment of incipient-stage fires. They are typically found in cabinets and are intended for trained occupant use and are spaced according to the hose length. The hose length and connection spacing are intended for all spaces of the building.
- *Class III* indicates a standpipe system that combines both Class I and II connections directly connected to a water supply and is for the use of in-house personnel capable of furnishing effective water discharge during the more advanced stages of fire in the interior of workplaces. Many times these connections will include a 2.5-inch reducer to a 1.5-inch connection.

When a standpipe system *control valve* is located within a stairwell, the maximum length of hose should not exceed 100 feet. If the control valve is located in areas other than the stairwell, the length of hose should not exceed 75 feet. Code requires that fire hoses on Class II and Class III standpipe systems be equipped with a shutoff-type nozzle.

9.8.3 Hand-Held Fire Extinguishers

There are several different classifications of fire extinguishers, each of which extinguishes specific types of fire (Figure 9.33). Newer fire extinguishers use a picture/labeling system to designate which types of fires they are to be used on, whereas older fire extinguishers are labeled with colored geometrical shapes

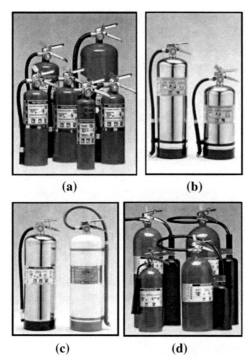

FIGURE 9.33   Various types of fires extinguishers are in common use: (a) MP, HD, and DC series multipurpose dry chemical; (b) WC-series wet chemical, 2-1/2 and 6L; (c) PW 2-1/2 and WM 2-1/2 series water mist; and (d) CD series carbon dioxide. *(Source: Larsen's Manu-facturing Co.)*

with letter designations (Figure 9.34). These labels indicate suitability for use as Class A, B, and C fire extinguishers.

Classification of Hand-Held Extinguishers

The U.S. Department of Labor says, "Portable fire extinguishers are classified to indicate their ability to handle specific classes and sizes of fires. Labels on extinguishers indicate the class and relative size of fire that they can be expected to handle." These classifications are as follows:

Class A fire extinguishers are designed to put out fires caused by organic solids and ordinary combustibles such as wood, textiles, paper, some plastic, and rubber. The numerical rating for this class of fire extinguisher refers to the amount of water the fire extinguisher holds and the amount of fire it will extinguish. To extinguish a Class A fire, extinguishers utilize either the heat-absorbing effects of water or the coating effects of certain dry chemicals. Class A fire extinguishers should be clearly marked with a triangle containing the letter "A." If in color, the triangle should be green.

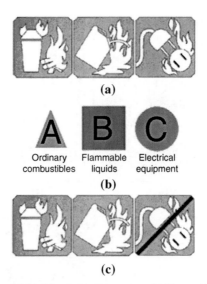

FIGURE 9.34 (a) New- and (b) old-style labeling systems. (c) Diagonal line through picture indicates extinguisher should not be used for an electrical fire.

Class B fire extinguishers are used to put out fires involving flammable and combustible liquids and gases. They work by starving the fire of oxygen and interrupting the fire chain by inhibiting the release of combustible vapors. Class B fires include gasoline, oil, and paraffin. The numerical rating for this class of fire extinguisher states the approximate number of square feet of a flammable liquid fire that a nonexpert can expect to extinguish. This includes all hydrocarbon- and alcohol-based liquids and gases that will support combustion. Class B fire extinguishers should be clearly marked with a square containing the letter "B." If in color, the square should be red.

Class C fire extinguishers are most effective for use on fires that involve live electrical equipment where a nonconducting material is required. This class of fire extinguishers does not have a numerical rating, but the presence of the letter "C" indicates that the extinguishing agent is nonconductive. Class C fire extinguishers should be clearly marked by a circle containing the letter "C." If in color, the circle should be blue.

Class D fire extinguishers are special types designed and approved for specific combustible materials (metals), such as magnesium, titanium, zirconium, potassium, and sodium, that require an extinguishing medium that does not react with the burning metal. Class D fire extinguishers should be clearly marked by a five-point painted star containing the letter "D." If in color, the star should be yellow. These extinguishers generally have no rating and are not given a multipurpose rating for use on other types of fires.

Class K fire extinguishers are effective for fighting fires involving cooking fats, grease, oils, and so forth, in commercial cooking environments. These fire extinguishers work on the principle of saponification, which takes place when alkaline mixtures such as potassium acetate, potassium citrate, or potassium carbonate are applied to burning cooking oil or fat. The alkaline mixture combined with the fatty acid creates soapy foam on the surface, which holds in the vapors and steam and extinguishes the fire. Class K fire extinguishers should be clearly marked with the letter "K."

Labeling

If a multipurpose extinguisher is being used and in order for users to be able to quickly identify the classification of a fire extinguisher in the event of an emergency, each unit should be clearly labeled. The approved marking system combines pictographs of both recommended and unacceptable extinguisher types on a single identification label. Many extinguishers available today can be used on different types of fires and will be labeled with more than one designator, for example, A-B, B-C, or A-B-C. It should also be noted that British standards and classifications differ slightly from American standards and classifications.

Types of Fire Extinguishers

There are several types of fire extinguishers, the most important being the following:

Dry chemical extinguishers come in a variety of types and are usually rated for multipurpose use (class A, B, and C fires). They are filled with a foam or powder extinguishing agent and use a compressed, nonflammable gas as a propellant. One advantage a dry chemical extinguisher has over a CO_2 extinguisher is that it leaves a nonflammable substance on the extinguished material, reducing the likelihood of re-ignition.

Water or APW (air-pressurized water) extinguishers are filled with water and pressurized with oxygen. AWP extinguishers should only be used on Class A fires (ordinary combustibles) and never on grease, electrical, or Class D fires—the flames will only spread and likely make the fire bigger.

Carbon dioxide (CO_2) extinguishers contain carbon dioxide, a nonflammable gas, and are highly pressurized. They are most effective on Class B and C (liquids and electrical) fires. Since the gas disperses quickly, these extinguishers are only effective from 3 to 8 feet. The carbon dioxide is stored as a compressed liquid in the extinguisher; as it expands, it cools the surrounding air. The cooling will often cause ice to form around the "horn," where the gas is expelled from the extinguisher. However, they don't work very well on Class A fires because they may not be able to displace enough oxygen to put the fire out, causing it to re-ignite. The advantage of CO_2 extinguishers have over dry chemical extinguishers is that they don't leave a harmful residue and may therefore be a good choice for an electrical fire on a computer or other favorite electronic device such as a stereo or TV.

Halon extinguishers contain a gas that interrupts the chemical reaction that takes place when fuels burn. Halon is an odorless, colorless gas that can cause asphyxiation, and halon extinguishers have a limited range, usually 4 to 6 feet. An advantage of halon is that it is clean, leaving no corrosive or abrasive residue after release. This minimizes cleanup, which makes halon more suitable for valuable electrical equipment, computer rooms, telecommunication areas, theaters, and so forth. However, pressurized fire suppression system cylinders can be hazardous and if not handled properly are capable of violent discharge. Moreover, the cylinder can act as projectile, potentially causing injury or death. Halon has been banned from new production, except for military use, since January 1, 1994 because its properties contribute to ozone depletion and long atmospheric lifetime, usually 400 years. However, Halon reuse is still permitted in the United States.

NFPA Code 10 addresses all issues pertaining to portable fire extinguishers and contains clear, widely accepted rules for distribution and placement, maintenance, operation, inspection, testing, and recharging. Recognized as a first line of defense against fires, portable extinguishers, when maintained and operated properly on a small containable fire, can prevent the fire from spreading beyond its point of origin. NFPA Code 10 requires owners of extinguishers to have monthly inspections performed and to maintain records of the inspections.

9.8.4   Smoke and Heat Detection Systems

Smoke and heat detection systems play a pivotal role in green buildings. Kate Houghton, director of marketing for Kidde Fire Systems, says:

By detecting a fire quickly and accurately (i.e., by not sacrificing speed or causing false alarms) and providing early warning notification, a fire-detection system can limit the emission of toxic products created by combustion, as well as global-warming gases produced by the fire itself. These environmental effects often are overlooked, but undoubtedly occur in all fire scenarios. Therefore, reducing the likelihood of a fire is an important part of designing a green building.

A *smoke detector* or *smoke alarm* is a device that detects smoke and issues an alarm to alert nearby people of the threat of a potential fire.

Smoke alarms that are properly installed and maintained play a critical role in reducing fire deaths and injuries. A household smoke detector will typically be mounted in a disk-shaped plastic enclosure about 150 mm in diameter and 25 mm thick, but the shape can vary by manufacturer (Figure 9.35). Because smoke rises, most detectors are mounted on the ceiling or on a wall near the ceiling. It is imperative that smoke detectors are regularly maintained and checked that they operate properly. This will ensure early warning to allow emergency response well before a fire causes serious damage. It is not uncommon for modern

FIGURE 9.35 Ceiling-mounted smoke detector. *(Drawing courtesy of Scott Easton.)*

types of systems to detect smoldering cables or overheating circuit boards. Smoke detectors are typically powered by one or more batteries, but some can be connected directly to a building's wiring. Often the smoke detectors that are directly connected to the main wiring system also have a battery as a power supply backup in case the facility's wiring goes out. Batteries should be checked and replaced periodically to ensure appropriate protection. Early detection can save lives and help limit damage and downtime. Laws governing the installation of smoke detectors may differ from one jurisdiction to another.

Most smoke detectors work either by optical detection or by ionization, and in some cases both detection methods are used to increase sensitivity to smoke. A complete fire protection system will typically include spot smoke detectors that can signal a fire control panel to deploy a fire suppression system. Smoke detectors can either operate alone or be interconnected to cause all detectors in an area to sound an alarm if one is triggered, or be integrated into a fire alarm or security system. Smoke detectors with flashing lights are also available for the deaf or hearing impaired. A smoke detector cannot detect carbon monoxide to prevent carbon monoxide poisoning unless it comes with an integrated carbon monoxide detector.

Aspirating smoke detectors (ASD) can detect combustion at the early stages, and are 1000 times more sensitive than conventional smoke detectors, giving early warning to building occupants and owners. An ASD generally consists of a central detection unit that sucks air through a network of pipes to detect smoke, and in most cases requires a fan unit to draw in a representative sample of air from the protected area through its network of pipes. Although ASDs are extremely sensitive, and are capable of detecting smoke before it is even visible to the human eye, their use is not recommended in environments that are unstable due to the wide range of particle sizes that are detected.

Optical smoke detectors are light sensors that include a light source (infrared LED), a lens to collimate the light into a beam like a laser, and a photodiode or other photoelectric sensor at right angles to the beam as a light detector. Under normal conditions (i.e., in the absence of smoke) the sensor device detects no light signal and therefore produces no output. The source and the sensor device are arranged so that there is no direct "line of sight" between them. When smoke enters the optical chamber into the path of the light beam, some light is scattered by the smoke particles, and some of the scattered light is detected by the sensor; the alarm is set off by the increased input of light into the sensor.

Projected beam detectors are employed mainly in large interior spaces, such as gymnasia and auditoria. A unit on the wall transmits a beam, which is either received by a receiver or reflected back via a mirror. When the beam is less visible to the "eye" of the sensor, it sends an alarm signal to the fire alarm control panel. Optical smoke detectors are generally quick in detecting slow-burning, smoky fires.

Ionization detectors are sometimes known as ionization chamber smoke detectors (ICSDs). They are capable of quickly sensing flaming fires that produce little smoke. This detector employs a radioactive material to ionize the air in a sensing chamber; the presence of smoke affects the flow of the ions between a pair of electrodes, which triggers the alarm. While over 80% of the smoke detectors in American homes are of this type, and although ionization detectors are less expensive than optical detectors, they are frequently rejected for projects seeking LEED certification for environmental reasons. The majority of residential models are self-contained units that operate on a 9-volt battery, but construction codes in some parts of the country now require installations in new homes to be connected to the house wiring, with a battery backup in case of a power failure.

Heat detectors can detect heat and can be either electrical or mechanical in operation. Most heat detectors are designed to trigger alarms and notification systems before smoke even becomes a factor. The most common types of heat detectors are thermocouple and the electro-pneumatic, both of which respond to changes in ambient temperature. Typically, if the ambient temperature rises above a predetermined threshold, an alarm signal is triggered.

Good detection has many benefits (beyond triggering the alarm system), the main one being that in many cases there is a chance to extinguish a small, early blaze with a fire extinguisher. Also, intelligent smoke detectors can differentiate between different alarm thresholds. These systems typically have remote detectors located throughout the facility that are connected to a central alarm system.

9.8.5   Fire Doors

Fire doors are made of fire-resistant material that can be closed to prevent the spread of fire and are designed to provide extra fire-spread protection for certain areas of a building (Table 9.4). The fire rating classification of the wall into which a door is installed dictates the required fire rating of the door. The location of the wall in the building and prevailing building codes establish the wall's fire rating. Fire doors are normally installed in staircases from corridors or rooms or cross-corridor partitions to laboratories, plant rooms, workshops, storerooms, machine rooms, service ducts, and kitchens, as well as to defined fire compartments. They are also employed in circulation areas that extend the escape route from the stair to a final exit or to a place of safety, and entrances and lobbies; at routes leading onto external fire escapes; and in corridors that are protected from adjoining accommodation by fire-resisting construction.

According to the NFPA, doors are rated with respect to the number of hours they can be expected to withstand fire before burning through. There are 20-, 30-, 45-, 60-, and 90-minute fire doors as well as 2-hour and 4-hour fire doors that are certified by an approved laboratory such as Underwriters Laboratories (UL). Because fire doors are rated physical fire barriers that protect wall openings from the spread of fire, they are required to provide automatic closing in the event of fire detection. Fire doors should usually be kept closed at all times, although some are designed to stay open under normal circumstances and to close automatically or manually in the event of a fire. Fire door release devices

TABLE 9.4 Interior Fire-Rated Doors and Glass Lites Classification.

Class	Fire Rating	Location and Use	Glass Lite Size Allowed		
			Area	Height	Width
A	3 hr	Fire walls separating buildings or various fire areas within a building; 3–4 hr walls	None	None	None
B	1½ hrs (HM) 1 hr (other)	Vertical shafts and enclosures such as stairwells, elevators, and garbage chutes; 2 hr walls	100 in	33 in	10–12 in
B	1 hr	Vertical shafts in low-rise buildings and discharge corridors; 1–1½ hr walls	100 in	33 in	10–12 in
C	¾ hr	Exit access corridors and exitway enclosures; 1 hr walls	1296 in	54 in	54 in
N/A	20 min. (⅓ hr)	Exit access corridors and room partitions; 1 hr walls	No limit	No limit	No limit

are electro-mechanical devices that enable automatic-closing fire doors to respond to alarm signals from detection devices such as smoke detectors, heat detectors, and central alarm systems. This permits closing the door before high temperatures melt the fusible link. Fusible links should always be used as backup to the releasing device.

Fire-rated door assemblies comply with NFPA Code 80 and are listed and labeled by UL, for the fire ratings indicated, based on testing according to NFPA Code 252. Assemblies must be factory-welded or come complete with factory-installed mechanical joints and must not require job fabrication on site.

Exit Routes

As defined by OSHA, an exit route is a continuous and unobstructed path of exit travel from any point within a workplace to a place of safety. All buildings require fire exits, which enable users to exit safely in the event of an emergency. Well-designed emergency exit signs are necessary for emergency exits to be effective. In the United States, fire escape signs often display the word "EXIT" in large, well-lit, green or red letters. An exit route must be permanent and must be separated by fire-resistant materials. Construction materials used to separate an exit from other parts of the workplace must have a one-hour fire resistance rating if the exit connects three or fewer stories and a two-hour fire resistance rating if the exit connects four or more stories.

Unless otherwise stipulated by code, at least two exit routes must be provided in a workplace to permit prompt evacuation of employees and other building occupants during an emergency. The exit routes must be located as far away as practical from each other so that if one exit route is blocked by fire or smoke, employees can evacuate using the second exit route. More than two exit routes must be available in a workplace if the number of employees, the size of the building, its occupancy, or the arrangement of the workplace is such that all employees would not be able to evacuate safely. Likewise, a single exit route is permitted where the number of employees, the size of the building, its occupancy, or the arrangement of the workplace is such that all employees would be able to evacuate safely.

Exit routes must be free and unobstructed, and must be arranged so that employees are not required to travel toward a high-hazard area, unless the path of travel is appropriately shielded by suitable partitions or other physical barriers. No materials or equipment may be placed, either permanently or temporarily, along the exit route. The exit access must not go through a room that can be locked, such as a bathroom, to reach an exit or exit discharge, nor may it lead into a dead-end corridor. Where the exit route is not substantially level, it is necessary to have stairs or ramps.

OSHA requirements stipulate that each exit discharge must lead directly to the exterior or to a street, walkway, refuge area, public way, or open space with access to the outside. These must be large enough to accommodate the building occupants likely to use the exit route. Exit stairs that continue beyond the level on which the exit discharge is located must be interrupted at that level by doors, partitions, or other effective means that clearly indicate the direction of travel leading to the exit discharge.

It is important to note that exit doors must not be locked from the inside, and each doorway or passage along an exit access that could be mistaken for an exit (e.g., a closet) must be marked "Not an Exit" or similar designation, or be identified by a sign indicating its actual use. Furthermore, exit route doors must be free of decorations or signs that obscure their visibility, and employees must be able to readily open them from the inside at all times without keys, tools, or special knowledge. A device, such as a panic bar that locks only from the outside, is permitted on exit discharge doors. Exit route doors may be locked from the inside only in mental, penal, or correctional facilities, and then only if supervisory personnel are continuously on duty and the employer has a plan to remove occupants from the facility during an emergency.

Where a fall hazard exists in the use of an outdoor exit route, it must have guardrails to protect unenclosed sides. If snow or ice is likely to accumulate along the route, it must be covered, unless it can be demonstrated that any snow or ice accumulation will be removed before it presents a slipping hazard. Also, the outdoor exit route must be reasonably straight and have smooth, solid, substantially level walkways, and must not have a dead-end that is longer than 20 feet (6.2 meters). To protect people and property during building fires requires the employment of three essential design elements:

- Alarms to provide early warnings
- Automatic sprinklers or other suppression systems
- Fireproof compartments to contain flames and smoke

These elements work together to give occupants time to escape and firefighters time to arrive. Eliminating any one of the three—detection, suppression, or compartmentation—would compromise the integrity of the building.

Compartmentation

Building regulations in most jurisdictions stipulate that large buildings need to be divided into compartments and that these compartments must be maintained should a fire occur. In order to do this there are a range of fire-stopping products and methods available offering between 30 and 240 minutes fire compartmentation protection for construction movement joints and service penetrations. A fire compartment can therefore be defined as a space within a building extending over one or several floors that is enclosed by separating members such that the fire spread beyond it is prevented during the relevant fire exposure. Fire compartments are sometimes referred to as *fire zones*. Compartmentation is critical to preventing a fire from spreading into large spaces or into the whole building. It involves the specification of fire-rated walls and floors sealed with firestop systems, fire doors, fire dampers, and the like. But to be effective, the walls, floor, and ceiling need to contain flames and smoke within the compartment. These components must also provide sufficient insulation to prevent excessive heat radiation outside the compartment.

The division of the building into discrete fire zones offers perhaps the most effective means of limiting fire damage. Compartmentation techniques are designed to contain the fire within the zone of origin by limiting vertical and horizontal fire spread. They also provide at least some protection for the rest of the building and its occupants even if first aid firefighting measures are used and fail. Also, they provide protection for inventory and business operations and delay the spread of fire prior to the arrival of the fire department. But determining the required fire resistance for a compartment depends largely on its intended purpose and on the expected fire. Either the separating members enclosing the compartment should resist the maximum expected fire or they should contain the fire until occupants are evacuated.

The load-bearing elements in the compartment must always resist the complete fire process or be classified to a certain resistance measured in terms of periods of time equal to or longer than the requirement of the separating members. The most important elements to be upgraded are the doors, floors, and walls; penetrations through floors and walls; and cavity barriers in the roof spaces. Halls and landings should typically be separated from staircases to prevent a fire from traveling vertically up or down the stairwell to other floors. However, creation of new lobbies can have an unacceptable negative impact on the character of a fine historic interior. To be effective therefore, compartmentation needs to be correctly planned and implemented.

The main function of fire stopping is to stop the spread of fire between floors of a building. Flame-retardant material is installed around floor openings designed to contain conduit and piping. A firestop is a product that, when properly installed, impedes the passage of fire, smoke, and toxic gases from one side of a fire-rated wall or floor assembly to another. Typical firestop products include sealants, sprays, mechanical devices (e.g., a firestop collar), foam blocks, or pillows. These products are installed primarily in two applications: around penetrations that are made in fire-resistive construction for the passage of pipes, cables, or HVAC systems; and where two assemblies meet, forming an expansion joint such as the top of a wall, curtain wall (edge of slab), or floor-to-floor joints. Typical opening types include:

- Electrical through-penetrations
- Mechanical through-penetrations
- Structural through-penetrations
- Nonpenetrated openings (e.g., openings for future use)
- Re-entries of existing firestops
- Control or sway joints within fire-resistant wall or floor assemblies
- Junctions between fire-resistant wall or floor assemblies
- "Head-of-wall" (HOW) joints, where non-load-bearing wall assemblies meet floor assemblies

Compliance with all applicable laws and regulations is the owner's responsibility, including the adopted and enforced fire code within a specific jurisdiction. Fire codes govern the construction, protection, and occupancy details that affect the fire safety of buildings throughout their life span. Numerous fire codes have been adopted throughout the United States, the vast majority of which are similar and based on one of the model codes available today or in the past. One requirement in all of these model codes is that fire safety features incorporated into a building at the time of its construction must be maintained throughout a building's life. This requires any fire-resistant construction to be maintained (Figure 9.36).

Alarm and Notification Systems

Fire alarm systems are essential to any facility, particularly in large buildings where there may be visitors or personnel who are unfamiliar with their surroundings. Bruce Johnson, director of Fire Service Activities with the International Code Council, says:

Fire alarm systems and smoke alarms are life safety systems that save countless lives each year, both civilians and firefighters. The International Residential Code requires interconnected, hardwired smoke alarms in all new construction (Section R313) and the International Building Code and International Fire Code (Section 907.2) call for manual or automatic fire alarm systems in most commercial buildings with high life occupancy or other hazards. In addition to new construction, the International Fire Code also has provisions for fire alarm systems and smoke alarms in existing structures (Section 907.3).

Fire alarms alert building occupants of a fire *and* alert emergency responders (police and fire) through a central station link to initiate appropriate responses.

Fire alarm control panels (FACP), or fire alarm control units (FACU), comprise electric panels that function as the controlling components of a fire alarm system (Figure 9.37). The FACP panel receives information from environmental sensors designed to detect any changes associated with fire. It also monitors their operational integrity and provides for automatic control of equipment and transmission of information necessary to prepare the facility for fire based on a predetermined sequence. An FACP panel may also supply electrical energy to operate any associated sensor, control, transmitter, or relay. There are currently four basic types of FACP on the market: coded panels, conventional panels, addressable panels, and multiplex systems.

Mass notification systems (MNSs) are invaluable in the protection of a wide range of facilities, and use both audible and visible means to distribute potential life-saving messages. An MNS is much more than an alarm system. By using the technologies based on fire alarm codes and standards, fire system manufacturers are able to produce a robust life safety and security system.

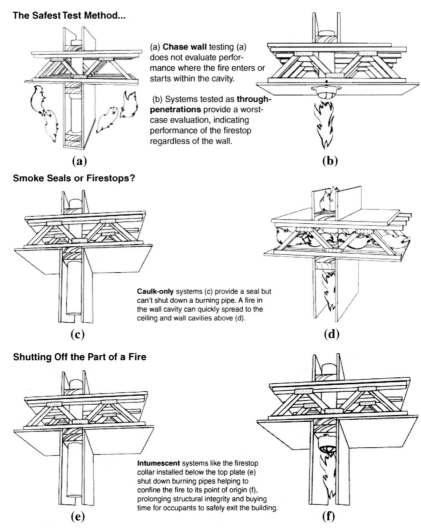

The Safest Test Method...

(a) **Chase wall** testing (a) does not evaluate performance where the fire enters or starts within the cavity.

(b) Systems tested as **through-penetrations** provide a worst-case evaluation, indicating performance of the firestop regardless of the wall.

(a) **(b)**

Smoke Seals or Firestops?

Caulk-only systems (c) provide a seal but can't shut down a burning pipe. A fire in the wall cavity can quickly spread to the ceiling and wall cavities above (d).

(c) **(d)**

Shutting Off the Part of a Fire

Intumescent systems like the firestop collar installed below the top plate (e) shut down burning pipes helping to confine the fire to its point of origin (f), prolonging structural integrity and buying time for occupants to safely exit the building.

(e) **(f)**

FIGURE 9.36 Various fire-stopping systems used in construction.

The impact of increasingly sophisticated technology has had a significant impact on today's alarm systems. For example, they now have the ability to provide more information to the fire department and first responders. In many cases, they can do more than just indicate that there has been an alarm in the building; they can be directed by the kind of alarm and where the alarm is. Moreover, many modern systems now include speakers that provide alerts in place of (or in addition to) traditional bell-type alarms. These speakers also can be used in emergencies other than fires to instruct and inform occupants of the situation. Such voice-actuated systems can include prerecorded or live

FIGURE 9.37 Siemens MXL fire alarm control panel (*top*) and graphic annunciator (*bottom*). These devices were installed in Potomac Hall at James Madison University. *(Photograph by Ben Schumin, 2003, Wikipedia.)*

messages that play in the event of fire or another emergency. Typical pre-recorded messages tell occupants that an alarm has been sounded and that they should remain in their designated area for further instruction. Building management can then manually use the system to deliver additional information and prepare occupants for an evacuation, if necessary. Alert systems can also close fire doors, recall elevators, and interface and monitor installed suppression systems such as sprinklers. It should be noted that when fire alarm systems are properly installed and maintained, they perform very well. But when they are not properly installed and maintained, the public and fire service may be subject to unnecessary "false alarms" that put everyone at risk.

Alarm systems can also connect with a building's ventilation, smoke management, and stairwell pressurization systems—all of which are critical to life safety. Again, these features are dependent on the building in which the particular system is installed. In addition, the integration of MNS and fire alarm control systems is a growing and positive trend that will hopefully continue and be applied in more types of facilities and multibuilding properties, including schools, high-rise buildings, mass transit hubs, and even public gathering places such as churches, theaters, and restaurants.

An annunciator is basically a unit containing two or more indicator lamps, alpha-numeric displays, or other equivalent means in which each indication provides status information about a circuit, condition, or location. An annunciator panel is sometimes employed to monitor the status of the different areas in a designated fire zone, theft protection, and control of a facility's alarm devices. There may be several fire zones in a building. Each fire zone is clearly marked on the panel, which identifies the different zones and their specific security status. Should a fire occur, an indicator light flashes on the panel and identifies the fire's location. For example, the light on the panel might indicate that a fire has occurred in Fire Zone 4. This information allows the fire department to quickly locate the fire.

9.8.6   Codes and Standards

One of the most important objectives of any design must be code compliance. There are a number of relevant national codes that relate to green building fire protection systems that are published by the NFPA. It should be noted that fire codes can vary substantially from one jurisdiction to another, and while these codes are not mandatory in all jurisdictions, they should nevertheless be adhered to whenever possible because they provide maximum safety for property and personnel and can help guide system design and installation:

> **NFPA 72, National Fire Alarm Code.** Governs the design, installation, operation, and maintenance of fire detection and alarm systems. It includes requirements for detector spacing, occupant notification, and control panel functionality.
>
> **NFPA 750, Standard on Water Mist Fire Protection Systems**. Governs classification of water mist systems and incorporates requirements for their design, installation, operation, and maintenance.
>
> **NFPA 2001, Standard on Clean Agent Fire Extinguishing Systems.** Governs the design, installation, operation, and maintenance of clean-agent systems and includes requirements for assessing design concentrations, safe personnel exposure levels, and system discharge times. It also stipulates that an agent be included in the EPA's Significant New Alternatives Policy list.

Finally, when building green today, there are numerous fire protection options that can be employed. Careful consideration of a building and its anticipated hazards help determine which areas require protection. Because of recent advances in technology, fire detection and suppression systems can now adequately support and sustain a modern green building philosophy. The methodical selection of a clean-agent or water-mist system can also help contribute to LEED certification credits for building owners and developers.

Chapter 10

Green Design and Building Economics

10.1 INTRODUCTION

Over the past decade, popular attention to "green" building has increased dramatically, to the extent that the built environment and sustainability have become closely intertwined. Moreover, building "green" provides us with an opportunity to use our resources more efficiently while creating buildings that enhance human health, build a better environment, and provide cost savings. However, the current financial crisis and the global economic recession continue to put an enormous strain on the nation's construction industry. Peter Morris, principal of the global construction consultancy Davis Langdon, believes that the dramatic reduction in construction activity is encouraging increased competition among bidders and lower escalation pressure on projects so that, in many projects, cost trends have become negative, leading to moderate construction price deflation. However, one of the biggest causes of concern, according to Morris, is contractor financing and working capital. Many contractors are finding it increasingly difficult to maintain adequate cash flow for their operations, and none have the resources for significant expansion of working capital. This has caused considerable concern in the construction industry and has obliged many bidders to be more cautious and judicious in project selection, and to focus on projects that have sound cash flows.

In recent years, broad awareness of the significant benefits that green buildings offer has increased. For example, the U.S. military, including the Air Force and Navy, now require that their new buildings be certified by the Leadership in Energy and Environmental Design (LEED®) Rating System. This may be partly because they recognize the linkage between wasteful energy consumption and the exposure of military forces to confrontation related to oil resources. As Boston Mayor Thomas M. Menino put it, "High-performance green building is good for your wallet. It is good for the environment. And it is good for people." As costs fall, the appealing financial performance of existing green buildings becomes clearer. In a 2006 survey of developers by McGraw-Hill Construction, respondents reported that they expected to see occupancy rates for green buildings 3.5% higher than market norms and rent levels increasing by 3%.

In addition, operating costs are estimated to be 8 to 9% lower. These numbers are receiving the attention of developers and investors, which is driving the growth of eco-construction today.

According to CalRecycle, "a green building may cost more up front, but saves through lower operating costs over the life of the building. The green building approach applies a project life cycle cost analysis for determining the appropriate upfront expenditure. This analytical method calculates costs over the useful life of the asset." Also, "these and other cost savings can only be fully realized when they are incorporated at the project's conceptual design phase with the assistance of an integrated team of professionals. The integrated systems approach ensures that the building is designed as one system rather than a collection of stand-alone systems."

During the past decade, we have witnessed a rapid emergence of eco-construction that reflects the building industry's growing confidence that the extra costs of building green are a good investment. Although the upfront costs of building green may be higher than those for conventional construction, that premium is shrinking. Precise benefits (e.g., reduced energy bills and reduced potable water consumption) can easily be computed, while others (e.g., positive impacts on occupant health or security) are usually much more difficult to quantify.

Incisive Media's "2008 Green Survey: Existing Buildings" found that nearly 70% of commercial building projects in the United States have already incorporated some kind of energy-monitoring system. It also found that energy conservation is the most widely implemented green program in commercial buildings, followed by recycling and water conservation. Moreover, approximately 65% of building owners who have incorporated green features claim that their investments have already resulted in a positive return. This return is expected to improve even more as the market for green materials and design expertise grows and matures. In this respect, Taryn Holowka, director of marketing and communications with the U.S. Green Building Council (USGBC™), says that "the supply of materials and services is going up and the price is coming down."

Turner Construction's "2008 Green Building Market Barometer," reported that approximately 84% of respondents believed that their green buildings have resulted in lower energy costs, and 68% claimed lower overall operating costs. Likewise, "75% of executives said that recent developments in the credit markets would not make their companies less likely to construct Green buildings." In fact, the survey revealed that 83% would be "extremely" or "very" likely to seek LEED certification for buildings they are planning to build within the next three years. Executives reported that green buildings generally have better financial performance than non-green buildings, especially in the following sectors:

- Higher building values (72%)
- Higher asking rents (65%)
- Greater return on investment (52%)
- Higher occupancy rates (49%)

This is confirmed by Jerry Jackson of Texas A&M University, who says,

A growing body of empirical literature indicates that LEED- and ENERGY STAR-certified buildings do command higher rents and greater occupancy rates relative to conventional buildings. For example, rent premium estimates from four recent studies using the CoStar national real estate database range from 4.4 to 51%. Occupancy premiums range from 4.2 to 17.9%. Each of these studies attempted to control for other factors such as building age.

Many commercial building owners now offer education programs to assist tenants in carrying out green programs in their space, reflecting a growing understanding of the significance of environmental awareness among employees and customers in addition to greater use of green materials and systems application. Davis Langdon conducted a comparative study in 2006 in which the construction costs of 221 buildings were analyzed. It was found that 83 buildings were constructed with the intent of achieving LEED certification whereas 138 lacked any sustainable design intentions. The study revealed that a majority of the buildings analyzed were able to achieve LEED certification without increased funding. In another investigation conducted by Davis Langdon of a diverse range of studies, it was found that the average construction cost premium required to achieve a moderate level of green features, equivalent to a LEED Silver certification, was roughly between 1 and 2%. However, what is particularly interesting is that in half or more of the green projects in these studies there was a zero increase in construction costs.

Yet even with increased awareness of the benefits of sustainable design, property owners and developers are sometimes slow off the mark to embrace green building practices. As Jerry Jackson (2009) points out,

Considering this information from the developer's perspective would seem to make the choice of sustainable versus conventional project development a rather easy choice. However, developer views of sustainable building projects are considerably less enthusiastic. A recent survey by Building Design and Construction in August 2007 found that while 94% of respondents thought the trend in sustainable building projects was "growing," 78% thought sustainable design added "significantly to first costs." Thirty-two percent of respondents estimated additional costs to be from 6- to 10%, while 41% estimated sustainable construction premiums to be 11% or greater.

This appears to be the sentiment of CB Richard Ellis's "Green Downtown Office Markets: A Future Reality," which depicts the general progress of the green building movement.

The report scrutinizes the obstacles preventing broad-based acceptance of sustainable design in office construction. Perhaps the main obstacle to embracing design sustainability is the perception of initial outlay compared to long-term benefits. This is in spite of an increasing number of studies similar to the one conducted by Davis Langdon in 2006 that clearly show no significant difference in average costs for green as compared to conventional buildings. Another hurdle that requires addressing is the lack of sufficient data on development, construction costs, and time needed to recoup investments. Nevertheless, a recent CB Richard Ellis (CBRE) white paper cites preliminary studies showing that building a property to basic LEED certification standards can be achieved with zero additional cost. According to CBRE's "Who Pays for Green? The Economics of Sustainable Buildings" (EMEA Research, 2009), "However, building a greener building—designed to achieve one of the higher standards of accreditation—is likely to add somewhere between 5% and 7.5% to construction costs."

There seems to be a lack of interest in research on green building in the United States. Such research currently constitutes an estimated $193 million per year or roughly 0.2% of federal research funds. This approximates a mere 0.02% of the estimated $1 trillion value of annual U.S. building construction, despite the fact that the construction industry represents approximately 9% of GDP. It is unfortunate that the construction industry can currently manage to reinvest only 0.6% of sales in research. This is markedly less than the average for other U.S. industries and for private-sector construction research investments in other industrialized countries.

Various green organizations strongly suggest that, unless we move decisively toward increasing and improving green building practices, we will soon likely be confronted with a dramatic adverse impact of the built environment on human and environmental health. Building operations today are estimated to account for 38% of U.S. carbon dioxide emissions, 71% of electricity use, and 40% of total energy use. If the energy required in the manufacture of building materials and in building construction is included, this number goes up to an estimated 48%. Moreover, buildings consume roughly 12% of the country's water in addition to rapidly increasing amounts of land.

Construction and remodeling account for 3 billion tons, or 40%, of the raw material used globally each year, which has a negative impact on human health. In fact, up to 30% of new and remodeled buildings may experience acute indoor air quality (IAQ) problems such as sick building syndrome (SBS). But, as is the case with most projects, determining building strategies early in the design process and sticking with them can result in the most efficient cost models for building. Implementing a goal-setting session at the beginning of each project to determine appropriate strategies and cost and time investments can result in lower sustainable design construction costs.

An appropriate analysis of green construction costs can be achieved through several methods. For example, it is possible to use either the LEED Rating System or the Green Globes® Rating System as a benchmark for success. Higher levels of certification may carry increased costs, but empirical market data suggest that LEED's "Certified" and "Silver" levels and one or two Green Globes "Globes" carry little or no premium over traditional building costs for most building types. Specialty project types, such as health care or research, often have program criteria and specific needs that are at odds with sustainable design, primarily as it relates to energy use and environmental constraints.

10.2   COSTS AND BENEFITS OF GREEN DESIGN

Even with the global economy in the state that it is, most people support policies that protect the environment, but many concerned developers and investors really want to know at what cost and how building green will benefit their investment's financial viability (See Figure 10.1.) In this respect, Peter Morris (2007) says,

Clearly there can be no single, across the-board answer to the question, "what does green cost?" On the other hand, any astute design or construction professional recognizes that it is not difficult to estimate the costs to go green for a specific project. Furthermore, when green building concepts and features are incorporated early in the design process, it greatly increases the ability to construct a certified green building at a cost comparable to a code compliant one. This means that it is possible today to construct green buildings or buildings that meet the U.S. Green Building Council's LEED third-party certification process with minimal increase in initial costs.

For LEED certification, there are studies that suggest that conventionally constructed buildings can often qualify for 12 or more LEED points by current building standards and their inherent design qualities. In many cases, between 15 and 20 additional points can be achieved for little to no additional cost, qualifying most buildings for the minimum rating. As mentioned previously, some studies have concluded that the cost of achieving Silver certification varies between 2 and 6% above traditional construction. To achieve the higher certification, Gold or Platinum, can add significantly to the cost of the project, primarily because of the costs of increasingly efficient technologies for water conservation and energy performance. These estimates are from LEED V2, and we have yet to determine how the latest version, LEED V3, will affect them. We have also yet to determine the effect of the new International Green Construction Code (IgCC). Global construction consulting firm

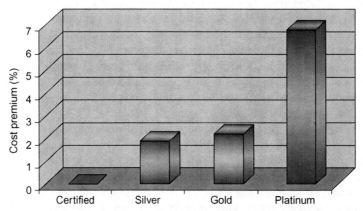

FIGURE 10.1   Estimated extra costs for building green using the various LEED Rating Systems. *(Source: USGBC.)*

Davis Langdon suggests that to be successful in building green and to keep the costs of sustainable design under control, three critical factors must be understood and implemented.

- Clear goals are critical for managing the cost. It is insufficient to simply say that we want the project to be green; the values should be defined as early in the design process as possible.
- Once the sustainability goals have been defined, it is essential to integrate them into the design and to integrate the design team so that the building elements can work together to achieve those goals. Buildings can no longer be broken down and designed as an assemblage of isolated components. This is the major difference between traditional building techniques and sustainable design.
- Integrating the construction team into the project team is critical. Many sustainable design features can be defeated or diminished by poor construction practices. Such problems can be eliminated by engaging the construction team, including subcontractors and site operatives, in the design and procurement process.

"Greening Buildings and Communities: Costs and Benefits," an important study resulting from the largest international research of its kind, is based on extensive financial and technical analysis of 150 green buildings across the United States and in 10 countries. It provides the most detailed findings to date on the costs and financial benefits of building green. The study found that the benefits of building green consistently outweigh any potential cost premium. The main conclusions arrived at include the following:

- Most green buildings cost 0 to 4% more than conventional buildings, with the largest concentration of reported "green premiums" between 0 and 1%. Green premiums increase with the level of greenness, but most LEED buildings, up through the Gold level, can be built for the same cost as for conventional buildings. This stands in contrast to a common misperception that green buildings are much more expensive than conventional buildings.
- Energy savings alone make green building cost-effective, outweighing the initial cost premium. The present value of 20 years of energy savings in a typical green office ranges from $7 per square foot (Certified) to $14 per square foot (Platinum)—more than the average additional cost of $3 to $8 per square foot for building green.
- Green building design goals are associated with improved health and enhanced student and worker performance. Health and productivity benefits remain a major motivating factor for green building owners, but are difficult to quantify. Occupant surveys generally demonstrate greater comfort and productivity in green buildings.
- Green buildings create jobs by shifting spending from fossil fuel-based energy to domestic energy efficiency, construction, renewable energy, and

other green jobs. A typical green office creates roughly one-third of a permanent job per year, equal to $1 per square foot of value in increased employment compared to a similar non-green building.

- Green buildings are seeing increased market value (higher sales/rental rates, increased occupancy, and lower turnover) compared to comparable conventional buildings. CoStar, for example, reports an average increased sales price from building green of more than $20 per square foot, providing a strong incentive to build green even for speculative builders.
- For roughly 50% of green buildings in the study's data set, the initial "green premium" is paid back by energy and water savings in five or fewer years. Significant health and productivity benefits mean that over 90% of green buildings pay back an initial investment in at most a five-year period.
- Green community design (e.g., LEED-ND) provides distinct benefits to owners, residents and municipalities, including reduced infrastructure costs, transportation and health savings, and increased property values. Green communities and neighborhoods have a greater diversity of uses, housing types, job types, and transportation options, and they appear to better retain value in a market downturn than conventional sprawl.
- Annual gas savings in walkable communities can be as much as $1000 per household. Annual health savings (from increased physical activity) can be more than $200 per household. CO_2 emissions can be reduced by 10 to 25%.
- Upfront infrastructure development costs in conservation developments can be reduced by 25%, or approximately $10,000 per home.
- Religious and faith groups build green for ethical and moral reasons. Financial benefits are not the main motivating factor for many places of worship, religious educational institutions, and faith-based nonprofits. A survey of faith groups building green found that the financial cost-effectiveness of green building makes it a practical way to enact the ethical/moral imperative to care for the Earth and communities. Building green has also been found to energize and galvanize faith communities.

Even when green building upfront costs exceed what was originally estimated, usually because of inefficient planning and execution, they can be quickly recouped through lower operating costs over the life of the building.

10.2.1 Economic Benefits of Green Building

It cannot be overemphasized that, to achieve maximum cost savings, green design strategies need to be incorporated into the project's conceptual design phase by an integrated team of professionals. Using an integrated systems approach ensures that the building is designed in a holistic manner as one system rather than a number of standalone systems, as is normal with conventional construction. The challenge here is that not all green benefits are easy to quantify. For example, how do you measure improved occupant health, comfort, and productivity, or pollution reduction? This is why these benefits are excluded from cost analyses. It is therefore prudent to set aside a small portion of the building budget (e.g., as a contingency) to cover differential costs associated with less tangible green building benefits or to cover the cost of researching and analyzing green options. Even when in difficult times, many green building measures can be incorporated into a project with minimal or zero increased upfront costs and yet yield substantial savings and other benefits (Figure 10.2) over the life of the facility.

No matter how interested an owner or developer is in green building and sustainability, the bottom line remains: What does "green" cost? Typical translation: Does it cost more? This then raises the question: More than what? For example, is it more than what the building would have cost without the sustainable design features or is it more than available funds? The answers to these questions have until recently been largely elusive because of a lack of hard data. During recent years, however, we have seen various organizations conduct considerable research into green building and sustainability costs. We now have substantial data on building costs that allow us to compare the costs of comparable green and traditional nonsustainable buildings.

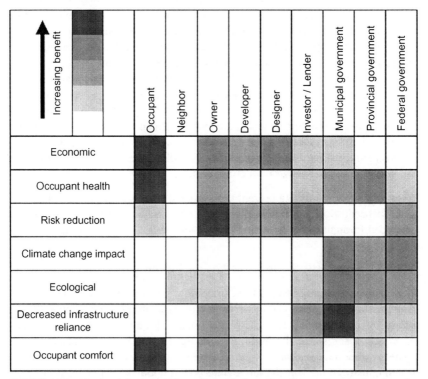

FIGURE 10.2 Matrix of Green Building stakeholder benefits. *(Source: "A Business Case for Green Buildings in Canada," presented to Industry Canada by Lucuik et al., Morrison Hershfield, 2005.)*

To adequately assess sustainable design and how it relates to construction costs, it is imperative to analyze the costs and benefits using a holistic approach. This basically means including evaluation of operations and maintenance costs, user productivity and health, design and documentation fees, and the like. This is largely because empirical experience demonstrates that construction cost implications have the greatest impact on fundamental decisions about sustainable design. Helping teams to understand the actual construction costs of real projects that are achieving green, and providing a methodology that allows them to manage these costs, can go a long way toward determining whether or not green is the answer.

Green construction is helped by the fact that the cost of green design has dropped significantly in the last few years as the number of green buildings has increased. The trend of declining costs, resulting from increased experience in green construction, has manifested itself in a number of U.S. states. It can be concluded that many projects are able to achieve sustainable design either within the initial budget or with minimal supplemental funding. This suggests that developers should continue to find ways to incorporate project goals and values, regardless of budget, by making choices. However, every building project is unique and should be considered as such because there are no one-size-fits-all solutions; benchmarking with other comparable projects can be valuable and informative, but not predictive. Any cost estimate for sustainable design for a specific building must be specific to that building, its objectives, and its particular circumstances and attributes.

In a recent study, Greg Kats of Capital E summarized some of the financial benefits of going green (see Table 10.1). His conclusion is that financial benefits

TABLE 10.1 Summary of Findings of Green Financial Benefits

Category	20-year Net Present Value
Energy savings	$5.80
Emissions savings	$1.20
Water savings	$0.50
Operations and maintenance savings	$8.50
Productivity and health value	$36.90 to $55.30
Subtotal	$52.90 to $71.30
Average extra cost	(−$3.00 to −$5.00)
Total 20-year net benefit	**$50 to $65**

Note: Numbers based on costs per square foot. The financial benefits of going green are related mainly to productivity. (Source: Adapted from Capital E Analytics, "The Costs and Financial Benefits of Green Buildings," 2003)

for building green are estimated to be between $50 and $70 per square foot in a LEED-certified building; this is more than 10 times the additional cost. These financial benefits come in the form of lower energy, waste, and water costs, lower environmental and emissions costs, and lower operational and mainte-nance costs; lower absenteeism; increased productivity and health; higher retail sales; and easier reconfiguration of space, resulting in less downtime and lower costs. According to Katz, cost estimates based on a sample of 33 office and school buildings suggest only 0.6% greater costs for LEED Certified, 1.9% for Silver, 2.2% for Gold, and 6.8% for Platinum levels. Although these esti-mates are for direct costs, they nevertheless closely reflect estimates provided by the USGBC. What is perhaps surprising is that, other than LEED, few studies have looked at rating systems such as Green Globes. The principal motivators that influence the long-term value of green building appear to be increasing energy costs (75%), government regulations/tax incentives (40%), and global influences (26%)

The lack of adequate clear, credible data pertaining to development, con-struction costs, and time required to recoup investment is the most obvious reason for the industry's sometimes lethargic acceptance of green construction, which is why education has become the most important tool in promoting green strategies. The main obstacles revealed in the Kats study include

- Too multidisciplinary—41%
- Not convinced of increased return on investment (ROI)—37%
- Lack of understanding of benefits—26%
- Lack of service providers—20%
- Too difficult—17%
- Greenwashing—16%
- Lack of shareholder support—10%

Most building owners and developers have one main objective, to build a pro-ject and then sell it. In this sense, they construct or revamp an office building and lease its space with the hope or calculation of selling the asset within a three- to five-year time frame in order to repay debts and ensure a profit. The speed with which this process is completed affects the amount of profit gen-erated on the sale of the building. Uninformed developers who are under the false perception that green construction costs more can trigger fears since they are already concerned about the costs of short-term debt and conventional build-ing materials. However, according to Davis Langdon, typical benefits for build-ing owners and tenants include

- Ability to command higher lease rates
- Reduced risk of building obsolescence
- Potential higher occupancy rates
- Higher future capital value
- Less need for refurbishment in the future

- Higher demand from institutional investors
- Lower operating costs
- Mandatory for government tenants
- Lower tenant turnover
- Enhanced occupant comfort and health
- Improved IAQ, which increases employee productivity and satisfaction

10.2.2   Cost Considerations of Green Design

There are two categories of costs for green buildings: direct capital and direct operating. These are explained next.

Direct Capital Costs

Direct capital costs are those associated with the original design and construction of the building; they generally include interest during construction (IDC). A general misperception held by some building stakeholders is that the capital costs of constructing green buildings are significantly higher than those of conventional buildings; however, many in the green building field believe that green buildings actually cost less than conventional buildings, or at least cost no more. Empirical evidence shows that savings are achieved by downsizing systems through better design and eliminating unnecessary systems, which offset any increased costs caused by more advanced systems. Capital and operational costs are normally relatively easy to measure because the required data are readily available and quantifiable. Productivity effects, on the other hand, are difficult to quantify yet nevertheless important because of their potential impact. There are other indirect and external effects that can be wide-reaching and that may prove difficult to quantify.

Direct Operating Costs

Direct operating costs include all applicable expenditures required to operate and maintain a building over its full life. Included are the total costs of energy use, water use, insurance, maintenance, waste, property taxes, and so forth. The primary costs are those associated with heating, cooling, and maintenance activities such as painting and roof repairs and replacement, as well as less obvious items such as churn (the costs of reconfiguring space and services to accommodate occupant moves). All costs relating to major renovations, cyclical renewal, and residual value or demolition are excluded from this category.

Insurance is essentially a direct operating cost (and is discussed in greater detail in Chapter 16). Green buildings have many tangible benefits that reduce or mitigate a variety of risks and should be reflected in insurance rates for the building. Likewise, the fact that green buildings generally provide a healthier environment for occupants should be reflected in health insurance premiums.

Indeed, the general attributes of green buildings (e.g., natural light, off-grid electricity, and commissioning) should reduce a broad range of liabilities, and general site locations potentially reduce risks of property loss due to natural disasters. Furthermore, a fully integrated building design will typically reduce the risk of inappropriate systems or materials use, which could have a positive impact on other insurable risks. Insurance companies sometimes offer premium reductions for certain green features, such as commissioning or reduced reliance on fossil fuel–based heating. The list of premium reductions will undoubtedly increase with further education and awareness and as the broad range of benefits are more fully recognized and understood. In any case, prior to taking out a policy, it is advisable to consult an insurance agent or attorney.

The churn rate reflects the frequency with which building occupants are moved, either internally or externally, including occupants who move but remain within a company and those who leave a company and are replaced. It has been found that, because of increased occupant comfort and satisfaction, green buildings typically have lower churn rates than conventional buildings.

10.2.3   Increased Productivity

The positive effect on productivity is one of the many benefits that green buildings offer, as shown by numerous studies. However, because these studies are often broadly based and rarely focus on unique green building attributes, they need to be supplemented by thorough, accurate, and statistically sound research to fully comprehend the effects of green buildings on occupant productivity, performance, and sales. The difficulty of properly attributing such gains as reduced absenteeism and staff turnover rates appears to have made the benefits of "green" the exception rather than the rule. It seems prudent, however, to include any productivity gains attributable to a green building in the life-cycle cost analysis, particularly for an owner-occupied building. Key features of green buildings relating to increased productivity are possible because of controllable ventilation, temperature, and lighting systems; daylighting and views; natural and mechanical ventilation; pollution-free environments; and vegetation. It is not always clear why these features produce improved productivity, although studies show that healthier employees are typically happier employees, which means increased worker satisfaction, improved morale, increased productivity, and reduced absenteeism.

The Lawrence Berkeley National Laboratory recently conducted a study on the potential ramifications of building green and concluded that improvements to indoor environments such as commonly found in green buildings can help reduce health care costs and work losses from communicable respiratory diseases by 9 to 20%; from reduced allergies and asthma from 18 to 25%; and from nonspecific health and discomfort effects from 20 to 50%. Hannah Carmalt, a project analyst with Energy Market Innovations, notes that, "the most intuitive explanation is that productivity increases due to better occupant health and

therefore decreased absenteeism. When workers are less stressed, less con-
gested, or do not have headaches, they are more likely to perform better."
High-performance buildings have many potential benefits including increased
market value, lower operating and maintenance costs, improved occupancy for
commercial buildings, and increased employee satisfaction and productivity for
owner-occupied buildings

A study by W. H. Fisk and A. H. Rosenfeld (Fisk and Rosenfeld, 1998)
concluded that green buildings add $20 to 160 billion in increased worker pro-
ductivity annually. This is due to the fact that LEED-certified buildings yield
significant productivity and health benefits, such as heightened employee
productivity and satisfaction, fewer sick days, and fewer turnovers. Other independent studies have shown that better climate control and improved air quality
can increase employee productivity by an annual average of 11 to 15%. Still, it
is necessary to define the particular elements of green buildings that are directly
related to productivity. Sound control, for example, while recognized to
increase productivity, is often excluded from green-related studies, mainly
because it is not considered to be a particularly green feature. Likewise, the
presence of biological pollutants (e.g., molds) is associated with decreased productivity, yet this is also excluded from green studies because typical green
buildings do not automatically eliminate such pollutants even though they
reduce them because of improved ventilation. However, it should be noted that
in commercial and institutional buildings, payroll costs generally significantly
overshadow all other costs, including those involved in a building's design,
construction, and operation.

In today's world, occupant control of the indoor environment has become
one of the most significant elements of green buildings that affect productivity
and thermal comfort; control is required over temperature, ventilation, and
lighting. Green buildings usually incorporate this feature because it can notice-
ably decrease energy use by ensuring that areas are not heated, cooled, or lit
more than necessary. These measures are decisive in maintaining energy efficiency and occupant satisfaction within a building. Increased productivity is
often associated with increased emotional well-being, as various studies have
clearly shown. One such study conducted by the Heschong Mahone Group
(HMG) in 2003 found that higher test scores in daylight classrooms were
achieved because students were happier. It also found that when teachers were
able to control the amount of daylighting in classrooms, students appeared to
progress 19 to 20% faster than students in classrooms that lacked daylighting
control. Similar studies on office settings clearly showed a significant rise in
productivity when there was individual control over temperature, lighting,
and ventilation.

A view to the outdoors is another common feature of green buildings associated with productivity in offices. HMG's 2003 study confirmed correlations
between productivity and access to outdoor views: Test scores were generally
10 to 15% higher and call center performance increased by 7 to 12%. This

reinforces HMG's 1999 study of classrooms in the Capistrano School District, which found that children progressed 15% faster in math and 23% faster in reading in the classrooms with the largest windows.

Ventilation can be of paramount importance because it allows fresh air to cycle through the building and removes stale or pollutant air from the interior. Germs, molds, and various volatile organic compounds (VOCs), such as those emitted by paints, carpets, and adhesives, can often be found in buildings lacking adequate ventilation, and they can cause SBS. Typical symptoms include inflammation, asthma, and allergic reaction. Ventilation also plays a critical role in worker productivity, as evidenced by the extensive research that has been conducted to address these issues. This is why it is imperative to minimize the use of toxic materials inside a building. Many of the products used in conventional office buildings, such as carpets and copying machines, contain toxic materials. Minimizing them decreases potential hazards associated with their use and disposal.

Furthermore, because these materials are known to leak pollutants into the indoor air, proper ventilation is required to avoid adverse impacts on worker productivity. From the preceding, however, it becomes evident that productivity can be impacted by many factors, all of which can influence the bottom line. Moreover, it has been shown that there is less staff turnover when employees are satisfied, which in turn helps improves a firm's overall productivity. Less time spent on job training allows more time to be spent being productive. Staff retention is one of the decisive reasons that many firms make the decision to green their office buildings in today's competitive world.

Committing to a market value for occupant productivity gains, and having them accurately reflected in the business case at the decision-making point, is not easy in the case of speculative or leased facilities. Even so, there is now enough data and evidence quantifying these gains to support taking them into account on some basis. And while an owner of a leased facility may not financially benefit directly from increased user productivity, there can be indirect benefits in the form of increased rental fees and occupancy rates. For the majority of commercial buildings, the use of a conservative estimate for the potential reduction in salary costs and for productivity gains will loom large in any calculation.

10.2.4 Improved Tenant/Employee Health

Superior air quality is one of the principal attributes of green buildings. The "green" attributes normally include features such as abundant natural light, access to views, and effective noise control. Each of these qualities is for the benefit of building occupants, making a building a better place in which to work and live. Building occupants are increasingly seeking green building features, such as superior air quality, control of air temperatures, and views. The Urban Land Institute (ULI) and the Building Owners and Managers Association

(BOMA) jointly conducted a 1999 survey that found that occupants rated air temperature (95%) and air quality (94%) most crucial in terms of tenant comfort. The study also determined that 75% of buildings did not have the option or capability to adjust these features and that many individuals were willing to pay higher rents to obtain it. These features were the only ones considered "most important" and on the list of those with which tenants were least satisfied. The survey also determined that the principal reasons that tenants moved out were heating or cooling problems.

Natural light, clean air, and thermal comfort are required for health and productivity, and provide an enjoyable living and work environment. The credible studies demonstrating an intrinsic connection between green building strategies and occupant health and well-being are endless. William Fisk, in a 2002 study, "How IEQ Affects Health, Productivity," estimated that 16 to 37 million cases of colds and flu could be avoided by improving indoor environmental quality (IEQ). This translates into a $6 to $14 billion annual savings in the United States. IEQ also reduces SBS (a condition whereby occupants become temporarily ill) by 20 to 50%, resulting in annual savings of $10 to $30 billion.

10.2.5   Better Recruitment and Retention

Harris Interactive conducted a national survey that concluded that more than a third of U.S. workers would be more inclined to work for companies with strong green credentials, and highlighted the growing influence of environmental issues surrounding staff recruitment and retention. These findings are confirmed by Timothy R. Johnson, a principal at GCA International, who says, "Firms that focus on the growing market sector of green building and sustainable design will be more attractive to top-notch candidates for reasons of recognized workload availability, progressive growth, and employment stability." This is clear evidence that providing a healthy and pleasant work environment increases employee satisfaction, productivity, and retention. It also clearly shows that such an environment increases a company's ability to compete for the most qualified employees as well as for business.

Statistical data and other evidence leave no doubt that high-performance green buildings can increase a company's ability to recruit and retain employees because of good air quality, abundant amounts of natural light, and better circulated heat and air conditioning, all of which help provide a more pleasant, healthier, and more productive place to work. With this in mind, it is surprising that a willingness to join and remain with an organization is often overlooked when considering how green buildings affect employees. The economics of employee retention is important and should be seriously considered, particularly since one estimate puts the cost of losing a single good employee at roughly $50,000 to $150,000. Many organizations experience a 10 to 20% annual turnover, some of it from employees they would have wanted to retain. In a workforce of, say, 100 people, turnover at this level implies 10 to 20 people

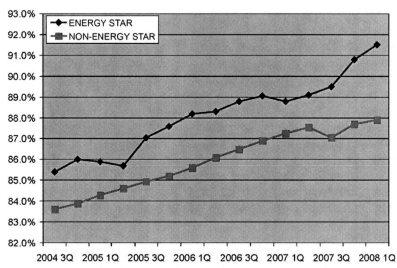

FIGURE 10.3   Occupancy rates for two building types by quarter through 2008. *(Source: "Does Green Pay Off?" by Norm Miller, Jay Spivey, and Andy Florance.)*

leaving per year. In some cases people decide to leave because of poor physical and working environments. Figure 10.3 compares the difference in occupancy rates between ENERGY STAR® and non-ENERGY STAR buildings.

10.2.6   Increased Property Values and Marketability

The number of reliable studies on the intrinsic relationship between property values and green buildings is rather limited, even though this is an important consideration that should be quantified and included in any economic calculations. There are many factors that will or may increase property values for green buildings. Indeed, enhancement of property value is a key factor for speculative developers who fail to directly achieve operating cost and productivity savings. It is particularly relevant to speculative developers who intend to either sell or lease a new building, although it can also have a bearing on the decision process in general, including for developers who intend to occupy a building while maintaining an eye on the market value of this asset. Many in the real estate industry unfortunately remain oblivious to the real benefits of green buildings because of a lack of adequate education in sustainability, and therefore they cannot fully convey these benefits to prospective purchasers or tenants.

Jerry Yudelson (2008), a well-known building green proponent, discovered that increased annual energy savings promote higher building values and cites as an example a 75,000-square-foot building that saves $37,500 per year in energy costs versus a comparable building built to code. (This saving might result from saving 50 cents per square foot per year.) At a capitalization rate of 6%, which is typical today in commercial real estate, green building standards add $625,000 ($8.33 per square foot) to the value of a building. This means that for a small upfront investment, an owner can reap benefits that typically offer a rate of return exceeding 20% with a payback of three years or fewer.

The fact that high-performance buildings can offer building owners many important benefits ranging from higher market value to more satisfied and productive employee occupants is not always apparent. The primary reason for this is that most of the benefits accrue to tenants, who usually need proof before they are willing to pay the cost of investments that are perceived to help them be more productive or to save money. It is only very recently, mainly because of increased awareness, that tenants have started to fully appreciate the benefits of cleaner air, more natural lighting, and flexible spaces that can be modified as and when required.

All available evidence points to the fact that green buildings, particularly green buildings with good natural lighting, can have a dramatic effect on property value and sales of commercial buildings. Furthermore, several U.S. studies report that there is a sound economic basis for green buildings, but only when operational costs are included in the equation. More specifically, whole-building studies conclude that net present values (NETs) for green as opposed to conventional buildings range from $50 to $400 per square foot ($540 to $4300 per square meter). NET depends on a building's length of life (e.g., 20 to 60 years) and the degree to which it incorporates green strategies.

One of the main conclusions from these studies is that, generally, NET increases as the greenness of the building increases. A CoStar study found that LEED buildings can command rent premiums of about $11.24 per square foot versus their conventional peers, in addition to a 3.8% increase in occupancy rate. It also found that rental rates in ENERGY STAR buildings can boast a $2.38-per-square-foot premium versus comparable non-ENERGY STAR buildings, in addition to a occupancy rate 3.6% higher. What is perhaps more remarkable, and what may prove to be a trend signaling greater attention from institutional investors, is that LEED buildings are commanding a surprising $171 more per square foot than their conventional counterparts, and ENERGY STAR buildings are commanding an average of $61 more. This is extraordinary because most leasing arrangements, particularly in the office/commercial sector, provide little incentive to undertake changes that might be construed as beneficial to the environment. For example, leases often have fixed rates with no regard to energy or water consumption, even though the lessees have control over most energy- and water-consuming devices.

10.2.7   Miscellaneous Indirect Benefits

The numerous indirect benefits of green building include improved image, risk reduction, future proofing, and self-reliance. These and similar benefits may be captured by investors and should not be discarded in decision economics considerations. Although they may be difficult to quantify, and in some cases even intangible, they should nevertheless be factored into the business case because they are intrinsically connected to sustainable design and can significantly impact the value of a green building.

Enhanced Image

One of the key messages conveyed by sustainable buildings is concern for the environment, which is why they can be used as a public relations tool. Moreover, even if we disregard their financial benefits, green buildings are generally perceived by the public as modern, dynamic, and altruistic. A green building serves as a physical and permanent message about the commitment of an organization to environmental stewardship and accountability. It can therefore provide a strong symbolic message of an owner's commitment to sustainability. Some of the benefits that companies can enjoy from these perceptions include employee pride, satisfaction, and well-being, which often translate into reduced turnover, better recruitment, and improved morale. These powerful images can be a motivational factor in a company's decision to pursue occupancy in a green facility.

Reduced Risk

Commenting on the recent downturn in the U.S. economy, Peter Morris of Davis Langdon Consultants comments, "Risk remains a serious concern for construction projects. Delay and cancellation of projects, even projects under construction, is a growing trend." Morris also proposes a key theme for project owners in the current market turmoil which is that, "the successful adoption of a competitive procurement strategy, in order to secure lower costs, will depend on active steps by the owner and the project team to ensure that the contractor is in a position to provide a realistic and binding bid and that contractors' bidding costs are minimized."

Employing green building principles can mitigate many potential perceived risks. In this regard, the Environmental Protection Agency (EPA) currently classifies IAQ as one of the top five environmental health risks. Moreover, increased litigation is evidenced with regard to mold-related problems. SBS and building-related illness (BRI) are major concerns and often end up being resolved in the courts. Business owners and operators are increasingly facing legal action from building tenants who blame the building for their health problems. The main cause of SBS and BRI is poor building design and/or construction, particularly of the building envelope and mechanical systems. Green buildings should emphasize and promote not only safe but exceptional air quality, and no recognized green building should ever have to suffer from SBS or BRI.

Future Proofing

Green buildings are inherently efficient and safe, and as such help to ensure that they will not be at a competitive disadvantage in the future. Davis Langdon, an AECOM company, sums it up very well: "Going green is 'future-proofing' your asset." This is largely because of a number of potential risks that are significantly mitigated in green buildings, including the following:

- Energy conservation protects against future energy price increases.
- Occupants of green buildings are generally more comfortable and contented, so it can be assumed that they will generally be less likely to be litigious.
- Water conservation shields against water fee increases.
- A documented effort to build or occupy a healthy green building demonstrates a level of due diligence that could stand as an important defense against future litigation or changes in legislation, even when faced with currently unknown problems.

Self-Reliance

Green buildings often incorporate natural lighting, ventilation, and internal energy and water generation, which makes them less likely to rely on external grids and less likely to be affected by grid-related problems or failures such as blackouts, water shortages, or contaminated water. This element is increasingly important globally because of the growing potential risk of terrorism. Local self-reliance is steadily moving to prime time, and the Institute for Local Self-Reliance (ILSR) continues to develop cutting-edge solutions to the problems facing communities around the world.

Five American communities that have taken steps toward energy independence follow. They show that at the local level energy independence has become a realistic possibility as communities around the United States explore available renewable resources and the technology necessary to harness them. The five U.S. towns listed are but a few of the many around the nation creating models for clean energy production and self-reliance.

- Rock Port, Missouri, a small town with a population of 1400, has become the first community in the nation to be completely powered by wind.
- Greensburg, Kansas, is rebuilding itself as a "model green town" after being hit by a disastrous tornado, and is now expecting to provide enough power using alternative energy methods to meet all of its energy needs in the foreseeable future.
- Reynolds, Indiana, another small Midwestern town with a population of 540, was chosen to execute the state government's "BioTown USA" experiment. The plan is to power the town with a range of locally available biomass.
- In San José, California, the city council recently gave the city manager the authority to negotiate the terms of "an organics-to-energy bio-gas facility."
- Warrenton, Virginia, like San José, is taking a "trash into treasure" approach. Mayor George Fitch has spearheaded an effort to build a "biorefinery," and reduce the town's greenhouse-gas emissions 25% by 2015.

10.2.8   External Economic Effects

It is not easy to give a precise definition of external effects. They generally consist of a project's costs or benefits that accrue to society and are not readily captured by the private investor. Examples are reduced reliance on infrastructure such as sewers and roads, reduced greenhouse gases, and reduced health costs. The extent to which these benefits can be factored into a business case relies on the extent to which they can be converted from the external to the internal side of the ledger. This constitutes a vital factor in any assessment of the costs and benefits of green buildings. The costs of vegetated roofs, for example, are borne by the developer or investor, while much of the benefit, such as reduced heat island effects and reduced stormwater runoff, accrues at a broader societal level. Where the investor is a government agency, or where a private developer is compensated for including features that produce benefits at a societal level, the business case can encompass a much broader range of effects. For example, the state of Oregon offers tax incentives for green building, thereby providing a direct business case payoff to the investor. Another example is Arlington, Virginia, which allows higher ratios of floor space to land coverage for green buildings.

Green has become a buzzword, and journalists everywhere are writing about a "green economy," "green technology," and even "green jobs." Many manufacturers are jumping on the green bandwagon and increasingly claiming to be "green," while others are trying to measure the effect of "green" technology on the job market. "Green" encourages job creation, partly because green building technologies are often labor-intensive rather than material- or technology-intensive. For example, there are significant environmental impacts associated with the transportation of materials in the construction industry. In addition, by promoting the use of local and regional materials, local and regional job creation is encouraged and promoted.

Buildings can be singled out as having the largest indirect environmental impact on human health. Other, perhaps less critical, impacts (e.g., damage to ecosystems, crops, and structures/monuments), as well as resource depletion, should also be considered even though they do not have a large associated indirect cost relative to human health. Infrastructure costs such as water use and disposal are typically paid by governments and are rarely cost effective or even cost neutral, and in many instances governments are required to heavily subsidize water use and treatment. On the other hand, external environmental costs consist mainly of pollutants in the form of emissions to air, water, and land, and the general degradation of the ambient environment.

Furthermore, green building can have economic ramifications and provide export opportunities on a much broader scale as a result of increased international recognition and related export sales. The 2005 Environmental Sustainability Index, jointly prepared by Yale and Columbia, benchmarked the ability of nations to protect the environment by integrating data sets of natural resource endowments, pollution levels, and environmental management efforts into a smaller set of indicators of environmental sustainability. The United States unexpectedly ranked only 45th of the countries in this index.

10.3 LIFE-CYCLE COSTING

Life-cycle costing (LCC) combines capital and operating costs to determine the net economic effect of an investment, and evaluates the economic performance of additional investments that may be required for green buildings. It is based on discounting future costs and benefits to dollars of a specific reference year, referred to as present value (PV) dollars. LCC makes it feasible to intelligibly quantify costs and benefits and compare alternatives based on the same economic criterion or reference dollar. The World Business Council for Sustainable Development (WBCSD) recently came out with a study that suggests that key players in real estate and construction often misjudge the costs and benefits of green buildings. Peter Morris of Davis Langdon says,

Perhaps a measure of the success of the LEED system, which was developed to provide a common basis for measurement, is the recent proliferation of alternative systems, each seeking to address some perceived imbalance or inadequacy of the LEED system, such as the amount of paperwork, the lack of weighting of credits, or the lack of focus on specific issues. Among these alternative measures are broad-based approaches, such as Green Globes, and more narrowly focused measures, such as calculations of a building's carbon footprint or measurements of a building's energy efficiency (the ENERGY STAR® rating). All these systems are valid measures of sustainable design, but each reflects a different mix of environmental values, and each will have a different cost impact.

10.3.1 Initial/First Costs

Construction projects typically have initial or upfront costs that may include capital investment in land acquisition, construction, or renovation and in equipment needed to operate a facility. Land acquisition costs are normally included in the initial cost estimate if they differ among design alternatives. A typical example is comparing the cost of renovating an existing facility against new construction on purchased land.

The assumed increase in first costs is the most cited reason for not incorporating green elements into a building design. Some aspects of design have few or no first costs, including site orientation and window and overhang placement. Other sustainable systems that incorporate additional costs in the design phase (e.g., an insulated shell) can be offset, for example, by the reduced cost of a smaller mechanical system. Material costs can be reduced during a project's construction phase by the use of dimensional planning and other material efficiency strategies. Such strategies can reduce the amount of building material needed and cut construction costs, but they require forethought by designers to ensure a building that creates less construction waste solely through its dimensions and structural design. An example of dimensional planning is designing rooms of 4-foot multiples, since wallboard and plywood sheets come in 4- and 8-foot lengths. Another example is designing a room in 6- or 12-foot multiples to correspond with the length of carpet and linoleum rolls.

10.3.2 Life-Cycle Cost Analysis

The method of evaluating a project's total relevant cost over time, product, or measure is known as life-cycle cost analysis (LCCA). LCCA takes into consideration all costs—including first costs such as capital investment, purchase, and installation; and future costs such as energy, operation, maintenance, capital replacement, financing; and any resale, salvage, or disposal—over the life of the project or product. It is thus an engineering economic analysis tool for comparing the relative merits of competing project alternatives.

George Paul Demos, estimating engineer at Colorado Department of Transportation (CDOT), notes:

The first component in an LCC equation is cost. There are two major cost categories by which projects are to be evaluated in an LCCA: initial expenses and future expenses. Initial expenses are all costs incurred prior to occupation of the facility. Future expenses are all costs incurred after occupation of the facility. Defining the exact costs of each expense category can be somewhat difficult at the time of the LCC study. However, through the use of reasonable, consistent, and well-documented assumptions, a credible LCCA can be prepared.

According to Demos, the following are considered to be major steps that are essential to performing a proper cost analysis:

- Establish objectives.
- Identify constraints and specify assumptions.
- Define the base case and identify alternatives.
- Set the analysis period.
- Define the level of effort for screening alternatives.
- Analyze traffic effects.
- Estimate benefits and costs relative to the base case.
- Evaluate risk.
- Compare net benefits and rank alternatives.
- Make recommendations.

Sieglinde Fuller of the National Institute of Standards and Technology (NIST) comments:

LCCA is especially useful when project alternatives that fulfill the same performance requirements, but differ with respect to initial costs and operating costs, have to be compared in order to select the one that maximizes net savings. For example, LCCA will help determine whether the incorporation of a high-performance HVAC or glazing system, which may increase initial cost but result in dramatically reduced operating and maintenance costs, is cost-effective or not.

However when it comes to budget allocation, LCCA is not beneficial.

While there is general consensus that a life-cycle approach is valid, most building stakeholders prefer to focus on minimizing direct costs or, at best, on short time-frame payback. Many developers, building owners, and other stakeholders hold the view that basing opinions on anything other than reducing direct costs is fiscally irresponsible; in reality, the opposite is often the case. The lack of adoption is largely due to a corporate structure that typically dissociates direct and operating costs. It also results from the lack of a mandate that constructors reduce operating costs, although they are mandated to reduce construction costs. This unfortunate reality is also evidenced by owner/developers, who oversee construction of buildings for their own use.

LCCA's primary objective is to calculate the overall costs of project alternatives and to select the design that safeguards the facility's ability to provide the lowest overall cost of ownership in line with its quality and function. It should be performed early in the design process to allow for any needed design refinements or modifications before finalization to optimize life-cycle cost. Likewise, it is important to ensure that the design complies with the IgCC now in effect. Another very important challenge for LCCA (or any economic evaluation method for that matter) is to evaluate the economic effects of alternative designs for buildings and building systems and to quantify these effects as dollar amounts. LCCA is especially suited to the evaluation of design alternatives that satisfy a required performance level, but that may have differing investment, operating, maintenance, or repair costs, and possibly different life spans.

Although lowest life-cycle cost provides a straightforward and easily interpreted measure of economic evaluation, there are other commonly used methods such as net savings (or net benefits), savings-to-investment ratio (or savings benefit-to-cost ratio), internal rate of return, and payback period. Fuller sees these as being consistent with the lowest LCC measure of evaluation if they use the same parameters and length of study period. Almost identical approaches can achieve cost-effective choices for building-related projects—whether called cost estimating, value engineering, or economic analysis. After all costs are identified by year and amount and discounted to present value, they are added to arrive at total LLCs for each alternative, including the following:

- Initial design and construction costs
- Maintenance, repair, and replacement costs
- Other environmental or social costs/benefits (impacts on transportation, solid waste, water, energy, infrastructure, worker productivity, and outdoor air emissions, etc.)
- Operating costs of energy, water/sewage, waste, recycling, and other utilities

Appropriate adjustments should be made to all dollar values expended or received over time on a comparable basis; this is necessary for the valid assessment of a project's life-cycle costs and benefits. Time adjustment is required because a dollar today will not have an equivalent value in the future. Supplementary measures, however, are considered to be relative—that is, they are computed for an alternative relative to a base case. According to Sieglinde Fuller,

Supplementary measures of economic evaluation are Net Savings (NS), Savings-to-Investment Ratio (SIR), Adjusted Internal Rate of Return (AIRR), and Simple Payback (SPB) or Discounted Payback (DPB). They are sometimes needed to meet specific regulatory requirements. For example, the FEMP LCC rules (10 CFR 436A) require the use of either the SIR or AIRR for ranking independent projects competing for limited funding. Some federal programs require a Payback Period to be computed as a screening measure in project evaluation. NS, SIR, and AIRR are consistent with the lowest LCC of an alternative if computed and applied correctly, with the same time-adjusted input values and assumptions. Payback measures, either SPB or DPB, are only consistent with LCCA if they are calculated over the entire study period, not only for the years of the payback period.

A holistic or integrated approach through active, deliberate, and full collaboration among all players is the most likely way to achieve successful green buildings. Building-related investments typically involve a great deal of uncertainty relating to costs and potential savings. An LCCA greatly increases the likelihood of deciding on a project that can save money in the long run. However, this does not alleviate all of the potential uncertainty associated with the LCC results, mainly because LCCAs are typically conducted early in the design process, when only estimates of costs and savings are available rather than specific dollar amounts. This uncertainty in input values means that actual results may differ from estimated results. The LCCA can be applied to any capital investment decision, and is particularly relevant when high initial costs are traded for reduced future cost obligations.

A 2007 study by Davis Langdon updating an earlier study states,

It is clear from the substantial weight of evidence in the marketplace that reasonable levels of sustainable design can be incorporated into most building types at little or no additional cost. In addition, sustainable materials and systems are becoming more affordable, sustainable design elements are becoming widely accepted in the mainstream of project design, and building owners and tenants are beginning to demand and value those features.

Likewise, Ashley Katz, a communications coordinator for the USGBC, says, "Costs associated with building commissioning, energy modeling and additional professional services typically turn out to be a risk mitigation strategy for owners. While these aspects might add on to the project budget, they will end up saving projects money in the long run, and are also best practices for building design and construction."

10.4 TAX BENEFITS AND INCENTIVES

State and local governments throughout the country are drafting new green building regulations to take advantage of incoming stimulus funding. The IgCC reinforces this sustainability trend. For example, the tax benefits available to businesses through the Energy Policy Act (EPAct) of 2005 have been extended through 2013. Also, as a follow-up effort to encourage environmentally friendly construction and energy savings, many states now have various tax incentive programs for green building. New York and Oregon offer state tax credits, while other states, such as Nevada, offer abatements on property and sales tax. Oregon's credits vary and are based on building area and LEED certification level—for example, at the Platinum level a 100,000-square foot building can expect to receive a net present value (NPV) tax credit of up to $2 per square foot that is transferrable from public or nonprofit entities to private companies (e.g., contractors or benefactors), making it more attractive than a credit that applies only to private owners. Also, DOE approved appliance rebate programs for certain states and territories as of November 17, 2011.

New York State has various sales tax exemptions, property tax abatements, and personal tax credits and incentives to encourage residential energy-saving measures, onsite renewable generation, solar and wind renewable energy systems, and use of alternative fuel for space heating and hot-water heating. However, for onsite generation systems (e.g., wind, solar, biomass, and fuel cells) to be eligible, they must be grid-connected and net-metered. New York also offers a tax credit for builders who meet energy goals and use environmentally preferable materials that can be up to $3.75 per square foot for interior work and $7.50 per square foot for exterior work. Builders can apply this credit against their state tax bill. To qualify, a building needs to be certified by a licensed architect or engineer and meet specific requirements for energy use, water use, IAQ, waste disposal, and materials selection. This means that the energy used in new buildings must not exceed 65% of that allowed under the New York energy code; in rehabilitated buildings, energy use cannot exceed 75%.

In 2005, Nevada passed a poorly considered green building incentive package in an effort to spur private green development. The state offered a property tax abatement of up to 35% for up to 10 years to projects that achieve LEED Silver certification. Thus, if the property tax represented 1% of value, it could be worth as much as 5% of the building cost, which translates to much more than the actual cost of achieving LEED Silver on a large project. This package encouraged a large number of Nevada projects to pursue LEED certification, including the $7 billion, 17 million-square-foot City Center in Las Vegas, which is one of the world's largest private development projects to date. The hastily written legislation forced the 2007 session of the Legislature to rethink and modify the program because it created an enormous financial crisis for the state. Nevada also provides for sales-tax abatement for green materials used in LEED Silver-certified buildings. South Carolina introduced a program of tax incentives that meet certain Green Globes or LEED standards for energy efficiency.

When appraising currently existing incentive programs, it should be noted that New York, Oregon, and Maryland preceded Nevada and have used their state income tax code as the primary tool to promote green building. In addition, many jurisdictions have created their own unique programs. Virginia followed the Nevada model by allowing property tax abatements at the local level. New Mexico used the income tax credit approach, and Hawaii tried a new idea, requiring a green building to receive priority processing during governmental approval reviews.

There are many federal tax incentives available for green building. One is EPAct 2005, which offers two major tax incentives for differing aspects of green buildings: a tax credit of 30% for use of solar thermal and electric systems; and a tax deduction of up to $1.80 per square foot for projects that reduce energy use for lighting, HVAC, and water-heating systems by at least 50% compared with the 2001 baseline standard. These tax deductions may be taken by the design team leader (typically the architect) when applied to government projects. Consumers should always check the various state websites for the latest updates. Some of the more prevalent federal tax incentives are listed in Table 10.2.

TABLE 10.2 Building Green Tax Incentives

Who Is Eligible?	Incentive
Consumers	Credits for home energy improvements such as windows, insulation, and envelope and duct sealing Credits for efficient air conditioners, heat pumps, gas or oil furnaces, and furnace fans; in new or existing homes, credits for efficient gas, oil, or electric heat pumps and water heaters Credits for qualified solar water heating and photovoltaic systems, and small wind and geothermal heat pump systems
Businesses	Deductions for new or renovated buildings that save 50% or more of projected annual energy costs for heating, cooling, and lighting compared to model national standards Partial deductions for efficiency improvements to individual lighting, HVAC and water heating, and envelope systems Investment tax credits for combined heat and power systems Credits for qualified solar water heating and photovoltaic systems and for certain solar lighting systems Credits for qualifying microturbines (power-producing systems that typically run on natural gas run small to medium-size commercial buildings)
Builders and Manufacturers	Credits for homes that exceed national model energy codes by 50%, subject to certification Small credits for manufactured homes that exceed national model codes by 30% or that meet ENERGY STAR standards Credits for high-efficiency refrigerators, clothes washers, and dishwashers (special consumer promotions may be available for qualifying products)

It is obvious that tax credits can provide significant savings by reducing the amount of income tax owed, unlike a deduction, which reduces the amount of income that is taxable. The reader should always check online for the latest tax incentive updates, as many new programs are continually initiated and older programs expire. For example, the Federal Solar Tax Credit has been extended for eight years. Also, on February 17, 2009, President Obama signed into law the American Recovery and Reinvestment Act (ARRA), which created new incentives for solar energy, modified existing incentives, and provides billions of dollars in funding for renewable energy projects. With ARRA, the United States can in the coming years become the largest solar market in the world. For additional details on tax incentives and credits visit http:// energytaxincentives.org, www.dsireusa.org/incentives, http://seia.org, www. energy.gov/taxbreaks.htm, and www.aceee.org/energy/index.htm.

10.5   OTHER GREEN BUILDING COSTS

Other green building costs include operational expenses for energy, water, and other utilities, which depend to a large extent on consumption, current rates, and price projections. However, since energy and, to a lesser extent, water consumption, building configuration, and the building envelope are interdependent, energy and water costs are usually assessed for the building as a whole rather than for its individual systems or components. Sometimes the latest, greenest technology has yet to be approved and may cause both delays and additional costs.

10.5.1   Operational Energy and Water Costs

Accurate forecasts of energy costs during the preliminary design phase of a project are rarely simple. Assumptions have to be made regarding use profiles, occupancy rates, and schedules, all of which can have a dramatic impact on energy consumption. Several computer programs currently on the market, such as Energy-10 and eQuest, can provide information regarding assumptions about a facility's consumption of energy. Alternatively the information and data can come from engineering analysis. Software packages, such as ENERGY PLUS (DOE), DOE-2.1E, and BLAST, are excellent, but they require more detailed input that is not normally available until later, when the design is more fully developed. It is important to determine prior to program selection whether annual, monthly, or hourly energy consumption estimates are required and whether a program is capable of adequately tracking savings in energy consumption even when design changes take place or when different efficiency levels are simulated. Figure 10.4 shows typical costs incurred by an HVAC system over its expected useful life of 30 years.

Estimates of energy use in conventional and green buildings vary, but the consensus is that green buildings on average use 30% less energy than conventional buildings do, which is why energy is a substantial and widely recognized

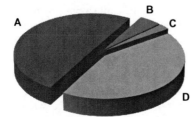

A: Energy cost, 50.0%

B: Maintenance cost, 40.7%

C: Replacement cost, 2.3%

D: HVAC first cost, 43.0%

FIGURE 10.4 Typical costs (percent) incurred by an HVAC system over its 30-year useful life. *(Source: Washington State Department of General Administration, 2012.)*

cost of building operations that can be reduced through energy efficiency and related measures that are part of green design. A detailed survey of 60 LEED-rated buildings showed that green buildings, when compared to conventional buildings, reaffirm the following conclusions:

- On average more energy efficient by approximately 25 to 30%
- More likely to generate renewable energy onsite
- Lower electricity peak consumption
- More likely to purchase grid power generated that is from renewable energy sources

Energy savings in sustainable buildings come primarily from reduced electricity purchases and secondarily from reduced peak energy demand. On average, green buildings are estimated to be 28% more efficient than conventional buildings and to generate 2% of their power onsite from photovoltaics (PV). The financial benefits from a 30% reduction in consumption at an electricity price of $0.08/kilowatt adds up to about $0.30 per square foot annually with a 20-year NPV of over $5 per square foot, which is equal to or more than the average additional cost associated with building green. Jerry Yudelson, author of *The Green Building Revolution*, says:

Many green buildings are designed to use 25- to 40-percent less energy than current codes require; some buildings achieve even higher efficiency levels. Translated to an operating cost of $1.60 to $2.50 per square foot for electricity (the most common energy source for building), this energy savings could reduce utility operating costs by 40 cents to $1 per square foot per year. Often, these savings are achieved for an added investment of just $1 to $3 per square foot. With building costs reaching $150 to $300 per square foot, many developers and building owners are seeing that it is a wise business decision to invest 1 to 2% of capital cost to secure long-term savings, particularly with a payback of less than three years. In an 80,000-sq-ft building, the owner's savings translates into $32,000 to $80,000 per year, year after year, at today's prices.

Environmental and health costs associated with air pollution caused by non-renewable electric power generation and onsite fossil fuel use are generally excluded when making investment decisions. Table 10.3 highlights the reduced energy used in green buildings as compared with conventional buildings.

TABLE 10.3 Energy Use in Green Buildings Compared
with Conventional Buildings

	Certified	Silver	Gold	Average
Energy efficiency (above standard code)	18%	30%	37%	28%
Onsite renewable energy	0%	0%	4%	2%
Green power	10%	0%	7%	6%
Total	**28%**	**30%**	**48%**	**36%**

10.5.2 Operation, Repair, and Maintenance Costs

Sustainability studies have shown that over their lifetime, LEED-certified buildings typically cost less and are easier to operate and maintain than conventional buildings. This puts them in a position to command higher lease rates than conventional buildings in their markets. However, operation, maintenance, and repair (OM&R) costs and nonfuel operating costs are often more difficult to estimate than other building expenditures. Operating schedules and maintenance standards vary from one building to the next; this variation is significant even when the buildings are of the same type and age. This is why it is important, when estimating these costs, to use common sense and good judgment. Published estimating guides and supplier quotes can sometimes provide relevant information. Some estimating guides derive their cost data from databases such as RSMeans and BOMA, which typically report, for example, average owning and operating costs per square foot, number of square feet, number of stories, building age, and geographic location.

Typically, a green building can recoup any added costs within the first year or two of its life cycle once it becomes operational. Studies show that green buildings typically use 30 to 50% less energy and 40% less water than their conventional counterparts, yielding significant operational cost savings. The New Buildings Institute (NBI) recently released a study indicating that new buildings certified by LEED on average perform 25 to 30% better than buildings that are not LEED-certified in energy use. The study also suggests that buildings achieving Gold and Platinum LEED certification have average energy savings approaching 50%.

10.5.3 Durability and Replacement Costs

One of the more significant benefits of building green is the incorporation of durable materials that prolong the life of a building and its systems; such buildings enjoy lower system replacement and material costs. Replacing a roof, flooring, HVAC system, or the whole building itself results in the highest cost to the environment and to the owner's bottom line. While many of these features also reduce operating costs, an owner's commitment to proactive maintenance is the key to keeping systems working well into their prime.

The number and timing of capital replacements of building systems are based to a large extent on the estimated life of a system and the length of the study period. It is expected that the same sources providing the cost estimates for the initial investments will provide estimates of replacement costs and expected useful lives. Likewise, a good starting point for estimating future replacement costs is to use these costs as of the base date. The LCCA method is designed to escalate base-year amounts to their future time of occurrence. The term *residual value* of a system or component is sometimes mentioned; this basically represents the value the system or component will have after being depreciated—its remaining value at the end of the study period—or at the time it is replaced during the study period. According to Sieglinde Fuller, residual values can be based on value in place, resale value, salvage value, or scrap value, net of any selling, conversion, or disposal costs. By using simple rule-of-thumb calculations, the residual value of a system with a remaining useful life can be determined by linearly prorating its initial costs.

10.5.4 Finance Charges and Other Costs

For federal projects, finance charges and taxes do not normally apply, although finance charges and other payments do apply if a project is financed through an energy savings performance contract (ESPC) or a utility energy services contract (UESC). These charges are normally included in the contract payments negotiated with the energy service company (ESCO) or the utility.

Nonmonetary benefits or costs are associated with project-related issues for which there is no meaningful way of assigning a dollar value. Despite efforts to develop quantitative measures of benefits, there are situations that simply do not lend themselves to such an analysis. For example, projects may provide certain benefits such as improved quality of the working environment and preservation of cultural and historical resources. By their nature, these benefits are external to the LCCA and difficult to assess, but if they are considered significant, they should be taken into account in the final investment decision and included in a life-cycle cost analysis and in the project documentation.

To formalize the inclusion of nonmonetary costs or benefits in decision-making, the analytical hierarchy process (AHP) can be used. This is one of a set of multiattribute decision analysis (MADA) methods that can be employed when considering qualitative and quantitative nonmonetary attributes in addition to common economic evaluation measures when evaluating project alternatives. The "ASTM E 1765: Standard Practice for Applying Analytical

Hierarchy Process (AHP) to Multiattribute Decision Analysis of Investments Related to Buildings and Building Systems," published by AS International, presents a general procedure for calculating and interpreting AHP scores for a project's total overall desirability when making building-related capital investment decisions. An excellent source of information for estimating productivity costs is the *Whole Building Design Guide (WBDG)* Productive Branch.

10.6   ECONOMIC ANALYSIS TOOLS AND METHODS

The federal government is the nation's largest owner and operator of built facilities. During the energy crisis of the 1970s followed by the 1980s crisis, it was faced with increasing initial construction costs and ongoing operational and maintenance expenses. As a result, facility planners and designers decided to use economic analysis to evaluate alternative construction materials, assemblies, and building services with the goal of lowering costs. In today's difficult economic climate, building owners wishing to reduce expenses or increase profits are again employing economic analysis to improve their decision making in planning, designing, and constructing a building. Moreover, federal, state, and municipal entities have all enacted legislative mandates requiring the use of building economic analysis to determine the most economically efficient or cost-effective choice among building alternatives. Figure 10.5 illustrates the general steps taken in an economic analysis.

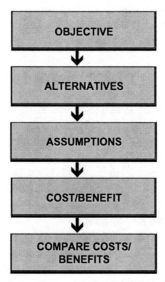

FIGURE 10.5   Economic analysis. *(Source: Adapted from Whole Building Design Guide.)*

10.6.1 Present-Value Analysis

Present-value analysis is based on the simple concept that the value of a dollar profit today is greater than the value of a dollar profit next year. How much greater is determined by the "discount rate," as in how much of a discount you would expect if you were buying a dollar's worth of next year's profit. The discount rate used in NPV calculation is usually the cost of debt, also known as the weighted average cost of debt. NPV allows decision makers to compare various alternatives on a similar time scale by converting the various options to current dollar figures. A project is generally considered acceptable if the net present value is positive over its expected life.

Take as an example a building owner considering changing the building's lighting from traditional incandescent to fluorescent bulbs. The initial investment to change the lights themselves is estimated to be $40,000. After the initial investment, the cost to operate the lighting system is estimated to be $2000, but the change will yield $15,000 in savings each year, thus producing an annual cash flow of $13,000 every year after the initial investment. For the sake of simplicity, a discount rate of 10% is assumed, and it is calculated that the lighting system will be used over a five-year period. This scenario produces the following NPV calculations:

$$t=0 \text{ NPV} = (-40,000)/(1+0.10) \ 0 = -40,000.00$$
$$t=1 \text{ NPV} = (13,000)/(1.10) \ 1 = \$11,818.18$$
$$t=2 \text{ NPV} = (13,000)/(1.10) \ 2 = \$10,743.80$$
$$t=3 \text{ NPV} = (13,000)/(1.10) \ 3 = \$ 9,767.09$$
$$t=4 \text{ NPV} = (13,000)/(1.10) \ 4 = \$8,879.17$$
$$t=5 \text{ NPV} = (13,000)/(1.10) \ 5 = \$8,071.98$$

Based on this information, the total NPV over the lifetime of the project comes to $9,280.22.

The value of discounting is that it adjusts costs and benefits to a common point in time. Thus, to be able to add and compare cash flows incurred at different times during the life cycle of a building, they need to be made time-equivalent. The LCC method converts them to present values by discounting them to a common point in time, which is usually the base date. To some extent, the selection of the discount rate is dependent on the use to which it will be put. The interest rate used for discounting essentially represents the investor's minimum acceptable rate of return.

The "Federal Discount Rate FY 2012" in Principles and Guidelines states, "Discounting is to be used to convert future monetary values to present values. Calculate present values using the discount rate established annually for the formulation and economic evaluation of plans for water and related land resources." The discount rate for federal energy and water conservation projects is determined annually by DOE's Federal Energy Management Program (FEMP);

for federal projects not primarily concerned with energy or water conservation, the discount rate is determined by the Office of Management and Budget (OMB). These discount rates do not include the general rate of inflation but rather represent real discounts. In OMB and FEMP studies, annually recurring cash flows such as operational costs are normally discounted from the end of the year in which they are incurred. Under MILCON guidelines, they are typically discounted from the middle of the year. All single amounts such as replacement costs and residual values are discounted from their dates of occurrence.

The length of the study period begins with the base date, which is the date to which all cash flows are discounted. The study period includes any planning, construction, and implementation periods as well as the service or occupancy period. It remains unchanged for all considered alternatives. The service period, however, essentially begins when the completed building is occupied or when a system is put into service. This is the period over which operational costs and benefits are evaluated. In FEMP analyses, the service period cannot exceed 25 years. The contract period in ESPC and UESC projects lies within the study period, starting when the project is formally accepted, energy savings begin to accrue, and contract payments begin to come due. The contract period generally ends when the loan is paid off.

According to Sieglinde Fuller, "[LCCA] is particularly suitable for the evaluation of building design alternatives that satisfy a required level of building performance but may have different initial investment costs, different operating and maintenance and repair costs, and possibly different lives." However, LCCA can be applied to any capital investment decision in which relatively higher initial costs are traded for reduced future obligations. Also according to Fuller, LCCA is an approach that provides a much better assessment of the long-term cost effectiveness of a project than alternative economic methods that mainly focus on first costs or on operating costs in the short run. Fuller also says that LCCA can be performed at various levels of complexity, but its scope can vary from a "back-of-the-envelope" calculation to a detailed analysis with thoroughly researched input data, supplementary measures of economic evaluation, complex uncertainty assessment, and extensive documentation.

An important attribute of life-cycle cost analysis is that it can be performed in either constant or current dollars. Both methods produce identical present-value life-cycle costs. However, a constant-dollar analysis does not include the general rate of inflation, which means that it has the advantage of not requiring an estimate of the rate of inflation for the years in the study period. A current-dollar analysis, on the other hand, does include the rate of general inflation in all dollar amounts, discount rates, and price escalation rates. Constant-dollar analysis is generally recommended for federal projects except those financed by the private sector, such as through the ESPC and the UESC. There are several alternative financing analyses available that are usually performed in current dollars if the analyst wants to compare contract payments with actual year-to-year operational or energy cost savings.

10.6.2   Sensitivity Analysis

Sensitivity analysis is a technique recommended by FEMP for energy and water conservation projects. It is typically employed to investigate the robustness of an analysis that includes some form of mathematical modeling. Critical assumptions should be varied and net present value and other outcomes recomputed to determine how sensitive outcomes are to changes in assumptions. The assumptions that deserve the greatest attention rely on the dominant benefit and cost elements and the areas of greatest uncertainty for the program being analyzed. In general, a sensitivity analysis is used for estimates of benefits and costs, discount rate, general inflation rate, and distributional assumptions. Models used in the analysis should be well documented and, where possible, available for independent review.

10.6.3   Break-Even Analysis

Break-even analysis is useful for tracking a business' cash flow. It focuses on the relationship between fixed cost, variable cost, and profit, and is mostly used by decision makers who want to know the maximum cost of an input that will allow the project to still break even, or, conversely, the minimum benefit a project can produce and still cover the cost of the investment. To perform a break-even analysis, benefits and costs are set equal, all variables are specified, and the break-even variable is solved mathematically. Since we are dealing with cash flow, depreciation, which is a noncash expense, is subtracted from operating expenses. The variables needed to compute a break-even analysis for a particular project include (1) gross profit margin, (2) operating expenses (less depreciation), and (3) annual debt service (total monthly debt payments for the year).

10.6.4   Computer Estimating Programs

Computer programs can considerably reduce the time and effort spent on an LCCA by performing the computations and documenting the study. Many LCCA software programs are available, all of which can be found on the Internet. For example, the DOE Building Technologies website has information on more than 200 software tools. Some software is free and downloadable. A few of the more widely used applications follow.

- Isograph's AvSim + and RCMCost for system availability simulation and reliability-centered maintenance, and Weibull Analysis and Life Cycle Costing modules.
- Economic Analysis Package (ECONPACK) for Windows, a comprehensive economic analysis program incorporating calculations, documentation, and reporting capabilities. ECONPACK is structured to permit use by noneconomists in preparing complete, properly documented economic analyses in support of DOD funding requests. It was developed by the Army Corps

of Engineers, and its analytic capabilities are reportedly generic, providing standardized methodologies and calculations for a broad range of capital investments such as hospitals, family housing, information systems, utility plants, maintenance facilities, commercially financed facilities, and equipment.

- Building Life-Cycle Cost (BLCC), an FEMP program that analyzes capital investments in buildings. BLCC calculates life-cycle costs, net savings, savings-to-investment ratio, internal rate of return, and payback period for federal energy and water conservation and renewable energy projects that are agency-funded or alternatively financed. Version 5.3-09, developed by the NIST, performs economic analyses by evaluating the relative cost-effectiveness of alternative buildings and building-related systems or components. Typically, BLCC is used to evaluate alternative designs that have higher initial costs but lower operating costs over their life than do lowest-initial-cost designs.
- Life-Cycle Cost in Design (WinLCCID) was originally developed for MILCON analyses by the Construction Engineering Research Laboratory of the Army Corps of Engineers. It is a life-cycle costing tool for evaluating and ranking design alternatives for new and existing buildings and for "what-if" analyses based on variables such as present and future costs and/or maintenance and repair costs.
- ENERGY-10® is a cost-estimating program tool that helps architects, builders, and engineers to rapidly (within 20 minutes) identify the most cost-effective energy-saving measures for designing a low-energy building. Using site-specific climate data, different combinations of materials, systems, and orientations can be shown to yield lesser or greater results based on energy use, comparative costs, and reduced emissions. Use of this software at the early phases of a design can reportedly result in energy savings of 40 to 70%, with little or no increase in construction cost. ENERGY-10 is available through the Sustainable Buildings Industry Council (SBIC).
- Success Estimator Estimating and Cost Management System, a cost-estimating tool available from U.S. Cost, Inc., gives estimators, project managers, and owners real-time, simultaneous access to their cost data and estimating projects from any Internet-connected computer.

Relevant Codes and Standards

Many codes and standards are relevant to green building, including
- International Green Construction Code (IgCC), 2012
- AS E2432—Standard Guide for the General Principles of Sustainability Relative to Building, 2011
- Circular No. A-94 Revised—Guidelines and Discount Rates for Benefit-Cost Analysis of Federal Programs, February 2011
- 10 CFR 436 Subpart A—Federal Energy Management and Planning Programs, Methodology and Procedures for Life-Cycle Cost Analyses

- Energy Policy Act of 2005
- Executive Order 13123—Greening the Government through Efficient Energy Management, DOE Guidance on Life-Cycle Cost Analysis Required by Executive Order 13123
- Executive Order 13423—Strengthening Federal Environmental, Energy, and Transportation Management
- Facilities Standard for the Public Buildings Service, P100 (GSA)—Chapter 1.8: Life-Cycle Costing
- *Standards on Building Economics*, 6th ed. ASTM, 2007
- *Sustainable Building Technical Manual*, DOE/EPA, 1996
- NAVFAC P-442, *Economic Analysis Handbook National Facilities Engineering Command, October 1993*
- Tri-Services Memorandum of Agreement (MOA) on "Criteria/Standards for Economic Analyses/Life-Cycle Costing for MILCON Design," 1991

Chapter 11

Green Project Commissioning

11.1 INTRODUCTION

The building commissioning industry is relatively new and continues to grow. Over the last decade, it has grown from a disparate group of researchers and engineers into an established network of professionals and is increasingly being embraced by public and private organizations such as the American Society of Heating, Refrigerating and Air-Conditioning Engineers (ASHRAE), the Building Commissioning Association™ (BCA), and the National Conference on Building Commissioning (NCBC). Also, for the first time California Building Standards Commission (CBSC) has adopted the first statewide green building code, which took effect in January 2011.

The first "national green building code," which was approved in November 2011 by the International Green Construction Code (IgCC), further emphasizes the importance of commissioning. This is because building commissioning is an important quality-assurance service in the building industry and offers improved project delivery results, and because it undoubtedly enhances a building's value to its owner. This is partly why more and more engineering firms currently consider commissioning services as a core business component. Recent studies clearly demonstrate that commissioning has been found to be the single-most cost-effective strategy for reducing energy consumption, costs, and greenhouse-gas emissions in buildings today. That is why it is being integrated into the construction process to ensure that owners and investors get good buildings for their investments.

Building commissioning (Cx) is one way to reduce risk for new construction projects or major capital improvements (e.g., renovations), and it is a comprehensive way to assess and tune up the performance of existing buildings. Many building owners are now demanding higher performance in buildings from their architects, engineers, and contractors. This increased momentum for commissioning is coming from energy and environmental policy makers as well as the private sector, and is increasingly resonating with building owners' interest in greening their properties and those seeking LEED® or Green Globes® Certification. *ASHRAE Guideline 0, The Commissioning Process*, defines commissioning as "a quality-oriented process for achieving, verifying, and documenting that the performance of facilities, systems, and assemblies meets defined objectives and criteria."

Modern buildings increasingly contain sophisticated conservation and environmental control technologies that, to function properly, require careful supervision of installations, testing and calibration, and operator training. Many modern sustainable (and conventional) buildings may possess high-technology electrical or air-conditioning systems, or employ certain sustainable features, that require specialized attention to ensure that they operate as designed. Successful project commissioning can be very helpful in reducing operating and maintenance costs, as well as in extending the useful life of equipment. It also helps fulfill LEED Certification requirements.

Commissioning provides better planning, coordination, and communication between various stakeholders, resulting in fewer Change Orders, shorter punch lists, and fewer callbacks. In addition, the commissioning of new construction projects can reduce construction delays, ensure that the correct equipment is properly installed, increase productivity, and provide healthier occupant conditions (thereby reducing employee absenteeism). Once the project is completed, it is important that all the as-built information and operating and maintenance information be passed on to the owners and operating staff. Research has shown that returns for these commissioning services often pay for themselves in energy savings within a year of the project being completed.

As mentioned earlier, building commissioning is a fairly recent concept that includes what was historically referred to as "testing, adjusting, and balancing." But in today's high-technology world, it goes much further, acquiring additional importance when complex mechanical and electrical systems are involved, and there is a need to ensure that these systems operate as intended in order to achieve the energy savings and improved building environment that justify their installation. Cx is also crucial to achieving optimum performance when special building features are installed to generate renewable energy, recycle waste, or reduce other environmental impacts. Furthermore, commissioning practices should be specially tailored to address the size and complexity of the building, its systems, and its components in order to verify performance and to confirm that all requirements are met as per the construction documents and specifications. In addition to verification of the installation and performance of systems, the commissioning process ultimately culminates in the production of a commissioning report for the owner that will be helpful should problems arise in the future.

The incorporation of total building commissioning (TBCx) in the design, construction, and operation of a new building can help eliminate any potentially frustrating failures of essential systems that do not operate according to specifications or as intended. The level of commissioning applied should be appropriate to the complexity of the project and its systems, to the owner's need for assurances, and to the budget and time available. For example, HVAC commissioning costs vary but are usually in the range of 1 to 4% of the value of the mechanical contract. It is prudent in this case to request several quotations.

A recent study entitled "Building Commissioning: A Golden Opportunity for Reducing Energy Costs and Greenhouse-Gas Emissions" by Evan Mills (2009) responds to an apparent lack of end-user confidence and understanding regarding the nature and level of energy savings that can be achieved through commissioning. The report tackles this issue head on primarily by assembling numerous case studies and previously unpublished data, in addition to incoporating performance benchmarks using standardized assumptions. Some of its key findings:

- Median commissioning costs were found to be between $0.30 and $1.16 per square foot for existing buildings and new construction, respectively, and 0.4% of total construction costs for new buildings.
- Median whole-building energy savings are between 16% and 13%.
- Median payback times are between 1.1 and 4.2 years.
- Median benefit-cost ratios are between 4:5 and 1:1.
- Cash-on-cash returns are between 91% and 23%.
- Very considerable reductions in greenhouse-gas emissions were achieved at a negative cost of $-\$110$ and $-\$25$/tonne CO_2-equivalent.
- High-tech buildings are particularly cost effective, saving large amounts of energy and emissions due to their energy intensiveness.
- Projects employing a comprehensive approach to commissioning attain nearly twice the overall median level of savings and five times the savings of projects with a constrained approach.
- Nonenergy benefits are extensive and often offset part or all of the commissioning cost.
- Limited multiyear post-commissioning data indicate that savings often persist for a period of at least 5 years.
- Uniformly applying median whole-building energy savings value to the stock of U.S. nonresidential buildings will yield an energy-savings potential of $30 billion by the year 2030 and annual greenhouse-gas emissions reductions of about 340 megatons of CO_2 each year. An industry equipped to deliver these benefits will have a sales volume of $4 billion per year and support approximately 24,000 jobs.

It is important to note that, while commissioning of building systems varies from one project to another, most projects generally entail equipment startup and HVAC systems, electrical, plumbing, communications, security, and fire management systems and their controls and calibration. Large or complex projects may include other systems and components. Cx usually begins with checking the documentation and design intent for reference. Performance testing of components is conducted when first arriving on the job site and again after installation is complete. The final step of commissioning is usually providing maintenance training and operation and maintenance (O&M) manuals.

Surprisingly, according to the *Whole Building Design Guide (WBDG)*, there are currently no building code requirements at a national level that mandate commissioning. However, studies repeatedly show that proper commissioning is cost effective and benefits all new construction and renovation building programs. Furthermore, recent case studies of private-sector facilities have concluded that the Cx process can significantly improve building energy performance by 8 to 30%. Formal commissioning of complex building types with highly integrated building systems can reap dramatic benefits. Indeed, the WBDG says that some government agencies (e.g., the General Services Administration, the U.S. Naval Facilities Engineering Command, and the U.S. Army Corps of Engineers) have adopted formal requirements, standards, or criteria for commissioning of their capital construction projects, but that the level of commissioning used depends on a number of factors, including available project funds.

11.2   FUNDAMENTAL COMMISSIONING BASICS: WHAT IS COMMISSIONING?

Building commissioning (Cx) is an all-inclusive process of systematic quality assurance that verifies building systems to be designed, installed, tested, and capable of being operated and maintained to perform interactively according to the design intent and the owner's operational needs. For new construction, the process ideally begins at a project's inception (i.e., the beginning of the design process) and continues through construction startup, inspection, testing, balancing, acceptance, training, and an agreed-on warranty period (i.e., occupancy and operations). It therefore encompasses all the necessary planning, delivery, verification, and risk management of critical functions performed in, or by, facilities. Cx also accomplishes higher energy efficiency, improved environmental health and occupant safety, and improved indoor air quality by making sure the building components are working correctly and that the plans are implemented with the greatest efficiency. It basically confirms that all systems are efficient and cost effective, that their installation has been adequately documented according to requirements written into the Contract Documents, and that the operators are adequately trained.

As a quality-assurance-based process, commissioning is intended to deliver preventive and predictive maintenance plans, tailored operating manuals, and training procedures for users to follow. Its principal function is therefore to ensure that the various systems such as HVAC&R and associated controls, domestic hot-water, lighting control, renewable energy (PV, wind, solar, etc.), and other energy-using building systems meet the owner's performance requirements, and perform and operate as intended and at optimal efficiency (Figure 11.1). For example, commissioning can:

- Ensure that a new building begins its life with systems at optimal productivity, which increases the likelihood that the building will maintain this level of performance throughout its useful life.
- Restore an existing building to high productivity.
- Ensure that building renovations and equipment upgrades function as designed.

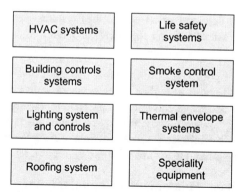

FIGURE 11.1   General scope for building commissioning and the major systems that typically need it.

11.3   BUILDING COMMISSIONING OBJECTIVES, BENEFITS, AND COSTS

1.3.1   Why Commision?

Most research clearly shows that the vast majority of building energy systems fail to function at their full potential. Irrespective of how carefully a building is designed, if the systems, equipment, and materials are not installed and operating as intended, the building will not perform well. Poor communication of design intent, inadequate equipment capacity, inferior equipment installation, insufficient maintenance, and improper system operation all adversely impact energy cost savings. And perhaps now more than ever, effective operation requires subsystems and components that work effectively and reliably with a building staff that has the knowledge and resources to operate and maintain them. This is sometimes difficult in today's competitive construction environment, where building owners and project team members have become increasingly cost-conscious and seldom allocate an adequate budget for quality-assurance processes. The nature of deficiencies frequently found in noncommissioned energy projects vary considerably and include

- Airflow problems
- Underutilized energy-management systems for optimum comfort and efficient operation
- Short cycling of HVAC equipment leading to premature failures
- Inadequate documentation of project installation/operational requirements during the warranty period
- Inappropriate heating and cooling sequence of operation
- Erroneous lighting and equipment schedules
- Erroneous calibration of controls and sensors
- Improperly installed or missing equipment

- Malfunctioning economizers (free cooling) systems
- Inadequate or lack of training for O&M personnel
- O&M manuals that are not specific to installed equipment

The outcome of poorly performing buildings can cause system and equipment problems that result in higher than necessary utility bills. Furthermore, any unexpected or excessive equipment repair and replacements due to premature failures cost the owner money. Having good indoor environmental quality (IEQ) helps mitigate employee absenteeism, tenant complaints, and turnover. It also minimizes the potential for lawsuits and expensive retrofits.

11.3.2   Benefits of Fundamental Commissioning

The intent of fundamental commissioning, according to LEED, is "to verify that the project's energy-related systems are installed and calibrated to perform according to the Owner's Project Requirements basis of design and construction documents. Benefits of commissioning include reduced energy use, lower operating costs, fewer contractor callbacks, better building documentation, improved occupant productivity, and verification that systems perform in accordance with the Owner's Project Requirements." Building commissioning offers many benefits that can be achieved no matter when the process begins, but the earlier it takes place, the greater the potential benefits will be. Since all modern building systems are integrated, a deficiency in one or more components can adversely impact the operation and performance of the others. Rectifying these deficiencies can therefore result in numerous benefits:

- Improved energy efficiency, which generally means lower utility bills (Figure 11.2)
- Improved occupant comfort and workplace performance
- Improved functioning of systems and equipment, which reduces design problems
- Reduced requests for information (RFIs) and Change Orders
- Provides faster and smoother equipment startup due to systematic equipment and control testing procedures
- Increased owner satisfaction
- Increased occupant safety
- Significant life-cycle extension of equipment/systems
- Enhanced environmental/health conditions by improving quality of indoor environment
- Improved building system/equipment reliability and maintainability
- Improved building documentation
- Shortened occupancy transition period and reduced post-occupancy corrective work
- Increased value as a result of better-quality construction

From the preceding list, we see that commissioning generally facilitates the delivery of a project that provides a safe and healthy facility, optimizes energy

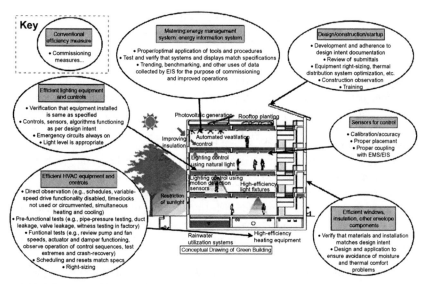

FIGURE 11.2 Illustration showing the relationships between commissioning and energy-efficiency measures. *(Source: From report prepared for the California Energy Commission Public Interest Energy Research by Mills, 2009.)*

use, reduces operating costs, ensures adequate O&M staff orientation and training, and improves installed building systems documentation. In addition, commissioning benefits owners through improved energy efficiency, improved workplace performance due to better IAQ, reduced threats and risks, and prevention of business losses. Some industry sources estimate that on average the operating costs of a commissioned building are between 8 and 20% below those of a noncommissioned building. Meanwhile, the cost of not commissioning is equal to the costs of correcting deficiencies plus those of inefficient operation. Commissioning is even more crucial for mission-critical facilities, as the significant cost of not commissioning can be measured in terms of downtime and interruption of required facility use, which can be quite substantial.

The most significant benefit of Cx is the result that come from better building control, which extends equipment life, improves operating efficiency through avoidance of frequent equipment cycling, and improves thermal comfort and IAQ. General building performance is also enhanced through commissioning by improved coordination of systems. Furthermore, it provides the owner with additional in-house knowledge for optimizing equipment, system, and control efficiencies, which helps to minimize occupant complaints and employee absenteeism and increase staff retention. Although difficult to quantify, it is estimated that the health and productivity benefits of a building with good IAQ are likely to be worth approximately five times the energy and operating cost savings of a building with poor IAQ.

Another advantage of commissioning is that it provides better upfront performance accountability since problem prevention is invariably less expensive than problem correction. Upfront performance accountability and quality control allow frequent comparison of consistent project construction with project design, thus providing rapid feedback to design professionals on the design's dynamic performance. Proper commissioning also considerably reduces the risk of liability from equipment failure or environmental hazards. In fact, companies typically use Cx on projects to ensure peak performance, which positively impact the bottom line and business continuity. Likewise, manufacturers find the commissioning process essential because of the high levels of environmental control and occupational safety required in their processes. Most government projects employ commissioning because mission-critical facilities support essential public infrastructures. While it is evident that projects with special performance needs, in particular, require proper commissioning, in fact all projects, if they are to perform satisfactorily, require some level of commissioning.

A number of factors are driving up the demand for commissioning of modern facilities, including performance needs and the desire to obtain green certification through programs such as LEED and Green Globes, which been developed to improve energy efficiency and environmental performance in buildings. Fundamental commissioning of building energy systems considered is a prerequisite for LEED V3 Certification (although enhanced commissioning can achieve two possible credits); it is also a requirement in Green Globes. Buildings certified by these rating systems are likely to include highly efficient power and lighting systems as well as photovoltaic and active/passive solar technologies, which, from an owner's perspective such sophisticated building technologies, should be accompanied by strict construction quality-assurance and performance verification measurements. Commissioning generally provides this. The new IgCC national green building code is quite significant, as it requires extensive pre- and post-occupancy commissioning and education of building owners and maintenance employees. Moreover, to comply with the new code, every project is required to choose an additional "elective," which pushes the envelope for the developer even further.

11.3.3   Goals and Objectives of Commissioning

A study prepared in October 2003 for a group of more than 40 California government agencies concluded that investing in green construction pays for itself 10 times over. Conducted by the Capital E Group at Lawrence Berkeley National Laboratory, with input from a number of state agencies, the study reflects the most definitive cost benefit analysis of green building to date.

Although the commissioning process is sometimes misinterpreted to mean focusing solely on testing at completion of the Construction Phase, in reality it is primarily a collaborative process for planning, delivering, and operating buildings that work as intended. ASHRAE defines commissioning as "the process of ensuring that systems are designed, installed, functionally tested, and capable of being operated and maintained to perform in conformity with the design intent... Commissioning begins with planning and includes design, construction, start-up, acceptance and training, and can be applied throughout the life of the building." This definition accurately depicts commissioning as a holistic process that spans from pre-design planning to post-construction operation, consisting basically of a checks-and-balances process that ensures a building's systems perform as intended. According to the WBDG, the main parts of commissioning are to:

1. Define and document requirements at the commencement of each phase and appropriate updates throughout the process.
2. Establish and document commissioning process tasks and responsibilities for subsequent phase delivery team members.
3. Verify and document compliance as each phase is completed.
4. Deliver construction projects that meet the owner's needs at the time of completion.
5. Verify that O&M personnel and occupants are properly trained.
6. Maintain facility performance across its life cycle

New construction project commissioning typically goes through the pre-design and design phases to establish a project owner's needs, goals, scope, and design solutions. The evaluation of proposed designs and constructed work can only be made by comparison with objective criteria and measures, which can be found in well-documented project requirements. Project development is a continuous learning process where building performance decisions undergo continuous refinement over the project's life cycle. The main commissioning activities supporting this principle include

- Comprehending key program goals and objectives
- Comprehending the needs of special building types
- Determining key threats, risks, and consequences
- Critical analysis of systems to facilitate achieving goals
- Conducting important commissioning programming activities

11.3.4 Factors Affecting the Cost of Commissioning

Commissioning costs vary with each project and depend on several factors such as the project's size, complexity, and the scope of the commissioning process. For this reason, commissioning costs are difficult to accurately estimate. There is no standard convention for determining what is included in the total cost of commissioning. Because it is difficult to define precisely, the cost of commissioning is most often presented as a range of potential rather than specific dollar amounts. But no matter how it is defined, that cost of commissioning generally accounts for only a small percentage of the overall construction budget and an even smaller percentage of the building's life-cycle costs. Some of the many factors that can impact the overall cost of commissioning include

- Size and type of building
- Number, type, and complexity of systems to be commissioned and the sample rate of like systems and equipment
- Phase at which commissioning begins (e.g. design, construction, or post-construction)
- Degree to which the commissioning authority (CxA) actively performs testing (as opposed to passively observing it)
- New construction or renovation
- Required deliverables (design intent document, commissioning plan, commissioning specification, O&M manuals, training plans, final report, etc.
- Commissioning process protocol: Does it include documenting and witnessing all equipment pre-startup and startup activities, pre-functional test procedures, functional test procedures, spot checks, etc.?
- Extent to which operators assist in testing (including future building operators in testing can help reduce the time required by the CxA).
- Costs allocation (e.g., commissioning consultant's fees, increased contractor bids, increased designer fees, O&M personnel time)
- Tools available (e.g., installed sensors, meters, trend logs)

11.3.5   Long-Term Cost Implications of Commissioning

The potential for long-term cost savings from building commissioning is considerable and, theoretically, can induce owners to perform it with payback as a major consideration. It is not surprising that studies show that commissioning costs per square foot tend to be higher for more complex buildings such as hospitals; therefore, as a result of their relatively high-energy intensity, commissioning payback has also been found to be lowest in these building types. For existing buildings, the median whole-building energy cost savings associated with Cx was found to be about 15%. Some of the potential long-range cost benefits of conducting an effective commissioning process are outlined here:

- Buildings that are properly documented are easier and less time-consuming to maintain, which translates into significantly lower operating and maintenance costs.
- A commissioned building is generally more energy-efficient and is therefore likely to consume less energy than if it had not been commissioned.
- Where IAQ controls have been commissioned and are operating properly, tenants and employees have been found to be more productive, have less absenteeism, and be less likely to contract "sick building syndrome" (SBS) or "building-related illness" (BRI).
- For specialized (e.g., industrial and research) facilities, the value of lost processes, experiments, and/or collections can be far greater than the cost of

commissioning or a potential product loss caused by improper control or malfunction of those systems.

11.4 PLANNING THE COMMISSIONING PROCESS

Since each building is in many ways unique, it is necessary for the commissioning process to be adapted to meet the specific needs of individual building projects. To get its full benefit, a plan should provide guidance in the execution of the commissioning process, preferably early in the design process. It is also very important to establish a clear method for sharing information at the earliest stages. Likewise, the plan should contain a process for identifying planning delivery team member roles, responsibilities, and tasks for the various project phases and activities. These include development and approval of commissioning plans, oversight of review and acceptance procedures, documentation compliance, commissioning schedules, and testing and inspection plans.

Planning must include identification of special testing needs for unique or innovative assemblies and measures that will ensure appropriate O&M training. This forms part of the Bid and Contract Documents and is binding on the contractor; it also outlines many of the contractor's responsibilities, procedures, and tasks throughout the Cx process. The specifications take precedence over the commissioning plan, included in which should be a full description of the functional performance testing (FPT) to be performed during the Acceptance Phase and culminating with staff training and warranty monitoring.

Normally, the commissioning process culminates in a final, complete commissioning report that is prepared and submitted to the owners along with drawings and relevant equipment manuals. This report should contain all documentation pertaining to the Cx process, procedures, and testing results, in addition to records of any deficiencies and accepted corrections. System commissioning requires specialized knowledge, which is why it is usually conducted by a mechanical consultant with appropriate experience and training. This consultant should be hired by and responsible directly to the project's owner and be independent of the mechanical consultant firm and general contractor. Where very large or complex projects are involved, it may be necessary to designate a special commissioning coordinator.

The architect or designer of record (DOR) is normally assigned responsibility for overseeing commissioning. In cases where total building commissioning is requested, this typically includes additional essential building systems such as exterior walls, plumbing, acoustics, and roofing. Having these systems commissioned can provide many advantages, including reducing moisture penetration and infiltration noise problems, and contributing to the building's energy and resource efficiency, in addition to facilitating occupant productivity. One proposed structure of the commissioning plan is shown in Table 11.1. Bear in mind that all information it contains must be project-specific.

TABLE 11.1 Proposed Commissioning Plan Structure

Element	Description
Introduction	Purpose and general summary of the plan
General project information	Overview of the project, emphasizing key project information and delivery methods
Scope	Building assemblies, systems, subsystems, and equipment to be commissioned
Team contacts	Team member contact information
Plan and protocols	Communication channels to be used throughout the project
Process	Details of tasks to be accomplished during planning, design, construction, and occupancy stages, with associated roles and responsibilities
Documentation	Documents required to identify expectations, track conditions/decisions, and validate/certify performance
Schedule	Specific sequences of events and relative time frames, dates, and durations

Source: U.S. General Services Adminstration, The Building Commissioning Guide, 2005.

11.4.1 Documentation: Compliance and Acceptance

As previously mentioned, commissioning serves as a general record of the owner's expectations for project performance during the project delivery process. It is a conscious team effort to document the continuity of the project as it progresses from one phase to the next. In the earliest phases (i.e., planning and development), we see the establishment of planning and programming documents that begin to define an owner's requirements, goals, and standards for building performance. Appropriately documenting project delivery in its entirety produces a chronological perspective that outlines and clarifies the iterative process of determining agreed-to project requirements at each phase of the development process. Commissioning documentation therefore becomes the road map for criteria to be met by facilities when they are put in service; it verifies that designed and installed systems meet the specified standards.

After the building is occupied, commissioning documentation ensures that the building can be maintained, retuned, or renovated to meet future needs. The OPRs are documented at project initiation and record compliance, acceptance, and operations throughout the facility's life cycle. These include:

A. *The Contractor is required to deliver to the Commissioning Authority one copy of the following as specified in the Cx Plan and other sections of the specifications and Contract Documents:*

    1. Shop drawings and product data relating to systems or equipment to be commissioned. The CxA shall review and incorporate any comments via the designated design engineer

    2. Start-up check lists along with the manufacturers start up procedures for installed equipment. CxA will review, assist, and recommend approval if appropriate

    3. Provide all System Test reports. CxA will review and compile prior to FPT

    4. Completed Equipment Start-up certification forms in addition to the manufacturer's field or factory performance and start up test documentation. CxA will review prior to FPT

    5. Completed Test and Balance Reports. CxA will review prior to FPT

    6. Equipment and other Warrantees

    7. Proposed Training Plans

    8. O&M Information per the requirements of the Cx Plan, Division 1 requirements

    9. Record Drawings

B. Record Drawings: Contractor is to maintain at the site an updated set of record or "as-built" documents reflecting actual conditions of installed systems.

The following is a list of commissioning activities and documentation adapted from a checklist provided by the U.S. Department of Energy, Office of Energy Efficiency & Renewable Energy:

1. **Owner's requirements:** List and describe the owner's requirements and basis-of-design intent with performance criteria.
2. **Commissioning plan:** Create the plan as early in the Design Phase as possible, including the management strategy, and list all features and systems to be commissioned.
3. **Bid Documents:** Integrate commissioning requirements in the construction Bid and Contract Documents. Designate Construction Specifications Institute (CSI) Construction Specification Section 01810, Division 1, for general commissioning requirements. Use the unassigned Sections 01811 through 01819 to address requirements specific to individual systems. Notify the mechanical and electrical subcontractors of Division 15 and 16 commissioning requirements in Sections 15995 and 16995.
4. **Functional performance test procedures and checklists:** Develop functional performance test procedures or performance criteria verification checklists for each element identified in the commissioning plan.
5. **Commissioning Report:** Complete a final commissioning report and submit it to the project owner. The report should summarize all the tasks, findings, and documentation of the commissioning process and should address the actual performance of building systems in reference to the design documents. This report should identify each component, piece of equipment, system, or feature, including the results of installation observation, startup and checkout, operation sampling, FPT, and performance criteria verification. All test reports by various subcontractors, manufacturers, and controlling authorities should be incorporated into the final report.

6. **Training:** Assemble written verification that training was conducted for appropriate personnel on all commissioned features and systems.

7. **Operation and maintenance manuals:** Review O&M manuals for completeness, including instructions for installation, maintenance, replacement, and startup; replacement sources; parts lists; special tools; performance data; and warranty details.

8. **Recommissioning management manual:** Develop an indexed recommissioning management manual containing guidelines for establishing and tracking benchmarks for whole-building energy use and equipment efficiencies, recommendations for sensor recalibration frequency, list of all user-adjustable set points and reset schedules, and list of diagnostic tools.

9. **Acceptance Phase:** This is not strictly a separate phase of the building delivery process, but it is during this period that the facility and its systems and equipment are inspected, tested, verified, and accepted, including performance testing of equipment and systems, fire system verification, final punch list development, code official inspections, Certificate of Occupancy, and the like. Additionally, it is during this phase that most of the formal training occurs, which generally includes requirements after the Construction Phase is substantially complete and the building is occupied. The architect/engineer (A/E) and contractor now finalize the "as-built" or record documentation. The end of this phase is marked by an "Approved Functional Completion" document.

Most of this section of the *Whole Building Design Guide* is based on the commissioning process recommended in *ASHRAE Guideline 0-2005 & Total Building Commissioning*. It is strongly recommended that project teams follow the process outlined in this guideline or that found in the *Total Building Commissioning Process (TBCxP)*. *Guideline 0* has been adopted by both ASHRAE and the National Institute of Building Sciences (NIBS) and does not focus on specific systems or assemblies, but adheres to a standard process that can be used to commission any building system critical to a building's functioning.

The NIBS Total Building Commissioning Program™ is currently working with industry organizations to develop a set of 11 (eventually 18) commissioning guidelines for various systems and assemblies related to total building commissioning. The Acceptance Phase is of particular importance to innovative and unique (e.g., sustainable) buildings. Sometimes this phase may include training and the development of the system manuals. The scheduling and clearness of Acceptance Phase tasks are very important because they provide information on what was delivered and information the owner can use to facilitate successful operation and maintenance of all commissioned building components and systems.

11.5   COMMISSIONING AUTHORITY (COMMISSIONING SERVICE PROVIDER)

One of the first and most important decisions that a building owner needs to make is selecting the commissioning authority (CxA), because that entity heads up the commissioning team, facilitates and is responsible for the entire commissioning process. The CxA (sometimes referred to as the commissioning service provider or commissioning agent) is an objective, independent advocate for the building owner. While many stakeholders have crucial roles to play in the Cx process, the CxA is key. The commissioning authority consists of a team of senior specialists who direct and oversee the commissioning process. According to LEED, "the Commissioning Authority (CxA) is typically a third-party advocate for the owner, and LEED requires the CxA for the project to be independent of the design team. The CxA should focus on the process and have a strong background encompassing design, construction, operation, and quality process control."

The CxA should be retained early in the project's programming phase and will have various responsibilities, including the following:

- Reviewing component and equipment submittals by contractors, the various systems to be commissioned, and the contractor's pre-functional/startup checklists
- Generally providing technical and procedural oversight during the different phases of the project
- Conducting functional performance testing during construction
- Validating and testing, adjusting, and balancing (TAB)
- Directing the project's functional acceptance testing and warranty phases

Moreover, the CxA is also required to review and provide support in training, as-built documentation, O&M manuals, and handover of the facility to the project owner. After completion of the project, the Warranty Phase begins; the CxA is expected to periodically monitor the facility during this phase in order to optimize it with the actual occupancy.

11.5.1 Commissioning Authority's Responsibilities

The commissioning authority organizes and leads the commissioning team and essentially guides the process for all commissioned systems. During the preconstruction phase, this includes peer review of design submittals, ensuring that owner's expectations are adequately documented in the OPRs, producing the systems manual, and reviewing all contractor submittals (see sidebar). It should

Responsibilities of the Phases

Construction Phase
- Organize, chair, and prepare meeting minutes for Cx meetings.
- Review and assist in documenting all commissioning requirements to be included in the specifications. Prepare and update the commissioning plan as work progresses.

Continued

Responsibilities of the Phases—cont'd

- Review relevant project documentation such as shop drawings, TAB reports, product data, record drawings, and O&M data for compliance and to ensure system functionality.
- Develop, maintain, and approve all review documents, startup checklists, and issue logs (e.g., design, submittal, construction, site visit, and startup reviews; functional testing reports; O&M manuals to ensure completeness and applicability; and training and end-of-warranty reviews).
- Observe construction and attend progress meetings, as required, to observe progress and assist other parties, facilitate the Cx process, and help expedite completion. Although the CxA is responsible for collecting and compiling all checklists and test and data forms, it does not direct work or approve/accept materials, systems, or equipment.
- Monitor installation and periodically inspect systems and equipment during the first year of occupancy, conduct any deferred testing, and serve as a resource for the building staff.
- Prepare and submit the Final Commissioning Report.
- Compile O&M information and systems overview and format O&M manuals.
- Witness selected tests startups and equipment training.

Acceptance Phase
- Verify, test, adjust, and balance by spot-checking all TAB reports, control component calibration, and equipment performance certifications (test 100% of key systems or a sample percentage).
- Analyze all trend logs.
- Test all equipment and systems to ensure correct operation and functioning as per specifications, including failure and safety modes.
- Review training plans and coordinate training activities between the O&M staff and contractors/vendors to ensure appropriate staff training.
- Record commissioning procedures and provide the Cx report with testing documentation.
- Verify that contractors/vendors provide proper O&M material (fan curves, pump curves, operating parameters, etc.), not just the equipment-mounting information.
- Follow through to ensure that all commissioning issues are resolved.

Warranty Phase
- Discuss with building users any problem areas that have developed after building acceptance and verify corrections made by appropriate subcontractors.
- Assist the owner and facility staff in developing reports and documents and requests for services to resolve outstanding problems and issues with the contractor and the designers.
- Provide follow-up training to O&M staff, especially for new staff that was assigned to the building and did not previously receive vendor training.
- Check building performance and conduct seasonal and other deferred testing on systems as required by the specifications.
- Make suggestions for improvements and identify areas that may come under warranty.

be noted that the construction contractor is not responsible for delivering the design intent, nor is the A/E responsible for installation.

11.5.2   The Commissioning Team

According to the Building Commissioning Guidelines developed by Energy Design Resources,

Commissioning is a team process in which members of the project team each play defined roles. The commissioning team often includes the building owner or project manager, commissioning provider, design professionals, general contractor, subcontractors and manufacturer's representatives. For LEED projects, the LEED coordinator should also be a member of the commissioning team.

Futhermore, "the team may also include facility staff, testing specialists or utility representatives. It is important to remember that the commissioning team does not manage the design and construction of the project. It merely promotes communication among team members to identify and resolve issues in a collegial and systematic fashion."

The individuals on the commissioning team, through coordinated actions, are responsible for implementing the process and are led by the CxA. All traditional parties to the design and construction process are vital to the commissioning process and have roles to play as part of the CxA team. Those roles generally provide an extra focus to the team's efforts and in some cases delineate required assignments and rules that are normally included in the traditional process but often ignored or poorly executed. It is important for all commissioning team members to be involved as early as possible in the project to allow the valuable input of their knowledge and experience into the design and to allow them to become active participants in the initial checkout and acceptance of the facility.

The initial step in the commissioning process is for the CxA to develop a commissioning plan (preferably at project inception phase) and then identify and lay out the composition of the commissioning team. This should be followed by a scoping meeting, which all team members are required to attend. The purpose of this meeting is largely to outline the roles and responsibilities of team members and to describe the commissioning process and its scheduling. It is necessary that the commissioning team work as a cohesive unit so that all of the steps in the Cx process are completed and the facility objectives are met.

In general, roles and responsibilities within the commissioning team do not change. The owner/user is normally responsible for clearly communicating the facility's needs and for explaining the design and functional intent. The architect/engineer is responsible for designing a facility that accomplishes what the user wants and that is in compliance with all regulations and accepted practices. Construction contractors, subcontractors, and vendors are responsible for supplying and installing the facility in accordance with the Contract Documents.

The size and number of members that comprise the commissioning team will vary depending on project size, type, and complexity. However, in many cases, the team will include:

- Project/facility owner
- Commissioning agent (CxA)
- Project manager
- Users
- Operating personnel
- Architect/engineer
- Structural, mechanical, electrical, LEED/sustainability, elevator, fire protection, seismic, and other technical experts
- Construction manager agent (CMa)
- Construction contractor and subcontractors

General descriptions of the commissioning team roles and responsibilities, according to the Army Corps of Engineers LEED Commissioning Plan Template; it is based on Platform Environment Control Interface (PECI) model commissioning plans, as follows:

- *Commissioning Authority:* Coordinates the Cx process; develops and updates the Cx plan, assists, reviews, and approves incorporation of commissioning requirements into construction documents; writes or approves tests; oversees and documents performance tests; and develops the commissioning report.
- *Professional Engineer:* Facilitates the Cx process; coordinates between the general contractor and the CxA; approves test plans and signs off on performance; performs construction observation; approves O&M manuals (design−bid−build contracts).
- *General Contractor:* Facilitates the Cx process, ensures that subcontractors perform their responsibilities, and integrates Cx into the construction process and schedule.
- *Subcontractors:* Demonstrate proper system performance.
- *Designer of Record:* Develops and updates basis of design, incorporates commissioning requirements in construction documents, observes construction, approves O&M manuals (design−build contracts) and assists in resolving problems.
- *Project Manager:* Facilitates and supports the Cx process.
- *Manufacturers/Vendors:* Equipment manufacturers and vendors provide documentation to facilitate the commissioning work and perform contracted startup.

11.5.3   Commissioning Authority Qualifications and Certification

Although the building industry at times remains divided on which party should be the commissioning authority, an independent CxA is strongly recommended—in other words, neither the contractor, the A/E, nor the CM. The CxA should be motivated solely by the needs of the owner and the facility user and should not be a competitor of the A/E or contractor. Individual CxA team members should be highly specialized in the types of facilities and systems to be commissioned (LEED, for example, has specific requirements in this respect), and the more complex the project, the more experience required of the CxA. Selecting a CxA team with directly relevant experience is of particular importance on projects with special or mission-critical needs such as hospitals or labs. Because of the level of technical oversight that is expected, the CxA should be a Certified Commissioning Professional (CCP) or a Licensed Professional Engineer or have applicable experience in the specialized systems/facilities being installed, in addition to extensive experience in design, optimization, remediation, and acceptance testing of applicable systems as well as building manual preparation and training.

Today, we find that many building projects now require performance certifications such as LEED, Green Globes, ENERGY STAR®, OSHA, and others. To obtain certification, requirements have to be determined during the planning and design phases so that commissioning for certifications can be included in the OPRs and commissioning plans. Several organizations, including The American Institute of Architects (AIA), are formulating new programs and training and Contract Documents to assist their members in offering building commissioning as an additional service to their clients. The Building Commissioning Association has created the CCP program to raise professional standards and provide a means of certification in the building commissioning industry. To earn CCP certification, individuals are required to complete an application form that is reviewed by the Building Commissioning Certification Board, and pass a two-hour written examination. Likewise, ASHRAE now offers an exam for Cx process management professionals with the intention of helping building owners and others find qualified people to lead the commissioning process.

The growing complexity of today's building designs and equipment has resulted in a greater emphasis on the building commissioning process as a quality assurance measure. The cost ramifications for delayed occupancy and the benefits of early detection of design and installation faults on their own provide more than adequate economic justification for the majority of commissioning projects. Commissioning can employ various methods that focus on building systems and assemblies and can be readily customized to suit specific project needs. However, whatever the commissioning approach and system focus, a clear articulation of performance expectations; rigorous planning and execution; and comprehensive project testing, operational training, and documentation is crucial to achieving success at the end of the day.

Commissioning and retro-commissioning (RCx) a building should not be seen merely as means to save energy or reduce the payback period of investments; among other things, they help the environment, produce healthier buildings, improve the economic performance of a building, and increase productivity. Additionally, if a building is seeking LEED Certification for new construction or for an existing building, commissioning will invariably be required. For the latest updates to the changes that have been made to commissioning requirements, owners should look at the latest version of LEED (currently V3) released by the USGBC.

11.6   THE COMMISSIONING PROCESS

In most cases, formal commissioning has now become a necessity because the majority of modern buildings incorporate complex and digitally controlled HVAC systems and integrated natural ventilation systems; some, especially if they are "green," incorporate renewable energy, onsite water treatment, occupancy sensor lighting controls, and other high-technology innovations. The building commissioning process is generally interwoven with overall project delivery; however, it is not usually requested for projects with minimal mechanical or electrical complexity such as typical residential projects.

11.6.1   Commissioning Process

The following is from Section 01 91 00: General Commissioning Requirements.[1] It is "Guidance for designers and specifiers, with suggested language to be modified and incorporated into project specifications," and provides a brief overview of the typical commissioning tasks during construction and the general order in which they occur:

1. Commissioning during construction begins with an initial commissioning meeting conducted by the CxA where the commissioning process is reviewed with the commissioning team members.
2. Additional meetings will be required throughout construction, scheduled by the CxA with necessary parties attending, to plan, coordinate, schedule future activities and resolve problems.
3. Equipment documentation is distributed by the A/E to the CxA during the normal submittal process, including detailed startup procedures.
4. The CxA works with the Contractor in each discipline in developing startup plans and startup documentation formats, including providing the contractor with construction checklists to be completed during the installation and startup process.
5. In general, the checkout and performance verification proceeds from simple to complex; from component level to equipment to systems and intersystem levels with construction checklists being completed before functional testing occurs.
6. The contractors, under their own direction, will execute and document the completion of construction checklists and perform startup and initial checkout. The CxA documents that the checklists and startup were completed according to the approved plans. This may include the CxA witnessing startup of selected equipment.
7. The CxA develops specific equipment and system functional performance test procedures.
8. The functional test procedures are reviewed with the A/E, CxA, and contractors.

---

1 Source: BuildingGreen, Inc. 2007.

9. The functional testing and procedures are executed by the Contractors under the direction of, and documented by, the CxA.
10. During initial functional tests and for critical equipment, the Engineer will witness the testing.
11. Items of noncompliance in material, installation, or setup are corrected at the contractor's expense and the system is retested.
12. The CxA reviews the O&M documentation for completeness.
13. The project will not be considered substantially complete until the conclusion of commissioning functional testing procedures, as defined in the Commissioning Plan.
14. The CxA reviews and coordinates the training provided by the Contractors and verifies that it was completed.
15. Deferred testing is conducted as specified or required.

Commissioning is an integral prerequisite in LEED Certification. For its New Construction, Commercial Interiors, Schools, and Core and Shell categories, LEED has two commissioning components: (1) fundamental commissioning, a prerequisite (i.e., obligatory); and (2) enhanced commissioning, which receives up to two possible credits but is not a prerequisite.

For optimum results, commissioning should start at the pre-design stage and take place through all phases of the building project. A commissioning agent should also be designated as early as possible in the project—again, ideally during pre-design. While it is beneficial to employ a third-party CxA to provide a more comprehensive design and construction review, it is nevertheless acceptable to appoint a qualified member of the design team as the CxA, providing there is no conflict of interest. The CxA is required to serve as an objective advocate of the owner, direct the commissioning process, and deliver to the owner final recommendations regarding the performance of commissioned building systems.

Also, the CxA is expected to lead the commissioning process and to introduce standards and strategies early in the design process. Additionally, the CxA should ensure the implementation of selected measures by clearly stating all requirements in the construction documents. After completion of construction, the CxA verifies that all systems and equipment meet minimum requirements as per the Contract Documents and are operating as designed and intended. The CxA should also provide guidance on how to operate the building at maximum efficiency.

11.6.2   Fundamental Commissioning

As noted earlier, the LEED intent of fundamental commissioning is "to verify that the building's energy-related systems are installed, calibrated, and perform according to the Owner's Project Requirements, basis of design and construction documents." As mentioned, fundamental commissioning is a prerequisite for

LEED Certification (unlike Enhanced Commissioning which is a credit and not a prerequisite) and is required for both new construction and major retrofits, as well as for medium or large energy management control systems that incorporate more than 50 control points. It is especially important for large or very complex mechanical or electrical systems are in place or where the onsite renewable-energy generation systems (e.g., solar hot water heaters or photovoltaic arrays) and should also be considered when innovative water conservation strategies, such as composting toilets or graywater irrigation systems, are installed.

If LEED Certification is to be pursued, the commissioning team will need to implement the following commissioning activities:

- The owner or project team must designate an individual as the CxA to lead, review, and oversee the commissioning process activities until completion. This individual should be independent of the project's design or construction management unless the project is smaller than 50,000 square feet.
- The designated CxA should have documented commissioning authority experience with at least two building projects, and ideally meet the minimum qualifications of having an appropriate level of experience in energy systems design, installation, and operation, as well as commissioning planning and process management. LEED recommends that a designated CxA have hands-on field experience with energy systems performance, startup, balancing, and troubleshooting, and knowledge about energy systems, automation control, testing, operation, and maintenance procedures.
- The CxA should clearly document and review the OPRs and the basis-of-design (BOD) for the building's energy-related systems (usually developed by the A/E). Updates to these documents must be made during design and construction by the design team. The commissioning process does not absolve or diminish the contractor's obligation to meet the contract requirements.

Design Phase commission for both fundamental and enhanced commissioning is intended to achieve a number of specific objectives, including the following:

- Ensuring that the Owner's Project Requirements—design and operational intent—are clearly documented and fully understood. The OPRs detail the functional requirements of the different building systems from the owner's perspective and should be fully measurable and verifiable. They include facility use, occupant comfort, and project success. Where the owner lacks sufficient experience to formally document these requirements, the CxA may facilitate the process by conducting a workshop on OPR development.
- Verifying and ensuring that the OPR recommendations are communicated to the design team during the design process so they can develop a BOD document that appropriately describes system configurations and control sequences that will be put in place to meet the OPRs and avoid later modifications to the contract.
- Ensuring that the commissioning process for the Construction Phase is appropriately reflected in the construction documents (CDs). The CxA

conducts design reviews in the context of the BOD, preferably be able to perform an initial review prior to 50% completion of CDs. The CxA is also required to provide the architect with specifications for incorporation into the Contract Documents. All tasks to be performed during commissioning are described in the commissioning plan developed by the CxA.

Prior to completion of the design process, the CxA develops a Construction Phase commissioning plan and reports results, findings, and recommendations directly to the owner. However, the owner and the design team are responsible for updating their respective documents. Construction Phase commissioning for both fundamental and enhanced commissioning is intended to achieve a number of Contract Document objectives, including:

- Commissioning requirements and OPRs should from the outset be incorporated into the CDs. During and immediately prior to the Construction Phase, the CxA may review contractor submittals related to the systems to be commissioned.
- In developing and implementing a proper commissioning plan, the CxA typically establishes protocols for functional performance testing based on project specifics and the sequence of operations developed by the controls engineer and the Contract Documents. Teamwork and accountability should be strongly encouraged.
- The CxA should hold a kick-off meeting with the contractors and other stakeholders.
- Verification and documentation should be provided showing that the installation and performance of energy-consuming equipment and systems meet the OPRs and the BOD. After equipment startup, the CxA should conduct periodic pre-functional checks of installation progress to make sure that system mounting allows easy and safe O&M access to ensure proper maintenance over the life of the building.
- Verification and documentation should be provided showing that all equipment and systems in place are installed according to the manufacturers' recommendations and to industry-accepted minimum standards. Once the equipment is fully installed, the CxA will conduct functional performance testing to evaluate performance for all sequences of operation. Because some functional testing can only be conducted in certain seasons, the commissioning process usually extends beyond the completion date of the construction.
- A Final Commissioning Report should be completed that includes an executive summary, a list of participants and roles, a brief building description, an outline of commissioning and testing scope, and an overall description of testing and verification methods used during commissioning.
- The CxA verifies that O&M documentation left onsite is complete. Moreover, after completion of the commissioning process, the CxA's final report may include the preparation of an O&M manual for the project.

- Verify that training of the owner's operating personnel is adequate to operate and maintain all equipment and systems, and to maintain master Cx "issue log" throughout construction.

For existing buildings, continuous maintenance is very important, especially since building systems over the years tend to become less efficient, mostly because of changing occupant needs, building renovations, and obsolete systems that end up causing occupant discomfort and complaints. Unless these problems are appropriately addressed, such as by investing in a commissioning process, a facility's operating costs will dramatically increase, making it less attractive to new and existing tenants. Although commissioning needs may differ from project to project, commissioning the building envelope systems, power distribution, domestic water heating, ductwork, and any hydronic piping system is strongly recommended for any project. The many advantages of commissioning has been found to typically pay for itself in less than a year. Figure 11.3 shows a sample checklist for a completed project.

11.6.3   Enhanced Commissioning

For a LEED credit (2 points), enhanced commissioning is required in addition to the fundamental commissioning prerequisite. The intent of enhanced commissioning, according to LEED, is "to begin the commissioning process early in the design process and execute additional activities after systems performance verification is completed." For the Commercial Interiors category, for example, the intent is to verify and ensure that the tenant space is designed, constructed, and calibrated to operate as intended, which requires commissioning process activities in addition to the fundamental commissioning prerequisite requirements, as stated in the relevant *LEED Reference Guide* (e.g., "Green Building Design and Construction," 2009). As stated by LEED, the duties of the CxA are as follows:

1. Prior to the end of design development and commencement of the construction documents phase, a CxA independent of the firms represented on the design and construction team, must be designated to lead, review and oversee the completion of all commissioning process activities. Although it is preferable that the CxA be contracted by the Owner, for enhanced commissioning the CxA may also be contracted through design firms or construction management firms not holding construction contracts. This person can be an employee or consultant of the owner, although this requirement has no deviation for project size. Furthermore to meet LEED requirements this person must:

   - have documented commissioning authority experience in at least 2 building projects
   - be independent of the project's design and construction management

- not be an employee of the design firm, though the individual may be contracted through them
- not be an employee of, or contracted through, a contractor or CM of the construction project

2. The CxA must report all results, findings and recommendations directly to the owner.

HVAC Equipment and System

- ☐ Variable Speed Drives
- ☐ Hydronic Piping Systems
- ☐ HVAC Pumps
- ☐ Boilers
- ☐ Chemical Treatment Systems
- ☐ Air Cooled Condensing Units
- ☐ Makeup Air Systems
- ☐ Air-Handling Units
- ☐ Underfloor Air Distribution
- ☐ Centrifugal Fans
- ☐ Ductwork
- ☐ Fire/Smoke Dampers
- ☐ Automatic Temperature Controls
- ☐ Laboratory Fume Hoods
- ☐ Testing, Adjusting, and Balancing
- ☐ Building/Space Pressurization
- ☐ Ceiling Radiant Heating
- ☐ Underfloor Radiant Heating

Electrical Equipment and System

- ☐ Power Distribution Systems

- ☐ Lighting Control Systems
- ☐ Lighting Control Programs
- ☐ Engine Generators
- ☐ Transfer Switches
- ☐ Switchboard
- ☐ Panelboards
- ☐ Grounding
- ☐ Fire Alarm and Interface Items with HVAC
- ☐ Renewable Energy Systems
- ☐ Security System

Plumbing System

- ☐ Domestic Water Heater
- ☐ Air Compressor and Dryer
- ☐ Stormwater Oil/Grit Separators

Building Envelope

- ☐ Building Insulation Installation
- ☐ Building Roof Installation Methods
- ☐ Door and Window Installation Methods
- ☐ Water Infiltration/Shell Drainage Plan

FIGURE 11.3 Detailed checklist of systems to be commissioned on completion of a project. *(Source: Adapted from BuildingGreen.com, Section 01 91 00 of the General Commissioning Requirements.)*

3. The CxA must conduct a minimum of one commissioning design review of the OPRs' BOD, and design documents prior to the mid-construction documents phase, and back-check the review comments in the subsequent design submission.

4. The CxA must review contractor submittals and confirm that they comply with the OPRs and BOD for systems being commissioned. This review must be conducted in parallel with the review of the architect or engineer of record and submitted to the design team and the owner.

5. The CxA or other members of the project team are required to develop a systems manual that provides future operating staff with the necessary information to understand and optimally operate the commissioned systems. For Commercial Interiors, the manual must contain the information required for re-commissioning the tenant space's energy-related systems.

6. The task of verifying that the requirements for training operating personnel and building occupants have been completed may be performed by either the CxA or other members of the project team.

7. The CxA must be involved in reviewing the operation of the building with O&M staff and occupants and having a plan in place for resolving outstanding commissioning-related issues within 10 months after substantial completion. For Commercial Interiors, there must also be a contract in place to review tenant space operation for O&M staff and occupants.

11.6.4 Retro-Commissioning: Commissioning for Existing Buildings

Although building commissioning has become a critically important aspect of new construction projects and is used primarily to ensure that all installed systems perform as intended, the reality is that most existing buildings have never been subjected to commissioning or quality-assurance process and, not unexpectedly, have been found to be performing well below their intended design potential. RCx can address problems that have occurred during design or construction or those that may have developed throughout the building's life. In 1996, the Lawrence Berkeley National Laboratory (LBNL) conducted a study (Association of State Energy Research Technology Internships and DOE) of 60 different types of buildings that confirms this inefficiency, showing that

- Fifty percent, or more, had control problems.
- Forty percent had HVAC equipment problems.
- Fifteen percent had missing equipment.
- Twenty-five percent had building automation systems (BASs) with economizers, variable frequency drives (VFDs), and advanced applications that were simply not operating correctly.

The term retro-commissioning simply refers to the commissioning of existing buildings not previously commissioned and usually focuses on energy-using equipment (e.g., mechanical, lighting, and related controls), with the objectives being to reduce energy waste, obtain energy cost savings for the owner, and identify and fix existing problems, using diagnostic testing and O&M tune-up activities. The Building Commissioning Association defines RCx, or commissioning for existing buildings, as "a systematic process for investigating, analyzing, and optimizing the performance of building systems by improving their operation and maintenance to ensure their continued performance over time... [to] make the building systems perform interactively to meet the owner's current facility requirements." It is important to first determine how the installed systems are designed to operate, measure and monitor their operation, and then prepare a prioritized list of the operating opportunities of the varous systems.

The RCx process basically reviews the functionality of equipment and systems installed and optimizes how they work together to facilitate the reduction of energy waste, increase comfort, and improve building operation. RCx may also be required to address issues such as modifications to system components; function/space changes from the original design intent; failure to operate according to designed benchmarks; and complaints regarding IAQ, temperature, BRI, SBS, and the like. Figure 11.4 shows a CEE hospital whose owner decided to recommission (ReCx). The first floor was built in 1981; the upper floors were added in 1982. The owner's objective was to reduce operating costs while maintaining or improving IAQ and comfort.

FIGURE 11.4 This CEE hospital, a nine-story, 600,000-square-foot acute care facility in Minneapolis, recently underwent recommissioning. For this project, CEE partnered with the Energy Systems Lab at Texas A&M University, which is considered to be the most experienced and effective health-care facility recommissioning provider in the United States.
(Source: Center for Energy and Environment.)

According to the Center for Energy and Environment,

Opportunities for recommissioning measures were identified through field measurements carried out by highly-skilled and experienced engineers to quickly zero in on sub-optimal central-system operating strategies that waste energy. Major recommissioning opportunities identified included:

1. *Calibration of control system instrumentation;*
2. *Resetting supply air temperature set point;*
3. *Resetting duct static pressure set point;*
4. *Replacing bad inlet guide vanes with VFDs;*
5. *Calibration of VAV terminal boxes;*
6. *Improving economizer operation;*
7. *Optimizing the chiller and chilled water pump operation;*
8. *Performing hot water and chilled water balance;*
9. *Optimizing heating water temperature reset schedule and on/off sequence;*
10. *Reducing outside air flow;*
11. *Calibration of thermostats;*
12. *Performing air balance;*
13. *Determining the minimum outside air damper position; and*
14. *Repairing linked flex ducts and leaky reheat control valves.*

ReCx is a systematic process used to diagnose, identify, and correct performance problems in existing buildings that might otherwise prevent key central-system functions from being fully implemented; it ensures that these systems continue to operate optimally for the life of the facility. The key goals of an ReCx program typically include:

- Optimization of energy consumption
- Reduction of energy use
- Identification of chronic maintenance problems
- Improvements in building comfort, IAQ, lighting, and so forth.

All operating improvements made should be recorded, and the building operator should be trained on how to sustain efficient operation as well as implement capital improvements. RCx continues to witness increasing prominence as a cost-effective strategy for improving energy performance and helping to make the building's systems perform interactively in a manner that addresses the owner's current and anticipated facility requirements.

Recommissioning (ReCx) applies mainly to buildings previously commissioned or retro-commissioned, for which the original commissioning process documentation shows that the building systems performed as intended at one point in time. The intent of recommissioning, therefore, is to help ensure that the benefits of the initial commissioning or retro Cx process remain valid. The need for recommissioning depends on several factors, among them changes in function and use of the facility, quality and schedule of preventive maintenance activities, and frequency of operational problems. In some cases, ongoing commissioning may be necessary to resolve operating problems, improve comfort, optimize energy use, and identify energy and operational retrofits for existing buildings. Periodic ReCx may also take place when a building previously commissioned undergoes another commissioning process to help keep it operating optimally.

11.6.5 Warranty Phase

Following construction completion and turnover, a building goes into the hands of the owner and operators. However, although the project may be considered complete, some commissioning tasks from the initial commissioning contract continue throughout the typical one-year warranty period.

Systems Performance Monitoring

Systems performance monitoring includes early occupancy of the building and continues through the Warranty Phase, during which the commissioning agent, representatives of the owner, the A/E team, and the contractor's team verify that ongoing system performance is in line with specifications. This is achieved by repeating selected systems' functional performance tests and by reviewing energy bills and other performance-related documentation. The CxA prepares a report to the owner stipulating any issues and/or confirming that the systems are functioning as designed and intended. Then the contractor updates and finalizes all documentation, reflecting actual conditions at the end of the warranty period, and prepares final modifications to the O&M manuals and as-builts resulting from testing.

According to the GSA's *Building Commissioning Guide*, the CxA is responsible for delivering a Final Commissioning Report during the post-construction period. As stated by the GSA in the "Final Commissioning Report," the following must, at a minimum, be included:

- *A statement that systems have been completed in accordance with the Contract Documents and that the systems are performing in accordance with the final Owner's Project Requirements document.*
- *Identification and discussion of any substitutions, compromises or variances between the final design intent, Contract Documents and as-built conditions.*
- *Description of components and systems that exceed Owner's Project Requirements and those [that] do not meet the requirements and why.*
- *Summary of all issues resolved and unresolved and any recommendations for resolution.*
- *Post-construction activities and results including deferred and seasonal testing results, test data reports and additional training documentation.*
- *Lessons learned for future commissioning project efforts.*
- *Recommendations for changes to GSA standard test protocols and/or facility design standards (i.e., GSA P-100, etc.).*

The importance of the Final Commissioning Report is that it serves as a pivotal reference and benchmark document for any future recommissioning of the facility.

Deferred Seasonal Testing

Sometimes testing is delayed because of site or equipment conditions or inclement weather, but is to be completed during the warranty period. Likewise, functional testing is performed on systems that cannot be tested during the project's Acceptance Phase because of seasonal load issues that prevent reasonable testing or because operation of certain integrated systems is dependent on seasonal building loads. Functional performance testing is scheduled in the opposite season from the one in which it was initially conducted to confirm successful operation of the integrated systems under the building loads specified. These tests may also be necessary to demonstrate the performance of the occupied building where insufficient internal loads prevented the CxA from adequately challenging the systems during initial testing.

The requirements for deferred and seasonal testing need to be clearly defined in the Contract Documents to avoid confusion, as the CxA and some contractor personnel will be required to return to the site after the project is completed—more specifically, a few months prior to expiration of the contractor's one-year warranty—to confirm that all facility systems are operating as planned and to interview facility staff and assist them in addressing outstanding performance problems or warranty issues, particularly before the warranty period expires. It is advisable to put aside money for the execution of this activity in addition to the traditionally withheld warranty amounts.

Post-Occupancy Testing

In performing testing during the post-occupancy phase, the CxA or a representative must take care not to void any equipment warranties. The building owner should require that the contractor and subcontractors provide the commissioning provider with a full set of warranty conditions for all equipment to be commissioned, because some warranty provisions require that the installing contractor perform the testing under the supervision of the commissioning provider. To meet warranty requirements, the CxA should, in all cases:

- Revisit the site once operators have become acquainted with their systems and have additional questions about them that may not have been fully understood during training.
- Provide follow-up training to O&M staff, especially if new staff have been assigned and so did not receive vendor training.
- Speak to building users to identify any problem areas that may have developed after building acceptance, and review equipment malfunctions or system operation issues to determine whether corrective work under warranty coverage should be requested and, if so, by which subcontractor or supplier.

- Provide assistance to maintenance personnel in documenting occupant complaints to accurately determine whether real equipment and/or system problems are the cause.
- Perform appropriate seasonal testing and check building performance.
- Provide the owner with adequate support to enable resolution of issues with the contractor and designers.
- Conduct an eleventh-month walk-through to seek out any system problems prior to expiration of the warranty.

We find that the high-performance building movement of recent years, in addition to various energy rating prerequisites, has brought building commissioning well into the mainstream. Moreover, with the increasing complexity of mechanical systems and the continuous development of new technologies, the process of total building commissioning (TBCx) has taken on an increasingly important role. As mentioned earlier, Cx entails more than commissioning typical systems such as HVAC; it includes other elements such as lighting systems and controls, as well as building envelope and fenestration to ensure a building's optimum performance. It is advisable for owners to consider recommissioning their facilities periodically to ensure that equipment performance levels are maintained as originally intended.

We also find that during these challenging times, particularly since the 9/11 terrorist attacks, designers and owners have a greater focus and urgency regarding occupant safety in public facilities; this in turn has created a need to deliver and commission facilities with enhanced builiding safety measures. The trend toward increased security has become a global issue and is likely to increase the standard of care required in the design and operation of all forms of new construction and existing buildings.

Relevant Resources, Codes, and Standards

- AIA B211™–2004 Standard Form of Architect's Services: Commissioning. Fixed scope of services requiring architect to develop a commissioning plan, a design intent document, and commissioning specifications, based on owner's identification of systems to be commissioned
- *ASHRAE Guideline 0-2005: The Commissioning Process*, ASHRAE, 2005. Industry-accepted model commissioning guide
- *ASHRAE Guideline 1.1: HVAC&R Technical Requirements for the Commissioning Process*, ASHRAE, 2008
- The International Green Construction Code (IgCC), in effect as of March 2012
- The California Building Standards Commission ("CalGreen") 2010 standard
- *NIBS Guideline 3-2006: Exterior Enclosure Technical Requirements for the Commissioning Process*, National Institute of Building Sciences, 2006
- *The Building Commissioning Guide*, GSA, 2005
- Building Commissioning Association, www.bcxa.org
- Model Commissioning Plan and Guide Specifications, Version 2.05, PECI, February 1998—Available from PECI, 921 SW, Washington, DC; and Suite 312, Portland, OR 97205; email peci@peci.org

- "No Operator Left Behind: Effective Methods of Training Building Operators," Bradley Brooks, in *Proceedings of the 2007 National Conference on Building Commissioning*; www.peci.org/ncbc/proceedings/2007/Brooks_NCBC2007.pdf
- LEED Commissioning for New and Existing Buildings, Wilkinson, HPAC Engineering, 2008; http://hpac.com/mag/popular/leed_commissioning_buildings/
- Continuous Commissioning Guidebook for Federal Managers, DOE, 2002; www1.eere.energy.gov/femp/pdfs/ccg02_introductory.pdf

Chapter 12

Project Cost Analysis

12.1 INTRODUCTION

Many project owners and developers who want to develop green/sustainable projects find themselves immersed in painstaking analysis and rigorous planning, which really means that a comprehensive compilation of construction-related cost information is pivotal for assessing the merits of a proposed construction undertaking. Fortunately, there are many software packages on the market with a wide selection of attributes, such as "Projectmates" by Systemates, which provides a simple web-based interface for submitting requests and managing project budgets and contracts, and has capabilities ranging from document management and scheduling to financial budgeting and Change Order management. Projectmates software contains more than 40 different modules for various types of construction projects that can save time and increase accountability among the various project participants. Furthermore, while owners may choose to delegate responsibilities to other professionals on the project team, such decisions are left to their discretion and control.

It is indeed fortunate for project managers and estimators that there is such a wide range of computer-aided cost estimation software available, ranging in sophistication from simple spreadsheet calculation software to integrated systems involving design and price negotiation over the Internet. Such software involves costs for purchase, maintenance, training, and hardware, so the user will experience significant benefits. Cost estimates may be prepared more rapidly and with less effort with these products. Professor Chris Hendrickson, co-director of the Green Design Institute, lists some of the more common features of computer-aided cost-estimation software:

- *Databases for unit cost items such as worker wage rates, equipment rental, and material prices. These databases can be used for any cost estimate required. If these rates change, cost estimates can be rapidly recomputed after the databases are updated.*
- *Databases of expected productivity for different components types, equipment, and construction processes.*

- *Import utilities from computer-aided design software for automatic quantity take-off of components. Alternatively, special user interfaces may exist to enter geometric descriptions of components to allow automatic quantity take-off.*
- *Export utilities to send estimates to cost control and scheduling software. This is very helpful to begin the management of costs during construction.*
- *Version control to allow simulation of different construction processes or design changes for tracking changes in expected costs.*
- *Provisions for manual review, override, and editing of any cost element resulting from the cost estimation system.*
- *Flexible reporting formats, including electronic reporting rather than simply printing cost estimates on paper.*
- *Archives of past projects to allow rapid cost-estimate updating or modification for similar designs.*

There are many challenges facing the construction industry as it continues to be riddled with procurement problems arising from an impractical division between the design and construction process, a lack of organization among subcontractors, strained relationships between design professionals and construction team members on the integrated project team, antiquated design and construction methods, preventable delays, and an inferior end product. However, in recent years the construction industry has demonstrated some improvements in technologies and methods in which clients have been the driving force for change leading to the development of improved and more sophisticated services (e.g., project management, facilities management, better control of cash flow, and alternate procurement methods). This need for change originated from the desire for a more competitive and efficient industry, including a stronger emphasis on education programs that help to merge construction and design to emphasize multiskilled trades.

One of the leading "green economists" is the project management consulting firm Davis Langdon whose study with colleagues, "Costing Green: A Comprehensive Cost Database and Budgeting Methodology" (Langdon et al., 2004), compared the square-foot construction costs of 61 buildings pursuing Leadership in Energy and Environmental Design (LEED®) certification to those of similar buildings without green objectives. Taking into consideration climate, location, and other variables, the study came to the conclusion that for many sustainable projects, aiming for LEED certification resulted in little or no impact on the general budget. Another point worth noting is that construction lending is basically real estate lending, and the lender should therefore be aware that the primary security for the loan is the real estate to be developed and that for the loan to be repaid the development has to be completed. Experience has shown that few real estate borrowers are able to repay a construction loan from the assets listed in their financial statements. The borrower's professional and financial capabilities are key elements in the loan determination and should be thoroughly examined before finalizing any loan commitments.

Professor Hendrickson comments: "The costs of a constructed facility to the owner include both the initial capital cost and the subsequent operation and maintenance costs. Each of these major cost categories consists of a number of cost components." Therefore, to be able to determine the capital cost for a construction project, it is necessary to estimate the expenses related to the initial erection of the facility, which according to Hendrickson, include

- *Land acquisition, including assembly, holding, and improvement*
- *Planning and feasibility studies*
- *Architectural and engineering design*
- *Construction, including materials, equipment, and labor*
- *Field supervision of construction*
- *Construction financing*
- *Insurance and taxes during construction*
- *Owner's general office overhead*
- *Equipment and furnishings not included in construction*
- *Inspection and testing*

In addition to the preceding, the project owner must consider the operation and maintenance cost of the project over its life cycle, for which Hendrickson includes the following expenses:

- *Land rent, if applicable*
- *Operating staff*
- *Labor and material for maintenance and repairs*
- *Periodic renovations*
- *Insurance and taxes*
- *Financing costs*
- *Utilities*
- *Owner's other expenses*

Fully appreciating the significance of each of these cost elements will depend largely on the project's type, size, and location, as well as its management organization, among many other considerations. As far as the owner is concerned, the ultimate objective is to achieve the lowest possible overall project cost while at the same time meeting the project's specified quality and investment objectives.

To arrive at an accurate estimate of the total project cost (TPC), one must include all *hard* and *soft* costs to reflect the building being complete and usable as intended. (Hard and soft costs are discussed in a later section.) Costs include construction costs, construction contingency, architect and/or engineer fee, project contingency, owner services, and administrative fees. Depending on the general conditions and Contract Document requirements, the total project cost may also include infrastructure, furniture and equipment, voice/data, instructional technology, moving, and custodial equipment. As for construction costs, this is basically the fee charged by a general contractor or construction firm to manage a project.

Construction costs per gross square foot vary from state to state, whether the project is new construction or renovation, and the type, size, and complexity of the project (e.g., office, educational, hospital, residential). Hendrickson correctly points out the important thing is for:

Design professionals and construction managers to realize that while the construction cost may be the single largest component of the capital cost, other cost components are not insignificant. For example, land acquisition costs are a major expenditure for building construction in high-density urban areas, and construction financing costs can reach the same order of magnitude as the construction cost in large projects such as the construction of nuclear power plants.

From the owner's perspective, it is equally important to estimate the corresponding operation and maintenance cost of each alternative for a proposed facility in order to analyze the life cycle costs. The large expenditures needed for facility maintenance, especially for publicly owned infrastructure, are reminders of the neglect in the past to consider fully the implications of operation and maintenance cost in the design stage.

In the vast majority of cases, construction budgets contain a contingency clause for unexpected cost overruns that may occur during construction. Although this contingency amount may be included within each cost item, it is preferably included as a single category, namely a construction contingency, which is normally a percentage of the project's estimated cost. This contingency amount is based on several factors, including the complexity and size of the project and whether it is new construction or renovation. For example, for large new construction projects, the contingency is generally about 5% of the total cost, whereas for renovations, it may be roughly 7%. Likewise, for small interior projects, it may be as high as 10%. Any remaining contingency amounts can be released on substantial completion of the project. The general fee for the developer, builder, or construction manager (CM) is usually released as a direct percentage of the value of the subcontractual work completed to date. Figure 12.1 is a sample project cost breakdown showing the main elements to be considered when preparing a budget estimate for the project.

12.2   BUDGET DEVELOPMENT AND REQUIREMENTS

Whether the project in question is large university complex or a small single-family residence, it is important to set a budget estimate for it. This is basically setting out a financial plan to design and build the particular project and estimating the costs needed to complete it. Thus, regardless of whether the project to be constructed is large or small, a prudent developer will certainly find it necessary to create a budget for it. The primary purpose of a budget is to understand and control costs and cost overruns. Cost overruns are mitigated by the inclusion of appropriate contingencies in the budget estimate to cover Change Orders and the like, and these contingency allowances are disbursed as the project proceeds to cover the additional costs. Figure 12.2 contains a graph that shows the relationship between contingencies and the direct cost budget.

COST BREAKDOWN

ITEMS	DESCRIPTION	BUDGET AMOUNT	COSTS PAID	COSTS TO BE PAID
0100	**GENERAL CONDITIONS**			
0101	Architecture and Design			
0102	Permits and Fees			
0104	Testing and Special Inspections			
0105	Soil Engineering			
0106	Structural Engineering			
0107	Survey and Topography			
0108	Tile 24 Compliance			
0109	Insurance and Bonds			
0110	Temporary Fencing and Security			
0111	Supervision			
0115	Temporary Sanitation			
0150	Temporary Utilities			
0199	Contingency (10% of hard costs)			
0200	**SITE WORK**			
0210	Clearing and Grubbing			
0211	Demolition			
0221	Site Grading			
0222	Excavation and Backfill			
0224	Erosion Control			
0255	Permanent Utilities			
0260	Paving			
0262	Curb and Gutters			
0263	Walks			
0271	Fences and Gates			
0280	Sewer Connections—Septic Tank			
0300	**CONCRETE**			
0310	Formwork			
0320	Steel Reinforcement			
0330	Concrete Foundation			
0331	Concrete Slabs and Patios			
0335	Concrete Driveways			

FIGURE 12.1 An example of a project cost breakdown and budget estimate. To estimate the total project cost (TPC), it is necessary to include all hard and soft costs to fully execute the building as intended. TPCs include construction costs, contingencies, architect/engineer fee, owner services, and administrative fees. Depending on the general conditions and Contract Document requirements, TPC may also include costs for infrastructure, furniture and equipment, instructional technology, moving, and others.

Continued

0400	MASONRY			
0410	Stucco			
0421	Fireplace Masonry			
0422	Concrete Blocks			
0425	Fireplace Facing/Stone Mantel			
0442	Marble			
0445	Exterior Stone Veneer			
0500	**METALS**			
0510	Rough Frame Hardware			
0512	Structural Steel			
0570	Ornamental Stairs and Rails			
0600	**WOODS AND PLASTICS**			
0610	Rough Carpentry Labor			
0611	Framing/Sheathing Materials			
0619	Wood Trusses			
0620	Finish Carpentry Labor			
0622	Interior Millwork and Trim			
0640	Cabinetry			
0643	Wood Stairs and Railings			
0700	**THERMAL AND MOISTURE PROTECTION**			
0710	Waterproofing			
0720	Insulation			
0731	Roof—Composition Shingle			
0732	Roof—Tile			
0740	Exterior Siding			
0750	Roof—Membrane Build-up			
0760	Flashing and Sheetmetal			
0780	Skylights			
0800	**DOORS, WINDOWS, AND GLASS**			
0820	Exterior Doors			
0821	Interior Doors			
0830	Sliding Glass Doors			
0850	Metal Windows			
0860	Wood Windows			
0870	Finish Hardware			
0872	Garage Doors and Operators			
0883	Mirrors			

FIGURE 12.1—Cont'd

0900	**FINISHES**			
0925	Gypsum Drywall			
0930	Tile Countertop			
0934	Marble Countertop			
0938	Formica Countertop			
0960	Wood Flooring			
0961	Wood Mantels			
0965	Linoleum			
0968	Carpet			
0970	Tile Floor			
0990	Exterior Paint			
0991	Interior Paint			
0995	Wallpaper			
1000	**SPECIALTIES**			
1017	Tub and Shower Enclosures			
1030	Prefabricated Fireplace			
1040	Signage			
1080	Toilet and Bath Accessories			
1100	**EQUIPMENT**			
1105	Vacuum System			
1140	Kitchen Appliances			
1200	**FURNISHINGS**			
1250	Window Coverings			
1400	**CONVEYING SYSTEMS**			
1420	Elevators			
1500	**MECHANICAL**			
1540	Plumbing—Rough			
1545	Plumbing—Finish			
1546	Plumbing Fixtures			
1550	Fire Protection			
1580	Heating and Air Conditioning			
1600	**ELECTRIC**			
1610	Rough—Electrical			
1614	Finish—Electrical			
1650	Lighting Fixtures			
1670	Telephone and Prewire			
1676	Intercom and Prewire			
1678	Television and Prewire			

FIGURE 12.1—Cont'd

Continued

2000	MISCELLANEOUS			
	Cleanup			
3000	**OVERHEAD AND PROFITS**			
	Overhead			
	Profits			
	SUBTOTAL			
	Loan Cost			
	Insurance, Compensation, P.L. & P.D., Social Security & Unemployment Insurance			
	Builders Overhead and Profit %			
	TOTAL COST OF CONSTRUCTION			
	LAND COST			
	TOTAL			

I certify that to the best of my knowledge the above is a true and correct statement of the estimate cost of this job.

Signed: _____ Date: _____

FIGURE 12.1—Cont'd

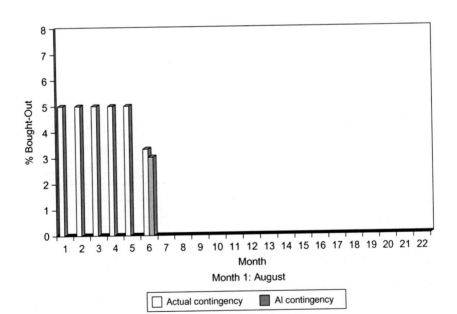

FIGURE 12.2   Contingency as a percentage of direct cost budget.

Among the more significant challenges that consultants and contractors face is keeping the project within the specified budget. Randy White, CEO of White Hutchinson Leisure & Learning Group in Kansas City, says,

Many a project gets into serious trouble when, for whatever reason, the project can't be developed within the budget. Usually, by the time the problem is discovered, it's too late to increase the budget, as financing has already been secured. So to keep the project within budget, critical features end up being compromised, such as the theming, finishes, and the quality of the materials, furniture and equipment, the things that really matter the most to creating the guest experience. Or certain attractions are eliminated, so the project never performs as originally planned and projections are never achieved. In fact, such last-minute deletions and changes can seriously threaten a project's very long-term survival.

For obtaining a construction loan, it is particularly important to have a budget for the project. Thus, prior to calculating the required construction loan amount, a basic budget is needed, the main components of which include

- *Hard costs:* Direct costs associated with the labor and materials used for actual physical construction
- *Soft costs:* Indirect or "offsite" costs not directly related to labor or materials for construction (e.g., architectural plans, engineering and permit fees)
- *Closing costs:* All costs associated with origination and closing the construction loan (e.g., title cost, loan fees, discount fees, insurance, appraisals, closing fees)
- Land acquisition costs
- Inspection fees
- Reserves (estimated interest on the loan during construction and contingency reserve for unforeseen expenses and cost overruns)
- Possible equipment, furnishings, and other unforeseen necessities

Because many projects are constrained by limited funding, there is a strong need to have a budget to initially define funding requirements. The project manager develops the budget based on the cost estimates calculated at the beginning of each project phase and refined once there is more accurate information defining the project's scope. Refining the budget occurs through studies and analysis in the design development process. When owners try to fix the budget too early in the project life cycle, they may be surprised by the significant increases in the budget over what was initially set forth. Randy White says, "This reoccurring problem is often caused by the nature of the design process. Design proceeds from general to specific and from conceptual to detailed. Accordingly, there is limited ability to accurately predict construction costs at the onset when initial project planning takes place and accurate costs are needed as part of the business plan to secure financing." With respect to project cost overruns, White says,

Cost overruns are also caused by the traditional design-bid-build process. First the project is designed, and then a contractor is selected by either competitive bid or negotiation to build the project. This process precludes value engineering until the project is already designed. So by the time the bid comes in over budget, the only way to reduce costs is to make major compromises in finishes, quality or components.

The project manager typically has the critical task of developing and tracking the project budget. The PM first develops a project budget in the early feasibility phase and continues to refine it throughout the project phases until the project is "bought out" by the general contractor prior to the start of the construction phase. All the elements of the budget should be clearly defined and fine-tuned throughout each phase. Specialized estimating software is often used to create, develop, monitor, and track budgets. In the development of a budget, there are certain logical steps that should be followed:

Step 1. Know precisely how much money is available for the project when attempting to develop a project budget. This should include all costs from project initiation through award of a construction contract to completion. At various points in the different stages, more detail, specificity, and definition are developed, and these estimates become more certain and realistic.

Step 2. Determine the mandatory or vital expenses for project success. Each project has certain essential requirements, and these should be given priority in the budget. For example, incorporating certain green features to achieve LEED certification is very important, which means that this should be high on the budget's priority list.

Step 3. Collect preliminary estimates from several companies and contractors. Once you identify the key elements of the project budget, you can start requesting estimates from area businesses to determine which offers the best price or value. In any case, preconceived cost indices can often be unrealistic and misleading. A new building project in an urban setting, for example, is far more costly than on a green field site. Construction managers usually have a good understanding of true market conditions and pricing in specific regions because their livelihood depends on it. Should your project budget not align with the project's expectations, then one or both will need realignment.

Step 4. Once the final budget is determined, clearly spell out, write down, and distribute it to all team members to ensure that everyone is on the same page. In the final analysis, this will depend largely on the type of contract entered into between the owner and the contractor—for example, whether it is design–bid–build, design–build, or cost-plus.

It is inappropriate to rely totally on "program" and/or "preliminary" estimates for setting a final project budget because it is far too early in the design/construction process. Until the time when a fairly accurate budget estimate can be developed, the project remains too conceptual in terms of scope and program size to accurately estimate final costs. After the architect completes the schematic design, the project's scope of work is more clearly defined to the extent that a realistic budget estimate can be arrived at to provide effective discipline and direction. However, this is still insufficient to bid the project and will not be sufficient until the Contract Documents are completed, including all plans and specifications. The following are the different phases or milestones a project must go through from design concept to completion and occupancy:

- Project initiation program budget estimate
- Planning/programming preliminary project estimate

- Design (conceptual design, schematic design) budget estimate
- Contract Documents (drawings, project manual, etc.)
- Bidding and awarding contract—project estimate
- Construction, commissioning
- Occupancy

Figure 12.3 is an example of a preliminary construction budget for a typical project (new construction)—a government agency is considering a new facility to allow expansion of its activities. It should be noted that these costs can vary depending on the region of the country where the project is located and local conditions. Another example of a simple project budget is shown in Figure 12.4, which has been adapted from "Charter School Facilities—A Resource Guide on Development and Financing," NCB Development Corp. and Charter Friends National Network (2010). It shows the typical components of a project budget, including the purchase and renovation of a building as opposed to new construction.

12.3   PROJECT BUYOUT AND BID SHOPPING

Project buyout and bid shopping are two concepts familiar to most construction professionals. The term *buyout*, as it relates to construction project mobilization, basically refers to procuring materials and equipment and arranging subcontracts. It is the time interval between the pre-construction and construction phases of a project and is among the most critical first steps in achieving a project's overall profitability. Even before breaking ground, making or losing money may be determined by how well the project is bought out.

Also, it is during buyout that purchase orders and subcontracts are issued. This includes selection of both suppliers and subcontractors and finalizing their purchase orders and contracts. Unfortunately, buyouts are necessary because, often due to time constraints during the bidding phase, complete, meticulous analyses of bids from subcontractors may not have been carried out. Figure 12.5 shows the budget percentage of buyout of trade contract cost versus time on a $15 million project. Bid shopping, on the other hand, while legal, is considered to be unethical because details of a bid are revealed to a competitor in an effort to achieve an overall lower bid for the project.

The buyout process starts during the tender preparation stage, as the contractor solicits and assesses offers in the process of assembling the cost estimate. Should a contractor's proposal be successful and the contractor awarded the contract, the next step is negotiating the contract. Material procurement and subcontracting are typically the two distinct parts of the buyout process discussed during contract negotiations, even though both may be the responsibility of a single department or individual. Project buyout is an ethical and necessary practice, conducted during pre-construction, that enables a general contractor to clarify scopes of work and to streamline specific activities. As construction professionals more fully understand the ethical issues separating the

SAMPLE CONSTRUCTION BUDGET

This is an example of a project budget for a government agency that proposes to construct a new facility in which to expand its activities. It should be noted that these costs will vary depending on what region of the country the project is located.

Expenses
Hard Construction Costs (8,000 SF @ $97/SF)

Foundation, Framing, Drywall, Flooring, Roofing	$581,000
Plumbing, Electrical, Security System	80,000
Fixtures, Furnishings, and Equipment	50,000
HVAC	27,000
Landscaping	19,000
Site Work	18,000
Subtotal Hard Costs	**$775,000**
Land Acquisitions	175,000

Soft Construction Costs

Architect and Engineers	31,000
Fees	4,000
Subtotal Soft Construction Costs	**$35,000**
Contingency	10,000
Total Expenses	**$995,000**

Revenues

Individual Contributions	$295,000
ABC Foundation	125,000
Government Grant	75,000
Corporate Donations	100,000
DEF Foundation	85,000
Other Foundations	65,000
XYZ Corporation (in-kind)	50,000
Other corporate donations	25,000
Fundraising Events	15,000
ABC Corporation	50,000
XYZ Foundation (pending)	50,000
To Be Raised from Other Sources	60,000
Total Revenues	**$995,000**

Notes:

1. Hard construction costs include any costs that cannot be physically moved, in other words site work, renovations or construction work, plumbing, electrical, landscaping, parking lot, demolition, flooring, roofing, HVAC, wiring, fire and security alarms, playgrounds, fixtures, appliances, etc. that become a permanent part of the site.
2. Soft construction costs include fees, surveys, permits, architect and engineer fees, and so on.
3. Contingencies are usually between 5 and 10% of construction costs, depending on the size and complexity of the project.

FIGURE 12.3   Sample construction budget.

USES OF FUNDS:

Acquisition of building	$250,000
Construction/renovation Costs (hard costs)	
Demolition of old walls	75,000
Electrical	65,000
Plumbing	80,000
Heating/ventilation	40,000
Roof	50,000
Drywall and painting	140,000
Carpet	35,000
Windows	40,000
Fixtures and Fit-out	55,000
Site work	20,000
Total Construction:	600,000
Hard Cost Contingency (15%)	90,000
Total Acquisition & Construction:	940,000
Legal Fees	10,000
Appraisal	5,000
Architect	30,000
Project Manager	10,000
Engineering	5,000
Insurance during construction	3,000
Closing Costs	5,000
Financing fees (loan origination fee etc.)	7,000
Interest during construction	35,000
Inspection fees	5,000
Environmental studies	12,500
Accountant	5,000
Security	8,000
Bonding	6,000
Total:	146,500
Soft Cost Contingency (5%)	7,325
Grand Total:	**$1,093,825**

SOURCES OF FUNDS:

Start-Up Grant	$150,000
Donations	238,325
Loan	705,500
Grand Total:	**$1,093,825**

Notes:
1. Hard costs contingencies are anything related to the building structure or its materials. Large contingency budgeted is due to extensive nature of renovations. On average renovation projects have a 7–10% contingency and for new construction it is usually about 5%.
2. Soft costs are related to all other costs, including architectural, financing, inspection and legal fees. These costs are normally about 5% of the total project costs.

FIGURE 12.4   Simple project budget.

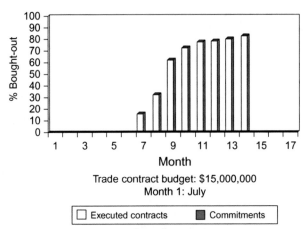

FIGURE 12.5   Percentage of buyout (executed and committed) of a budget's trade contract.

unacceptable practice of bid shopping from the ethical practice of project buy-out, the efficiency and quality of estimating and subsequent project management are improved. Estimators are normally required to bear the responsibility of obtaining bids and performing project buyout while maintaining high ethical standards.

The purpose of a buyout estimate is therefore to order materials once the project becomes viable for the contractor or subcontractor. It differs from a bid estimate in that the latter is a detailed estimate in order to bid a project. A typical example of a bid estimate versus a buyout estimate would be metal studs for drywall. During the bid period, it is good enough to know the total linear footage of studs by size and gauge. A buyout, however, requires greater detail: for example, in addition to the bid information, the lengths for each application. However, in mechanical and electrical scopes of work, the bid estimate and buyout estimates are very similar.

Project buyout takes place between the award of a bid to the general contractor and the issuing of subcontracts and purchase orders. While bid shopping is not illegal, it is considered an unethical practice where details of a bid are revealed to a competitor in an effort to solicit an overall lower bid. A better understanding of ethical versus acceptable construction practices can help construction professionals identify the basic differences between bid shopping and project buyout while at the same time steering clear of unethical practices and maintaining the ability to remain competitive.

In an article, "Contrasting Bid Shopping and Project Buyout," authors Cody Andreasen, Mark Lords, and Kevin R. Miller of Brigham Young University state,

The justification, for some contractors, when arguing in favor of bid shopping is, that if a bid is revealed to another subcontractor, then a lower bid may be forthcoming which may translate into a lower overall bid on the project, benefit the owner, and thus increase the likelihood of being awarding the project. It can be argued that this is no different than shopping for a car or bartering for goods in a foreign country. However, in the auto industry there is an expectation that pricing will be disclosed to other dealers in the buying process. That expectation does not exist in the construction industry. A construction project is something yet to be built. It is not an existing product and any changes in the cost typically will affect the quality or schedule of the project. Therefore, the owner is not receiving the same product if bid shopping occurs.

A technique often used to prevent bid shopping is for a subcontractor to submit a bid at the last minute, thus preventing the general contractor from receiving the bid by shopping it.

By and large, bid shopping occurs because most subcontractors being shopped believe that if they do not reduce their price, they won't get the job. Additionally, if business is slow, subcontractors may be willing to accept lower profit margins just to keep their crews busy even when it means they will only break even on the project. Sometimes, the general contractor will induce a subcontractor with additional work on being awarded the job. Although it may appear that the owner is the principal beneficiary of bid shopping by receiving a lower price for the project, usually one finds that the result is a lower-quality project in addition to greater risks and warranty problems down the road. According to Andreasen and colleagues,

[The main benefits of] project buyout [is that it] allows a period of time for the contractor to ensure that each scope of work is covered by only one subcontractor. Occasionally a contractor finds that two subcontractors submitted bids for an overlapping scope of work. Since both subcontractors do not need to perform the work, the general contractor will determine which subcontractor will perform the work for the overlapping scopes. The subcontractor that doesn't perform the overlapping work will generally provide a credit to the general contractor for the reduction in their scope of work. If the opposite is found and there has been work that was assumed to be included in a subcontractor's scope of work, but was not included in the bid, the general contractor generally negotiates with a subcontractor to have the work included in their subcontract and that negotiation may increase the contract amount for the subcontractor.

Another instance where changes could be made during the project buyout process results when a subcontractor anticipated a different project schedule than the general contractor. As a result, the subcontractor may not have sufficient crews to complete the job in the time frame or manner desired by the general contractor. In this case, the general contractor may elect to use a different subcontractor in order to maintain the project schedule.

Darin C. Zwick and Kevin R. Miller, authors of the article "Project Buyout," emphasize the importance of completing the buyout process as early as possible:

By completing buyout early, future delays are avoided in the event that a given scope of work is difficult to buyout due to conflicts with subcontractors or suppliers. It also protects the project from price escalation.

Other tasks that occur during the buyout process by the general contractor include checking the following items to ensure that the subcontractor can perform the work for the project:

- *Insurance and liability coverage,*
- *Evidence of state Workman's Compensation coverage,*

- *Evidence of proper local and state subcontractor licenses,*
- *Evidence of proper bonding requirements if required.*

The expiration dates of the previous items need to be verified to prevent lapses of coverage while the subcontractor is working on the project.

Another consideration that companies need to examine during the buyout process is the financial stability of the subcontractors and vendors. During economic downturns, companies may declare bankruptcy, leaving the general contractors in a precarious situation.

The project buyout specialist's main duties are to focus on awarding scopes of work to subcontractors and to act as liaison between field operations and subcontractors while keeping the contract amount within the budget. This is particularly important because it relates to disputes and problems that often exceed field management's ability to solve in a timely manner. During the pre-construction phase, this includes technical support to both the design–build and estimating departments. In addition, the buyout specialist is responsible for the hiring of new subcontractors for all projects, including cost control, corporate and contract compliance, quality control, and customer satisfaction. He or she is also responsible for the quality and completeness of project buyout and small-business utilization, as well as customer relations and client satisfaction in all areas within the firm.

Team building and a holistic approach are necessary if maximum support for and from each team member is to be achieved. This includes setting and monitoring goals and collaborating with purchasing and operational goal setting. There are currently several proprietary buyout software packages on the market that can save time and reduce effort by automating bid solicitation. Buyout software is also an important tool for determining where the project is in the buyout process at any particular moment, and it can establish percent complete, minimize exposure, and rapidly show how actual prices compare with estimated costs. Moreover, it can offer access to standard cost codes, categories, and tax groups stored in various applications such as accounts payable. One example of such software is from Sage Software (www.sagecre.com), which offers a package called "Buyout" (Figure 12.6) that has the following features:

- Builds a worksheet of material and subcontract items to be bought out automatically by reading the estimating file.
- Combines multiple estimates into a single worksheet, an important feature for contractors who receive price discounts based on volume purchases.
- Creates one-time items in the Buyout item window.
- Views items the way you want to see them: by work breakdown structure (WBS), location, phase, material class, and so forth.
- Groups materials or subcontract items for ease in obtaining prices. Creates quote sheets and assigns material items and subcontract items to them.
- Assigns multiple vendors and subcontractors to quote sheets.

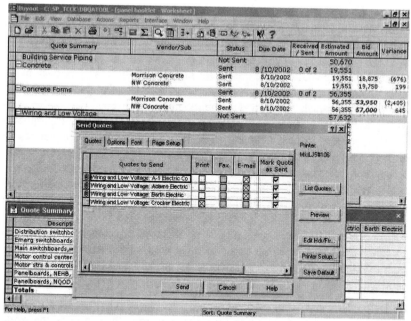

FIGURE 12.6   Sage Software's Buyout software.

- Uses prices from its standard price database for items in the quote sheet.
- Automatically submits requests for quotes and sends purchase orders via email, fax, or hard copy.
- Splits items out of one quote for the creation of new quote sheets.
- Uses the Summary quote sheet to organize vendors' or subcontractors' quotes from low to high.
- Saves prices from the quote sheet to a standard price database.
- Changes prices for any item and updates the estimating database with pricing from the price database.
- Generates requests for quotes (RFQs) and purchase orders (POs) directly from the price database and issues them automatically via email, fax, or hard copy.
- Updates estimates with its database prices, revised quantities, and vendor/ subcontractor selections.

12.4   GENERAL CONDITIONS AND SUPPLEMENTAL CONDITIONS

General conditions are considered among the most important documents in the project manual because they set forth and define the rights and responsibilities of the different parties, particularly the owner and contractor, in the construction process as well as the specific terms of the contract. They also specify and define the surety bond provider; the design professional's role, authority, and responsibilities; and the requirements governing the various parties' business and legal relationships. These conditions are "general" and can apply to almost any project. It is vital that the contractor knows exactly what they contain.

It should be noted that many trade and professional organizations have developed their own standard documents and general conditions. The most widely used may be those published by the American Institute of Architects (AIA), specifically AIA Document A201. This document has been well tested in the courts and is familiar to most contractors. The ConsensusDOCS 200, Standard Agreement and General Conditions between Owner and Contractor, is also widely used. Likewise, ConsensusDOCS 410, Agreement and General Conditions between Owner and Design-Builder (Cost of Work Plus Fee with Guaranteed Maximum Price [GMP]) is sometimes used.

There are a number of standard clauses that typically appear in the general conditions. The following are some of them with a brief description of each. Bear in mind that they may vary depending on the type of general conditions contract.

> **Definitions and General Provisions.** This clause provides definitions for the purpose of the Contract Documents relevant to the contracts, the work, and the drawings and specifications. It also clarifies the ownership, use, and overall intent of the Contract Documents.
>
> **Owner Responsibilities.** This clause defines the information and services that the owner is required to supply. It also defines the owner's rights and responsibilities and the owner's right to stop or carry out the work.
>
> **Contractor General Obligations and Responsibilities.** This clause lays out the obligations of the contractor regarding construction procedures and site operations, employees, labor and materials, warranty, taxes, permits, fees and notices, schedules, samples and product data, and cleanup. It reads as follows:
>
> *The Contractor is required to execute and complete the Works and remedy any defects therein in strict accordance with the Contract, with due care and diligence and to the satisfaction of the Architect, and shall provide all labor, including the supervision thereof, materials, and all other things, whether of a temporary or permanent nature. The Contractor shall also take full responsibility for the adequacy, stability and safety of all site operations and methods of construction, but the Contractor shall not be responsible, unless expressly stated otherwise in the Contract, for the design or specification of the Permanent Works or of any Temporary Works prepared by the Architect.*
>
> **Administration of the Contract.** This clause delineates the duties, responsibilities, and authority of the architect for the administration of the contract—specifically, the architect's responsibility for conducting periodic site inspections and issuing periodic reports to the owner or lender. It also deals with issuing modifications in drawings and technical specifications and

assisting the contractor in the preparation of Change Orders and other contract modifications, as well as assisting in inspections, signing certificates of completion, and making recommendations on acceptance of work completed under the contract. The architect is required to review detailed drawings and shop drawings, price breakdowns, and progress payments estimates, as well as requests for additional time and methods for handling claims and disputes.

Pre-Construction Conference and Notice to Proceed. This clause deals with the procedures for conducting a pre-construction conference to acquaint the different parties with one another. For example: "Within ten calendar days (or as stated in the Contract Documents) of contract execution, and prior to the commencement of work, the Contractor shall attend a pre-construction conference with representatives of the Owner, Architect, and other interested parties and stakeholders." This clause also requires that the contractor begin work only after receipt of a written Notice to Proceed from the owner or designee. The contractor may not begin work prior to receiving this notice.

Availability and Use of Utility Services. This section deals with the availability of utility services: "The Project Owner shall ensure that all reasonably required amounts of utilities are available to the Contractor from existing outlets and supplies, as specified in the contract. Unless otherwise provided in the contract, the amount of each utility service consumed shall be charged to or paid for by the Contractor at prevailing rates charged to the Owner."

Assignment and Subcontracting. This clause deals with the assignment and awarding of subcontracts by the general contractor for portions of the work. It states,

The Contractor shall not, except after obtaining prior written approval of the Project Owner, assign, transfer, pledge or make other disposition of the Contract or any part thereof or of any of the Contractor's rights, claims or obligations under the Contract. In the event the Contractor requires the services of subcontractors, the Contractor shall also obtain prior written approval of the Owner for all such subcontractors. The approval of the Owner does not relieve the Contractor of any of his obligations under the Contract, and the terms of any subcontract shall be subject to and be in conformity with the provisions of the Contract.

Construction by Owner or Others. This clause deals with the owner's right to perform some of the construction work with his or her own forces or to award separate contracts to other parties besides the general contractor. It reads:

The Contractor shall in accordance with the requirements of the Architect/Project Manager and the Contract afford all reasonable opportunities for carrying out portions of the work by the Owner or to any other contractors employed by the Owner and their workmen or the Owner's workmen who may be employed in the execution on or near the Site of any work not included in the Contract or of any Contract which the Owner may enter into in connection with or ancillary to the Works.

Permits and Codes. This section states,

*The Contractor shall give all notices and comply with all applicable laws, ordi-
nances, codes, rules and regulations. Before installing the work, the Contractor shall
examine the drawings and the specifications for compliance with applicable codes
and regulations bearing on the work and shall immediately report any discrepancy it
may discover to the Architect/Project Manager.*

Change Orders. This clause explains how changes are authorized and processed according to the relevant clauses of the contract. Changes Orders are
one of the areas of greatest contention between owners and contractors. Generally, according to this clause, "The Architect may instruct the Contractor,
with the approval of the Owner and by means of Change Orders, all variations in quantity or quality of the Works, in whole or in part, that are deemed
necessary by the Architect. Processing of Change Orders shall be governed
by appropriate clauses of the General Conditions."

Construction Progress Schedule. Time is always a pivotal factor on any
project. Project schedules govern project startup, progress, and anticipated
completion dates. This clause addresses issues associated with delays and
extensions of time. For example, "Schedules shall take the form of a progress chart of suitable scale to indicate appropriately the percentage of work
scheduled for completion by any given date during the construction period."

Progress Payments. This section specifies how applications for progress
payments are to be processed. It reads:

*The Owner/Lender shall make progress payments approximately every 30 days as the
work proceeds, on estimates of work completed and which meets the standards of quality established under the contract, as approved by the Project Manager/Architect.
Before the first progress payment can be processed under this contract, the Contractor
shall furnish a breakdown of the total contract price showing the amount included
therein for each principal category of the work, which shall substantiate the payment
amount requested in order to provide a basis for determining progress payments.*

The clause also deals with the withholding of payments and failure to pay.

Protection of Persons and Property. This clause addresses issues relating
to safety of property and people. It deals specifically with the handling of
hazardous materials and emergencies, as well as overall safety programs
and requirements. For example,

*The Contractor shall (unless stated otherwise in the Contract) indemnify, hold and save
harmless and defend at his own expense the Project Owner, its officers, agents, and
employees from and against all suits, claims, demands, proceedings, and liability of
any nature or kind, including costs and expenses, for injuries or damages to any person
or any property which may arise out of or in consequence of acts or omissions of the
Contractor or its agents, employees, or subcontractors in the execution of the contract.*

Insurance of Works and Bonds. This clause deals with insurance (including liability insurance) and bonding requirements of the various parties and
the coverage period stipulated. That is, "The Defects Liability Period for

loss or damage arising from a cause occurring prior to the commencement of the Defects Liability Period and for any loss or damage experienced by the Contractor in the course of any operations carried out for the purpose of complying with the Contract obligations."

Examination of Work before Covering Up. This clause has to do with acceptance of the work by the architect (as the owner's agent) and stipulates how and when the contractor will uncover and/or correct any work deemed unacceptable. Accordingly,

No work shall be covered up or put out of view without the prior approval of the Project Manager/Architect. The Contractor shall afford full opportunity for the Project Manager/Architect to examine and measure any work which is about to be covered up or put out of view and to examine foundations before permanent work is placed thereon. The Contractor shall give due notice to the Project Manager/Architect whenever any such work or foundations is ready for examination and shall without unreasonable delay advise the Contractor accordingly to attend for the purpose of examining and measuring such work.

Clearance of Site on Substantial Completion. This clause states, "Upon the substantial completion of the Works the Contractor shall clear away and remove from the Site all rubbish, constructional plant, surplus materials, and temporary works so as to leave the whole of the Site and Works clean and in a workmanlike condition to the satisfaction of the Project Manager/Architect."

As-Built Drawings. The term "as-built drawings," as used in this clause, refers to "drawings submitted by the Contractor or subcontractor at any tier to show the construction of a particular structure or Work as actually completed under the Contract. 'As-built drawings' shall be synonymous with 'Record drawings.'"

Miscellaneous Provisions. This clause deals with various matters such as liquidated damages, taxation, disputes, prohibition against liens, warranty of construction, energy efficiency and other green issues, and waiver of consequential damages.

Termination/Suspension of the Contract. Either party has the right to terminate the contract under certain conditions. These conditions are clarified in this clause. For example,

The Contractor shall on the written order of the Architect/Project Manager suspend the progress of the Works or any part thereof for such time or times and in such manner as required by the Architect/Project Manager and shall, during such suspension, properly protect and secure the Works as specified by the Architect/Project Manager. The Project Owner should be notified and written approval sought for any suspension of work in excess of three (3) days.

12.4.1   Supplemental Conditions

Supplemental conditions are special conditions, also known as supplementary general conditions, special provisions, or particular conditions. They typically amend or supplement the standard general conditions and other provisions of the Contract Documents as indicated. Supplemental conditions normally deal with matters that are project-specific and beyond the scope of the general conditions. They may either add to or modify them. Examples of project-specific information that may appear in the supplemental conditions include

- Safety requirements
- Contractor's bond requirements
- Bonus payment information
- Defects liability period
- Cost fluctuation adjustments
- Progress payment retainage
- Services provided by the owner
- Temporary facilities provided by the owner
- Owner-provided materials

12.5 CONTINGENCIES AND ALLOWANCES

Contingencies are generally necessary to cover unknown, unforeseen, and/or unanticipated conditions or circumstances that are not possible to adequately evaluate or determine from the information on hand at the time the cost estimate is prepared. Contingency allocations specifically relate to project uncertainties of the current known and defined project scope that may arise; they are not a prediction of future project scope or schedule changes. The amount of contingency allocated relates to the amount of assessed risk and should not be reduced without appropriate supporting justification. Furthermore, a contingency amount in the cost estimate mitigates the impact of cost increases inherent in an overly optimistic estimate and provides the opportunity for earlier discussion of how to address potentially adverse circumstances.

Contingencies in a project budget represent the degree of risk in the estimate and are traditionally calculated as an across-the-board percentage addition to the base estimate, typically based on initial estimates, previous experience, and historical data (see Chapter 4 for additional information). This approach is arbitrary and has serious flaws because it is generally illogically arrived at and therefore often not appropriate for the project at hand. Moreover, it is difficult for an estimator to justify or defend. A percentage addition results in a single-figure prediction of estimated cost that often does not reflect reality; nor does it encourage creativity in estimating. Examples of typical contingency types normally found in budgets that should be considered when major projects are involved are the following:

- A *construction contingency* is basically used to cover cost growth during construction. It is a percentage of construction cost held by the PM to resolve issues during construction, which is why it should not be used until the project is in the construction phase. This contingency will be higher for renovations in older buildings, for buildings with complicated site conditions, or for complex projects. It is contained in the contractor's guaranteed maximum price (GMP), but the PM must approve use of any contingency funds prior to being committed by the general contractor.

- A *design contingency* is for changes or modifications during the design process for such factors as incomplete scope definition and inaccuracy of estimating methods and data. Design contingency amounts are based on the amount of design completed and are a percentage of construction cost held to represent the completeness of the design. This design contingency is understandably higher during the early phase of the project's design. As the design is completed and the scope of work is more defined, it is reduced until it becomes zero in the cost element at the completion of the permit phase.
- A *project contingency* is a percentage of project cost retained for risks in individual project costs such as professional fees, hazardous materials abatement (e.g., asbestos), communications wiring, and the like. Money allocated as a contingency in the project budget should not be utilized for additional scope or other changes to the project once the design is completed.
- A *program contingency* is optional and may be employed to cover scope or program changes requested by the user group or owner. An alternative is to have the general contractor carry an allowance line item in the GMP contract.

Various other contingencies cover areas or items that may show a high potential for risk and change such as environmental mitigation, utilities, and highly specialized designs.

12.5.1   Construction Contingencies

A construction contingency is essentially a predetermined amount or a percentage of the contract budgeted for unexpected or unforeseen changes or design shortfalls identified after construction has commenced. It is a helpful tool that financially prepares owners for budget escalations. When underwriting a commercial construction loan request, it is prudent to analyze the four major elements of total construction cost: land cost, hard costs, soft costs, and contingency reserve. Since there are always cost overruns in almost any commercial construction project, a contingency reserve builds a cushion into the budget to cover them. If managed properly, a contingency reserve can provide a safeguard for the designer, contractor, and owner to complete the project within the allocated budget. And while there is no specific formula for computing it, many underwriters feel comfortable using 5% of the construction estimate/bid for new construction (although in complicated projects the contingency can be as high as 10%) and 7% for remodeling/renovation projects. Land costs are not included because these are usually known in advance and are fixed. It is unlikely that there will be a cost overrun connected to the purchase of the land itself.

Construction project budgets typically include a construction contingency to allow the project to proceed with minimal interruption for small or insignificant (nonscope) changes or cost overruns. The typical construction contract includes a specific completion date or a specific number of working days to complete, and the contractor can be required to pay liquidated damages if the work is not completed within this period. At the same time, the contractor is entitled to proceed with the work without undue interruption. To minimize delays due to external causes, the client must be capable of implementing minor (i.e., nonscope) changes without causing any administrative delay.

Since it is almost impossible to produce a perfect set of construction documents, it is highly likely that there will be some errors and/or omissions. Whatever the case, changes always occur on construction projects. The owner must therefore ensure that an appropriate contingency is included to cover the costs of any changes in the scope of a project, such as upgrades, additional equipment, or perhaps enlargement of the building footprint. Moreover, financing costs may change with the market. A small contingency may be sufficient to cover final documentation of drawings by designers, but a construction contingency of roughly 3% to cover changes in market conditions and estimate variances should be planned.

One of the objectives of contingency planning is to determine a confidence value by means of a percentage in potential cost and schedule growth. The contingency value is an indicator of the level or degree of project development, and typically the less defined a project, the higher that value. Issues, such as scope definition and quality assurance, have a significant impact on confidence, risks, and resulting contingency development. In determining a contingency value, consideration must be given to the details and information available at each stage of planning, design, and construction for which a cost estimate is being prepared. While most construction budgets contain an allowance for contingencies or unexpected costs during construction, this allowance may be included either within each cost item or as a single category in the construction contingency.

Another aspect important to the owner's contingency is accounting for risk. Risk is created when some aspects of the project are unknown or when certain project elements are likely to cause concern. Generally, the estimated contingency amount to be retained is based on historical experience and the anticipated difficulty of a particular construction project. For example, one construction firm places estimates of the anticipated cost into five specific categories:

- Design development changes
- Schedule adjustments
- General administration changes (e.g., wage rates)
- Different site conditions than those expected
- Third-party requirements imposed during construction (e.g., new permits)

Any contingent amounts not disbursed during construction can be released toward the end of construction to the project owner or used to add additional project elements. Costs related to a construction project are typically divided into two essential components: *hard costs* and *soft costs*.

Hard Costs

Hard costs are generally considered to be by far the largest portion of allocated expenses in a construction budget and generally consist of all of the costs for physical items and visible improvements (i.e., actual construction costs incurred). These line items include site preparation (grading/excavating), concrete, framing, electrical, carpentry, roofing, and landscaping. Hard costs have been described as the bricks-and-mortar expenses. In some cases, they may include land, but that particular cost is usually separated out in order to find out the actual construction expenses.

Soft Costs

Soft costs are typically indirect or "offsite" costs not directly related to labor or materials for construction. They are nonphysical expenses and involve all other fees involved in the completion of the project. Soft costs include transfer taxes; origination points; mortgage insurance (if applicable); overhead; attorney fees; professional fees; permits; title insurance; appraisal fee; testing; hazard insurance; marketing; construction insurance; and the like. Another primary soft cost category, if applicable, is fixtures, furnishings, and equipment (FF&E). Soft costs are generally estimated as a percentage of the total project budget during the planning stages. And as the planning and design of a project progresses, the soft cost contingency percentage can be increased or decreased.

There are numerous software packages on the market that facilitate calculation of the total cost of a commercial construction project. To achieve accurate results, one must include hard costs, soft costs, and land costs, as well as the contingency reserve, which for new construction is generally about 5% of total project cost. Design professionals can establish a project cost estimate by using several methods. One approach is to use estimates whose development is based on project parameters and major cost elements or on an analysis of historical bid data, actual cost, or a combination of the two. But whatever method is used, in the final analysis special care must be taken to ensure that the capital cost estimate is complete and realistic, not overly optimistic. Underestimation of project construction and related costs is one of the more common problems faced in the economic analysis and budgeting of a project. Contingency funding is a fiscal planning tool that helps manage the risk of cost escalations and cover potential cost estimate shortfalls. Inclusion of a contingency amount in the cost estimate mitigates the adverse impact of cost increases inherent in an overly optimistic estimate and provides an opportunity for an earlier discussion of how potential circumstances can be addressed.

It is strongly advised that large projects have an overall management contingency. Such a contingency is usually a "standalone" amount of the cost estimate that is managed by an executive and used for a broad array of uncertainties and potential risks. Some of the "Project Oversight Management" contingency allowance will be disbursed to manage costs, manage the approved budget and schedule deviations, address adverse impacts caused by modifications, and analyze or implement initiatives to address or mitigate potential cost overruns or schedule delays.

The transfer of costs to and from contingency and allowance line items needs to be administered and tracked carefully to allow decision makers to take appropriate action. Cost transfers should correspond to the major component type of cost escalation. Thus, if a proposed work is clearly outside of a well-defined scope but is found to be essential to the project's well-being and can be readily justified, then a management decision can be made to disburse payment for the added work or Change Order from either the management contingency or another appropriate contingency. On the other hand, if there are distinct fees or FF&E issues that have contingencies, careful tracking of these contingencies can help the PM and management better analyze potential cost overruns. The rationale for supporting contingency transfers should be noted and incorporated into all relevant reporting. This is to allow a periodic comparison analysis of available contingency amounts to establish contingency usage rates. This analysis will alert project managers if potential problems exist as well as confirm if a reasonable and sufficient amount of contingency remains to keep the project within the latest approved budget.

Construction cost estimates should not be presented as a lump-sum total but rather as the sum of costs for each major element of the project. Contingency allowances can be clearly identified as individual line items associated with each major element. This allows the PM and reviewers of future updates to track where and how project costs are changing and how they may impact project completion. This can be achieved by providing information on reoccurring patterns and reasons behind cost escalation. Contingencies are normally disclosed as a dollar value or as a percentage of the major element costs depending on the format used.

12.5.2   Budget Allowances

Budget allowances are somewhat similar to contingencies in that their purpose is to reserve funds for circumstances that are ill-defined and thus more prevalent in the earlier design phases of a project, when the uncertainties are most evident. For example, a construction budget estimate may include an allowance for green building energy enhancements to achieve LEED Silver or Gold certification, site preparation, demolition, and so forth. But unlike contingencies, allowances are usually identifiable single items/issues and are placed in budgets as individual line items.

Certain allowances may also be carried by the general contractor after attaining the approval of the PM and provided they do not exceed their budgets or estimates to cover such items that they believe may arise (based on prior experience). Furthermore, as an optional allowance, the General Contractor may also carry a contingency to cover scope or program changes that the User Group/Owner may request during construction. This Allowance is an agreed upon amount between the Contractor and the PM and must be approved by the PM prior to being committed by the Contractor. The PM can carry Allowances in any of the Cost and Time Summary (CATS) categories for questionable or additive alternate construction and nonconstruction items as necessary. On the subject of allowances, the Chicago law firm of Sabo & Zahn states,

An allowance is a line item in a construction budget that serves as a placeholder during the bidding and initial construction contract phase. It is used when a particular item to be used in the construction has not been picked or completely specified. For example, if the carpeting has not been selected at the time of bidding, rather than delay the bidding, an allowance for the carpeting can be used. Normally, in this situation the total amount of carpeting to be used is known. If, for instance, the house will have 300 yards of carpeting, with a $50 per square yard allowance, the contractor will include a carpet allowance of $15,000 in the bid. This allowance will cover the cost of the materials as well as the cost of installation. The contractor will also have its overhead and profit included in the proper category. At some later date, the owner will pick the actual carpeting. If the actual cost for that carpeting is $60 per yard, then the contractor will be entitled to a change order for the increased cost—in this example, $3,000. On the other hand, if the actual cost of the carpeting turns out to be only $40, then the change order will reflect a deduct of $3,000.

The key to properly administering allowances is to account for them by proper change orders. At the time that the actual material is selected and approved by the owner, a change order must be issued and signed. This change order must indicate that the allowance for that item is being deleted, with a credit to the contract for the allowance amount, with a corresponding increase in the construction cost in the amount of the actual cost. In our example with a $60 carpet cost, the allowance of $15,000 would be credited to the owner and the $18,000 actual cost would be added to the contract, for a net increase of $3,000.

However, the best practice is ensure that everything is clearly identified and specified prior to solicitation of bids rather than provide any allowances.

As a follow up, the allowance section of AIA Document A201-1997 states that, "The Contractor shall include in the Contract Sum all allowances stated in the Contract Documents. Items covered by allowances shall be supplied for such amounts and by such persons or entities as the Owner may direct, but the Contractor shall not be required to employ persons or entities to whom the Contractor has reasonable objection." According to the American Institute of Architect's Document A201-1997,

Unless otherwise provided in the Contract Documents:

1. *Allowances shall cover the cost to the Contractor of materials and equipment delivered at the site and all required taxes, less applicable trade discounts.*
2. *Contractor's costs for unloading and handling at the site, labor, installation costs, overhead, profit and other expenses contemplated for stated allowance amounts shall be included in the Contract Sum but not in the allowances.*

3. *Whenever costs are more than or less than allowances, the Contract Sum shall be adjusted accordingly by Change Order. The amount of the Change Order shall reflect (1) the difference between actual costs and the allowances under Clause 3.8.2.1 and (2) changes in Contractor's costs under Clause 3.8.2.2.*

It is important to note that AIA A201-1997 stipulates that "materials and equipment under an allowance shall be selected by the Owner within sufficient time to avoid causing delay to the Work." This means that if the additional time caused by the delay is significant, the contractor may be entitled to additional compensation.

12.6   GREEN PROJECT COST MANAGEMENT

To achieve success, we need to align project management with caring for the environment, which basically requires merging relatively new environmentally friendly approaches with traditional project management methodologies. The best results are usually achieved when a project's activities are fully integrated and costs are managed collaboratively. The integrated project team should typically be engaged, at the earliest phases of design, in target costing, value management, and risk management. Owners and developers are sometimes tempted to place a guaranteed maximum price on the project before the design stage is complete, but this should be resisted to ensure quality and functionality for the building owner or stakeholder. If the project owner comes under pressure to seek a fixed price at an earlier stage of the process, it is prudent to agree on an incentive scheme for the sharing of benefits. It goes without saying that the owner should have a clear understanding of actual construction costs, both hard and soft. Likewise, the owner and project manager must be able to identify and differentiate between underlying costs and risk allowances, in addition to being able to distinguish between profit and overhead margins.

12.6.1   Successful Cost Management Procedures

The project manager is generally responsible for management of the running and overall cost of the project and reports regularly to the owner (or lender, depending on the Contract Documents). One of the PM's responsibilities is regularly reviewing designs as they develop and providing advice on costs to the integrated project team, as well as obtaining feedback from it. This continuous cost oversight is of particular benefit in assessing individual decisions and is especially useful on large and complex projects. It may also prove useful to schedule periodic formal assessments of the whole project, as budgetary estimates, at each project phase (Figure 12.7). The PM's roles and responsibilities and limits of authority should be clearly agreed on at the start of the project, so that everyone knows exactly what he or she is empowered to do in managing project costs and cost overruns.

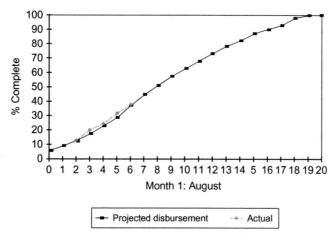

FIGURE 12.7 The cumulative disbursements shown are scheduled as a percentage of total direct cost versus time, and are based on the current project budget and 20.5-month construction period. The CM's projection of disbursements generally follows a realistic S curve. The project's cumulative net direct cost disbursements to date are shown to be roughly in line with initial budget estimates.

According to the U.K. Office of Government Commerce, the main ingredients of successful project cost management are:

- To manage the base estimate and risk allowance
- To operate change control procedures
- To produce cost reports, estimates, and forecasts; the PM is directly responsible for understanding and reporting the cost consequences of any decisions and for initiating corrective actions, if necessary
- To maintain an up-to-date estimated outturn cost and cash flow
- To manage expenditure of the risk allowance
- To initiate action to avoid overspend
- To issue a monthly financial status report

Furthermore, the cost management objectives during the construction phase include delivery of the project at the appropriate capital cost using the value criteria established at the project's inception and ensuring that, throughout the project, comprehensive and accurate accounts are kept of all transactions, payments, and changes. Again, according to the U.K. Office of Government Commerce, the chief areas that cost management teams should consider during the design and execution of a construction project are as follows:

- Identifying elements and components to be included in the project and constricting expenditure accordingly.
- Defining the project program from inception to completion.
- Making sure that designs meet the scope and budget of the project and delivering quality as appropriate and as conforms to the brief.
- Checking that orders are properly authorized.

- Certifying that the contracts provide full and proper control and that all incurred costs are as authorized. All materials are to be appropriately specified to meet the project's scope and design criteria and that materials can be procured effectively.
- Monitoring all expenditure relating to risks to ensure that it is appropriately allocated from the risk allowance and properly authorized. Also, monitoring use of the risk allowance to assess the impact on overall outturn cost (i.e., costs of actual expenditure, accruals, and estimated costs to complete the work).
- Maintaining strict planning and control of both commitments and expenditure within budgets to help prevent any unexpected cost over-/underruns. All transactions are to be properly recorded and authorized and where appropriate, decisions are justified.

12.6.2 Risk Allowance Management

The construction business can be very risky for both the owner and contractor; this is mainly due to the plethora of potential risks that can be caused by unexpected and noncontrollable issues (including those relating to incorporating green features), which is why risk allowances are necessary. Risk allowances should be managed by the party in the best position to manage the risks—usually the project owner or someone representing the owner—with the advice and support of the project manager.

What risk allowance management essentially consists of is a procedure to move costs out of the risk allowance column into the base estimate for the project work, either as risks materialize or as actions are taken to manage them. Formal procedures are required for controlling quality, cost overruns, project delays, and Change Orders. Risk allowances should not be disbursed unless the identified risks to which they relate actually occur. When risks not previously identified come up, they should be treated as Change Orders to the project. Likewise, risks that materialize but have insufficient risk allowance allocated for them should be treated as Change Orders (variation orders).

Figure 12.8 is a graph designed to assist the PM and owner in monitoring key project indicators; it presents an overall picture of how the project is progressing. It should be noted that the graph is to be supplemented with notes for each key element—for example:

Site Work: The site has been cleared. The north waterline connection, and two (2) north and two (2) south sanitary connections have been installed and stubbed through the foundation wall. Electrical connections to the temporary switchgear have been made. A temporary concrete sidewalk has been placed along Washington Avenue.

Potential risks are always preferably defined by allocating specific costs to them, as opposed to just inflating the total cost to compensate for inadequate early planning. A risk-allocated cost contingency is normally needed and included in the total project cost estimate to help mitigate potentially significant

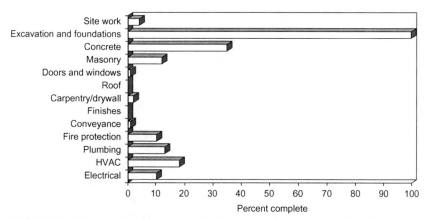

FIGURE 12.8   This graph is designed to assist the project manager and owner monitor the key elements of the project as well as present an overall picture of how it is progressing.

risks. Risk management and contingency funding are particularly useful for mitigating risks that cause cost escalations and overruns during the course of a project's execution.

In preparing the initial project budget, it is strongly recommended that a risk assessment on the entire project be performed in order to identify and quantify the potential risk areas and types. This will help mitigate uncertainties and help create a conservative cost expectation. Risk assessments should also be performed on a regular basis throughout the project's execution to update contingency amounts. Examples of risk assessment areas that may cause concern include failure to perform, heterogeneous or irregular site conditions, utility impacts, hazardous materials, environmental impacts, third-party concerns, and so forth. When quantifying risk as a contingency amount, expectation of occurrence, severity, and anticipated dollar value are variables that may be considered and utilized. After all known risk mitigation, the budget's cost estimate contingency allowance levels should reflect the actual amount of remaining risk associated with the project's major cost elements. An overall management contingency can also be included to cover unknown, unanticipated risks.

Risks and risk allowances should normally be reviewed and evaluated on a regular basis, particularly when formal estimates are prepared, from the design and construction phases through substantial completion and occupancy. The introduction of changes after the briefing and outline design stages are complete should be avoided as much as possible. Change Orders can be minimized by ensuring that, from the start of the project, the Contract Documents are as clear, complete, and comprehensive as possible and that they have been approved by stakeholders. This may require early meetings with planning authorities to discuss their requirements and to ensure that the designs are adequately developed and coordinated before construction begins. For renovation of existing buildings, the type of risks may differ slightly and may require site investigations or condition surveys.

12.6.3   Cost Planning

Over the years, we have witnessed many ups and downs in the construction industry. During economic downturns, the industry experiences more than its fair share of bankruptcies. Many of these bankruptcies could have been prevented had the project owner and project manager taken adequate precautions. The primary causes of bankruptcies have been shown to be inadequate cash resources and failure to convince creditors and the main project lender (if the project is financed) that this inadequacy is only temporary. The need to forecast cash requirements and a project's expected cash flow (i.e., transfer of money into or out of the firm) is important for the project to succeed, particularly if there are cost overruns, economic recession, and the like. Cash flow planning can take many forms, but is necessary as there will always be a time lag between an entitlement to receive a payment for work executed and actually receiving it.

In an element-designed cost plan, the estimate is broken down into several components that can then be compared with later estimates or with actual costs as the project progresses. In applying this approach, each element or item is treated as a distinct cost center, although money can be still be transferred between elements as long as a reasonable balance between elements and the overall target budget are maintained. It must be emphasized that green buildings require intensive planning to ensure optimal results, but any additional effort is usually worth it if you consider that operating costs will be substantially reduced over the life of the facility.

The initial cost plan is often based on unsubstantiated estimates, which nevertheless provide a fair basis for determining the validity of future assessments. The project manager can control costs through ongoing reviews of estimates for each cost center against its target budget. Figure 12.9 is a practical example of a detailed cost estimate with comments. It is interesting to note some of the buyouts and line items that appear low or acceptable. As the project design continues to develop and is priced, variances in cost from the initial cost plan are noted and recorded. A decision must then be taken as to whether that item can be authorized with a corresponding increase in cost, which then requires an equal reduction elsewhere, or whether the item needs to be redesigned to keep within the proposed budget. Furthermore, if a lender is involved, the consultant or PM needs to check the lender's policy to determine if funding of contract deposits is permitted on such subcontracts as structural steel, precast panels, and curtain wall systems. In most cases, the lender's policy is to refrain from funding deposits.

Most often, the owner's designated representative has overall responsibility for general management of the project, including the estimated cost, and therefore has to be satisfied that suitable methods are in place for controlling the project's cost. Where the design process requires the allocation of a significant amount of money, such costs should be appropriately assessed against the budget amount and properly authorized. To facilitate matters, the owner's

Site Area (Acres): 26,310   **Building Area (SFG):** 358,000

Item of Work	Borrower's Budget ($)	$/SFG	$/ACRE	% Total	Comments
Mass Grading	0	0.00	0	0.00%	By previous land owner
Fine Grade and Spoil Removal	240,000	0.67	9,122	0.95%	OK-Bought Out
Site Concrete	230,000	0.64	8,742	0.91%	OK-Bought Out
Asphalt Paving, Striping and Signage	460,000	1.28	17,483	1.83%	Seems Low
Landscaping Allowance	75,000	0.21	2,851	0.30%	Seems Low-Allowance
Site Irrigation Allowance	25,000	0.07	950	0.10%	Allowance
Parking Equipment	60,000	0.17	2,280	0.24%	Seems High
Site Plumbing	334,500	0.93	12,713	1.33%	OK-Bought Out
Site Electrical	93,250	0.26	3,544	0.37%	Low-Buy Out Loss
Fencing with Gate for Secure Parking	15,000	0.04	570	0.06%	Acceptable
Total Site Work	**1,532,750**	**4.28**	**58,255**	**6.09%**	**Acceptable**
Building Excavation, Stone for Fill	263,000	0.73	N/A	1.04%	OK-Bought Out
Building Concrete	1,841,000	5.14	N/A	7.31%	OK-Bought Out
Precast Concrete	800,000	2.23	N/A	3.18%	Low/OK-Bought Out
Caulking of Precast and Windows	70,000	0.20	N/A	0.28%	Acceptable
Masonry	20,000	0.06	N/A	0.08%	Acceptable
Structural Steel and Metal Decking	2,585,000	7.22	N/A	10.26%	Low/OK-Bought Out
Utility Court Steel Doors	10,000	0.03	N/A	0.04%	Acceptable
Steel Stairs and Misc. Metals	300,000	0.84	N/A	1.19%	OK-Bought Out
Steel Precast Support at Tall Entries	50,000	0.14	N/A	0.20%	Unknown Scope
Spray on Fireproofing	362,000	1.01	N/A	1.44%	Acceptable
Foundation and Basement Waterproofing	35,000	0.10	N/A	0.14%	Acceptable
Misc. and Sanitary Caulking	9,150	0.03	N/A	0.04%	Acceptable
Metal Penthouse Siding and Louvers	255,100	0.71	N/A	1.01%	Seems High
60 Mil EPDM Roofing	249,432	0.70	N/A	0.99%	Acceptable
Windows, Entrances, Glass and Glazing	864,000	2.41	N/A	3.43%	OK-Bought Out
Mirrors	5,000	0.01	N/A	0.02%	OK-Bought Out
Overhead Dock Doors	10,000	0.03	N/A	0.04%	Acceptable
Drywall and Metal Studs	1,433,583	4.00	N/A	5.69%	Acceptable
Acoustical Ceilings	469,063	1.31	N/A	1.86%	Acceptable
Carpeting	470,000	1.31	N/A	1.87%	Seems Low
Lobby Floors Allowance	42,000	0.12	N/A	0.17%	Allowance
Ceramic Floor and Wall tile	75,000	0.21	N/A	0.30%	Seems low
VCT	90,000	0.25	N/A	0.36%	Acceptable
High Pressure P-Lam Flooring	25,000	0.07	N/A	0.10%	Acceptable
Painting and Vinyl Wall Covering	360,000	1.01	N/A	1.43%	Acceptable
Toilet Partitions	30,000	0.08	N/A	0.12%	Seems Low
Flag Pole	3,500	0.01	N/A	0.01%	Acceptable
Install Interior Signage	10,000	0.03	N/A	0.04%	Acceptable
Fire Extinguishers and Cabinets Allowance	20,000	0.06	N/A	0.08%	Allowance
Dock Equipment	15,000	0.04	N/A	0.06%	Acceptable
Access Flooring	145,000	0.41	N/A	0.58%	Acceptable
Window Blinds and Draperies	64,000	0.18	N/A	0.25%	Unknown Scope
Operable Walls	70,000	0.20	N/A	0.28%	Acceptable
Carpentry	124,500	0.35	N/A	0.49%	Acceptable
P-Lam Vanity Tops	24,000	0.07	N/A	0.10%	Acceptable
Hollow Metal Doors & Frames Allowance	79,000	0.22	N/A	0.31%	Allowance
Wood Doors Allowance	86,000	0.24	N/A	0.34%	Allowance
Hardware Allowance	149,500	0.42	N/A	0.59%	Allowance
Lobby Features Allowance	20,000	0.06	N/A	0.08%	Allowance
Toilet Accessories	35,957	0.10	N/A	0.14%	Acceptable
Building Directory and Floor Directories	7,500	0.02	N/A	0.03%	Acceptable
Misc. Counter Tops	500	0.00	N/A	0.00%	Unknown Scope
Closet Shelves and Rods	1,500	0.00	N/A	0.01%	Acceptable
Chair Rail	500	0.00	N/A	0.00%	Acceptable
T.V. Brackets	500	0.00	N/A	0.00%	Acceptable
Sound Absorbing Wall Panels	35,000	0.10	N/A	0.14%	Acceptable
NovaWall Wall System	21,000	0.06	N/A	0.08%	Unknown Scope
Elevators	795,000	2.22	N/A	3.16%	Acceptable
Plumbing	865,000	2.42	N/A	3.43%	OK-Bought Out
Fire Protection (no fire pump)	392,000	1.09	N/A	1.56%	OK-Bought Out
HVAC	3,310,000	9.25	N/A	13.14%	OK-Bought Out
Electrical	3,555,750	9.93	N/A	14.12%	OK-Bought Out
Total Building	**20,554,035**	**57.41**	**N/A**	**81.61%**	**Compares Low**
Total Trade Costs Site and Building	**22,086,785**	**61.69**	**N/A**	**87.69%**	**Compares Low**
General Conditions	1,618,000	4.52	N/A	6.42%	Acceptable
General Liability Insurance	150,215	0.42	N/A	0.60%	Acceptable
Subcontractor Bond Costs	221,000	0.62	N/A	0.88%	Acceptable
GC's Fee	710,000	1.98	N/A	2.82%	Acceptable
GC's Construction Contingency	400,000	1.12	N/A	1.59%	Acceptable
Total General	**3,099,215**	**8.66**	**N/A**	**12.31%**	**Acceptable**
Total Direct Cost Budget	**25,186,000**	**70.35**	**N/A**	**100.00%**	**Needs Borrower Contingency**

FIGURE 12.9 Project budget review for an office building. This building has a total building area of 358,000 square feet and a total cost budget of $25,286,000. It is interesting to note the various line item comments in the end column.

representative will frequently delegate appropriate financial authority for design development decisions to the integrated project team. For particularly large or complex projects, the owner may decide to change the delegated levels for each cost center. The payment process is normally managed in the same manner as the design/construction process. All payments should be made as per the contract agreement and on time. Payments for Change Orders, provisional sums, and the like, should be discharged after formal approval is given and as the work is carried out.

Note that, once the construction process begins, any instructions issued to the integrated project team requesting a change through a formal Change Order can have a pronounced impact on the project's cost and possibly other impacts such as time delay. This is why the project team should have specific procedures and protocols for issuance of instructions and information that ensure any issued instructions are within assigned authority and that, before any instruction or information is transmitted, the costs of proposed Change Orders are properly estimated and their impact fully evaluated. Any issued Change Order instruction should be fully sustainable in terms of value for money and overall positive impact on the project. Furthermore, adequate and continuous monitoring of the total costs of all issued instructions is necessary, and where costs are determined to be outside the delegated authority, specific approval is required.

Payments were discussed in Chapter 4. Normally, the client, as the contracting party, is legally obligated to make all payments to the integrated project team, including interim and final payments as per the contract. These payments are usually made at various stages of the work in progress or by application from the general contractor (usually at monthly intervals) after inspection and assessment of the value of work in place. The client's lender (e.g., a bank) should be updated constantly on the project's progress—whether satisfactory or otherwise—by means of regular reports, memos, and cash flow and budget forecasts.

It should be noted that some lenders have a policy of not permitting funding for mobilization but only for actual work in place, while others permit funding for certain trades. For this reason, the terms of the contract should be verified prior to commencement of construction. The contract may also include clauses that allow the project manager under certain circumstances to claim additional payments as specified in the contract's general conditions. Any justification for additional payments may be the result of risks that are essentially considered to be the client's under the contract, such as a change in scope or additional work, or to be caused by the client's failure to comply with its contract obligations, which may be caused by a disruption to the project's scheduled program because of modifications, delivery delays, and the like.

Chapter 13

Green Specifications and Documentation

13.1 INTRODUCTION

Working drawings and specifications are the primary documents used by a contractor to bid and execute a project. Specifications are precisely written documents that go with the construction documents and describe materials as well as installation methods. They describe the project to be constructed, supplementing drawings and forming part of the contract, and describe qualities of materials, their methods of manufacture and their installation, and workmanship and mode of construction. They also provide other information not shown in the drawings, including a description of the final result. Many designers have considerable difficulty preparing a competent set of standard building specifications, partly because it demands that they shift gears, using a different medium to express design content: written instead of drawn. They also propel the designer into the technical realm of materials not normally dealt with on a daily basis and which the designer may not be up to speed on.

Specifications should complement drawings, not overlap or duplicate them, and normally prescribe the quality standards of construction expected on the project. They indicate the procedure by means of which it may be determined whether requirements are satisfied. Because specifications are an integral part of the Contract Documents, they are considered to be legal documents, and should therefore be comprehensive, accurate, and clear. Specification writing has two principal objectives: to define the scope of work and to act as a set of instructions. Defining the scope of work is at the core of specification writing. The required quality of the product and services must be clearly communicated to bidders and the party executing the contract, and must ensure that the completed project conforms to this specified quality. Projects now generally incorporate specifications in a project manual that is issued as part of the contract package along with drawings, bidding requirements, and other contract conditions. The specification writer should ensure that the requirements are compatible with the methods to be employed and that the methods selected in one specification are compatible with those selected in another.

A primary function of project specifications is to deliver detailed information regarding materials and methods of work for a particular construction project. They cover various components relating to the project, including general conditions, scope of work, quality of materials, and standards of workmanship. The drawings, collectively with the project specifications, define the project in detail and clearly delineate exactly how it is to be constructed. The project drawings and specifications are an integral part of the Contract Documents and are inseparable. They reflect what the project specifications are unlikely to cover; the project specifications outline what the drawings are unlikely to portray. Specifications are also sometimes used to clarify details that are not adequately covered by the drawings and notes. Project specifications always take precedence over the drawings, should the drawings conflict with them.

The Construction Specifications Institute (CSI) MasterFormat® is the most widely used standard for organizing specifications for building projects in the United States and Canada. Its format of organization is widely recognized. CSI is a nationwide organization composed of architects, engineers, manufacturers' representatives, contractors, and other interested parties who closely collaborated to develop this format. Its specification standards are noted in MasterFormat, which in 2004 was expanded from 16 to 50 divisions as described later in this chapter. It should be noted that the 1995 edition of the format is no longer be supported by CSI. The MasterFormat 2011 Update, produced jointly by the CSI and Construction Specifications Canada (CSC), replaces all previous editions.

In recent years we have witnessed a fundamental change in specification writing due to technology and green-related practices, which have had a tremendous impact on the construction industry and on the general way we conduct our business. Examples of this are specification production and reproduction, which in a few short years have progressed tremendously. Master systems are now commercially available in electronic form that a specifier can simply load into the computer and get instant access to drawing checklists and explanation sheets. After editing the relevant sections, a printout can be made with an audit trail that informs and records what has been deleted and what decisions remain undetermined.

The CSI comments,

Construction projects use many different kinds of delivery methods, products, and installation methods, but one thing is common to all—the need for effective teamwork by the many parties involved to ensure the correct and timely completion of work. The successful completion of projects requires effective communication amongst the people involved, and that in turn requires easy access to essential project information. Efficient information retrieval is only possible when a standard filing system is used by everyone. MasterFormat provides such a standard filing and retrieval scheme that can be used throughout the construction industry.

Green building specifications can be easily incorporated into CSI MasterFormat in three general ways: (1) environmental protection procedures, (2) green building materials, and (3) practical application of environmental specifications.

13.2   DO WE NEED SPECIFICATIONS?

Construction specifications are necessary mainly because drawings alone typically fail to define the qualitative issues of a scheme. Well-executed specifications form the written portion of the Contract Documents that govern the project. Design decisions are continuously made as drawings develop from schematic sketches to detailed design to construction documents. Drawings are intended to depict the general configuration and layout of a design, including its size, shape, and dimensions. They inform the contractor of the quantities of materials needed, their placement, and their general relationship to each other. Technical specifications are a critical component of the Contract Documents as they reflect the design intent and describe in detail the quality and character of materials and the standards to which the materials and their installation must conform, in addition to other issues that are more appropriately represented in written rather than graphic form. The bottom line is that no matter how beautiful a designer's concept is, it is difficult to envisage the project as properly executed without clear, concise, accurate, and easily understood Contract Documents that include well-written specifications.

While it is true that construction drawings may contain all the information about a structure that can be presented graphically, they nevertheless omit information that the contractor must have but which is not adaptable to graphic presentation. Information that is in this category includes quality-related criteria for materials, specified standards of workmanship, prescribed construction methods, and the like. In the event of a discrepancy between the drawings and the specifications, the specifications must be considered the final authority. For most projects, the specifications consist of a list of 50 divisions that usually starts with a section on general conditions. These are the rules of the job and provide instructions for what to do in any anticipated situations on the project. The general conditions start with a general description of the building, including type of foundation, types of windows, character of framing, utilities to be installed, and so on. This is followed by definitions of terms used in the specs and then certain routine declarations of responsibility and other project-related issues.

According to Douglas D. Harding, a California attorney, "Every project manager should be intimately aware of the general conditions to the project as part of his/her project administration effort." Furthermore, without actual knowledge of the general conditions, contractors and subcontractors take an unacceptable risk that may ultimately cause ruin. In this respect Harding lists a number of important general condition clauses that can directly impact project success if they are not given adequate consideration:

- *Progress Payments:* When are they due? Is there a "condition precedent" clause?
- *Retention:* How much and when is it due?

- *Change Orders:* Overhead and profit, time extensions, inclusions
- *Delay:* Notice and time impact analysis
- *Scheduling:* Who has scheduling responsibilities? What kind of schedule is required? Is the subcontractor required to complete and maintain a schedule?
- *Order of Precedence:* What is the order? Do specs rule drawings or do drawings rule specs?
- *Notice:* How many days after a delay to give notice? How is notice to be delivered: verbally, by mail, by registered mail?

It should be evident that even well-drawn construction drawings are unable to adequately reveal all aspects of a construction project because many aspects cannot be shown graphically. An example of this is trying to describe in a drawing the quality of workmanship required for the installation of electrical equipment or who bears responsibility for supplying the materials. This can be done only with extensive notes. For the majority of projects, the standard procedure is to supplement construction drawings with written descriptions that define and limit the materials and fabrication according to the intent of the engineer or designer. Thus the specifications are an important part of the project because they eliminate possible misinterpretation and ensure positive control.

Time and cost restraints, tend to discourage individuals (or small firms) from writing a completely new set of specifications for each project. Because of this and other issues, specifiers often have to turn to alternative solutions. In this respect, the superiority of master systems over traditional written specifications is overwhelming. Moreover, because of liability issues, specifiers generally feel more comfortable relying on specifications that have repeatedly proven themselves. Typical advantages of employing master systems include accuracy, correctness of language and format for ease of preparation, and the many sources, extensive product databases, and reference material available, from which a complete set of specifications can be compiled for each new project.

Master spec systems are also referred to when modifications are implemented to fit particular conditions of a given job or new specifications are incorporated. They contain guide specifications for many materials that are constantly updated, which allows the specifier to edit out unnecessary text rather than generate new information for each project. In November of 2009, CSI launched GreenFormat™, an online database that organizes sustainable product attributes. GreenFormat states:

[It] is a web-based CSI format that allows manufacturers to accurately report the sustainability properties of their products. It provides designers, constructors and building operators with basic information to help meet "green" requirements. . . Manufacturers report the attributes of their products through a comprehensive, online questionnaire. Their entries are then displayed through www.greenformat.com, where designers, constructors and building operators can search for products that fit their projects.

Some of the more popular sources for specification material are listed next, many of which can be retrieved from the Internet and public libraries:

- Master specifications (MasterSpec®, MasterFormat, SpecText®, BSD SpecLink®-E, EZ-Spec™, 20-20 CAP Studio, and many others)
- Local and national codes and ordinances
- Federal specifications—for example, SpecsIntact, used by the General Services Administration (GSA), NASA, and the U.S. Naval Facilities Engineering Command (NAFVAC)
- National standards organizations, such as the American National Standards Institute (ANSI), National Institute of Building Sciences (NIBS), National Fire Protection Association (NFPA), National Institute of Standards and Technology (NIST), and the Association for Contract Textiles
- Industry associations (Fire Equipment Manufacturers' Association, American Plywood Association, Brick Industry Association, etc.)
- Testing societies—for example, American Society for Testing and Materials (ASTM), American Society for Nondestructive Testing, Underwriters Laboratories (UL)
- Manufacturers' catalogs (Sweets Catalog, MANU-SPEC™, SPEC-DATA®)
- Industry-related publications (e.g., *Construction Specifier*, *Architecture*, the online *Green Magazine*, *Interior Design*, *Architectural Lighting*, *Architectural Record*)
- Books on relevant subjects
- Information from files of previously written specifications

In addition, there are numerous firms providing online specification-writing services that have emerged in recent years. These services can easily be found on the Internet.

13.3 SPECIFICATION TYPES AND CATEGORIES

In preparing a specification document, the specifier has to determine early in the process which format or method is to be used to communicate the desired design intent to the contractor. There are two broad categories of specifications, *closed* and *open*, and most products can be specified by either. Within these categories are four generic types of industry-standard specification:

- Descriptive
- Reference
- Performance
- Proprietary

The type chosen depends on several factors, discussed next.

13.3.1 Closed Specification

A closed (also called prescriptive or restrictive) specification limits a product to a single manufacturer or a few brand-identified types or models and prohibits substitutions. This type of specification is more often used in the private sector in cases where specifiers feel more comfortable resorting to a specific propriety product with which they are familiar, and which will meet the specific criteria of the project. However, it should be noted that this procedure (particularly when only one product is named) is not competitive, and rarely attracts the most favorable price for the owner. Also, while the closed specification is common in private construction work, it is generally prohibited by the Code for public projects and is required by law to be bid under open specifications. An open specification allows products of any manufacturer to be used if the product meets the specified requirements.

The closed proprietary specification method is considered the easiest form to write but the most restrictive in application, because it names a specific manufacturer's product. It generally establishes a narrower definition of acceptable quality than do performance or reference standard methods, and gives the designer complete control over what is installed. The specification can also be transformed into an open proprietary specification in which multiple manufacturers or products are named or alternatives solicited by adding the phrase "or equal." This would increase potential competition and encourage a lower installation price from potential vendors. There are instances where a multiple choice may not be appropriate, as for example in a renovation project where a specific brick is required for repairs to an existing brick façade.

13.3.2 Open Specification

Also called Performance or Nonrestrictive, an open specification gives the contractor some choice in how to achieve the desired results and is the type required by California's Public Contract Code. Proprietary specifications may also be used as open specifications but with the addition of the "or equal" clause, which allows the contractor to consider other products for bid if they are shown to be equal in performance and specifications. Due to the ambiguity surrounding this clause, and the disagreements it often perpetuates, specifiers generally shy away from incorporating it into the proprietary specifications.

Descriptive Specification

A descriptive specification is a method of open specification that is gaining popularity and is sometimes referred to as prescriptive specifications. As the name implies, this type of specification describes in detail the requirements for the material or product and the workmanship required for its fabrication and installation without providing a trade name. Government agencies sometimes stipulate this type of specification to allow greater competition among product manufacturers. Descriptive specifications are more difficult to write than proprietary ones because the specifier is required to include all the product's relevant physical characteristics in the specification bearing in mind the specifier has already decided that the specified product meets functional needs. For an individual product, proprietary, performance, and descriptive specifying techniques may be used.

Reference Standards

Reference standards specify standards such as ASTM, ASHRAE, State of California, Federal, and others. The various manufacturers must meet these standards, which basically describe a material, product, or process referencing a recognized industry standard or test method as the basis for the specification. It is often used to specify generic materials such as Portland cement or clear glass. Thus, in specifying gypsum wallboard for example, the specification can state that all gypsum wallboard products shall meet the requirements of ASTM C36. It is worth mentioning that a number of construction industry members have voiced the opinion that specifications should not only make references to the applicable standards, but they should also quote the relevant parts of the referenced standards.

With the reference standard specification the product is described in detail so that the specifier is relieved of the necessity to repeat the requirements but can instead refer to the recognized industry standard. In employing a reference standard, the specifier should not only possess a copy of that standard, but should also know what is required by the standard, including choices that may be contained therein, and which should be enforced by all suppliers. This type of specification is generally short and fairly straightforward and easy to write. In addition, the use of reference standard specifications reduces a firm's liability and the possibility for errors.

Performance Specifications

Performance specifications have been developed over recent years for many types of construction operations. Rather than specifying the required construction process, performance specifications establish the performance requirements of the finished facility without dictating the methods by which the end results are to be achieved. The precise method by which this performance is obtained is left to the construction contractor. This gives the greatest latitude to contractors because it allows them to use any material or system that meets the required performance criteria, provided the results can be verified by measurement, tests, or other acceptable methods. Performance specifications are difficult to write; the specifier needs to know all the criteria for a product or system, determine an appropriate method for testing compliance, and write a clear and lucid document. This requires sufficient data to be provided to ensure that the product can be adequately demonstrated. Performance specifications are primarily used in cases where a specifier wants to inspire new ways of achieving a particular result in specifying complex systems.

Proprietary (Product) Specifications

Proprietary (product) specifications often use a combination of methods to convey the designer's intent. They are normally written by referencing specific products by manufacturer and brand or model name, and apply to materials and equipment. For example, a proprietary specification for a terracotta tile includes the product or products selected by the specifier, a description of the size and design, and the ASTM standard, grade, and type required. Proprietary specs are distinguished from prescriptive specs in that the physical characteristics are inferred rather than explicitly stated. For an individual product, proprietary, performance, and descriptive specs may be used. Proprietary specifications can be made "open" by adding the phrase "or equal."

13.4   DEVELOPING THE PROJECT MANUAL

The CSI developed the first standard format for organizing construction information in 1963, which later became known as MasterFormat. In 1964 the American Institute of Architects (AIA) developed the concept of a project manual primarily to meet the pressing need for a consistent arrangement of building construction specifications. A project manual consists of an assemblage of documents related to project construction, and is employed to guide the construction process. It typically includes bidding requirements (contract forms, bonds, certificates, etc.), sample documents, contract conditions, and technical specifications, which together with the drawings, constitute the Contract Documents.

The project manual has gained general acceptance in the industry and is greatly preferred to the traditional method of organizing the Contract Documents, which was a matter of individual preference by the design firm producing them, resulting in a wide diversity of organization around the country that became very confusing. As design firms and contractors became increasingly nationwide in their operations, the project manual concept continued to develop; while it may differ depending on the size and type of project, a typical project manual may include, but not be limited to the following:

- *General project information:* These include
  - Title page with names and addresses of all parties responsible for the project (owners, architects, civil engineers, mechanical engineers, electrical engineers, and structural engineers) in addition to a statement of compliance by the architect or engineer of record
  - Table of contents
  - Schedule of drawings
- *Bidding requirements:* These apply where contracts are awarded through the bidding process and include

- The invitation to bid and advertisement for bids
- Instructions to bidders (prequalification forms, bid forms, information available to bidders, date and time of bid opening, notice of pre-bid conference, etc.)
- *Contract forms:* These include
  - *Sample forms:* public entity crime form, owner/contractor agreement, performance and payment bond, Change Order, bid form (may require the general contractor's license number, subcontractors list and license numbers, etc.)
  - *Bonding requirements:* labor and materials payment bonds, required on projects costing above a certain amount
  - *Insurance requirements:*
    - Workers' compensation and employer's liability
    - Public liability (personal injury, bodily injury, and property damage)
    - Products and completed operations liability
    - Owner's protective liability
    - Business automobile liability (for owned, nonowned, and hired vehicles)
    - Property all-risks coverage (to 100% of the value at risk subject to acceptable deductibles)
  - *Contract condition:* General conditions of the contract such as AIA Form 201 or similar preprinted forms; supplementary conditions include anything not covered in the general conditions, such as addenda (changes before contract signing), and Change Orders (changes after contract signing); also, general conditions and supplementary conditions, including but not limited to
    - Deductive alternates (required if bidding is on a project where funds may be reverted and rebidding is not possible in the remaining time available, and when the client wants the option to negotiate with the apparent low bidder
    - Notice of time limit and method of payment to the contractor, including final payment
    - Time limit for completion of construction
    - Penalty to the contractor for failure to comply with time limits
    - Federal wage rates and hourly scales where applicable (not required for projects financed by local or state funds)
    - Provision for who should pay for standard tests of concrete, plumbing, electrical, steel, etc., as required by industry standards
    - At client's discretion, incentives for early completion of the project
  - *Technical specifications:* Written technical requirements concerning building materials, components, systems, and equipment shown on the drawings regarding standards, workmanship quality, performance of related services, and stipulated results to be achieved by construction methods (Figure 13.1)

MASTERSPEC® SMALL PROJECT™ 2005

COMBINED TABLE OF CONTENTS - MASTERFORMAT 2004 (Section Text Only)

© 2005 The American Institute of Architects

Issue Date	Sect. No.	SECTION TITLE	SECTION DESCRIPTION
DIVISION 01 - GENERAL REQUIREMENTS			
2003	011000	SUMMARY	Summary of the Work, Owner-furnished products, use of premises, and work restrictions.
2003	012000	PRICE AND PAYMENT PROCEDURES	Allowances, alternates, unit prices, contract modification procedures, and payment procedures.
2005	013000	ADMINISTRATIVE REQUIREMENTS	Project management and coordination, submittal procedures, delegated design, and Contractor's construction schedule.
2003	014000	QUALITY REQUIREMENTS	Testing and inspecting procedures.
2005	014200	REFERENCES	Abbreviations, acronyms, and trade names referenced in MASTERSPEC SMALL PROJECT.
2003	015000	TEMPORARY FACILITIES AND CONTROL	Temporary utilities and facilities for support, security, and protection.
2003	016000	PRODUCT REQUIREMENTS	Product selection and handling and product substitutions.
2003	017000	EXECUTION AND CLOSEOUT REQUIREMENTS	Examination and preparation, cutting and patching, installation, and closeout requirements.
DIVISION 02 - EXISTING CONDITIONS			
2003	024119	SELECTIVE STRUCTURE DEMOLITION	Demolition and removal of selected portions of buildings and site elements.
DIVISION 03 - CONCRETE			
2003	033000	CAST-IN-PLACE CONCRETE	General building and structural applications.
2005	033713	SHOTCRETE	Pneumatically projected mortar and concrete.
2005	034100	PRECAST STRUCTURAL CONCRETE	Conventional precast units.
2005	034500	PRECAST ARCHITECTURAL CONCRETE	Exposed surface units.
2003	034713	TILT-UP CONCRETE	Wall panels.
DIVISION 04 - MASONRY			
2003	042000	UNIT MASONRY	General applications, walls, partitions.
2005	042300	GLASS UNIT MASONRY	Glass block.
2005	044300	STONE MASONRY	Stone veneer laid in mortar.
2003	047200	CAST STONE MASONRY	Architectural units set in mortar.
DIVISION 05 - METALS			
2005	051200	STRUCTURAL STEEL FRAMING	Framing systems.
2005	052100	STEEL JOIST FRAMING	Standard SJI units.
2005	053100	STEEL DECKING	Roof, floor, composite types.
2005	054000	COLD-FORMED METAL FRAMING	Load-bearing and curtain-wall studs; floor, ceiling, and roof joists.
2003	055000	METAL FABRICATIONS	Iron, steel, stainless steel, and aluminum items (not sheet metal).
2003	055100	METAL STAIRS	Steel; with pan, abrasive-coated, and floor plate treads and tube railings.
2003	055200	METAL RAILINGS	Metal railings, including glass panels and wood rails.
DIVISION 06 - WOOD, PLASTICS, AND COMPOSITES			
2005	061000	ROUGH CARPENTRY	Framing, sheathing, subflooring, etc.
2005	061053	MISCELLANEOUS ROUGH CARPENTRY	Rough carpentry for minor applications.
2005	061600	SHEATHING	Wall and roof sheathing, subflooring, underlayment, and related products.
2005	061753	SHOP-FABRICATED WOOD TRUSSES	Metal-plate-connected members.
2003	061800	GLUED-LAMINATED CONSTRUCTION	Glued-laminated beams, arches, and columns.
2005	062000	FINISH CARPENTRY	Exterior and interior trim, siding, paneling, shelving, and stairs.
2005	064013	EXTERIOR ARCHITECTURAL WOODWORK	Trim, door frames, shutters, and ornamental items.
2005	064023	INTERIOR ARCHITECTURAL WOODWORK	Trim, custom cabinets, counter tops, flush paneling, and stairwork and rails.
2003	066113	CULTURED MARBLE FABRICATIONS	Vanity tops, shower walls, and tub surrounds.

> **MASTERSPEC SMALL PROJECT SPECIFICATIONS TABLE OF CONTENTS - COMBINED - Page 1 of 6**

Continued

FIGURE 13.1   MasterSpec® Small Project six-page specifications Combined Table of Contents. *(Source: American Institute of Architects; see www.masterspec.com/masterspec/pdfs-os04/TOC%20SP%20Combined%20Complete.pdf.)*

MASTERSPEC® SMALL PROJECT™ 2005
COMBINED TABLE OF CONTENTS - MASTERFORMAT 2004 (Section Text Only)
© 2005 The American Institute of Architects

MASTERSPEC SMALL PROJECT SPECIFICATIONS TABLE OF CONTENTS - COMBINED - Page 2 of 6

FIGURE 13.1—Cont'd

MASTERSPEC® SMALL PROJECT™ 2005
COMBINED TABLE OF CONTENTS - MASTERFORMAT 2004 (Section Text Only)
© 2005 The American Institute of Architects

MASTERSPEC SMALL PROJECT SPECIFICATIONS TABLE OF CONTENTS - COMBINED - Page 3 of 6

FIGURE 13.1—Cont'd

MASTERSPEC® SMALL PROJECT™ 2005

COMBINED TABLE OF CONTENTS - MASTERFORMAT 2004 (Section Text Only)
© 2005 The American Institute of Architects

Issue Date	Sect. No.	SECTION TITLE	SECTION DESCRIPTION
2005	124813	ENTRANCE FLOOR MATS AND FRAMES	Recessed and surface-applied flexible floor mats and frames.

DIVISION 13 - SPECIAL CONSTRUCTION

2005	132416	SAUNAS	Modular and precut saunas; includes heaters and accessories.
2005	133419	METAL BUILDING SYSTEMS	Systems consisting of structural framing, roofing and siding panels, and standard components.

DIVISION 14 - CONVEYING EQUIPMENT

2005	142400	HYDRAULIC ELEVATORS	Preengineered hydraulic and roped-hydraulic units.
2005	142600	LIMITED-USE/LIMITED-APPLICATION ELEVATORS	Limited-use/limited-application elevators.
2005	144200	WHEEL CHAIR LIFTS	Vertical and inclined types, and stairway chairlifts.

DIVISION 21 - FIRE SUPPRESSION

2005	210500	COMMON WORK RESULTS FOR FIRE SUPPRESSION	Motors, hangers and supports, vibration isolation and seismic restraints, and meters and gages.
2003	211000	WATER-BASED FIRE-SUPPRESSION SYSTEMS	Wet-pipe sprinklers and standpipes.

DIVISION 22 - PLUMBING

2005	220500	COMMON WORK RESULTS FOR PLUMBING	Motors, hangers and supports, vibration isolation and seismic restraints, and meters and gages.
2005	220523	GENERAL-DUTY VALVES FOR PLUMBING PIPING	Gate, globe, check, and ball valves.
2005	220700	PLUMBING INSULATION	Piping and equipment insulation.
2005	221113	FACILITY WATER DISTRIBUTION PIPING	Domestic and fire-protection water utility services outside the building.
2005	221116	DOMESTIC WATER PIPING	Potable-water piping and specialties inside the building.
2003	221123	DOMESTIC WATER PUMPS	Horizontal and vertical in-line pumps.
2003	221313	FACILITY SANITARY SEWERS	Sanitary sewerage and underground structures outside the building.
2005	221316	SANITARY WASTE AND VENT PIPING	Soil, waste, and vent piping and specialties inside the building.
2005	221353	FACILITY SEPTIC TANKS	Tank, distribution box, and drainage pipe.
2005	221413	FACILITY STORM DRAINAGE PIPING	Stormwater piping and specialties inside the building.
2003	221429	SUMP PUMPS	Wet-pit mounted, simplex packaged units and submersible sump pumps.
2003	223100	DOMESTIC WATER SOFTENERS	Fully-automatic, pressure-type, household water softener.
2005	224000	PLUMBING FIXTURES	Fixtures, carriers, faucets, and trim.

DIVISION 23 - HEATING VENTILATING AND AIR CONDITIONING

2005	230500	COMMON WORK RESULTS FOR HVAC	Motors, hangers and supports, vibration isolation and seismic restraints, and meters and gages.
2005	230523	GENERAL-DUTY VALVES FOR HVAC PIPING	Gate, globe, check, and ball valves.
2003	230593	TESTING, ADJUSTING, AND BALANCING FOR HVAC	AABC and NEBB certified testing and balancing.
2005	230700	HVAC INSULATION	Pipe, duct, and equipment insulation.
2005	230900	INSTRUMENTATION AND CONTROL FOR HVAC	Electric/electronic controls and sequences for HVAC systems and equipment.
2005	231113	FACILITY FUEL-OIL PIPING	Piping, specialty valves, and transfer pumps.
2005	231123	FACILITY NATURAL-GAS PIPING	Natural-gas piping and specialties.
2005	231126	FACILITY LIQUEFIED-PETROLEUM GAS PIPING	LP-gas piping and specialties.
2005	232113	HYDRONIC PIPING	Heating and cooling water piping and condensate drain piping.
2003	232123	HYDRONIC PUMPS	Base mounted and inline; close coupled and separately coupled pumps.
2005	232300	REFRIGERANT PIPING	Piping, specialties, and refrigerant inside the building.
2003	233100	HVAC DUCTS AND CASINGS	Metal and fibrous ducts and accessories.
2005	233423	HVAC POWER VENTILATORS	Roof and wall mounted centrifugal fans, and ceiling-mounted and inline ventilators.
2003	233600	AIR TERMINAL UNITS	Fan-powered and shutoff, single-duct units.
2003	233713	DIFFUSERS, REGISTERS, AND GRILLES	Diffusers, registers, and grilles.
2005	235100	BREECHINGS, CHIMNEYS, AND STACKS	Gas vents, chimneys, and grease ducts.
2005	235213	ELECTRIC BOILERS	Electric, hot-water boilers.

MASTERSPEC SMALL PROJECT SPECIFICATIONS TAB LE OF CONTENTS - COMBINED - Page 4 of 6

FIGURE 13.1—Cont'd

MASTERSPEC® SMALL PROJECT™ 2005
COMBINED TABLE OF CONTENTS - MASTERFORMAT 2004 (Section Text Only)
© 2005 The American Institute of Architects

FIGURE 13.1—Cont'd

MASTERSPEC® SMALL PROJECT™ 2005
COMBINED TABLE OF CONTENTS - MASTERFORMAT 2004 (Section Text Only)
© 2005 The American Institute of Architects

FIGURE 13.1—Cont'd

Being legal, specification language must be written in a clear, precise, and unambiguous manner in order to communicate the intended concept. In this respect, a convention has developed over the years as to the information that should be shown on the drawings and information that should more appropriately be in the specifications. Drawings should depict information that can be most aptly and effectively expressed graphically. This includes dimensions, sizes, proportions, gauges, arrangements, locations, and interrelationships. In addition, drawings express quantity, whereas specifications normally describe quality, and they denote type (e.g., wood), whereas specifications denote species (e.g., oak). Well-written specifications, on the other hand, are essentially based on a number of broad general principles as outlined here:

- They should transmit information only that lends itself to the written word (e.g., standards, descriptions, procedures, guarantees, names).
- They should be clear, concise, and technically correct.
- They should avoid ambiguous words that could lead to misinterpretation.
- They should be written using simple words in short, easy to understand sentences.
- They should use technically correct terms, and avoid slang or "field" words.
- They should avoid fielding conflicting requirements.
- They should avoid repeating requirements stated elsewhere in the contract.

Confusion may arise in some cases when there are exceptions to these principles. For example, building departments in most municipalities only accept drawings with applications for building permits, and refuse to accept a project manual with specifications. Additionally, all data demonstrating building code compliance must be indicated on the drawings. However, the repetition of identical data on the specifications and the drawings exposes the documents to potential errors and inconsistency. To achieve better communication, the specifier should:

- Avoid specifying standards that cannot be measured or phrases that are subject to wide interpretation.
- Acquire a thorough understanding of the most current standards and test methods referred to and the sections applicable to the project. Use accepted standards to specify quality of materials or workmanship required, such as "Portland Cement: Conform to ASTM C150, Type I or Type II, low alkali. Maximum total alkali shall not exceed 0.6 percent."
- Avoid specifications that are impossible for the contractor to execute.
- Use clear, simple, direct statements, concise terms, and correct grammar and punctuation. Avoid words or phrases that are ambiguous and imply a choice that may not be intended.
- Be impartial in designating responsibility. Avoid exculpatory clauses such as "the General Contractor shall be totally responsible for all . . .," which try to shift responsibility.

- Describe only one important idea per paragraph to make reading and comprehension easier, editing and modifying at a later date. Specifications should be as short and concise as possible, omitting words like *all*, *the*, *an*, and *a*.
- Capitalize the following:
  - Contract Documents: Specifications, Working Drawings, Contract, Clause, Section, Supplementary Conditions
  - Major parties to the contract—for example, Contractor, Client, Owner, Architect
  - Specific rooms in the building—for example, Living Room, Kitchen, Office
  - Grade of materials—for example, No.1 Douglas Fir, FAS White Oak
  - All proper names
- Avoid underlining anything in a specification, as this implies that the remaining material can be ignored.
- Use the terms *shall* and *will* correctly. *Shall* designates a command: "The Contractor shall . . ." *Will* implies choice: "The Owner or Architect will . . ."

It is imperative that the specifications and construction drawings be fully coordinated as they complement each other. Moreover, they should not contain conflicting requirements, errors, omissions, or duplications. The following sidebar shows a summary of project manual requirements for a new construction project.

Summary of General Project Manual Requirements*

Contact list
Location map/site plan/building plans/elevations (reduced scale)
Borrowers Loan Agreement (BLA)
Architect/engineer (A/E) agreement (design services)
CM agreement
Construction agreement
Consultant services agreement
Additional service billings
Project Analysis Report (PAR)
Project Status Report (PSR) template
Borrower's draw requests
Construction schedule
GC/CM Applications for Payment (current and log)
Log of Change Order and/or pending Change Orders
Change Orders
Request for information (RFI) log
Submittal log
Buyout/subcontractor log
Vendor log
Allowances

* To be edited as required based on nature of project.

13.5   ORGANIZATION AND FORMAT OF SPECIFICATIONS

Over the decades, many revisions and expansions to MasterFormat have occurred resulting from changes in the construction industry. The Construction Specifications Institute and Construction Specifications Canada originally created the 16-division MasterFormat in 1963. Today, this format is widely used in the United States and Canada for preparing construction specifications for nonresidential building projects. MasterFormat has become the standard for titling and arranging construction project manuals containing bidding requirements, contracting requirements, and specifications. Since its inception, the CSI struggled to standardize the specification numbering system and the format of the sections, and produced a modified MasterFormat in 1995.

During recent years the CSI actively sought to further improve MasterFormat by adding new divisions to the system. In a concerted effort to address rapidly evolving computer and communications technology, a modified MasterFormat was introduced in 2004 that included a significant expansion and reorganization of the project manual division numbers. In the new MasterFormat edition, division numbers are increased from 16 to 50, of which 13 are left blank to provide for future revisions and to allow construction products and technology to further evolve (Figure 13.2). The revised numbering system allows for more than 100 times the number of subjects at the same level in the old numbering system.

MasterFormat is a well-structured system employed by specifiers for organizing information into project manuals, for organizing cost data, for filing product information and other technical data, for identifying drawing objects, and for presenting construction market data. Although construction projects use a number of delivery methods, products, and installation methods, effective communication among the people involved on a project is crucial to achieving successful completion. MasterFormat facilitates standard filing and retrieval schemes throughout the construction industry; without a standard filing system familiar to all users, information retrieval would be almost impossible.

The new MasterFormat standard provides a master list of divisions, and section numbers and titles within each division, for organizing information about a facility's construction requirements and associated activities, A full explanation of the titles used in MasterFormat is provided along with a general description of the coverage for each title. A keyword index of requirements, products, and activities is also provided to help users find appropriate numbers and titles for construction subjects. The current MasterFormat groups, subgroups, and divisions consist essentially of 50 specification divisions, which are described in the following sections.

PROCUREMENT AND CONTRACTING REQUIREMENTS GROUP:

- Division 00   — Procurement and Contracting Requirements

SPECIFICATIONS GROUP

General Requirements Subgroup

- Division 01   — General Requirements

Facility Construction Subgroup

- Division 02   — Existing Conditions

- Division 03   — Concrete

- Division 04   — Masonry

- Division 05   — Metals

- Division 06   — Wood, Plastics, and Composites

- Division 07   — Thermal and Moisture Protection

- Division 08   — Openings

- Division 09   — Finishes

- Division 10   — Specialties

- Division 11   — Equipment

- Division 12   — Furnishings

- Division 13   — Special Construction

- Division 14   — Conveying Equipment

- Division 15   — RESERVED FOR FUTURE EXPANSION

- Division 16   — RESERVED FOR FUTURE EXPANSION

Continued

FIGURE 13.2   The revised 2004 MasterFormat edition has replaced the 1995 edition. *(Source: Construction Specification Institute, Inc.)*

- Division 17 — RESERVED FOR FUTURE EXPANSION

- Division 18 — RESERVED FOR FUTURE EXPANSION

- Division 19 — RESERVED FOR FUTURE EXPANSION

Facility Services Subgroup

- Division 20 — RESERVED FOR FUTURE EXPANSION

- Division 21 — Fire Suppression

- Division 22 — Plumbing

- Division 23 — Heating, Ventilating, and Air Conditioning

- Division 24 — RESERVED FOR FUTURE EXPANSION

- Division 25 — Integrated Automation

- Division 26 — Electrical

- Division 27 — Communications

- Division 28 — Electronic Safety and Security

- Division 29 — RESERVED FOR FUTURE EXPANSION

Site and Infrastructure Subgroup

- Division 30 — RESERVED FOR FUTURE EXPANSION

- Division 31 — Earthwork

- Division 32 — Exterior Improvements

- Division 33 — Utilities

- Division 34 — Transportation

- Division 35 — Waterways and Marine Construction

- Division 36 — RESERVED FOR FUTURE EXPANSION

FIGURE 13.2—Cont'd

- Division 37 — RESERVED FOR FUTURE EXPANSION

- Division 38 — RESERVED FOR FUTURE EXPANSION

- Division 39 — RESERVED FOR FUTURE EXPANSION

Process Equipment Subgroup

- Division 40 — Process Integration

- Division 41 — Material Processing and Handling Equipment

- Division 42 — Process Heating, Cooling, and Drying Equipment

- Division 43 — Process Gas and Liquid Handling, Purification, and Storage Equipment

- Division 44 — Pollution Control Equipment

- Division 45 — Industry-Specific Manufacturing Equipment

- Division 46 — RESERVED FOR FUTURE EXPANSION

- Division 47 — RESERVED FOR FUTURE EXPANSION

- Division 48 — Electrical Power Generation

- Division 49 — RESERVED FOR FUTURE EXPANSION

FIGURE 13.2—Cont'd

13.5.1   Specification Section Format

The specification section provides a uniform standard for arranging specification text in a project manual's sections using a three-part format, and reduces the chance of omissions or duplications. According to the CSI,

Rather than grouping administrative, product requirements and execution requirements under each product separately, SectionFormat provides a uniform approach to organizing specification text within each section. SectionFormat is based upon the principle that a section should be organized by grouping the administrative requirements, product requirements, and execution requirements for each product together.

Thus each specification section covers a particular trade or subtrade (e.g., drywall, carpet, ceiling tiles). Furthermore, each section is divided into three basic parts, each of which contains the specifications about a particular aspect of each trade or subtrade (Figure 13.3). The updates are intended to reflect changes in the industry related to advancements in information technology and electronic publishing.

According to CSI, "PageFormat™ offers a recommended arrangement of text on a specification page within a project manual, by providing a framework for consistently formatting and designating articles, paragraphs and subparagraphs.

Product Specification

Franchise: Hilton	Project: Evolution '05 Prototype
Brand: Hilton Garden Inn	Location: N/A

Item Code: CA-11 Guestrooms
Item: Guestroom

Issue Date: 07/22/2005
Print Date: 08/25/2006

Description: Enhanced loop - tip sheer

Pattern Name: Custom

Color: Custom

Pat Design No: Custom   **Aprvd Strike-Off No:** M45104

Dr No: N/A   **Reference No:** N/A

<u>**Dimensions**</u>

Carpet Width: 12 ft 0 in

Ptrn Repeat: Width: N/A Length: N/A

<u>**Specs**</u>

Contents: 100% solution dyed nylon

Dye Method: Solution dyed

Guage: 1/10   **Pile Height:** .250"   **Oz Wt-SY:** 32   **Tot Oz Wt-SY:** 74.10

Prm Backing: Woven Polypropylene

Sec Backing: ActionBack

Carpet Pad: Pad to be Hartex Super pad. Fiber: Polyester; Weight: 40 oz; Density: 8.8; Thickness: 0.25-0.375; Breaking Strength: MD 32 lbs, CMD 31 lbs, ASTM 2262; Sound Absorption: NAC .61 ASTM C423-77; Thermal conduct: 2.02 ASTM C518-76.

Installation: Stretch-in

Locations: Guestroom

Fire Safety: Material must meet NFPA-253 or ASTM E-648 as required by local and state fire codes for use in hotels and other public areas.

Notes:
1. Carpet installation shall comply with all carpet manufacturer standards.
2. Certification of compliance required verifying that carpet meets or exceeds all local and state fire codes.
3. Manufacturer to provide maintenance instructions to owner.

Special Notes: 1. Templeton approved strike-off number: 45104.

Quantity: Quantity to be verified by installer before purchasing.

Vendors:

Templeton Carpet, 1900 Willowdale Road NW, Dalton, GA, 30720
Ph: 706-275-8665 Fax: 706-275-0687
Contact: Kelly Barker, Templeton Carpet Ph: 281-807-7927 Fax: 281-807-7158

Changes to this specification must be approved by the Designer.   CA-11

FIGURE 13.3   The Product Specification sheet shown is for a carpet for the guest bedrooms in a Hilton Garden Inn. The vendor is Templeton Carpet. *(Source: Paradigm Design Group.)*

It also includes guidance for page numbers and margins." Recent updates to Page-Format focus on making documents easier to read without limiting specific technology or processing methods. These updates offer greater freedom to use sophisticated publishing and electronic media techniques on a wider variety of display devices.

Part 1: General

Part 1 of the specification outlines the general requirements for the section and describes the administrative, procedural, and temporary requirements unique to it. This part is an extension and amplification of subjects covered in Division 1. In general, Part 1 outlines quality control requirements, delivery and job condition requirements, trades with which this section needs to be coordinated, and submittals required for review prior to ordering, fabricating, or installing material for that section. It generally consists of the following:

1. *Description and Scope:* This Article should include administrative and procedural requirements specific to the Section. It should also specify the scope of the work and the interrelationships between work in this and the other sections. It should also include a list of important generic types of products, work, and specified requirements, as well as the following:
   a. *Products supplied but not installed:* List products that are only supplied under this Section but whose installation is specified in other sections.
   b. *Products installed but not supplied:* List products that are only installed under this Section but furnished under other sections.
   c. *Allowances:* List products and work included in the section that are covered by quantity allowances or cash allowances. Do not include cash amounts. Descriptions of items should be included in Part 2 or Part 3.
   d. *Unit prices:* Include statements relating to the products and work covered by unit prices and the method to be used for measuring the quantities.
   e. *Measurement procedures:* State the method to be used for measurement of quantities. Complete technical information for products and types of work should be specified in the appropriate articles of Part 2 and Part 3.
   f. *Payment procedures:* Describe the payment procedures to be used for measurement of quantities used in unit price work. Complete technical information for products and types of work should be incorporated in the appropriate articles of Part 2 and Part 3.
   g. *References:* List standards referenced elsewhere in the Section, complete with designations and titles. Industry standards and associations may be identified here. This Article does not require compliance with standards but merely a listing of those used.
   h. *Definitions:* Define unusual terms not explained in the Contracting Requirements and that are utilized in unique ways not included in standard references.
   i. *Alternates:* Check whether the acceptability of alternates is detailed in the General Requirements.
2. *Quality Assurances:* These include prerequisites, standards, limitations, and criteria that establish an acceptable level of quality for products and workmanship. To achieve this, the following are required for consideration:
   a. *Qualifications:* List statements of qualifications of consultants, contractors, subcontractors, manufacturers, fabricators, installers, and applicators of products and completed work.

b. *Regulatory requirements:* Describe obligations for compliance with specific code requirements for contractor-designed items and for public authorities and regulatory agencies, including product environmental requirements.

c. *Certifications:* Include statements to certify compliance with specific requirements.

d. *Field samples:* Include statements to establish standards used to evaluate the work with the assistance of field samples. These are physical examples representing finishes, coatings, or a material finish such as wood, brick, or concrete.

e. *Mock-ups:* Include statements to establish standards by which the work will be evaluated by the use of full mock-ups. These are full-size assemblies erected for construction review, testing, operation, coordination of specified work, and training of the trades.

f. *Pre-installation meetings:* Determine requirements for meetings to coordinate products and techniques, and to sequence related work for sensitive and complex items.

3. *Submittals:* Include but do not limit to requests for certain types of documentary data and affirmations of the manufacturer or contractor to be furnished as per the contract. Includes requests for specific types of product data and shop drawings for review, as well as submittal of product samples and other relevant information, including warranties, test and field reports, environmental certifications, maintenance information, installation instructions, and specifics for closeout submittals.

4. *Product Handling, Delivery, and Storage:* To furnish instructions for various activities, to include:

a. *Packing, shipping, handling and unloading:* Specify requirements for packing, shipping, handling, and unloading that pertain to products, materials, equipment, and components specified in the section.

b. *Acceptance at site:* Describe the conditions of acceptance of items at the project site. This normally applies to owner-provided products.

c. *Storage and protection:* Outline special measures including temperature control that are needed to prevent damage to specific products prior to application or installation.

d. *Waste management and disposal:* Affirm any special measures required to minimize waste and dispose of waste for specific products.

5. *Project and Site Conditions:* Determine physical or environmental conditions or criteria that are to be in place prior to installation, including temperature control, humidity, ventilation, and illumination required to achieve proper installation or application. Statements that reference documents where information may be found pertaining to such items as existing structures or geophysical reports. For example, all wall tiling should be completed prior to cabinet installation.

6. *Sequencing and Scheduling:* This is required where timing is critical and where tasks and/or scheduling need to be coordinated and follow a specific sequence.

7. *Maintenance:* List items to be supplied by the contractor to the owner for future maintenance and repair. Delineate provisions for maintenance services applicable to critical systems, equipment, and landscaping.

8. *Warranties:* Terms and conditions of special or warranty or bonds covering the conformance and performance of the work should be spelled out and the Owner should be provided with copies.

9. *System Startup, Owner's Instructions, Commissioning:* List applicable requirements for the startup of the various systems. Include requirements for the instruction of the owner's personnel in the operation of equipment and systems. State requirements for Commissioning of applicable systems to ensure installation and operation are in full compliance with design criteria.

10. *System Description:* Describe performance or design requirements and functional requirements of a complete system. Limit descriptions to composite and operational properties to the extent necessary to link multiple components of a system together, and to interface with other systems.

Part 2: Products

Part 2 describes the materials, products, equipment, fabrications, components, mixes, systems, and assemblies being specified and that are required for the project. It also details the standards to which the materials or products must conform to fulfill the specifications and similar concerns. Materials and products are included with the quality level required. Included in the itemized subsections are the following:

1. *Manufacturers:* This section is used when writing a proprietary specification, and will include a list approved manufacturers. The Section should be coordinated with the Product Options and Substitutions Section. Names of manufacturers may be supplemented by the addition of brand names, model numbers, or other product designations.

2. *Materials, Furnishings, and Equipment:* A list should be provided of materials to be used. If writing descriptive or performance specifications, detail the performance criteria for materials, furnishings, and equipment. Describe the function, operation, and other specific requirements of equipment. This article may be omitted and the materials included with the description of a particular manufactured unit, equipment, component, or accessory. Environmental concerns such as toxicity, recycled content, and recyclability can be addressed here.

3. *Manufactured Units:* Fully describe the complete manufactured unit, such as standard catalog items.

4. *Components:* Describe the specific components of a system, manufactured unit, or type of equipment.
5. *Accessories:* Describe requirements for secondary items that aid and assist specified primary products or are necessary for preparation or installation of those items. This Article should not include basic options available for manufactured units and equipment.
6. *Mixes:* This Section specifies the procedures and proportions of materials to be used when site-mixing a particular product. This Article relates mainly to materials such as mortar and plaster.
7. *Fabrication:* Describe manufacturing, shop fabrication, shop assembly of equipment and components, and construction details. Specify allowable variations from specified requirements.
8. *Finishes:* Describe shop or factory finishing here.
9. *Source Quality Control:* Indicate requirements for quality control at off-site fabrication plants.
   a. *Tests, inspection:* Describe tests and inspections of products that are required at the source, that is, plant, factory, mill, or shop.
   b. *Verification of performance:* State requirements for procedures and methods for verification of performance or compliance with specified criteria before items leave the shop or plant.
10. *Existing Products:* Create a list of characteristics of assemblies, components, products, or materials that must match existing work, including matching material, finish, style, or dimensions. Specify compatibility between new and existing in-place products.

Part 3: Execution

Part 3 describes the quality of work—the standards and requirements specified for the installation of products and materials. Site-built assemblies and site-manufactured products and systems are included. This part specifies basic onsite work and includes provisions for incorporating products in the project. It also describes the conditions under which the products are to be installed, the protection required, and the closeout and post-installation cleanup and protection procedures. The subheadings in this section include:

1. *Inspection:* Define what the Contractor is required to do, for example to the subsurface, prior to installation. Sample wording may include "The moisture content of the concrete should meet manufacturer's specifications, prior to installation of the flooring material."
2. *Preparation:* Specify actions required to prepare the surface, area, or site to incorporate the primary products of the Section. Also stipulate the improvements to be made prior to installation, application, or erection of primary products. Describe protection methods for existing work.

3. *Installation, Construction, and Performance:* The specific requirements for each finish should be specified, as well as the quality of work to be achieved, and includes
   a. *Special techniques:* Describe special procedures for incorporating products, which may include spacings, patterns, or unique treatments.
   b. *Interface with other work:* Include descriptions specific to compatibility and transition to other materials. This may include incorporating accessories, anchorage, as well any special separation or bonding.
   c. *Sequences of operations:* Describe the required sequences of operation for each system or piece of equipment.
   d. *Site tolerances:* State allowable variations in application thicknesses or from indicated locations.
4. *Field Quality Control:* State quality control requirements for onsite activities and installed materials, manufactured units, equipment, components, and accessories. Specify the tests and inspection procedures to be used to determine the quality of the finished work.
5. *Protection:* Where special protection is necessary for a particular installation, such as marble flooring, this Section must be included. Provisions for protecting the work after installation but prior to acceptance by the owner should be cited.
6. *Adjust:* Describe final requirements to prepare installed products to perform properly.
7. *Clean:* Describe in detail the cleaning requirements necessary for the installed products.
8. *Schedules:* To be used only if deemed necessary. When schedules are included indicate item/element/product/equipment and location.
9. *Demonstration:* State installer or manufacturer requirements to demonstrate or to train the owner's personnel in the operation and maintenance of equipment.

13.6 GREENING SPECIFICATIONS

The term "greening specifications" seems to be increasingly used in the industry. But when greening your specs—for example, by introducing resource- or energy-efficient methods or materials, you should ask two important questions:

- Should the specifications call out or emphasize new green attributes to ensure that bidders take particular notice, or should the modified specs be quietly woven in so that contractors and subcontractors do not use them to make a bigger (and more expensive) deal out of the changes than they reasonably should?

- Should the "green" additions be placed throughout the body of the specs, or, should they be gathered into one place where everyone can see them and where they are easy to update over time?

There are many opinions regarding these questions and considerable merit for each of the answers to them. Numerous articles on green specifications are available on the Internet, and it would be prudent to research them. There are also several green organizations such as GreenFormat (a web-based CSI format). Here are some comments that may be helpful in greening your project:

- Integrate the green specs into your conventional specs without putting a particular emphasis on them.
- Establish the purpose of going green in your general or introductory specs.
- Use a program that has high-performance scopes of work that integrate the various trade contractors' work with pre- and post-work checklists and performance-based verifications.
- Take advantage of available green specification software programs to generate your specs.

For guidance on these early sections of your specs, see the Green*Spec*® "Introduction to Guideline Specifications." According to BuildingGreen.com:

These guideline specifications are designed to be modified as needed for new development, retrofits, and maintenance. They are organized into three Division 01 sections:

- *01 81 09 Testing for Indoor Air Quality*
- *01 81 13 Sustainable Design Requirements*
- *01 91 00 General Commissioning Requirements*

Together, these three sections provide an overview of sustainable design requirements that might be appropriate in a wide variety of projects. When these sections are used in actual project specifications, specific requirements must be inserted throughout the construction documents to ensure compliance with the sustainable design intent.

In many specifications we find the phrase "or equivalent" or "approved equal" indicating a substitute for the original product. This is particularly important in green specifications because of the large amount of misinformation and "greenwashing" that plagues the market. It is important to be precise and very clear on about the attributes the determination of equivalency is based on. Some of the many green attributes that may be important in a green product specification include recycled-content, durability, reusability, emissions, biodegradability, and the like. It is not fair to the owner, the contractor, the design consultant, or the employees/occupants not to make clear from the outset what constitutes a fair substitution.

The CSI offers a sample outline to illustrate its three-part SectionFormat with respect to environmental product specifications (see sidebar).

CSI Green Product Specification Outline

Part 1. General
1 Environmental Requirements
   1. List applicable environmental standards, regulations, and requirements.
   2. Include VOC requirements.
   3. List recycled content requirements.
   4. Identify reuse, recycling, and salvaging methods.
   5. Reference Division 1, Environmental Procedures for Construction.
      i) VOCs or chemicals to avoid.
      ii) General environmental procedures.
      iii) Reuse, recycling, or salvaging requirements.
      iv) Healthful building maintenance.

Part 2. Products
2 Specific Environmental Product Attributes
   1. Product contains no xxxx chemicals (list and identify).
   2. Product contains xx percent recycled content:
      i) Identify post-industrial recycled content.
      ii) Identify post-consumer recycled content.
   3. Product is recyclable after useful life.
   4. Product is certified by an independent third party.
      i) Recycled content
      ii) Sustainably harvested
   5. Product is durable (list warranty).
   6. Product is moisture-resistant (if applicable).
   7. Include any other environmental attributes.

Part 3. Execution
3 Environmental Procedures
   1. Address environmental installation of materials.
   2. Include protection of materials.
   3. Identify environmental methods of cleanup.
   4. Include recycling of scrap during construction.
   5. Reference Division 1 Environmental Procedures.

The primary purpose of sample specifications is to supplement rather than replace standard specifications. Model green specs are designed to be edited, adapted, and incorporated into the standard specifications of building projects, and generally augment standard specifications by providing additional environmental information, such as sustainable building criteria, definitions, and performance requirements.

13.7 COMPUTERIZED SPECIFICATION-WRITING SYSTEMS

With technology evolving at such a rapid pace, it should not be surprising that the era of manual preparation of building specifications is rapidly disappearing into the annals of history. Indeed, over the past decade, a number of firms have developed various versions of automated specification-writing systems, and many now offer these services online to architects, interior designers, engineers, and others. Moreover, while CAD has revolutionized the drawing process, many architects and designers no longer possess the knowledge and expertise for, or interest in, writing specifications for their projects. Computer software has reduced the need for some of the traditional skills shaped by years of specification-writing experience. Computer resources offer practitioners greater efficiency and the ability to deal with continually expanding compilations of information, and they provide almost immediate access to information from thousands of applicable electronic databases. This is coupled with an increasingly complex construction industry, changing methods of procurement, and the tremendous pressure on architects and designers to prepare contract documentation, including specifications of the highest quality and in little time.

A further complication is the fact that the complexity of new commercial construction requires continually updated knowledge, particularly regarding the adoption of new "green" codes (e.g., The International Green Construction Code [IgCC]) and more eco-friendly products. The changing nature of the industry, with its new energy-efficient products, laws, environmental regulations, techniques for assembling products, and building industry practices, cannot be adequately presented in educational courses for engineers and architects. Specialists are therefore urgently needed to manage and update information.

Today, there are an increasing number of automated CAD specification packages on the market. Of the better known is Building Systems Design's (BSD) SpecLink®-E, an automated master guide and specification management system with built-in intelligence to help you significantly speed up editing tasks and reduce production time while minimizing errors and omissions. Combined with the industry's most comprehensive and up-to-date master database, SpecLink-E enables you to accelerate specification development with tremendous accuracy and integrity.

SpecLink-E includes specification sections for use in construction documents, short-form specifications, and design criteria documents. Furthermore, it allows images of up to 100 K in size to be inserted into the text in any of the following formats: Windows bitmap (.bmp), Windows metafile (.wmf), portable network graphics (.png), JPEG file interchange format (.jpg, .jpeg), and graphics interchange format (.gif). Hyperlinks to websites can also be inserted into the text. SpecLink-E uses master guide specifications in CSI 3-part format, and has a database of over 780 master specification sections and more than 120,000 data links that automatically include related requirements and exclude incompatible options as you select specification text (Figure 13.4). Building Systems Design also offers PerSpective® early design performance specifications organized in CSI's UniFormat, which is the industry's first commercially available database of performance-based specifications.

Another well-known program comes from InterSpec, LLC; it uses a proprietary technology for construction document management solutions and services built on its patented e-SPECS® specification management program (Figure 13.5).

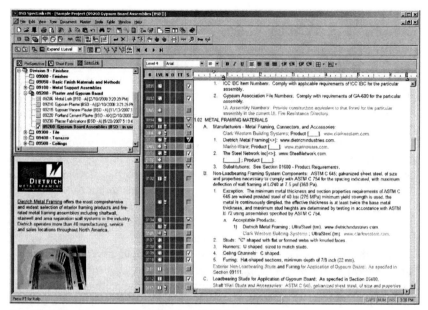

FIGURE 13.4 Screenshot of the BSD SpecLink-E summary catalog listing and computer screen printout. *(Source: Building Systems Design, Inc.)*

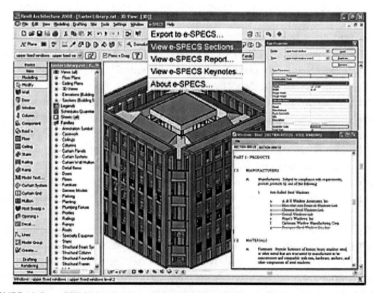

FIGURE 13.5 e-SPECS integrates specifications with CAD and BIM applications. It can extract all the project requirements from the Revit model facilitating redlining and commenting in collaboration with the team members. *(Source: InterSpec.)*

e-SPECS software automates specification writing by extracting product and material requirements directly from the project drawings; it also connects a large database of specifications to an electronic architectural drawing of the project. e-SPECS integrates directly with Autodesk's AutoCAD, AutoCAD Architecture, AutoCAD MEP, and all Revit®-based applications, in addition to supporting all libraries of MasterSpec. InterSpec has a do-it-yourself program designed for architects and designers on small projects that can be very helpful for small firms.

For architects and engineers who spend many hours on every project preparing construction specifications, e-SPECS software saves time and money while ensuring that the specifications agree with the construction documents. Like other automated systems it enables design professionals to increase their productivity and reduce their costs. Also, by linking the architect's CAD drawings to the master guide specifications, there is no longer the need to mail or deliver large blueprint drawings to the spec writer.

With these automated systems, the designer can input all required information in the earliest phases of the project, before any drawings are even available, and almost instantly obtain an outline or preliminary specification. InterSpec has announced a major upgrade to e-SPECS and recently released e-SPECS Version 7.0, which includes Revit Phase Support, Office Master Updater, and enhanced publishing functionality. GreenWizard and InterSpec have formed a collaborative partnership so that future integrated tools from InterSpec and GreenWizard will help solve issues of green and LEED-oriented materials' specification, including product discovery and evaluation.

SpecsIntact is another automated software system for preparing standardized facility construction specifications; it is used in construction projects worldwide. SpecsIntact was initially developed by the National Aeronautics and Space Administration (NASA) to assist architects, engineers, specification writers, project managers, construction managers, and other professionals doing business with NASA, the NAVFAC, and the U.S. Army Corps of Engineers (USACE). According to its developers, the system provides quality assurance reports and automated functions that reduce the time required to complete project specifications. The Unified Facilities Guide Specifications (UFGS) Master employed by SpecsIntact is divided into functional divisions according the CSI format, with each one containing related specification sections. The principal elements in the sections are annotated using SpecsIntact's application of the eXtensible Markup Language (XML), a tagging scheme that provides the intelligence used in automatically processing these sections.

20-20 CAP Studio is an integrated package of applications that automate the design and specification process, and that contains two base applications, 20-20 CAP Designer and 20-20 CAP Worksheet. CAP Designer is a CAD-based design tool that operates within AutoCAD. CAP Worksheet is a power specification tool for product pricing, specification, and estimating. The program can import complete large-scale space plans and layouts into CAP Worksheet for full specification, discounting, and order entry. It specializes in furniture specification.

ARCOM recently announced its "revolutionary" Altarix, a new productivity tool in its SpecWare® suite of specification software. According to ARCOM,

Altarix is easy and intuitive to use, with a sleek interface that is more than just a document processor. Altarix provides standard functions like the ability to:

- *Add, delete, and edit specification text*
- *Select options and automatically add punctuation*
- *Globally format all sections in a project*
- *Globally set headers and footers for a project*

But Altarix goes beyond these basic capabilities and provides tools specially designed for specifiers, allowing them to:

- *Add project notes and automatically track their status and resolution*
- *Insert specification sections into a project from another project*
- *Easily jump from one location in a specification section to another location*
- *Track completion progress for each specification section*

Specifiers have complete control over their documents and can change the specifications as fast as clients change their mind. Specifiers can also produce complete project manuals with the Altarix program.

The continuously evolving technology of specification writing is transforming the way architects and interior designers prepare specifications for construction projects. The main advantage of automated systems is that they provide greater accuracy in less time and at a lower cost. These systems also eliminate or minimize costly construction modifications caused by omissions, discrepancies, or improper quality controls. A firm's proprietary interactive online editing systems can be integrated into specification development over the Internet with secure password access. A completed specification manual can readily be delivered online for client downloading, and can be easily printed and bound or presented on a CD-ROM. For smaller design firms that lack the necessary resources, outsourcing may be worth considering as the most effective way to proceed on a project.

13.8 LIABILITY ISSUES

Several international associations exist in order to facilitate legal, technical, and political issues of international concern to the construction industry. All professionals, including architects and engineers, are expected to exercise reasonable care and skill in the implementation and execution of the various aspects of their work. The level of professional performance should be consistent with that normally provided by any qualified practitioner under similar circumstances. However, this does not imply that projects will be executed to 100% perfection at all times.

The law relating to professional responsibility and liability has in recent years become very active and has assumed unprecedented urgency. The parameters of risk and exposure have expanded dramatically in professional practice so that, under current law, if a professional designer enters into a contractual agreement and specifies a subsystem of a commercial or institutional space, he or she assumes responsibility for the satisfactory performance of that system. Another emerging area that is causing significant concern is a professional designer's liability to third parties who have no connection with the contract for claims of negligence or design errors that allegedly lead to injury of persons using the building. The legal bases for the majority of liability suits often overlap, but generally include professional negligence, breach of contract, implied warranty or misrepresentation, joint and several liability, and liability without fault for design defects. Moreover, if designers fail to reject defective work by a contractor or supplier, they may now be considered professionally negligent and in breach of contract.

Product liability (i.e., building product performance) is another area of exposure in which the architect is held responsible for damages caused by faulty materials and components, and sometimes for the cost of their replacement. This places a heavy emphasis on appropriate selection and specification of building products with long records of satisfactory performance and thus often discourages the use of new materials (e.g., green products), and methods. Product liability is primarily concerned with negligence, which is discussed in Chapter 16. It especially affects manufacturers, retailers, wholesalers, and distributors. Furthermore, with the upsurge of green delivery systems, designers and specifiers are increasingly finding themselves involved in product liability suits. The best way to minimize product liability is to specify products that are manufactured for their intended use and have been adequately tested.

Finally, design professionals can more effectively protect themselves from liability suits by working within their area of expertise, using concise contracts and specifications, complying with codes and regulations, using reputable and licensed contractors, maintaining accurate records, retaining legal counsel, and ensuring that adequate and appropriate liability insurance is in place.

Referenced Standards

- ASHRAE—American Society of Heating, Refrigerating and Air-Conditioning Engineers
  - 62.1-2007: Ventilation for Acceptable Indoor Air Quality
  - 90.1-2004: Energy Standard for Buildings Except Low-Rise Residential Buildings
- ASTM—American Society for Testing and Materials International
  - C518-04—Test Method for Steady-State Thermal Transmission Properties by Means of the Heat Flow Meter Apparatus

- C1371-04a—Standard Test Method for Determination of Emittance of Materials Near Room Temperature Using Portable Emissometers
- C1549-04—Standard Test Method for Determination of Solar Reflectance Near Ambient Temperature Using a Portable Solar Reflectometer
- D6400-04—Standard Specification for Compostable Plastics
- D6868-03—Standard Specification for Biodegradable Plastics Used as Coatings on Paper and Other Conpostable Substrates
- D7081-05—Standard Specification for Non-Floating Biodegradable Plastics in the Marine Environment
- E408-71 (2008)—Standard Test Methods for Total Normal Emittance of Surfaces Using Inspection-Meter Techniques
- E1333-96: Test Method for Determining Formaldehyde Concentrations in Air and Emissions Rates from Wood Products Using a Large Chamber
- E1980-01—Standard Practice for Calculating Solar Reflectance Index of Horizontal and Low-Sloped Opaque Surfaces
- BIFMA—Business and Institutional Furniture Manufacturers Association
- CRI—Carpet and Rug Institute
  - Green Label Plus Testing Program (carpet and carpet adhesive)
  - Green Label Testing Program (Carpet Pad)
- CRRC—Cool Roof Rating Council
  - 1—Product Rating Program Manual
- CRS—Center for Resource Solutions
  - green-e Product Certification Requirements (for renewable energy)
- CSA—Canadian Standards Association
  - 2810: Life Cycle Impact Assessment: Pulp and Paper Production Phase
- EPAct—Energy Policy Act (EPAct) of 1992 (water fixture maximum flow rates)
- EPA—Environmental Protection Agency
  - Comprehensive Procurement Guide
  - ENERGY STAR® Rating System
  - National VOC Emission Standard
  - Priority PBT List
  - Reference Test Method 24—Surface Coatings
- FSC—Forest Stewardship Council's Principles and Criteria
- GREENGUARD Environmental Institute
- GREENGUARD Certification Standards for Low Emitting Products for the Indoor Environment
- GREENGUARD Product Emission Standard for Children and Schools
- Green Seal
  - GC-03—Anti-Corrosive Paints
  - GC-09—Residential Central Air-Conditioning Systems
  - GC-12—Occupancy Sensors
  - GC-13—Split-Ductless Air-Source Heat Pumps (GC-13)
  - GC-15—Residential Central Air Source Heat Pumps
  - GS-05—Compact Fluorescent Lamps
  - GS-11—Paints
  - GS-13—Windows (GS-13)
  - GS-14—Window Films (GS-14)
  - GS-31—Electric Chillers
  - GS-32—Photovoltaic Modules

- GS-36—Commercial Adhesives, October 19, 2000
  - GS-37—Industrial and Institutional Cleaners: Green Seal Environmental Standard for General-Purpose, Bathroom, Glass, and Carpet Cleaners Used for Industrial and Institutional Purposes
  - GS-40—Industrial and Institutional Floor Care Products: Finishes and Compatible Strippers Used for Industrial and Institutional Purposes
  - GS-43—Recycled-Content Latex Paint, August 1 2006
- IDA—International Dark-Sky Association
- ISO—International Organization for Standardization
  - 14001:1996—Environmental Management Systems – Specification with Guidance for Use
  - 14021:1999—Environmental Labels and Declarations – Self-Declared Environmental Claims Type II Environmental Labeling
  - 14024:1999—Environmental Labels and Declarations, Type I Environmental Labeling—Principles and Procedures
  - 14025:2006—Environmental Labels and Declarations, Type III Environmental Declarations—Principles and Procedures
  - 14040:2006—Environmental Management, Life Cycle Assessment: Principles and Framework
  - 21930—Sustainability in Building Construction, Environmental Declaration of Building Products
- NFRC—National Fenestration Rating Council
  - 100-04—Procedure for Determining Fenestration Product Thermal Properties
  - 200-04—Procedure for Determining Fenestration Product Solar Heat Gain Coefficients at Normal Incidence
  - 300-04—Procedures for Determining Solar Optical Properties of Simple Fenestration Products
  - 400-04—Procedure for Determining Fenestration Product Air Leakage
  - 500-04—Procedure for Determining Fenestration Product Condensation Resistance Values
- PEFC—Programme for the Endorsement of Forest Certification
- RFCI—Resilient Floor Covering Institute
- FloorScore®—Testing program certified by SCS to comply with the VOC emissions criteria of the CA CHPS Section 01350 emissions standard
- SCS—Scientific Certification Systems
- SFI—Sustainable Forestry Initiative
- SGS—SGS Group
- NSF—National Science Foundation
  - 140—Sustainable Carpet Assessment
- SmartWood
- SCAQMD—South Coast Air Quality Management District
  - South Coast Rule #1113, Architectural Coatings
  - South Coast Rule #1168, October 3, 2003, Adhesive/Sealant VOC Limits, State of California
- CHPS—The Collaborative for High-Performance Schools
  - Section 01350—Special Environmental Requirements (emissions testing requirements)

- OEHHA—Office of Environmental Health Hazard Assessment
- California Proposition 65 Chemicals—List of chemicals known by California EPA to cause cancer
- UL—Underwriters Laboratories, Inc.
- USDA—U.S. Department of Agriculture
  - Bio-based Compliant Program
- U.S. Energy Policy Act 1992

Chapter 14

Types of Building Contract Agreements

14.1 INTRODUCTION

Participants in a construction project generally have to sign an American Institute of Architects (AIA) contract, a ConsensusDOCS contract, or something similar. In some cases the owner or lender may prefer to use its own. Whatever the case, this contract sets out the terms for the parties and outlines the role of the project architect. The AIA contract will most likely be more favorable to the architect's interests than to those of the other parties to the contract, which is why some owners and contractors choose an alternative such as ConsensusDOCS or that offered by the Engineers Joint Contract Documents Committee® (EJCDC). Choosing the most appropriate contract for your purpose is important, as the construction process is often governed by contracts that involve complex relationships in several tiers. Thus, as indicated, there are many contract types, with some favoring the client, others the builder. In any case, contracts should be taken seriously and carefully scrutinized. They are usually categorized according to type of payment but can be tailored to incorporate common elements from several different contract types.

As just mentioned, one alternative to AIA contracts is ConsensusDOCS, developed by construction industry organizations that represent the interests of building owners, contractors, and surety providers. Organizations that represent the interests of design professionals did not play a major role in developing ConsensusDOCS contracts. Consequently, a ConsensusDOCS contract does not set specific standards for the quality of design work. Another difference between ConsensusDOCS and AIA contracts is that in the former, the role of the architect in communications between the building owner and the contractor is reduced.

Another alternative to AIA contracts is the EJCDC Contract Documents. The EJCDC has a suite of contracts relating to construction projects. Like ConsensusDOCS, contracts, they downplay the role of the architect in managing the project.

It is of great importance for the contractor to know how to bid a construction job if there is to be any chance at ever turning a profit. In today's competitive marketplace, the most effect approach to bidding a construction job is to have an experienced estimator on hand who can use a computer estimating program to come up with the best (and lowest) possible cost. Construction bid software can be readily purchased by a general contractor and installed on most computers, or, increasingly, it can be web-based. Its use is no longer a luxury, but has become an absolute necessity for successful bidding, and its benefits are very substantial. Such programs can develop budgets and establish cost baselines, as well as allow general contractors to keep track of the financial status of a project on a daily or hourly basis rather than have it done manually by an employee. Furthermore, all budgetary information is stored in one location and is easily accessed as opposed to using the traditional filing system, and the likelihood of errors and omissions is far less than it would be if the computations were done by hand.

14.2   BIDDING PROCESS AND BUILDING CONTRACT TYPES

There is no set way to bid construction jobs, but a tried and true method is developing the lowest bid based on an accurate cost estimate. However, it is vital to know how to bid as this can make the difference between success and bankruptcy for a contractor. Normally, construction bidding consists of submitting a proposal to carry out a described residential or commercial construction project for an agreed sum and within certain parameters (generally according to the Contract Documents). Bidding for a project can occur at the construction manager, general contractor, or subcontractor level, and contractors can submit their bids for the total cost of a construction tender to the project owner, developer, or consultant. A decision is then made to award the contract, taking into account factors such as price, contractor qualifications, and time to complete the project.

Construction contracts need to be carefully drafted and managed to avoid possible exposure to financial penalties and the risk of turning a potentially profitable project into an unprofitable one. As for bidding on federal projects, Federal Acquisition Regulation (FAR) 28.102 stipulates that all construction projects over $100,000 be subject to the Miller Act, which requires performance and payment bonds. Performance bonds represent a promise of surety to the government that once the contract is awarded, the contractor will perform its obligations on it. But whether federal or private, for the contract to be lucrative, it must minimize or eliminate risk factors as much as possible.

There are a number of project delivery systems, the most common of which were discussed in greater detail in Chapter 3. The three primary systems are

- Traditional design−bid−build (Figure 14.1)
- Design−build (Figure 14.2)
- Construction management—adviser or at-risk (Figure 14.3)

Each of these delivery systems has distinct advantages and disadvantages depending on the project, and all can be used to successfully plan, design, and bring a given construction project to successful completion.

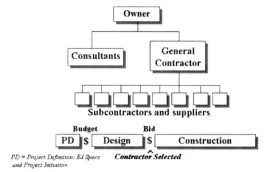

FIGURE 14.1 Structure and schedule of the traditional design−bid−build delivery system.

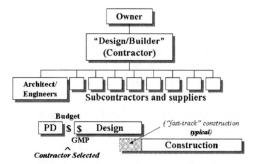

FIGURE 14.2 Structure and schedule of the design−build delivery system.

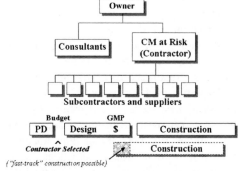

FIGURE 14.3 Structure and schedule of the Construction Manager at Risk delivery system. *(Source: Drawings by Mary K. Crites, AIA. 2007; www.maricopa.edu/facilitiesplanning/docs/delivery_methods.pdf.)*

14.2.1 Design−Bid−Build

Design−bid − build (DBB) is the traditional method of project delivery and has been the most widely used since ancient times. It is also the one with which project owners are most familiar. DBB is a linear process where one task follows completion of another with no overlap. It commences with an owner selecting an architect to prepare construction documents. Most often the architect will release these documents either publicly to any general contractors or to a select, prequalified group that are invited to bid what they believe the total cost of construction will be. This bid is inclusive of various other bids from subcontractors for each specific trade, and the general contractor's fee is generally built in. The majority of government contracts are required to bid competitively using this method. Contractors bid the project exactly as it is designed and the lowest responsible, responsive bidder is awarded the work. The design consultant team is selected separately and reports directly to the owner. A DBB contract is most suited for less complicated projects that are budget-sensitive but not necessarily schedule-sensitive or subject to change. The owner can define and control the design through the architectural consultant.

14.2.2   Design–Build

The design–build project delivery system differs from traditional DBB, in that the design and construction aspects of the project are contracted with a single entity. On this, Mark Cladny of On Site Systems, Inc., comments, "The design–build approach has been touted as the contracting method of choice for the future, eliminating construction disputes and hastening the design and construction process," and this may be so. The system essentially focuses on combining the design, permit, and construction schedules in a manner that streamlines the traditional DBB environment. A guaranteed maximum price (GMP) is typically provided by the design-builder early in the project, based on design criteria prepared by the owner and a moderately developed design created by the architect. The design-builder is typically either the general contractor or design professional (architect or engineer). This system minimizes project risk for an owner and reduces the delivery schedule by overlapping the design and construction phases. If the design-builder is the contractor, any required design professionals will typically be retained directly by the contractor.

As outlined in Chapter 3, there are potential problems associated with the design–build delivery system. For example, cost estimating for a design–build project can be difficult when, say, design documents are preliminary and liable to change over the course of the project. To address this situation, design–build contracts are often written to allow for unexpected situations without penalizing either the design-builder or the owner. Organizations such as the Design–Build Institute of America provide standardized contracts for design-builders to use, although it is not unusual for the design-builder to provide its own contractual documents, particularly for well-established firms and firms that have constructed similar projects.

14.2.3   Construction Manager as Constructor

In the manager-as-constructor delivery system, a construction manager (CM) is hired prior to completion of the Design Phase to act as project coordinator and general contractor. As discussed in Chapter 3, this allows the CM to work directly with the architect or project team and circumvent potential design issues prior to completion of the construction documents. Once the tender documents have been completed, the construction manager-adviser invites contractors or subcontractors to bid for the various divisions of work for the project.

14.3 BID SOLICITATION AND TYPES OF BUILDING CONTRACTS

The bid solicitation process requires making published construction data (tender documents) readily available to interested parties, such as construction managers, contractors, and the public. Project owners release prepared project details, including Contract Documents and specifications, in the form of tender documents (normally for a fee) to interested general contractors, subcontractors, and other parties to solicit bids. This is done through services that are usually subscription-based or that charge a flat fee. It is difficult to structure a formula for quoting an exact fee largely because it is extremely uncommon for two properties to be exactly alike in terms of the variables that need to be considered. Moreover, it becomes a guessing game trying to quote a fee on a cost-per-square-foot basis without having all the information upfront.

The owner is responsible for determining the type of contract to be used on a particular project. This is determined by several factors such as the type of project and the amount of risk the owner is willing to accept. The following are several basic types of construction contracts currently in use:

- Lump sum
- Cost-plus
- Guaranteed maximum price
- Unit price
- Project management
- Labor
- Negotiated

14.3.1 Lump Sum

The lump sum contract is typically used with the design−bid−build method of project procurement and often in contracts where the contractor agrees to supply all labor and materials for a fixed sum. This type of contract (also sometimes called "stipulated sum" or "fixed fee") is suitable if the scope and schedule of the project are sufficiently defined to allow estimation of project costs. However, if the building contractor miscalculates, there is nowhere to go except to absorb the cost. The lump sum is the most widely used contract in building construction, and it is generally more favorable for the owner because the cost of the project is known from the outset.

In this contract type the cost of management is placed squarely on the shoulders of the contractor, and adds responsibility in the pre-contract assessment carried out by the contractor for ensuring that all potential contingencies are covered as agreed extras. It does provide a degree of certainty for both parties because the contract clearly spells out what is involved. Once the contract is signed, both parties (the owner and contractor) must live up to its terms, and any modifications that follow will be considered "Change Orders" and will add to the cost and possible delays. A lump sum agreement normally protects the project owner from unethical contractors hoping to take advantage.

According to Mark Cladny of On Site Systems (http://www.onsitesystems. net/methods.html), many factors determine whether the lump sum method is suited to a project:

1. *Is there a requirement to have lump sum competitive bidding due to public or internal policy concerns? Do these policies preclude other contracting methods?*
2. *Have plans and specifications been developed by an architect and related consultants that exactly depict the project requirements? With the competitive nature of the architectural industry, many architects and designers have come to rely on previously generated details and "canned" specifications, without thoroughly modifying them to depict the current project.*
3. *[Have] sufficient time and resources been allocated to produce and final design prior to the start of construction?*
4. *Does the owner or architect possess the expertise to issue partial documents and coordinate the design and construction process to allow concurrency of design and construction activities by the issuance of bid packages?*
5. *Have the budgets and financing allowed consideration for the expected amount of scope and non-scope changes?* Scope changes *are ones regarded as different from the bidding documents due to any number of issues, including:*
   - *Unexpected or nondepicted site conditions.*
     - *Changes to the building's characteristic due to program changes or user requirements.*
     - *Changes in technology from the design phase to building turn over.*

Non-scope changes *could include:*

- *Claims for delay due to lack of design coordination, inadequate or nondescript documents.*
- *Ambiguity regarding sub-system design responsibility.*
  - *Interference with other contractors or public agencies.*
  - *Interference with existing tenants or occupants.*

Table 14.1 summarizes the advantages and the disadvantages of lump sum contracts.

Building contractors normally add 10 to 15% to the expected cost of the project to account for unforeseen contingencies. This amount may be increased based on the builder's assessment of risk and partly depends on the type of project and the special circumstances surrounding it, and by agreement can either be included in the total cost of the contract or treated as a maximum contingency fund. This means that, provided no surprises eventuate, the owner will not be

TABLE 14.1 Evaluating the Lump Sum Contract

Advantages	Low financial risk to owner.
	Lump sum contract price not subject to adjustment on basis of contractor's cost experience in performing contract (unless agreed by both parties)
	Places maximum risk and full responsibility for all costs and resulting profit or loss on contractor
	Because the price is fixed, unforeseen contingencies or variations in material or labor prices will not affect owner
	Costs known at outset
	Contractor selection is relatively easy
	Provides degree of certainty for both parties because it clearly spells out expectations and terms
	Maximum incentive for contractor to control costs and perform effectively and impose minimum administrative burden on contracting parties
Disadvantages	Architect not involved because this contract is an agreement between owner and contractor for final fixed-price sum; because architect does exercise traditional role, the quality of work cannot be checked and not controlled by an expert
	Difficult and costly to make changes
	Specifications may not be clear, so contractor can use alternative/inferior brands of materials; only minimum specifications usually provided
	Early project start not possible due to need to complete design prior to bidding process
	One can expect considerable ambiguity in specifications, measurements, mode of payment, and so on
	Although contract is executed on fixed-price basis, contractor may claim extras by giving different reasons since specifications, measurements, and so on are not precise
	Contractor requests money in advance from owner, and proceeds with work at own pace; in some cases, he or she may deliberately hold up work toward the end to extract maximum money from owner, leaving owner feeling helpless because money is tied up with contractor

liable for this sum. However, if cost overruns are associated with the work that are due not to errors or omissions in the design but instead to the contractor's poor performance, or even the weather, the contractor must absorb the loss with no additional requests for compensation from the owner. In practice, sometimes when costs exceed estimates, disputes may arise over the scope of work or attempts to substitute less expensive materials for those specified. In addition, if the contractor is able, through superior performance, to achieve cost savings, these benefits go solely to him or her.

A lump sum contract can also be used in conjunction with an award-fee incentive and performance incentives, providing that the award fee or incentive is based solely on factors other than cost. As for the stipulated sum contract, a section may be included that stipulates certain unit price items. Unit price is often used for items that have indefinite quantities, such as pier depth. A fixed price is established for each unit of work.

Lump Sum Contract with Price Adjustment

The lump sum contract with price adjustment provides for an upward or downward revision of the stated contract price on the occurrence of specified contingencies. It can be used when there is serious doubt regarding the market's stability or labor conditions that may exist during an extended period of contract performance, and contingency allowances that would otherwise be included in the contract price can be identified and covered separately in the contract. It is important here to ensure that contingency allowances not be duplicated by inclusion in both the base price and the adjustment requested by the contractor in the economic price adjustment clause.

14.3.2 Cost-Plus Contract

Cost-plus is a contract agreement wherein the owner agrees to pay the cost of all labor and materials on a cost-plus basis; that is, the owner reimburses the contractor for all costs associated with the contract in addition to a fee covering the contractor's profit and nonreimbursable overhead costs. It is somewhat like a labor contract, except that here the contractor buys the materials and provides the labor and is reimbursed accordingly. Cost-plus contracts can be either costs plus percentage of costs, under which the fee is an agreed percentage of the "costs," or costs plus fixed fee, where the fee is independent of the contractor's "costs." With this type of contract, it is important to define exactly what is included in "costs" (e.g., are soft costs such as supervision and overhead reimbursable?). Because of the discretionary and subjective nature of a cost-plus building contract, it ends up being the best for the contractor and the riskiest for the investor or building owner. The owner may suddenly discover that the project ends up being twice what was initially agreed to. Care must be taken when using this type of agreement, as it is frequently open to abuse.

Contractors prefer cost-plus bidding because it relieves them of sticking to a set price and guarantees them a profit regardless of project cost. However, this bid model must not be allowed to be used by building contractors as a haven for poor estimating. The various additional costs should be discussed with the contractor at the time the bid is being reviewed. Although surprises are not uncommon in construction, they can be kept to a minimum by a careful drafting of the contract. Cost-plus is typically used for projects where the scope of work is indeterminate or very uncertain and the kinds of labor, material, and equipment needed are difficult to ascertain. It is also used for very specialized projects and where the scope is difficult to define or when time is the most important factor and construction is required to start before completion of the contract

---

TABLE 14.2 Evaluating the Cost-Plus Contract

Advantages	Provides contractor with an additional amount at project completion, so maximizes contractor's incentive to control costs and perform effectively and on schedule
	Similar advantages to those that are part of labor contracts
Disadvantages	Quality of work cannot be checked or controlled by an expert because architect no longer has a role to play
	Difficult to verify whether contractor's records are accurate, which is vital since contractor's reimbursement is based on records of workers employed and materials purchased

---

documents. Under this arrangement, complete records of all time and materials used by the contractor on the work must be accurately maintained. Table 14.2 shows an evaluation of the cost-plus contract.

14.3.3   Guaranteed Maximum Price

Guaranteed maximum price (GMP) contracts are becoming more popular as a means to minimize risk, avoid claims, and integrate the diverse interests of a complex project. They are based on competitive bids for each trade subcontract, but the contractor/administrator charges an additional fee for taking on the risk of the guarantee. The GMP type of contract is essentially a variation of the cost-plus contract but should not be confused with it. It is more suited for projects where the scope is well defined, and is especially suited for turnkey projects, in which the contractor is reimbursed for actual costs incurred for labor, materials, equipment, subcontractors, overhead, and profit up to an agreed maximum fixed amount. The contractor also warrants that the project will be constructed in accordance with the Contract Documents and that the cost of the project to the owner will not exceed the agreed total maximum price. Costs over the maximum are borne by the contractor and any savings below the maximum price will revert to the owner.

It is important that the owner, from the outset, clearly state whether the maximum price reflects the total costs (fee excluded) or the total costs the owner pays, including the fee. Normally the owner pays the fee and the contractor pays the costs in excess of the maximum. The contractor is typically allocated a contingency amount for construction changes that are within the design intent of the project. However, changes that exceed the design intent require approval by all stakeholders. Here the owner tends to play an active role throughout the entire process. The whole issue of cost is manageable when the savings are shared rather than negotiated from an adversarial position. Another advantage of GMP occurs when work must start ahead of final drawings. There are often issues that delay completion of the drawings and therefore the start of construction. The GMP format allows owners to minimize risk when proceeding with work ahead of final drawings. Penalty and incentive clauses are often included in the agreement relating to costs, schedule, and quality performance. This type of contract is sometimes preferred when a design is less than 100% complete.

14.3.4   Unit Price Contract

With a unit price contract, the work to be performed is broken into various segments, usually by construction trade, and a fixed price is established for each unit of work. The contractor is paid as the contract proceeds by requiring that the actual quantities of work completed be measured and multiplied by the pre-agreed per-unit price. The final price of the project depends on the final quantities of the items used to carry out the work, which may eventually vary from what was initially estimated. This allows the terms of this type of contract to accommodate some flexibility in price adjustment, and means that the agreed value may be subject to amendment if the actual volume is reduced or exceeds the original negotiated terms and price. Tender estimates provided by contractors are based on specifications and estimated quantities supplied by the owner. However, during and after the work, the price is based on actual, not estimated, quantities completed. For the contractor this removes some of the risk in the bidding process because payment is based on actual quantities and not a lump sum amount. The contractor's unit price must cover both direct and indirect costs, overheads, contingencies, and profit. For this reason, the owner usually provides fixed quantities for contractors to use as the basis of their unit price costing. For example, painting is typically done on a square-foot basis. However, when additional work is required, a separate invoice should be presented (Figure 14.4).

Unit price contracts are seldom used for an entire major construction project, but they are frequently used for agreements with subcontractors, and for maintenance and repair work. They are also typically suitable for projects where quantities are ill defined and therefore cannot be accurately measured before the project starts, such as highway construction. Thus, the owner can provide quantities for excavation, pipelaying, and backfill. The contractor can quote a dollar amount per cubic yard for soil excavation, a dollar amount per linear foot of piping laid, and a dollar amount per cubic yard of backfill installed, and come up with a total bid based on the quantities that the owner provides. The project's final price will not be known with certainty until the project is completed. Additionally, it is prudent for the owner or the owner's representative to "track" actual quantities by some method of measuring, such as counting truckloads of materials or weighing steel. The relative advantages and disadvantages of the unit price contract are given in Table 14.3.

BILLING FORM
ADDITIONAL WORK PERFORMED

EMPLOYEE NAME:

DATE:

PROJECT NO.

JOB NAME:

FOR PERIOD UP TO AND INCLUDING: _____

DESCRIPTION OF WORK PERFORMED, INCLUDING WHO WAS INVOLVED, DATES, ETC.

HOURS: _____

EXPENSES: _____

_____

_____

APPROVED: **SIGNATURE:**

FIGURE 14.4   Example billing form used for additional work completed.

14.3.5   Project Management Contract

Most construction projects have one individual in charge of coordinating the various trades to make sure no step is missed and to make sure the work is properly executed. This requires a considerable amount of expertise. In the project management contract, the architect agrees to manage the contract, as defined by the scope of the agreement, for a specified duration of time for monetary consideration. To complete a project successfully requires good communication between the client and the project manager. This contract type addresses many of the most common causes of conflict in such agreements, which can be short- or long-term. Table 14.4 lists the advantages and disadvantages of the project management contract.

TABLE 14.3 Evaluating the Unit Price Contract

Advantages	Architect involved because he/she provides quantities of each item (in bill of quantities) and negotiates unit prices with contractor.
	With the unit price contract, owner makes payments to contractor only after architect has verified measurements onsite and certified contractor's bills for payment.
	Owner pays only for volume of work done onsite and not for anything extra.
	There is assurance (because architect involved) that quality of the work will be according to specifications and Contract Documents.
	"Typical" drawings can be used for competitive bidding.
	Contractor initially invests own money for starting work, so owner does not give contractor a large advance.
	Selection of contractor is easier because it is generally considered to be the most scientific and most suitable for construction projects where different types of items, but not quantities, can be accurately identified in Contract Documents.
	Contractor is safeguarded against any potential contingencies or variations in labor or material rates.
	Degree of built-in flexibility in scope, and quantities easily adjustable.
Disadvantages	For contractor: initially has to invest own money; while one of most preferred in building contracts, it is not unusual to combine a unit price contract for parts of a project with other types of contracts such as lump sum.
	Final cost not known at outset (because bills of quantities at bid time are only estimates); additional site staff is needed to measure, control, and report on completed units.

TABLE 14.4 Evaluating the Project Management Contract

Advantages	Clients can focus on core objectives while architect (project manager) looks after management of project and related issues, so deadlines are met, quality is maintained, and costs are controlled and within budget.
	Project manager coordinates with all agencies, including consultants, contractor, and suppliers to ensure construction proceeds as planned.
Disadvantages	Clients hesitate to go in for project management contract partly because they have to pay extra, in addition to the fees paid to the architect.
	On the other hand, many of clients will opt for project management because it saves them from a lot of headache and because it allows them to concentrate on their work as the project proceeds. In the long run, the clients actually save since the project is completed on time and costs are controlled.

TABLE 14.5 Evaluating the Labor Contract

Advantages	Some owners prefer this type of contract because by buying all material she or he is able to pocket much of contractor's profit.
	This type of contract ensures that the owner can choose and buy materials and can be sure of brand and quality that will be used in the construction.
Disadvantages	Dismisses role of architect who no longer has a role to play, so quality of work cannot be checked or controlled by expert.
	Owner can be misled on quality of sand, bricks, etc., because of lack of knowledge/experience with materials.
	Much stress and tension resulting from organizing on-time supply of materials, etc., as work progresses.
	Casual nature of employment, inferior labor status, lack of job security, and poor economic conditions.
	Contractor can often strike better bargain when negotiating with suppliers and vendors because owner is one-time client whereas contractor normally has the advantage of being a regular client.
	Strong possibility of pilferage of material stored at site.
	Work delays because laborers, masons, etc., can fail to turn up, sometimes because of the lure of work on another site.
	Workers are generally paid on a daily basis, so an unscrupulous labor contractor may purposely go slow in the hope of getting paid more.

14.3.6 Labor Contract

In a labor contract, the contractor only supplies the labor while the owner buys and supplies all the material required for project execution. The system of employing contract labor is prevalent in many industries including construction, and involves skilled and semiskilled jobs. The worker is considered to be employed as contract labor when he or she is hired in connection with the work by or through a contractor. Contract workers are classified as indirect employees: persons who are hired, supervised, and paid by a contractor who, in turn, is compensated by the project owner. Contract labor has to be employed for specific work and for a definite duration. Table 14.5 shows an review of the labor contract in terms of advantages and disadvantages.

14.3.7 Negotiated Contract

A negotiated contract is not dissimilar to the design–bid–build method in that the project's design and construction are performed by different firms. Sometimes a negotiated contract is used in lieu of the tendering process, especially when an owner has had previous experience with a particular contractor. That contractor may be invited to submit a proposal or offer to the owner or the owner's representative based on the Contract Documents. This is then followed by negotiations regarding price, scope of work, time to execute the project, and other contractual issues. Once an agreement is reached, the contract is signed. Negotiations may form part of a tendering process. After evaluating the submitted tenders, a short list of top-ranking firms is created; negotiations regarding work content, risk, liability issues, and contract-related issues follow. If the owner decides on negotiating with the top-ranking firms after the tenders are opened and analyzed, this must be made clear to all via the invitation to tender to ensure impartiality.

The latest design−build documents use one agreement for both design and construction, with three possible methods of payment available to the parties:

- Stipulated sum
- Cost-plus fee with guaranteed maximum price
- Cost-plus fee without guaranteed maximum price

14.4 AMERICAN INSTITUTE OF ARCHITECTS CONTRACT DOCUMENTS

According to its website, the American Institute of Architects (AIA) offers more than 100 forms and contracts that: "define the relationships and terms involved in design and construction projects." The AIA prepared and developed these contracts with the consensus of owners, contractors, attorneys, architects, engineers, design professionals, and others. The result is a comprehensive collection of contracts and forms that are widely recognized as the industry standard. The AIA organizes this collection using two basic methods: (1) by families based on types of projects or particular project delivery methods, and (2) by series based on use of the document. For the series method, the AIA Contract Documents are divided into six alphanumeric series according to use or purpose (see text box that follows). Current AIA documents by series are listed in Table 14.6. The AIA advises users to exercise independent judgment and to use legal counsel when deciding which documents are appropriate for a particular project.

TABLE 14.6 Current AIA Document Series

Series	Document Type	Document Numbers
A	Owner/contractor agreements	A101, A102, A103, A105, A107, A132, A133, A134, A141, A142, A151, A195, A201, A232, A251, A295, A305, A310, A312, A401, A441, A503, A533, A701, A751
B	Owner/architect agreements	B101, B102, B103, B104, B105, B106, B107, B108, B109, B132, B142, B143, B144ARCH-CM, B152, B153, B161, B162, B188, B195, B201, B202, B203, B204, B205, B206, B207, B209, B210, B211, B212, B214, B252, B253, B305, B503, B509, B727,

---

TABLE 14.6   Cont'd

Series	Document Type	Document Numbers
C	Other agreements	C101, C106, C132, C191, C195, C196, C197, C198, C199, C401, C441, C727
D	Miscellaneous documents	D101, D200, D503
E	Exhibits	E201, E202
F	[reserved]	[reserved]
G	Contract administration and project management forms	G601, G602, G612, G701, G701S, G701CMa, G702, G702S, G703, G703S, G704, G704CMa, G704DB, G705, G706, G706A, G707, G707A, G709, G710, G711, G712, G714, G714CMa, G715, G716, G732, G736, G737, G801, G802, G803, G804, G806, G807, G808, G808A, G809, G810

AIA Contract Documents (Series A through D and G)[1]

A-Series: Owner/Contractor Agreements

A101–2007, Standard Form of Agreement between Owner and Contractor where the Basis of Payment Is a Stipulated Sum

AIA Document A101™–2007 is a standard form of agreement between owner and contractor for use where the basis of payment is a stipulated sum (fixed price). A101 adopts by reference, and is designed for use with, AIA Document A201™–2007, General Conditions of the Contract for Construction. A101 is suitable for large or complex projects. For projects of a more limited scope, use of AIA Document A107™–2007, Agreement between Owner and Contractor for a Project of Limited Scope, should be considered. For even smaller projects, consider AIA Document A105™–2007, Agreement between Owner and Contractor for a Residential or Small Commercial Project. *Note:* A101–2007 replaces A101–1997 (expired 2009).

A102–2007 (formerly A111–1997), Standard Form of Agreement between Owner and Contractor where the Basis of Payment Is the Cost of the Work Plus a Fee with a Guaranteed Maximum Price

This standard form of agreement between owner and contractor is appropriate for use on large projects requiring a guaranteed maximum price, when the basis of payment to the contractor is the cost of the work plus a fee. AIA Document A102™–2007 is not intended for use in competitive bidding. AIA Document A102–2007 adopts by reference and is intended for use with AIA Document A201™–2007, General Conditions of the Contract for Construction. NOTE: A102–2007 replaces A111–1997 (expired 2009).

---

1. Excerpted from the AIA website (www.aia.org/contractdocs/AIAS076742), Spring 2012.

A103–2007 (formerly A114–2001), Standard Form of Agreement between Owner and Contractor where the Basis of Payment Is the Cost of the Work Plus a Fee without a Guaranteed Maximum Price

AIA Document A103™–2007 is appropriate for use on large projects when the basis of payment to the contractor is the cost of the work plus a fee, and the cost is not fully known at the commencement of construction. AIA Document A103–2007 is not intended for use in competitive bidding. A103–2007 adopts by reference, and is intended for use with, AIA Document A201™–2007, General Conditions of the Contract for Construction. NOTE: A103–2007 replaces A114–2001 (expired 2009).

A105–2007 (formerly A105–1993 and A205–1993), Standard Form of Agreement between Owner and Contractor for a Residential or Small Commercial Project

AIA Document A105™–2007 is a stand-alone agreement with its own general conditions; it replaces A105–1993 and A205–1993 (expired 2009). AIA Document A105–2007 is for use on a project that is modest in size and brief in duration, and where payment to the contractor is based on a stipulated sum (fixed price). For larger and more complex projects, other AIA agreements are more suitable, such as AIA Document A107™–2007, Standard Form of Agreement between Owner and Contractor for a Project of Limited Scope. AIA Documents A105–2007 and B105™–2007, Standard Form of Agreement between Owner and Architect for a Residential or Small Commercial Project, comprise the Small Projects family of documents. Although A105 and B105 share some similarities with other agreements, the Small Projects family should NOT be used in tandem with agreements in other document families without careful side-by-side comparison of contents.

A107–2007, Standard Form of Agreement between Owner and Contractor for a Project of Limited Scope

AIA Document A107™–2007 is a stand-alone agreement with its own internal general conditions and is intended for use on construction projects of limited scope. It is intended for use on medium-to-large sized projects where payment is based on either a stipulated sum or the cost of the work plus a fee, with or without a guaranteed maximum price. Parties using AIA Document A107–2007 will also use A107 Exhibit A, if using a cost-plus payment method. AIA Document B104™–2007, Standard Form of Agreement between Owner and Architect for a Project of Limited Scope, coordinates with A107–2007 and incorporates it by reference.

For more complex projects, parties should consider using one of the following other owner/contractor agreements: AIA Document A101™–2007, A102™–2007 or A103™–2007. These agreements are written for a stipulated sum, cost of the work with a guaranteed maximum price, and cost of the work without a guaranteed maximum price, respectively. Each of them incorporates by reference AIA Document A201™–2007, General Conditions of the Contract for Construction. For single family residential projects, or smaller and less complex commercial projects, parties may wish to consider AIA Document A105™–2007, Agreement between Owner and Contractor for a Residential or Small Commercial Project. NOTE: A107–2007 replaces A107–1997 (expired 2009).

A132–2009 (formerly A101CMa–1992), Standard Form of Agreement between Owner and Contractor, Construction Manager as Adviser Edition

AIA Document A132™–2009 is a standard form of agreement between owner and contractor for use on projects where the basis of payment is either a stipulated sum (fixed price) or cost of the work plus a fee, with or without a guaranteed maximum price. In addition to the contractor and the architect, a construction manager assists the owner in an advisory capacity during design and construction.

The document has been prepared for use with AIA Documents A232™–2009, General Conditions of the Contract for Construction, Construction Manager as Adviser Edition; B132™–2009, Standard Form of Agreement between Owner and Architect, Construction Manager as Adviser Edition; and C132™–2009, Standard Form of Agreement between Owner and Construction Manager as Adviser. This integrated set of documents is appropriate for use on projects where the construction manager only serves in the capacity of an adviser to the owner, rather than as constructor (the latter relationship being represented in AIA Documents A133™–2009 and A134™–2009). NOTE: A132–2009 replaces A101CMa–1992 (expired 2010).

A133–2009 (formerly A121CMc–2003), Standard Form of Agreement between Owner and Construction Manager as Constructor where the Basis of Payment Is the Cost of the Work Plus a Fee with a Guaranteed Maximum Price

AIA Document A133™–2009 is intended for use on projects where a construction manager, in addition to serving as adviser to the owner, assumes financial responsibility for construction of the project. The construction manager provides the owner with a guaranteed maximum price proposal, which the owner may accept, reject, or negotiate. Upon the owner's acceptance of the proposal by execution of an amendment, the construction manager becomes contractually bound to provide labor and materials for the project and to complete construction at or below the guaranteed maximum price. The document divides the construction manager's services into two phases: the preconstruction phase and the construction phase, portions of which may proceed concurrently in order to fast track the process. AIA Document A133–2009 is coordinated for use with AIA Documents A201™– 2007, General Conditions of the Contract for Construction, and B103™–2007, Standard Form of Agreement between Owner and Architect for a Large or Complex Project. A133–2009 replaces A121CMc–2003 (expired 2010).

Caution: To avoid confusion and ambiguity, do not use this construction management document with any other AIA construction management document.

A134–2009 (formerly A131CMc–2003), Standard Form of Agreement between Owner and Construction Manager as Constructor where the Basis of Payment Is the Cost of the Work Plus a Fee without a Guarantee Maximum Price

Similar to AIA Document A133™–2009, AIA Document A134™–2009 is intended for use when the owner seeks a construction manager who will take on responsibility for providing the means and methods of construction. However, in AIA Document A134–2009 the construction manager does not provide a guaranteed maximum price (GMP). A134–2009 employs the cost-plus-a-fee method, wherein the owner can monitor cost through periodic review of a control estimate that is revised as the project proceeds.

The agreement divides the construction manager's services into two phases: the preconstruction phase and the construction phase, portions of which may proceed concurrently in order to fast track the process. A134–2009 is coordinated for use with AIA Documents A201™–2007, General Conditions of the Contract for Construction, and B103™–2007, Standard Form of Agreement between Owner and Architect for a Large or Complex Project. A134–2009 replaces A131CMc–2003 (expired 2010).

Caution: To avoid confusion and ambiguity, do not use this construction management document with any other AIA construction management document.

A141–2004, Agreement between Owner and Design-Builder

AIA Document A141™–2004 replaces A191–1996 (expired) and consists of the agreement and three exhibits: Exhibit A, Terms and Conditions; Exhibit B, Determination of the Cost of the Work; and Exhibit C, Insurance and Bonds. Exhibit B is not applicable if the parties select to use a stipulated sum. AIA Document A141–2004 obligates the design-builder to execute fully the work required by the design-build documents, which include A141 with its attached exhibits, the project criteria and the design-builder's proposal, including any revisions to those documents accepted by the owner, supplementary and other conditions, addenda and modifications. The Agreement requires the parties to select the payment type from three choices: (1) Stipulated Sum, (2) cost of the work plus design-builder's fee, and (3) cost of the work plus design-builder's fee with a guaranteed maximum price. A141–2004 with its attached exhibits forms the nucleus of the design-build contract. Because A141 includes its own terms and conditions, it does not use AIA Document A201™.

A142–2004, Agreement between Design-Builder and Contractor

AIA Document A142™–2004 replaces AIA Document A491™–1996 (expired) and consists of the agreement and five exhibits: Exhibit A, Terms and Conditions; Exhibit B, Preconstruction Services; Exhibit C, Contractor's Scope of Work; Exhibit D, Determination of the Cost of the Work; and Exhibit E, Insurance and Bonds. Unlike AIA Document B491–1996, AIA Document A142–2004 does not rely on AIA Document A201™ for its general conditions of the contract. A142–2004 contains its own terms and conditions.

A142–2004 obligates the contractor to perform the work in accordance with the contract documents, which include A142 with its attached exhibits, supplementary a other conditions, drawings, specifications, addenda, and modifications. Like AIA Document A141™–2004, AIA Document A142–2004 requires the parties to select the payment type from three choices: (1) Stipulated Sum, (2) Cost of the work Plus Design-Builder's Fee, and (3) Cost of the Work Plus Design-Builder's fee with a Guaranteed Maximum Price.

A151–2007 (formerly A175ID–2003), Standard Form of Agreement between Owner and Vendor for Furniture, Furnishings and Equipment where the basis of payment is a Stipulated Sum

AIA Document A151™–2007 is intended for use as the contract between owner and vendor for furniture, furnishings, and equipment (FF&E) where the basis of payment is a stipulated sum (fixed price) agreed to at the time of contracting. AIA Document A151–2007 adopts by reference and is intended for use with AIA Document A251™–2007, General Conditions of the Contract for Furniture, Furnishings and Equipment. It may be used in any arrangement between the owner and the contractor where the cost of FF&E has been determined in advance, either through bidding or negotiation. NOTE: A151–2007 replaces A175ID–2003 (expired 2009).

A195–2008, Standard Form of Agreement between Owner and Contractor for Integrated Project Delivery

AIA Document A195™–2008 is a standard form of agreement between owner and contractor for a project that utilizes integrated project delivery (IPD). AIA Document A195–2008 primarily provides only the business terms and conditions unique to the agreement between the owner and contractor, such as compensation details and licensing of instruments of service. A195 does not include the specific scope of the contractor's work; rather, it incorporates by reference AIA Document A295™–2008, General Conditions of the Contract for Integrated Project Delivery, which sets forth the contractor's duties and obligations for each of the six phases of the project, along with the duties and obligations of the owner and architect. Under A195–2008, the contractor provides a guaranteed maximum price. For that purpose, the agreement includes a guaranteed maximum price amendment at Exhibit A.

A201–2007, General Conditions of the Contract for Construction

The general conditions are an integral part of the contract for construction for a large project and they are incorporated by reference into the owner/contractor agreement. They set forth the rights, responsibilities, and relationships of the owner, contractor, and architect. Though not a party to the contract for construction between owner and contractor, the architect participates in the preparation of the contract documents and performs construction phase duties and responsibilities described in detail in the general conditions. AIA Document A201™–2007 is adopted by reference in owner/architect, owner/contractor, and contractor/subcontractor agreements in the Conventional (A201) family of documents; thus, it is often called the "keystone" document. NOTE: A201–2007 replaces A201–1997 (expired 2009).

A232–2009 (formerly A201CMa–1992), General Conditions of the Contract for Construction, Construction Manager as Adviser Edition

AIA Document A232™–2009 sets forth the rights, responsibilities, and relationships of the owner, contractor, construction manager and architect. A232–2009 is adopted by reference in owner/architect, owner/contractor, and owner/construction manager agreements in the CMa family of documents. Under A232–2009, the construction manager serves as an independent adviser to the owner, who enters into a contract with a general contractor or multiple contracts with prime trade contractors. NOTE: A232–2009 replaces A201CMa–1992 (expired 2010).

CAUTION: Do not use A232–2009 in combination with agreements where the construction manager takes on the role of constructor, such as in AIA Document A133™–2009 or A134™–2009.

A251–2007 (formerly A275ID–2003), General Conditions of the Contract for Furniture, Furnishings and Equipment

AIA Document A251™–2007 provides general conditions for the AIA Document A151™–2007, Standard Form Agreement between Owner and Vendor for Furniture, Furnishings and Equipment where the basis of payment is a Stipulated Sum. AIA Document A251–2007 sets forth the duties of the owner, architect, and vendor, just as AIA Document A201™–2007, General Conditions of the Contract for Construction, does for building construction projects. Because the Uniform Commercial Code (UCC) governs the sale of goods and has been adopted in nearly every jurisdiction, A251–2007 recognizes the commercial standards set forth in Article 2 of the UCC, and uses certain standard UCC terms and definitions. A251 was renumbered in 2007 and was modified, as applicable, to coordinate with AIA Document A201–2007. NOTE: A251–2007 replaces A275ID–2003 (expired 2009).

A295–2008, General Conditions of the Contract for Integrated Project Delivery

AIA Document A295™–2008, provides the terms and conditions for AIA Documents A195™–2008 Standard Form of Agreement between Owner and Contractor for Integrated Project Delivery, and B195™–2008, Standard Form of Agreement between Owner and Architect for Integrated Project Delivery, both of which incorporate AIA Document A295–2008 by reference. Those agreements provide primarily only business terms and rely on A295–2008 for the architect's services, the contractor's pre-construction services, and the conditions of construction. A295 not only establishes the duties of the owner, architect, and contractor, but also sets forth in detail how they will work together through each phase of the project: conceptualization, criteria design, detailed design, implementation documents, construction, and closeout. A295 requires that the parties utilize building information modeling.

A305–1986, Contractor's Qualification Statement

An owner preparing to request bids or to award a contract for a construction project often requires a means of verifying the background, references, and financial stability of any contractor being considered. These factors, along with the time frame for construction, are important for an owner to investigate. Using AIA Document A305™–1986, the contractor may provide a sworn, notarized statement and appropriate attachments to elaborate on important aspects of the contractor's qualifications.

A310–2010, Bid Bond

AIA Document A310™–2010, a simple, one-page form, establishes the maximum penal amount that may be due to the owner if the selected bidder fails to execute the contract or fails to provide any required performance and payment bonds. NOTE: A310–2010 replaces AIA Document A310™–1970, which expired on December 31, 2011.

A312–2010, Performance Bond and Payment Bond

AIA Document A312™–2010 incorporates two bonds—one covering the contractor's performance, and the other covering the contractor's obligations to pay subcontractors and others for material and labor. In addition, AIA Document A312–2010 obligates the surety to act responsively to the owner's requests for discussions aimed at anticipating or preventing a contractor's default. NOTE: A312–2010 replaces AIA Document A312™–1984, which expired on December 31, 2011.

A401–2007, Standard Form of Agreement between Contractor and Subcontractor

AIA Document A401™–2007 establishes the contractual relationship between the contractor and subcontractor. It sets forth the responsibilities of both parties and lists their respective obligations, which are written to parallel AIA Document A201™–2007, General Conditions of the Contract for Construction, which A401–2007 incorporates by reference. AIA Document A401 may be modified for use as an agreement between the subcontractor and a sub-subcontractor, and must be modified if used where AIA Document A107™–2007 or A105™–2007 serves as the owner/contractor agreement. NOTE: A401–1997 replaces A401–1997 (expired 2009).

A441–2008, Standard Form of Agreement between Contractor and Subcontractor for a Design-Build Project

AIA Document A441™–2008 is a fixed price agreement that establishes the contractual relationship between the contractor and subcontractor in a design build project. AIA Document A441–2008 incorporates by reference the terms and conditions of AIA Document A142™–2004, Standard Form of Agreement between Design-Builder and Contractor, and was written to ensure consistency with the AIA 2004 Design-Build family of documents. Because subcontractors are often required to provide professional services on a design-build project, A441 provides for that possibility.

A503–2007 (formerly A511–1999), Guide for Supplementary Conditions
AIA Document A503™–2007 is not an agreement, but is a guide containing model provisions for modifying and supplementing AIA Document A201™–2007, General Conditions of the Contract for Construction. It provides model language with explanatory notes to assist users in adapting AIA Document A201–2007 to specific circumstances. A201–2007, as a standard form document, cannot cover all the particulars of a project. Thus, AIA Document A503–2007 is provided to assist A201–2007 users either in modifying it, or developing a separate supplementary conditions document to attach to it. NOTE: A503–2007 replaces A511–1999 (expired 2009).

A533–2009 (formerly A511CMa–1993), Guide for Supplementary Conditions, Construction Manager as Adviser Edition
Similar to AIA Document A503™–2007, AIA Document A533™–2009 is a guide for amending or supplementing the general conditions document, AIA Document A232™–2009. AIA Documents A533–2009 and A232–2009 should only be employed on projects where the construction manager is serving in the capacity of adviser to the owner and not in situations where the construction manager is also the constructor (CMc document-based relationships). Like A503–2007, this document contains suggested language for supplementary conditions, along with notes on appropriate usage. NOTE: A533–2009 replaces A511CMa–1993 (expired 2010).

A701–1997, Instructions to Bidders
AIA Document A701™–1997 is used when competitive bids are to be solicited for construction of the project. Coordinated with AIA Document A201™, General Conditions of the Contract for Construction, and its related documents, AIA Document A701–1997 provides instructions on procedures, including bonding requirements, for bidders to follow in preparing and submitting their bids. Specific instructions or special requirements, such as the amount and type of bonding, are to be attached to, or inserted into, A701.

A751–2007 (formerly A775ID–2003), Invitation and Instructions for Quotation for Furniture, Furnishings and Equipment
AIA Document A751™–2007 provides (1) the Invitation for Quotation for Furniture, Furnishings and Equipment (FF&E) and (2) Instructions for Quotation for Furniture, Furnishings and Equipment. These two documents define the owner's requirements for a vendor to provide a complete quotation for the work. The purchase of FF&E is governed by the Uniform Commercial Code (UCC), and AIA Document A751–2007 was developed to coordinate with the provisions of the UCC. NOTE: A751–2007 replaces A775ID–2003 (expired 2009).

B-Series: Owner/Architect Agreements
B101–2007 Standard Form of Agreement between Owner and Architect

B102–2007 Standard Form of Agreement between Owner and Architect without a Predefined Scope of Architect's Services

B103–2007 Standard Form of Agreement between Owner and Architect for a Large or Complex Project

B104–2007 Standard Form of Agreement between Owner and Architect for a Project of Limited Scope

B105–2007 Standard Form of Agreement between Owner and Architect for a Residential or Small Commercial Project

B106–2010 Standard Form of Agreement between Owner and Architect for Pro Bono Services

B107–2010 Standard Form of Agreement between Developer-Builder and Architect for Prototype(s) for Single Family Residential Project

B108–2009 Standard Form of Agreement between Owner and Architect for a Federally Funded or Federally Insured Project

B109–2010 Standard Form of Agreement between Owner and Architect for a Multi-Family Residential or Mixed Use Residential Project

B132–2009 Standard Form of Agreement between Owner and Architect, Construction Manager as Adviser Edition

B142–2004 Standard Form of Agreement between Owner and Consultant where the Owner contemplates using the design-build method of project delivery

B143–2004 Standard Form of Agreement between Design-Builder and Architect

B144ARCH-CM–1993 Standard Form of Amendment for the Agreement between Owner and Architect where the Architect Provides Construction Management Services as an Adviser to the Owner

B152–2007 Standard Form of Agreement between Owner and Architect for Architectural Interior Design Services

B153–2007 Standard Form of Agreement between Owner and Architect for Furniture, Furnishings and Equipment Design Services

B161–2002 Standard Form of Agreement between Client and Consultant for use where the Project is located outside the United States

B162–2002 Abbreviated Form of Agreement between Client and Consultant for use where the Project is located outside the United States

B188–1996 * Standard Form of Agreement between Owner and Architect for Limited Architectural Services for Housing Projects

B195–2008 Standard Form of Agreement between Owner and Architect for Integrated Project Delivery

B201–2007 Standard Form of Architect's Services: Design and Construction Contract Administration

B202–2009 Standard Form of Architect's Services: Programming

B203–2007 Standard Form of Architect's Services: Site Evaluation and Planning

B204–2007 Standard Form of Architect's Services: Value Analysis, for use where the Owner employs a Value Analysis Consultant

B205–2007 Standard Form of Architect's Services: Historic Preservation

B206–2007 Standard Form of Architect's Services: Security Evaluation and Planning

B207–2008 Standard Form of Architect's Services: On-Site Project Representation

B209–2007 Standard Form of Architect's Services: Construction Contract Administration, for use where the Owner has retained another Architect for Design Services

B210–2007 Standard Form of Architect's Services: Facility Support

B211–2007 Standard Form of Architect's Services: Commissioning

B212–2010 Standard Form of Architect's Services: Regional or Urban Planning

B214–2007 Standard Form of Architect's Services: LEED® Certification

B252–2007 Standard Form of Architect's Services: Architectural Interior Design

B253–2007 Standard Form of Architect's Services: Furniture, Furnishings and Equipment Design

B305–1993 Architect's Qualification Statement

B503–2007 Guide for Amendments to AIA Owner-Architect Agreements

B509–2010 Guide for Supplementary Conditions to AIA Document B109™–2010 for use on Condominium Projects

B727–1988* Standard Form of Agreement between Owner and Architect for Special Services

C-Series: Other Agreements

C101–1993 (formerly C801–1993), Joint Venture Agreement for Professional Services
AIA Document C101™–1993 is intended for use by two or more parties to provide for their mutual rights and obligations in forming a joint venture. It is intended that the joint venture, once established, will enter into an agreement with the owner to provide professional services. The parties may be all architects, all engineers, a combination of architects and engineers, or another combination of professionals. The document provides a choice between two methods of joint venture operation. The "division of compensation" method assumes that services provided and the compensation received will be divided among the parties in the proportions agreed to at the outset of the project. Each party's profitability is then dependent on individual performance of pre-assigned tasks and is not directly tied to that of the other parties. The "division of profit and loss" method is based on each party performing work and billing the joint venture at cost plus a nominal amount for overhead. The ultimate profit or loss of the joint venture is divided between or among the parties at completion of the project, based on their respective interests. NOTE: AIA Document C101–1993 was renumbered in 2007, but its content remains the same as C801–1993 (expired May 2009).

C106–2007, Digital Data Licensing Agreement
AIA Document C106™–2007 serves as a licensing agreement between two parties who otherwise have no existing licensing agreement for the use and transmission of digital data, including instruments of service. AIA Document C106–2007 defines digital data as information, communications, drawings, or designs created or stored for a specific project in digital form. AIA C106 allows one party to (1) grant another party a limited nonexclusive license to use digital data on a specific project, (2) set forth procedures for transmitting the digital data, and (3) place restrictions on the license granted. In addition, C106 allows the party transmitting digital data to collect a licensing fee for the recipient's use of the digital data.

C132–2009 (formerly B801CMa–1992), Standard Form of Agreement between Owner and Construction Manager as Adviser
AIA Document C132™–2009 provides the agreement between the owner and the construction manager, a single entity who is separate and independent from the architect and the contractor, and who acts solely as an adviser (CMa) to the owner

---

* AIA Documents B188–1986 and B727–1988 expire on April 30, 2012.

throughout the course of the project. AIA Document C132–2009 is coordinated for use with AIA Document B132™–2009, Standard Form of Agreement between Owner and Architect, Construction Manager as Adviser Edition.

Both C132–2009 and B132–2009 are based on the premise that there will be a separate construction contractor or multiple prime contractors whose contract(s) with the owner will be jointly administered by the architect and the construction manager under AIA Document A232™–2009. AIA Document C132–2009 is not coordinated with, and should not be used with, documents where the construction manager acts as the constructor for the project, such as in AIA Document A133™–2009 or A134™–2009. NOTE: C132–2009 replaces B801CMa–1992 (expired December 2010).

C191–2009, Standard Form Multi-Party Agreement for Integrated Project Delivery

AIA Document C191™–2009 is a standard form multi-party agreement through which the owner, architect, contractor, and perhaps other key project participants execute a single agreement for the design, construction, and commissioning of a Project. AIA Document C191–2009 provides the framework for a collaborative environment in which the parties operate in furtherance of cost and performance goals that the parties jointly establish. The non-owner parties are compensated on a cost-of-the-work basis. The compensation model is also goal-oriented, and provides incentives for collaboration in design and construction of the project. Primary management of the project is the responsibility of the Project Management Team, comprised of one representative from each of the parties. The Project Executive Team, also comprised of one representative from each of the parties, provides a second level of project oversight and issue resolution. The conflict resolution process is intended to foster quick and effective resolution of problems as they arise. This collaborative process has the potential to result in a high-quality project for the owner, and substantial monetary and intangible rewards for the other parties.

C195–2008, Standard Form Single Purpose Entity Agreement for Integrated Project Delivery

AIA Document C195™–2008 is a standard form single purpose entity (SPE) agreement through which the owner, architect, construction manager, and perhaps other key project participants, each become members of a limited liability company. The sole purpose of the company is to design and construct a project utilizing the principles of integrated project delivery (IPD) established in Integrated Project Delivery: A Guide. AIA Document C195–2008 provides the framework for a collaborative environment in which the company operates in furtherance of cost and performance goals that the members jointly establish. To obtain project funding, the company enters into a separate agreement with the owner. To design and construct the project, the company enters into separate agreements with the architect, construction manager, other non-owner members, and with non-member consultants and contractors. The compensation model in the non-owner member agreements is goal-oriented and provides incentives for collaboration in design and construction of the project, and for the quick and effective resolution of problems as they arise. This highly collaborative process has the potential to result in a high-quality project for the owner, and substantial monetary and intangible rewards for the other members.

C196–2008, Standard Form of Agreement between Single Purpose Entity and Owner for Integrated Project Delivery

AIA Document C196™–2008 is a standard form of agreement between a single purpose entity ("the SPE") and a project owner, called the owner member.

AIA Document C196–2008 is intended for use on a project where the project participants have formed the SPE utilizing AIA Document C195™–2008, Standard Form Single Purpose Entity Agreement for Integrated Project Delivery. This AIA Document is coordinated with AIA Document C195–2008 in order to implement the principles of integrated project delivery, including the accomplishment of mutually-agreed goals. C196 provides the terms under which the owner member will fund the SPE in exchange for the design and construction of the project. The SPE provides for the design and construction of the project through separate agreements with other members, including an architect and construction manager, utilizing AIA Document C197™–2008, Standard Form of Agreement between Single Purpose Entity and Non-Owner Member for Integrated Project Delivery. The SPE may also enter into agreements with non-member design consultants, specialty trade contractors, vendors, and suppliers.

C197–2008, Standard Form of Agreement between Single Purpose Entity and Non-Owner Member for Integrated Project Delivery

AIA Document C197™–2008 is a standard form of agreement between a single purpose entity ("the SPE") and members of the SPE that do not own the project, called non-owner members. AIA Document C197–2008 is intended for use on a project where the parties have formed the SPE utilizing AIA Document C195™–2008, Standard Form Single Purpose Entity Agreement for Integrated Project Delivery. C197–2008 is coordinated with C195–2008 in order to implement the principles of integrated project delivery, including the accomplishment of mutually-agreed goals. All members of the SPE, other than the project owner, will execute C197–2008.

AIA Document C197–2008 provides the terms under which the non-owner members provide services to the SPE to complete the design and construction of the project. The specific services the non-owner members are required to perform are set forth in the Integrated Scope of Services Matrix, which is part of the C195–2008 Target Cost Amendment and is incorporated into the executed C197–2008. In exchange for the non-owner members' services, the non-owner members are paid the direct and indirect costs they incur in providing services. Additionally, C197 allows for the non-owner members to receive profit through incentive compensation and goal achievement compensation.

C198–2010, Standard Form of Agreement between Single Purpose Entity and Consultant for Integrated Project Delivery

AIA Document C198™–2010 is a standard form of agreement between a single purpose entity ("the SPE") and a consultant. AIA Document C198–2010 is intended for use on a project where the parties have formed the SPE utilizing AIA Document C195™–2008, Standard Form Single Purpose Entity Agreement for Integrated Project Delivery. C198–2010 is coordinated with C195–2008 in order to implement the principles of integrated project delivery. The specific services the consultant is required to perform are set forth within the document as well as the Integrated Scope of Services Matrix, which is part of the C195–2008 Target Cost Amendment. In addition to traditional compensation for services, C198–2010 allows for the consultant to receive additional profit through incentive compensation and goal achievement compensation.

C199–2010, Standard Form of Agreement between Single Purpose Entity and Contractor for Integrated Project Delivery

AIA Document C199™–2010 is a standard form of agreement between a single purpose entity ("the SPE") and a contractor. AIA Document C199–2010 is intended for use on a project where the parties have formed the SPE utilizing AIA Document C195™–2008, Standard Form Single Purpose Entity Agreement for Integrated Project Delivery. C199–2010 is intended to be a flexible document. C199 can be used for a contractor that only provides construction services, or it can also be used for a contractor that will provide both pre-construction and construction services. C199 is not intended for use in competitive bidding and relies on an agreed to contract sum, which can be either a stipulated sum (fixed price) or cost of the work plus a fee, with a guaranteed maximum price. In addition to compensation for the contract sum, C199 allows for the contractor to receive additional profit through incentive compensation and goal achievement compensation.

C401–2007 (formerly C141–1997), Standard Form of Agreement between Architect and Consultant

AIA Document C401™–2007 is a standard form of agreement between the architect and the consultant providing services to the architect. AIA Document C401–2007 is suitable for use with all types of consultants, including consulting architects. This document may be used with a variety of compensation methods. C401–2007 assumes and incorporates by reference a preexisting owner/architect agreement known as the "prime agreement." AIA Documents B101™–2007, B103™–2007, B104™–2007, B105™–2007, and B152™–2007 are the documents most frequently used to establish the prime agreement. C401–2007 was modified in 2007 to be shorter and more flexible by "flowing down" the provisions of the prime agreement, except as specifically stated in C401. NOTE: C401–2007 replaces C141–1997 (expired May 2009).

C441–2008, Standard Form of Agreement between Architect and Consultant for a Design-Build Project

AIA Document C441™–2008 establishes the contractual relationship between the architect and a consultant providing services to the architect on a design-build project. AIA Document C441–2008 is suitable for use with all types of consultants, including consulting architects and may be used with a variety of compensation methods. C441 assumes and incorporates by reference a preexisting prime agreement between design-builder and architect. C441–2008 was written to ensure consistency with AIA Document B143™–2004, Standard Form of Agreement between Design-Builder and Architect, and with other documents in the AIA 2004 Design-Build family of documents.

C727–1992, Standard Form of Agreement between Architect and Consultant for Special Services

AIA Document C727™–1992 provides only the terms and conditions of the agreement between the architect and the consultant—the description of services is left entirely to the parties, and must be inserted in the agreement or attached in an exhibit. It is often used for planning, feasibility studies, post-occupancy studies, and other services that require specialized descriptions.

D-Series: Miscellaneous Documents

D101–1995, Methods of Calculating Areas and Volumes of Buildings

This document establishes definitions for methods of calculating the architectural area and volume of buildings. AIA Document D101™–1995 also covers interstitial space and office, retail, and residential areas.

D200–1995, Project Checklist

The project checklist is a convenient listing of tasks a practitioner may perform on a given project. This checklist will assist the architect in recognizing required tasks and in locating the data necessary to fulfill assigned responsibilities. By providing space for notes on actions taken, assignment of tasks, and time frames for completion, AIA Document D200™–1995 may also serve as a permanent record of the owner's, contractor's, and architect's actions and decisions.

D503–2011, Guide for Sustainable Projects, including Agreement Amendments and Supplementary Conditions

AIA Document D503™–2011 is not an agreement, but is a guide that discusses the roles and responsibilities faced by Owners, Architects, and Contractors on sustainable design and construction projects. D503 also contains model provisions for modifying or supplementing the following AIA Contract Documents: A201™–2007, General Conditions of the Contract for Construction; A101™–2007, Standard Form of Agreement between Owner and Contractor where the basis of payment Is a Stipulated Sum; and B101™–2007, Standard Form of Agreement between Owner and Architect. D503 provides model language with explanatory notes to assist users in adapting those documents for use on a sustainable project. A201–2007, A101–2007, and B101–2007, as standard form documents, cannot address all of the unique requirements and risks of sustainable design and construction. Thus, AIA Document D503–2011 is provided to assist users either in modifying those documents, or developing separate supplementary conditions documents to attach to them.

E-Series: Exhibits

E201–2007, Digital Data Protocol Exhibit

AIA Document E201™–2007 is not a stand-alone document, but must be attached as an exhibit to an existing agreement, such as the AIA Document B101™–2007, Standard Form of Agreement between Owner and Architect, or A101™–2007, Agreement between Owner and Contractor. Its purpose is to establish the procedures the parties agree to follow with respect to the transmission or exchange of digital data, including instruments of service. AIA Document E201–2007 defines digital data as information, communications, drawings, or designs created or stored for a specific project in digital form. E201 does not create a separate license to use digital data, because AIA documents for design or construction, to which E201 would be attached, already include those provisions. Parties not covered under such agreements should consider executing AIA Document C106™–2007, Digital Data Licensing Agreement.

E202–2008, Building Information Modeling Protocol Exhibit

AIA Document E202™–2008 is a practical tool for managing the use of building information modeling (BIM) across a project. It establishes the requirements for model content at five progressive levels of development, and the authorized uses of the model content at each level of development. Through a table the parties complete for each project, AIA Document E202–2008 assigns authorship of each model element by project phase. E202 defines the extent to which model users may rely on model content, clarifies model ownership, sets forth BIM standards and file formats, and provides the scope of responsibility for model management from the beginning to the end of the project. Though written primarily to support a project using integrated project delivery (IPD), E202 may also be used on projects delivered by more traditional methods. E202 is not a stand-alone document, but must be attached as an exhibit to an existing agreement for design services, construction, or material. NOTE: E202–2008 is available in AIA Contract Documents software, but is not available in print.

G-Series: Contract Administration and Project Management Forms

G601–1994, Request for Proposal—Land Survey

AIA Document G601™–1994 allows owners to request proposals from a number of surveyors based on information deemed necessary by the owner and architect. G601–1994 allows owners to create a request for proposal through checking appropriate boxes and filling in project specifics, thus avoiding the costs associated with requesting unnecessary information. G601–1994 may be executed to form the agreement between the owner and the land surveyor once an understanding is reached.

G602–1993, Request for Proposal—Geotechnical Services

Similar in structure and format to AIA Document G601™–1994, AIA Document G602™–1993 can form the agreement between the owner and the geotechnical engineer. It allows the owner to tailor the proposal request to address the specific needs of the project. In consultation with the architect, the owner establishes the parameters of service required and evaluates submissions based on criteria such as time, cost, and overall responsiveness to the terms set forth in the request for proposal. When an acceptable submission is selected, the owner signs the document in triplicate, returning one copy to the engineer and one to the architect, thus forming the agreement between owner and geotechnical engineer.

G612–2001, Owner's Instructions to the Architect

AIA Document G612™–2001 is a questionnaire, drafted to elicit information from the owner regarding the nature of the construction contract. AIA Document G612–2001 is divided into three parts: Part A relates to contracts, Part B relates to insurance and bonds, and Part C deals with bidding procedures. The order of the parts follows the project's chronological sequence to match the points in time when the information will be needed. Because many of the items relating to the contract will have some bearing on the development of construction documents, it is important to place Part A in the owner's hands at the earliest possible phase of the project. The owner's responses to Part A will lead to a selection of the appropriate delivery method and contract forms, including the general conditions. Part B naturally follows after selection of the general conditions because insurance and bonding information is dependent on the type of general conditions chosen. Answers to Part C will follow as the contract documents are further developed.

G701–2001, Change Order

AIA Document G701™–2001 is for implementing changes in the work agreed to by the owner, contractor, and architect. Execution of a completed AIA Document G701–2001 indicates agreement on all the terms of the change, including any changes in the contract sum (or guaranteed maximum price) and contract time. The form provides space for the signatures of the owner, architect, and contractor, and for a complete description of the change.

G701S–2001, Change Order, Subcontractor Variation

AIA Document G701S™–2001 modifies AIA Document G701™–2001 for use by subcontractors. Modifications to G701–2001 are shown as tracked changes revisions—that is, additional material is underlined; deleted material is crossed out. NOTE: G701S–2001 is not available in print, but is available in AIA Contract Documents software and on the AIA Documents-on-Demand™ website.

G701CMa–1992, Change Order, Construction Manager-Adviser Edition

AIA Document G701™CMa–1992 is for implementing changes in the work agreed to by the owner, contractor, construction manager adviser, and architect. Execution of a completed AIA Document G701™–2001 indicates agreement on all the terms of the change, including any changes in the Contract Sum (or Guaranteed Maximum Price) and Contract Time. It provides space for the signatures of the owner, contractor, construction manager adviser, and architect, and for a complete description of the change. The major difference between AIA Documents G701CMa–1992 and G701–2001 is that the signature of the construction manager adviser, along with those of the owner, architect, and contractor, is required to validate the change order.

G702–1992, Application and Certificate for Payment

AIA Documents G702™–1992, Application and Certificate for Payment, and G703™–1992, Continuation Sheet, provide convenient and complete forms on which the contractor can apply for payment and the architect can certify that payment is due. The forms require the contractor to show the status of the contract sum to date, including the total dollar amount of the work completed and stored to date, the amount of retainage (if any), the total of previous payments, a summary of change orders, and the amount of current payment requested. AIA Document G703–1992 breaks the contract sum into portions of the work in accordance with a schedule of values prepared by the contractor as required by the general conditions. NOTE: The AIA does not publish a standard schedule of values form.

AIA Document G702–1992 serves as both the contractor's application and the architect's certification. Its use can expedite payment and reduce the possibility of error. If the application is properly completed and acceptable to the architect, the architect's signature certifies to the owner that a payment in the amount indicated is due to the contractor. The form also allows the architect to certify an amount different than the amount applied for, with explanation provided by the architect.

G702S–1992, Application and Certificate for Payment, Subcontractor Variation

AIA Document G702S™–1992 modifies AIA Document G702™–1992 for use by subcontractors. Modifications to G702–1992 are shown as tracked changes revisions—that is, additional material is underlined; deleted material is crossed out. NOTE: G702S–1992 is not available in print, but is available in AIA Contract Documents software and on the AIA Documents-on-Demand™ website.

G703–1992, Continuation Sheet

AIA Documents G702™–1992, Application and Certificate for Payment, and G703™–1992, Continuation Sheet, provide convenient and complete forms on which the contractor can apply for payment and the architect can certify that payment is due. The forms require the contractor to show the status of the contract sum to date, including the total dollar amount of the work completed and stored to date, the amount of retainage (if any), the total of previous payments, a summary of change orders, and the amount of current payment requested. AIA Document G703–1992 breaks the contract sum into portions of the work in accordance with a schedule of values prepared by the contractor as required by the general conditions. NOTE: The AIA does not publish a standard schedule of values form.

G703S–1992, Continuation Sheet, Subcontractor Variation

AIA Document G703S™–1992 modifies AIA Document G703™–1992 for use by subcontractors. Modifications to G703–1992 are shown as tracked changes revisions—that is, additional material is underlined; deleted material is crossed out. NOTE: G701S–1992 is not available in print, but is available in AIA Contract Documents software and on the AIA Documents-on-Demand™ website.

G704–2000, Certificate of Substantial Completion

AIA Document G704™–2000 is a standard form for recording the date of substantial completion of the work or a designated portion thereof. The contractor prepares a list of items to be completed or corrected, and the architect verifies and amends this list. If the architect finds that the work is substantially complete, the form is prepared for acceptance by the contractor and the owner, and the list of items to be completed or corrected is attached. In AIA Document G704–2000' the parties agree on the time allowed for completion or correction of the items, the date when the owner will occupy the work or designated portion thereof, and a description of responsibilities for maintenance, heat, utilities, and insurance.

G704CMa–1992, Certificate of Substantial Completion, Construction Manager-Adviser Edition

AIA Document G704™CMa–1992 serves the same purpose as AIA Document G704™–2000, except that this document expands responsibility for certification of substantial completion to include both the architect and the construction manager.

G704DB–2004, Acknowledgement of Substantial Completion of a Design-Build Project

Because of the nature of design-build contracting, the project owner assumes many of the construction contract administration duties performed by the architect in a traditional project. Because there is not an architect to certify substantial completion, AIA Document G704™DB–2004 requires the owner to inspect the project to determine whether the work is substantially complete in accordance with the design-build documents and to acknowledge the date when it occurs. AIA Document G704DB–2004 is a variation of AIA Document G704™–2000 and provides a standard form for the owner to acknowledge the date of substantial completion.

G705–2001 (formerly G805–2001), List of Subcontractors

AIA Document G705™–2001 is a form for listing subcontractors and others proposed to be employed on a project as required by the bidding documents. It is to be filled out by the contractor and returned to the architect for submission to the owner. NOTE: AIA Document G705–2001 was renumbered in 2007, but its content remains the same as in AIA Document G805™–2001 (expired May 31, 2009).

G706–1994, Contractor's Affidavit of Payment of Debts and Claims

The contractor submits this affidavit with the final request for payment, stating that all payrolls, bills for materials and equipment, and other indebtedness connected with the work for which the owner might be responsible has been paid or otherwise satisfied. AIA Document G706™–1994 requires the contractor to list any indebtedness or known claims in connection with the construction contract that have not been paid or otherwise satisfied. The contractor may also be required to furnish a lien bond or indemnity bond to protect the owner with respect to each exception.

G706A–1994, Contractor's Affidavit of Release of Liens

AIA Document G706A™–1994 supports AIA Document G706™–1994 in the event that the owner requires a sworn statement of the contractor stating that all releases or waivers of liens have been received. In such event, it is normal for the contractor to submit AIA Documents G706–1994 and G706A–1994 along with attached releases or waivers of liens for the contractor, all subcontractors, and others who may have lien rights against the owner's property. The contractor is required to list any exceptions to the sworn statement provided in G706A–1994, and may be required to furnish to the owner a lien bond or indemnity bond to protect the owner with respect to such exceptions.

G707–1994, Consent of Surety to Final Payment

AIA Document G707™–1994 is intended for use as a companion to AIA Document G706™–1994, Contractor's Affidavit of Payment of Debts and Claims, on construction projects where the contractor is required to furnish a bond. By obtaining the surety's approval of final payment to the contractor and its agreement that final payment will not relieve the surety of any of its obligations, the owner may preserve its rights under the bond.

G707A–1994, Consent of Surety to Final Reduction in or Partial Release of Retainage

This is a standard form for use when a surety company is involved and the owner/contractor agreement contains a clause whereby retainage is reduced during the course of the construction project. When duly executed, AIA Document G707A™–1994 assures the owner that such reduction or partial release of retainage does not relieve the surety of its obligations.

G709–2001, Work Changes Proposal Request

This form is used to obtain price quotations required in the negotiation of change orders. AIA Document G709™–2001 is not a change order or a direction to proceed with the work. It is simply a request to the contractor for information related to a proposed change in the construction contract. AIA Document G709–2001 provides a clear and concise means of initiating the process for changes in the work.

G710–1992, Architect's Supplemental Instructions

AIA Document G710™–1992 is used by the architect to issue additional instructions or interpretations or to order minor changes in the work. It is intended to assist the architect in performing its obligations as interpreter of the contract documents in accordance with the owner/architect agreement and the general conditions of the contract for construction. AIA Document G710–1992 should not be used to change the contract sum or contract time. It is intended to help the architect perform its services with respect to minor changes not involving adjustment in the contract sum or contract time. Such minor changes are authorized under Section 7.4 of AIA Document A201™–2007.

G711–1972, Architect's Field Report

The architect's project representative can use this standard form to maintain a concise record of site visits or, in the case of a full-time project representative, a daily log of construction activities.

G712–1972, Shop Drawing and Sample Record

AIA Document G712™–1972 is a standard form by which the architect can log and monitor shop drawings and samples. The form allows the architect to document receipt of the contractor's submittals, subsequent referrals of the submittals to the architect's consultants, action taken, and the date returned to the contractor. AIA Document G712–1972 can also serve as a permanent record of the chronology of the submittal process.

G714–2007, Construction Change Directive

AIA Document G714™–2007 is a directive for changes in the Work for use where the owner and contractor have not reached an agreement on proposed changes in the contract sum or contract time. AIA Document G714–2007 was developed as a directive for changes in the work which, if not expeditiously implemented, might delay the project. Upon receipt of a completed G714–2007, the contractor must promptly proceed with the change in the work described therein. NOTE: G714–2007 replaces AIA Document G714™–2001 (expired May 31, 2009).

G714CMa–1992, Construction Change Directive, Construction Manager-Adviser Edition

AIA Document G714™CMa–1992 serves the same purpose as AIA Document G714™–2007, except that this document expands responsibility for signing construction change directives to include both the architect and the construction manager.

G715–1991, Supplemental Attachment for ACORD Certificate of Insurance 25-S

AIA Document G715™–1997 is intended for use in adopting ACORD Form 25-S to certify the coverage required of contractors under AIA Document A201™–2007, General Conditions of the Contract for Construction. Since the ACORD certificate does not have space to show all the coverages required in AIA Document A201–2007, the Supplemental Attachment form should be completed, signed by the contractor's insurance representative, and attached to the ACORD certificate.

G716–2004, Request for Information (RFI)

AIA Document G716™–2004 provides a standard form for an owner, architect, and contractor to request further information from each other during construction. The form asks the requesting party to list the relevant drawing, specification, or submittal reviewed in attempting to find the information. Neither the request nor the response received provides authorization for work that increases the cost or time of the project.

G732–2009 (formerly G702CMa–1992), Application and Certificate for Payment, Construction Manager as Adviser Edition

AIA Document G732™–2009 serves the same purposes as AIA Document G702™–1992. The standard form AIA Document G703™–1992, Continuation Sheet, is appropriate for use with G732–2009. NOTE: G732–2009 replaces AIA Document G702™CMa–1992 (expired December 31, 2010).

G736–2009 (formerly G722CMa–1992), Project Application and Project Certificate for Payment, Construction Manager as Adviser Edition

Use AIA Document G736™–2009 with AIA Document G737™–2009, Summary of Contractors' Applications for Payment. These forms are designed for a project where a construction manager is employed as an adviser to the owner, but not as a constructor, and where multiple contractors have separate, direct agreements with the owner.

Each contractor submits separate AIA Documents G732™–2009 and G703™–1992, payment application forms, to the construction manager-adviser, who collects and compiles them to complete G736–2009. AIA Document G737–2009 serves as a summary of the contractors' applications with totals being transferred to AIA Document G736–2009. The construction manager-adviser can then sign G736, have it notarized, and submit it along with the G737 to the architect. Both the architect and the construction manager must certify the payment amount. NOTE: G736–2009 replaces AIA Document G722™CMa–1992 (expired December 31, 2010).

G737–2009 (formerly G723CMa–1992), Summary of Contractors' Applications for Payment, Construction Manager as Adviser Edition
Use AIA Document G736™–2009 with AIA Document G737™–2009, Summary of Contractors' Applications for Payment. These forms are designed for a project where a construction manager is employed as an adviser to the owner, but not as a constructor, and where multiple contractors have separate, direct agreements with the owner.

Each contractor submits separate AIA Documents G703™–1992 and G732™–2009, payment application forms, to the construction manager-adviser, who collects and compiles them to complete AIA Document G736–2009. AIA Document G737–2009 serves as a summary of the contractors' applications with totals being transferred to G736. The construction manager-adviser can then sign G736, have it notarized, and submit it along with the G737 to the architect. Both the architect and the construction manager must certify the payment amount. NOTE: G737–2009 replaces AIA Document G723™CMa–1992 (expired December 31, 2010).

G801–2007 (formerly G605–2000), Notification of Amendment to the Professional Services Agreement
AIA Document G801™–2007 is intended to be used by an architect when notifying an owner of a proposed amendment to the AIA's owner/architect agreements, such as AIA Document B101™–2007. NOTE: G801–2007 replaces AIA Document G605™–2000 (expired May 31, 2009).

G802–2007 (formerly G606–2000), Amendment to the Professional Services Agreement
AIA Document G802™–2007 is intended to be used by an architect when amending the professional services provisions in the AIA's owner/architect agreements, such as AIA Document B101™–2007. NOTE: G802–2007 replaces AIA Document G606™–2000 (expired May 31, 2009).

G803–2007 (formerly G607–2000), Amendment to the Consultant Services Agreement
AIA Document G803™–2007 is intended for use by an architect or consultant when amending the professional services provisions in the AIA's architect-consultant agreement, AIA Document C401™–2007. NOTE: G803–2007 replaces AIA Document G607™–2000 (expired May 31, 2009).

G804–2001, Register of Bid Documents
AIA Document G804™–2001 serves as a log for bid documents while they are in the possession of contractors, subcontractors, and suppliers during the bidding process. The form allows tracking by bidder of documents issued, deposits received,

and documents and deposits returned. AIA Document G804–2001 is particularly useful as a single point of reference when parties interested in the project call for information during the bidding process.

G806–2001, Project Parameters Worksheet

AIA Document G806™–2001 is an administrative form intended to help maintain a single standard list of project parameters including project objectives, owner's program, project delivery method, legal parameters, and financial parameters.

G807–2001, Project Team Directory

AIA Document G807™–2001 is used as a single point of reference for basic information about project team members including the owner, architect's consultants, contractor, and other entities. AIA Document G807–2001 differs from AIA Document G808™–2001, Project Data, which contains only data about the project and project site. G807–2001 should be carefully checked against the owner/architect agreement so that specific requirements as to personnel representing the owner and those involved with the architect in providing services are in conformance with the agreement.

G808–2001, Project Data

AIA Document G808™–2001 is used for recording information about approvals and zoning and building code issues gathered in the course of providing professional services. AIA Document G808–2001 should be completed piece by piece as a project progresses and periodically reviewed to ensure information relevance. The attached worksheet, AIA Document G808A™–2001, Construction Classification Worksheet, can be used to supplement the G808–2001.

G808A–2001, Construction Classification Worksheet

AIA Document G808A™–2001, Construction Classification Worksheet, can be used to supplement AIA Document G808™–2001, which is used for recording information about approvals and zoning and building code issues gathered in the course of providing professional services. AIA Document G808–2001 should be completed piece by piece as a project progresses and periodically reviewed to ensure information relevance. AIA Document G808A–2001 can help a design team work through the range of code compliance combinations available before choosing a final compliance strategy.

G809–2001, Project Abstract

AIA Document G809™–2001 establishes a brief, uniform description of project data to be used in the tabulation of architect marketing information and firm statistics. The intent is to provide a single sheet summary where information can be sorted, compiled, and summarized to present a firm's experience. Information compiled in AIA Document G809–2001 can support planning for similar projects and answer questions pertaining to past work.

G810–2001, Transmittal Letter

AIA Document G810™–2001 allows for the orderly flow of information between parties involved in the design and construction phase of a project. It serves as a written record of the exchange of project information and acts as a checklist reminding the sender to tell the recipient what exactly is being sent, how the material is being sent, and why it is being sent.

14.5 CONSENSUSDOCS CONTRACT DOCUMENTS

ConsensusDOCS® publishes a comprehensive catalog of more than 90 documents covering most Contract Document needs. These standard contracts were developed by a coalition of 35 leading construction industry associations and members from stakeholders in the design and the construction industry. ConsensusDOCS contracts incorporate best practices and fairly allocate risk to help reduce costly contingencies and adversarial negotiations. They are regularly updated to keep pace with changes in construction law and industry practice.

The catalog addresses all major project delivery methods and provides coordinated administrative forms. Table 14.7 lists current and scheduled future ConsensusDOCS documents. AIA Contract Documents are divided into six alphanumeric series according to document use of purpose (see text box that follows the table).[2]

TABLE 14.7 ConsensusDOCS Documents

Current Contract Documents series, www.consensusdocs.org	200 Series: General Contracting Documents
	300 Series: Collaborative Documents
	400 Series: Design–Build Documents
	500 Series: Construction Management Contracts
	700 Series: Subcontracting Documents
	800 Series: Program Management Documents
Future documents scheduled	Time Extension Addendum
	Schedule Specification
	Purchase Agreement
	Construction Manager at Risk Short Form
	Land Survey Agreement
	Teaming Agreement for Design–Build Projects
	Subcontractor Qualifications
	Joint Venture Agreement
	Joint Venture Agreement for Design–Build
	Federal Design-Builder and Design Professional Agreement
	Subconsultant Agreement Between an Architect and Engineer
	Owner and Geotechnical Consultant Agreement

---

2. Excerpted from *ConsensusDOCS Contracts Catalog* (www.consensusdocs.org/catalog/), Spring 2012.

ConsensusDOCS General Contracting Documents (200 Series)

Current Contract Documents

ConsensusDOCS 200: Agreement and General Conditions between Owner and Constructor (Lump Sum)
An integrated agreement and general conditions document between the Owner and contractor performing work on a lump sum basis. Appropriate for use in competitive bid environments or in situations requiring a negotiated lump sum contract. View 200 Guidebook Comments. View 200 Index.

ConsensusDOCS 200.1: Time and Price Impacted Materials
Provides a method for establishing the market price of a construction commodity and for calculating a price adjustment for that commodity if it has an extraordinary cost increase or decrease. View 200.1 Guidebook Comments.

ConsensusDOCS 200.2: Electronic Communications Protocol Addendum
Helps the parties determine acceptable formats and technology for electronic communications, including Building Information Modeling (BIM). This groundbreaking document allows for communications management and consistency throughout the project. View 200.2 Guidebook Comments.

ConsensusDOCS 202: Change Order
Used to formalize changes in the work and adjustments to contract time and price.

ConsensusDOCS 203: Interim Directed Change
A unilateral order issued by the Owner in the absence of agreement on price and time for changes in the work.

ConsensusDOCS 204: Request for Information
Used by contractors or Subcontractors to request information or instructions.

ConsensusDOCS 205: Short Form Agreement between Owner and Constructor (Lump Sum)
This convenient, short-form agreement and general conditions document is premised on concepts and language found in ConsensusDOCS 200.

ConsensusDOCS 220: Contractor's Qualification Statement for Engineered Construction
May be used as a generic pre-qualification statement or a contract-specific qualification statement. Includes Schedules A-C regarding current/past projects and key personnel.

ConsensusDOCS 221: Contractor's Statement of Qualifications for a Specific Project
Helps Owners to assess the qualifications of a contractor. Includes Schedules A-C regarding current/past projects and key personnel.

ConsensusDOCS 222: Architect-Engineer's Statement of Qualifications for a Specific Project
Helps Owners to assess the qualifications of an architect-engineer.

ConsensusDOCS 235: Short Form Agreement between Owner and Contractor (Cost of Work)
This convenient, short-form agreement and general conditions document is premised on concepts and language found in the ConsensusDOCS 510.

ConsensusDOCS 240: Agreement between Owner and Design Professional
Coordinated for use with ConsensusDOCS 200 series (Owner-contractor documents), this agreement is used between the Owner and the architect-engineer performing a full range of design and administrative services for the project. View 240 Guidebook Comments. View 240 Index.

ConsensusDOCS 245: Short Form Agreement between Owner and Design Professional
Describes the relationship between the Owner and the architect-engineer and places most transaction-specific information at the front, and addresses services from schematic design through construction contract administration.

ConsensusDOCS 260: Performance Bond
Developed with the assistance of organizations representing the surety industry, this standardized performance bond form is coordinated for use with Consensus-DOCS 200 and 500 documents.

ConsensusDOCS 261: Payment Bond
Developed with the assistance of organizations representing the surety industry, this standardized payment bond form is coordinated for use with ConsensusDOCS 200 and 500 documents.

ConsensusDOCS 262: Bid Bond
Developed with the assistance of organizations representing the surety industry, this standardized bid bond form is coordinated for use with ConsensusDOCS 200 and 500 documents.

ConsensusDOCS 263: Warranty Bond
Used for the correction of a defect in the Work during a one-year Correction of Work period. Provisions addressing general conditions and Surety obligation are provided in this document.

ConsensusDOCS 270: Instructions to Bidders on Private Work
Used for bid submission and award, it provides information about pre-bid procedure, including obtaining bidding documents and additional information prior to opening of bids, and the examination of bidding documents and worksite.

ConsensusDOCS 280: Certificate of Substantial Completion
Establishes the date of substantial completion of the work or a designated portion thereof.

ConsensusDOCS 281: Certificate of Final Completion
Establishes the date of final completion of the work.

ConsensusDOCS 290: Guidelines for Obtaining Financial Owner Information
Helps the contractor identify the type of information that should be requested of the Owner, and why this information is important.

ConsensusDOCS 290.1: Owner Financial Questionnaire
contractors and Subcontractors use this form to request specific information about the Owner's legal structure, ownership of the land, construction financing, and insurance matters.

ConsensusDOCS 291: Application for Payment (Guaranteed Maximum Price, GMP)
Facilitates the calculation and documentation of progress payments.

ConsensusDOCS 292: Application for Payment (Lump Sum)
Facilitates the calculation and documentation of progress payments.

ConsensusDOCS 293: Schedule of Values
Provides a breakdown of the cost of elements of the work and should be used with the ConsensusDOCS Application for Payment forms ConsensusDOCS 291 and 292.

ConsensusDOCS 907: Equipment Lease
Offered as either a one-page agreement or two-page general conditions, this lease agreement is accompanied by an instruction sheet on assumptions in the document which may require modification on items in the standard form requiring completion.

Collaborative Documents (300 Series)

ConsensusDOCS 300: Tri-Party Collaborative Agreement
This is the first standard IPD agreement published in the United States. The Owner, Designer, and Constructor all sign the same agreement. This agreement incorporates Lean principles and is also known as a relational contract. A core team at both the project management and project development levels is created to make consensus-based project decisions to increase project efficiency and results. View 300 Guidebook Comments. I Read "IPD for Public and Private Owners."

ConsensusDOCS 301: Building Information Modeling Addendum
The first standard Contract Document that globally addresses legal issues and administration associated with utilizing Building Information Modeling (BIM), it is intended to be used as an identical contract addendum for all project participants inputting information into a BIM Model. It also includes a BIM Execution Plan, which allows the parties to determine the level for which BIM model(s) may be relied on legally. View 301 Guidebook Comments.

ConsensusDOCS 310: Green Building Addendum
Another industry first—appropriate for use on projects with green building elements, particularly those seeking a third-party green building rating certification such as LEED. It provides a contractual mechanism to identify clear objectives, and assign roles and responsibilities to achieve green goals. The parties designate a Green Building Facilitator (GBF) to coordinate or implement identified objectives, which can be a project participant or consultant. It contemplates that such services will be included in the underlying agreement with the project participant or in a separate agreement with a GBF. View 310 Guidebook Comments. I Read an article on ConsensusDOCS 310.

Design-Build Documents (400 Series)

ConsensusDOCS 400: Preliminary Agreement between Owner and Designer-Builder
Intended to be used in conjunction with ConsensusDOCS 410 or 415 to take the project through schematic design only.

ConsensusDOCS 410: Agreement and General Conditions between Owner and Design-Builder (Cost of Work Plus Fee with Guaranteed Maximum Price (GMP)
May be used as a follow-up document to ConsensusDOCS 400 or as a stand-alone document that addresses the entire design-build process. View 410 Guidebook Comments.

ConsensusDOCS 415: Agreement and General Conditions between Owner and Design-Builder (Lump Sum Based on the Owner's Program Including Schematic Design Documents)
Unlike the ConsensusDOCS 410, this document cannot be used as a stand-alone document to address the entire design-build process. It is intended to be a follow-up document to ConsensusDOCS 400, assuming that the owner's program or other project information includes schematic design documents.

ConsensusDOCS 420: Agreement between Design-Builder and Design Professional
Delineates the respective rights and responsibilities of the Design-Builder and the architect-engineer.

ConsensusDOCS 421: Statement of Qualifications
Provides information to Owners to assess the qualifications of a Designer-Builder.

ConsensusDOCS 450: Agreement between Design-Builder and Subcontractor (Design-Builder Assumes Risk of Owner Payment)
Intended for use where the Subcontractor has not been retained to provide substantial portions of the design for the project, and payment to the Subcontractor is not conditioned on the Design-Builder having received payment from the Owner.

ConsensusDOCS 460: Agreement between Design-Builder and Design–Build Subcontractor (Subcontractor Provides a Guaranteed Maximum Price and Design-Builder Assumes Risk of Owner Payment)
Intended for use where the Subcontractor is retained by the Design-Builder early in the design phase, basically providing the same design and construction services as the Design-Builder provides the Owner under ConsensusDOCS 410 and 415. Construction is performed based on cost of the work, plus a fee, up to the GMP. Payment to the Subcontractor is not conditioned on the Design-Builder having received payment from the Owner for subcontract work satisfactorily performed.

ConsensusDOCS 470: Performance Bond (Surety Is Liable for Design Costs of Work)
Bond between the Surety and the Designer-Builder where the Surety is liable for the design costs of the work. Provisions addressing Surety obligations, limited liability for design, and dispute resolution are provided in this document. Space is provided to fill in the bond sum and names of Owner (Obligee), Designer-Builder (Principal), Surety, Surety Representative and Project.

ConsensusDOCS 471: Performance Bond (Surety Is Not Liable for Design Services)
Bond between the Surety and the Designer-Builder where the Surety is not liable for the design costs of the work. Provisions addressing Surety obligations, on liability of design, and dispute resolution are included.

ConsensusDOCS 472: Payment Bond (Surety Is Liable for Design Costs of Work)
Bond between the Surety and the Designer-Builder where the Surety is liable for the design costs of the work. Provisions addressing Surety obligations, on liability of design, and dispute resolution are included.

ConsensusDOCS 473: Payment Bond (Surety Is Not Liable for Design Services)
Bond between the Surety and the Designer-Builder where the Surety is not liable for the design costs of the work. Provisions addressing Surety obligations, on liability of design, and dispute resolution are included.

ConsensusDOCS 481: Certificate of Substantial Completion
Establishes the date of substantial completion of the work.

ConsensusDOCS 482: Certificate of Final Completion
Establishes the date of final completion of the work.

ConsensusDOCS 491: Application for Payment (Cost of Work and a Guaranteed Maximum Price (GMP) Has Been Established)
Used with the ConsensusDOCS 410 and provides for notarization.

ConsensusDOCS 492: Application for Payment (Lump Sum)
Used with the ConsensusDOCS 415 and provides for notarization.

ConsensusDOCS 495: Change Order for Cost Plus with Guaranteed Maximum Price (GMP) Design-Build Contracts
Used with the ConsensusDOCS 410 and requires signatures of the Designer-Builder and the Owner.

ConsensusDOCS 496: Change Order for Lump Sum Design-Build Contracts
Used with the ConsensusDOCS 415 and requires signatures of the Designer-Builder and the Owner.

Construction Management contracts (500 Series)

ConsensusDOCS 500: Agreement and General Conditions between Owner and Construction Manager (CM Is At-Risk)
An integrated agreement and general conditions document, the ConsensusDOCS 500 also provides an option for pre-construction services, such as providing estimates of the Project, reviewing drawings and specifications for constructability problems, creating schedules for procurement of long lead items, and developing Trade contractor interest in the Project. It may be used in a variety of negotiated contract situations in which the Owner desires a comprehensive set of pre-construction and/or construction services from the Construction Manager and seeks the assurance of an overall project cost ceiling. View 500 Guidebook Comments.

ConsensusDOCS 510: Agreement and General Conditions between Owner and Construction Manager (Cost of Work with Option for Preconstruction Services)
Intended to form an integrated agreement and general conditions document between the Owner and the Construction Manager performing work on a cost of the work plus a fee basis without a GMP. It also provides an option for the contractor to provide pre-construction services similar to the ConsensusDOCS 500. It may be used in a variety of negotiated contract situations in which the Owner desires a comprehensive set of pre-construction and/or construction services from the contractor, and it may be particularly applicable in situations where project variables, such as a well-defined scope of the work, may be unknown at the time of contract execution. With pre-construction services added, this document becomes the equivalent of a Construction Manager at Risk (CM@R) agreement.

ConsensusDOCS 525: Change Order/Construction Manager Fee Adjustment
This form is for projects built under the Construction Management method of contracting.

Subcontracting Documents (700 Series)

ConsensusDOCS 703: Purchase Agreement for Noncommodity Goods
Standard purchase agreement between a Constructor and an equipment manufacturer for noncommodity goods, which may include some installation labor.

ConsensusDOCS 705: Invitation to Bid/Subbid Proposal
Used for Subcontractors to describe the scope of work covered in their bids.

ConsensusDOCS 706: Performance Bond
This bond can be requested by a contractor from a Subcontractor to guarantee the Subcontractor's performance.

ConsensusDOCS 707: Payment Bond
This bond form can be requested by a contractor from a Subcontractor to guarantee that the Subcontractor will pay laborers and material suppliers.

ConsensusDOCS 710: Application for Payment
Provides a standardized format for Subcontractor's requests for payment.

ConsensusDOCS 721: Statement of Qualifications
Used by the Subcontractor to provide information, such as personnel qualifications, industry references, performance history and safety record, to the contractor who is assessing the Subcontractor's qualifications to work on a specific project.

ConsensusDOCS 725: Agreement between Subcontractor and Subsubcontractor
The first and only standard agreement, this simplified form is for use between a Subcontractor and a Subsubcontractor and is suited to the generally less complex relationship between these two parties.

Exhibit E
Insurance requirements to 725 Standard Subsubcontractor Agreement.

ConsensusDOCS 750: Agreement between Contractor and Subcontractor
(contractor Assumes Risk of Owner Payment)
This document is intended to be generally compatible with ConsensusDOCS 200 or other agreements. An indemnity agreement is also included. View 750 Guidebook Comments.

ConsensusDOCS 750.1: Rider between Contractor and Subcontractor for Material Storage at Subcontractor's Site
Governs the storage of specific materials and equipment at a Subcontractor's yard, and sets a standard agreement for storage that will ensure the minimum precautions and coverages are agreed on (and purchased if they are not covered in the Builder's Risk Policy for the Project). It may be attached as a rider to ConsensusDOCS 750.

ConsensusDOCS 751: Short Form Agreement between Contractor and Subcontractor (Contractor Assumes Risk of Owner Payment)
This convenient subcontract form places all negotiated points and project-specific terms at the beginning of the document. The contractor assumes the risk of Owner nonpayment. An indemnity agreement is also included.

ConsensusDOCS 752: Subcontract for Use on Federal Construction
The first and only standard subcontract agreement for federal projects that is compliant with the contracting requirements and practices found in the 2009 Federal Acquisition Regulation (FAR). View 752 Guidebook Comments. I Read an article on ConsensusDOCS 752.

ConsensusDOCS 760: Bid or Proposal Bond
Used when a bid or proposal bond is required.

ConsensusDOCS 781: Certificate of Substantial Completion
Establishes the date of substantial completion of the work.

ConsensusDOCS 782: Certificate of Final Completion
Establishes the date of final completion of the work.

ConsensusDOCS 790: Subcontractor Request for Information (RFI)
Used by Subcontractors to request information or instructions.

ConsensusDOCS 795: Change Order
Formalizes changes in the work and makes adjustment to subcontract time and price.

ConsensusDOCS 796: Interim Directed Change
A unilateral order issued by the contractor in absence of agreement on price and time for changes in the Subcontractor's work.

Program Management Documents (800 Series)
ConsensusDOCS 800: Program Management Agreement and General Conditions between Owner and Program Manager
The contractual configuration is of a "pure/agent program manager," not at risk, either with all design and construction contracts signed by the Owner or the Program Manager signing the contracts as the agent of the Owner. The Program Manager can be seen as replacing the Owner's facilities staff and may oversee a project delivery accomplished under a variety of methods (e.g., design-bid-build or design-build) for each discrete project or site. This contract provides a scope of services presented in a matrix to be used as a menu for the parties to assign duties.

ConsensusDOCS 801: Construction Management Agreement between Owner and Construction Manager (Construction Manager Is Owner's Agent and Owner Enters into All Trade Contractor Agreements)
May be used with the construction management process when the Owner awards all the trade contracts.

ConsensusDOCS 802: Agreement between Owner and Trade contractor (Construction Manager is Owner's Agent)
Describes the legal relationship between the Owner and each Trade contractor, who becomes prime to the Owner. This document is compatible with the ConsensusDOCS 801.

ConsensusDOCS 803: Agreement between Owner and Architect/Engineer (Construction Manager Acting as Agent Has Been Retained by Owner)
Developed expressly to coordinate with ConsensusDOCS' other Construction Management agency forms, specifically ConsensusDOCS 801 and 802.

ConsensusDOCS 810: Agreement between Owner and Owner's Representative
Agreement between an Owner and a person/entity acting as an independent contractor, who shall serve as the Owner's authorized representative for a specific project, assuming that the Owner will retain both an architect-engineer and a contractor.

ConsensusDOCS 812: Interim Directed Change
Issued by the Owner to the Trade contractor in the absence of agreement on price and time for changes in the trade contract work.

ConsensusDOCS 813: Change Order
Used to formalize changes in the trade contract work and make adjustment to time and price.

ConsensusDOCS 814: Certificate of Substantial Completion
Establishes the date of final completion of the work.

ConsensusDOCS 815: Certificate of Final Completion
Establishes the date of final completion of the work.

14.6 ENGINEERS JOINT CONTRACT DOCUMENTS COMMITTEE CONTRACT DOCUMENTS

The Engineers Joint Contract Documents Committee (EJCDC) is the third organization that has developed standard documents representing the latest and best thinking involved in engineering design and construction projects. The committee is made up of the National Society of Professional Engineers' (NSPE) Professional Engineers in Private Practice group, the American Counsel of Engineering Companies, the American Society of Civil Engineers, and the Associated General Contractors, and its members include representatives of more than 15 professional engineering design, construction, owner, legal, and risk management organizations. Table 14.8 lists some of the documents that

TABLE 14.8 EJCDC Documents

Construction	Standard General Conditions of the Construction Contract (C-700)
	Bid Bond; Damages Form (C-435)
	Bid Bond; Penal Sum Form (C-430)
	Certificate of Substantial Completion (C-625)
	Change Order (C-941)
	Construction Payment Bond (C-615)
	Construction Performance Bond (C-610)
	Construction-Related Documents Set (C-990)
	Construction-Related Documents Set (C-990)
	Contractor's Application for Payment (C-620)
	Engineer's Request for Instructions on Bonds and Insurance for Construction (C-051)
	Field Order (C-942)
	Guide to the Preparation of Supplementary Conditions (C-800)
	Narrative Guide to the 2007 EJCDC Construction Documents (C-001)
	Notice of Award, Download (C-510)
	Notice to Proceed (C-550)
	Owner's Instructions Concerning Bonds and Insurance for Construction (C-052)
	Owner's Instructions Regarding Bidding Procedures (C-050)
	Suggested Bid Form for Construction Contracts (C-410)
	Suggested Form of Agreement between Owner and Contractor; Cost-Plus (C-525)

Continued

TABLE 14.8   EJCDC Documents—Cont'd

	Suggested Form of Agreement between Owner and Contractor; Stipulated Price (C-520)
	Suggested Instructions to Bidders for Construction Contracts (C-200)
	Work Change Directive (C-940)
Contract Document Sets	Owner–Engineer Documents Set (E-990)
	Construction-Related Documents Set (C-990)
	Design–Build Document Set (D-990)
	Engineer–Subconsultant Agreements Set (E-991)
	Environmental Remediation Set (R-990)
	Full Design–Bid–Build Document Set (A-990)
	Owner–Engineer Documents Set (E-990)
	Procurement Agreements Set (P-990)

can be downloaded for various prices. For other EJCDC documents, visit www.nspe.org/ejcdc/index.html.

While there are many organizations that produce their own versions of contract documents, it is always advisable to consult an attorney to ensure that a particular contract meets your specific needs.

Chapter 15

Green Business Development

15.1   INTRODUCTION

"Green building" has been in the spotlight for several years now, and if you are interested in starting a construction business, you should certainly consider making it a green construction business. But starting a successful green company is a serious business and does not happen by accident. Like other types of businesses, it requires careful planning to start and succeed. It requires, among other things, the ability to manage the business with all its complexities on a day-to-day basis and taking into account information relating to market analysis, planning, accounting and bookkeeping, advertising, targeting the market, analyzing the competition, and more. Having said that, however, companies that are presently embracing green building appear to be among the few in building construction and design that are succeeding. This may be because green buildings generally incorporate nontoxic building materials and products (i.e., green), provide healthier spaces, use recycled building components, are more energy efficient, and so on.

The deterring factor in all of this is the rapid downturn in the economy and the fact that the construction industry has been at a virtual standstill. Many professionals have suddenly found themselves unemployed and seeking employment for the first time in years. These professionals have been forced into retirement or have had to abandon the safety of an organization that regularly delivered their paycheck. This has nudged many who may be at their peak to reevaluate their future prospects and employment strategies and to consider becoming their own boss. The obvious concerns that dominate this strategy are job satisfaction, location, and stress, in addition to cash flow, health insurance, and retirement funding. Nevertheless, if you are seriously thinking about starting a new construction business, there are several things to consider prior to making too many plans.

Planning has become even more imperative with some green builders being left reeling by the economic recession and the dismal construction market, forcing some customers to adopt a tight budget—owners/developers are feeling less inclined to splurge on green materials and products. The good news is that there are numerous incentives for building green, such as the 2009 American Recovery and Reinvestment Act (ARRA), which offers homeowners tax credits to encourage them to make their residences more energy efficient. Furthermore, there are new directives for the design and construction of greener buildings, which combine to create significant opportunities as well as challenges for the construction industry. But before taking the plunge, you must know where to start. In recent years, many new green building codes have come into effect that need to be understood and complied with. You will also need to position your construction company as a certified "green builder," which will require time, research, and investment.

Still, independence has many attractions and advantages, not least of all being your own boss, having flexible hours, and having greater control of your future. Whether as a green general contractor or professional consultant, you or your accountant bill clients for services rendered. If new to contracting, you may initially decide to subcontract all or most of the work to "specialty" contractors who typically will bill you on a monthly basis. This may reduce the initial overhead, but whether you are new in the contracting business or are an established contractor with solid, loyal clients, concentrating on the basic elements of the business is essential for survival and growth.

Generally speaking, the individual proprietorship is the form of entity used by most small businesses at startup. However, a partnership may be the best way to go if additional capital or expertise is needed. But the freedom of independence often comes with a heavy price tag, not least of which is the initial loss of security. By being independent, one may suddenly breathe an illusion of freedom, but the question that soon needs answering is: Where is the next dollar coming from? Family members, in particular, need to be mentally prepared for the reality of being unemployed as well as the challenges that starting a new business bring to the table.

15.2   THE OFFICE: HOME-BASED VERSUS BRICKS AND MORTAR

After taking the various factors into consideration, a determination is made to incorporate, so now the decision must be made whether to start searching for office space or to work from home. This is a decision that is influenced by many considerations such as available resources, whether foot traffic is necessary, number of staff needed, whether to work full- or part-time, whether the business will be web-based or not, and many others. If the business is to be home-based, it will require easy and preferably separate access such as a separate walk-out basement. Many competitive startup businesses are initially home-based, which has several advantages, particularly with the new technology available that has become part of our culture.

The available technology includes the Internet, instant messaging, video conferencing, and other innovative workflow tools that make effective telecommuting a reality. Moreover, working from home obviously saves time that would otherwise be spent traveling to and from an office. Some of the other upsides of a home-based office include less risk and startup costs, allowing you to test the waters without excessive expenditure. Likewise, you can outsource tasks (e.g., managing accounts, public relations, website management). There are a number of significant disadvantages, however, related to working from home, such as being constantly distracted, particularly with a large family, children, and so on. Moreover, meeting clients and subcontractors at home can sometimes be awkward and not present the professional impression desired. Additionally, you need to ensure that adequate parking space is provided and zoning ordinances are not infringed.

Should the decision be made not to work from home, then appropriate office space will be needed. The cost of rental space is mainly determined by the size and location of what's being rented. When a suitable office is found, have a lease prepared in the name of the corporation rather than in your name; this will minimize liability exposure should the business not succeed. Also, while people can work in a tight space for short periods of time, particularly during the startup phase of an operation, it will be difficult to maintain productivity and retain employees over longer periods unless they are comfortable and appropriate space is allocated. An additional incentive is to make the office space as "green" as possible, sending the right message to visiting clients and others that you practice what you preach.

One of the upsides to having a bricks-and-mortar office is that a physical location causes fewer distractions and may even attract walk-in traffic (e.g., by noticing the sign). It also reflects more professionally on the firm and transmits an air of confidence and efficiency to clients and potential clients. The main downside is a greater risk factor and more startup costs. It also requires a greater full-time commitment upfront to get the office ready for business, as well as to hire some staff (e.g., an assistant). To minimize travel time, the location should not be too far from your residence and adequate parking space is always a definite plus, particularly if many visitors and workers are expected. It may be wise initially to avoid taking out a long lease, particularly if there are lingering doubts about the possible success of the new venture. It may be possible to enter into a month-to-month rental agreement at first. However, if you find yourself tied into a long lease, check the agreement to see whether there is a sublease clause that allows for leasing the premises should you decide to close the business.

Prior to viewing potential office space, determine what you can afford and what your budget allocation is; take into account not only the rent but also items such as furniture and utilities. Also, when calculating space needs, make sure you understand the difference between "gross" square feet and "usable" square feet. For example, usable square feet typically means the area available for things like workstations, and it generally consists of the total or "gross" square feet less areas occupied by lobbies, restrooms, kitchens, and so on. Thus, when inspecting a prospective office, you may wish to ensure that it meets current needs and can accommodate future expansion.

As a potential tenant, you should also check whether the lease is a net-net lease requiring you to pay all expenses, including utilities, lighting, signs, taxes, insurance, maintenance, and garbage collection, and if so, whether this is acceptable. Space requirements will vary according to individual needs, the allocated budget, and the type and size of the business. Most startups need less space than well-established ones. Typically new businesses require approximately 100 square feet per workstation; this does not include space for aisles, equipment, and other shared areas. Another important consideration is the placing of a professional-looking sign by the front entrance or in another prominent location.

15.3 CREATING A SUCCESSFUL BUSINESS PLAN

Most business advisors and experienced entrepreneurs generally agree that it is wise to develop a business plan prior to starting a business and marketing your construction services. A well-structured plan can certainly help you move forward and make the right decisions, and help make your business successful. The business plan that you will need to put together is basically a written document describing the business, its objectives, and its strategies, as well as the market it is in and its financial forecasts. It will also detail precisely what services are being offered, who are the services' proposed clients, who is the competition, and the proposed method to advertise and promote your services during the first year of business and beyond. Not all business plans, however, are the same, nor do they need the same level of detail. You might develop a fairly simple plan first and then expand and elaborate on it as you prepare to approach bankers or investors. Having a business plan is essential because it will help generate interest from potential lenders, prospective employees, and strategic partners. As an operating tool, it can help manage the business and effectively work toward its success.

For getting started, it is possible to develop a plan in several stages that meet your real business needs. In writing the business plan, it should be kept simple, concise, and neatly formatted, and preferably in a Microsoft® Word format with attached or embedded spreadsheets in Microsoft Excel. Fancy graphics, "padding," and flowery language are unnecessary and should be avoided. It is possible to use business planning software to prepare your business plan, if so desired, although it may lack the flexibility to accurately convey all of the features and potential of your new business. However, business planning software has the advantage of a logical step-by-step approach, and generally formats the plan for you. Plus, unless you have sufficient startup capital to finance setting up the new business on your own (e.g., for signage, office equipment, payroll, rent, utilities), you will probably need to deal with bank loans or investors or both, and for that you will need a more extensive business plan.

For business startups in particular, proper planning is one of the keys to success and its importance cannot be overemphasized. Putting together a business plan, including the research and thought even before beginning to write, forces one to take a serious, objective, and unemotional view of the entire business project. The business plan will invariably assist in identifying areas of strengths and weaknesses. But to be truly effective and convincing, it must show, among other things, the marketing strategy to be employed.

The whole idea of having a business plan is to communicate ideas to others while providing the basis for a financial proposal. Research shows that setting up a new business is fraught with difficulties and challenges, and that over half of all new businesses fail within the first 10 years. The main cause of failure is lack of planning and lack of adequate financing. As previously mentioned, finding startup capital for a new business will not be easy, which is why owners are initially expected to use their own funds or a bank loan linked to income or security other than the business (e.g., a home equity loan) or, as a last resort, to borrow from friends or relatives.

A business plan is designed to serve several functions in addition to securing external funding. For example, it helps in measuring success within the business; for new businesses it is often used to ensure that the various aspects of running the business have been researched and adequately thought out, thereby avoiding unexpected surprises. However, it must be said that a business plan is typically required by lenders (i.e., banks) when applying for financing; it can help convince banks or potential investors that you are worthy of receiving financial assistance for the new venture, especially if you can provide a professional-looking basic sales and expense forecast, leading to high profits and minimal loss. The principal components of a business plan are shown in the sidebar.

Business Plan Components

Introduction. This should primarily consist of a brief but comprehensive summary of how the company was formed, what type of business it is (e.g., green construction), and the people linked to it.

Mission and Vision. This generally reflects the objectives, aspirations, and direction of the company's business, as well as its expected goals and achievements. A mission statement generally outlines both short- and long-term goals and strategies.

Management. Even if the new business is a one-person operation, a key ingredient for business success is the strength of your management skills. When the business consists of more than one person, describe the management team with short biographies of principals and key personnel who will be instrumental to its success. Include each team member's role, background, position and responsibilities, and why he or she is specifically qualified for his or her role.

Services Offered. Outline in detail the type of services to be offered (e.g., green building, sustainability consulting, remodeling and alterations, permitting, site preparation, carpentry, concrete foundations, painting, plumbing and utilities installation, exterior renovations), the market for these services, and how you will fit them into this market. Include drawings, specifications,

Continued

Business Plan Components—cont'd

previous projects executed, and anything else that would enhance your presentation. It is also important to highlight any special skills, factors, and qualities that give your firm an edge over the competition.

Financial Plan. Here you need to include financial statements; this is a critical part of the business plan and condenses your strategies and assumptions into the cost of setting up the business and its expected profits. This is the section lenders and investors will be most interested in to evaluate your financial prospects. The financial section should clearly show projections for the first few years of business (depending on the lender's requirements) and may contain formal records of the business' financial activities, including

- Written statement of key business assumptions
- Twelve-month profit and loss projection
- One-year cash flow projection
- Income statement
- Projected balance sheet and break-even point
- Personal financial statement
- Report on cash management

Executive Summary. Although the executive summary appears as the first part of the plan, it is not typically written until the whole document is complete. It basically summarizes the most important information and aspects of your business plan and normally does not consist of more than one or two pages. The executive summary outlines information relating to the services offered by the business, its key people, why there is a need for this company, the current market, the competition, and the strategies that will be employed. Because lenders and investors are often very busy, they normally will not spend more than a few minutes reviewing a business plan before deciding whether they should read it in detail or move on to another plan. When it is decided to read any part of the business plan, it is typically the executive summary, which is why it is so imperative for it to be both appealing and convincing; it needs to be able to capture the investor's attention and imagination. The executive summary, therefore, is considered to be the most important part of the business plan, and it will determine whether the remaining pages will be read or not.

15.4 STARTUP BASICS

You will incur many expenses long before you even start operating your business. It is important to estimate these expenses accurately, and then try to plan where to get the needed capital. People often underestimate startup costs and begin their business in a haphazard and unplanned way. Without adequate funding, the business would be almost impossible to establish, operate, and made to be successful. Inadequate funding, being "undercapitalized," is one of the primary reasons many small businesses fail within their first year of operation. Perhaps because of lack of experience, new business owners also frequently fail to include a contingency amount to meet unforeseen expenses and, consequently, fail to secure adequate financing to carry their business through the period before it reaches a break-even status and starts to show a profit.

Most experts recommend that startup funding be adequate to cover operating expenses for six months to a year to allow the business to find customers and get established. This is because many startups are likely to end up spending more money than originally planned. However, it is not possible to determine the amount of financing needed without detailed cost projections in hand. Some experts suggest a two-part process: Develop an accurate estimate of your startup costs and put together a projection of operating expenses for at least the first six months of operation. Performing these two exercises will present a clearer picture of the business and help identify potential problems needing resolution, thereby ensuring success.

15.4.1   Startup Costs

It is rarely easy to figure out what the startup costs will be for a new business, mainly because you have a moving target that is easy to underestimate and frequently subject to change. Startup costs reflect expenses incurred prior to commencing with the business plan, usually before the first month. It is no secret that many new companies incur initial costs for legal work, logo design, brochures, and other miscellaneous items. Using a startup worksheet to plan initial financing will help gather the necessary information to set up initial business balances and prepare a preliminary estimate of expenses (Figure 15.1). Needless to say, estimating the amount of capital needed to start a business requires careful analysis of a number of factors. A list of realistic one-time costs for opening your doors will be needed that should include all necessary furniture, fixtures, and equipment. The list should also include, but not be limited to, the full price, down payment, or cash price of items; if purchased on an installment plan, you need the amount of each monthly payment for each product.

First-year expenses should appear in the profit and loss statement and expenses incurred before that must appear as startup costs. Once the initial estimate of cash needed to start is determined, it is possible to calculate how much money is actually available, or can be made available, to help set up the business; if this proves inadequate, then a decision must be made on where the remaining startup money required can be found.

15.4.2   Employees and Required Forms

Your accountant should be the first person to consult on whether you should hire yourself or others as full- or part-time employees. This is because you may need to register with the appropriate state agencies or obtain worker's compensation insurance or unemployment insurance (or both). Numerous major firms now allow (or prefer) some of their employees to work from home and only come into the office once a week, say, or as required; this may be suitable for accountants or estimators.

Start-up Costs Estimates: The first step is to put together a list of realistic expenses of one-time costs for opening your doors. Such a list would include what furniture, fixtures and equipment is needed, as well as the cost, cash price, down payment if purchased on an installment plan and the amount of each monthly or periodic payment. Record them in the costs table below:

Down Payment	$_____
Amount of each payment	$_____

The furniture, fixtures and equipment required may include such things as desks, moveable partitions, storage shelves, file cabinets, tables, safe, special lighting, and signs. Other start-up costs are shown next.

TYPICAL START-UP COSTS ITEMS TO BE PAID ONLY ONCE

Furniture, Fixtures, & Equipment:

Interior decorating	$_____
Installation of fixtures and equipment	$_____
Starting inventory	$_____
Deposits with public utilities	$_____
Legal and other professional fees	$_____
Licenses and permits	$_____
Advertising and opening promotion	$_____
Advance on lease	$_____
Other miscellaneous cash requirement	$_____
TOTAL ESTIMATED CASH NEEDED TO START =	$_____

ESTIMATED MONTHLY EXPENSES

Salary of owner-manager	$_____
All other salaries and wages	$_____
Payroll taxes and expense	$_____
Rent or lease	$_____
Advertising	$_____
Delivery expense	$_____
Office supplies	$_____
Telephone	$_____
Other utilities	$_____
Insurance	$_____
Property taxes	$_____
Interest expense	$_____
Repairs and maintenance	$_____
Legal and accounting	$_____
Miscellaneous	$_____
TOTAL ESTIMATED MONTHLY EXPENSES =	$_____
Multiply by 4 (4 months)	$_____
Add: Total Cash needed to start above	$_____
TOTAL ESTIMATED CASH NEEDED	$_____

FIGURE 15.1   A draft startup worksheet is used to produce a preliminary cost estimate and to plan an initial financial strategy for a new business venture.

In larger establishments, it takes many hours of hard work to prepare and file the various payroll reports and other necessary government forms; this puts a heavy burden on anyone trying to keep up with everything on his or her own. When the business has grown sufficiently to allow the hiring of qualified employees and/or managers, take the opportunity to do so—hire them. Having qualified and well-trained personnel can significantly improve a company's performance and help expand it. Consult your accountant when the decision is made to hire new employees to determine what type of personnel files will be needed for each. Typically, the minimum forms needed are an I-9 form, IRS form W-4, and the state equivalent form for employee income tax withholding. If using independent sub-contractors, they should sign IRS form W-9. Again, consult with your accountant as to whether state law requires subcontractors to be included on the firm's tax policy.

Expense Reports

Almost all larger firms have developed standardized, digitized expense report forms for their employees so that they can request reimbursement for business expenses. Even with a startup business, it is vital to monitor expenditures, and a standard form may be the best way to do so because it makes bookkeeping easier. When the expense form is not standardized, it should nevertheless be neatly typed and organized, identifying each location, project name and number, and applicable dates, with all original receipts and supporting documentation attached in date order. It should be handed over to accounting to process and record as soon as possible.

15.4.3 Utilities, Equipment, and Furnishings

When leasing a new office, utility, equipment, and furnishing expenses are a necessary overhead. Advance deposits, especially for new businesses, are often required when signing up for power, gas, water, and sewer. Also, once the decision is made to establish your own business and an office has been leased (if you are not working from home), request a phone number (consider an 800 number) as well as a domain name for your new business' website (more about this later in this chapter). When you get the phone number, find out when the next issue of the telephone directory is to be published and the deadline for a listing so that you can include a display ad in the yellow pages under the classification that best describes the company's services (compare to what the competition has).

Equipment

Businesses invariably differ in the type of equipment they need, but it should be "green" if possible, as this will reflect well on the company. It may also be prudent for a startup business to preserve cash for inventories or working capital by initially purchasing good used fixtures and equipment at a much lower price. Be sure to obtain more than one quote for the equipment you need. With the recent changes in the income tax laws, you will have to do extra analysis to determine if a lease program or a direct purchase is the best way to proceed. Whether to buy or lease depends on several factors that an accountant can advise you on. For new companies that want to keep their initial startup costs to a minimum, it may be smarter to lease as much as possible, especially electronic equipment, computers, copiers, printers, telephone systems, and certain other products, because of continuous advances taking place in these fields. All cash down payments for equipment purchased on contract should be appropriately recorded.

Furnishings

Whenever possible, get office furnishings (desks, credenzas, file cabinets, bookcases, chairs, end tables, lamps) that are eco-friendly and be sure to record their cost so that you can deduct it from your taxes. When paying in cash, enter the full retail price and, if payment is by installments, note the down payment as a startup cost.

Decorating or Remodeling

If the office you are moving into needs some redecoration or reconfiguration, make an estimate of what the total cost will be and try to negotiate with the landlord to pay for it, or to deduct it from the base rent. Also, talk to suppliers from whom you plan to purchase materials and other services and record these expenses. It is unlikely that you will consider undertaking major work unless you are contemplating leasing for an extended period.

15.4.4 Suppliers

For a new business, many suppliers may be reluctant to ship their goods without some sort of assurance that they will be paid. It may help to have some good credit references; this is why it is important to have a rapport with your banker because she or he can provide acceptable references to suppliers. Identify key suppliers and determine whether they need convincing that you are honest, hard-working, and in for the long haul, and that your business is expected to be solid and have a good chance for success. Some suppliers may request C.O.D. payments during the early stages of getting started; take this fact into consideration when preparing your financial planning and startup estimates. Once you have become established with suppliers, send your financial data to Dun & Bradstreet so that you can be listed in their files. Most U.S. firms recognize D&B as very reliable for obtaining correct credit information about a registered company.

15.4.5 Accounting and Bookkeeping

For every type of business, it is important to set up a good accounting and recordkeeping system and to learn as much as possible about what taxes your new company is responsible for paying. Company documents and tax and corporate filings are generally required to be kept for three years, including a list of all owners and addresses and copies of all formation documents, financial statements, annual reports, and company amendments or changes.

Being a new business owner, you may decide to do the recordkeeping yourself, but if you do, it is advisable to engage the assistance of an accountant to help set up the books based on the simple method outlined here. Moreover, if possible, let the accountant "keep the books" for the first few months until you feel comfortable taking them over. After a short time, you or an employee will likely be in a position to do the accounting. Whatever the case, use a separate checkbook and bank account for your business so as to avoid commingling private and business accounts. Normally a simple "cash" accounting system requires records of original entry; a "general journal" to record extraneous transactions; and a "general ledger" to which accounts from the three records are posted at the end of each month. This system can be readily converted to an accrual accounting method by journalizing accounts receivable, payable, accruals, and so on. The balance sheet and income statement can be completed fairly easily once these entries are posted.

Consult your accountant regarding preparing financial statements and reversing the accruals to be equipped for the following month's entries. It is now possible to enter the gross payroll, payroll deductions, and the net amount in the check register. Where employees are involved, it is expedient to give them a pay stub itemizing all the facts while maintaining supplementary payroll sheets with all information about each employee. Having individual payroll records and control accounts in the general ledger provides the necessary information to complete the various payroll tax reports and returns as they become due. At the end of each annual accounting period, all the information for filing tax returns will be available for your accountant and/or bookkeeper to go through and submit the final returns to the IRS.

Accounting files should be stored on computers (with backups on CDs or portable hard drives) instead of in file cabinets; this also makes it easier to email and make offsite backup copies when traveling. Reviewing documents onscreen rather than printing them out also helps the environment, as does sending emails instead of paper letters. Software, such as GreenPrint, is also available and helps eliminate blank pages from documents before printing and can convert to PDF for paperless document sharing.

15.4.6 Miscellaneous Expenses

There are numerous other startup expenses that need to be taken into consideration when estimating the amount of cash that will be needed for a new business, including both business and personal living expenses. Thus, if you are leaving a salaried position to start your own business, include in your expense projection an estimate of the expenses you and your household will incur for the months it will take to get the business going. At this point it probably makes sense to review certain categories (e.g., equipment, office supplies, advertising/promotion) with cost control in mind. If it appears that your estimated startup costs are greater than originally anticipated, it may be time to review and reevaluate your list of projected expenses and decide whether it is more practical to purchase or lease used office equipment or furnishings than buying new. The classified ads may be a good place to start your search; they can lead you to bankruptcy auctions, house sales, and furniture resellers in addition to individual items. To ensure that you are on the right path, do not shy from asking your attorney or accountant for referrals to business owners who have relevant experience in evaluating startup costs.

Additional information and advice on startup can be found on the U.S. Department of Commerce Minority Business Development Agency website (www.mbda.gov). Articles can be found here that discuss the amount of money needed to start a new business. The site includes helpful checklists and provides referrals to other information resources. The U.S. Small Business Administration (www.sba.gov) was created specifically to assist and counsel small businesses. Its publication, "Small Business Startup Kit," includes a checklist for calculating costs. The SBA's online Women's Business Center (www.sba.gov/content/womens-business-centers) includes a helpful section on evaluating startup costs for new businesses and starting a contracting business.

15.5 CREATING AN IMAGE AND MARKETING A NEW BUSINESS

Creating a distinct corporate image is absolutely vital for corporations in today's market. It is not surprising that every company wants to have a favorable image in the global marketplace. A corporate image distinguishes your company from its competitors and provides a picture of it to potential customers and the general public. Moreover, a corporate image builds confidence and credibility, helping your target audience understand you and your firm better, because it reflects your principles, beliefs, and productivity, and increases trust. As discussed earlier in this chapter, starting a new business can be risky, but your chances of success significantly increase with proper planning, including having a distinctive corporate image. The following sections describe some of the steps needed to succeed in setting up a new business venture, whether it is a green contracting business or professional green consultancy.

15.5.1 The Company's Image

A new business must create a good company image and business identity that reflect confidence and efficiency. Usually, this means hiring a professional to design a corporate logo, business card, letterhead, and promotional material. The logo should be simple and not easy to forget; it represents the visual image of your company and will be used in a variety of applications. Moreover, an attractive and professionally created logo and letterhead can go a long way to sending clients an image of confidence and trustworthiness while reducing their perception of risk, making it easier to command a premium price for your services. A good logo also says who you are, how you are different from your competition, and why a client should do business with you. The need to have a good logo cannot be overemphasized.

15.5.2 Advertising and Promotion

With any new business, it is important to get the word out so that customers start coming through your door or to your home page. To do that you will need to research your target audience and develop a marketing message that resonates with them. Some small new businesses start their operation with a grand-opening announcement, in addition to press releases to the local press and relevant business publications. Circulars can be printed and distributed to potential clients or placed in a newspaper to be distributed to subscribers. The dollar cost of planned advertising and marketing announcing the launch of a new business should be recorded and reflected in the budget and should include the cost of all promotional items, including flyers, brochures, phone calls, and signs. The competition's ads should be studied along with their websites.

15.5.3 Marketing

Prior to marketing your services, adequate research is required to get all the facts, just as with writing the business plan. This will help you formulate a successful marketing strategy that will target your ideal customer and thus be much more methodical and effective. Research will also facilitate the development of professionally designed brochures and other marketing materials by determining who your target audience is and what its preferences are. Because you are selling a specialized service, it is imperative to know how to market it. To do this successfully several key questions need to be answered, including

- Who are the typical customers?
- Are the services you have to offer what your customers want?
- What is the budget of targeted customers and how much are they willing to pay? (This will obviously vary depending on the project.)
- Why should potential customers prefer you over the competition?
- What media type will best reach your target audience and have the most impact?

After answering these questions satisfactorily, you will be in position to start developing and implementing a successful marketing strategy. Figure 15.2 is a typical letter to get the word out and let customers and potential clients know you are now open for business.

Mr. John Doe   November 24, 2011
President
XYZ Developers Inc.
1070 East Market St.,
Leesburg, VA 20176

(Tel)   (703) 777 1234
(Fax)   (703) 777 2345
Email: sdoe@XYZ-developers.com

Re: Green Building Services

Dear Mr. Doe:

I am taking this opportunity to apprise you of green building services that we offer to
property developers and investors.

ABC Green Building International has recently been formed to provide green
construction services. Although ABC is a newly formed company, its principals have
over twenty-five years of experience in design, construction, and sustainable
practices. Our specialty is green construction and we have a number of LEED
Accredited Professionals on our staff to ensure improved occupant health, protection
of ecosystems, and reducing energy consumption in our projects. For further
information and an overview of our services, please view our website at: www.abc-
greenbuilding.com. We would be delighted to discuss our services with you and bid
on any upcoming projects.

We are able to travel anywhere within the United States to provide services to meet
your requirements.

Please take a few days to look things over, and then I'll call you to set up a time to
discuss your requirements.

Sincerely,

Sam Kubba, AIA, RIBA, Ph.D., LEED AP
Principal
ABC GREEN BUILDING INTERNATIONAL

SAK/bs
enclosures

cc: General Files

ABC/PROMO/LETTER/John Doe/Promo.doc

FIGURE 15.2   A promotional letter can be sent to potential clients offering green construction services now that your company is open and ready for business. Printed promotional material should
accompany the letter.

Time Management

The better organized you are, the more efficient you are and the less time will be wasted. Good organization can be facilitated by the appointment of an assistant or office manager to deal with the operational aspects of the business and to make it as automated and efficient as possible. This will allow you to concentrate on the business aspects, including marketing. It will also free you from having to follow up on normal day-to-day issues (e.g., processing orders, paying utility and other bills).

15.6 TRACKING AND IDENTIFYING SOURCES FOR LEADS

There are several ways to identify potential sources of project leads, depending to some extent on whether your business is essentially a one-person operation or one with several employees. These include the following:

- Send out flyers, brochures, emails, and so on, to potential clients and advertise your services in the local press.
- Check specialized construction search engines such as "bidclerk" (www. bidclerk.com) or "buildingonline" (www.buildingonline.com). Check the Internet for others. They can provide excellent leads for construction projects that are coming up for bid in your area.
- Browse the Internet, particularly broker sites displaying vacant land. All major real estate firms typically have websites, and some of these have client lists to build up potential customer confidence. These lists can be researched to see which, if any, names are worth following up on. Inquire as to whether it is possible to get information regarding potential buyers so that you can send them promotional information.
- Visit neighborhood commercial real estate agents to see what commercial properties, including vacant land, are currently on the market. Some properties may require renovation. A list of all possible leads should be made and followed up with letters and brochures offering your company's services.
- Find possible leads by driving around your area or around areas that appear "ripe" for development. Many of the clients will be lenders (e.g., banks, lending institutions) who will provide financing to property developers or individuals wanting to build a custom home or commercial building (Figure 15.3). Make a list using the yellow pages, the Internet, and the public library of these institutions and send them promotional material.

15.7 THE IMPORTANCE OF SELLING YOURSELF

In your business, the ability to convince potential clients that you are the best person for the job is critical. You want to make a good impression, be viewed as genuine, and be taken seriously as well. Despite the need to sell yourself to others, for many people this may not come naturally. People are often introverts

(a)

(b)

FIGURE 15.3 (a) An office building under construction in Arlington, Virginia, has a building sign showing the name of the lending institution and the general contractor. Send promotional information to the bank so that your firm's name is on their contractors' list; also send a letter to the construction company to seek subcontracting work. (b) Construction under way in Bowie Town Center, Maryland, offers opportunities for subcontracting work.

or shy, or lack a sense of self-worth. These are just a few of the obstacles that can potentially get in the way. How one dresses is another element that can get in the way. To dress for success you have to "dress" the part. The following sections discuss several keys to successful dressing and self-presentation.

15.7.1 Dress for Success

To be successful, it is important to dress appropriately for corporate environments. For example, you do not want to show up in tennis shoes and jeans at an executive meeting, or in a skirt and heels for a construction site walk-through. For general meetings, appropriate dress normally means that men should wear dress slacks, a clean button-up shirt with tie, and a blazer or suit jacket. A jacket may not be necessary in some situations—for example, if you are inspecting a building site during the summer. Obviously, it is not necessary to wear a business suit when you are working outdoors. However, in the final analysis this will depend on the individual situation, the environment, and the audience for which you are dressing. Another important aspect of proper business attire for a man is to be clean-shaven because scruffiness is unprofessional. To further enhance your image and complete your ensemble, hair should be clean and well-groomed and cologne or aftershave should be subtle, not overpowering.

Larger organizations may have a business casual dress code for employees to follow, whatever that means. Business casual dress is not as formal as wearing professional business clothes—suits are acceptable but not necessary. Dressing professionally but comfortably is the point, and outfits should have a relaxed, comfortable appearance while still looking neat and smart. Examples of business casual are cotton trousers and khakis for men; combine these with a collared shirt to create a professional but relaxed appearance. But even for a business with a casual dress code, you should still dress professionally, especially when having face-to-face meeting with clients or customers.

This also applies to women who become confused sometimes about what is appropriate and what is not. In this they are not alone, as even women in executive positions sometimes admit to not knowing which styles best suit their bodies. Great basics for most work environments include collared shirts and pencil skirts or good slacks. If you are still unsure of what to wear to work, observe some of the professional, successful women in your industry because they can offer appropriate examples of what is acceptable in your particular environment (Figure 15.4); Figure 15.5 highlights the dress policy of one government department.

While proper clothing might seem logical and common sense for some, it does not come naturally to everyone. It is a well-known fact that first impressions can significantly impact how a person is ultimately perceived, and this is why proper dress is so important. A person's appearance, therefore, is a powerful form of communication. When used properly, it can be an effective tool for

(a)

FIGURE 15.4   The photos in (a) and (b) show appropriate professional business attire; those in (c) show inappropriate business clothes.

Continued

portraying confidence, trust, and ability. Regardless of the occasion, however, dressing appropriately is one of the easiest ways to impress a potential client. It may not guarantee that you get this job, but it may help you get the next one. A note of warning: Clothing and accessories should not attract so much attention as to distract from a meeting's real purpose. Also, when attending a business meeting with a client, bank manager, or whoever, start off with a firm handshake and follow up with eye contact.

15.7.2   Introductions, Correspondence, and Meetings

When attending an event where you are likely to meet potential clients (e.g., a conference, seminar, or even dinner), be sure to carry business cards and perhaps some literature about your company. Also try to portray an air of confidence; this will give you the appearance of an accomplished professional. Be cool, calm, and collected, and, most important, *think* before you speak. Be organized and prepared and have the necessary knowledge to answer any questions you may be asked, and show customers that you can execute the job successfully.

(b)

(c)

FIGURE 15.4—cont'd

Correspondence

Correspondence is increasingly an online function, which means less paper is being used. Likewise, business files are now being kept on computers instead of in file cabinets, making it easier to make offsite backup copies or take them with you when you move to new offices. Documents can be reviewed onscreen

A. POLICY

   1. Generally. All City employees shall comply with all applicable and appropriate standards of dress, personal appearance, neatness and cleanliness. This policy describes the City's general expectations for employees regarding personal appearance and dress. Department directors are responsible for the enforcement of this policy. All departments shall comply with this policy unless the department has a stricter standard.

   2. Dress Code. The standard of dress for employees who are not required to wear a uniform is "business casual." The purpose of this standard is to allow employees to work comfortably while at the same time project a professional image. Because not all casual dress is suitable for the office, this policy will set out clothing considered appropriate to wear to work. This is a general overview. Examples of appropriate and inappropriate attire are included. Neither list is all-inclusive; both are subject to change. No dress code can cover all contingencies so employees are required to exercise sound judgment in the choice of clothing to wear to work.

     a. General Guidelines.

        i. Clothing should be pressed and never wrinkled (unless the material is wrinkled by design);

       ii. Seams must be finished;

      iii. Clothing containing the City of Surprise name and/or logo is encouraged;

      iv. Torn, dirty, patched, or frayed clothing is unacceptable;

       v. Clothing containing words, terms, or pictures that may be offensive to other employees and the general public is unacceptable;

      vi. Clothing that reveals an employee's back, chest, stomach or underwear is unacceptable.

     b. Examples of Acceptable Business Casual Dress.

        i. Cotton slacks (similar to Dockers©), wool pants, flannel pants, dressy Capri's, and attractive synthetic dress pants;

       ii. Casual dresses or skirts (dresses and skirts should be at an appropriate length - mid-thigh or below);

      iii. Casual shirts, dress shirts, sweaters, tops, golf or polo shirts, and turtlenecks; suit and sport jackets generally;

      iv. Loafers, clogs, conservative athletic shoes, boots, flats, dress heels, and leather deck-type shoes and sandals are acceptable; closed toe and closed heel shoes are required in certain operational areas;

       v. Perfume and cologne should be used with restraint, as some people are allergic to chemicals in perfume and make-up.

     c. Examples of Unacceptable Business Casual Dress.

        i. Jeans, sweatpants, exercise pants, shorts, bib overalls, leggings, and any spandex or other form-fitting pants;

       ii. Mini-skirts, skorts, sun dresses, beach dresses, and spaghetti strap dresses;

      iii. Tank tops, midriff tops, shirts with potentially offensive words, terms, logos, pictures, cartoons, or slogans, halter-tops, tops with bare shoulders, sweatshirts, and t-shirts (unless worn under a blouse, shirt, jacket, or dress);

      iv. Beach/pool style flip-flops and slippers;

       v. Hats are not appropriate in the office. Head covers that are required for religious tradition are permitted.

     d. Casual Days. Fridays, and other days as determined by the department director are declared as "casual" days. On casual days, jeans (clean, non-wrinkled, with finished seams), sweatshirt and t-shirts (that do not contain words, terms, or pictures that may be offensive to other employees and the general public), are permitted in addition to all other acceptable business casual attire.

     e. Department Directors and Above. While persons in positions of department director and above ("managers") are covered by the City's "business casual" dress code, managers are expected to be attired in "business formal" clothing (suit, jacket and pants, or dress paired with appropriate accessories) for City Council meetings, City Council workshops, formal meetings in the office, and formal events in or outside the office. Managers are expected to know or ascertain all situations when more formal dress is expected and dress accordingly.

   3. Appearance. The maintenance of high standards of personal cleanliness and appearance by all City employees is essential to creating and maintaining a favorable public image. Employees are expected to observe proper habits of personal grooming and hygiene at all times.

   4. Exceptions. Exceptions to this policy may be requested and will be granted for bona fide religious and philosophical reasons.

FIGURE 15.5   Dress policy for a government department.

rather than having to be printed out. Electronic instead of paper communication is far more efficient and cost effective. However, there are various other ways to communicate (e.g., via telephone or in person depending largely on one's personality). Introverts tend to prefer email because it is efficient and avoids direct contact; extroverts, on the other hand, usually prefer direct communication. Before sending an email or letter, make sure that it is being sent to the correct person. Also, all correspondence should be reviewed for accuracy—spellcheck should always be used before transmitting.

General Meetings

Successful meetings are usually the result of good organization and adequate preparation. This is a time when you will meet clients, investors, executives, and others to discuss relevant topics (e.g., client projects, marketing strategies, financing). Be prepared with questions and issues you want to cover and anticipate in advance what your objectives are. During business meetings, be careful to stick to proper meeting etiquette, as this is an arena where poor etiquette can reflect negatively on you and your firm. Correct meeting etiquette automatically improves your chances of success and communicates comfort and trust to everyone involved, including colleagues, clients, and customers. In today's business world, it is these people who can greatly impact your firm's ability to succeed and flourish.

Informal Meetings

Informal meetings are generally relaxed and may not necessarily take place in an office or meeting room. Nevertheless, a sense of professionalism and good business etiquette are still required. In this respect, punctuality is always important. In addition, the purpose of the meeting should be clearly outlined to the proposed attendees. Failing to relay the proper information is poor business etiquette and may cause embarrassment and prevent the meeting from achieving its objectives. Normally, the person calling the meeting is the most senior person or the person with the most direct or urgent interest in the topic at hand. He or she may also be responsible for determining (through consultation) the meeting's time, place, and agenda. Generally, someone is appointed (usually through the chair) to take minutes that can later be typed and distributed to all attendees for future reference.

Formal Meetings

It is unlikely that the owner of a startup firm will attend many formal meetings; nevertheless, it is important to have a clear understanding of required etiquette. As a professional, you should dress appropriately and be punctual; mobile phones should be switched off. It is imperative to be well prepared and any reports or other information to be used should be handed out prior to the meeting, allowing adequate time to review. If you are unsure regarding the seating pattern, you should ask. During formal meetings, follow these rules:

- When speaking, always address the chair unless it is clear that no one else is doing so.
- When discussions are under way, allow more senior figures to contribute first.
- Acknowledge any opening remarks with a brief recognition of the chair and other participants.
- Refrain from interrupting a speaker, even if you strongly disagree. Note what has been said and come to it when appropriate with the chair's permission.
- Do not divulge confidential information regarding a meeting. This is considered very unethical and a serious breach.

Other factors that can add to a new company's chances of achieving success, and that should be carefully considered, include the following:

- *Creating a Network:* Having a good network is almost synonymous with business success. Although a lucrative contract may sometimes be the result of a single contact, it takes a strong network to generate a continuous stream of remunerative projects.
- *Communication Skills:* Senior- and executive-level professionals are expected to have excellent verbal skills, since this competency is a primary determinant for moving up the corporate ladder. However, writing skills can be a major challenge to those who depend on others to put pen to paper, especially since consulting and construction projects often require some form of written report. Publishing quality articles that attract the attention of potential clients and the industry is another, cost-effective approach to spreading the word. However, face-to-face contact remains the most effective form of communicating and potentially gives the best returns. The downside is that it is time-consuming and can be expensive.
- *Hard Work:* The chances of a new business succeeding without putting in the hours and the effort are virtually zero. There is obviously some flexibility in work hours when you become your own boss, but this is no eight-to-five job and hard work and effort will definitely be needed to build a business. Sometimes a startup business is lucky and immediately falls into client work and so becomes complacent. Others may find themselves straddling the fence and may have not made a wholehearted commitment to the business and continue to look for a suitable position. Not being fully committed prevents you from aggressively building a presence, marketing your firm's services, and obtaining a web domain and building a website. Not feeling fully committed costs you money and ultimately reflects badly on you.
- *Marketing Skills:* It is necessary to both identify your target market and develop a detailed strategy that gives you a competitive edge and draws customers to you and your company rather than to the competition. To succeed

you must be willing to engage in relentless self-promotion to bring in needed new business. Seek out specific target markets that need your services and are willing to pay for it. Also develop a list of the main competition in your field and do an honest appraisal of its strengths and weaknesses; then contemplate how you can successfully compete with the available resources.

- *Grassroots:* This is another affordable method of marketing. It consists of taking advantage of available resources to spread the word about your service and entails distributing your marketing material at local businesses, churches, chambers of commerce, and community centers. It also includes networking to connect with potential customers and strategic partners. Joining a chamber is important because it can facilitate building your network and provide an ecosystem portal to members who may be looking for business; it can also be sources of services.

- *Financial Security:* Being financial secure is key to succeeding at being your own boss in a new business. You must have the ability to survive the dry, difficult periods that could easily last a year or more. If survival seems difficult under such circumstances, it may be prudent to reconsider the decision to be an independent business person. If, however, you feel exceptionally strong in one area, such as building systems, but are very poor in, say, marketing, it may be possible and wise to team up with others who can compensate for these weaknesses and help you succeed.

- *People Skills:* Dictating orders to employees within the firm is not the same as dealing with clients—this is much more complicated and takes skill. For example, contractors have to respond to a multitude of personalities with little or no background information on clients' likes and dislikes. Bullying techniques and intimidation, which some bosses seem to thrive on inside companies, fail to get much of a welcoming response from clients and potential customers. Moreover, independent contractors may find themselves quickly dropped if their performance is not up to "snuff." While possessing great people skills is a great asset and may bring in the work, it will not necessarily help you retain it or get repeat business; this can only be achieved by hard, persistent effort.

- *Self-Direction:* Some people are unable to work on their own initiative and experience great difficulty in performing in an unstructured environment. Independence can certainly be freeing and exhilarating, but it can also be intimidating and lonely without daily, face-to-face interaction, especially if such individuals have never been their own boss. This is especially true of those who work out of a home office instead of in rented space.

15.8   FORMS, LICENSES AND PERMITS, INSURANCE, AND BANKING

Having made the decision to start a new business, it has become your responsibility to understand and comply with government laws and regulations that apply to it. Prior to incorporating and registering a new firm, there are several bureaucratic and legal hurdles that must be overcome. These laws are designed to protect you, your customer, and your employees. You may now be required to obtain a number of licenses and permits from federal, state, and local governments before you can open for business. Licensing and permit requirements for small businesses vary from one jurisdiction to another, so you will need to contact your state and local governments to determine which permits, licenses, and other specific obligations are required. Before doing so, however, a decision must be made on the proposed business name and its legal structure.

15.8.1   Name and Legal Structure

It is always wise to consult your attorney and accountant before deciding on a legal structure because there is no universally "right" structure for all businesses. Choosing the right one depends on your specific needs. Since there are advantages and disadvantages to each type of business structure, it is important to understand the various options available before setting up your company. There are four forms of business ownership to choose from: (1) Sole Proprietorship, (2) Partnership (general or limited), (3) Limited Liability Company (LLC), and (4) Corporation or S corporation.

Many new small-business owners seem to prefer Sole Proprietorship, perhaps because it is the least complicated and simplest form of business organization to set up. An individual proprietorship is basically owned and operated by one person, and apart from local business licenses there are minimal government fees and paperwork. On the other hand, there are also considerable risks that need to be considered (e.g., the vulnerability to creditors of your personal assets and other liabilities such as lawsuits). In addition, you may not be able to take advantage of certain tax breaks reserved for more formal business structures such as corporations or limited liability companies. Moreover, as a Sole Proprietorship, your company name is not protected, which means that there is nothing to prevent another company from incorporating under it. This is why it is wise to work closely with an attorney and avoid many of the potential pitfalls and challenges that setting up a new business may face.

Partnerships and Sole Proprietorships are similar in that they are easy to set up and maintain and require no government fees or annual state paperwork. This may also be the way to go if you feel you need additional capital or expertise. A disadvantage with a Partnership entity is that you and your partners are each held fully responsible for all company debts. Thus, if any of the partners defaults on a company loan, creditors can still go after you personally to satisfy the entire debt. This includes your personal bank accounts, property holdings, and other assets. Furthermore, just as in a Sole Proprietorship, your company name is not protected and can be used by any other business.

The standard for many of today's businesses is incorporating, largely because it shields you and the company from personal liability. Creditors are prevented from going after personal assets to make up for any company short-falls should your business hit hard times. In addition, the corporate business structure offers significant tax savings, company name protection, and increased opportunities for raising capital. If you decide to incorporate, you need to choose a C corp. or S corp. setup to take advantage of the various tax options available (consult with an accountant). However, unlike a Sole Proprietorship, corporations require some initial setup fees and perhaps a certain amount of regular maintenance.

Setting up as a C corporation only makes sense if you have a significant amount of startup capital and feel ready for the big time, or if you are wanting to sell shares of stock in your business. This is unlikely to apply to the vast majority of startups. A good alternative is the S corporation, which avoids the double taxation of a C corporation. This form provides a tax-efficient way to structure your business if you expect losses in the short term, allowing individual shareholders to report losses on their tax returns rather than pay the C corporation's double taxation. Before making a final determination, consult with an attorney and an accountant and check with your Secretary of State (most states are now online). It should be noted that running your business as a corporation has some serious disadvantages, especially for new small businesses, including strict laws and higher state income taxes in some states; these are in addition to increased legal work and heavier accounting and tax reporting requirements. Moreover, closing down a corporation is often difficult.

For many new entrepreneurs, choosing a business structure basically comes down to liability protection, tax savings, and convenience. This is why many startups today prefer forming a Limited Liability Company since this type of entity requires fewer formalities and less ongoing paperwork than a corporation, while maintaining the same personal liability protection and tax flexibility. Just as with a corporation, the business name is protected and the company is shielded from creditors and other liabilities such as lawsuits. Likewise, with an LLC minimal company records are required to be kept. Many professionals consider the LLC to combine the best aspects of incorporation with the tax advantages of Partnership while omitting much of the red tape that accompanies both. In the end, only you can decide which form will serve you best.

15.8.2 Federal Employer Identification Number

An Employer Identification Number (EIN) is a unique nine-digit number assigned by the IRS to business entities operating in the United States for the purposes of identification. The EIN is required for almost all types of businesses and acts as your business identifier on all types of registrations and documents; most banks will not let you set up a business checking account or apply for a loan without it. Apply for an EIN as soon as possible (if a corporation, after receiving your corporate charter). You can do so by going to the IRS and downloading Form SS-4. Once this is filled out, call (866) 816-2065 for your EIN. Once you get the number, download and fill out Form 2553 (S-election) if you want to avoid double taxation on your company's earnings.

You are strongly advised to mail this form via certified mail, return receipt requested, because the IRS frequently tends to misplace this important tax election and the burden of proof is solely on you to prove you sent it within the appropriate time. If the number is used for identification rather than employment tax reporting, it is usually called a Tax payer Identification Number (TIN). Businesses considered proprietorships do not need an EIN; in this case the owner/operator's Social Security Number (SSN) is used on tax documents. Should you chose to form an LLC, you will need to decide how you prefer to be taxed (e.g., as a Sole Proprietorship, Partnership, S corporation, or C corporation) and use IRS Form 8832 to document your decision.

Whatever your business, you will need a fictitious business name permit, also called "dba" or "doing business as" permit. In choosing a business name, it is generally good practice to choose one that best describes your product or service to make the public better aware of just what the firm has to offer. Apply for a fictitious business name with your state or county offices if you plan on going into business under a name other than your own. Banks will also require a certificate or resolution pertaining to your fictitious name at the time you apply for a company bank account.

15.8.3 Licenses and Permits

Now that you have made the decision to start a new business, the next step is to obtain a number of licenses and permits from federal, state, and local governments. A contractor must have the appropriate license prior to entering into and performing a construction contract. The purpose is to regulate the industry for the protection of the public, including homeowners, commercial project owners, and even public project owners. Licensing and permitting requirements for small businesses may vary from one jurisdiction to another, so you must contact your state and local governments to determine any specific requirements prior to setting up a new business. Keeping this in mind, the following sections describe some of the different federal, state, and local licenses and permits that may be required prior to opening for business. Note that the impact of performing construction work without a proper license varies from one state to another, but in many states an unlicensed contractor cannot enforce a construction contract against a project owner based on statute and case law. However, the expiration of a previously valid license is a common exception to this.

Business Operation License

A business operation license grants a company the authority to do business within a city and/or county and can be obtained from the city (for a fee) in which the business will be operating or from the local county if the business is located outside city limits. A business license is required by most cities or counties, even when operating from home. If you plan to initially run the business from home, you should first carefully investigate the zoning ordinances in your area. Some residential neighborhoods have strict regulations that prohibit use of a home for business purposes.

After you file a license application, the city planning or zoning department will check to ensure that the location is zoned for the intended purpose and that there are sufficient parking spaces to meet code requirements. If the area is not zoned for your type of business, a variance or conditional-use permit will be needed before permission to operate is granted. This normally can be achieved by presenting your case before the city's planning commission. Getting a variance is usually quite straightforward as long as you can show the commission that your business in its proposed location will not adversely impact the neighborhood. However, in many areas, attitudes toward doing business from home are gradually changing and becoming more supportive, making it less difficult to obtain a variance for a home-based business.

Occupational Licenses

It is often easy to overlook the need for certain licenses and permits prior to opening for business. Many types of new businesses will require an occupational license through state or local licensing agencies. Such businesses include real estate brokers, building contractors, those in the engineering profession, electricians, plumbers, insurance agents, and many others. Moreover, in many states and jurisdictions, occupational licenses will not be granted to conduct business unless relevant state examinations are passed. Your state government can be contacted for a list of occupations that require licensing and passing of exams, or you can check on the Internet.

Signage Permit

Numerous cities and jurisdictions have sign ordinances that restrict the size, location, and sometimes the lighting and type of sign that can be installed outside a business. To avoid costly mistakes, regulations need to be checked to see whether any signage restrictions are imposed by your city or county. Written approval of the landlord (if renting a house or apartment) should be secured before going to the expense of having a sign designed and installed.

Other Licenses and Permits

Many kinds of interstate activities regarding license and permit requirements are controlled by federal regulations, but in most cases this is not a cause for concern. However, a few types of businesses do require federal licensing, including investment advisory services. Check with the Federal Trade Commission website to see whether your business requires a federal license. The same types of permits and licenses required by cities are typically required by counties. If your business is outside the city or town's jurisdiction, these permits will apply to you. County regulations are generally not as strict as those of adjoining cities. Localities may have individual variations, or they may require additional permits or licenses (e.g., zoning, fire, alarm), so both the city and county need to be contacted once you have your basic business information, business address, and tax ID number.

15.8.4   Insurance

The importance of having adequate insurance cannot be overstated, especially for general contractors and professionals. This is discussed in greater detail in Chapter 16 (Litigation and Liability Issues). Premiums are usually high, especially for business liability, but no general contractor or consultant can operate with peace of mind without full coverage. There are many types of insurance for businesses but these are usually packaged as "General Business Insurance" or a "Business Owner's Policy." A good insurance agent can be helpful as to the types of insurance you will need and the coverage available (e.g., general liability, health, fire, property, burglary, company vehicles, worker's compensation, business interruption, malpractice). It is advisable to seek estimates from two or three different agents. It is also imperative to have adequate liability insurance, and anyone contemplating offering contracting services is strongly advised to consult with an attorney. If you have employees and plan to offer them health insurance, talk to your agent about the upfront fee and record the premium payment you will need to make before opening your business. Health insurance costs are among the most important concerns facing small-business owners today.

15.8.5   Banking

You will not be able to open a business account for your firm without a valid tax ID number (TIN). Once you have this, you can use it to open a business checking account. Find a bank that is convenient and where you feel comfortable with the bank manager. She or he can be one of your best references as well as a source for advice and help on financial matters. Even if you are not interested in qualifying for a loan yet, banks can provide numerous other financial services fundamental to your business, including business checking accounts, business credit cards, and even a credit reference. Added to this, banks have great contacts in the community and can be an excellent source of business leads; this is why having a good relationship with your bank manager is of paramount importance. It may also be useful to develop a line of credit so that it will be there should you need it further down the road. Make sure you maintain business and personal finances separately from the beginning to avoid commingling and thereby complicating bookkeeping and tax returns.

Likewise, you cannot establish a bank account without a Federal ID number or SSN along with your Certificate of Assumed (fictitious) Business Name. If you are incorporated, you may be required to provide a copy of the minutes and a corporate resolution authorizing the account. It is better to visit the bank you wish to open an account with and discuss with the bank manager what the specific requirements are for a business checking account and to see whether you feel comfortable with him or her. Requirements vary from bank to bank; some are fairly simple while others are extremely complex. The more important issue here is establishing a rapport and empathy early on with the bank manager and making sure that she or he understands your potential needs and is willing to give you a bank reference and other things you may need.

15.9   TAXES, STRATEGIES, AND INCENTIVES

The ultimate goal of a business tax plan is to minimize your business' tax bill. This makes it an important ingredient for a successful business. Whether it is capitalizing on business deductions, Section 179 depreciation, home office write-offs for the self-employed, tax deductions for business vehicles, business travel, rental property depreciation, or finding other tax-friendly ways to run your business, a good accountant is indispensible to get the best results in applying small-business tax deduction strategies. A necessary requirement for all business startups is to submit applications for federal and state ID numbers and to request "Business Start Up" application forms from the IRS and from the State Tax Commission. After these are sent in, you will be notified of your number and get a packet of information. Following this, you will periodically receive depository forms, quarterly report forms, W-2 s, W-4-As, estimated tax forms, and other relevant material.

Depending on the type of entity you form and the size of your new business, you will encounter various payroll expenses for taxes—that is, FICA (Social Security), SUE (state unemployment), FUE (federal unemployment)—and Workers' Compensation (WC) and/or State Disability Insurance (SDI). If you are a Sole Proprietor or a Partnership, you will be required to file and pay federal estimated tax reports each quarter based on estimated annual income. Partnerships file an annual information return, and each partner's share of profits is included in their individual personal income tax return. Corporations are also required to file returns for estimated taxes. Your accountant is the person who should be taking care of this.

15.9.1 Tax Deductions and Write-Offs

Maximize what you can deduct according to Section 179 of the IRS Tax Code and discover what you can write off by knowing what constitutes a legitimate business expense. For this, a proficient accountant will be needed to prepare your returns. Tax consultant David Wetzel says, "Proper planning will result in you getting all the deductions you deserve. Poor planning raises a red flag with the IRS. Sloppy looking returns and indications of poor recordkeeping will earn you a trip to see your friendly IRS agent." Here are some possible tax deductions, but they may change from time to time:

- Rent typically can be deducted for a rented office as a business expense. With a home-based office, the business must be located in a separate room within the home. Ideally, it is located in a walk-out basement with a separate entrance. To claim office-in-home expenses, you need to calculate the square foot percentage of your home office in relation to the total square footage; then apply this percentage for deductions for utilities, mortgage or rent, insurance, Internet service, and so on, to arrive at the final deduction.
- Utilities include water, electricity, and phones; they normally can be deducted for outside rented offices. With home-office settings, it is better to install a second phone line for your business. This is the safest approach for taking phone deductions on your business taxes. Check with your accountant for maximum deductions.
- Furniture purchases can be deducted, but amounts will vary; reportedly, you can now deduct 100% of all office furniture costs without having to depreciate them over several years. Check with your accountant.
- All supplies purchased for your office can be deducted. It is important to keep receipts.
- Website building and maintenance for a business can be written off as a business expense.
- Computer equipment, either new or recently purchased, can be deducted 100% without having to depreciate it.
- Computer software purchased for business use is 100% deductible.
- Business travel and other expenses can be claimed based on the actual mileage the vehicle is used for business. Check with your accountant to ascertain the mileage rate at the time.
- Entertainment expenses that are legitimately for client entertaining, such as business lunches, can be recouped. Check with your accountant.
- Insurance that small-business owners carry can generally be deducted at 100% of the premiums, providing they do not exceed your business' net profits. Check with your accountant regarding requirements and stipulations.

- Travel, where the "primary reason" for a trip is business-related by a sole proprietor, partner, or LLC member, can be written off; this is for all transportation costs within the United States. However, no deductions can be made where the primary motivation is not business-related (i.e., for a vacation). Nevertheless, by mixing a few vacation days into your business trip, you can legally deduct all your transportation costs, cab fare, and so on.

15.10   THE INTERNET AND ONLINE MARKETING

One of the Internet's greatest attributes is that it has leveled the playing field when it comes to competing with large corporations. Not so long ago, having a website for your new business was considered a luxury, whereas today it has become an absolute necessity, so much so that few businesses can thrive without an online component. One reason is that the Internet has created enormous marketing opportunities to reach previously unimaginable numbers of people around the globe. Your site is accessible to people who might not otherwise have access to your business, and if it is well designed, a small business can project the image and professionalism of a much larger company. Having a business website has therefore become a high priority because it not only is a great marketing tool but also allows you to develop your green building services and enables rapidly launching of successive marketing campaigns.

The importance of a website should not be underestimated; it is a specialized tool that lets you reach many new clients. Even with a small, newly created operation, when it comes to benefiting from a website, size is almost irrelevant. Whether your firm is a one-person entity or a 1000-employee corporation, without a website you are merely losing business to other companies that have one. It is necessary to update the site regularly with additional information about your company, its projects, and what is happening in your field, and to make modifications as needed.

In today's competitive world, the Internet is considered one of the best ways to generate high-quality new business opportunities, mainly as a result of its ease of use, speed, ability to target the audience you seek, and affordability; plus, it does not sleep. There is, however, a steep learning curve to successful Internet marketing at both strategic and tactical levels; what works today may not work tomorrow, taking into account the very nature of the Internet marketplace. Also, taking full advantage of the Internet means more than just creating a website and waiting for potential clients to find it—it must be part of an overall Internet marketing strategy. Property developers or persons in search of a green building contractor should be able to quickly find your photo and background information on the site. At a minimum, list previous jobs and relevant career highlights. A well-thought-out strategy will help guide all the other decisions you will make over the months and years ahead.

The new company's website should be viewed as a platform that features your available services to clients and potential customers around the world. But, even before setting one up, it is imperative for you to have a service provider. This is a high priority for general communication and for sending promotional material; in fact, most clients consider email availability vital and find it burdensome and inefficient to have to communicate everything by any other means. When choosing an email address, it should be simple and professional, and preferably reflect the domain name after the website is established and the domain name registered. The firm's email signature should be provided, including complete contact information and an active link to the website. Visitors to a site often prefer to make the first contact via email, either because they prefer keeping it impersonal or because it is easier for them to articulate what they are looking for quickly. This is why a "Contact Us" button/page and/or the footer on the home page are typical locations for listing corporate mailing and email addresses and other relevant contact information.

15.10.1 Importance of a Website in a Competitive World

The public expects most businesses and organizations to have a website—one that can supply, at the very least, basic information about the services offered and a means of getting in contact. In fact, today a website has become nearly as essential as a telephone, fax machine, or printed brochure—maybe even more so. Not having one of these tools can put your new company at a serious disadvantage. Furthermore, having a website has many advantages, such as marketing your company's services to the world. It is important to consider what information you want potential customers to gather when visiting the site. For example, a website can inform consumers and end users about the benefits of green contracting and how it can add to their bottom line. Well-designed websites usually serve a number of functions, including:

- Providing important information about the firm and its services
- Leading to answers about any outstanding issues customers may have
- Providing downloadable files such as brochures, research findings, templates, and other relevant information
- Making it easy for visitors to contact you
- Steering inquiries from potential customers to relevant links
- Motivating users to visit the site and come back for more information
- Selling construction services and products (e.g., green products) online
- Providing clients and customers with more efficient service
- Helping to recruit staff
- Enabling customers to provide feedback

Having a well-designed company website can be a great asset and brings great benefits to a firm by providing clients and users with better access to its services.

In addition, it can help in resolving clients' issues quickly and satisfactorily, allowing you and your staff to focus more time and energy on other pressing tasks. The list of potential services that can be offered via a website is quite substantial. Yet builders and professionals sometimes fail to comprehend a site's proper function and therefore fail to make appropriate decisions regarding content and form.

Taking a current "green" website as an example, the one you create should include green construction projects that the firm has executed, green building costs, and other facts and figures, as well as any awards that the firm has received and contact information. The firm's phone number should be displayed prominently on the home page. It can also include information about the firm's mission, structure, and responsibilities. In addition, a good website can be used to draw attention to upcoming events such as green industry conferences, new products, and other time-sensitive information. Whatever it contains, however, the website should concentrate on appealing to your target audience—that is, clients and potential customers.

In many cases, general contractors use their websites to communicate with subcontractors, consultants, clients, and other project team members to explain or ask questions regarding bidding guidelines, building schedules, variation orders, and so on. Furthermore, they may provide downloadable forms, building fact sheets, and equipment procedures for field workers, subcontractors, and/or manufacturers who can access them remotely from an offsite location. Confidential information can be password-protected, so only authorized individuals can access it by logging in.

15.10.2   Planning a Website

Perhaps the first clarification required to proceed with building a website is to decide whether to hire a professional to design it or whether to do it yourself because you are on a tight budget and feel sufficiently competent. Whatever you decide, remember that the website is an important means of communication with your target audience; thus, one of the first steps in planning the site must be to decide on content. This should be about what customers need or want, not about the company itself. Ben Seigel, principal of the web design firm Versa Studio, says,

Planning is essential for most businesses and organizations. In practice, many people fail to plan their websites. Sometimes the ever-busy, dynamic nature of running a business is to blame; there are so many operational demands that proper time is not allotted to projects. But this often happens because people fail to recognize that planning for the Web is just as important as planning for anything else in a business.

It is important, therefore, to contemplate what potential site visitors will want to know or see when logging in, making sure that the image you want the site to convey to visitors actually contributes to that end. Furthermore, the website should be organized so that prospects can easily identify the firm's areas of expertise. In this phase, you really need to research your market and determine the main competitors and their main strengths and weaknesses. Periodic searches of competitors' websites can be helpful and should include other general contractors and green building websites; the types of services being offered to visitors, and how they are marketing themselves online. It may also be prudent for a new company starting from scratch to conduct a quick survey to learn precisely what services may be required. To get started on building a website for your new business, a few requirements need to be satisfied:

- Deciding on a domain name (URL, uniform resource locator; web address) and registering it
- Deciding on a web-hosting service
- Finding appropriate web authoring software or services to design the website

Registering a Domain Name and Setting Up the Website

Prior to setting up a commercial website (as opposed to a personal one), a name for the company must be decided on. Once this is determined, a search needs to be conducted to see whether that name is available—that is, not taken by someone else. Numerous companies let you register a domain name online using a search tool to ensure that the name chosen is not reserved or already in use. Once the name's availability is confirmed, you can register it immediately (for a small fee) and formally make it your web address. The extension ".com," is the most common, followed by ".net," ".us," ".gov," and others because the majority of users are familiar with these. A domain is essentially the name and address of your website, all in one. For example, if your company name is ABC Green Contracting, your web address may be *www.ABCgreencontracting. com*. Domain registration is inexpensive, and the name can usually be registered with the same hosting company that provides the webspace (i.e., space on a computer owned by a hosting company). It is usually best to choose a domain name that is simple and easy to remember, and that best describes the business. Choosing a good domain name is important because it provides extra branding for your site and makes it easier for people to remember the URL.

Web Hosting

It is not surprising that most startup companies do not want to own or invest in a server—a powerful computer that is always online has the capacity to store an entire firm's website files, as well as the content and operations of its network. It is therefore necessary to find and hire a knowledgeable web host for this purpose. Numerous hosting companies can be found online, but some research is needed to find the one that will best serve your needs. A host generally will accept your site into its computers and securely store your files and data, ensuring that they will be available to you and your customers every day, 24/7.

The space provided on the web for your firm by the hosting company is set up so that whenever someone types your firm's domain name into the browser, she or he is automatically connected to your website. The prudent thing to do is to spend some time online researching topics (e.g., "domain hosts") to find a host that meets your needs and budget. Of note, green web hosting providers are popping up more frequently as the eco-movement becomes widespread. Here are some possible domain hosts to start with: www.NetworkSolutions.com, www.greengeeks.com, www.fatcow.com, www.justhost.com, www.super greenhosting.com, www.GoDaddy.com, and www.greenteahosting.com.

After establishing an account with a hosting service, instructions will be provided on how to upload your website onto its server. This is normally achieved using File Transfer Protocol (FTP) utility software. Many hosting services also have file upload options on the control panel you use to manage your site. The hosting service chosen should preferably allow you to build your site online using drag-and-drop, fill-in-the-blank templates, which are simple to apply and typically do not require much technical knowledge. This type of service is offered by Yahoo! Small Business, StartLogic, and ValueWeb, among others.

15.10.3   Content and Components

The fact that the Internet is constantly evolving requires the administrator to constantly monitor it and to regularly refresh, update, and add new content. This helps increase visibility in the search engines and gives customers a reason to continuously come back to your site.

Basic Content

The website's principal function should be attracting visitors and converting them into clients using various means, including the articulation of services and/or products offered. This is often stated in a mission or vision statement. Since it is unlikely that many, if any, Internet users actually know you, it is extremely important that what you say on your website captures the visitor's attention, establishes credibility, provokes interest, and motivates her or him to action. This presents you with the opportunity to project the kind of image desired, in addition to being able to highlight any particular aspect of the organization you want noticed. The most important component here is obviously the target visitor, whose special needs you must cater to, and who should be taken into consideration and included in website planning and design.

It is obvious that no customer should ever have to work to get the information he or she is looking for. Therefore, use the minimum needed to enhance your central message and to tell it simply and clearly in an attractive format.

It is also important to ensure that the corporate image you wish to portray matches that of the image displayed in other formats and media. Once a theme is decided on, it must be consistent and you must stick to it. In addition, unless you are considering hiring a professional designer to manage all photos and other graphics on your site, you will need a graphics program to do it yourself (except, of course, if the hosting company has its own tools). However, basic digital photo and graphics editors can be found online for free, although sophisticated top-end programs (e.g., Adobe Photoshop) can run into hundreds of dollars. The editing software chosen should be capable of resizing and cropping images; resolving color and contrast issues; setting resolution to control how sharp the images are; and saving the images using appropriate color modes and formats. Check out the many graphics programs on the market before making a decision on which to purchase.

In today's business environment, creating a company website is not terribly difficult and a basic one can be built in several hours. For beginners, a simple method for building your own site is to download a template that attracts you (using "Save As" on your browser), then edit it with an editing program. The Internet is full of suitable templates so that you are not simply imitating another site.

It should be emphasized here that such imitation is illegal; however, a free online template is acceptable, which is simple and easily downloaded. Once this is done, explore your options for upgrading and customization. Your favorite website builder can be used to complete the design and customization, after which you can immediately publish to the web. There are several popular web design software packages that can be used (e.g., Adobe Dreamweaver; Microsoft Expression Web, which replaced FrontPage; and NetObjects Fusion 11). However, before making a final decision about which design software to get, make sure it has the ability to design both in HTML and drag-and-drop. It would be wise to do a little research on the software packages you are considering before buying.

During the process of building your website, check for bugs in the system such as spelling and grammar mistakes, which will reflect negatively on the site owner and its administrator. This should be done one page at a time and in a thorough manner. Once a page is error-free and everything is working as intended, move on to the next page; repeat the process until the website has been checked completely. Also, when looking for bugs, ensure that all the links work–that is, they take users where they are supposed to go. Additionally, page navigation should be kept consistent across the site, and generally all pages should be printable.

Figure 15.6 shows two online template designs that have been edited. In Figure 15.6(a), the contracting buttons' links are on the left side of the page whereas in the architecture template they are at the top of the page. Both easily navigate to where you want to go (e.g., to About Us, Projects, Services, Firm Principals and Qualifications). The final layout depends largely on personal

FIGURE 15.6   (a) Edited contracting home page template, available free from www.free websitetemplates.com, showing the link buttons on the *left* of the page. (b) Architecture home page template, from www.free websitetemplates.com, showing the link buttons on the *upper* part of the page.

preference as to how the page is composed and designed. However, for a professional-looking website, it is almost always best to use a design professional. The only exception may be when the website owner is proficient in web design and has significant graphic design background.

Adding New Content

As the firm grows and develops, and with emerging opportunities and challenges, your website will need to keep pace with new content, which may require

- Adding profiles of new senior staff to the "About Us" page
- Adding new service lines, awards, and anything else of note to the home page with appropriate links
- Adding photos and text regarding newly completed projects (e.g., "Gallery of Completed Projects")
- Testing for bugs and ensuring that affected pages are optimized for peak performance whenever new content is added website
- Updating all links and the site map after changes

15.10.4   Pinpoint the Audience and Determine Interests

Pinpointing a target audience depends very much on the research you have conducted and the answers you have come up with. The primary purpose here is to know potential clients so well that you answer any questions they might have before they ask, then make it easy for them to "buy" the concepts you are selling. Clients include existing customers, potential customers, people interested in your specialty area (who may never have heard of you), organizations, individuals, different groups, and so on. If you are a green builder/contractor, your targeted groups might look a little like this:

- Property owners and facility managers
- Lenders
- Investors
- Designers, architects, and engineers
- Subcontractors
- Green product manufacturers

Take into account the need to prioritize. Your potential and existing clients need to learn why you are qualified to do what you do, and why your firm can offer a better service than the competition. In fact, your new website will be designed to sell the company's services as no one else can. Thus, if your objective is to promote an image of a sustainable contractor specializing in green building, building owners and investors may take a higher priority than general users. This should be reflected in the structure of your website and in the weighting that you give to each aspect in site navigation.

Potential clients must know that you are to be trusted. The majority of viewers may be unfamiliar with your company, which is why it is so important to continuously hammer this message home. There are a number of ways to reinforce your message (e.g., company logo, university crest). If you have projects that have won an award, make sure everyone who visits the site is aware of this. It is advisable to place the company logo on every page, in a consistent location and in the same size, so that whichever page a user visits first, the logo will be proudly displayed.

Converting Visitors to Clients

You must understand your audience so that you can tailor your website to suit their preferences. It is also vital that you articulate what your site is for; if you do not know, you cannot expect that viewers will. Also, the time available to make a serious impression on visitors is very limited (probably 15 seconds or less) before they are gone. As the window of opportunity to impress and sell your services is quite small, avoid the use of huge flashy graphics that take forever to load because most visitors do not have the patience to wait for them to appear and will therefore flee. Likewise, easy and unobstructed navigation is essential to a successful design. If the path you lay out for potential customers is twisted and difficult to follow, they will get lost and you may end up losing them. A successful site, therefore, is not necessarily the most attractive one (although it helps), nor is it one consisting of the latest web technology. It does not rely on the total number of site visitors, but rather relies on how many visitors return and are converted to clients and how much business it can generate. Implementing the preceding suggestions will certainly help push your website to the top of visitors' lists of those to visit when they want information about your topic.

A successful website needs to be continuously refreshed and periodically updated. If it is not continuously updated, or carries out-of-date information, it reflects negatively on the firm. Decide on who is to be the Webmaster—the person who will be responsible for updating the site, introducing new content, monitoring security (e.g., creating firewalls), and so on. The real key to long-term business success on the Internet is continuous maintenance of your website to meet the needs of clients and potential customers.

Company Blogs

Blogs have become quite popular in recent years and are usually created to enhance a business website and drive more traffic to it by bolstering its credibility. They are often used to report company and industry events, comment on relevant news stories, and let people know when new products or services are expected. A blog may also attract professionals inside your industry and possibly calls from the press. It can be housed on- or offsite, with its own URL. If housed with its own URL, care should be taken not to distract from the main website.

Attracting Traffic

One of the necessary ingredients for success is ensuring that potential customers know not only that the website exists but also that they are able to access it without difficulty. Your business may offer the best products or services on the web, but it does not mean a thing if potential customers cannot find them. Additionally, the website and email address should be clearly mentioned on all letterheads, brochures, cards, and advertising material. Regular promotional campaigns and strategies are helpful in driving traffic to your website. They are limited only by your creativity, imagination, and marketing strategy.

Registering your site with as many search engines as possible, but especially with the leading search engines (e.g., Google, Yahoo!, Bing, and MSN) is an excellent way to get your site noticed and will help bring your company to the fore whenever users ask these search engines to scan the Internet for your kind of services. This is necessary for the website to succeed, even though it may appear at times to be a very challenging task. It typically takes at least a few months for a website to generate responses and become recognized. This is the time it normally takes the big search engines to index a new one. However, the popularity of a site, and the speed at which it becomes recognized, really depends on how it is promoted and the services it offers. In addition, it is helpful to register with online directories, such as the Green*Spec*® Directory, to promote your firm. This is another excellent way to increase online recognition and visibility and drive qualified prospects to your website. It should be noted that many contractors and green associations maintain online service directories.

Search engines evaluate your site basically by using what is known as a "web crawler" that reads the "meta tags" in the header of your HTML pages. This is why it is very helpful to include a title and description tag on your site, as well as appropriate keywords. One of the concepts behind a high search engine ranking is "keyword density," which basically means that the web pages should include key terms that drive searches. Useful "buzzwords" for green contractors may include "green," "green builder," "sustainable," "contractor," and other key industry terms. Being able to drive visitors to your website is not the main objective. The real objective is to be able to convert site visitors to clients by making it easy for new prospects to learn about your services and by encouraging them to contact you.

15.10.5   Website Security

Security is considered one of the most important concerns with any website. The Webmaster needs to be vigilant about the security of your site's content, your network, and your customers' private information by creating firewalls, installing the latest security software programs, and so on. There are many security software programs on the market (e.g., CA Internet Security, Norton™ 360, Norton Internet Security, McAfee). It is also important that whoever is responsible for keeping the site up to date be able to write and update the pages as needed while keeping abreast of new technologies to keep content fresh and interesting.

Last, but perhaps not least, the website should adopt a "best practices" approach. Successful business websites are based on doing simple things well, and each of the objectives set out here must be clearly defined. Although, obviously, other things need thinking through when developing your website, many of these relate more to personal objectives than anything else.

Chapter 16

Building Green Litigation and Liability Issues

16.1 INTRODUCTION

The majority of liability scenarios in green building and sustainable construction do not differ from those in regular building and are therefore already addressed by existing laws, which courts can apply to green project defects. Nevertheless, green design strategies and systems can present unique challenges in possible risk and liability that need to be well understood to minimize the potential for failure or underperformance, and thus the litigation that may result. We have witnessed a sharp increase in litigation, much of it related to green buildings, due partly to the lack of proper preparation for risk and liability, causing developers to face unforeseen problems in the form of job site interruptions, negative bottom-line impacts, and red tape. This increasing litigation is forcing the industry to change in a manner that actually has the potential to improve the quality of construction projects.

The dramatic rise in litigation related to building has also given an increased urgency to finding lawyers, forensic experts, and consultants that specialize in green building and construction. Also, because most green building laws involve some form of green incentive, there exists an inherent potential liability for design professionals (architects, engineers, etc.), contractors, and other project participants should the owner not achieve the promised "green" certification. This newly emerging risk of green liability for design professionals when a Leadership in Energy and Environmental Design (LEED®) project fails to reach the proposed level of certification is very concerning, and perhaps more so because it often remains uncovered in the standard language of errors and omissions insurance, which excludes guarantees.

Additionally, insufficient precedence (although this is changing) means there remains a lack of relevant data on this issue and the associated risks it implies. This has resulted in some insurers' resistance to moving quickly to provide a professional liability insurance that deals with the unique issue of building green. It is therefore not surprising that lawsuits are becoming a major problem to all concerned, although, partly because of the skyrocketing costs of traditional litigation, many construction disputes tend to be settled before they go to court.

In spite of more than a decade of research and litigation and billions of dollars in insurance claims and lost productivity, issues such as sick building syndrome (SBS), building-related illness (BRI), and indoor air quality (IAQ) remain prevalent in the commercial buildings sector. Most of these efforts have been reactive—that is, a problem is reported and its cause identified, and then it is fixed. While this may be the least expensive, and perhaps most practical, solution, it is failing to halt the rising tide of costly, debilitating, and otherwise avertable solutions that owners and industry are experiencing.

Mold-related claims have dramatically increased in recent years, and developers and contractors are increasingly being held responsible for damages as a result of mold. Ted Bumgardner, vice president of the San Diego-based construction consultant Gafcon Inc., says, "Mold has grown into a big business." Building occupants are suing property owners, owners are suing contractors, and contractors are suing design consultants and product manufacturers. Many insurance companies are struggling to find ways to address mold claims and, in most cases, are now entirely excluding mold from their coverage. The insurance industry is further responding by changing policy language, claims-handling procedures, and loss reserving, while continuing to try to manage the regulators. Some architects and engineers, in an effort to reduce their professional liability, have also started to react to this problem and have eliminated the term "supervision" from their contractual responsibility during construction, replacing it with the term "observation" or "inspection" to more correctly describe their services.

Green building projects are generally required to adhere to zoning and building code requirements. Projects that fail to do so can adversely impact the consultant and expose him or her to multimillion-dollar litigation. Many U.S. cities now require implementation of a green building standard through municipal zoning. Boston is the first city in the nation to stipulate this with a municipal zoning code that requires all large-scale projects to meet LEED certification standards.

The construction industry generally has had to deal with higher premiums for all types of insurance, but, following the World Trade Center attacks on 9/11, those costs have skyrocketed beyond all expectations. Legal claims for all types of building envelope failures continue to rise and are typically made against developers, contractors, property management corporations, architects, engineers, building trades, and government authorities among others.

Recent surveys clearly indicate a growing urgency to make buildings more eco-friendly and energy efficient; in fact, significant progress has been achieved with the help of government incentives. Although the green building movement is having a very significant and positive impact on the construction industry as a whole, aspects such as risk of liability and "greenwashing"—misleading environmental benefits of a product or service—are causing considerable disquiet and need to be addressed. But while green buildings are generally more efficient users of energy and materials, resulting in reduced safety factors for the different building systems, they nevertheless sometimes incorporate nonstandard materials and systems that are subject to increased risk of failure. To minimize these risks, qualified and experienced design consultants should be employed to ensure that the design process is correctly implemented. Moreover, the advent of building information modeling (BIM) technology promises to be a great asset in helping to improve the overall design of systems for construction projects and thus in lessening potential risks.

In this respect, Ward Hubbell, president of the Green Building Initiative (GBI), says,

One of our most pressing issues is the fact that some buildings designed to be green fail to live up to expectations. And in business, as we all know, where there are failed expectations there are lawsuits. All practices and/or products that could possibly result in a firm's exposure and liability should be clearly identified. The good news is that this period of increased legal action, or the threat thereof, will in fact motivate the kind of clarity and measurement that both reduces liability risks and results in better buildings.

This is why clear and precise documentation is particularly important in rated green buildings to ensure that performance goals and specified objectives are met.

Also important is consideration of allocating performance responsibility and coordinating obligations, and of the consequences when the specified result or objectives are not achieved, including unique insurance, indemnity, liability, and consequential damage. In addition to contractors and subcontractors, design professionals are highly vulnerable to claims from clients, owners, and users. This vulnerability is partly due to the failure of some consultants, particularly architects and engineers, to understand the challenges that professionalism imposes. Design professionals confront daily complex legal issues in their practice; they need to be above reproach in every aspect of their dealings with others and in the management of the firm. Moreover, they may need to concentrate more than most professionals on maintaining good relationships with colleagues and coworkers to meet the many potential challenges they will face. Doing so will help minimize claims and also help to attract potential clients.

More often than not, litigation is usually the result of a general breakdown in understanding between the parties involved, either in the interpretation of the Contract Documents or in the working communications between the various parties on the construction site. When the conflicting parties are incapable of reaching agreement, the courts become the final resort for resolving the situation and are called upon to adjudicate based on case law. In the construction industry, problems and failures are most often the result of defects caused by faulty design, gross negligence, or poor execution, or by deterioration—a natural process unless it is excessive or rapid.

Carl de Stefanis, president of Inspection & Valuation International, Inc. (IVI), a prominent construction consulting and due diligence firm, says that claims against firms providing due diligence services are an increasing concern. He states that roughly 80% of these claims are building envelope–related, including roofs, exterior insulation and finish system (EIFS), windows, masonry, and so on. Not far behind are claims relating to building codes and, in recent years, mold- and water-related property claims have also become a major concern for the insurance industry. Insurance companies are now grappling with the challenges of how to address these issues, and some carriers have decided to exclude mold from their coverage entirely (Figure 16.1).

Acronyms and phrases, such as indoor air quality (IAQ), indoor environmental quality (IEQ), sick building syndrome (SBS), and building-related illness (BRI), are tossed around so arbitrarily that building owners and managers are encouraged to just shrug them off. This is surprising since all indications are that the number of commercial buildings with poor IAQ and the

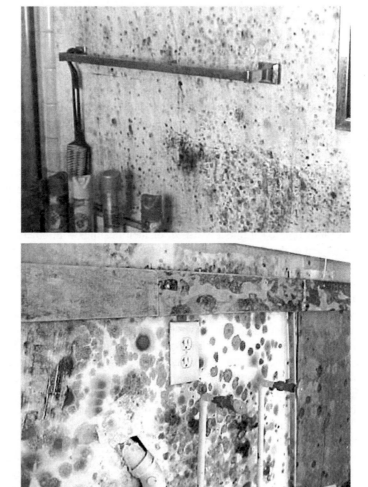

FIGURE 16.1 Heavy mold infestation caused by water penetration left unattended. *(Source: Servpro Industries, Inc.)*

corresponding increases in litigation over its consequences are quite significant. These increases are bound to impact insurance carriers, which pay many of the costs of health care and the costs arising from general commercial liability. In cases where an action is brought against a professional, such as an architect or engineer, an expert in the same discipline will likely be required to give an opinion as to whether negligence was a factor in the design, execution, or performance of duty. The right expert witness can make the difference between winning and losing a case. However, in the majority of cases, the investigation will involve much more than expert opinion; for example, laboratory and other tests may be recommended to help determine the cause of failure. The role that experts are required to play will therefore vary depending on the case in question.

Legal services involving architectural and engineering experts are invariably required for a variety of issues, including structural failure assessments, building envelope investigations, exposure reconstruction, assessments involving mold growth, and construction defect evaluations. There are also occasions when consultants may be asked to reconstruct events that took place years ago. The expert consultant's conclusions will typically be used by the client to evaluate a claim's strength and used as an evidentiary tool in ensuing dispute-resolution proceedings. But in addition to consulting and expert testimony services for both defendants and plaintiffs, the expert may be required to perform case evaluations, assist with settlements, and provide advice on litigation avoidance, as well as alternative dispute resolution (ADR).

As pointed out earlier, we find that, even in today's litigious environment, the vast majority of disputes in the construction industry reach resolution before they ever see the inside of a courtroom. Most are resolved through some form of settlement or ADR, and in various jurisdictions, courts are now making it mandatory for some parties to resort to dispute resolution of some type, usually mediation, before allowing a case to be tried in court. ADR's rising popularity is in part due to the tremendous overload of court cases and in part due to the knowledge that it is usually less expensive than traditional litigation. Added to this is the fact that parties sometimes feel a need to have greater control over the selection of those who will decide their dispute, in addition to the desire for confidentiality in the negotiations.

16.2 LIABILITY ISSUES

The public often hears of the many benefits of pursuing green or sustainable building design and construction, as this emerging market segment continues to increase its market share of the construction industry. Many owners and developers are embracing modern technologies, design elements, and operational models, even though they bring with them new liability risks. Because of the extreme complexity of liability issues, however, it is not possible to adequately address the many concerns and legalities that may arise with regard to liability here; builders, manufacturers, and designers are strongly advised to consult their attorneys and professional liability insurance carriers and agents for advice on these matters.

Although building owners and managers are rarely expected to guarantee the safety or well-being of their tenants, visitors, and guests, they are required to exercise reasonable care to protect them from foreseeable events. It should also be noted that there have been major changes in the civil False Claims Act and how it is enforced. These changes have resulted in increased compliance and enforcement risks for all construction companies conducting business with government agencies, including projects funded with federal or state money. This increased concern with liability is partly a result of the enormous upsurge in interest in green buildings, which has seen many misconceptions and exaggerations put forth by owners, designers, manufacturers, and distributors.

"Greenwashing" is a pejorative term often used within the industry for many misconceptions; it is derived from the term "whitewashing" and was coined by environmental activists to describe efforts by corporations to portray themselves as environmentally responsible in order to mask environmental wrongdoings. The term can be applied to building materials, systems, buildings, companies, and so forth. Indeed, eco-labeling is often found to be unreliable because of a lack of independent validation by third parties. Furthermore, greenwashing has the potential to ultimately discredit the entire green building industry, in addition to being the source of numerous lawsuits, because the ultimate declared goals of green buildings are not achieved through use of the following categories. Greenwashing claims can be categorized into two basic groups: (1) green materials and products and (2) green performance.

16.2.1   Green Material and Products Claims

Because of the lack of a precise definition of what constitutes a green building, material or system, providers will frequently find a material or product with limited green properties and then market those properties, and the material or product, as being "green." As an example, a material that uses a highly recycled content might also contain excessive amounts of urea-formaldehyde because of the way it was produced; thus, even though the material's overall impact on the environment may be negative, it is nevertheless erroneously marketed as green. Such false claims also occur when material or system providers base their claims on unreliable and inaccurate information. As green products and the green building industry become better understood and increasingly become part of the mainstream, and as processes (e.g., life-cycle analyses) are increasingly understood and adopted, these risks are likely to begin to decline. In addition, employing reliable material rating systems should help reduce the plethora of false claims that currently plague the market both in the United States and abroad.

16.2.2   Green Performance Claims

A phenomenon that is often cited within the green building industry is the misrepresentation of a person's or a building stakeholder company's knowledge and expertise regarding building green. When building owners and other stakeholders rely on this professed expertise, the result can be a dismal failure of the building to achieve its stated goals. In considering a building's operational performance, Ward Hubbell, president of the Green Building Initiative, notes,

There is an expectation that green buildings will, in addition to reducing environmental impacts, offer lower energy and water costs, less maintenance and other long-term benefits to the building owner. However, while the design may incorporate a wide range of green features, there are, of course, a tremendous number of variables between a building's design and occupancy that can impact operational performance. These potential areas of misunderstanding can be mitigated by following good business practices that facilitate clear communication and common expectations between building owner, designer, and rating organizations.

To avoid potential disagreement between building owners, designers, and builders in the interpretation of what constitutes a successful green building, it is extremely important for building owners and stakeholders to explicitly delineate and communicate their thoughts at the start of a project. Issues and problems tend to compound when the parties are relatively new and lack an understanding of the concepts of green building. This becomes even more unsettling when it is discovered that a lawsuit was recently filed that even challenges the claims of a very prominent green building effort—the LEED certification program. However, a building's failure to achieve a promised level of green building certification can be problematic for the building owner and/or developer in that it can impact the building's ability to qualify for a tax incentive or grant on which the owner may have relied to assist in offsetting the project's initial costs. In the case of public buildings, new laws are emerging in many jurisdictions that require public buildings to have green certification.

Herbert Leon MacDonell, author of *The Evidence Never Lies*, rightly says, "You can lead a jury to the truth, but you can't make them believe it"; this is why good field notes and photographs are imperative because they form the basis of solid documentation. Because field notes provide firsthand recorded observations of a failure or claim, they are indispensable and irreplaceable. In addition to being accurate and articulate, they should be written in a manner that is clear, neat, legible, and self-explanatory. Likewise, photographs are usually required to provide a visual record and are imperative for forensic investigations, particularly with issues such as mold or failures. Photographs should be of the best quality (preferably high-quality digital) and taken from different positions and viewpoints so as to get a comprehensive view of the scene in question. They should also be sufficiently annotated and filed appropriately. Digital photographs can be stored on a computer hard drive or on CDs. Whether it is a failure- or performance-related issue, it may be advisable for the forensic/consultant expert to supplement the project's documentation or scene with videography. This will depend largely on the circumstances prevailing at the time, the documentation already in hand, and whether it is deemed necessary for additional clarification.

It is a well-known fact in the construction industry that assigning culpability for green disputes usually boils down to a matter of negligence, ignorance, or incompetence. U.S. courts typically require qualified experts to testify to the standard of care applicable to the case in dispute, and that qualified experts testify to the professional's performance as measured by that applicable standard. It should be noted, however, that since building contractors are not legally considered "professionals" with respect to making independent evaluations and judgments based on learning and skill, the principle of standard of care may not be applied to them. Builders are nevertheless held to a "duty to perform," meaning that the provisions of their contracts require strictly following and executing project plans, specifications, and Contract Documents, which normally consist of four essential parts in combination: drawings, specifications, agreements, and conditions.

16.2.3　Alternative Dispute Resolution

Alternative dispute resolution typically refers to one of several processes used to resolve disputes between litigating parties. These include mediation, arbitration, negotiation, and collaborative law; conciliation and litigation are sometimes considered additional categories. As stated before, courts are increasingly requiring parties to use some type of ADR, most often mediation, before permitting their cases to be heard. In past AIA agreements, standard dispute resolution provisions called for nonbinding mediation as a condition precedent to binding arbitration. In the AIA's new design—build documents, this requirement for other forms of dispute resolution remains, although the new documents now offer the parties three methods to choose from: binding arbitration, litigation, and a third method to be decided on by the parties themselves. If the parties fail to choose a binding method of dispute resolution, litigation becomes the selection by default. It should be noted that the terms and conditions in the AIA's 1996 family of design—build documents incorporated AIA A201, General Conditions of the Contract for Construction, but AIA Document A141, Agreement between Owner and Design—Builder, no longer incorporates the A201 form; A141 now contains its own general conditions as Exhibit A to the agreement. The same is true for AIA Document A142, Agreement between Design-Builder and Contractor.

Although traditionally the architect served as the owner's representative and handled much of the project administration, including the Certificate of Substantial Completion of the project, some of these tasks have been taken over by the BIM manager. Likewise, on a design—build project, the administrative functions of the architect, who is part of the design—build team, are significantly different. Recognizing this, the AIA has prepared a new form entitled "G704/DB™−2004, Acknowledgment of Substantial Completion of a Design–Build Project," which requires the owner to inspect the project and acknowledge the date when substantial completion occurs. This form is a variation of G704−2000, Certificate of Substantial Completion, which was used previously.

16.3   FAILURE, NEGLIGENCE, AND STANDARD OF CARE

Failures are rarely the result of a single factor, but most often are the result of interrelated multiple factors. Thus, when an accident or failure takes place, it is important for an investigation team to be put in place, its emphasis and main function being finding the root causes of the failure. Even when all parties agree that the failure was from a single technical cause, responsibility may still not be accepted as clear-cut, especially given the general complexity of today's project delivery systems. As new high-technology methods of project delivery continue to be introduced, we will see a corresponding evolution in legal liability interpretation. Liability issues are further complicated and impacted by the fact that architects, engineers, and builders all have obligations to third parties, who are not included in the project contracts. In fact, typically the third parties are the ones to submit injury and property damage claims in the event of physical, product, or performance failures. The legal responsibility of design professionals clearly extends beyond the party with whom they have contracted, to any other party who may be injured by an alleged act of negligence and standard of care.

The most important factors that can impact the legal issues and interests associated with construction failures are (1) the type of failure, (2) the cause of failure, and (3) the parties impacted by the failure and their interests in it. While each of these is important, the strategy employed to effectively represent the party's legal interests should be based mainly on the facts relevant to the failure. These facts can usually be determined through documentation, interviews, and public records, and they represent the baseline from which any legal analysis commences. This is one of the reasons that it is crucial for the legal and technical teams to be interwoven from the outset of any investigation. Also at the outset, it is important to determine the ultimate objective for conducting a legal analysis relating to the failure. For example, it could be to determine the cause of the failure, what steps if any to be taken to avoid causing additional harm or damage, and who is legally responsible.

Most often in these types of cases, a methodical team approach, using both legal and forensic expertise, is adopted to help achieve the best possible outcome. The overall objectives are interdependent, and their individual importance may vary at different points in the investigation process. Moreover, the legal concerns surrounding a construction failure will vary with the roles and responsibilities of the different parties. The main interest of project owners is to preserve their rights against the potentially responsible parties and to have the project remedied and returned to its pre-failure condition. Design professionals may be concerned about their legal exposure to injured parties. Although the various parties have different interests, there are a number of actions that they need to take when a failure occurs. The initial steps typically include the formation of an investigative team; that team will develop a plan of action and deal with public agencies and the media, as well as protect confidentiality.

The elements of the investigative team will depend on a number of factors, such as the type and size of the failure in question. Furthermore, the team's leader should be a senior-level member of the client organization; such a person is likely to have the members' respect and support. All persons involved should take an objective view, during the investigation, of the cause of failure. The investigative team should include independently retained, qualified consultants and experts to assist in the investigation and in the analysis of the various components of the project. Depending on the magnitude and type of failure, additions to the team may include an attorney, as lead counsel to assess the viability of legal claims and defenses, and one or more forensic experts to study the technical causes of the failure. Likewise, the team should include persons who have personal knowledge of the project, particularly concerning work relating to the failure in question, which may be subject to litigation.

A significant number of all major disputes and litigations are related to construction failures, which is why, when a structural failure does occur, the contractor should be prepared to take action. The first steps following a failure are critical to conducting the investigation as well as to preventing further potential damage or loss of life. Moreover, the actions taken immediately following a failure can have a tremendous impact on the outcome of any subsequent technical investigation. This is because much of the evidence associated with a collapse or structural failure is often of a perishable nature and needs to be preserved and protected. The key factor in avoiding or mitigating the impact of conflicts is having the appropriate mechanisms in place to manage related issues, and acting proactively before something becomes an issue.

An investigation cannot succeed in its objectives of determining the cause of failure and correctly attributing responsibility if evidence is tampered with in any way; the existing failure scene, its condition, and other circumstances on the site will play a crucial role in determining the most important factors that caused the failure and thus directly impact an investigation's outcome. Moreover, this evidence is a major element in assisting in the development of hypotheses and theories of the failure's cause, which is why the site should be protected immediately, completely documented, and appropriately recorded. This process can be greatly facilitated if the owner and contractor cooperate with the investigation team. Even though disputes in the construction industry are fairly common, they are not inevitable and can be prevented with proper care and due diligence.

16.3.1   Procedural Issues

A typical failure is demonstrated by the partial roof collapse in July 2007 of a luxury high-rise building in Greenwood Village, Colorado, in which 14 construction workers from Beck Construction and Concrete Express of Dallas were injured. Rick Palese of Everest Development says that concrete for a flat roof was being poured by Concrete Express onto a metal sheet that sagged and became detached, pinning one person (Figure 16.2). The permits show that the building was to have 261 units. The plans also called for restaurants, shops, a spa, and a theater. The cause of the failure was under investigation. Beck Residential was the project lead contractor.

Document Collection

Compiling a project's documents is one of the earliest and highest-priority actions that need to be taken when a failure occurs. The primary sources for these documents are obviously the design and engineering consultants, the contractors, and the owner. In addition to the Contract Documents (i.e., construction drawings: architectural, structural, electrical, and specifications), the principal documents typically required to conduct a preliminary evaluation include:

- Shop drawings and assembly drawings
- Other contractor submittals

FIGURE 16.2   Fourteen people were injured when a section of the roof of the "The Landmark," a $140 million high-rise building, collapsed onto the 13th and 12th floors as concrete was being poured for the building's roof. *(Source: CBS Broadcasting Inc.)*

- Change Orders, warranties, and so on
- Test reports
- Boring logs
- Construction-monitoring photographs and reports
- Engineer-of-record calculations
- Relevant correspondence

Interviews

It is important for an investigative team, once it has been commissioned, to investigate the defect, system failure, or collapse. Appointing a person to be in charge immediately facilitates efforts to conduct interviews with eyewitnesses and other persons (e.g., project personnel who may be able to provide useful information regarding the investigation), and it may assist with focusing on and determining the probable causes of the failure. It is important to conduct the interviews as soon after a reported incident as possible but especially prior to its "contamination" and while recollections are still fresh. The specific information to be sought from interviews depends largely on the type of failure, the circumstances of the failure, and the interviewees' knowledge and expertise.

Experts' Cooperation

It is in the interests of all parties during an investigation for experts to cooperate with each other and those with whom there may have common interests. This allows them to pool resources and avoid unnecessary duplication of effort. Other areas of potential collaboration include possible destructive testing, identification of debris and relevant components, as well as the sharing of interview information. However, while sharing basic information and fostering cooperation is often desirable and may greatly facilitate an investigation, in situations where there is a possibility of litigation, the client or the client's attorney should always be consulted to ensure that any sharing does not inadvertently compromise the client's interests.

The Preliminary Assessment

From the initial investigation of possible failure scenarios and possible contributing factors, a preliminary evaluation should emerge. Once the input of the various consultants, eyewitnesses, and staff members is considered and recorded, there is a need for follow up by conducting preliminary structural analyses and tests to try to determine the viability of initial hypotheses as well as to identify other possibilities that may have triggered the failure; these analyses and tests can include excessive occupancy loading or environmental factors (e.g., excessive snow load on the roof), strong winds, and so on—anything that might exceed the building envelope's design capabilities to withstand.

The expert should fully understand the strategies and objectives of both the client and the attorney before offering advice regarding photographs, testing, or measurements, which must be taken as soon as possible to preserve evidence. The expert must also confer with the client regarding analytical procedures, obtain approval to implement them, and identify important documents that will be required. Whatever the case, forensic investigations should be conducted on the assumption that the consultant will be required to offer sworn testimony in court about the investigation and the conclusions reached. As court procedure rules vary from one jurisdiction to another, it is prudent for the client's attorney to provide the expert with specific guidance.

Negligence

Lawsuits based on negligence are considered the most common civil action in tort law. They occur when a person fails to exercise the standard of care that a reasonable, prudent person would have exercised in a similar circumstance (sometimes called a lack of "due care"). Very often the bottom line in assigning culpability for failure issues or construction disputes is narrowed down to a result of negligence, ignorance, or incompetence. Tort law in the United States is generally defined by states rather than federal government. Negligence cases generally get to federal court through a diversity of jurisdictions, even though they will typically be tried with some state's negligence law as the basis for a decision. It should also be noted that the state law of negligence is usually common rather than statutory, with the effect that what is determined to be a lack of due care will differ from one jurisdiction to another. Moreover, in a negligence suit, the plaintiff has the burden of proving that the defendant failed to act as a reasonable person would have acted under the circumstances. The court will be expected to instruct the jury as to the standard of conduct required of the defendant.

"Gross negligence" is when a person or party shows unrestrained disregard of consequences; that is, where ordinary care is not taken in circumstances under which injury or grave damage is likely to result. A determination of gross negligence is a legal conclusion that can only be arrived at by a court of law. The distinction between ordinary negligence and gross negligence amounts to a rule of policy that a failure to exercise due care in such situations as where the risk of harm is great will give rise to legal consequences harsher than those arising from negligence in less hazardous situations. Negligence is also said to occur when something is omitted that ought to be done.

With respect to design professionals, evidence of negligence is often found in the preparation of Contract Documents—for example, where there is a lack of coordination between construction drawings and site conditions or evident discrepancies between building plans or specifications and shop drawings. When a design professional can be shown to be negligent, his or her license to practice may be temporarily or permanently revoked. However, while project designers and architects are required to possess and apply the same degree of skill, knowledge, and ability as other members of their profession, are required to exercise a standard of care, and are expected to apply their best judgment in executing an assignment, they cannot assume or guarantee that a perfect set of plans or Contract Documents will be provided or that the outcome will always achieve all of the owner's or consultant's objectives. It is therefore important that design professionals not undertake projects that clearly exceed their technical abilities or exceed those of the personnel available to work on them. Only experienced, competent, and qualified staff should be assigned to a task. Junior and inexperienced personnel must be carefully supervised by fully qualified professionals. Outside consultants may be required to supplement the firm's own capabilities to achieve optimum results.

Standard of Care and Duty to Perform

The doctrine of reasonable standard of care is basically that one who undertakes to render services in the practice of a profession is required to exercise the skill and knowledge that members of that profession (whether architect, engineer, contractor, etc.) normally possess. Here, design professionals often cite a California Supreme Court case. The Court in that case essentially noted that services of experts are sought because of their special skill. They have a duty to exercise the ordinary skill and competence of members of their profession, and a failure to discharge that duty will subject them to liability for negligence. The Court added, however, that those who hire such professionals should not expect infallibility but only reasonable care and competence. Whether the design professional has violated the standard of care is usually determined by a jury in which the conflicting testimony of experts is heard and acted on. With respect to the design professional, key points to remember relating to standard of care include the following:

- Possess learning and skill that is ordinarily possessed by the profession (at each stage of their work—for example, preparing drawings and specifications; overseeing the bidding process; approving shop drawings, equipment cut sheets, and contractor payment requests; and observing the work to guard the owner against construction defects).
- Exercise the care and skill ordinarily possessed by reputable members of the profession practicing in the same or similar locality and under similar circumstances.
- Use reasonable diligence and best judgment in the execution of the project.
- Achieve the objective—that is, accomplish the purpose for which the design professional was employed.

With respect to contractors, California courts have consistently held that the standard of care applicable to negligence claims against a contractor is that of a licensed contractor under similar circumstances. Moreover, expert testimony is required as to the standard of care itself, as well as to a defendant's compliance with it. Therefore, any reference to what a contractor defendant "should" have done using the standard of a reasonable person would not only be irrelevant but would be prejudicial and improper. According to Raymond T. Mellon, senior partner with the law firm Zetlin & De Chiara LLP,

It is imperative to note that the standard of care is kinetic and continually evolving. Both events and technology can and do affect and change the standard of care. Recent technological changes in the last 20 years have mandated revisions to building construction and safety. For example, many building codes now require various types of computerized fire safety devices, smoke detectors, strobe lights and other safety features not available 20 years ago. While complying with statutory requirements for safety in building design is straightforward, the important issue raised by 9/11 is what design changes and technological advances must be incorporated into a building in the absence of statutory mandates. The fact that a building code in a particular municipality has not yet been amended to include new safety features or technological advances, does not, by itself, provide a safe haven from liability for damages incurred by a terrorist attack.

Expert Witnesses

The investigation and testimony of an expert consultant or witness can have an enormous impact on reputations and livelihoods, even though this service is essentially required to determine the cause of, assign responsibility for, and prevent a repeat occurrence of a failure. For this and other reasons, it is imperative that forensic experts understand the accepted standards that professional consultants and contractors are expected to meet. For example, is the standard-of-care bar raised when the designer/consultant or contractor is a LEED Accredited Professional? The fact that a consultant makes a mistake, and that the mistake causes injury or damage, is normally insufficient to prove professional liability. For there to be professional liability, it must be proven that the services offered were professionally negligent, meaning that they fell below the expected standard of care of the profession. While it may be obvious that design professionals today have a duty to practice sustainable design, that standard may or may not rise to the level of promising the achievement of a specific LEED certification level.

R. T. Ratay, a forensic engineering expert, defines the standard of care simply as being "that level and quality of service ordinarily provided by other normally competent practitioners of good standing in that field, when providing similar services with reasonable diligence and best judgment in the same locality at the same time and under similar circumstances." To determine this, qualified experts are brought in and required to testify in U.S. courts to the standard of care applicable to the case on trial; these experts are also required to testify to the professional's performance as measured by that standard. Building contractors are exempted from the principle of standard of care because they are not deemed "professionals" in the sense of making independent evaluations and judgments based on learning and skill. However, they are held to a "duty to perform," which basically means that they must strictly adhere to the plans, specifications, Contract Documents, and provisions of their contracts. Also, although the expert is generally hired by one of the parties, it is important that he or she always remain objective and neutral.

Of note, on December 1, 2010, Rule 26 of the Federal Rules of Civil Procedure was amended to limit discovery that may be obtained from a party's testifying expert. The newly amended rule provides added protection because it now limits disclosures to actual "material of a factual nature." It should be noted that the trial systems may differ from country to country. For example, the U.S. trial system differs significantly from the German trial system—not only in many of its details but also in its fundamentals. In the United States, judges most often have only a passive role, whereas in Germany the judge usually takes a very active role, generally controlling the proceedings, examining the witnesses, and in the end typically acting as the decision maker. There are other differences (e.g., the lack of pretrial discovery) in addition to the decisive role of German court experts.

16.4 ALTERNATIVE DISPUTE RESOLUTION VERSUS TRADITIONAL LITIGATION

The Nationwide Academy for Dispute Resolution (U.K.) Ltd. (NADR, 2000), says,

A crucial distinction between litigation and ADR is that whilst many legal practitioners engage in ADR processes, there is no legal or professional requirement for either the ADR practitioner or for party representatives at ADR processes to be legally qualified or to be members of legal professions such as the bar or the law society. Many of those who engage in ADR practice are first and foremost experts in particular fields such as architects, builders, civil engineers, mariners, scientists and social workers, albeit with a thorough understanding of ADR processes and some knowledge and understanding of law. In-house legal experts in large corporate organizations can take part in the entire ADR process without engaging professional lawyers thus cutting costs further, both in terms of time lost through communicating with the professionals and in respect of legal fees and costs.

16.4.1 Alternative Dispute Resolution Techniques

The downturn in the economy has provided strong encouragement for professionals and businesses, both small and large, to find new and innovative ways to minimize expenses. It follows that with the rising costs associated with traditional litigation and other issues, many businesses are now trying to avoid traditional litigation—either turning to ADR or negotiating a pre-litigation settlement to resolve construction disputes. It is also evident that there has been a significant surge in alternative dispute resolution popularity in recent years, especially with state and federal courts. The popularity of ADR is partly due to dissatisfaction in the industry with the current state of traditional civil litigation. This has invariably provided further impetus to the development of various ADR techniques, even though they have been around for many years. In fact, the American Arbitration Association (AAA), a public service, not-for-profit organization, has been the leading advocate of ADR since 1926.

ADR typically encompasses a variety of processes and techniques for resolving disputes without litigation (i.e., that fall outside of the judicial process). Most individuals and corporations try to avoid being involved in lawsuits because formal litigation can entail lengthy delays, high costs, unwanted publicity, and ill will. After a decision has been rendered, appeals may be filed that will extend the time required to reach a final result still further. Most ADR techniques, on the other hand, are typically fast, conclusive, and inexpensive.

Research shows that ADR techniques (e.g., arbitration, mediation, negotiation, and other out-of-court settlement procedures) are now widely employed in the vast majority of construction disputes. They are especially useful where minor defects or failures are involved (Figure 16.3). These techniques are essentially based on the premise that disputes can best be resolved through negotiation or mediation immediately after a conflict comes to light rather than through tedious, costly, and time-consuming traditional civil litigation. In such cases, the only elements governing a quick resolution are the eagerness of the parties to end the dispute and the complexity of the cases to be resolved. There are occasions when one of the parties to a dispute will insist on litigation; in these circumstances, it is usually because certain legal precedents have been shown to be favorable to that party.

FIGURE 16.3   Common failures that are not excessive in magnitude can perhaps be more readily resolved through ADR procedures.

TABLE 16.1 Primary Advantages and Disadvantages of Different Forms of Resolution

Process	Advantages	Disadvantages
Negotiation/ Assisted Negotiation	Parties have control Confidential	No structure Entrenched bargaining positions likely
Mediation	Structured Skilled mediator helps avoid entrenched positions Control and resolution lie with parties Helps maintain future commercial relationship for parties Costs less than litigation Quick result Confidential	No decision if parties do not agree A resolution may not be reached
Arbitration	Structured Can be quick, timetable controlled by parties Costs may be less than litigation Confidential	Parties do not have control Imposed decision May jeopardize future relationship of parties
Litigation (court action)	Structured	Timetable controlled by court Costs may be significant Parties do not have control Imposed decision May jeopardize future relationship of parties Long waiting times On public record (no confidentiality)

Generally, settlement discussions are the best, least bruising, most private, and least expensive way of resolving construction disputes. For example, most forms of ADR are not open to public scrutiny, unlike disputes settled in court. Moreover, hearings and awards can remain private and confidential, which helps to preserve positive working relationships. There are different methods of resolving construction disputes in the United States as Table 16.1 clearly shows. ADR techniques have proven to be viable, cost-effective alternatives to litigation. According to the American Arbitration Association,

Arbitration is the submission of a dispute to one or more impartial persons for a final and binding decision, known as an "award." Awards are made in writing and are generally final and binding on the parties in the case. Mediation, on the other hand, is a process in which an impartial third party facilitates communication and negotiation and promotes voluntary decision making by the parties to the dispute. This process can be effective for resolving disputes prior to arbitration or litigation.

Many jurisdictions in the United States have now made it mandatory to use ADR methods and will accept a case only if ADR methods have failed to resolve the dispute in question. It should be noted that most attorneys advise clients to be cautious in choosing ADR methods over traditional litigation (especially when it is in their interest) and, when ADR is decided on, encourage them to choose voluntary and nonbinding methods whenever possible.

Arbitration

Arbitration is becoming the favored practice for resolving construction disputes because for many it is considered to be a low-cost and time-efficient alternative to court litigation. It is also one of the oldest and more common forms of ADR. There are two basic types of arbitration: (1) binding, which means the parties must follow the arbitrator's decision and courts will enforce it; and (2) nonbinding, which means that either party is free to reject the arbitrator's decision and take the dispute to court as if the arbitration had never taken place. Although arbitration includes participation of the parties typically on a voluntary basis, it can only be legally enforceable and binding if agreed to by the parties beforehand. For the arbitrator's decision to be binding, a separate written agreement needs to be in place certifying that the arbitration agreement has been read, understood, and accepted. Here is an example of a simple arbitration clause:

All claims and disputes arising from or in connection with this Contract are to be settled by binding arbitration in the state of (insert state in which parties agree to arbitrate) or another location mutually agreed on by the parties. The arbitral award is final and binding on both parties *and may be confirmed in a court of competent jurisdiction.*

Traditionally, the AAA has been the forum for resolution of arbitration cases, and it has developed detailed rules governing the arbitration process. Copies of its Dispute Resolution Procedures Governing Arbitration and Mediation, as amended through January 1, 2003, may be obtained from the American Arbitration Association, as well as from the AAA's website (www.adr.org).

When using arbitration, a third party is agreed on; that person acts as a private judge and is authorized to render a final decision. The feuding parties control the range of issues to be resolved by arbitration, the scope of the relief to be awarded, and many of the procedural aspects of the process. The contract between the owner and builder or designer may very well include an arbitration clause, although even if there isn't one, a disputant party can still choose to take the case to arbitration. This may be why most residential construction contracts now contain arbitration clauses; likewise, arbitration provisions in commercial construction contracts are becoming increasingly common. Nevertheless, it is extremely prudent to obtain legal advice before proceeding with arbitration. The main attraction of arbitration is that it costs less than a trial and has the added advantage of often having a speedy outcome, as well as relative privacy of the parties. Also, because the parties control the process, they can enjoy tremendous flexibility.

Once a decision to go to arbitration has been made, all the facts need to be sorted out and the pertinent issues in dispute established. A decision also must be made about who the potential witnesses for each side are to be, as well as who the arbitrator is to be and the venue for the arbitration hearing. In most initial settlement discussions, expert consultants do not usually appear, although the parties may rely on their expertise in formulating their respective positions. However, they do usually participate in mediations and are often the primary presenters of their clients' technical positions and are relied on to provide continuous input throughout the ongoing arguments. It is highly unlikely that the arbitration process can proceed without the use of expert witnesses, partly because arbitrators often probe deeply into technical matters of a case, and also to achieve for the parties credible and authoritative presentations of their technical positions in the case. However, with regard to litigation, an expert's testimony at a trial involving architecture, engineering, and construction is usually required for the purpose of elucidating to the judge or jury the technical aspects of the case; plus, both of the parties want a judge or jury to hear their technical positions as expressed by their own experts as well as hear challenges to opposing experts' opinions and conclusions.

Mediation

Mediation is an ADR nonbinding process by which an impartial person, the mediator, facilitates negotiations between parties to a dispute to promote reconciliation, settlement, or understanding. The main objective is therefore to assist the parties in voluntarily reaching an acceptable resolution of issues in dispute with the aid of a neutral third party mediator. The mediator's role is solely advisory, in that he or she may offer suggestions on ways of resolving the dispute but in no way tries to impose a resolution on the parties. Mediation proceedings are confidential and private. Also, there are no hard-and-fast rules of procedure, so the mediator and the parties can seek out the most effective and efficient method to resolve the conflict. Mediation has several important advantages and fewer disadvantages than the other options, which is why it is gaining popularity as the ADR method of choice for resolving disputes. Among the advantages of mediation are lower costs, efficiency, the ability to preserve business relationships, and potential speed and flexibility in resolving the dispute.

A nominated mediator must have the approval of all involved parties. The approval process includes his or her disclosure of any past, present, or future relationships with the participants in the mediation. Mediators often have special training that allows them to assist parties in identifying the real issues that separate them while fully comprehending what is in their best interest; they help the parties reach an agreement on some or all of the disputed issues in order to provide certainty and clarity. For this reason, the mediator should have expertise in construction and construction claims in addition to being well versed in construction law. A mediator, like a facilitator, makes primarily procedural suggestions regarding how parties can reach agreement, because the final objective is reaching a compromise. Disputing parties often find themselves having to forgo some of what they consider as their legal rights in the matter.

Some mediators try to set the stage for bargaining by making minimal procedural suggestions and intervening in the negotiations only to avoid or overcome a deadlock. Others may get much more personally involved in forging the details of a resolution. Mediation has been found to be particularly useful in highly polarized disputes where the parties have either been unable to initiate a productive dialogue or have been talking but appear to have reached a seemingly insurmountable impasse. If it is determined that the parties are unable to reach an amicable agreement, the case will have to go to court. This will incur significant costs (e.g., legal expenses) as well as considerable time and energy. But even if mediation fails, the parties may still make an effort to settle just to avoid the expensive court process, without either admitting fault. In some countries such as the United Kingdom, ADR is synonymous with what in other countries is generally referred to as mediation.

Unassisted and Assisted Negotiation

Negotiation is a time-proven approach to resolving disputes between feuding parties through discussions and mutual agreements. In an unassisted negotiation, the parties attempt to reach a settlement without involvement of outside parties. This process is essentially voluntary and there is no third party to facilitate or impose a resolution. Negotiation is one of the most fundamental ADR methods, offering parties maximum control over the process. As with any endeavor, however, it can be effective or unproductive. To be successful, the parties should at the outset identify both the issues on which they differ and possible settlement options. They should also disclose their respective needs and interests to enable them to negotiate acceptable agreement terms and conditions.

The most appropriate and successful approaches to negotiations are those in which negotiators conduct discussions that focus on the common interests of the parties and not on the parties' relative power or positions. A construction industry negotiation variant is the "step negotiation," which is a multitiered process that is sometimes used when the information that the parties have in place is, for example, incomplete, or employed as a mechanism to break a deadlock. But while no specific mechanism or formula mandates the settlement negotiations, many contracts today contain a "good faith" negotiation clause requiring any dissatisfaction with performance under a contract to be communicated to the other party, usually in writing. The final objective of negotiations is that each party end up in a better position than if it had not negotiated.

In an assisted negotiation, a third party (outsider) to a dispute is agreed to; she or he brings the parties together and, to varying degrees, helps them resolve their disagreements. Here the decisions remain in the hands of the parties themselves, and the function of the third party is mainly to help them negotiate a mutually acceptable agreement. In some cases, the third party may suggest a particular settlement. This model is increasingly being used internationally in construction contracts.

16.4.2   Traditional Litigation

Litigation is the traditional form of dispute resolution and is based on taking action through the courts. In traditional litigation, a judge typically listens to arguments and expert witnesses on the interpretation of relevant law as applied to the particular dispute, and then makes a determination as to who wins and who loses. This process can be complex and drag on for years and can also be extremely costly. Traditional litigation necessitates the observance of certain protocols regarding rules of evidence and procedures for reports, pretrial discovery, interrogatories, depositions, direct/redirect and cross/re-cross examination, and so forth. It is helpful for the consultant, building owner, or product manufacturer to have a basic understanding of the different stages and procedural details of civil lawsuits. Litigation procedures are generally governed by statute in each jurisdiction. Federal and civil trials are normally governed by the *Federal Rules of Civil Procedure* (FRCP). Although each state has its own rules of civil procedure, in many states these are similar to the FRCP. Moreover, many courts have their own local rules of procedure that supplement federal or state rules. Traditional litigation lawsuits are normally divided into two stages: pretrial and trial.

Under U.S. law, discovery is the lawsuit's pretrial phase in which the parties are identified and the issues in dispute are clarified. In addition, each party is given an opportunity to learn about the other party's witnesses and potential evidence. Each party can request documents and other evidence from the other parties or can use a subpoena or other discovery devices to compel the production of evidence and depositions. Many states are adopting discovery procedures that are more or less based on the federal system; in some cases, the federal model is closely adhered to while in others it is not. Some states take a totally different approach to discovery.

When a lawsuit is filed against an individual or firm, the plaintiff must first "serve" the defendant with a "summons" and "complaint." In most jurisdictions, traditional litigation starts with the service of a summons. The complaint consists of the plaintiff's factual and legal allegations against an individual or firm. The summons is basically a document that notifies the individual party or firm that it is required to appear in the lawsuit and file a response to the complaint. This usually takes place sometime after a failure or deficiency has actually occurred. In federal court, the suit starts when the summons and complaint are filed with the court prior to service on the defendant. This gives the attorneys for the defendant an opportunity to reply to the complaint and possibly file a counterclaim. If a counterclaim is filed, the plaintiff's attorney is expected to answer it. Either way, the defendant must respond to the plaintiff's complaint by denying or admitting the allegations. Should the defendant fail to respond to the plaintiff's allegations, it may be considered an admission of guilt by the defendant.

Sometimes the defendant may decide to use a third-party practice to bring an outside party into the suit—a person or firm that may be liable to the defendant if the defendant is liable to the plaintiff. Thus, a defendant owner may serve a third-party complaint on a contractor or consultant designer claiming that should the owner be found liable to the plaintiff, the contractor or consultant designer would be liable to the owner to the extent that the liability was caused by defective construction or design. Discovery, inspection, and disclosure techniques are frequently used by attorneys to reveal details of the adverse party's claim; they also allow a party to inspect an adversary's files. This can be initiated by the attorney of the defendant, plaintiff, or third party by service of a document demand on the adversary to present certain documents that the party wishes to inspect (e.g., minutes of job meetings, photographs, tests, correspondence).

Interrogatories, which are part of discovery, consist of written questions formulated by one party and served on the other; they are required to be answered in writing and may only be served on a party to the lawsuit. Interrogatories are intended to cover issues that help prove or disprove the presence of material fact. Where the issues are technically complex, experts may be brought in to assist and clarify the issues. The party receiving the interrogatories is required to answer in writing, and under oath, within a specified time period (30 days under FRCP 33).

Depositions are oral questions, with answers given as testimony under oath; they may be taken from witnesses as well as relevant parties. Normally, they take place prior to trial and consist of in-depth questions of a party or witness by the attorneys of the various parties involved in the case. Depositions mainly consist of a cross-examination under oath recorded verbatim, but in the absence of judge and jury. They often assume cardinal importance at a trial and can be used for any relevant purpose, including the discrediting of trial testimony—that is, *impeaching*, or throwing doubt on, a witness' credibility.

Pretrial conferences are often employed by judges and attorneys to encourage settling. Thus, prior to trial the attorneys for both sides often meet with the trial judge in a pretrial conference to review the evidence and clarify the issues in dispute, and to determine whether a settlement is possible. If a settlement is not possible, a date for trial commencement is set and the attorneys and judge then decide how the trial will be conducted and what types of evidence are to be admissible. This is followed by jury selection (called *voir dire*), which depends largely on jurisdiction. It can be accomplished either by the attorneys themselves or by the judge, depending on the rules constraining the court where the case is being heard.

Once jury selection is complete, the trial can begin with opening statements by the attorneys of the plaintiff and defendant. These statements typically outline the strategy that will be used by the respective attorneys to prove their case.

When the lawyers have finished their opening statements, the plaintiff's and defendant's witnesses are called to the stand individually to be examined and cross-examined by the adversary's attorneys. Closing arguments by the attorneys are then made both attorneys, followed by the judge's instructions to the jury on the laws applicable to the facts of the case. Finally, the jury is allowed to deliberate to make a judgment. During deliberations, the jury has an opportunity to inspect all the relevant testimony and documents entered into evidence prior to reaching a final verdict. Once the jury has reached a decision, it issues its verdict in favor of one of the parties.

16.4.3 Professional Ethics and Confidentiality

Most of today's professional specialty fields have their own rules of professional conduct, adherence to which is monitored by the memberships' leaders. Thus, the vast majority of contractors and professional architects or engineers continue to be governed by the ethical obligations of their professional organizations and associations, most of which have a code of ethics. Typical examples are those of the American Institute of Architects (AIA) and the American Society of Civil Engineers.

Lawyers are fiduciaries of their clients, and one of their fiduciary obligations is to protect the confidentiality of clients' information. It is important that experts and professionals in civil litigation who are or were retained by an attorney on a client's behalf likewise always protect client confidentiality, after being made aware of this duty, by treating all information obtained from the client, directly or through the attorney, in complete confidence. This obligation arises from the fact that the attorney hired the expert/professional to act for the client and as the attorney is sworn by his or her code of conduct to maintain and preserve the client's confidences, so must the expert/professional be sworn. It should be noted that litigants do not normally file discovery documents in civil litigation unless a matter is to go to trial. Such documents and recordings should not be released to third parties without the attorney's permission, although a release may waive the privilege.

In the legal profession, there is a fiduciary responsibility and duty of loyalty owed to a client that prohibits an attorney (or law firm) from representing any other party with interests that conflict with those of a current client (an exception would be made when all affected parties give written consent). And although the law pertaining to conflict of interest is very complicated, a potential conflict obviously exists when an attorney attempts to oppose a party she or he previously represented. A potential conflict of interest exists whenever such prior representation may influence the attorney's loyalty to a current client in any way. In fact, attorneys, experts, and professionals may be privy to customer lists and other proprietary business information, hold government secret classification status, and be committed to keeping the information confidential. All such prior confidential commitments could come into conflict when an expert or professional agrees to act as a consultant and possibly testify. Even the appearance of impropriety should be avoided at all costs by attorneys, experts, and professionals.

Whenever a potential conflict does exist, it is the duty of the professional to disclose its nature to the attorney and to allow him or her to determine whether the professional should continue to function on the client's behalf. As a technical specialist, a consulting expert has a professional obligation to uncover all relevant facts relating to the issue being investigated, whether it is in the client's best interest or otherwise. In addition, if a construction failure should occur, each of the involved parties (owner, contractor, designer, consultant, etc.) will typically retain an expert consultant to investigate the causes of and responsibilities for the failure. If later deemed necessary, the expert may be required to provide technical support in the litigation of potential claims. Experts and consultants can play critically important roles in litigation, and the evidence of an expert witness can have a dramatic impact on a case and, in many cases, determine its success or failure. Bear in mind that the service performed and role played by a particular expert will vary from case to case.

The special knowledge that experts have by virtue of their skill, experience, training, and/or education goes beyond the normal experience of the general public, to the extent that others may officially (and legally) rely on their opinion on such matters. This acknowledged expertise in their chosen subject often gives their testimony considerable moral authority in the eyes of the court, and their written reports—in which their view of the evidence is both concise and clear and able to find for one side over the other—may facilitate a settlement. Furthermore, the expert witness is formally allowed to offer an opinion as testimony in court without having been a witness to any occurrence relating to the lawsuit. It is important, too, that the expert witness be articulate and able to present highly technical matters relating to architecture and engineering in language that can be easily comprehended by the nonexpert. In the resolution of construction-related claims, the expert becomes a valuable and necessary professional.

16.5 INSURANCE PROGRAMS AND REQUIREMENTS

The extremely complex field of liability makes it almost impossible to adequately address the endless concerns and legal matters that may arise with regard to liability issues here; this is why builders, manufacturers, and designers are strongly advised to consult their attorneys and professional liability insurance carriers for advice. Although building owners and managers are not expected to guarantee the safety or well-being of their tenants, visitors, and guests, they are required to exercise reasonable care to protect them from foreseeable events. The number of liability lawsuits filed against U.S. companies has increased so dramatically over the last decade that it has become a major concern to all involved in the construction industry.

Attitudes toward insurance vary, and few fully understand it. To put it simply, an insurance policy consists of a legal contract expressed in relatively complex legal terms. It promises to provide compensation for something (e.g., a potential loss) that may never happen, although in the construction industry it can and often does happen. Indeed, contractors lacking adequate insurance risk the success of their business and even potential bankruptcy. Independent and expert advice is imperative. Insurance should be considered as crucial to the continued operation of any business, as few companies and proprietors in the building trades have sufficient capital to meet adversity with their own resources. While many inherent risks may be inevitable in most businesses, in building construction (including green) projects properly arranged and appropriate insurance can remove or mitigate many of the risks for a small cost compared to the potential devastating liabilities that can ensue by not having it.

With most things that are new, there are challenges to be faced, and certainly going green is no exception. From an insurance perspective, Zurich Insurance says, "What matters the most is the claims process. It's important to ask, are there people in place who know how to order and rebuild to green standards?" Zurich Insurance goes on to say, "What we see now is a growing number of insurance carriers providing different coverage forms for green buildings, and the question is how much real coverage they are providing. There are obviously some unique exposures associated with these types of facilities, and the key issue will be how the market addresses those exposures with coverage." There are several types of insurance policies common in the construction industry that contractors and subcontractors should have to protect themselves and the owner/developer from financial loss, whether in the form of property damage, bodily injury, or both. The most common include:

- Commercial general liability (CGL)
- Builder's Risk
- Errors and omissions (E&O)
- Workers' compensation insurance
- Professional liability

16.5.1   Commercial General Liability

For construction companies, liability is one of the most important types of insurance to carry. It is intended to protect the policyholder and other named insured for claims instituted by third parties resulting from damages or injuries sustained by persons or property onsite due to faulty workmanship as a result of the contractor's operations or negligence. In most cases, the construction contract will specify that subcontractors name the general contractor as an additional insured on their respective CGL policies. Both general contractors and subcontractors typically hold CGL policies that cover personal injury and property damage on either an occurrence or claims-made basis. Also, CGL policies contain a general aggregate limit that states the maximum that the insurer will pay out during the policy period for damages resulting from bodily injury, property damage, and personal and advertising injury. If for some reason the contractor does not have adequate insurance, the owner/developer can be held financially responsible.

Obtaining general liability insurance has not been easy for many contractors in recent years mainly because of the construction defect litigation crisis. To address this crisis, some in the insurance industry have started restricting coverage in their CGL policies by placing numerous exclusions in the "fine print," resulting in lack of coverage for many types of lawsuits. These exclusions typically include but are not limited to EIFS, mold, lead, products-completed operations, prior completed work, damage to work performed by subcontractors on the contractor's behalf, subsidence, contractual liability limitation, independent contractor exclusion, roofing operations exclusions, and so on. It is imperative for stakeholders to fully understand the coverage implications of these exclusions if they are contained in the CGL policy and to be able to decide on what action is to be taken, if any.

An attorney and past-partner with Howrey LLP, Seth Lamden, says "After nearly three decades of litigation and scores of decisions from nearly every jurisdiction, courts still do not agree on whether a commercial general liability policy protects a general contractor against claims for property damage occurring after the construction project is complete when the damage is caused by the defective work of a subcontractor." Lamden goes on to say:

The battle still rages on as to whether a general contractor should be entitled to coverage for liability resulting from damage to the completed project caused by the defective work of a subcontractor. A close reading of the standard form CGL policy reveals that the policy should provide such coverage. Until courts agree on this issue, however, general contractors should be aware of the insurance law in their jurisdiction regarding how CGL policies are interpreted in the construction context to make sure that they truly are protected from the main risk they face on construction projects.

16.5.2 Builder's Risk

Builder's Risk policies differ from CGL policies in that they focus on losses incurred by the policyholder as a result of damage to the project during construction, rather than on losses incurred by third parties. They are generally purchased on a project-specific basis and usually indemnify against damage and losses due to fire, theft, wind, hail, lightning, explosion, and similar forces. Faulty workmanship and construction defects are generally excluded from coverage, as are earthquakes, floods, acts of war, or intentional acts of the owner.

Builder's Risk insurance is designed to insure construction projects and covers buildings and other structures while being built, including building materials and equipment that will become part of the building or structure, whether

FIGURE 16.4 There were more than 1500 people in this luxury department store in Seoul, South Korea, when a section of the five-story building collapsed in June 1995. More than 500 people were killed as a result. The building was constructed using steel-reinforced concrete pillars. The collapse was blamed on faulty construction.

being temporarily stored, on a delivery truck, or merely sitting on the project lot waiting to be installed. Such a policy will usually be in force from the start of construction to acceptance of the completed project (Figure 16.4). It is typically purchased by the project owner, although the general contractor may purchase it if it is stipulated as a condition of the contract. In most Builder's Risk policies, there are fewer exclusions; many even cover flood, earthquake, and testing. In addition, they provide broader transit and off-premises coverage. Nevertheless, a potential problem may arise with a separate Builder's Risk policy: securing permanent coverage when the policy expires. Builder's Risk policies typically contain a provision stating when the coverage will expire, but this provision will vary according to the policy. Such provisions may create potential problems should permanent commercial property insurance not be put in place in a timely manner.

Furthermore, the owner always has the option to delegate responsibility for obtaining the required Builder's Risk insurance to the general contractor. This is not an uncommon practice with larger contractors because they are generally more familiar with the market and the type and amount of coverage needed, which is why they prefer to have a degree of control over who will ultimately be insuring the project. It is important to maintain sufficient Builder's Risk insurance, and insurance consultants need to be familiar with the construction contract, particularly clauses pertaining to responsibility for procuring adequate and appropriate insurance. It is incumbent on all stakeholders to know not only what coverage is required by the construction contract but also what coverage is available in the marketplace so that the proposed insurance policy provides appropriate protection against the many potential challenges and exposures that may arise.

16.5.3   Errors and Omissions

Errors and omissions policies are basically malpractice insurance and protect a professional from claims of malpractice in the execution of his or her work relating to an error or omission in providing services that can lead to a lawsuit. Errors or omissions can occur in almost any transaction, although many owners tend to expect and believe that the architect will produce a perfect set of design and Contract Documents. Design professionals know this is not possible, yet hesitate to discuss the potential of errors and omissions with the client. It is in the interest of the design professional that this discussion take place.

E&O coverage extends to the payment of defense costs, court costs, and costs resulting from any judgments up to the policy limit. This insurance may exclude negligent acts other than errors and omissions ("mistakes") and is most often obtained by consultants, brokers, and various agents. Because liability for a design professional's negligence claims are not addressed in CGL policies, the E&O policy becomes the professional's primary protection tool. The professional categories covered include architects, engineers, accountants, and attorneys for alleged professional errors and omissions that may amount to negligence. Errors and omissions insurance protects a company from claims when a client holds the company responsible for errors or for the failure of executed work to perform as promised in the contract. The prudent administrator or consultant therefore tries to provide a "safety net" to ensure that mistakes are caught and corrected before causing major problems. Checklists and construction reviews are very helpful in this respect.

Owners should watch out for gaps in coverage, which are common in E&O policies. A gap, or lapse, may be caused by a professional's not renewing a policy on the same day it is to expire. Some carriers will not allow professionals to back-date their coverage to the expiration date without a valid and acceptable explanation (e.g., a natural disaster or personal medical issue that prevented on-time renewal) in addition to a signed warranty letter to the carrier that states that the professional is unaware of any pending claims. For example, if the effective date of insurance is 01/01/2012 and coverage is to expire on 01/01/2013, and the professional fails to renew the E&O coverage on or before 01/01/2013, he or she may have to enroll with a gap in coverage, resulting in a loss of coverage of prior acts. This means that the professional will have no coverage for any business placed "prior" to the new effective date. Some carriers may allow a 30- to 45-day grace period, although it is not uncommon for them not to do this. Perhaps the most important thing the administrator can do to control professional liability is to educate the client and reach a clear understanding that errors will occur and that by working together such errors can be corrected in a timely manner.

16.5.4   Workers' Compensation

Workers' compensation generally provides compensation and medical care for employees who are injured in the course of employment, in exchange for mandatory relinquishment of their right to sue their employers for negligence. In the United States, this insurance is not required by every employer and, furthermore, in some states it may not be required for small businesses that employ less than three to five employees. Nevertheless, in today's litigious environment, having such coverage is strongly recommended if at all possible. The alternative may mean being set up for a massive legal bill or unlimited personal and/or corporate financial liability, including punitive damages that could run in the millions. In this type of plan, general damages for pain and suffering and punitive damages for employer negligence are generally not included.

16.5.5   Professional Liability

Within this environment of increasing litigation, the private practice of a design professional or contractor can be particularly vulnerable, not least because of the damage that even an unfounded lawsuit can do to a reputation and financial stability. Business insurance is especially imperative because it offers protection to both the owner and the business with coverage for claims related to allegations of negligent activities or failure to use reasonable care. To minimize potential risks of litigation there are several steps that can be taken. The following are five major areas of interaction between the consultant/administrator and other members of the project team where the most can be done to protect from liability:

- Professionalism
- Interpersonal relationships
- Business procedures
- Technical procedures
- Professional liability insurance

Of note, some insurance policies go further than the standard professional liability coverage, which in most cases does not cover defamation (libel and slander), breach of contract or warranty, security, personal injury, intellectual property, and the like. Coverage can often be added to provide indemnity "for any civil liability." The operative clause of a civil liability policy is long and normally consists of an extensive list of exclusions so that liabilities, such as employers' and public liability, are not covered under this policy because they are the subject of other forms of insurance. Whatever the case, it is imperative that before signing a policy, it is well understood, preferably with the assistance of an attorney.

16.5.6   Standard Documents and Related Issues

Construction projects in the United States often involve the use of standard documents published by organizations such as the American Institute of Architects or the Associated General Contractors of America (AGC), to name only two. Some project owners nevertheless prefer to prepare their own documents, using a combination of standard and other forms. In both the AIA and AGC documents, project owners are required to provide Builder's Risk insurance covering the interests of all those involved in the project. This policy should provide the following:

- "All-risk" or comparable coverage
- Coverage for property at the job site and material stored offsite or in transit
- Coverage for all parties/stakeholders to the contract—for example, owner, lenders, contractors, subcontractors, architect/engineers
- Permission for waivers of subrogation among the parties
- Coverage for the duration of the project

The construction contract itself, although not a party, remains a crucial element of the construction project. Contractors' equipment insurance should provide coverage for construction equipment (e.g., forklifts, bulldozers, mobile tools). The project owner can also add the construction project to its regular commercial property policy, ensuring that there is compliance with the preceding criteria, or alternatively purchase separate builder's risk coverage. This approach has the advantage of significantly broader coverage than that provided by standard commercial property insurance. In some Contract Documents (e.g., AIA), should the owner decide not to purchase Builder's Risk insurance, the company must inform the general contractor in writing prior to commencement of the project. This gives the contractor the option of obtaining the necessary insurance, including protecting the interests of all parties, and charging this coverage to the owner.

The Joint Contracts Tribunal (JCT) in the United Kingdom has a Standard Building Contract, which is considered to be one of the most common standard contracts used to procure building work; it is updated regularly to take into account changes in legislation and industry practice and relevant court decisions from litigation. For example, the JCT 2005 contract looks very different from the JCT 1998 contract, even though the insurance provisions have not been significantly altered. Today most U.K. commercial building work is carried out under a standard form of contract, with or without amendments. Notably, the JCT has recently launched a new set of publications, the JCT 2011 Tracked Change Documents, which highlight the differences between JCT 2005 and the latest 2011 edition.

There are a number of factors that need be recognized and taken into consideration when taking out an insurance policy for a building project:

- Several parties have insurable interests in the overall construction project.
- Materials and equipment can normally be located on or off the job site, and at different times may belong to the owner, general contractor, or subcontractors. Some material and equipment may be owned by suppliers, but these individuals or entities are not considered to be subcontractors. Because their interests are rarely covered by construction policies, they have to be specifically added.

- To make it easier to purchase insurance and to avoid potential gaps in coverage, one of the parties to the contract will usually be required to assume responsibility for insuring the project on behalf of all parties.
- Responsibilities of the various stakeholders, including obtaining insurance, are generally clearly articulated and delineated in the construction contract.

In building construction, losses may be the result of many factors, including:

- Negligence, lack of skill and care, and failings in design, specification, workmanship, or materials
- The building construction process—that is, materials, workmanship, security, heat work, health and safety procedures, and so on
- Environmental factors, including weather, ground conditions, and proximity to hazards
- The inability to complete a contract on time, resulting in financial losses (whether due to insured perils or insolvency of the contractor)
- Close onsite proximity of various contractors, subcontractors, and professionals, leading to a significant public liability risk in addition to the employer's liability exposures resulting from carrying, lifting, working at height, confined spaces, collapse, dropping, toppling, and so on

If the general contractor sustains a loss due to the owner's failure to obtain or maintain coverage without notifying the general contractor, the owner will then have to assume responsibility for all reasonable damages the general contractor sustains. To meet contractual insurance obligations and requirements, the insurance consultant may be required to negotiate modifications in coverage to comply with the contract. In cases where it is not possible to obtain specific coverage for, say, flood or earthquake, modifications to the construction contract may be necessary and such requirements deleted.

16.5.7 Influence of Green Features on Insurance in the United States

With green building and the green industry moving deeper into the mainstream, many insurance carriers have started to take notice. Still, professional liability typically reflects how the insurance industry as a whole has yet to fully comprehend the changing demands of green building and to respond with appropriate insurance policies. A number of insurers (e.g., Chubb, Zurich, Lexington) are realizing the new demands of green building and finding that they are at the vanguard of the insurance industry. They possessed the vision to respond rapidly to the growing needs and demands of policyholders, while also understanding the risk reduction and economic value associated with green building construction.

In this respect, Peter Thompson, a vice president at Chubb & Son, says, "For years, Chubb has been a leader in insurance protection for green buildings through its commercial property insurance policies." The Chubb Group of Insurance Companies also says that commercial property owners now have a cost-effective way to repair damaged property using "green" materials and environment-friendly construction techniques. According to the Chubb Group, a new endorsement, "Customarq Classic Package Insurance," offers commercial property owners a "green upgrade" option

... to help pay to repair damaged covered property using environmentally-friendly materials, low-impact construction processes and efficient heating and cooling technology after an insured loss. Chubb will pay the difference between repairing the property to its pre-loss condition and repairing it using green materials and construction techniques, subject to the applicable limit of insurance. The option is available for property that did meet, or no longer met, green certification standards in effect at the time of loss or damage.

"As businesses look to help protect the environment and improve their operations, Chubb continues to innovate around our customers' desire to go green," says Bill Puleo, worldwide mono-line property manager for Chubb Commercial Insurance. "Whether businesses wish to enhance their buildings with green features, achieve LEED certification, or build green from the outset, the property insurance we provide can be tailored accordingly."

Another example of forward thinking is Lexington Insurance, a Chartis company, which provides property–casualty and general insurance. Lexington announced in August 2009 the introduction of Upgrade to Green® Builder's Risk, which provides coverage that supports green building construction and renovation projects registered with LEED or Green Globes®. Upgrade to Green is available as an endorsement to Lexington's Completed Value Builder's Risk policy and extends coverage to address the risks green buildings face in three key areas of construction: project management and administration, site ecological impacts, and consumption of resources. According to a Lexington Insurance 2009 press release,

In the event of a covered loss to a green building, Upgrade to Green provides coverage for the fees of qualified professionals associated with the building's design and restoration, as well as the costs of re-commissioning building systems and replacing vegetative roofs. Additionally, the product responds to changes to the relevant rating system criteria or loss of anticipated rating points as a result of a covered loss. For example, Upgrade to Green provides coverage if the new rating criteria require more points to achieve the anticipated certification level, or if the criteria to achieve the same number of points are more stringent and costly. It also provides extra remuneration for reconstruction or rework to secure new rating points if, as a result of covered damage, anticipated rating points are lost and no longer available.

This is emphasized by Liz Carmody, Lexington senior vice president, who says, "We are committed to supporting sustainable development through innovative green products and services ... Upgrade to Green, which is part of our Ecosurance® portfolio of green products, addresses the unique risks that green construction and renovation presents to property owners." In the event of a covered loss to a green building, Upgrade to Green provides coverage for the fees of professionals qualified in the building's design and restoration; it also covers the costs of recommissioning building systems and replacing vegetative roofs. Furthermore, this coverage is designed to respond to changes in relevant rating system criteria or loss of anticipated rating points as a result of a covered loss.

Zurich Insurance says that stakeholders have to look at potential exposures not usually covered by standard Builder's Risk and property policies but that are possibly unique to green building. According to Mike Halvey, head of real estate at Zurich North America Commercial, these are the three most important elements in "green" coverage[1]:

1. **Betterment of the rating system.** *As the green movement continues to grow, the criteria by which a building can become certified by the U.S. Green Building Council may change. This can adversely affect real estate owners and developers because they may have designed their respective buildings using previous rating criteria. Zurich's Better Green coverage for builder's risk allows you to rebuild your building to the new criteria so that you can still achieve your original desired certified status such as LEED Silver, Gold or Platinum. With the property green coverage, if you lose your LEED certification that qualified you for government tax incentives, utility cost credits, reduced loan rates, or other financial incentives as the result of the loss of or damage to your property, the policy pays for the actual loss sustained up to the aggregate limit on the endorsement.*

2. **Debris recycling.** *Most builder's risk and property policies provide for debris removal. However, some of the LEED criteria require that debris be recycled either on the site or taken to a recycling center, which may be more costly than utilizing a landfill. Zurich's endorsement offers broader removal coverage providing for debris recycling and any associated additional costs.*

3. **LEED-accredited professional and building commissioning expenses.** *The LEED Green Building Rating System requires that qualified engineers help you with redesign or oversee the repair, rebuilding, or replacement of your building. As these professionals are not always on a contractor's staff, Zurich's endorsement would provide coverage for expenses incurred to hire a third-party engineer to assist. As a benefit to customers, Zurich has risk engineers with LEED-accredited professional designations who can also provide recommendations to ensure that a project is being built to the green standards. Zurich will pay for these expenses and losses up to the aggregate limit provided for under the endorsement.*

And, according to J. R. Steele, an attorney and LEED Accredited Professional,[2]

Since green construction, and especially large-scale green construction, is a fairly new phenomenon in the United States, there is very little legal analysis regarding

---

[1]Excerpted from "Green Buildings Insurance" online article (www.zurichna.com/zna/REALESTATE/greenbuildingsinsurancearticle.htm).

[2]This excerpt is from "Green Construction: Initiatives and Legal Issues Surrounding the Trend" in *Business Law Today*, 17(2), 2007.

green building disputes. Oftentimes, the problem faced by green building contractors, owners, and design professionals—especially those new to green construction—is that they fail to recognize that there are differences between a "normal" construction project and a green building project. Consequently, parties to green building contracts often rely on standard contracts that do not necessarily address the risks unique to green building projects. Failing to recognize those risks within the contract creates the potential for disputes and litigation down the road. Therefore, one of the keys to a successful green building project is to recognize and deal with the risks of that particular project.

Steele cites the following as examples of some of the unique risks parties to green construction projects often fail to address:

- *Defining which party is responsible for administrating the LEED-certification process, which can be a time-consuming*
- *Defining who is responsible if the project fails to achieve LEED certification and what sort of damages flow from such a failure*
- *Confirming that there is adequate insurance coverage, including professional liability insurance for design professionals, that takes into account the green nature of the project*
- *Checking warranty and guaranty language to confirm that new green construction procedures, or installation materials and/or techniques, do not void the warranty or guaranty for a product*
- *Dealing with long-term performance goals and length of warranty issues*
- *Determining whether any intellectual property infringements will result from using new green techniques or equipment and who is responsible for dealing with any infringement that may arise*
- *Investigating availability of green construction material and the replacement price for such material*
- *Recognizing in the project construction schedule the length of time and inspection process associated with LEED certification in the project construction schedule*

From the preceding examples, it is clear that there are various issues that parties participating in green building projects need to address. J. R. Steel emphasizes that in the end the parties that will most likely succeed in avoiding costly litigation are those that emphasize open communication, clearly define performance expectations, and fully examine the risks of a green project, as well as deal with these issues in their contract.

Indeed, the insurance industry as a whole needs to follow the lead of these "enlightened" insurance companies and embrace green building practices. In some cases, this may require the creation of new add-ons to current polices (which we have already started to see). However, what is most needed is a more institutional change in valuation and risk assessment that takes into account the sustainable quality of a building project, as well as its potential positive impact on the environment. Many of problems arise when building owners, designers, and builders differ in their interpretation of what constitutes a successful green building, particularly when building owners fail to explicitly communicate their thoughts at the start of the project. Issues of this kind are compounded when the parties are relatively new to the concepts of the green construction.

Two of the main areas that should be addressed are (1) a building's failure to achieve a promised level of green building certification and (2) a building's operational performance. Regarding operational performance, there is some expectation that green buildings will, in addition to reducing environmental impacts, reduce energy and water costs, require less maintenance, and provide other long-term benefits to the building's owner. The point to note, however, is that while a design may incorporate a wide range of green features, there are numerous considerations between design and occupancy that can invariably impact the building's operational performance.

Legal action may be brought against the building owner/developer, the builder, the professional consultant (e.g., architect or engineer), or the product manufacturer. Sometimes an expert may be used in the pretrial stages, possibly to give an affidavit supporting one or more issues of the case. The expert may also serve solely as an expert witness at trial or may play a combination of these roles. There are also times when an expert will serve solely as a consultant to the attorney and remain in the background.

One of the important principles governing insurance contracts is good faith. This basically implies that there is a duty of any entity taking out insurance to disclose all material facts and to expeditiously notify the insurer of any events that may lead to a claim. Indeed, the most common problems relating to insurance are failure by a party to make full disclosure of all material facts when taking out a policy and failure to promptly notify of possible claims. Insurers seek out any breaches of this duty to avoid liability. Also, an insured is required to have an "insurable interest" in the subject matter of the insurance; otherwise, the insurance policy is typically considered to be null and void. The insurable interest may pertain to an interest in the property or to a liability or potential liability, such as damages caused by negligence or breach of contract.

For this reason, the insured can most often recover only what has actually been lost and, based on the indemnity principle, is not permitted to legally make a profit from an insurance policy. Connected to this principle is the concept of subrogation, which basically allows the insurer to take over any claims that the insured might have in place against third parties and to receive any payments or compensation made to the insured by third parties. In many instances, insurance policies, construction contracts, and certificates include what is known as a "waiver of subrogation" (also known as a "transfer of rights of recovery"). This is essentially a process by which insurers are able to transfer risk and to limit the rights of recovery from another party on behalf of the insured. Attorneys often advise clients to have this waiver built into their policy so that there is no possibility that they are in breach of contract by failing to have it endorsed separately for every job requiring it. Whatever the case, it is strongly advised that, prior to finalizing any insurance policy, an attorney be consulted. Policies are frequently vague or inconsistently worded and may contain exclusions that limit their usefulness; therefore, parties taking out insurance should always carefully consider the policy's wording to ensure that it adequately serves the purpose.

Acronyms and Abbreviations

Organizations and Agencies

AAEE	American Academy of Environmental Engineers
ACEEE	American Council for an Energy Efficient Economy
ADAAG	Americans with Disabilities Act Architectural Guidelines
AFA	American Forestry Association
AIA	American Institute of Architects
ANSI	American National Standards Institute
APCA	Air Pollution Control Association
APM	Association for Project Managers
APPA	America Public Power Association
ASAE	American Society of Architectural Engineers
ASCE	American Society of Civil Engineers
ASHRAE	American Society of Heating, Refrigerating, and Air-Conditioning Engineers
ASID	American Society of Interior Designers
ASME	American Society of Mechanical Engineers
ASNT	American Society for Nondestructive Testing
ASPE	American Society of Plumbing Engineers
ASTM	American Society for Testing and Materials
ATBCB	Architectural and Transportation Barriers Compliance Board (Access Board)
AWEA	American Wind Energy Association
BAAQMD	Bay Area Air Quality Management District
BEA	U.S. Bureau of Economic Affairs
BBRS	Board for Building Regulations and Standards
BCA	Building Commissioning Association™
BCDC	Bay Conservation and Development Commission
BDRI	Building Diagnostics Research Institute
BEA	U.S. Bureau of Economic Affairs
BIFMA	Business and Institutional Furniture Manufacturers' Association
BLM	Bureau of Land Management
BOCA	Building Officials and Code Administrators
BOMA	Building Owners and Managers Association
BREEAM	Building Research Establishment Environmental Assessment Method
CalEPA	California Environmental Protection Agency
CalRecycle	California Department of Resources Recycling and Recovery

757

CARB	California Air Resources Board
CEC	California Energy Commission
CHPS	Collaborative for High-Performance Schools
CIBSE	Chartered Institution of Building Services Engineers
CIWMB	California Integrated Waste Management Board
CRI	Carpet and Rug Institute
CRS	Center for Resource Solutions
CSA	Canadian Standards Association
CSC	Construction Specifications Canada
CSI	Construction Specifications Institute; also Construction Standards Institute
CTBUH	Council on Tall Buildings and Urban Habitat
CUWCC	California Urban Water Conservation Council
DBIA	Design-Build Institute of America
DoD	U.S. Department of Defense
DOE	U.S. Department of Energy
DPW	Directorate of Public Works
DWR	Department of Water Resources (California)
EIA	Energy Information Administration
EJCDC	Engineers Joint Contract Documents Committee
EPA	U.S. Environmental Protection Agency
ERDC USACE	Engineer Research and Development Center, U.S. Army Corps of Engineers
ESI	European Standards Institute
FEMA	U.S. Federal Emergency Management Agency
FEMP	Federal Energy Management Program
FERC	Federal Energy Regulatory Commission
FSC	Forest Stewardship Council
FTC	Federal Trade Commission
GBA	Green Building Alliance
GBCI	Green Building Certification Institute
GEI	GREENGUARD Environmental Institute
GSA	General Services Administration
HUD	U.S. Department of Housing and Urban Development
IEA	International Energy Agency
IEEE	Institute of Electrical and Electronics Engineers
IESNA	Illuminating Engineering Society of North America
IFMA	International Facilities Management Association
IGBC	Indian Green Building Council
IH	Insurance Information Institute
ILSR	Institute for Local Self-Reliance
IPCC	Intergovernmental Panel on Climate Change
ISO	International Organization for Standardization
LADWP	Los Angeles Department of Water and Power
LBNL	Lawrence Berkeley National Laboratory
NAE	National Academy of Engineering
NAHB	National Association of Home Builders
NAHB RC	NAHB Research Center
NAS	National Academy of Sciences

NAVFAC	U.S. Naval Facilities Engineering Command
NBI	New Buildings Institute
NCARB	National Council of Architectural Registration Boards
NCBC	National Conference on Building Commissioning
NCPA	Northern California Power Agency
NEMA	National Electrical Manufacturers Association
NFPA	National Fire Protection Association
NFRC	National Fenestration Rating Council
NIBS	National Institute of Building Sciences
NIST	National Institute of Standards and Technology
OEE	Office of Energy Efficiency
OMB	Office of Management and Budget
OSHA	Occupational Safety and Health Administration (or Act)
OSWER	Office of Solid Waste and Emergency Response (EPA)
SBIC	Sustainable Buildings Industry Council
SBTF	Sustainable Building Task Force (California)
SCAQMD	South Coast Air Quality Management District
SCS	Scientific Certification Systems
SEC	Securities and Exchange Commission
SFI	Sustainable Forestry Initiative
SMACNA	Sheet Metal and Air-Conditioning Contractors' National Association
UL	Underwriters Laboratories
ULI	Urban Land Institute
UNEP	United Nations Environment Programme
USACE	U.S. Army Corps of Engineers
USAID	United States Agency for International Development
USDA	U.S. Department of Agriculture
USFS	U.S. Forest Service
USGBC	U.S. Green Building Council
USGS	United States Geological Survey
WBCSD	World Business Council for Sustainable Development
WCED	World Commission on Environment and Development
WHO	World Health Organization
WorldGBC	World Green Building Council

Standards, Legislation, and Programs

ADA	Americans with Disabilities Act
AHERA	Asbestos Hazard Emergency Response Act
ASHRAE 90.1	Building energy standard covering design, construction, operation, and maintenance
ASHRAE 52.2	Standardized method of testing building ventilation filters for removal efficiency by particle size
ASHRAE 55	Standard describing thermal and humidity conditions for human occupancy of buildings
ASHRAE 62	Standard defining minimum levels of ventilation performance for acceptable indoor air quality
ASHRAE 192	Standard for measuring air-change effectiveness

ASTM E408	Standard of inspection-meter test methods for normal emittance of surfaces
ASTM E903	Standard of integrated-sphered test method for solar absorptance, reflectance, and transmittance
BEEP	BOMA Energy Efficiency Program
CAA	U.S. Clean Air Act
CAAQS	California Ambient Air Quality Standards
CBC	California Building Code
CASBEE	Comprehensive Assessment System for Built Environment Efficiency
CERCLA	Comprehensive Environmental Response, Compensation, and Liability Act
CERL	Construction Engineering Research Laboratory (Army Corps of Engineers)
CFR	Code of Federal Regulations
CWA	U.S. Clean Water Act; also known as the FWPCA
DSIRE	Database of State Incentives for Renewables and Efficiency
EESA	Emergency Economic Stabilization Act of 2008
EISA	Energy Independence and Security Act of 2007
EPAct	Energy Policy Act of 2005
EPAct	Energy Policy Act of 1992
EPCA	Energy Policy and Conservation Act
FCAA	Federal Clean Air Act
FFHA	Federal Fair Housing Act
FOIA	Freedom of Information Act
FWPCA	Federal Water Pollution Control Act
GS	Green Seal
I-BEAM	IAQ Building Education and Assessment Model
IgCC	International Green Construction Code
IPMVP	International Performance Measurement and Verification Protocol
LEED	Leadership in Energy and Environmental Design
NAAQS	National Ambient Air Quality Standards
NAHBGreen	National Green Building Program
NEPA	National Environmental Policy Act of 1969
NFPC	National Fire Protection Code
NGBS	National Green Building Standard
PURPA	Public Utilities Regulatory Policy Act of 1978
RCRA	Resource Conservation and Recovery Act
TSCA	Toxic Substances Control Act
UBC	Uniform Building Code

Abbreviated General Terminology

A	area
A/V	audiovisual
AAQS	ambient air quality standards
AC	air-conditioning unit
AC	alternating current
ACH	air change per hour

ACM	asbestos-containing material
ACT AGE	actual age
AE	architect/engineer
AEI	advanced energy initiative
AEO	*Annual Energy Outlook* (DOE/EIA publication)
AF	acre-foot (of water)
AFC	application for certification
AFV	alternative-fueled vehicle
AFY	acre-feet per year
AGMBC LEED-NC	*Application Guide for Multiple Buildings and On-Campus Building Projects* (USGBC document)
AHM	acutely hazardous materials
AHP	analytical hierarchy process
AHU	air-handling unit
AIB	air infiltration barrier
AIRR	adjusted internal rate of return
AL	aluminum
amp/A	ampere
AOR	architect of record
AQMD	air quality management district
AQMP	air quality management plan
ASD	aspirating smoke detector
ATC	acoustical tile ceiling
BAS	building automation system
bcfd	billion cubic feet per day
BC	building code
BEES	Building for Environmental and Economic Sustainability Support
BG	biomass gasification
BIM	building information modeling
BIPV	building integrated photovoltaics
BiQ	building intelligence quotient
BL	building line
BLA	Building Loan Agreement
BM	benchmark
BMP	best management practice
BOD	basis of design; also beneficial occupancy date
BRI	building-related illness
BT	building technology
BTU	British thermal unit
BTUH	British thermal unit per hour
BUR	built-up roofing
CAA	Compliance Assurance Agreement
CAD	computer-aided design
CAFS	certified air filter specialist
CAPM	capital asset pricing model
CATS	cost and time summary
CAV	constant-volume air-handling unit
CBECS	Commercial Building Energy Consumption Survey
CCP	Certified Commissioning Professional

CD	construction division
CDVR	corrected design ventilation rate
CEERT	Coalition for Energy Efficiency and Renewable Technologies
CEU	continuing education unit
CFC	chlorofluorocarbons
CFM	cubic feet per minute
CFS	cubic feet per second
CIR	Credit Interpretation Ruling (USGBC)
CM-a	construction manager-advisor
CMBS	commercial mortgage-backed securities
CM	construction manager
CMU	concrete masonry unit
CO	carbon monoxide
CO	Certificate of Occupancy
CO_2	carbon dioxide
COC	chain-of-custody
COP	coefficient of performance
COS	center of standardization
COTS	commercial off-the-shelf
CPG	Comprehensive Procurement Guidelines
C.P.M.	Critical Path Method
CxA	commissioning authority
dB	decibel
DB	design–build
DBB	design–bid–build
DC	direct current
DEC	design energy cost
DHWH/WH	domestic hot water heater/water heater
DOC	determination of compliance
DOR	designer of record
DPB	discounted payback
DPTN	demountable partitions
DS	daylight sensing control; also disconnect switch, downspout
DW	drinking water; also drywall
E	each; also east, modulus of elasticity
E&C	engineering and construction
EAQ	environmental air quality
ECB	Engineering and Construction Bulletin
ECBEMS	energy management system
ECM	energy conservation measure
EEM	energy efficiency measure
EER	energy efficiency ratio
EFF AGE	effective age
EIFS	exterior insulation and finish system
EIR	Environmental Impact Report
EIS	Environmental Impact Statement
EL	easement line; also elbow
EL/ELEV	elevation
EMC	energy management controller

EMCS	energy monitoring and control system
EMF	electromagnetic field
EMP (LEED)	Energy Modeling Protocol
EMS	Environmental Management System
EO	executive order
EOR	engineer of record
EPDM	etheylic-propalene diene molomer
EQP/EQUIP	equipment
ERM	environmentally regulated material
ESA	environmental site assessment
ESC	erosion and sedimentation control plan
ESCO	energy service company
ESFR	early suppression fast response
ESP	energy service provider
ESPC	energy savings performance contract
ETS	environmental tobacco smoke
EUL	expected useful life
FAU	forced-air unit
FF&E	finishes, furniture (fixtures), and equipment
FFL	finished floor line
FIO	for information only
FIX	fixture
FM	Factory Mutual
FPT	functional performance testing
FS	full scale, full size; also federal specification
FTE	full-time equivalent; also full-time employee
FTG	footing
FY	fiscal year
GB	green building
GBTC	green building tax credit
GC	general contractor
GDP	gross domestic product
GEM	Global Environmental Method
GEP	good engineering practice
GF	glazing factor
GHG	greenhouse gas
GHP	geothermal heat pump
GIS	geographic information system
GMP	guaranteed maximum price
gpd	gallons per day
gpf	gallons per flush
gpm	gallons per minute
GRD/GD/G	Grade
GW	gigawatt
GWh	gigawatt (hour)
GWP	global warming potential
GYP	gypsum
GYP BD	gypsum board
H_2S	hydrogen sulfide

HAZMAT	hazardous material
HBA	homebuilder association
HBD/HDB	hardboard
HBN	healthy building network
HCFC	hydrochlorofluorocarbon
HET	high-efficiency toilet
HFC	hydrofluorocarbon
HFR	halogenated flame retardant
HP	horse power
HPB	high-performance building
HRA	health risk assessment
HT	height
HV	high voltage
HVAC	heating, ventilation, and air conditioning
HVAC&R	heating, ventilation, air conditioning, and refrigerants
HWD	hardwood
Hz	Hertz
IAQ	indoor air quality
IDC	interest during construction
IDG	installation design guide
IEPR	Integrated Energy Policy Report
IEQ	indoor environmental quality
in	inch/inches
INFO	information
INSUL	insulate/insulation
IPC	International Plumbing Code
IPLV	integrated part load value
IPMVP	International Performance Measurement and Verification Protocol
IRR	internal rate of return
ISO	independent system operator
ITC	investment tax credit
JCR	Job Cost Report
kg	kilogram
KIT	kitchen
km	kilometer
kV	kilovolt
kVA	kilovolt-ampere
KVAR	kilovolt-ampere reactive
kW	kilowatt
KWh	kilowatt (hour)
LAN	local area network
lav	lavatory
lb	pound
lb/h	pounds per hour
LCA	life-cycle assessment
LCC	life-cycle cost
LCCA	life-cycle cost analysis
LCGWP	life-cycle global warming potential
LCODP	life-cycle ozone depletion potential

LD BRG	load bearing
LE/FE	low-emission/fuel-efficient vehicle
LEED AP	LEED Accredited Professional
LEED-Homes	LEED tool for Homes
LEED-NC	LEED tool for New Construction and Major Renovations
LEED-ND	LEED tool for Neighborhood Development
LID	low-impact development
LL	live load
LPG	liquefied petroleum gas (propane and butane)
LQHC	low-quality hydrocarbon
LR	living room
LTV	loan-to-value
LV	low voltage
LVL	level
LW	lightweight
LZ	lighting zone
m	meter; also million, mega, milli, thousand
M&V	measurement and verification
M/F	male/female ratio
m/s	meters per second
MADA	multi attribute decision analysis
MAIN	maintenance
MAX	maximum
MCS	multiple chemical sensitivity
MDF	medium-density fiberboard
MEP	mechanical, electrical, and plumbing
MERV	minimum efficiency reporting value
MIC	microbiologically influenced corrosion
MMT	million metric tons
MNS	mass notification system
MOU	Memorandum of Understanding
MPPT	maximum power point tracking
MPR	minimum program requirement
MR	moisture-resistant
MSDS	material safety data sheet
MV	megavolt
MVA	megavolt-amperes
MW	megawatt
MWD	metropolitan water district
MWh	megawatt (hour)
N	north
NC	new construction
NCBC	National Conference on Building Commissioning
NEC	National Electrical Code
NES	national energy savings
NG	natural gas
NO	nitrogen oxide
NO$_2$	nitrogen dioxide
nom	nominal

NO_x	nitrogen oxides
NPDES	National Pollutant Discharge Elimination System
NPV	net present value
NS	net savings
NTS	not to scale
NZEB	net-zero energy building
O&M	operation and maintenance
O^3	ozone
OC (O/C)	on center
ODP	ozone depletion potential
OH/OVHD	overhead
OM&R	operation, maintenance, and repair
OPR	Owner's Project Requirements
OSA	outside air
OSB	oversight board
oz	ounce
PAH	polycyclic aromatic hydrocarbon
PAR	Project Analysis Report
PBD	particleboard
PBP	payback period
PCA	property condition assessment
PCB	polychlorinated biphenyl
PCC	precast concrete
PDT	project development team
PL	property line
PLF	pounds per lineal foot
PM	project manager
PML	probable maximum loss
PMO	project management oversight
POC	point of contact
PPM	parts per million
PPT	parts per thousand
PRM	performance rating method
PSF	pounds per square foot
PSI	pounds per square inch
PSR	Project Status Report
PTO	permit to operate
PU	per unit
PUC	Public Utilities Commission
PV	photovoltaic; also present value
PVC	polyvinyl chloride
QA	quality assurance
QC	quality control
QTY	quantity
R/RD	radius
RA	return air
RC	refrigerant charge
RD&D	research, development, and demonstration
REC	renewable energy certificate

REF	reference
REINF	reinforcement
REQ	requirement/required
REV	revision
RFI	request for information
RFP	request for proposal
RFQ	request for qualifications
RH	relative humidity
RTU	rooftop packaged unit
RUL	remaining useful life
S/SW	switch
SA	supply air
SAN	sanitary
SBS	sick building syndrome
SCH/SCHED	schedule
SD	smoke detector; also shop drawing, storm drain, supply duct
SEER	seasonal energy efficiency ratio
sq ft	square foot/feet
SFTWD	softwood
SHGC	solar heat gain coefficient
SIR	savings-to-investment ratio
SO$_2$	sulfur dioxide
SOG	slab-on-grade
SOW	scope of work
SO$_x$	oxides of sulfur
SPB	simple payback
SRI	solar reflectance index
STC	sound transmission coefficient
STD	standard
SYM	symbol; also symmetry/symmetrical
SYS	system
T	ton
TAC	toxic air contaminant
TBCx	total building commissioning
THK	thick/thickness
TL	total losses
TOG	total organic gases
TOPO	topography
TP	total phosphorous
TPD	tons per day
TPY	tons per year
TS	tensile strength
TSP	total suspended particulate matter
TSS	total suspended solids
TVOC	total volatile organic compounds
U/UR	urinal
UESC	utility energy service contract
UFGS	unified facilities guide specifications
UH	unit heater

UMCS	utility monitoring and control system
UPC	Uniform Plumbing Code
UST	underground storage tank
UTIL	utility
UV	ultraviolet radiation
V	volt
VAR	variable air volume
VAV	variable-volume air-handling unit
VB	vapor barrier
VENT	ventilation/ventilator
VMT	vehicle miles traveled
VOC	volatile organic compound
VOL	volume
VP	vent pipe
W	watt; also width, wide, west, wire
WB	wet bulb; also wood base
WBDG	*Whole Building Design Guide*
WC	water closet
WD	wood
WH/DHWH	water heater/domestic hot water heater
WP	waterproof; also weatherproof
WPM	waterproof membrane
WSTP	wastewater sewage treatment plant
WW/WTW	wall to wall
yd	yard
y/yr	year
ZEV	zero emissions vehicle

Glossary

Abatement A reduction in the degree or intensity of pollution or its elimination.

Absorption The process by which incident light energy is converted into another form of energy, usually heat.

Accessible The condition of a site, building, facility, or a portion thereof as being in compliance with accessibility guidelines.

Accessible Route A continuous unobstructed path connecting all accessible elements and spaces of a building or facility. Interior accessible routes may include corridors, floors, ramps, elevators, lifts, and clear floor space at fixtures. Exterior accessible routes may include parking access aisles, curb ramps, crosswalks at vehicular ways, walks, ramps, and lifts.

Acid Rain The precipitation of dilute solutions of strong mineral acids, formed of various industrial pollutants that mix in the earth's atmosphere. Sulfur dioxide and nitrogen oxides mix with naturally occurring oxygen and water vapor.

Acrylics A family of plastics used for fibers, rigid sheets, and paints.

Adaptability A design strategy that takes into account potentially different future functions in a space as needs evolve and change. Adaptable design should be considered a sustainable/ green building strategy because it minimizes the need for major renovations or the demolition of a structure to meet future needs.

Adapted Plant A plant that reliably grows in a given habitat with minimal attention from humans in the form of winter protection, pest protection, water irrigation, or fertilization once root systems are established in the soil. Adapted plants are considered to be low maintenance but not invasive.

Addendum A written or graphic instruction issued by the architect prior to execution of the contract, which modifies or interprets the bidding documents by additions, deletions, clarifications, or corrections. An addendum becomes part of the contract documents when the contract is executed.

Adhesive Any substance that is used to bond one surface to another by attachment, including adhesive bonding primers, adhesive primers, and any other primer.

Adobe A heavy clay soil used in many southwestern states to make sun-dried bricks.

Agency (1) The relationship between agent and principal. (2) An organization acting as agent. (3) An administrative subdivision of an organization, particularly in government.

Agent One authorized by another to act in his or her stead or behalf.

Agreement (1) A meeting of minds. (2) A legally enforceable promise or promises between two or among several persons. (3) On a construction project, the document stating the essential terms of the contract between owner and contractor which incorporates by reference the other contract documents. (4) The document setting forth the terms of the contract between the architect and owner or between the architect and a consultant. *Agreement* and *Contract* are often used interchangeably with no intended change in meaning.

Aggregate Fine, lightweight, coarse, or heavyweight grades of sand, vermiculite, perlite, or gravel added to cement to make concrete or plaster.

Air conditioning A process that simultaneously controls the temperature, moisture content, distribution, and quality of air.

Air filter A device designed to remove contaminants and pollutants from air passing through it.

Air-Handling Unit A mechanical unit for air conditioning or movement of air as in direct supply or exhaust of air within a structure.

Air Pollution The presence of contaminants or pollutant substances in the air that may be hazardous to human health or welfare, or that produces other harmful environmental effects.

Aligned Section A section view in which some internal features are revolved into or out of the viewing plane.

Allergen A substance capable of causing an allergic reaction because of an individual's sensitivity to it.

Alternating Current (AC) Electrical current that continually reverses its direction of flow. The frequency at which it reverses is measured in cycles per second, or Hertz (Hz). The magnitude of the current itself is measured in amperes (amps or A).

Alternative Energy Environmentally sound energy that is not extensively used in the United States, such as solar or wind energy (as opposed to fossil fuels).

Alternative Fuels Transportation fuels other than gasoline or diesel, including natural gas, methanol, and electricity.

Alternator A device for producing alternating current (AC) electricity, usually driven by a motor but also by other means, including water and wind power.

Ambient Lighting Lighting in an area from any source that produces general illumination, as opposed to task lighting.

Ambient Temperature The temperature of the surroundings.

American Bond A brickwork pattern consisting of five courses of stretchers followed by one bonding course of headers.

Ammeter A device used for measuring current flow at any point in an electrical circuit.

Ampere (A or Amp) A unit for measuring electric current (the flow of electrons). One amp is 1 coulomb passing in 1 second. One amp is produced by an electric force of 1 volt acting across a resistance of 1 ohm.

Analog The processing of data by continuously variable values.

Anemometer A device used to measure wind speed.

Angle of Incidence The angle between a surface and the direction of incident radiation; here the term applies to the aperture plane of a solar panel. Only minor reductions in power output within plus or minus 15 degrees.

Animal Dander Tiny scales of animal skin.

American National Standards Institute (ASTM) An umbrella organization that administers and coordinates the national voluntary consensus standards system. http://www.ansi.org/.

Approval Written or imprinted acknowledgement that materials, equipment, or methods of construction are acceptable for use in construction work; also, acceptance of a contractor's or owner's request or claim as valid.

Approved Equal Material, equipment, or methods approved by the architect, for use in the work, as being acceptable as an equivalent in essential attributes to the material, equipment, or method specified in the contract documents.

Arc A portion of the circumference of a circle.

Architect's Scale A scale used for expressing dimensions or measurements in feet and inches.

Array Here, a number of solar modules connected together in a single structure.

As-Built Drawings Record drawings completed by the contractor and turned over to the owner at the completion of a project, identifying any change or adjustments made to the conditions and dimensions of the work relative to the original plans and specifications.

Asphalt Shingles Shingles made of asphalt or tar-impregnated paper embedded with a mineral material; asphalt shingles are very fire resistant.

Assumed Liability Liability that arises from an agreement between parties, as opposed to liability that arises from common or statutory law.

ASTM International Formerly the American Society for Testing Materials, an organization that develops and publishes testing standards for materials and specifications used by industry. http://www.astm.org/

Authority Having Jurisdiction (AHJ) The governmental body responsible for the enforcement of any part of the standard codes, or the official or agency designated to exercise such a function and/or the architect.

Axial Load A weight distributed symmetrically to a supporting member, such as a column.

Axonometric Projection A set of three or more views in which the object appears to be rotated at an angle so that more than one side is seen

Backfill Any deleterious material (sand, gravel, etc.) used to fill an excavation.

Baffle A single opaque or translucent element used to diffuse or shield a surface from direct or unwanted light.

Ballast An electrical "starter" required by certain lamp types, especially fluorescents.

Balloon Framing A system of wood construction in which studs are continuous without an intermediate plate for the support of second-floor joists.

Baluster A vertical member that supports handrails or guardrails.

Balustrade A horizontal rail held up by a series of balusters.

Banister The part of a staircase that fits on top of the balusters.

Bar Chart A calendar that graphically illustrates a projected time allotment to achieve a specific function.

Base A trim or molding piece found at the interior intersection of the floor and the wall.

Beam A weight-supporting horizontal member.

Base Building The core (common areas) and shell of a building and its systems that typically are not subject to improvements to suit tenant requirements.

Base Flashing The flashing that covers the edges of a membrane. *See* Flashing.

Batten A narrow strip of wood used to cover a joint.

Batt Insulation An insulating material formed into sheets or rolls with a foil or paper backing, installed between framing members.

Bearing Wall A wall that supports any vertical loads in addition to its own weight.

Benchmark A point of known elevation from which surveyors can establish all grades.

Bill of Material A list of standard parts or raw materials needed to fabricate an item.

Bio-Based Derived from natural renewable resources such as corn, rice, or beets.

Biodegradable Composed primarily of naturally occurring elements that can be broken down and absorbed naturally into the ecosystem.

Biodiversity The tendency in ecosystems, when undisturbed, to support a great variety of species, forming a complex web of interaction. Human population pressure and resource consumption tend to dangerously reduce biodiversity; diverse communities are less subject to catastrophic disruption.

Blackwater Wastewater generated from toilet flushing, which has a higher nitrogen and fecal coliform level than gray water. Some jurisdictions include water from kitchen sinks or laundry facilities in the definition of blackwater.

Blistering A condition in which air or moisture is trapped beneath paint, creating bubbles that break into flaky particles and ragged edges.

Blocking The use of internal members to provide rigidity in floor and wall systems. Also used for fire draft stops.

Blueprints A set of documents containing all instructions necessary to manufacture a part. The key sections of a blueprint are the drawing, the dimensions, and the notes. Although traditional blueprints are blue, modern reproduction techniques now permit printing of black and white as well as colors.

Board Foot (B.F.) A unit of lumber measure equaling 144 cubic inches; the base unit (B.F.) is 1 inch thick and 12 inches square, or $1 \times 12 \times 12 = 144$ cubic inches.

Boiler A piece of equipment designed to heat water or generate steam.

Bond In masonry, the interlocking system of brick or block to be installed.

Boundary Survey A mathematically closed diagram of the complete peripheral boundary of a site, reflecting dimensions, compass bearings, and angles. The boundary survey should bear a licensed land surveyor's signed certification and may include a metes and bounds or other written description.

Breezeway A covered walkway with open sides between two parts of a structure.

Brick Pavers Special brick to be used on a floor surface.

British Thermal Unit (BTU) The amount of heat energy required to raise 1 pound of water from a temperature of $60°F$ to $61°F$ at 1 atmosphere pressure. One watt hour equals 3,413 BTU.

Building Codes Prevailing regulations, ordinances, or statutory requirements set forth by government agencies governing building construction practices and owner occupancy, adopted and administered for the protection of public health, life safety, and welfare. Building codes are interpreted to cover structural, HVAC, plumbing, electrical, life safety, and vertical transportation codes.

Building Density The total floor area of a building divided by the total area of the site (square feet per acre).

Building Envelope The enclosure of a building that protects the interior from outside elements—namely, the exterior walls, roof, and soffit areas.

Building Inspector A representative of a government authority employed to inspect construction for compliance with codes, regulations, and ordinances.

Building Line An imaginary line determined by zoning departments to specify the area of a lot on which a structure may be built (also known as a setback).

Building Permit A permit issued by an appropriate government authority allowing construction of a project in accordance with approved drawing and specifications.

Building-Related Illness A diagnosable disease or illness that can be traced to a specific pollutant or source within a building. *Also see* Sick Building Syndrome.

Building Systems Interacting or independent components or assemblies, forming single integrated units, that make up a building and its site work, such as pavement and flatwork, structural frame, roofing, exterior walls, plumbing, HVAC, and electrical.

Build-Out The interior construction and customization of a space (including services, space, and stuff) to meet the tenant's requirements; either new construction or renovation (also referred to as fit-out or fit-up).

Byproduct A material other than the intended product, generated as a result of an industrial process.

Caisson A below-grade concrete column for the support of beams or columns.

Callback A request by a project owner to the contractor to return to the job site to correct or redo some item of work.

Candela A common unit of light output from a source.

Cantilever A horizontal structural condition in which a member extends beyond a support, such as a roof overhang.

Capillary The action by which the surface of a liquid, where it is in contact with a solid, is elevated or depressed.

Carbon Footprint A measure of the overall contribution—by an individual, family, community, company, industry, product, or service—of carbon dioxide and other greenhouse gases in the atmosphere, which takes into account energy use, transportation methods, and other forms of carbon emission. A number of carbon calculators have been created to estimate carbon footprints, including one from the U.S. Environmental Protection Agency.

Carbon Sinks Carbon reservoirs and conditions that take in and store more carbon (carbon sequestration) than they release. Carbon sinks can serve to partially offset greenhouse gas emissions. Forests and oceans are common carbon sinks.

Carcinogens Substances that cause cancer in humans.

Casement A type of window hinged to swing outward.

Catch Basin A complete drain box made in various depths and sizes; water drains into a pit and then from it through a pipe connected to the drain box.

Caulk Any type of material used to seal walls, windows, and doors to keep out the weather.

Cavity Wall A masonry wall formed with an air space between each exterior face.

Cement Plaster Plaster comprising cement rather than gypsum.

Central HVAC System A system that produces a heating or cooling effect in a central location for subsequent distribution to satellite spaces that require conditioning.

Centrifugal A particular type of fluid-moving device that imparts energy to the fluid by high-velocity rotary motion through a channel; fluids enter the device along one axis and exit along another axis.

Certificate for Payment A statement from the architect to the owner confirming the amount of money due the contractor for work accomplished, for materials and equipment suitably stored, or both.

Certificate of Insurance A document issued by an authorized representative of an insurance company stating the types, amounts, and effective dates of coverage in force for a designated insured.

Certificate of Occupancy A document issued by a government authority certifying that all or a designated portion of a building complies with the provisions of applicable statutes and regulations, and permitting occupancy for its designated use.

Certificate of Substantial Completion A certificate prepared by the architect, on the basis of an inspection, stating that the work or a designated portion thereof is substantially complete, that establishes the date of substantial completion; states the responsibilities of the owner and the contractor for security, maintenance, heat, utilities, damage to the work, and insurance; and taxes the time within which the contractor shall complete the items listed.

Certified Wood A wood-based material used in building construction that is supplied from sources that comply with sustainable forestry practices, protecting trees, wildlife habitat, streams, and soil as determined by the Forest Stewardship Council or other recognized certifying organization.

Cesspool An underground catch basin for the collection and dispersal of sewage.

Chain of Custody A document that tracks the movement of a product from the point of harvest or extraction to the end user.

Change Order A written and signed document between the owner and the contractor authorizing a change in the work to be executed or an adjustment in the contract sum or time. The contract sum and time may be modified only by a change order. A change order may be in the form of additional compensation or time or in the form of less compensation or time (known as a deduction).

Checklist A list of items used to check drawings.

Chiller A piece of equipment designed to produce chilled water.

Circuit A continuous system of conductors providing a path for electricity.

Circuit breaker A device that acts as an automatic switch that shuts off power when it senses too much current.

Circumference The length of a line that forms a circle.

Clear Floor Space The minimum unobstructed floor or ground space required to accommodate a single, stationary wheelchair and occupant.

Clerestory A window or group of windows placed above the normal window height, often between two roof levels.

Coefficient of Utilization (CU) The ratio of light energy (lumens) from a source, calculated as received on the workplane, to the light energy emitted by the source alone.

Column A vertical weight-supporting member.

Combustion An oxidation process that releases heat; onsite combustion is a common heat source for buildings.

Commissioning (Cx) A systematic process to verify that building components and systems function as intended and required; systems may need to be recommissioned at intervals during a building's life cycle.

Common Use Interior and exterior rooms, spaces, or elements that are made available for the use of a restricted group of people (for example, occupants of a homeless shelter, the occupants of an office building, or guests of such occupants).

Component A fully functional portion of a building system, piece of equipment, or building element.

Composite Wood A product consisting of wood or plant particles or fibers bonded together by a synthetic resin or binder. Examples include plywood, particleboard, OSB, MDF, and composite door cores.

Composting Toilet A dry plumbing fixture that contains and treats human waste via a microbiological process.

Compressor A device designed to compress (i.e., increase the density) of a compressible fluid; a component used to compress a refrigerant; a component used to compress air.

Computer-Aided Drafting (CAD) A method by which engineering drawings may be produced on a computer.

Computer-Aided Manufacturing (CAM) A method by which a computer uses a design to guide a machine that produces parts.

Concrete Block A rectangular concrete form containing cells.

Condensation The process by which moisture in the air becomes water or ice on a surface (such as a window) whose temperature is lower than that of the air.

Condenser A device that condenses a refrigerant; an air-to-refrigerant or water-to-refrigerant heat exchanger; part of a vapor compression or absorption refrigeration cycle.

Conductor A material used to transfer, or conduct, electricity, often in the form of wires.

Conduit A pipe or elongated box used to house and protect electrical cables.

Conservation The preservation and renewal, when possible, of human and natural resources and the use, protection, and improvement of natural resources according to recognized principles that ensure their highest economic or social benefits.

Construction Documents All drawings, specifications, addenda, and other pertinent construction information associated with a specific construction project.

Contamination Intrusion of undesirable elements; the addition of foreign matter to a substance that reduces the value of that substance or interferes with its intended use.

Contingency Allowance A sum included in the project budget to cover unpredictable or unforeseen items of work, or changes in the work subsequently required by the owner.

Contour Line A line that represents the change in level from a given datum point.

Contract A legally enforceable promise or agreement between two or among several person. *See also* Agreement.

Convection The transfer of heat through the movement of a liquid or gas.

Cooling tower A piece of equipment designed to reject heat from a refrigeration cycle to the outside environment through an open-cycle evaporative process; an exterior heat rejection unit in a water-cooled refrigeration system.

Cornice The projecting or overhanging structural section of a roof.

Corrosion Dissolution and wearing away of metal caused by a chemical reaction such as between water and pipes or between chemicals and a metal surface.

Cost Appraisal Evaluation or estimate (preferably by a qualified professional appraiser) of the value, cost, utility, market for, or other attribute of land or a facility.

Cost Estimate A preliminary statement of approximate cost, determined by one of the following methods: (1) Area and volume—cost per square or cubic foot of the building. (2) Unit cost—cost of one unit multiplied by the number of units (for example, in a hospital, the cost of one patient unit multiplied by the total number of patient units). (3) Unit in place—cost in place of a unit, such as doors, cubic yards of concrete, and squares of roofing.

Coving The curving of a floor material against a wall to eliminate the open seam between floor and wall.

Cradle-to-Grave Analysis Analysis of the impact of a product from the beginning of its source-gathering processes, through the end of its useful life, to disposal of all waste products. *Cradle-to-cradle* is a related term signifying the recycling or reuse of materials at the end of their first useful life.

Crawl Space The area under a floor that is not fully excavated but excavated sufficiently to allow one to crawl under it to get to the electrical or plumbing devices.

Critical Path A schedule of tasks or sequences developed by the contractor, after carefully considering dependencies among construction tasks, to ensure that there are no delays on "critical" elements that would delay subsequent tasks and thereby delay project completion. The sequence of tasks that have no tolerance for delay, constitutes the critical path.

Cross-Section A slice through a portion of a building or member that depicts the various internal conditions of that portion or member.

Current The flow of electric charge in a conductor between two points having a difference in electrical potential (voltage), measured in amps.

Curtain Wall An exterior wall that provides no structural support.

Cut-Off Voltage The voltage levels at which the charge controller (regulator) disconnects the photovoltaic array from the battery, or the load from the battery.

Damper A device designed to regulate the flow of air in a distribution system.

Dangerous or Adverse Conditions Conditions that may pose a threat or possible injury to a field observer on a construction site, and that may require the use of special protective clothing, safety equipment, access equipment, or any other precautionary measures.

Date of Agreement The date stated in an agreement. If no date is stated, it could be the date on which the agreement is actually signed, if this is recorded, or it may be date established by the award.

Date of Commencement of the Work The date established in a notice to the contractor to proceed or, in the absence of this notice, the date of the owner/contractor agreement or such other date and may be established therein.

Date of Substantial Completion the date certified by the architect when the work or a designated portion thereof is sufficiently complete, in accordance with the contract documents, for the owner to occupy it or designated portion thereof for the use for which it is intended.

Datum Point Reference point.

Daylight Factor (DF) The ratio of daylight illumination at a given point on a given plane, from an obstructed sky of assumed or known illuminance distribution, to the light received on a horizontal plane from an unobstructed hemisphere of this sky, expressed as a percentage. Direct sunlight is excluded for both values of illumination. DF is the sum of the sky component, the external reflected component, and the internal reflected component. The interior plane is usually a horizontal workplane. If the sky condition is the CIE standard overcast condition, the DF remains constant regardless of absolute exterior illuminance.

Daylighting The controlled admission of natural light into a space by glazing, with the intent of reducing or eliminating electric lighting. By utilizing solar light, daylighting creates a stimulating and productive environment for building occupants.

Dead Load The weight of a structure and all its fixed components.

Decibel (dB) A unit of sound level or sound-pressure level that is 10 times the logarithm of the square of the sound pressure divided by the square of the reference pressure, or 20 micropascals.

Deconstruction The process of taking apart a structure with the primary goal of preserving the value of all useful building materials so that they may be reused or recycled.

Defective Work Work not conforming with contract requirements.

Deferred Maintenance Physical deficiencies that cannot be remedied with routine maintenance, normal operating maintenance, and so forth, excluding de minimis conditions that generally do not present a material physical deficiency to the subject property.

Design-Build Construction Construction for which an owner contracts with a prime, or main, contractor to provide all design and construction services for a project. Use of the design-build project delivery system has grown from 5% of U.S. construction in 1985 to 33% in 1999, and is projected to surpass low-bid construction in 2005. A design-build contract that is extended to include selection, procurement, and installation of all furnishings, furniture, and equipment is called a "turnkey" contract.

Details An enlarged drawing to show a structural aspect, an aesthetic consideration, or a solution to an environmental condition, or to express relationships among materials or building components.

Debt-to-Capital Ratio A measure of a company's financial leverage, calculated as long-term debt divided by long-term capital. Total debt includes all short-term and long-term obligations. Total capital includes all common stock, preferred stock, and long-term debt. Debt-to-capital ratio can provide a more accurate view of a company's long-term leverage and risk, since it considers long-term debt and capital only. By excluding short-term financing in its calculation, the ratio provides an investor with a more accurate picture of the capital structure a company will have if it owns a stock over a long period of time.

Diffuser A device designed to supply air to a space while providing a good mix of supply and room air and avoiding drafts; normally ceiling installed.

Digital The processing of data by numerical or discrete units.

Dimension Line A thin unbroken line (except in structural drafting) with each end terminating in an arrowhead, used to define the dimensions of an object. Dimensions are placed above the line, except in structural drawings where the line is broken and dimensions are placed in line breaks.

Direct Costs (Hard Costs) The aggregate costs of all labor, materials, equipment, and fixtures necessary for the completion of construction or improvements.

Direct Costs Loan; Indirect Costs Loan The portion of the loan amount applicable and equal to the sum of the loan budget amounts for direct costs and indirect costs, respectively, shown on the borrower's project cost statement.

Direct Current (DC) Electrical current that flows only in one direction, although it may vary in magnitude. Contrasts with alternating current.

Discount Factor The translation of expected benefits or costs in any given future year into present-value terms. The discount factor is equal to $1/(1 + i)t$, where i is the interest rate and t is the number of years from the date of initiation of the program or policy until the given future year.

Discount Rate The interest rate used in calculating the present value of expected yearly benefits and costs.

Disinfectant A chemical or physical process that kills pathogenic organisms in water, air, or on surfaces. Chlorine is often used to disinfect sewage treatment, water supplies, wells, and swimming pools.

Dormer A structure that projects from a sloping roof to form another roofed area. This new area is typically used to provide a surface on which to install a window.

Downcycling The recycling of a material in such a manner that much of its inherent value is lost.

Downspout A pipe connected to a gutter to conduct rainwater to the ground or sewer.

Drip Irrigation System An irrigation system that slowly applies water to the root system of plants to maximize transpiration while minimizing wasted water and topsoil runoff. Drip irrigation usually involves a network of pipes and valves that rest on the soil or underground at the root zone.

Drywall An interior wall covering installed in large sheets and made from gypsum board.

Duct Usually a sheet metal form used for the distribution of cool or warm air throughout a structure.

Due Diligence Here, a walk-through survey of and appropriate inquiries into the physical condition of a commercial property's improvements, usually in connection with a commercial real estate transaction. The scale and type of this survey or inquiry may vary for different properties and different purposes.

Dwelling Unit A single unit that provides a kitchen or food preparation area in addition to rooms and spaces for living, bathing, sleeping, and so forth. A dwelling unit can be single-family home or a townhouse used as a transient group home; an apartment building used as a shelter; guestrooms in a hotel that provide both sleeping and food preparation areas; and other similar facilities used on a transient basis. For purposes of these guidelines, use of the term "Dwelling Unit" does not imply that the unit is used as a residence.

Easement The right or privilege to have access to or through another's piece of property—for example, a utility easement.

Eave The portion of the roof that extends beyond the outside wall.

Ecological/Environmental Sustainability Maintenance of ecosystem components and functions for future generations.

Ecological Impact The impact that a human-caused or natural activity has on living organisms and their nonliving environment.

Ecosystem The interacting system of a biological community and its nonliving environmental surroundings; an ecological community together with its environment, functioning as a unit.

Egress A continuous and unobstructed way of exit travel from any point in a building or facility to a public way, comprising vertical and horizontal travel and possibly intervening room spaces, doorways, hallways, corridors, passageways, balconies, ramps, stairs, enclosures, lobbies, horizontal exits, courts, and yards. An accessible means of egress is one that complies with International Code Council guidelines, Section 1007, and does not include stairs, steps, or escalators. Areas of rescue assistance or evacuation elevators may be included as accessible means of egress.

Electric Current The flow of electrons, measured in amps.

Electrical Grid A network for electricity distribution across a large area.

Electricity The movement of electrons (subatomic particles), produced by a voltage, through a conductor.

Electrode An electrically conductive material, forming part of an electrical device, often used to lead current into or out of a liquid or gas. In a battery, electrodes are also known as plates.

Element An architectural or mechanical component of a building, facility, space, or site—for example, telephone, curb ramp, door, drinking fountain, seating, or water closet.

Embodied Energy The total energy that a product may be said to "contain," including all energy used in growing, extracting, and manufacturing the product, and the energy used to transport it to the point of use. The embodied energy of a structure or system includes the embodied energy of its components plus the energy used in its construction.

Emission The release or discharge of a substance into the environment, generally the release of gases or particulates into the air.

Energy Power consumed multiplied by duration of use. For example, the use of 1000 watts for 4 hours is 4000 watt hours.

Energy Conservation Thoughtful and frugal management of energy to conserve it. The result of such deliberate and planned conservation results in saving energy for future use.

Energy Efficiency A reduction in the amount of electricity and/or fuel to do the same work, typically without changing the quality of the services provided. Efficiency can be accomplished by utilizing high-efficiency appliances, better insulation, better building design, and mechanical improvements.

ENERGY STAR® Rating A designation given by the EPA and the U.S. Department of Energy (DOE) to appliances and products that exceed federal energy efficiency standards. The ENERGY STAR label helps consumers identify products that are energy-efficient and thus save money.

Engineer's Scale A scale used for dimensions that are in feet and decimal parts of a foot or when the scale ratio is a multiple of 10.

Environmental Tobacco Smoke (ETS) A mixture of smoke from the burning end of a cigarette, pipe, or cigar, and the smoke exhaled by a smoker (also referred to as secondhand or passive smoke). *See* Smoke-free Homes Program at www.epa.gov/smokefree.

Environmentally Friendly The degree to which a product does not harm the environment, including biosphere, soil, water, and air.

Epicenter The point of the earth's surface directly above the focus or hypocenter of an earthquake.

Expansion Joint A joint often installed in concrete construction to reduce cracking and to provide workable areas.

Expected Useful Life (EUL) The average number of years that an item, component, or system is estimated to function when installed new, assuming routine maintenance is practiced.

Exploded View A pictorial view of a device in a state of disassembly, showing the appearance and interrelationship of parts.

Extension Line A line used to visually connect the ends of a dimension line to the relevant feature on the part. Extension lines are solid and drawn perpendicular to the dimension line.

Façade The exterior covering of a structure.

Face of Stud (F.O.S.) The outside surface of the stud, used most often in dimensioning or as a point of reference.

Fascia A horizontal member located at the edge of a roof overhang.

Facility All or any portion of buildings, structures, site improvements, complexes, equipment, roads, walks, passageways, parking lots, or other real or personal property located on a site.

Felt Tar-impregnated paper used for water protection under roofing and siding materials and sometimes used under concrete slabs for moisture resistance.

Fiber Optics Optical, clear strands that transmit light without electrical current; sometimes used for outdoor lighting.

Fillet A concave internal corner in a metal component, usually a casting.

Filter A device designed to remove impurities from a fluid passing through it.

Final Completion Completion of work in accordance with the terms and conditions of the contract documents.

Final Inspection Final review of a project by the architect to determine final completion prior to issuance of the final certificate for payment.

Final Payment Payment made by the owner to the contractor, upon the architect's issuance of the final certificate for payment, of the entire unpaid balance of the contract sum as adjusted by change orders.

Finish Grade The soil elevation in its final state upon completion of construction.

Fire Barrier A continuous membrane, such as a wall, ceiling, or floor assembly, that is designed and constructed to a specified fire-resistant rating to hinder the spread of fire and smoke. The resistance rating is based on a time factor. Only fire-rated doors may be used in these barriers.

Fire Compartment of Fire Zone An enclosed space in a building that is separate from all other areas or sections by fire barriers having fire resistance ratings.

Fire Door A door between different types of construction that has been rated as able to withstand fire for a certain amount of time.

Fire Resistance Rating A classification, or hourly rating that in building codes is usually based on the fire endurance required. Fire resistance ratings are assigned by building codes for various types of construction and occupancies, and are usually given in half-hour increments.

Firestop Blocking placed between studs or other structural members to resist the spread of fire.

Firewall A type of fire barrier of noncombustible construction that subdivides a building or separates adjoining buildings to resist the spread of fire. A firewall has a fire resistance rating as prescribed in the National Building Council (NBC) and has the structural ability to remain intact under fire conditions for the required fire-rated time.

Flashing A thin, impervious sheet of material placed in construction to prevent water penetration or to direct the flow of water. Flashing is used especially at roof hips and valleys, roof penetrations, and joints between a roof and a vertical wall, and in masonry walls to direct the flow of water and moisture.

Floodplain Mostly level land along rivers and streams that may be submerged by floodwater. A 100-year floodplain is estimated to flooded once every 100 years.

Floor Joist A structural member for the support of floor loads.

Floor Plan A horizontal section taken at approximately eye level.

Flush Even, level, or aligned.

Flush-Out The operation of mechanical systems for a minimum of two weeks, using 100% outside air, upon completion of construction and prior to building occupancy to ensure safe indoor air quality.

Fly Ash Fine ash waste collected from the flue gases of coal combustion, smelting, or waste incineration.

Footcandle A common unit of illuminance used in the United States. The metric unit is the lux.

Footings Weight-bearing concrete construction elements poured in place in the earth to support a structure.

Footlambert The U.S. unit for luminance. The metric unit is the nit.

Formaldehyde A colorless, pungent, and irritating gas mainly used as a disinfectant and preservative and in synthesizing compounds such as resins.

Fossil Fuels Fuel derived from ancient organic remains such as peat, coal, crude oil, and natural gas.

Foundation Plan A drawing that graphically illustrates the location of various foundation members and conditions that are required for the support of a specific structure.

Frieze A decoration or ornament shaped to form a band around a structure.

Frost Line The depth at which frost penetrates the soil.

Fungi Any of a group of parasitic lower plants that lack chlorophyll, including molds and mildews.

Fuse A device that protects electrical equipment from short circuits. Fuses are made with metals designed to melt when the current passing through them is high enough. When the fuse melts, the electrical connection is broken, interrupting power to the circuit or device.

Galvanized steel Steel that has had zinc applied to its exterior surface to provide protection from rusting.

Gauge The thickness of metal or glass sheet material.

General Conditions When used by contractors; Construction project activities and their associated costs that are not usually assignable to a specific material installation or subcontract—for example, temporary electrical power. When used by everyone else: The contract document (often a standard form) that spells out the relationships between the parties to the contract. A construction example is AIA Document A201.

General Contract Any contract (together with all riders, addenda, and other instruments referred to therein) as contractor or any other person, which requires the general contractor or such other person to provide, or supervise or manage the procurement of, substantially all labor and material needed for completion of the improvements.

Generator A mechanical device used to generate DC electricity. Power is produced by coils of wire passing through magnetic fields inside the generator. Most AC-generating sets are also referred to as generators.

Geothermal Literally, heat from the earth—that is, energy obtained from the hot areas under the surface of the earth. Examples are geysers, molten rocks, and steam spouts.

Gigawatt (GW) A unit of measure of power equal to 1000 million watts.

Gigawatt-Hour (GWh) A unit of measure of energy. One gigawatt -hour is equal to 1 gigawatt used for 1 hour, or 1 m used for 1000 hours.

Girder A horizontal structural beam for the support of secondary members such as floor joists.

Glare The effect produced by luminance within the field of vision that is sufficiently greater than the luminance to which the eye is adapted. It can cause annoyance, discomfort, or loss of visual performance and visibility.

Global Warming An increase in the average temperature of the earth's surface, which usually precedes an increase in greenhouse gases; it is sometimes called the *greenhouse effect*. Greenhouse gases are released from burning gas, oil, coal, and wood.

Grading The moving of soil to elevate land at a construction site.

Graywater Wastewater that does not contain toilet wastes and can be reused for irrigation after simple filtration. Wastewater from kitchen sinks and dishwashers may not be considered gray water in all cases.

Green Building A building that uses energy, water, materials, and land in a manner that is much more efficient than in buildings that are built to code. Such environments are considered healthy for their occupants, cost-effective, and less expensive to operate and maintain.

Green Design A design, usually architectural, that conforms to environmentally sound principles of construction and material and energy use. A green building, for example, might use solar panels, skylights, and recycled building materials.

Green Energy A popular term for energy produced from renewable energy sources or, sometimes, from clean (low-emitting) energy sources.

Greenfield Land not previously developed beyond agriculture or forestry use.

Greenhouse Gas A gas in the atmosphere that traps some of the sun's heat and prevents it from escaping into space. Greenhouse gases are vital for making the Earth habitable, but their increasing presence contributes to climate change. They include water vapor, carbon dioxide, methane, nitrous oxide, and ozone.

Greenhouse Effect The rise in the temperature of air in the lower atmosphere due to heat trapped by greenhouse gases such as carbon dioxide, methane, nitrous oxide, chlorofluorocarbons, and ozone.

Green Power Electricity generated from renewable energy sources.

Greenwash Disinformation disseminated by an organization to present an environmentally responsible public image.

Grid An electrical utility distribution network.

Grid-Connected, or Grid-Tied An energy-producing system connected to the utility transmission grid.

Groundwater Fresh water beneath the Earth's surface, usually in aquifers, that supplies wells and springs and is a major source of drinking water.

Grout A mixture of cement, sand, and water used to fill joints in masonry and tile construction.

Guardrail A horizontal protective railing used around stairwells, balconies, and changes of floor elevation greater than 30 inches.

Halogen Lamp A type of incandescent globe made of quartz glass and a tungsten filament, which enables it to run at a much higher temperature than a conventional incandescent globe. Halogen's efficiency is greater than that of a normal incandescent lamp but not as great as that of a fluorescent light.

Harmonic Content Frequencies in the output waveform in addition to the primary frequency (usually 50 or 60 Hz). Energy in these harmonics is lost and can cause undue heating of the load.

Harvested Rainwater Captured rainwater used for indoor needs, irrigation, or both.

Hazardous Waste Byproducts of society that can pose a substantial or potential hazard to human health or the environment when improperly managed. Hazardous wastes possess at least one of four characteristics: ignitable, corrosive, reactive, or toxic, or appears on special EPA lists.

Head The top of a window or door frame.

Header A horizontal structural member spinning over openings, such a doors and windows, for the support of weight above the openings.

Header Course In masonry, a horizontal course of brick laid perpendicular to the wall face, used to tie together a double-wythe brick wall.

Heat Exchanger A device designed to efficiently transfer heat from one medium to another (for example, water-to-air, refrigerant-to-air, refrigerant-to-water, and steam-to-water).

Heat Island Effect Higher air and surface temperatures caused by solar absorption and re-emission from roads, buildings, and other structures.

Heat Pump A device that uses a reversible-cycle vapor compression refrigeration circuit to provide cooling and heating from the same unit (at different times).

Heat Recovery A process whereby heat is extracted from exhaust air before the air is dumped to the outside environment; the recovered heat is normally used to preheat incoming outside air. Recovery may be accomplished by heat recovery wheels or heat exchanger loops.

Hertz (Hz) A unit of measurement for frequency. Home mains power is normally 50 Hz in Europe and 60 Hz in the United States. The magnitude of the current is measured in amps.

High-Performance Green Building A building that creates a healthy indoor environment, with design features that conserve water and energy; efficiently use space, materials, and resources; and minimize construction waste.

Hydronic System A heating or cooling system that relies on the circulation of water for heat transfer. A typical example is a boiler with hot water circulated through radiators.

Illuminance The density of the luminous flux incident on a surface, expressed in footcandles or lux. *Illuminance* should not be confused with *illumination*, which is the process of illuminating or the state of being illuminated.

Impervious Surface A surface that promotes runoff of precipitation volumes to prevent infiltration into a subsurface.

Incandescent Lighting Electric lamps that are evacuated or filled with an inert gas and contain a filament (commonly tungsten). The filament emits visible light when heated to extreme temperatures by the passage of electric current through it.

Incident Light Light that shines on the surface of a photovoltaic cell or module.

Indemnification A contractual obligation by which one person or entity agrees to secure another against loss or damage from specified liabilities.

Indirect Cost Statement A statement by the borrower, in a form approved by the lender, of indirect costs incurred and to be incurred.

Indoor Air Pollution The presence of chemical, physical, or biological contaminants in indoor air.

Indoor Air Quality (IAQ) According to the EPA and the National Institute of Occupational Safety and Health: (1) introduction and distribution of adequate ventilation air; (2) control of airborne contaminants; and (3) maintenance of acceptable temperature and relative humidity. According to the American Society of Heating, Refrigeration, and Air-Conditioning Engineers (ASHRAE) Standard 62, 1989: "air in which there are no known contaminants at harmful concentrations as determined by cognizant authorities and with which a substantial majority (80% or more) of the people exposed do not express dissatisfaction."

Indoor Environmental Quality (IEQ) The evaluation of five primary elements—lighting, sound, thermal conditions, air pollutants, and surface pollutants—to provide an environment that is physically and psychologically healthy for a building's occupants.

Industrial Waste Unwanted materials produced in or eliminated from an industrial operation and categorized under a variety of headings, such as liquid wastes, sludge, solid wastes, and hazardous wastes.

Infill Site A site largely located within an existing community. For the purposes of LEED for Homes credits, an infill site is defined as having at least 75% of its perimeter bordering land that has been previously developed.

Inscribed Figure A figure that is completely enclosed by another figure.

Insolation The amount of sunlight reaching an area, usually expressed in watt-hours per square meter per day.

Inspection Examination of work completed or in progress to determine its conformance with the requirements of the contract documents. The Architect ordinarily makes only two inspections, one to determine substantial completion and the other to determine final completion. These inspections should be distinguished from the more general observations made by the Architect on visits to the site during the progress of the Work. The term *inspection* is also used to mean examination of the work by a public official, owner's representative, or others.

Insulation Any material capable of resisting thermal, sound, or electrical transmission.

Integrated Design Team The team of all individuals involved in a project from very early in the design process, including design professionals, owner's representatives, and the general contractor and subcontractors.

Internal Rate of Return The discount rate that sets the net present value of the stream of net benefits equal to zero. The internal rate of return may have multiple values when the stream of net benefits alternates from negative to positive more than once.

Inverter A device that converts DC power from a photovoltaic array/battery to AC power; used for both stand-alone and grid-connected systems.

Irradiance The solar power incident on a surface, usually expressed in kilowatts per square meter. Irradiance multiplied by time gives insolation.

Isometric Drawing A pictorial form in which the main lines are equal in dimension, normally drawn using 30- degree and 90-degree angles.

Jamb The side portion of a door, window, or any other opening.

Joist A horizontal beam used to support a ceiling.

Joule (J) The energy conveyed by 1 watt of power for 1 second; a unit of energy equal to 1/3600 kilowatt-hour.

Junction box A protective enclosure on a PV module where PV strings are electrically connected and where electrical protection devices such as diodes can be fitted.

Key Plan A reduced-scale plan for orientation purposes.

Kilowatt (kW) A unit of electrical power equaling 1000 watts.

Kilowatt-hour (kWh) The amount of energy that derives from a power of 1000 watts over a period of 1 hour; the kWh is a unit of energy. 1 kWh = 3600 kJ.

Landfill A cavity in the ground in which nonhazardous waste is accumulated and eventually covered with dirt and topsoil. Today's landfills are deemed sanitary and require special technology to eliminate methane gas.

Lattice A grille made by crisscrossing strips of material.

Ledger A structural framing member used to support ceiling and roof joists at the perimeter walls.

LEED® The acronym for Leadership in Energy and Environmental Design, which is a sustainable-design building certification system promulgated by the U.S. Green Building Council (USGBC). Also an accrediting program for sustainable design professionals (LEED AP) who have mastered the certification system. See www.usgbc.org/.

LEED Accredited Professional (LEED AP) The credential earned by candidates who passed LEED's accrediting exam between 2001 and June 2009.

Legend A description of any special or unusual marks, symbols, or line connections used in a drawing.

Liability Insurance Insurance that protects the insured against liability from injury to the insured, the insured's property, or another or another's property. In construction, the types of liability insurance are (1) completed operations insurance; (2) comprehensive general liability insurance; (3) contractor's liability insurance; (4) employer's liability insurance; (5) owner's liability insurance; (6) professional liability insurance; (7) property damage insurance; (8) public liability insurance; (9) special hazards insurance.

Lien A monetary claim on a property.

Life-Cycle Cost The sum of all costs of creation and operation of a facility over a period of time.

Life-Cycle Cost Analysis A technique to evaluate the economic consequences over a period of time of mutually exclusive project alternatives.

Light Pollution Waste light from building sites that produces glare, directed upward to the sky off the site.

Light-Shelf A horizontal element positioned above eye level to reflect daylight onto the ceiling.

Limit of Liability The maximum amount an insurance company agrees to pay in case of loss.

Lintel A load-bearing structural member supported at its ends, usually located over a door or window.

Live Load A temporary and changing load superimposed on structural components by the use and occupancy of the building, not including wind load, earthquake load, or dead load.

Load The electrical power being consumed at any given moment or averaged over a specified period. The load that an electricity-generating system supplies varies greatly with time of day and to some extent with season. In an electrical circuit, the load is any device or appliance that is using power.

Load-Bearing Wall A support wall that holds floor or roof loads in addition to its own weight.

Lumen (lm) The luminous flux emitted by a point source having a uniform luminous intensity of one candela.

Luminaire A complete electric lighting unit, including housing, lamp, and focusing and/or diffusing elements; informally referred to as a fixture.

Lux The International System (SI) unit of illumination. It is the illumination on a surface 1 square meter in area on which there is a uniformly distributed flux of 1 lumen.

Manifold A fitting that has several inlets or outlets to carry liquids or gases.

Masonry Opening The actual distance between masonry units where an opening occurs, not including the wood or steel framing around the opening.

Master Specification In construction, a resource specification section containing options for selection, usually created by a design firm, which once edited for a specific project becomes a contract specification.

Master Format The industry standard for organizing specifications and other construction information, published by CSI and Construction Specifications Canada. Formerly a 5-digit numbering system with 16 divisions, now a 6- or 8-digit numbering system with 49 divisions.

MasterSpec ® The subscription master guide specification library published by ARCOM and owned by the American Institute of Architects. http://www.specguy.com/www.masterspec.com.

Mastic An adhesive used to hold tiles in place; also an adhesive used to glue many types of materials in the building process.

Mechanical Drawing A scale drawing of a mechanical object.

Mechanics' Lien A lien on real property created by statute in all states in favor of persons supplying labor or materials for a building or structure for the value of the labor or materials.

In some jurisdictions a mechanic's lien exists for the value of professional services. Clear title to the property cannot be obtained until claims for labor, materials, or professional services are settled.

Megawatt (MW) A measurement of power equal to 1 million watts .

Megawatt-Hour (MWh) A measurement of power with respect to time (i.e., energy). One megawatt-hour is equal to 1 megawatt used for 1 hour, or 1 kilowatt used for 1000 hours.

Mesh A metal reinforcing material placed in concrete slabs and masonry walls to resist cracking.

Mezzanine, or Mezzanine Floor An intermediate floor level placed within a story, having occupiable space above and below its floor.

Module A system based on a single unit of measure.

Modulus of Elasticity (E) The degree of stiffness of a beam.

Moisture Barrier Typically, a plastic material used to prevent moisture vapor from penetrating into a structure.

Mortar A mixture of cement, sand, lime, and water that provides a bond for the joining of masonry units.

Mortgage The mortgage (s) to be made to the Lender to secure the note and any sums in addition to the loan amount advanced by the lender for completion of the improvements.

Multizone HVAC System A central all-air HVAC system that uses an individual supply air stream for each zone; warm and cool air are mixed at the air-handling unit to provide supply air appropriate to each zone; a multi-zone system requires several separate supply air ducts.

Native Vegetation Vegetation whose presence and survival in a specific region is not due to human intervention. Plants imported to a region by prehistoric peoples are sometimes considered native. Plants that are imported and then adapt to survive without human cultivation are referred to as naturalized.

Natural Ventilation The exchange or movement of air through a building by thermal, wind, or diffusion effects through doors, windows, or other intentional openings in buildings.

National Electric Code (NEC) Guidelines for all types of electrical installation to be followed when installing a photovoltaic system.

Negligence Failure to exercise due care under normal circumstances. Legal liability for the consequences of an act or omission frequently depends on whether or not there has been negligence.

Net metering The export of a home owner's surplus solar power (i.e., surplus to actual need) during the day to the electricity grid. This causes the home's electric meter to (physically) reverse direction and/or simply creates a financial credit on the home owner's electricity bill. (At night, the home owner draws from the electricity grid in the normal way).

Net Size The actual size of an object.

Noise Pollution Environmental pollution made up of harmful or annoying noise. The degree of pollution is usually measured in intensity, duration, and frequency. Examples include cars, airplanes, construction equipment, and traffic noise.

Noise Reduction Coefficient (NRC) The average of the sound absorption coefficient of the four octave bands, 250, 500, 1,000, and 2,000 Hertz, rounded to the nearest 0.05.

Nominal Discount Rate A discount rate that includes the rate of inflation.

Nominal Size The call-out size, which may not be the actual size of the item.

Nonbearing Wall A wall that supports no loads other than it own weight. Some building codes consider walls that support only ceiling loads to be nonbearing.

Nonconforming Work Work that does not fulfill the requirements of the contract documents.

Nonferrous Metal Metals containing no iron, such as copper and brass.

Nuclear Energy Energy or power produced by nuclear reaction (fusion or fission).

Oblique Drawing A pictorial form in which one view is an orthographic projection and the views of the sides have receding lines at an angle.

Occupiable Room A room or enclosed space designed for human occupancy in which individuals congregate for amusement, educational, or similar purposes, or in which occupants are engaged in labor and which is equipped with means of egress, light, and ventilation.

Off-Gassing A process of evaporation or chemical decomposition by which vapors are released from materials.

Ohm The resistance between two points of a conductor when a constant potential difference of 1 volt applied between these points produces a current of 1 amp in the conductor.

Ohm's Law A simple mathematical formula that allows each of voltage, current, or resistance to be calculated when the other two values are known. The formula is: $V = I \times R$, where V is the voltage, I is current, and R is resistance.

Opinion of Probable Costs Determination of probable costs, a preliminary budget, for a suggested remedy.

Operating Cost Any cost of the daily function of a facility.

Organic Compounds Chemicals that contain carbon. Volatile organic compounds (VOCs) vaporize at room temperature and pressure. They are found in many indoor sources, including common household products and building materials.

Orientation Position with respect to the cardinal directions: north, south, east, and west.

Orthographic Projection A view produced when projectors are perpendicular to the plane of the object, giving the effect of looking straight at one side.

Outlet An electrical receptacle that allows for current to be drawn from the system.

Ozone A naturally occurring, highly reactive gas containing triatomic oxygen formed by combination with oxygen in the presence of ultraviolet radiation. Ozone builds up in the lower atmosphere as smog; in the upper atmosphere it forms a protective layer that shields the earth from excessive exposure to damaging ultraviolet radiation.

Packaged Air Conditioner A self-contained air-conditioning unit designed to control air temperature, humidity, distribution, and quality.

Parapet A portion of wall extending above the roof level.

Partial Occupancy Occupancy by the owner of a portion of a project prior to final completion.

Particulate Small pieces of an airborne material, such as dusts, fumes, smokes, mists, and fogs; generally anything that is not a fiber and has an aspect ratio of 3 to 1.

Partition An interior wall.

Party Wall A wall dividing two adjoining spaces such as apartments or offices.

Passive Solar Home A home that utilizes part of the building as a solar collector, as opposed to an active solar home such one using a PV system.

Patent Defect A defect in materials and/or equipment of completed work that reasonably careful observation could have discovered, distinguished from a latent defect, which reasonable observation could not have discovered.

Pathogen A microorganism typically found in the intestinal tracts of mammals that can cause disease in other micro-organisms or in humans, animals, and plants. These may be bacteria, viruses, or parasites and are found in sewage, in runoff from animal farms or rural areas, and in water used for swimming. Fish and shellfish contaminated by pathogens, or the contaminated water itself, can cause serious illnesses.

Performance Bond A contractor's bond in which a surety guarantees to the owner that the work will be performed in accordance with the contract documents. Except where prohibited by statute, a performance bond is frequently combined with the labor and material payment bond.

Performance Specifications Written material containing minimum acceptable standards and actions necessary to complete a project.

Phase An impulse of alternating current. The number of phases depends on the generator windings. Most large generators produce a three-phase current that must be carried on at least three wires.

Photometer An instrument for measuring light.

Photovoltaic (PV) Any device that produces free electrons when exposed to light.

Photovoltaic (PV) Panel Often interchangeable with *PV module* (especially in single-module systems).

Photovoltaic System All the parts connected together that are required to produce solar electricity.

Pile A steel or wooden pole driven into the ground sufficiently to support the weight of a wall and building.

Pillar A pole or reinforced wall section used to support a floor and thus a building.

Planking Wood members having a minimum rectangular section of 1-1/2 inches to 3-1/2 inches in thickness, used for floor and roof systems.

Plan All final drawings, plans, and specifications prepared by the borrower, the borrower's architects, the general contractor, or the major subcontractors, and approved by the lender and the construction consultant. These describe and show the labor, materials, equipment, fixtures and furnishings necessary for the construction of the improvements, including all amendments and modifications thereof made by approved change orders (and showing minimum grade of finishes and furnishings for all areas of the improvements to be leased or sold in ready-for-occupancy conditions).

Plat A map or plan view of a lot showing principal features, boundaries, and location of structures.

Plenum An air space (above the ceiling) for transporting air from the HVAC system.

Plug Load All equipment that is plugged into the electrical system, such as task lights, computers, printers, and electrical appliances.

Polarity The direction of magnetism or the direction of flow of current.

Pollutant Generally, any substance introduced into the environment that adversely affects the usefulness of a resource or the health of humans, animals, or ecosystems.

Polyvinyl Chloride (PVC) A plastic material commonly used for pipe and plumbing fixtures and as an insulator on electrical cables. PVCs are toxic and are being replaced with alternatives made from more benign chemicals.

Post A vertical wood structural member generally 4 × 4 inches (100 mm) or larger.

Post-and-Beam Construction A type of wood framing using timber for structural support.

Postconsumer Materials/Waste Recovered materials that are diverted from municipal solid waste for the purposes of collection, recycling, and disposition.

Postconsumer Recycling Use of materials generated from residential or consumer waste for new or similar purposes—for example, converting office waste paper into corrugated boxes or newsprint.

Potable Water Water that is suitable for drinking, generally supplied by a municipal water system.

Power The basic unit of electricity equal to the product of current and voltage (in DC circuits)—the rate of doing work—expressed in watts. For example, a generator rated at 800 watts can provide that amount of power continuously. 1 watt = 1 joule/second.

Precast A concrete component that has been cast in a location other than the one in which it will be used.

Preconsumer Materials/Waste Materials generated in manufacturing and converting processes, such as scraps and trimmings, and cuttings, including materials from print overruns, over-issue publication runs, and obsolete inventories. Sometimes referred to as "postindustrial."

Present Value The current value of a past or future sum of money as a function of the time value of money.

Pressed Wood Products Materials used in building and furniture construction that are made from wood veneers, particles, or fibers bonded together with an adhesive under heat and pressure.

Primer The first coat of paint or glue when more than one coat will be applied.

Progress Payment Partial payment made during progress of the work on account of work completed and or materials suitably stored.

Progress Schedule A diagram, graph, or other pictorial or written schedule showing proposed and actual start and completion of the various elements of the work.

Project Cost The total cost of a project, including construction cost, professional compensation, land costs, furnishings and equipment, financing, and other charges.

Projection A technique for showing one or more sides of an object to give the impression of a solid object.

Project Manual The volume(s) prepared by the architect for a project, which may include bidding requirements, sample forms, and conditions of the contract and specifications.

Purlin A horizontal roof member laid perpendicular to rafters to limit deflections.

Quarry Tile An unglazed, machine-made tile.

Quick Set A fast-curing cement plaster.

Rafter A sloping or horizontal beam used to support a roof.

Radioactivity The spontaneous emission of matter or energy from the nucleus of an unstable atom; the emitted matter or energy is usually in the form of alpha or beta particles, gamma rays, or neutrons.

Radius A straight line from the center of a circle or sphere to its circumference or surface.

Radon (Rn), Radon Decay Product A radioactive gas formed in the decay of uranium; a radon decay product (also called daughter or progeny) can be breathed into the lungs, where it continues to release radiation as it decays.

Rainscreen A method of constructing walls in which the cladding is separated from a membrane by an airspace that allows pressure equalization to prevent rain from being forced in. Often used for high-rise buildings or for those in windy locations.

Rainwater Harvesting The collection, storage, and use of precipitation from a catchment area such as a roof.

Rapidly Renewable Material A material that is not depleted when used. Such materials are typically harvested from fast-growing sources and do not require unnecessary chemical support. Examples include bamboo, flax, wheat, wool, and certain types of wood.

Remote Area Power Supply (RAPS) A power generation system that provides electricity to remote and rural homes, usually incorporating power generated from renewable sources, such as solar panels and wind generators, as well as nonrenewable sources, such as gas-powered generators.

Readily Accessible Area An area of a subject property that is promptly made available for observation by the field observer at the time of a walk-through survey. Rapidly accessible areas do not require removal of materials or personal property, such as furniture, and are safely accessible in the opinion of the field observer.

Record Drawings Construction drawings revised to show significant changes made during the construction process, usually based on marked-up prints, drawings, and other data furnished by the contractor to the architect. This term is preferred over *As-Built Drawings*.

Rectifier A device that converts AC to DC, such as in a battery charger or converter.

Recycled Content The amount of pre- and post-consumer recovered material, usually expressed as a percentage.

Recycled Material Material that would otherwise be destined for disposal but is diverted or separated from the waste stream, reintroduced as material feedstock, and processed into marketed end products.

Reference Numbers Numbers on a drawing that refer the reader to another drawing for more detail or other information.

Reflectance The ratio of energy (light) bouncing away from a surface to the amount striking it, expressed as a percentage.

Refrigerant A heat transfer fluid in a refrigerating process, selected for its beneficial properties such as stability, low viscosity, high thermal capacity, and appropriate state change points.

Regionally Manufactured Material For purposes of this document, material that must be assembled as a finished product within a 500-mile radius of the project site. Onsite assembly, erection, or installation of finished components, as in structural steel, miscellaneous iron, or systems furniture, are not included in this definition.

Register An opening in a duct for the supply of heated or cooled air.

Regulator A device used to limit the current and voltage in a circuit, normally to allow the correct charging of batteries from power sources such as solar panels and wind generators.

Relative Humidity The amount of water vapor in the atmosphere compared to the maximum possible amount at the same temperature.

Release of Lien An instrument executed by a person or entity supplying labor, materials, or professional services on a construction project that releases that person's or entity's mechanic' lien against the project property.

Remaining Useful Life (RUL) A subjective estimate based on observations, average estimates of similar items, components, or systems, or a combination thereof, of the number of remaining years that an item, component, or system is estimated to be able to function in its intended use before replacement. RUL is affected by the initial quality of an item, component, or system, the quality of initial installation, the quality and amount of preventive maintenance, climatic conditions, extent of use, and so forth.

Renewable Energy Alternative energy that is produced from a renewable source.

Renewable Resource A resource that is capable of being restored or replenished (e.g., trees).

Requisition A statement prepared by the borrower in a form approved by the lender setting forth the amount of the loan advance requested.

Resistance (R) The property of a material that resists the flow of electric current when a potential difference is applied across it, measured in ohms.

Resistor An electronic component that restricts the flow of current in a circuit, sometimes used specifically to produce heat, such as in a water heater element.

Retainage A sum withheld from progress payments to the contractor in accordance with the terms of the owner/contractor agreement.

Retaining Wall A masonry wall supported at the top and bottom that is designed to resist soil loads.

R-Factor A unit of thermal resistance applied to the insulating value of a specific building material.

Return Air Air that has circulated through a building as supply air and returns to the HVAC system for additional conditioning or release from the building.

Residual Value The value of a building or building system at the end of a study period.

Reuse A strategy to return materials to active use in the same or related capacity.

Roof Drain A receptacle for removal of roof water.

Roof Pitch The ratio of total span to total rise, expressed as a fraction.

Rotation A view in which the object is apparently rotated or turned to reveal a different plane or aspect, all of which are shown within it.

Rough-In Preparation of a room or space for plumbing or electrical additions by running wires or piping for a future fixture.

Rough Opening A large opening made in a wall or roof frame to allow insertion of a door or window.

R-Value A unit that measures thermal resistance (the effectiveness of insulation); the higher the R-value, the better the insulation.

Salvaged Material Construction material recovered from existing buildings or construction sites and reused in other buildings. Common salvaged materials include structural beams and posts, flooring, doors, cabinetry, brick, and decorative items.

Sanitary Sewer A conduit or pipe carrying sanitary sewage.

Scale The relation between the measurement used on a drawing and the measurement of the object it represents. Also, a measuring device, such as a ruler, having special graduations.

Schedule of Values A statement furnished by the contractor to the architect reflecting the portions of the contract sum allocated to the work, and used as the basis for reviewing the contractor's applications for payment.

Schematic Diagram A diagram using graphic symbols to show how a circuit functions electrically.

Scratch Coat The first coat of stucco that is scratched to provide a good bond surface for the second coat.

Sealant Any material with adhesive properties formulated primarily to fill, seal, or water-proof gaps or joints between two surfaces. Examples are primers and caulks.

Section A view showing internal features as if the viewed object had been cut or sectioned

Seismicity The worldwide or local distribution of earthquakes in space and time. *Seismicity* is used as a general term for the number of earthquakes in a unit of time or for relative earthquake activity.

Septic Tank A tank in which sewage is decomposed by bacteria and dispersed by drain tiles.

Sheet Steel Flat steel weighing less than 5 pounds per square foot.

Shear Distribution The distribution of lateral forces along the height or width of a building.

Shear Wall Wall construction designed to withstand shear pressure caused by wind or earthquakes.

Shoring Temporary support made of metal or wood that supports other components.

Short-Term Costs Opinions of probable costs to remedy physical deficiencies, such as deferred maintenance, that may not warrant immediate attention but require repairs or replacements that should be undertaken on a priority basis in addition to routine preventive maintenance. Such opinions of probable costs may include costs for testing, exploratory probing, and further analysis should this be deemed warranted by the consultant. The performance of such additional services are beyond this book. Generally, the time frame for such repairs is within one to two years.

Sick Building Syndrome (SBS) A set of symptoms that affect building occupants during the time they spend in the building and that diminish or disappear when they exit the building. SBS symptoms cannot be traced to specific pollutants or sources within the building and may be localized to a specific room or zone Contrast with Building-Related Illness."

Sill A horizontal structural member supported by its ends.

Single-Line Diagram A pictorial form using single lines and graphic symbols to simplify a complex circuit or system.

Single Prime Contract The most common form of construction contract, in which the bidding documents are prepared by the architect/engineer for the owner and made available to a number of qualified bidders. The winning contractor enters into a series of subcontract agreements to complete the work under which a single entity provides design and construction services. Increasingly, owners are opting for a design–build contract.

Site A parcel of land bounded by a property line or a designated portion of a public right-of-way.

Site Improvements Landscaping and paving improvements for pedestrian and vehicular ways, outdoor lighting, recreational facilities, and the like.

Skylight A relatively horizontal, glazed roof aperture for the admission of daylight.

Slab-on-Grade Type of foundation construction for a structure with no basement or crawl space.

Smart Growth The management of a community's growth in such a way that land is developed according to ecological tenets such as minimizing dependence on auto transportation, reducing air pollution, and increasing infrastructure investment efficiency.

Solar Energy Energy from the sun.

Solar Heat Gain Coefficient Solar heat gain through a building's total window system relative to incident solar radiation.

Solar Module A device used to convert light from the sun directly into DC electricity by the photovoltaic effect, usually consisting of multiple solar cells bonded between glass and a backing material. A typical solar module is 100 watts of power output (although module power can range from 1 to 300 watts) with dimensions of 2×4 feet.

Solar Panel A device that collects energy from the sun and converts it into electricity or heat.

Solar Power Electricity generated by conversion of sunlight, either directly through photovoltaic panels or indirectly through solar-thermal processes.

Solar Reflectance (Albedo) The ratio of reflected solar energy to incoming solar energy over wavelengths of approximately 0.3 to 2.5 micrometers. A reflectance of 100% means that all energy striking a reflecting surface is reflected back into the atmosphere and none is absorbed by the surface.

Solar Reflectance Index (SRI) A measure of a material's ability to reject solar heat, as shown by a small temperature rise. SRI defined so that a standard black (reflectance 0.05, emittance 0.90) is 0 and a standard white (reflectance 0.80, emittance 0.90) is 100.

Special Conditions A section of the conditions of the contract, other than general and supplementary conditions, that describe conditions unique to a particular project.

Specifications A detailed, exact statement of particulars— especially statements prescribing materials and methods and quality of work—for a specific project. Specifications form part of the contract documents contained in the project manual and consist of written requirements for material, equipment, construction systems, standards, and workmanship.

Specific gravity The ratio of the weight of a solution to the weight of an equal volume of water at a specified temperature; used with reference to the sulfuric acid electrolyte solution in a lead acid battery as an indicator of battery state of charge. More recently referred to as *relative density*.

Stack Effect A rise of warm air that creates a positive pressure area at the top of a building and a negative pressure area at the bottom. This effect can overpower mechanical systems and disrupt building ventilation and air circulation.

Stakeholder Any party that might be affected by a company's policies and operations, including shareholders, customers, employees, suppliers, business partners, and surrounding communities.

Statute of Limitations A specific period of time in which legal action must be brought for alleged damage or injury or other legal relief. This period varies from state to state and depends on the type of legal action. Ordinarily, the statute of limitations commences with the occurrence of the damage or injury, or with discovery of the act resulting in the alleged damage or injury. In the construction industry, many jurisdictions define the period as commencing with the completion of work or of services performed in connection with it.

Storm Sewer A sewer used for conveying rainwater, surface water condensate, cooling water, or similar liquid wastes exclusive of sewage.

Stormwater Runoff Water volumes created during precipitation events that flow over surfaces into sewer systems or receiving waters.

Stucco A type of plaster made from Portland cement, sand, water, and a coloring agent that is applied to exterior walls.

Structural Frame The components or building systems that support the building's nonvariable forces or weights (dead loads) and variable forces or weights (live loads).

Stud A light vertical structure member, usually of wood or light structural steel, that serves as part of a wall to support moderate loads.

Subcontract An agreement between a prime contractor and a subcontractor to complete a portion of the work at the construction site.

Subcontractor A person or entity that has a direct or indirect contract with a subcontractor to perform any of the work at a construction site.

Substitution A material, product, or item of equipment offered in lieu of that specified.

Superintendent The contractor's representative at the site who is responsible for continuous field supervision, coordination, and completion of the work, and, unless another individual is designated in writing by the contractor to the owner and the architect, for the prevention of accidents.

Supervision Direction of the work by the contractor's personnel. Supervision is neither a duty nor a responsibility of the architect as part of professional services.

Surety Bond A legal instrument under which one party agrees to answer to another party for the debt, default, or failure to perform of a third party.

Surge An excessive amount of power drawn by an appliance when it is first switched on; an unexpected flow of excessive current, usually caused by excessive voltage, that can damage appliances and other electrical equipment.

Survey Observations made by a field observer during a walk-through survey to obtain information concerning the subject property's readily accessible and easily visible components or systems.

Sustainability A principle holding that the needs of present generations are to be met without compromising the needs of future generations. Also, achieving a balance among extraction and renewal and environmental inputs and outputs that causes no overall net environmental burden or deficit.

Sustainable Community A community that maintains its present levels of growth without damaging effects.

Symbol A stylized graphical representation of commonly used component parts shown in a drawing.

Synergy The action of two or more substances to achieve an effect that neither is individually capable of achieving. In toxicology, synergy exists when two exposures together (for example, asbestos and smoking) create far more risk than the combined individual exposures create.

System (a process) Interacting or interdependent components assembled to carry out one or more functions.

Task Lighting Light provided for a specific task, versus general or ambient lighting.

Tee A fitting, either cast or wrought, that has one side outlet at right angles to the run.

Temper To harden steel by heating and then sudden cooling by immersion in oil, water, or other coolant.

Template A piece of thin material used as a true-scale guide or as a model for reproducing various shapes.

Tensile Strength The maximum stretching of a piece of metal (such as rebar) before breaking, calculated in kilopounds (kps) equals 1000 pounds.

Termite Shield Sheet metal placed in or on a foundation wall to prevent termite intrusion.

Terrazzo A mixture of concrete, crushed stone, calcium shells, and/or glass, polished to a tile-like finish.

Thermal Comfort The appropriate combination of temperature, airflow, and humidity for occupant comfort in a building. Individually, an expression of satisfaction with the thermal environment; statistically, expression of satisfaction by at least 80% of occupants within a space.

Thermal Resistance (R) A unit of measure of a material's resistance to heat transfer. The formula for thermal resistance is $R = L/k$, where L is the thickness of the material in inches and k is the thermal conductivity of the material.

Thermostat An automatic device that controls the operation of HVAC equipment.

Third-Party Certification An independent and objective assessment of an organization's practices or system for chain of custody by an auditor who is independent of the party undergoing such assessment.

Three-Phase Power The combination of three alternating currents in a circuit, with their voltages displaced at 120 degrees or one-third of a cycle.

Timely Access Entry to the site that is provided to the consultant at the time of a site visit.

Timely Completion The completion of the work or a designated portion of it on or before the date required by a contract.

Title Insurer The issuer (s), approved by the interim lender and permanent lenders, of a title insurance policy or policies to insure a mortgage.

Tolerance The amount that a manufactured part may vary from its specified size.

Topographic Survey The configuration of a surface, including its relief and the locations of its natural and man-made features, usually recorded on a drawing showing surface variations by means of contour lines indicating height above or below a fixed datum.

Toxicity A measure of a material's ability to release poisonous or harmful particulates.

Toxic Waste Garbage or waste that can injure, poison, or harm living things and is sometimes life threatening.

Transformer A device that changes voltage from one level to another, or a device that transforms voltage levels to facilitate the transfer of power from the generating plant to the customer.

Transient Lodging A building, facility, or portion of either, excluding inpatient medical care facilities and residential facilities, that contains sleeping accommodations. Transient lodging may include, but is not limited to, resorts, group homes, hotels, motels, and dormitories.

Transistor A semiconductor device that switches or otherwise controls the flow of electricity.

Trap A fitting designed to provide a liquid seal that prevents the back passage of air without significantly affecting the flow of wastewater.

Triangulation A technique for making complex sheet metal forms using geometrical constructions to translate the dimensions from the drawing to the pattern.

Trimmer A piece of lumber, usually a two-by-four, that is shorter than the stud or rafter, used to fill in where the longer piece would have been normally used except for a window or door opening or some other opening in the roof, floor, or wall.

Truss A prefabricated sloped roof system incorporating top and bottom chords and bracing.

Turbulence Any deviation from parallel flow in a pipe due to rough inner wall surfaces, obstructions, and the like.

Underwriters Laboratories, Inc. (UL) A private testing and labeling organization that develops test standards for product compliance. UL standards appear throughout specifications, often in roofing requirements and always in equipment using or delivering electrical power (http://www.ul.com/).

Unfaced Insulation Insulation that does not have a facing or plastic membrane over one side.

Union Joint A pipe coupling, usually threaded, that permits disconnection without disturbing other sections.

Urea Formaldehyde A combination of urea and formaldehyde that is used in some glues and may emit formaldehyde at room temperature.

Utility Plan A floor plan of a structure showing locations of heating, electrical, plumbing, and other service system components.

Vacuum Any pressure less than that exerted by the atmosphere.

Valley The area of a roof where two sections come together to form a depression.

Valve A device that controls water flow in a distribution system. Common valve types include globe, gate, butterfly, and check.

Vapor Barrier *See* Moisture Barrier.

Vapor Compression Chiller Refrigeration equipment that generates chilled water via a mechanically driven process using a specialized heat transfer fluid as the refrigerant, and comprising four major components: compressor, condenser, expansion valve, and evaporator. The operating energy of a chiller is input as mechanical motion.

Variable Air Volume (VAV) HVAC System A central all-air HVAC system in which a single supply air stream and a terminal device at each zone provide appropriate thermal conditions by controlling the quantity of air supplied.

Vegetated Roof A roof that is partially or fully covered by vegetation, designed to counteract the heat island effect as well as to provide additional insulation and cooling during the summer.

Vehicular Way A route intended for vehicular traffic, such as a street, driveway, or parking lot.

Veneer A thin layer or sheet of wood.

Veneered Wall A single-thickness (1-wythe) masonry unit wall, with a backup wall of frame or other masonry, that is tied but not bonded to the backup wall.

Ventilation The exchange of air or the movement of air through a building. Ventilation may occur naturally, through doors and windows, or mechanically by motor-driven fans.

Vent Usually an opening in the eaves or soffit to allow the circulation of air over an insulated ceiling, typically covered with a piece of metal or screen.

Ventilation Rate The rate at which indoor air enters and leaves a building, expressed in one of two ways: number of changes of outdoor air per unit of time (air changes per hour, or "ach") and the rate at which a volume of outdoor air enters per unit of time (cubic feet per minute, cfm).

Vent Stack A system of pipes for air circulation that prevents water from being suctioned from the traps in the waste disposal system.

Vertical Pipe Any pipe or fitting installed in a vertical position or that makes an angle of not more than 45 degrees with the vertical.

View A drawing of a side or plane of an object as seen from one point.

Vision Glazing The portion of exterior windows above 2 feet 6 inches and below 7 feet 6 inches that permits a view to the exterior.

Volatile Organic Compound (VOC) A highly evaporative, carbon-based chemical substance that produces noxious fumes and is found in many paints, caulks, stains, and adhesives. VOCs are capable of entering the gas phase from either liquid or solid form.

Volt (E) or (V) The potential difference across a resistance of 1 Ohm when a current of 1 amp is flowing; the amount of work per unit charge in moving a charge from one place to another.

Voltage Drop The voltage lost along a length of wire or conductor due to the resistance of that conductor (also applies to resistors), calculated using Ohm's Law.

Voltage Protection A sensing circuit on an inverter that disconnects the unit from the battery if input voltage limits are exceeded.

Voltage Regulator A device that controls the operating voltage of a photovoltaic array.

Waiver of Lien An instrument by which a person or organization who has or may have a right of mechanic's lien against the property of another relinquishes such right.

Warranty Legally enforceable assurance of quality or performance of a product or work or of duration of satisfactory performance. *Warranty guarantee* and *guaranty* are substantially identical in meaning, although *Guarantee* (or *Guaranty*) is considered by some to indicate only duration of satisfactory performance or legally enforceable assurance furnished by a manufacturer or other third party. The Uniform Commercial Code provisions on sales (effective in all states except Louisiana) use *Warranty,* but recognize the continued use of *Guarantee* and *Guaranty.*

Waste Pipe A discharge pipe from any fixture, appliance, or appurtenance in connection with a plumbing system that does not contain fecal matter.

Waste Water Spent or used water from a home, farm, community, or industry that contains dissolved or suspended matter.

Water-Cement Ratio The ratio of the weight of water to the weight of cement.

Water Hammer Noise and vibration that develops in a piping system when a column of non-compressible liquid flowing through a pipe line at a given pressure and velocity is abruptly stopped.

Water Main A water supply pipe for public or community use.

Waterproofing Materials used to protect below- and on-grade construction from moisture penetration.

Water Table The level to which water will rise in a well (excluding artesian wells).

Watt (W) The unit of electrical power commonly used to measure electricity consumption of an appliance; the power developed when a current of 1 amp flows through a potential difference of 1 volt. One watt equals 1/746 of a horsepower and 1 joule per second.

Watt-Hour (Wh) A unit of energy equal to 1 watt of power used for 1 hour.

Wetland An area saturated by surface or ground water that contains vegetation adapted for life under these soil conditions. Marshes, swamps, and estuaries are examples. In stormwater management, the term refers to a shallow, vegetated, ponded area that serves to improve water quality and provide wildlife habitat.

Wind Lift (Wind Load) The force exerted by wind against a structure.

Windpower Power or energy derived from the wind (via windmills, sails, etc.).

Wiring (Connection) Diagram A pictorial form to show the individual connections within a unit and the physical arrangement of the components.

Working Drawings A set of drawings that provide the necessary details and dimensions to construct an object; they may include specifications.

Wythe A continuous masonry wall width.

Xeriscape™ A system of water-efficient choices in planting and irrigation The seven basic xeriscaping principles for conserving water and protecting the environment are (1) planning and design; (2) use of well-adapted plants; (3) soil analysis; (4) practical turf areas; (5) use of mulches; (6) appropriate maintenance; and (7) efficient irrigation.

Zenith Angle The angle between directly overhead and a line through the sun. The elevation angle of the sun above the horizon is 90 degrees minus the zenith angle.

Zinc A noncorrosive metal used for galvanizing other metals.

Zone Numbers Number/letter combinations on the border of a drawing that provide reference points indicating or locating specific points on the drawing.

Zoning Legal codes that restrict parts of cities or towns to particular uses, such as residential, commercial, industrial, and so forth.

Zoning Permit A permit issued by an appropriate government authority allowing the use of land for a specific purpose.

Bibliography

Allen, E., Iano, J., 2003. Fundamentals of Building Construction: Materials and Methods, fourth ed. John Wiley & Sons.

American Society for Testing and Materials (ASTM), 2001. Standard practice for applying analytical hierarchy process (AHP) to multiattribute decision analysis of investments related to buildings and building systems: ASTM E1765-11.

American Society for Testing and Materials (ASTM), 2006. Standard guide for property condition assessments: Baseline property condition assessment process: E2018-01.

American Society of Heating, Refrigeration, and Air-Conditioning Engineers (ASHRAE), 2008. Proposed standard 189P: Standard for the design of high-performance green buildings. Second public review, July.

Andreasen, C., Lords, M., Miller, K.R., 2009. Contrasting bid shopping and project buyout. In: Proceedings of the 45th Annual ASC Conference. Gainesville, FL, April 1–4.

Apgar, M., 1998. The alternative workplace: Changing where and how people work. Harvard Business Review (May–June), 121–135.

Armer, G.S.T., 2001. Monitoring and Assessment of Structures. Spon Press, an imprint of Taylor & Francis.

Arnold, C., 2006. Green movement sweeps U.S. construction industry. Broadcast on NPR, 2 July.

Bady, S., 2008. Green building programs more about bias than science, expert argues. Professional Builder.

Bennett, F.L., 2003. The Management of Construction: A Project Life Cycle Approach. Butterworth-Heinemann.

Bezdek, R.H., 2008. Green building: Balancing fact and fiction. Discussion moderated by S.E. Cannon and U.K. Vyas. Real Estate Issues 33 (2), 2–5.

Bonda, P., Sosnowchik, K., 2007. Sustainable Commercial Interiors. John Wiley & Sons.

Brundtland Commission, 1987. The Brundtland Report: Our common future. U.N. World Commission on Environment and Development.

Building Owners and Managers Association International, Urban Land Institute, 1999. What office tenants want. BOMA/ULI Office Tenant Survey Report.

Build It Green, 2009. New Home Construction: Green Building Guidelines; www.builditgreen.org.

Burr, A.C., 2008. CoStar green report: The big Skodowski. 30 January. CoStar Group.

Butters, F. (as participant), 2008. Green building: Balancing fact and fiction, discussion moderated by S.E. Cannon and U.K. Vyas. Real Estate Issues 33 (2), 10–11.

Calow, P., 1998. Handbook of Environmental Risk Assessment and Management. Blackwell Science Ltd.

Coggan, D.A., 2009. Intelligent building systems simply explained. http://www.automatedbuildings.com/news/jul99/articles/coggan/coggan.htm.

ConsensusDOCS, website: www.consensusdocs.org.

Craiger, P., Shenoi, S. (Eds.), 2007. Advances in Digital Forensics III. International Federation for Information Processing (IFIP). Springer.

Davis Langdon, 2004. Costing green: A comprehensive cost database and budgeting methodology.

Davis Langdon, 2007. The cost and benefit of achieving green buildings; www.davislangdon.com.

De Chiara, J., Panero, J., 2001. Time-Saver Standards for Interior Design and Space Planning. McGraw-Hill.

Deasy, C.M., 1985. Designing Places for People: A Handbook for Architects, Designers, and Facility Managers. Whitney Library of Design.

Del Percio, S., 2007. What's wrong with LEED? American City 14 (Spring); www.americancity.org.

DiLouie, C., 2006. Why do daylight harvesting projects succeed or fail? Lighting Controls Association (March).

Dorgan, C., Cox, R., Dorgan, C., 2002. The value of the commissioning process: Costs and benefits. Paper presented at US Green Building Council Conference. Farnsworth Group, Austin.

Dworkin, J.F., 1990. Waterproofing below grade. The Construction Specifier (March).

Edwards, S., 1986. Office systems—Designs for the contemporary workspace. PBC International Inc.

Environmental Building News, 2008. Energy dashboards: Using real-time feedback to influence behavior, Vol. 17 (12); see also other issues at www.buildinggreen.com/articles/VolumeTOC.cfm?Volume=17.

Environmental Information Administration (EIA), 2008. Annual energy outlook: Assumptions. Cited in Green Building Facts, U.S. Green Building Council (November).

Federal Emergency Management Agency (FEMA), 2004. FEMA 426, Reference manual to mitigate potential terrorist attacks in high occupancy buildings.

Fisk, W.J., 2002. How IEQ affects health, productivity. ASHRAE Journal (May).

Fisk, W.J., Rosenfeld, A.H., 1998. The indoor environment, productivity and health, and $$$. Strategic Planning for Energy and the Environment 17 (4), 53–57.

Fuller, S., 2010. Life-cycle cost analysis (LCCA). National Institute of Standards and Technology, (in WBDG).

General Services Administration (GSA), 2004. LEED cost study—Final report, submitted to the U.S. GBI. Steven Winter Associates, Inc. (October).

Glavinich, T.E., 2008. Contractor's Guide to Green Building Construction. John Wiley and Sons.

Goldsmith, S., 1999. Designing for the Disabled: The New Paradigm. Architectural Press.

Gore, A., 2006. The future is green. Vanity Fair Special Green Issue (May).

Gottfried, D., 2000. Sustainable Building Technical Manual. U.S. Green Building Council, U.S. Department of Energy, and U.S. Environmental Protection Agency.

Green Building Certification Institute (GBCI), 2009a. LEED® Green Associate Candidate Handbook. (May).

Green Building Certification Institute (GBCI), 2009b. LEED® Professional Accreditation Handbook. (May).

Green Building Initiative (GBI), 2009. Green Globes®, a nationally recognized alternative to LEED; www.thegbi.org.

Gunn, R.A., Burroughs, M.S., 1996. Work spaces that work: Designing high-performance offices. The Futurist (March–April), 19–24.

Haasl, T.P., Claridge, D., 2003. The cost effectiveness of commissioning. HPAC Engineering (October), 20–24.

Harmon, S.K., Kennon, K.E., 2001. The Codes Guidebook for Interiors, second ed. John Wiley & Sons.

Hartman, T., 2007. What does it really take to be an intelligent building? The Hartman Company (April).

Heerwagen, J., Zagreus, L., 2005. The human factors of sustainable building design: Post-occupancy evaluation of the Philip Merrill Environmental Center. U.S. Department of Energy, Annapolis, MD.

Hellier, C.J., 2001. Handbook of Nondestructive Evaluation. McGraw-Hill.

Heschong Mahone Group, 2003. Daylighting and productivity. Studies on behalf of the California Energy Commission's Public Interest Energy Research (PIER) program.

Hess-Kosa, K., 1997. Environmental Site Assessment Phase I, second ed. CRC Press.

Horman, M.J., Riley, D.R., Pulaski, M.H., Magent, C., Dahl, P., Lapinski, A.R., et al., 2005. Well built?: A forensic approach to the prevention, diagnosis and cure of building defects. RIBA Enterprises.

International Code Council (ICC), 2012a. International Green Construction Codes.

International Code Council (ICC), 2012b. International Building Code®.

Jackson, B.J., 2004. Construction Management JumpStart. Wiley Publishing, Inc.

Jackson, J., 2009. How risky are sustainable real estate projects? An evaluation of LEED and ENERGY STAR development options. JOSRE 1 (1).

Kats, G., 2007. The costs and benefits of green. Capital E Analytics.

Kats, G., 2008. Greening buildings and communities: Costs and benefits. Good Energies.

Kats, G., Alevantis, L., Berman, A., Mills, E., Perlman, J., 2003. The costs and financial benefits of green buildings: A report to California's Sustainable Building Task Force (SBTF).

Kennett, S., 2008. Making BREEAM robust. Building 10; www.building.co.uk.

Kibert, C.J., 2005. Sustainable Construction—Green Building Design and Delivery, first ed. John Wiley.

Koomen-Harmon, S., Kennon, K.E., 2001. The Codes Guidebook for Interiors, second ed. John Wiley & Sons.

Kubal, M.T., 2000. Construction Waterproofing Handbook. McGraw-Hill.

Kubba, S.A.A., 2003. Space Planning for Commercial & Residential Interiors. McGraw-Hill.

Kubba, S.A.A., 2007. Property Condition Assessments. McGraw-Hill.

Kubba, S.A.A., 2008. Architectural Forensics. McGraw-Hill.

Kubba, S.A.A., 2010a. Green Construction Project Management and Cost Oversight. Architectural Press.

Kubba, S.A.A., 2010b. LEED™ Practices, Certification, and Accreditation Handbook. Butterworth–Heinemann.

Lawrence Berkeley National Laboratory, 1996. The commissioning process: In search of a universal definition and application; labs21.lbl.gov/DPM/Assets/a2_sharpless.pdf.

Lechner, N., 2002. Chapter 16, Mechanical equipment for heating and cooling in E-book (December).

Levy, M., Salvadori, M.G., Woest, K. (Illustrator), 1994. Why Buildings Fall Down: How Structures Fail. W. W. Norton & Co Inc.

Lewis, B.T., Payant, R., 2001. Facility Inspection Field Manual: A Complete Condition Assessment Guide. McGraw-Hill.

Lippiatt, B.C., Norris, G.A., 1998. Selecting environmentally and economically balanced building materials. National Institute of Standards and Technology.

Liu, K., 2003. Engineering performance of rooftop gardens through field evaluation. National Research Council Canada (NRC).

Loveland, J., 2003. Daylight by design. LD+A (October).

Luhmann, T., 2007. Close Range Photogrammetry: Principles, Techniques and Applications. Wiley.

Luo, L., Harding, N.G., 2006. Delivering green buildings: Process improvements for sustainable construction. Journal of Green Building 1 (1), 123–140.

Lupton, M., Croly, C., 2004. Designing for daylight. Building Sustainable Design (February).

Macaluso, J., 2009. An overview of the LEED Rating System. Empire State Development. Green Construction Data (May).

Macdonald, S. (Ed.), 2002. Concrete Building Pathology. Blackwell Publishing Ltd.

Madsen, J.J., 2008. The realization of intelligent buildings. Buildings (March).

Mago, S., Syal, M., 2007. Impact of LEED®-NC projects on construction management practices. M.S. thesis, Michigan State University.

Markovitz, M., 2008. The differences between Green Globes and LEED. ProSales (August 27).

Martín, C., Foss, A., 2006. All that glitters isn't green and other thoughts on sustainable design. National Building Museum Blueprints 24 (4), 10–13.

Matthiessen, L.F., Morris, P., 2004. Costing green: A comprehensive cost database and budgeting methodology. Davis Langdon (July).

Matthiessen, L.F., Morris, P., 2007. The cost of green revisited. Davis Langdon.

May, S., 2009. Do green design strategies really cost more? Design Cost Data; www.dcd.com/insights/novdec_2006_10.html.

McClain, L.R., 2007. Design-Build Interoperability and Conceptual Design and Development of a Design-Build Management Control System. M.Sc thesis, Georgia Institute of Technology.

McGraw-Hill Construction, 2008a. Global green building trends. SmartMarket Report.

McGraw-Hill Construction, 2008b. The green home consumer: Driving demand for green homes. SmartMarket Report.

McGraw-Hill Construction, 2010. 2011 construction outlook. SmartMarket Report.

Mendler, S., Odell, W., 2000. The HOK Guidebook to Sustainable Design. John Wiley & Sons.

Mills, E., 2009. Building commissioning: A golden opportunity for reducing energy costs and greenhouse-gas emissions. California Energy Commission Public Interest Energy Research (PIER), July 21.

Morris, P., 2009. Market update—A guide to working in a recession. Davis Langdon.

Morris, P., Matthiessen, L.F., 2007. Cost of green revisited. Davis Langdon (July).

Muldavin, S., 2007. A strategic response to sustainable property investing. PREA Quarterly (Summer), 33–37; www.muldavin.com.

Murphy, B.L., Morrison, R.D., 2007. Introduction to Environmental Forensics, second ed. Academic Press.

Nadel, B.A., 2004. Building Security—Handbook for Architectural Planning and Design. McGraw-Hill.

National Institute of Building Sciences, The Whole Building Design Guide (WBDG). Available at www.wbdg.org.

Needy, K.L., Ries, R., Gokhan, N.M., Bilec, M., Rettura, B., 2004. Creating a framework to examine benefits of green building construction. In: Proceedings from the American Society for Engineering Management Conference. Alexandria, VA, October 20–23, pp. 719–724.

Peck, S.W., Callaghan, C., Kuhn, M.E., Bass, B., 1999. Greenbacks from green roofs: Forging a new industry in Canada. Canada Mortgage and Housing Corporation.

Piper, J.E., 2004. Handbook of Facility Assessment. Fairmont Press.

Poynter, D., 1997. Expert Witness Handbook: Tips and Techniques for the Litigation Consultant, second ed. Para Publishing.

Propst, R., 1968. The Office—A Facility Based on Change. The Business Press.

Prowler, D., 2008. Whole building design, revised and updated by S. Vierra, Steven Winter Associates, Inc.

Pulaski, M.H., Horman, M.J., Riley, D.R., 2006. Constructability practices to manage sustainable building knowledge. Journal of Architectural Engineering 12 (2).

Ratay, R.T., 2005. Structural Condition Assessment. John Wiley & Sons.

Reznikoff, S.C., 1989. Specifications for Commercial Interiors. Whitney Library of Design.

Rogers, E., Kostigen, T.M., 2007. The Green Book. Three Rivers Press.

Romm, J.J., 1994. Lean and Clean Management. Kodansha Amer Inc.

Sabo, W., Zahn, J.K., Attorneys at Law. Allowances; www.sabozahn.com/pdf/79.pdf.

Samaras, C., 2004. Sustainable development and the construction industry: Status and implications. Carnegie Mellon University.

Sampson, C.A., 2001. Estimating for Interior Designers. Watson-Guptill.

Sara, M.N., 2003. Site Assessment and Remediation Handbook, second ed. CRC Press.

Scheer, R., Woods, R., 2007. Is there green in going green? SBM (April).

Smith, W.D., Smith, L.H., 2001. McGraw-Hill On-Site Guide to Building Codes 2000: Commercial and Interiors. McGraw-Hill.

Starr, J., Nicolow, J., 2007. How water works for LEED. BNET, CBS Interactive Inc.

Steven Winter Associates, 1997. Accessible Housing by Design. McGraw-Hill.

Steven Winter Associates, 2004a. LEED® Applications Guide. Prepared for U.S. General Services Administration.

Steven Winter Associates, 2004b. LEED® Cost Study. U.S. General Services Administration.

Stodghill, A., 2008. LEED vs. Green Globes. It's the environment, stupid, August 27. Available at: http://itstheenvironmentstupid.blogspot.com/2008/08/leed-vs-green-globes.html.

Strong, K.F., Juliana, C.N., 2005. The basics of design–build. Construction Update (May). Gordon & Rees LLP.

Sullivan, P.J., Agardy, F.J., Traub, R.K., 2000. Practical Environmental Forensics: Process and Case Histories. Wiley.

Syal, M., 2005. Impact of LEED® projects on constructors. AGC Klinger Award proposal, Michigan State University.

Tilton, R., Jackson, H.J., Rigby, S.C., 1996. The Electronic Office: Procedures & Administration, eleventh ed. South-Western Publishing Company.

Turner Construction, 2008. Green Building Market Barometer; www.turnerconstruction.com/greenbuildings.

Turner, C., Frankel, M., 2008. Energy Performance of LEED for New Construction Buildings. New Buildings Institute (March 4).

United Nations, 1987. Report of the World Commission on Environment and Development. General Assembly Resolution 42/187, December 11.

U.S. Army Corps of Engineers (USACE), 2008. Army LEED Implementation Guide. USACE (January 15).

U.S. Department of Energy, Federal Energy Management Program (FEMP), 2003. The business case for sustainable design in federal facilities.

U.S. General Services Administration, 2005. Building Commissioning Guide.

U.S. Green Building Council (USGBC), 2012. LEED® AP Building Design & Construction Study Guide; www.usgbc.org.

U. S. Green Building Council, 2008. LEED 2009 for New Construction and Major Renovations.

U.S. Green Building Council, 2009a. LEED®—Green Building Rating System for New Construction and Major Renovations, version 2.2; www.usgbc.org.

U.S. Green Building Council, 2009b. LEED Reference Guide for Green Building Design and Construction, for the Design, Construction and Major Renovations of Commercial and Institutional Buildings, Including Core & Shell, and K-12 School Projects.

Vermont Green Building Network, 2009. Green building rating systems; www.vgbn.org.

Wilson, S.R., Dunn, S., 2007. Green downtown office markets: A future reality. CB Richard Ellis.

Woods, J.E., 2008a. Expanding the principles of performance to sustainable buildings. Focus on Green Building (Fall). Entrepreneur Media, Inc.

Woods, J.E. (as participant). 2008b. Green building: Balancing fact and fiction, discussion moderated by S.E. Cannon and U. K. Vyas. Real Estate Issues 33 (2), 10.

Yoders, J., 2008. Integrated Project Delivery using BIM. Building Design and Construction (April).

Yudelson, J., 2008a. The business case for green buildings. Yudelson Associates.

Yudelson, J., 2008b. The Green Building Revolution. Island Press.

Zelinsky, M., 1998. New Workplaces for New Workstyles. McGraw-Hill.

Zwick, D.C., Miller, K.R., 2004. Project buyout. Journal of Construction Engineering and Management (March/April), 245–248.

Index

A

Absorption chiller, 435
AC breaker panel, 462
AC induction motor, 447
Aerobic processing system, 375
AFF. *See* American Forest Foundation
Agency CM, 137
AIA. *See* American Institute of Architects
Air-conditioning systems
  ASHRAE's definition, 421
  central air-conditioning systems, 422–424
  packaged systems, 426
  split systems, 424–426
  vapor compression refrigeration, 421–422
  wall/window air-conditioning units, 422
Air filters, 437–439
Air-handling unit (AHU), 405
All-air systems, 426
Alternative dispute resolution, 726
  advantages and disadvantages, 736–737,
    736*t*
  AIA agreements, 726
  arbitration, 737–738
  design–build project, 726–727
  mediation, 738–739
  vs. traditional litigation
    AAA, 734–735
    court action, 740
    defendant, 740–741
    depositions, 741
    impeaching, 741
    interrogatories, 741
    jury judgment, 741–742
    minor defects/failures, 735
    Nationwide Academy for Dispute
      Resolution (UK) Ltd., 734
    popularity, 734–735
    professional ethics and confidentiality,
      742–743
    summons and complaint, 740–741
    voir dire, 741
  unassisted and assisted negotiation, 739–740
Alternative energy, 68, 115

Aluminum wiring, 445–446
American Arbitration Association (AAA),
  734–735
American Forest Foundation (AFF), 304
American Institute of Architects (AIA), 8, 50,
  143–144
  A-series, owner/contractor agreements,
    647–653
  B-series, owner/architect agreements,
    653–655
  collection methods, 646–667
  C-series, other agreements, 655–658
  design and construction projects, 646–667
  Document E202, 219–220
  D-series, miscellaneous documents,
    658–659
  E-series, exhibits, 659–660
  G702 Forms, 149–150
  G-series, contract administration and project
    management forms, 660–666
  project cost analysis, 578
American National Standards Institute (ANSI)
  standard, 88
American Recovery and Reinvestment Act
  (ARRA), 13, 32–33, 519
American Society of Heating, Refrigerating and
  Air-Conditioning Engineers (ASHRAE), 50,
  65, 315–316, 387–388, 401
American Society of Testing and Materials
  (ASTM), 599, 601, 602
Annual fuel utilization efficiency (AFUE), 412
Architectural and Transportation Barriers
  Compliance Board, 357–358
ARRA. *See* American Recovery and
  Reinvestment Act
Array mounting rack, 461
ASHRAE. *See* American Society of Heating,
  Refrigerating and Air-Conditioning Engineers
Aspirating smoke detectors (ASD), 483
ASTM. *See* American Society of Testing and
  Materials
At-risk CM, 137–138, 139

Note: Page numbers followed by *b* indicate boxes, *f* indicate figures and *t* indicate tables.

Lightning Source UK Ltd.
Milton Keynes UK
UKOW04n1829310314

229174UK00008B/207/P